THE ILLUSTRATED
ENCYCLOPEDIA OF THE
UNIVERSE

Editorial Direction: Sonya Newland
Design Direction: Helen Courtney
Picture Research: Robin Scagell and Anna Amari
Design: Helen Courtney and Dave Jones
Development Director: Ian Powling
Concept/Creative Director: Nick Wells

Thanks to: Myles Archibald, Jennifer Bishop, Peter Bond, Tom Cohn,
Chris Cooper, Bob Nirkind and Polly Willis

First published in the United States in 2001 by
Watson-Guptill Publications,
a division of BPI Communications, Inc.
770 Broadway
New York
NY 10003

Produced by Flame Tree Publishing, part of
The Foundry Creative Media Company Ltd,
Crabtree Hall, Crabtree Lane,
Fulham, London, SW6 6TY

ISBN 0-8230-2512-8

Printed in the United Kingdom

Library of Congress Control Number: 2001087745

First printing 2001

1 2 3 4 5 6 7 8 9 / 09 08 07 06 05 04 03 02

THE ILLUSTRATED
ENCYCLOPEDIA OF THE
UNIVERSE

FOREWORD BY SIR MARTIN REES

GENERAL EDITOR: IAN RIDPATH

Watson-Guptill Publications/New York

HOW TO USE THIS BOOK

ABOUT THE ENCYCLOPEDIA

The Encyclopedia comprises an 'Explore the Universe' contents sequence, which introduces some of the key facts and concepts discussed in the book together with a complete contents and key personalities list for each chapter. This is followed by eight chapters, each one dealing with a different aspect of the Universe: from the first notions of the heavens in The History of Astronomy, to the very latest achievements in Space Exploration.

Each chapter is divided into a series of main sections, beginning with an introduction to the subject and ending with a summary of the information contained within the chapter.

THE HISTORY OF ASTRONOMY
Historical coverage of astronomy from prehistory to the present day, including all the key events and people such as Galileo, Newton and Einstein.

THE LAWS OF PHYSICS
Coverage of all the fundamental principles of classical physics, including the laws governing motion, light, radiation and relativity.

IN SEARCH OF QUANTUM REALITY
Explanations of all the startling and thought-provoking concepts of quantum mechanics, the new area of science which provides the key to understanding the Universe.

THE UNIVERSE: PAST, PRESENT & FUTURE
A full investigation of all the groundbreaking ideas about the Universe, including the Big Bang and black holes, and a look at the possibilities for its future.

CONTENTS OF THE COSMOS
A comprehensive survey of what is contained within the observable Universe, from the types and properties of stars to the most distant galaxies and quasars.

OUR SOLAR SYSTEM
An in-depth discussion of all the components of our Solar System, beginning with the Sun, our nearest star, and looking at the characteristics of each of the planets.

WATCHING THE SKY
A full selection of charts for observation of the night sky, along with descriptions of equipment, techniques and practical projects for the keen astronomer.

SPACE EXPLORATION
A thorough survey of human space exploration and discovery, including missions, spacecraft and technology.

REFERENCE SECTION
Includes a detailed glossary, bibliography and web site directory. There are also listings of spacecraft, missions, mathematical concepts and celestial data as well as an extensive index.

THEMES

Six themes run through the book, the entries for which can be found in colour-coded boxes in fixed positions on each spread.

PEOPLE
Biographical profiles of the most significant people, including their major discoveries and contributions.

TIME
A series of entries focusing on specific time-related issues and developments, from the invention of time-keeping devices, time zones on Earth, through to wormholes and the possibility of time travel.

MILESTONES
Key events in the history of our discovery and understanding of the Universe, their implications and practical applications.

THEORIES & DEFINITIONS
Explanations of some of the more profound or significant theories and definitions governing scientific laws.

PRACTICAL
Experiments or activities that the amateur can engage in as well as descriptions of technology and hardware, how equipment works and professional procedures and techniques currently in use.

CONCEPTS & ISSUES
A look at extra-science concepts and issues relating to astronomy and cosmology, including philosophical, social, historical, political and cultural ideas and questions.

 In addition to these six theme boxes, statistical information can be found in data tables wherever appropriate.

A thematic entry is indicated by an active icon at the top of each spread. The page references lead to the previous and following examples of a particular theme.

page 255 ◀◁ THEORIES & DEFINITIONS ▷▶ page 283   ◀◁ THEORIES & DEFINITIONS ▷▶

ACTIVE   INACTIVE

USING THE ENCYCLOPEDIA

The information in the Encyclopedia can be accessed in a number of different ways.

> • The text can be read sequentially through the chapters to gain a holistic view of all the different aspects of a particular subject.

> • The themes listed on page 4 can be traced through the book using the icons at the top of each spread, which give page references to the previous and following boxes of the same theme.

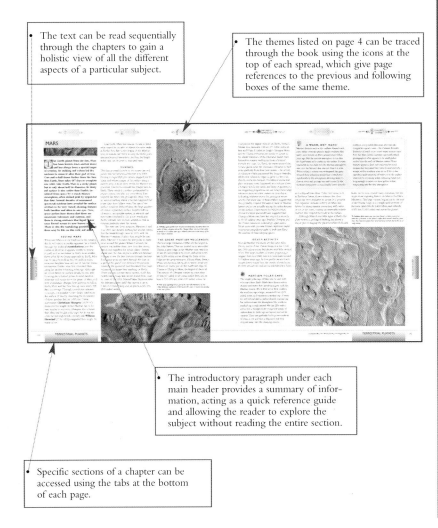

> • The introductory paragraph under each main header provides a summary of information, acting as a quick reference guide and allowing the reader to explore the subject without reading the entire section.

> • Specific sections of a chapter can be accessed using the tabs at the bottom of each page.

CROSS REFERENCING

A system of cross-referencing has been used whereby any terms or names that have an entry elsewhere in the encyclopedia are emboldened, with a page reference at the bottom of the spread.

- Terms: any scientific or astronomical terms that may require further explanation appear in bold the first time they appear in a chapter section; the cross reference at the bottom of the left-hand page leads to the appropriate entry in the glossary on pages 336 to 344.
- Names: people who have their own biographical entry in the 'People' theme appear in bold the first time they are mentioned in a chapter section. The cross-reference at the bottom of the right-hand page leads to the relevant thematic box.

Names and Dates
Full names and dates are given wherever possible the first time they appear in a chapter.

ABBREVIATIONS

A complete listing, including definitions, of all units of measurement can be found on pages 370–371 in the reference section. Listed below are some of the more common abbreviations used in the encyclopedia. Throughout the book, metric units are given first, followed by imperial units in brackets.

astronomical unit	AU
centimetres	cm
cubic metres	cubic m
cubic feet	cubic ft
degrees Celsius	°C
degrees Fahrenheit	°F
degrees Kelvin	K
feet	ft
grams	g
hours	h
inches	in
kilograms	kg
kilometres	km
kilometres per hour	km/h
kilometres per second	km/s
metres	m
miles per hour	mi/h
miles per second	mi/s
millimetres	mm
minutes	min
nanometres	nm
per cent	%
position angle	PA
radius	*r*
seconds	s
square centimetres	sq cm
square feet	sq ft
tonnes	t

STAR CHARTS

Each of the star charts contains a data box with information on some of the key celestial sights to spot in the relevant areas of the night sky. Some of these are visible to the naked eye, some need binoculars and others can only be detected with the aid of a telescope. The viewing method is indicated by the following icons:

Naked eye	Binoculars	Telescope

IMPORTANT NOTICE
All practical information in this book should be used with due care and attention. Viewing the Sun through any optical instrument can be dangerous and caution should be applied at all times. The authors and publishers cannot accept any responsibility for readers who ignore this advice.

CONTENTS

THE HISTORY OF ASTRONOMY ▷▷▶▶ 26

KEY TOPICS

Cosmologies and Cultures

ASTRONOMY is one of the crowning achievements of humankind – and surely one of the oldest. For over 5,000 years astronomers have searched the skies for answers to the most profound questions that can be asked: What is the design and purpose of the Universe? And where do we fit into it? Throughout history, cosmologies have reflected the concerns and beliefs of the cultures from which they sprang. With no evidence to dispute it, the assertion by Ptolemy that Earth was the centre of the Universe lay at the heart of astronomy for 1,400 years. From the ancient Egyptians making a connection between the pre-dawn rising of Sirius and the flooding of the Nile, to present-day speculation about what could have existed before the Big Bang, astronomy has challenged us at the deepest levels.

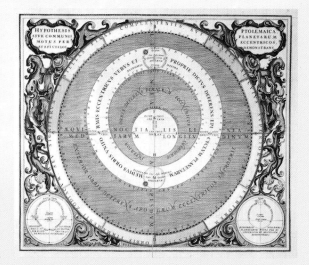

A Model for Science

ASTRONOMY has promoted the use of technology and the powerful techniques of methodical observation and recording. In every age, it embraced enquiry about great issues such as the infinite, philosophy and the relationships between celestial objects. In fact, astronomy can be seen as the model on which all modern sciences are based. Ancient astronomers puzzled over why celestial cycles could be regularly predicted when earthly elements and life seemed so chaotic. This pursuit of the logic of the spheres led to the study and classification that became the life sciences. The impulse to measure the position of the stars was fundamental to the development of geometry and physics. The observations of Eratosthenes led him to calculate the circumference of Earth.

Astronomical Instruments

INSTRUMENTS have been crucial to the measurement and description of the heavens. Astrolabes, quadrants and sundials were used to model the cosmos and test geometric constructs of epicycles and ellipses through the painstaking measurement of minutes and degrees. The design and accuracy of such instruments went relatively unchanged for centuries until Tycho Brahe refined the division of circles and kicked off a revolution in instrument design. The definitive breakthrough came when Galileo first used astronomy's most influential instrument: the telescope. These equipment upgrades were later used to clinch the Copernican theory and produce the data on which Newtonian gravitation was based. A succession of technologies – Herschel's huge reflectors, spectroscopes, photography and now telescopes in space – have continued the great tradition.

Milestones in Astronomy

PERHAPS the biggest watershed in astronomy came when the concept of celestial spheres moving around Earth was rejected in favour of the idea of planets orbiting around the Sun. Without this breakthrough, there would have been no Newtonian gravitation, the principles of which became crucial for the development of physics after 1700. Medieval astronomers had speculated about the infinite size of the cosmos, but it took the ever-increasing power of telescopes – from Galileo's first glimpses of the Milky Way to Herschel's sightings of nebulae – to give plausibility to this awesome idea. Discoveries in the twentieth century revealed an expanding Universe of Einsteinian warped space–time emanating from the incredibly hot and dense fireball of the Big Bang.

Amateur and Professional Astronomers

THROUGHOUT most of scientific history, astronomers were people who made a living doing other things. Greek astronomers, for instance, were librarians or taught students, whilst practitioners in medieval Europe and their counterparts in the Islamic world were usually university professors. Before the sixteenth century, most astronomers were occupied with refining the measurements of traditional tables. After the Copernican revolution, scientists began to do modern research, seeking evidence to prove that Earth either did or did not move. Tycho Brahe was one such astronomer who was paid full-time to do his work. An increasing emphasis on progressive technology and skills demanded new types of funding from governments and sponsoring agencies. The tradition of the wealthy self-funded amateur continued in nineteenth-century Britain, but by the time of Hubble, the cutting edge work was largely being done by professionals.

KEY TOPICS

Classical Physics

THIS CHAPTER focuses on what is sometimes called classical physics, most of which was already established before the end of the nineteenth century. In most cases, classical physics serves to define our everyday physical world. Also included in this chapter are Einstein's theories of relativity, and his descriptions of warped space and distorted time complete the classical view of physics – even though we do not experience the extremes of speed or gravity necessary to feel their effects. The newest area of physics, whose effects we also cannot experience because they are far too small, is called quantum mechanics. Its concepts are so radically different from classical physics that they are given a chapter all to themselves.

Laws Governing Motion

THE SCIENCE of physics is fundamentally about understanding the way in which objects move in response to forces acting upon them. The principles governing classical mechanics were set out by Isaac Newton, and for all normal purposes his laws of motion are still valid. Two other concepts are crucial to physics: energy and momentum. Energy is the ability of a body to do work, and takes many forms, including gravitational, kinetic, electromagnetic and nuclear. Momentum can be thought of as a body's quantity of motion. All laws of physics apply throughout the Universe, providing their calculations are within a frame of reference moving at a constant velocity. For more complex situations, physicists turn to the powerful laws of conservation.

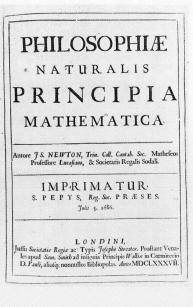

PHILOSOPHIÆ

NATURALIS

PRINCIPIA

MATHEMATICA

Autore JS. NEWTON, Trin. Coll. Cantab. Soc. Matheseos Professore Lucasiano, & Societatis Regalis Sodali.

IMPRIMATUR.

S. PEPYS, Reg. Soc. PRÆSES.

Julii 5. 1686.

LONDINI,

Jussu Societatis Regiæ ac Typis Josephi Streater. Prostant Venales apud Sam. Smith ad insignia Principis Walliæ in Cœmiterio D. Pauli, aliosq; nonnullos Bibliopolas. Anno MDCLXXXVII.

Gravitation

NEWTON'S amazing theory was that the same force which makes an apple fall from a tree also keeps the planets on their courses. His law of universal gravitation does not quite tell the whole story – it has been superseded by the deeper understanding of Einstein's general relativity – but it is good enough for everyday purposes. Essential for understanding gravitation are the terms mass and weight, which are often confused. Mass measures the amount of matter in a body. Weight is a measure of the gravitational force acting on that body. Every object that is moving freely under the influence of a gravitational force is, in fact, describing an orbit.

What is Matter?

EVERYTHING we see around us, from a speck of dust to a gigantic galaxy, is made up of atoms. Each of the chemical elements is composed of a different atom. These atoms can combine to form larger particles called molecules. The three familiar states of matter – solids, liquids and gases – can be understood in terms of the interaction between their atoms, molecules and the forces acting upon them. These changes of state are caused by heat which can be transferred through conduction, convection or radiation. All three types of heat transfer are important in the planetary sciences and astrophysics.

Electromagnetic Radiation

THE RANGE of different types of electromagnetic radiation is called the electromagnetic spectrum. It includes radio waves, infrared, ultraviolet and X-rays. The type of electro-magnetic radiation most familiar to us is visible light. The colour of light depends on its wavelength, although our perception of colour is largely subjective. The most intriguing thing about light is that it behaves both like a wave and a particle, depending on the type of experiment you choose to conduct. At the end of the nineteenth century, a clash between accepted theory and observation called the Ultraviolet Catastrophe heralded the revival of an old idea: perhaps light was made up of particles after all. Later, it led to the development of the new 'quantum' physics.

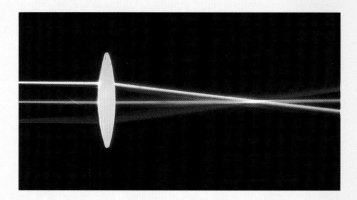

Einstein's Relativity

BY 1900, with a few exceptions, Newton's laws seemed to account for all mechanical phenomena and Maxwell's descriptions accounted for electromagnetic phenomena. However, a few puzzles remained unsolved. A little-known physicist named Albert Einstein came up with the answers which changed the field of physics forever. His theory of special relativity made astounding assertions: time is different for each observer; light has the same speed everywhere; mass and energy are two facets of the same phenomenon. Einstein's general theory of relativity went even further by stating that gravity is actually the curvature of space–time, inspiring the maxim, 'Matter tells space how to curve, space tells matter how to move'.

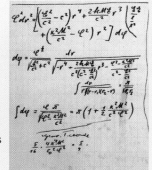

CONTENTS

KEY PERSONALITIES

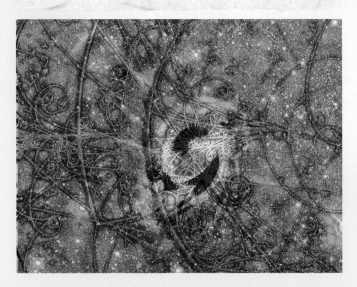

KEY TOPICS

Birth of the Quantum

AT THE end of the nineteenth century, physicists had a problem – the Ultraviolet Catastrophe. When they were measuring electromagnetic radiation using spectrographs, they found a mismatch between how objects called black bodies were actually emitting radiation and how the prevailing theory said they should be emitting it. In response to this problem, Max Planck came up with an idea: what if radiation did not flow continuously, but moved as a collection of lumps of energy he called quanta? This notion was so radical that even Planck himself did not wholly believe it for years. Quantum theory implies that matter gives up (and absorbs) energy in quantum jumps, rather like a ball loses potential energy rolling down a flight of stairs instead of down a slope.

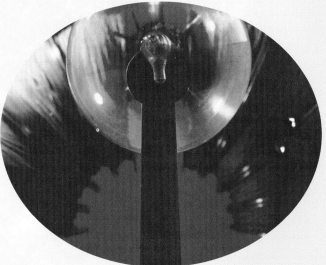

Light as Particles

THE WORK of Max Planck and his quanta was just the beginning. It was soon followed by experiments confirming that light (and all other types of electromagnetic radiation) could behave as separate particles rather than as a continuous wave. In 1905 Einstein successfully explained Heinrich Hertz's photoelectric effect in terms of particles of light. This was later confirmed by Arthur Compton, whose experiments with X-rays envisaged electrons and photons colliding and scattering like billiard balls.

Inside the Atom

THE IDEA of the atom dates back to the ancient Greeks, but our modern understanding of the atom and what is inside it was achieved through a succession of experiments and theories developed throughout the twentieth century. J. J. Thomson proposed a 'plum-pudding' model of the atom containing evenly distributed electrons. This was rejected in favour of the more commonly known visualization proposed by Niels Bohr – the 'planetary' model – in which electrons orbit around a nucleus. This model successfully explained the stability of the electron orbits but even this has now been superseded by more advanced theories. The nucleus – about 10,000 times smaller than the atom – is surrounded by so-called clouds of probability that help us to predict the location of the tiny electrons.

Uncertainty and Probability

ONE OF the weirdest concepts of quantum mechanics is the Uncertainty Principle, proposed by Werner Heisenberg. According to this principle, you can measure exactly where a particle is *or* you can exactly measure its momentum – but never both at the same time. The more accurately you know one measurement, the less you know about the other. In fact, scientists accept that they can only predict the probability of a particle having certain properties and dynamics. To do this, they use Irwin Schrödinger's equation, an important tool for predicting the way subatomic particles behave.

The Standard Model

BY THE end of the twentieth century, a host of new and exotic particle types had been discovered and classified. The matter particles called fermions are grouped into quarks and leptons. The four fundamental forces of nature – gravitational, electromagnetic, strong nuclear and weak nuclear – also have their own associated particles. These are called bosons. Physicists want to understand how all of these interact, and currently the Standard Model is the most comprehensive theory of matter on the quantum scale. It shows that three of the four forces of nature – electro-magnetic, strong nuclear and weak nuclear – can be understood in terms of quantum mechanics. The search is on to find a theory that will unify all the forces, including the fourth one: gravity.

Superstring Theory

THE FRONT runner for a 'theory of everything' that includes gravity is called superstring theory. It suggests that elementary particles are not point-like but are actually like tiny vibrating strings. The theory also calls for the Universe to have not four dimensions (including time) but 10. The reason we do not see these dimensions is that they are all curled up so infinitesimally small that we cannot detect them. The superstring theory is highly mathematical and is not so far supported by experimental evidence, but it is the best theory we have – for now.

CONTENTS

KEY PERSONALITIES

KEY TOPICS

Hubble's Expanding Universe

EDWIN Hubble made his startling discovery about the Universe in 1929. His work was based on the discoveries of another astronomer, Vesto Slipher, who ascertained that nearby stars, nebulae and galaxies were moving in relation to Earth. Some of them were moving towards us but most of them appeared to be receding. Hubble proposed what has since become known as Hubble's Law, which states that all distant galaxies are moving away from us, and the farther away they are, the faster they are moving. The implication was incredible and profound: the Universe is expanding, and at some point in the past it must have been squeezed into a very small region.

How Big is the Cosmos?

PERHAPS humans have always looked up at the night sky and mused about the age and size of the cosmos. For millennia these questions were embedded in myths and belief systems, but by the eighteenth century they were firmly established within the domain of science. When Messier published his catalogue of nebulae in 1781, many questions remained unanswered. What were nebulae? How far off were they? It was not until the early twentieth century that Hubble showed that one of the nebulae on Messier's list, M31 in Andromeda, was beyond the Milky Way. He went on to propose that distant galaxies were receding at incredible speeds, implying that the Universe could be infinite in extent. The Hubble Space Telescope has since revealed the beauty of galaxies much farther away than any previously studied.

The Big Bang

ALL THE evidence points to the fact that galaxies are racing apart from each other in an expanding Universe. Based on this observation, cosmologists now believe that the Universe was formed from a single cataclysmic event which has become known as the Big Bang. From this event emerged not only all the matter in the cosmos, but time and space as well. The study of high-energy collisions of particles within particle accelerators (pictured) helps physicists understand what occurred just after the Big Bang. The theory of the Big Bang is supported by the answer to a question that has puzzled scientists through the ages: Why is the night sky dark? The answer to what is called Olbers' paradox is explained by the fact that the Universe has a finite age.

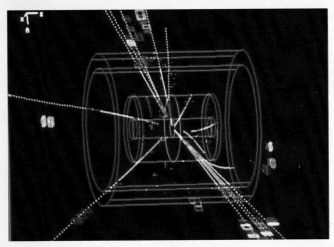

Cosmic Background Radiation

THE THEORY of the Big Bang suggests that all the matter in the Universe expanded from a single minuscule point around 13 billion years ago. The clinching evidence that the Universe started out this way was discovered by Penzias and Wilson in the 1960s. They detected a faint microwave glow covering the whole sky. This was subsequently recognized to be the very cool remnant of the intense hot radiation of the Big Bang. The glow seemed to be evenly diffused throughout the Universe, but this conflicted with theories of galaxy formation. Subsequently, the COBE satellite survey revealed that the density of the radiation was not smooth but rippled. This showed that regions within the primordial gas could have condensed to form galaxies.

A Cold and Dark Future?

IT IS currently believed that there is not enough matter in the Universe to prevent it from expanding forever. This implies that the Universe will have no end. But what will happen to everything in it? Since heat flows from hot to cold, everything in the cosmos will eventually reach the same temperature; the Universe will suffer 'heat death'. Stars will burn themselves out, ending as white dwarfs, neutron stars or black holes depending on their masses. Galaxies will collapse and become massive black holes. Black holes will lose mass and disappear in bursts of gamma rays. Even the protons that make up matter might decay at some point in the distant future.

Fundamental Questions

MANY deep cosmological questions remain unanswered. Where is the antimatter? The creation process of matter should theoretically produce equal amounts of antimatter. But evidence suggests that the Universe is predominantly composed of matter, presenting a theoretical puzzle yet to be solved. How did the Big Bang take place? A fuller explanation of the earliest time, just after the Big Bang – the so-called Planck era – awaits a quantum theory including gravity. What will be the ultimate fate of the Universe? The answer to this question requires greater knowledge of cosmological geometry and dark matter. The quest to find out continues. (Pictured: a galaxy containing a Type Ia supernova, used as a reference when measuring vast distances.)

CONTENTS

KEY PERSONALITIES

KEY TOPICS

Into the Cosmos

THERE ARE billions of galaxies out there and we happen to live within the outer suburbs of one of them – the Milky Way. Our planet, Earth, is illuminated by the Sun, itself an average star in middle age among countless billions of similar stars. Stars range from enormous red giants to extremely dense white dwarfs. Evidence now suggests that many stars have planets, although none has yet been found to support life. Some of the more exotic objects in the cosmic zoo include black holes and neutron stars, left over after the death of massive stars. It is likely that more incredible phenomena are still to be discovered, and yet we may never be able to catalogue fully the contents of the cosmos.

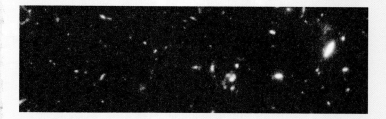

Lives of Stars

STARS ARE self-luminous balls of gas that start life as cold dense clouds. Throughout the cosmos, they are being born continuously in stellar nurseries such as the spectacular Eagle Nebula. They are usually formed in tightly knit clusters and many remain linked by gravity to make double or multiple systems. Stars appear different colours depending on their surface temperature and age. Young, hot stars are blue; as they age, they become cooler and redder. Much can be learned about a star by plotting its brightness and colour on a graph called the Hertzsprung–Russell diagram. This shows that most stars are found within a band called the main sequence.

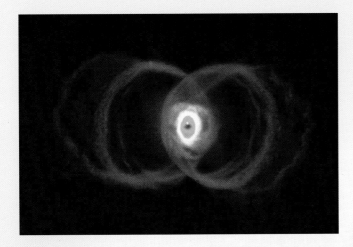

Star Death

STARS ARE powered by nuclear reactions, converting hydrogen into helium. When a star's nuclear furnace uses up its last stock of hydrogen, this signals the beginning of the end. The mass of the star determines what happens to it next. Stars like the Sun eventually swell into red giants and then turn into cooling white dwarfs. The more massive a star is, the quicker it burns out. It can collapse inwards, causing an explosion that results in a supernova. This then flings off its outer layers, leaving an incredibly dense, city-sized neutron star. The death of an even more massive star can result in the formation of a black hole, the strangest and most dense object in the cosmos.

Black Holes

THEY ARE perhaps the most amazing phenomena to be found in the Universe. You cannot see them through a telescope and yet they swallow entire stars. They are called black holes because light cannot escape them – but why is this? The answer, according to Einstein's theory of general relativity, is that the incredible density of a black hole causes space–time to curve in on itself. Around a black hole is a sphere called the event horizon and nothing beyond this point can reach the outside world.

It is widely believed that gargantuan black holes lie at the heart of many galaxies and that they provide the powerful radiation sources for active galactic nuclei.

The Milky Way

ON A clear moonless night it is possible to see a faint band of light spanning the sky from horizon to horizon. This is our Galaxy, the Milky Way. It comprises many billions of stars arranged in a flattened spiral disk. At its heart is a great bulge of stars lying about 25,000 light years away in the constellation Sagittarius. Surrounding the disk is a retinue of old stars, many grouped into globular clusters. Studies of these clusters have revealed the true size of our Galaxy and the Sun's position within it. Farther out is a massive halo of invisible dark matter.

Galaxies

DISTRIBUTED throughout the cosmos, galaxies are like islands in a vast empty ocean. They are collections of many billions of stars bound together by gravity and usually incorporate dust, gas and dark matter. They are immense and highly luminous. Yet even the brightest of them will only appear as a faint smudge through a small telescope because they are so distant. Early studies of galaxies focused on classifying them according to their shape, using Edwin Hubble's tuning-fork diagram. Although much more has been learned in the last half century about how these impressive structures are formed, an explanation of the mysterious dark matter around them is eagerly awaited.

CONTENTS

KEY PERSONALITIES

KEY TOPICS

The Sun

SITUATED in a spiral arm around two-thirds of the way out from the centre of the Milky Way is a medium-sized star. We call it the Sun. There are billions like it in the observable Universe, yet it is special to humans because we are so close to it. The Sun condensed from a large cloud of gas and dust about 4.6 billion years ago and formed a hot, dense core in which nuclear reactions began. Now about halfway through its life, the Sun provides a stable source of energy that sustains life on our planet.

Terrestrial Planets

THE FOUR planets that orbit closest to the Sun – Mercury, Venus, Earth and Mars – are the terrestrial planets, also known as the inner planets. They are all comparatively small, rocky worlds with an iron core, a partially molten mantle and a solid crust. Although they share these physical features, having evolved in a similar area near the Sun, we have learned a lot about how different the terrestrial planets are through robotic and ground-based investigations. Volcanoes and space debris impacts have produced diverse surface features and the planets' atmospheres are markedly different, varying from Mercury's negligible traces of gases to the dense carbon dioxide of Venus.

Our Planet: Earth

EARTH IS unique because it is the only planet on which we know life exists. The presence of liquid water over two-thirds of its surface and a habitable atmosphere are two key factors which make life possible. Earth is the largest of the terrestrial planets and lies third from the Sun after Mercury and Venus. It was formed about 4.6 billion years ago when particles from the gas and dust cloud that became the Sun clumped together. Earth has a dynamically changing surface, driven by active tectonic forces that cause volcanoes and earthquakes.

The Moon

THE MOON is about one-quarter of Earth's diameter and is our planet's only natural satellite. With surface features visible to the naked eye, it has been an object of wonder for millennia. The Moon is thought to have been formed from the impact on Earth of an object the size of Mars. Its closeness and relative bulk give it a powerful gravitational influence over the tides of Earth's oceans. The Moon appears as large as the Sun in the sky and can cause total eclipses by passing directly between the Sun and Earth.

Outer Planets

THE FOUR principal outer planets – Jupiter, Saturn, Uranus and Neptune – are also known as gas giants because they are much larger than the terrestrial planets and are composed mainly of hydrogen and helium. They lie in the outer Solar System beyond the asteroid belt. These four planets all have extensive and active atmospheres, with enormous storms that can last for years. Each gas giant has its own system of satellites and rings of ice and rock. Pluto, the ninth planet, lies so far beyond the gas giants in the outer Solar System that it reveals only vague surface markings to even our most sophisticated telescopes.

Asteroids, Comets and Meteorites

THE SOLAR SYSTEM contains a great assortment of smaller objects. Countless asteroids exist in the asteroid belt. They are typically irregularly shaped, tens of kilometres in diameter and are composed of rock and metal. Comets, made of frozen gases and dust, swarm in their billions in the Oort Cloud and Edgeworth–Kuiper belt. Some have orbits which take them near the Sun. As they approach the Sun, they heat up and develop tails of vaporized ice. Most of the debris entering Earth's atmosphere burns up, but larger fragments called meteorites can reach its surface. They can cause craters and, when found and studied, provide valuable information about the early Solar System.

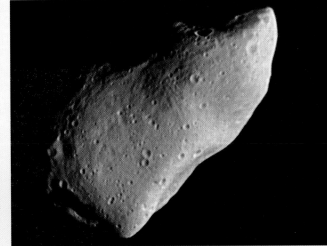

CONTENTS

KEY PERSONALITIES

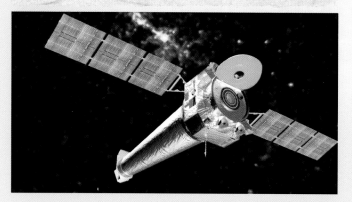

KEY TOPICS

The World of Astronomy

THE NIGHT sky is our window on the Universe. It contains innumerable diverse and wondrous objects: from gargantuan galaxies to dense black holes; from lonely, lifeless white dwarfs to stellar nurseries bursting with new stars. Professional astronomers now routinely study various types of radiation other than light and advances in technology such as the Hubble Space Telescope have extended the power of optical astronomy. But astronomy is not just for professionals – it is open to anyone. On a dark night away from city lights you, too, could turn your eyes to the sky and begin an enthralling journey through some of Nature's finest sights.

Observing Across the Spectrum

FOR CENTURIES humans have observed the light from celestial objects, but during the twentieth century, professional astronomers began to study radiation at many other parts of the electromagnetic spectrum. These range from long-wave, low-energy infrared and radio waves, to high-energy, short-wavelength X-rays and gamma rays. Observing objects in these different parts of the spectrum reveals information hidden to the optical telescope. Astronomical knowledge has increased dramatically with data from specialized equipment such as the Chandra X-ray Observatory, the Compton Gamma Ray Observatory and the 305-m (1,000-ft) radio dish at Arecibo in Puerto Rico.

Eyeing Up Equipment

YOU DO not need highly sophisticated equipment to observe much of what the night sky has to offer. In fact, many fascinating things can be seen with just the naked eye. An easily portable improvement is offered by binoculars. But to really start to explore the celestial wonders, you need a telescope. They come in two types: a refractor uses a lens to focus light and a reflector catches light with a concave mirror. The careful selection of mounts and eyepieces also contributes to successful astronomy. Computerized telescopes and charge-coupled devices (CCDs) are two of the new technologies to be found in the armoury of amateur astronomers.

Computers in Astronomy

AMATEUR astronomy is undergoing a revolution as more people use their home computers to pursue their interest. Thanks to the Internet, astronomy has now become a global village. Teachers and students can access remote telescopes such as the one on Mount Wilson (pictured) to conduct their own projects. Detailed star maps can be downloaded. Up-to-the-minute coverage of meteor showers and solar and lunar eclipses are relayed on the web. Updates on the Sun can be obtained by viewing live images transmitted from the Solar and Heliospheric Observatory (SOHO). Image-processing software makes it easy to enhance digital images from CCD cameras. And millions of computers are collaborating in the global SETI@home project that is analyzing signals from space in the search for extraterrestrial life.

Celestial Delights

THERE ARE five planets which can be seen easily with the naked eye at various times: Mercury, Venus, Mars, Jupiter and Saturn. You will need binoculars, at least, to see Uranus and Neptune. Pluto is barely perceptible even with a powerful telescope. Meteors can be glimpsed every night as dust burns up in the atmosphere. With a telescope you can often view comets, although they can be faint. Stars vary considerably in brightness and colour depending on their size, temperature and distance from Earth. Double stars can often simply appear close together from our line of sight, others are gravitationally bound together. Other objects of fascination are nebulae, clusters and the faint belt across the sky called the Milky Way.

Finding Your Way

AS AN aid to understanding where stars lie and how they move, think of stars as being on the inside of a sphere that rotates around Earth once every day. Although we no longer believe this to be literally true, the celestial sphere remains a useful concept for describing the sky. It allows us to plot the position of stars in a way comparable to longitude and latitude. The sky is divided into 88 constellation sections, many named by the ancient Greeks, whose boundaries are now laid down by the International Astronomical Union (IAU). The 20 Star Charts provided are the ideal starting-point for finding your way around the night sky.

CONTENTS

KEY PERSONALITIES

KEY TOPICS

Rockets

A GREEK engineer called Hero started it all with his reaction engine in AD 62. In the centuries that followed, gunpowder-fuelled rockets were being tried out on battlefields in Asia and Europe. The science of rocketry did not really take off until Robert Goddard launched his series of liquid-fuelled projectiles in the 1920s. Another pioneer was Wernher von Braun, whose V-2 rockets demonstrated their terrible destructive power in World War II. By the 1950s, rocket technology was well developed in the military sphere and its application for other purposes in space was spurred on by the Space Race.

The Space Race

THE GAME plan for the space race between Russia and USA began at the end of World War II. The advancing Russian army reached the labs of the German von Braun only to find he was already on his way to America. The Russians did, however, have their own rocket genius. He was Sergei Korolev and in 1957 he put Sputnik 1 into orbit, marking the beginning of the Space Age. Severely jolted, the Americans put their Explorer 1 up in 1958. After the Russians put a dog, Laika, into orbit, both Superpowers sent a succession of dogs and chimps aloft. The Russians won round three of the race by putting the first astronaut, Yuri Gagarin, briefly into orbit in 1961.

Satellites and Probes

BEFORE it was safe for humans to explore the Moon, much more needed to be understood: how to orbit it, how to land on it and what its surface was like. To answer these questions, the Americans and the Russians both launched numerous lunar probes with notable successes and spectacular failures. In the 1960s scientists turned their attention to Mars and Venus. In 1975 a Russian probe from the Venera series took the first pictures from the surface of another planet, Venus, and the follow-ing year two American probes called Viking landed on Mars to look for life. Other American probes have provided invaluable knowledge about the worlds in our Solar System out as far as Neptune.

Human Spaceflight

EVER since the Chinese pioneer Wan-Hu blew himself up trying to ride a rig of 47 gunpowder-driven rockets in 1500, humans have wondered whether it was possible to fly through space. We had to wait until the 1960s for the dream to become a reality, albeit only for a lucky elite of highly trained and dedicated specialists. There has been a string of great achievements. Yuri Gagarin became the first human to reach space in 1961. John Glenn was the first American to orbit Earth in 1962. Valentina Tereshkova became the world's first woman in space in 1963. The tradition continued with the crewing of the International Space Station (ISS) in 2000.

Missions to the Moon

AFTER the disaster of the American Apollo 1 fire in 1967, US public support for lunar missions plummeted and the Moon seemed farther away than ever. NASA was spurred on by the lunar orbiting achievements of the Russian Zond spacecraft and, in 1968, Apollo 8 took the first humans to orbit the Moon. After two more Apollo test flights, Apollo 11 landed safely on the Sea of Tranquility in 1969, allowing Neil Armstrong to step down a ladder and make his historic announcement: 'That's one small step for a man, one giant leap for mankind.'

Are We Alone?

UNABLE to accept that we might be the only intelligent life form in the Universe, we have sent out probes to other planets and built giant dishes to scan the skies. Life on Earth developed because of the presence of surface liquid water, and the search for life elsewhere has focused on this substance. It has been theorized that life on Earth might have begun with very tiny organisms and studies of Mars hold out the possibility of finding subsurface liquid water and perhaps the evidence of such microscopic life. Despite some false hopes and a lot of wishful thinking, we have yet to find any traces of life beyond our home planet. But the search goes on...

FOREWORD

We live at a special historical era — an era when the broad outlines of our cosmos are coming into focus. Just as the early mariners discovered the layout of the continents, so astronomers now have an outline map of our Universe.

The night sky is far more fascinating than the ancients could ever have envisioned. The Universe extends millions of times farther than the remotest stars we can see; it is the outcome of more than 10 billion years of evolution. Biologists trace life's history on Earth back to primitive organisms that lived more than three billion years ago. But we can now trace the Earth's history back to the birth of our Solar System in a spinning dusty disk, back to the formation of our Galaxy, back even to the first few seconds of a Big Bang.

Stars are not just 'points of light': they are 'fusion reactors' kept shining by nuclear energy; many have retinues of orbiting planets. Some of these planets will surely resemble our Earth in size and temperature. But will any of them harbour life — even intelligent? That is perhaps the number one problem in the whole of science.

A real understanding of our physical Universe — why it is expanding, why it contains atoms, and why it is governed by specific natural laws — still awaits a breakthrough. Theorists still seek new insights that will unify the two great ideas of the twentieth century: the quantum principle (governing the 'inner space' of atoms) and Einstein's relativity, which describes gravity, space and time.

We are poised between cosmos and microworld. It would take about as many human bodies to make up the Sun's mass as there are atoms in each of us. The chemistry of life depends on the properties of atoms — how they stick together to make the complex molecules in all living tissues. But to understand ourselves we must also understand the stars. All the oxygen and carbon atoms on Earth (and the rest of the periodic table) are nuclear waste from stars that exploded billions of years ago, before our Solar System formed.

Whenever I discuss my work with non-specialists, the topics most likely to strike a chord are the most basic and fundamental ones. What happened before the Big Bang? Is there life elsewhere? What causes gravity and mass? Is the Universe infinite? It's salutary to realize that, despite the amazing advances of recent years, none of these questions can be answered.

Probing the Universe is a world-wide enterprise, involving disparate technologies and imaginative ideas. This book describes, with stunning clarity and visual impact, the amazing recent advances in understanding particles, forces and the cosmos; but as the frontiers advance, their periphery gets longer. Exhilarating prospects lie ahead: we are still at the tentative beginnings of our cosmic conquest.

Sir Martin Rees
ASTRONOMER ROYAL

INTRODUCTION

This book will take you on a journey of exploration from the interior of atoms to the edge of the observable Universe. Along the way we will investigate stars and planets, and the clouds of gas and dust which spawn them; the galaxies of which they are a part; and, finally, the way in which astronomers currently believe the Universe and its contents came into being.

The story of astronomy is in many ways the story of human intellectual advance. Each step in our understanding of the Universe has widened humanity's horizons until, now, we can gaze on the relics of the very event in which the Universe appears to have been created – the Big Bang. In a mere four centuries, a blink of an eye in comparison to our evolutionary history, humans have progressed from assuming that the Earth was the centre of the Universe to recognizing that it is only a modest planet orbiting an average star, the Sun; that the Sun is just one member of a vast congregation known as the Galaxy; that there are countless other galaxies dotted throughout the Universe; and that the space between the galaxies is itself expanding, as though impelled by some enormous explosion over 10 billion years ago. And as astronomers have pushed back the boundaries of the known Universe, so the attitudes and expectations of humanity have grown accordingly. It is often asked why we should spend money on astronomy and space exploration or, indeed, any aspect of scientific research which does not promise an immediate practical benefit. Equally, one might ask what the benefits are of art galleries, museums or concert halls. The answer is, of course, that they are all cultural activities, designed to enrich our experience and help us learn more about ourselves and the world in which we live. Pursuing such activities is what makes us human.

Astronomy and cosmology, in particular, dwell on the greatest philosophical questions of all: How did the Universe begin? Where did we come from? Is there anyone else out there? Our generation is the first to be capable of answering these questions with any certainty, although our answers must always be regarded as open to revision in the light of future findings.

While telescopes have become ever-larger and more sensitive, new designs and manufacturing techniques have kept down their costs, and the use of electronic detectors in place of conventional photography has made them ever-more productive. With the introduction firstly of radio telescopes 50 years ago and now telescopes orbiting above the atmosphere to detect wavelengths that do not penetrate to the surface of the Earth, it has become possible to observe the Universe across the entire electromagnetic spectrum, from the very longest wavelengths (radio) to the shortest (X-rays and gamma rays). The resulting information on celestial objects and the processes at work within them is far more complete than was ever possible when observing in visible light alone.

Closer to home, newspaper headlines warn of issues that are of direct concern. There is much talk of global warming of the Earth, but little understanding of how much of it is due to slight changes in the energy output from the Sun. Until we fully understand the climatic variations

caused by natural effects we cannot properly assess our own impact on our planet. Another area of increasing concern is the threat from comets and asteroids. Its is now widely accepted that a mass extinction of living species, most notably the dinosaurs, was triggered 65 million years ago when a large asteroid hit Earth. Smaller impacts, capable of causing localized destruction such as of a city, happen more frequently. Only once we have fully surveyed our interplanetary environment can we assess the full risk we run from such wanderers, and decide how to deal with any of them that may be found to be on a collision course. Asteroids should also be regarded as a potential benefit as well as a risk – many of them may contain raw materials that can be extracted for use on Earth, in place of mining our own planet.

Everyone should have some understanding of the nature of the Universe in which we live. That may seem increasingly difficult as discoveries accumulate at an ever-increasing rate, but in many cases the acquisition of new knowledge helps clarify our picture of the Universe. Quasars, for example, were regarded as inexplicable when discovered in the 1960s. Now they are recognized as being the extremely bright centres of distant galaxies, the most extreme example of a range of activity occurring in the nuclei of galaxies. Pulsars, rapidly flashing radio sources, were another startling finding of the 1960s, but are now known to be one of the end products in the lives of certain stars. Similarly black holes, once the province of theorists and science-fiction writers, are now accepted as real objects, left behind by the deaths of the most massive stars. Enormous black holes are now thought to be the energy sources that heat up gas at the centres of quasars and other active galaxies. Most exciting of all, astronomers are now discovering planetary systems around numerous other stars.

On the other hand, as old mysteries are solved, new ones arise: Why do other planetary systems look different from our own? What is the origin of intense bursts of gamma rays from far off in space detected by orbiting satellites? What is the nature of the invisible dark matter that seems to make up the bulk of the mass of the Universe? Is the expansion of the Universe speeding up as the result of a mysterious cosmological force?

The best that a book such as this can hope to offer is a review of astronomical knowledge, and the physics that underpins it, as it stands at the start of a new century of discovery, certain in the knowledge that, before long, many of our ideas will seem incomplete or in some cases thoroughly naïve. What we have learned about our cosmic environment in the past few centuries, astounding though it is, amounts only to the first steps in a voyage of exploration that will continue as long as the human species remains in existence.

IAN RIDPATH

THE HISTORY OF ASTRONOMY

The development of astronomy over the past five millennia has been one of the greatest achievements of human civilization, for it has led humanity to wrestle with problems that rose above the bare necessities of daily survival. While it is true that some branches of astronomy have always been cultivated because of their practical usefulness – such as the recording of solar and lunar cycles for the development of reliable calendars, or the finding of the longitude at sea in the eighteenth century – the great breakthroughs in astronomy were essentially intellectual triumphs. They were so because they linked astronomy to other aspects of culture, such as religion, philosophy and education. The pursuit of new ideas in all these realms necessitated the cultivation of that creative thinking which is essential to the growth of complex civilizations. The first naming of the constellations by Sumerian shepherds some time before 3000 BC, the Greek theory of epicycles to explain the retrograde motions of the planets, and the emergence of Victorian astrophysics never put bread on the table, yet they were clear indications of how astronomy captivated the human mind when it came to finding answers to profound questions. This is still true today.

In this chapter various aspects of the history of astronomy will be examined across a 5,000-year time span. It will look at why particular astronomical concerns became important when they did, how theory and practice developed a dynamic, creative interrelationship, and how astronomy became a model for the other sciences.

The History of Astronomy
ANCIENT COSMOLOGIES

All cosmologies reflect the cultures that devise them, for they are solutions to questions asked by a particular people about the natural world. Because the near-eastern cultures of 2500 BC had only a limited geographical knowledge, and lived in lands dominated by approximately north- and south-flowing rivers, such as the Nile, Tigris and Euphrates, their cosmologies centred upon the particular importance of their own social

group, and a sky in which objects seemed to rise and set directly above their great waterways. The sky also provided the stage upon which a variety of deities controlled natural forces, and with which human beings had to come to terms. In all ancient cultures wind, water, light and disease were personified, and the greatest social fear was that their natural order would break down and chaos, itself a personified force of evil, would engulf the world. Religious ritual was invariably concerned with placating the capricious deities who could either sustain or destroy. In this respect, therefore, astronomy merged with natural history, medicine and public religion to form an interconnected body of knowledge.

▼ *A Chinese silk painting* showing the traditional dragon with the Moon. The animals depicted here are inhabitants of the Moon and the lady pictured riding the dragon was believed to have a palace on its silvery surface.

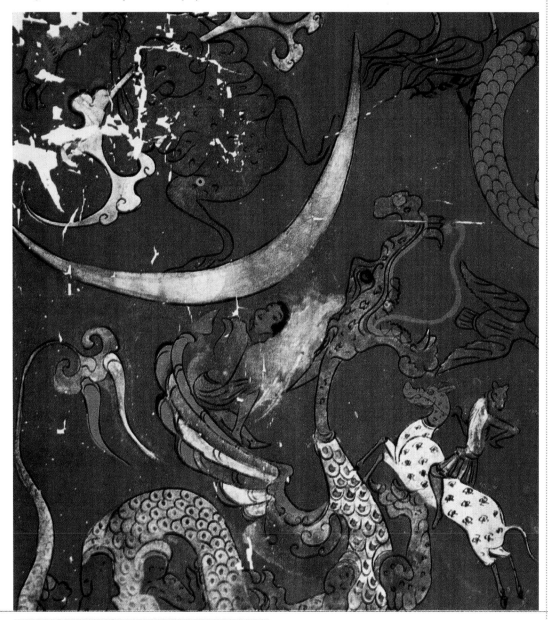

THE FOUR CORNERS OF THE WORLD

All the ancient near-eastern cultures saw the cosmos as being built in tiers, or levels. There was the Earth's surface, on which people lived, and above it a flat sky supported in some remote place by pillars at the 'four corners of the Earth'. Above the sky were the 'waters above the firmament' which, according to *Genesis*, broke forth to cause the Great Flood of Noah. Then there were the 'waters beneath the Earth', as well as the realms of the dead. This cosmology was shared, in its basic principles, by the ancient Egyptians, Babylonians, Sumerians and Jews, and its imagery can be found in the Old Testament. Astronomical bodies were seen as passing beneath the flat Earth at night. The Egyptians also saw the Sun God Ra entering the mouth of Nut, the Sky Goddess, who arched herself over the Earth as the Milky Way around the time of the spring **equinox**, and was reborn nine months later at the winter **solstice**. Only a broadening of geographical knowledge could undermine this cosmology, as observers in widely separated locations realized that phenomena which they all observed, such as **eclipses**, implied that both the Earth and the sky were curved.

⧖ THE 360° CIRCLE AND ITS POSSIBILITIES

Accurately determining the length of the year from early principles is not easy. The midday Sun stands at around the same **altitude** for several days in succession at the winter and summer solstices, making the exact solstice difficult to establish with simple shadow instruments. It was probably because of this that most early cultures worked on the basis of a 360-day year, and then came to add extra days after the error became obvious over a series of years. Of all the near-eastern peoples, however, it was the Assyrians and Babylonians, occupying the regions of modern Iran and Iraq, who made the first great innovations in astronomy and geometry. Driven by an official culture which demanded the observation and calculation of astronomical cycles for divinatory purposes, the Babylonians developed the 360° circle (each degree corresponding to a solar day) and exploited its perfect division into 60s. They timed observations with water clocks and discovered the 19-year cycle that underlies eclipse prediction. Babylonian astronomy and mathematics reached a peak around 500 BC, by which time it had developed a sophisticated theoretical foundation from which to analyze its long runs of observational data.

ANCIENT COSMOLOGIES

▲ *The ancient Egyptians associated their pantheon of gods with their views on the Universe. The Sky Goddess Nut (pictured) would arch her body over the Earth in the shape of what we now know to be the view of our own Galaxy, the Milky Way.*

THE BABYLONIANS

In ancient times, no one looked at the sky for what we would now call scientific purposes; they saw it instead as a tool for prophecy. The basic **constellation** figures probably date from before 3000 BC, and Mesopotamian tablets from 1100 BC depicting the Sun, Moon and **zodiac** figures still exist. While war, famine and disease made human life precarious, the heavens possessed a regularity which seemed to influence the Earth, through the solar seasons or lunar phases. While the Egyptians associated the pre-dawn, or heliacal, rising of Sirius with the flooding of the Nile, it was the Babylonians who first made regular records of celestial phenomena. Their dated observations of eclipses, and separations of Venus and other planets from the Sun, recorded on clay tablets, have proved invaluable to modern astronomers. Indeed, the Babylonians of around 1000 BC were the founders of systematic astronomy. They exploited the mistaken belief that there

THE ASTRONOMY OF ANCIENT BUILDINGS

Some of the oldest architectural structures in the world contain astronomical principles. Though good surveys of the Great Pyramid of Giza (built *c.* 2700 BC), revealing its accurate astronomical alignment, date from AD 1638, and those of Stonehenge from shortly after, much of our knowledge of Neolithic observatories really dates from after 1850. It is indeed remarkable that many Eurasian buildings erected before 1000 BC, and many Meso-American ones after about AD 500, were built on precisely laid-down axes. In addition to its north–south alignment, Egypt's Great Pyramid also contains finely cut shafts

were 360 days in the year to develop geometry, while their astronomical records, built up for the purposes of political and religious astrology, resulted in the basic concepts of scientific astronomy: observational data, mathematical analysis and verifiable predictions. Babylonian astronomy later influenced the Greeks.

HINDU ASTRONOMY

While Indian culture itself dates from earlier than 2000 BC, the basic elements of its astronomy are similar to those of Egypt and Mesopotamia. Before *c.* 1000 BC, the Hindus worked within a 360-day year, which presented problems in reconciling annual solar calendars with convenient Moon-phase cycles. The great developments in Hindu astronomy, however, took place at the same time as those of the Babylonians, and it is quite likely that Babylonian techniques of **intercalation** (the insertion of a period in the calendar to harmonize it with the solar year) and planetary table-making were conveyed to India by trade routes down the Tigris and Euphrates, and through the Persian Gulf. Following the development of Greek planetary theory, Indian astronomers began to make use of **epicycles** (circles whose centres move around the circumference of greater circles) in solving problems involving variable planetary speeds. The sophistication of Hindu astronomy by 300 BC is evidenced by the quality of their lunar tables and parallel developments in mathematics, which included the form of notation subsequently called 'Arabic' numbers. By this time, the modified Egyptian 365-day year had come into use, along with a properly intercalated calendrical cycle. By the Hellenistic, Roman and early Christian periods, trade began to flourish between the Mediterranean and India, and texts of Greek and Babylonian authorship began to be translated into Sanskrit.

which point to spiritually significant star positions. Such axes could, however, have been relatively easy to lay down from lines of noon shadows, observed over several years. The oldest Neolithic solar alignments in Europe are probably those of Stonehenge in the UK, and Newgrange in Ireland, both *c.* 2600 BC. The purpose behind these alignments, however, is impossible to know. Nor do we know how many alignments discovered by computer analyses from modern surveys were ever intended by the original builders, and improbable theories about these ancient observatories and their builders abound.

▲ *Neolithic structures such as this, Stonehenge in the UK, have posed many questions about ancient views of astronomy. The monoliths here are arranged along an axis that is aligned on the point of sunrise at the summer solstice.*

CHINA

China's astronomical culture also dates back to at least 2000 BC. It was the dominant astronomical culture of Japan and Korea. Chinese astronomy grew from its own roots, and while wrestling with many of the same problems as Babylonian astronomers – such as reconciling the lunar and solar calendars – developed independently from the cultures of India and the near East. The geographical barriers of the Himalayas and the Asian steppe probably facilitated this cultural separation. In China, astronomy and politics were intimately connected, as it was an emperor's duty to organize the heavens as well as the Earth. This meant that from a very early date, official sky watchers were employed to record the presence of **comets** or **supernovae**, while the ability to predict eclipses was of paramount importance to political stability. It was Chinese astronomers who recorded a 'guest star' in AD 1054, which we now know to have been the explosion of the Crab supernova. Perhaps because they accurately determined the year to be 365¼ days, and divided the **celestial equator** into 365¼°, the Chinese never developed that flexible coordinate geometry only possible with a 360° circle.

HISTORY OF ASTRONOMY

The History of Astronomy
GREEK ASTRONOMY

In the hands of the ancient Greeks, astronomy as a rational science took great leaps. The philosophical musings of men such as Thales and Pythagoras spawned a new creative thinking that developed into exact mathematical investigation. A determination to understand the Universe, and Earth's place within it, resulted in many milestone achievements: Aristotle's theories of the Four Elements; Hipparchus's discovery of the precession of the equinoxes; and the surprisingly accurate measurement of the Earth's circumference by Eratosthenes. The greatest figure, though, was Claudius Ptolemy, whose theories of a geocentric Universe provided a basis for belief that went unchallenged for 1,400 years.

THE ROOTS OF INTELLECTUAL FREEDOM

The flourishing of astronomy in Greece, after *c.* 500 BC, was an integral part of a much wider flowering of intellectual culture in the Greek world. Exactly what triggered this new wave of creativity – which laid down the foundations of modern western culture – has long been a subject of debate amongst scholars. For while the Greeks absorbed the 360° circle, some branches of mathematics, and systematic astronomical record-keeping from the Babylonians, it cannot be denied that the Greeks developed them into something vastly more cogent and influential. Lying at the heart of all Greek culture, be it medicine, architecture, politics or astronomy, was the assumption that the human and natural worlds were governed by rational principles. This would enable an investigator to unlock one puzzle after another, in an ordered sequence, to discover what no one had known previously.

While the Greeks believed in a pantheon of gods, by 400 BC philosophers in the tradition of Plato (427–347 BC) were speaking of one great, all-powerful creative force, or *logos* (in some ways similar to the Jewish Yahweh), whose reason underlay all ideas, including the human mind. And unlike the god-rulers of Egypt and Babylonia, Greek rulers were mortal men who could be over-thrown by free people.

ASTRONOMY AS A RATIONAL SCIENCE

The political and economic freedom enjoyed by the Greeks lay at the heart of the independent creativity from which their astronomy sprang. The leisured society of Greece produced the world's first professional academics, the Philosophers. Figures like Thales (*c.* 625–*c.* 547 BC) and Pythagoras (*c.* 580–*c.* 500 BC) now had the personal freedom to ask questions such as 'What is the substance out of which the Universe is made?' and 'Does the Universe have a central fire?'. Driven as they were by the concept of the rational structure of creation, there seemed no reason why solutions should not be found. The unification of these attitudes and intellectual freedoms with Babylonian mathematics caused a significant change of direction in human thought. In 585 BC Thales is said to have successfully predicted an **eclipse**, and some 40 years later Pythagoras began to explore the properties of circles and triangles, producing the theorem which still bears his name. By 480 BC, Parmenides and others were speaking of a spherical Earth, and realizing that the heavens were a sphere. And by 430 BC, the Athenian Meton, building upon the earlier work of Babylonian astronomers, had elucidated the 19-year eclipse cycle that is named after him.

CLAUDIUS PTOLEMY

The peak of Greek astronomical achievement is represented in the encyclopedic *Magna Syntaxis* ('Great Syntax') or, in Arabic, the *Almagest* of Ptolemy (fl. AD 127–141). Ptolemy lived, worked and observed in the Hellenized Egyptian city of Alexandria, and was culturally a Greek. His *Almagest* was clearly based on a knowledge of the works of numerous earlier Greek and Alexandrian astronomers, many of whose writings have since been lost. Using Hipparchus's earlier works as his basis, Ptolemy produced a definitive catalogue of 1,022 stars that fixed in cultural memory the basic 48 **constellations** which we still use. His greatest achievement, however, was the working out of a complex system of epicycles and perfect circles in which several centuries of recorded planetary observations

◀ *Ptolemy's summary* of Greek astronomical ideas was to dominate astronomical belief for 14 centuries, until Copernicus's idea of a heliocentric Universe gained credence.

▶ *The Ptolemaic System*
Ptolemy's celestial system had the Earth at its centre, with the Sun, Moon and planets moving around it in perfect circular orbits. To account for the observed movements of the planets, his cumbersome system specified that each planet also made a small circular motion (or epicycle) as it orbited the Earth, the centre of which – the deferent – described the path of the orbit.

were reconciled with the philosophy of uniform circular motions used in Greek geometry. Ptolemy's sometimes over-elaborate geometrical explanations nonetheless explained celestial phenomena as they were understood in AD 141, and firmly established a geocentric world view at the heart of astronomy for the next 1,400 years. Ptolemy's Earth-centred cosmology harmonized beautifully with the physics of Aristotle, provided quantities from which calendars could be calculated, and was used equally by Christian, Jewish, Arabic and Hindu astronomers.

SUN
SATURN
VENUS
MERCURY
MOON EARTH
MARS
JUPITER

PHILOSOPHY, MATHEMATICS AND ASTRONOMY

By 300 BC, two distinct intellectual traditions had emerged in Greek astronomy which in some ways prefigured the different approaches taken by modern scientists. The first – and oldest – tradition was philosophical and primarily concerned with the wider nature of the heavens. Anaximander of Miletus (611–547 BC), for instance, had tried to understand the fiery rotating nature of the planets, while Plato was to argue that the perfect heavens could only rotate in perfect circles at uniform speeds. The second tradition was mathematical, and concerned with finding geometrical representations

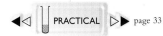

ERATOSTHENES MEASURES THE EARTH

By the time of Eratosthenes (*c.* 276–195 BC) no astronomer doubted the spherical nature of the Earth, but his measurement of its size was one of the triumphs of Greek geometry. Eratosthenes' technique hinged upon the geometry of com-

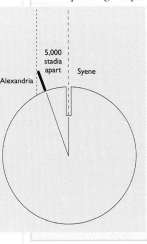

5,000 stadia apart
Alexandria | Syene

plementary angles within the circle, and as Librarian at the great Museum in Alexandria, Egypt, he had access to the most advanced scientific knowledge of the age. Eratosthenes knew that at midsummer, when the Sun touched the northern tropic, it shone right to the bottom of a deep well in the town of Syene, near the modern Aswan, and

hence was directly overhead. Yet in Alexandria, just over 7° to the north, the Sun cast shadows. Eratosthenes knew that as the rays of the Sun striking the Earth were parallel, the shadow angle cast in Alexandria must be the same as that produced by radial lines drawn from Syene and Alexandria to the centre of the Earth. As Syene and Alexandria were 5,000 Greek stadia apart, he deduced that they were one-fiftieth of a circle apart, so the circumference of the Earth must equal 250,000 stadia. Although there is some debate about the exact length of Eratosthenes' stadia, his figure produces an Earth circumference of about 47,000 km (29,000 miles). The modern value is 40,074 km (24,902 miles).

◄ *Eratosthenes' Experiment*
Eratosthenes found that at noon at midsummer, the Sun shone to the bottom of a deep well at Syene (Aswan). At that time in Alexandria, the Sun cast a shadow of just over 7°. Eratosthenes thereby calculated that Syene and Alexandria were one-fiftieth part of a circle apart. By measuring the distance on the ground, and multiplying it by 50, he obtained the circumference of the Earth: 250,000 stadia.

that could account for the **retrograde** loops and various motions of the planets. The most significant early contributor to this tradition was Eudoxus (407–355 BC), who tried to explain the planetary retrogrades by devising a purely geometric model which contained 27 spheres, representing the motion of the planets in relation to the **zodiac**, while rotating around the Earth. The path taken by a planet against what was then thought to be a fixed background of stars could be observed in this model as a figure-of-eight curve, which Eudoxus termed a *Hippopede* (Greek: 'Horsefetter'). Eudoxus' system was the ancestor of the **epicycle** and eccentric circles system later developed by Hipparchus (fl. 160–125 BC). Most Greek models of the cosmos assumed that the spheres rotated around the Earth, yet by 280 BC Aristarchus had devised a Sun-centred system, though it was irreconcilable with the physics of Aristotle.

▼ *A romanticized nineteenth-century picture showing Hipparchus using a large-aperture refracting telescope – an instrument which would not be invented for another 2,000 years.*

ARISTOTLE

Aristotle (384–322 BC) was the most influential and encyclopedic scientific writer of antiquity. Though a doctor by training, he wrote on everything from logic to cosmology. The comprehensiveness of his ideas ensured their survival for over two millennia. Through a mixture of precise observation of nature and careful deduction, Aristotle provided answers to many of the scientific problems of the day. Building on the ideas of Empedocles (d. *c.* 490 BC), Aristotle argued that all terrestrial change derived from the chaotic warring of the Four Elements. Earth was naturally heavy and stayed put at the centre of the Universe, while Water, Air and Fire raged about its surface. All astronomical bodies, however, were made of the same substance: the Quintessence, or

ANCIENT ASTRONOMICAL INSTRUMENTS

Astronomy, as opposed to theoretical geometry, needs practical observational data, and one of the many contributions of the Greeks to modern science was the concept of testing theory against observation. The key came from their exploitation of the Babylonian 360° circle, and Ptolemy's *Almagest* provides the best account of Greek astronomical instruments and techniques. In addition to measuring shadows, the Greeks set up fixed instruments to define and monitor celestial co-ordinates, such as the **meridian**, **zenith** and equinoxes. The **armillary sphere** probably began as a large graduated bronze ring set in the plane of

Fifth Element, which was not present on Earth. Because of their single-substance composition, there was no celestial conflict, as a result of which the heavens (which began at the sphere of the Moon) were changeless. Aristotle argued that this celestial constancy meant that **comets** or new stars must be 'meteorological' or airy bodies, confined to the atmosphere. Aristotle's physics gave a powerful coherence to all subsequent Greek science, while firmly anchoring it to the **geocentric** theory.

HIPPARCHUS: OBSERVER AND THEORETICIAN

Greek astronomy came to its zenith when its geometrical and philosophical developments became infused with the latest Babylonian computation techniques and planetary tables, in the time of Hipparchus. How Hipparchus, who did most of his work on the island of Rhodes, obtained his knowledge of the finest Babylonian works, we do not know. Perhaps he studied in Babylonia itself. One of his most remarkable achievements was the construction of the Hipparchian Diagram, from which he attempted to calculate the distances and respective sizes of the Moon and the Sun, from the ways in which the fast-moving (and hence closer) Moon exactly covers the slower-moving Sun during a total eclipse. Hipparchus calculated that the Moon's maximum distance was 67.5 earth radii, which is an astonishingly close approximation. He also made major refinements to the previous planetary theories of Apollonius, and used eccentric circles and epicycles. He is said to have discovered the **precession** of the **equinoxes**, and to have constructed the first star catalogue. As an observer, a theoretician, and a utilizer of Babylonian science, Hipparchus consolidated and redirected Greek astronomy.

the **celestial equator**, against which the Sun's seasonal positions could be recorded. This soon acquired meridian rings and a variety of sites for measuring planetary positions. Quadrants, with 90° scales, were used for measuring vertical angles. Ptolemy himself popularized an instrument consisting of three equidistant hinged 'Rulers' that formed a variety of triangles whereby the vertical angle of the Moon could be measured with great accuracy. Yet the technological triumph of Greek astronomy must be the geared bronze planetary calculating machine, found off the Greek island of Antikythera by divers in 1900, in a wreck dating from *c.* 80 BC. This Antikythera Mechanism is now preserved in the museum in Athens.

HISTORY OF ASTRONOMY

HISTORY OF ASTRONOMY

The History of Astronomy
ARABIC SCIENCE

The Islamic religion began in AD 622, when the prophet Mohammed exhorted the pagan Arabs of the Holy City of Mecca to 'surrender to the one God', or, in Arabic, 'Islam'. The faith was rapidly carried into Egypt, Mesopotamia (Iraq), and on to Syria. As they became more settled, Mohammed's followers came to absorb components of the cultures which they had overrun, and because the rules of the *Qur'an* (or *Koran*) for daily prayer and worship demanded accurate time and direction, astronomy became crucial. Islamic culture also came to include many peoples: Arabs, Persians, Mongols, Africans, Indians and Andalusian Spaniards, all of whom needed to use astronomy.

▲ *An Arabic work* on astronomy. In the eighth century AD, astronomical treatises began to be translated into Arabic from Greek, Sanskrit and other languages.

ULUGH BEG

Ever since the invasions of Genghis Khan in 1218, the Western Steppe and parts of Persia had been occupied by Mongols, many of whom had become Muslims. Ulugh Beg (1394–1449), prince and governor of Samarkand, was descended from these Mongols. He was an astronomer of brilliance who, in 1420, built a great three-storey drum-shaped observatory, 30 m (100 ft) high, and housing a 40-m (130-ft) radius masonry sextant set in the plane of the **meridian**. Astronomers ascended a staircase alongside the graduated edge of this sextant, each brass degree of which spanned 68 cm (27 in), and observed the Sun, Moon or stars through a small hole in the stone, at the geometrical centre of the instrument. Exact observation was made possible by attaching a finely graduated sliding pinhole eyepiece and scale between each degree. With this colossal sextant, and other instruments at Samarkand, Ulugh Beg determined the length of the solar year to within a minute of time. He also composed a *Zij* table which gave the precise positions of 1,018 stars. Tragically, Ulugh Beg was murdered by assassins employed by his own son, though his superb observations later became influential in Europe. Russian archeologists excavated his Samarkand observatory in 1908.

EARLY ARABIC ASTRONOMY

Around AD 350, the Christian St Ephrem had founded a school at Edessa, in Mesopotamia, where the writings of **Aristotle**, **Ptolemy** and other Greek scientists were translated into Syriac. Christian scholars were also active in Jundishapur, north of the Persian Gulf. After the new Muslim capital was built at Baghdad in AD 762, the Abbasid Caliphs began to collect Greek, Syriac and Indian manuscripts, and established the House of Wisdom in the new city. Here, Greek astronomy, medicine and other sciences were translated into Arabic. It was around this time that Hindu astronomy and the beginnings of mathematical techniques such as algebra began to be translated from Sanskrit into Arabic. In Baghdad, one of the most useful astronomical devices was to be developed: the *Zij*, or handbook of astronomical tables. Baghdad, therefore, became the first great centre of Muslim science, absorbing many cultural traditions and even employing scholars who were not Muslims. Astronomy, of course, was vital for several reasons. Firstly, the *Qur'an* encouraged the pious study of God's creation. Secondly, astronomy was necessary to establish prayer times and to find the direction of Mecca, towards which they had to pray. Thirdly, Islam needed a reliable calendar. As the Muslim world expanded, new centres of scientific excellence began to develop, most notably in Damascus, Cairo and southern Spain, between AD 900 and 1200.

ISLAMIC PLANETARY ASTRONOMY AND THE CALENDAR

The Islamic calendar is basically a lunar one, which means that observation of the new Moon – at its thinnest crescent – is a fundamental event for time reckoning. However, neither the lunar nor the solar cycles form neatly divisible numbers and from the very start, Muslims, like other peoples, faced the problem of reconciling the two. Although the prophet Mohammed ruled against **intercalation** on theological grounds, astronomers could not escape the problem; the result was that the calendar year was 11 days shorter than the solar year. Islam also required that the times of sunset, nightfall, dawn, noon and afternoon be precisely defined as the five times of prayer. In addition, it was necessary to devise reliable astronomical techniques to establish the direction of Mecca, for the orienting of mosques and for prayer. These last two religious requirements were scientifically exacting when one considers the great spread of **longitudes** traversed by the Islamic faith even in the Middle Ages. Arab astronomers, however, from their superior observations and computational techniques, were able to define physical constants to a new order of precision, while their reduced *Al-Manunkh* tables became our Almanacs.

altitude ▶ p. 336 epicycle ▶ p. 338 equant ▶ p. 338 intercalation ▶ p. 340 latitude ▶ p. 340 longitude ▶ p. 340 meridian ▶ p. 341 refraction ▶ p. 343 zenith ▶ p. 344

◀ *An Arabic shepherd* contemplates the vast expanse of space with stars and planets.

ALHAZEN AND OPTICS

Perhaps the most brilliant of all Arabic scientists was Alhazen (*c.* AD 965–1039), who, building on Aristotle's *Meteorologia*, did fundamental work in optics. Working as a doctor in Cairo, Alhazen was fascinated with vision. He dissected the eyes of cattle to reveal their optical structures, and went on to project images with a camera obscura, make simple lenses and to study the colours into which white light broke down when **refracted**. Despite this, it appears that Alhazen never questioned Aristotle's premise that light coming from astronomical bodies was pure and white, and decomposed into six colours only when contaminated by a passage through air, water or glass – an idea later overthrown by **Isaac Newton** (1642–1727). One of Alhazen's most influential studies concerned the refraction of astronomical bodies by the atmosphere. Why do the Sun and Moon look distorted on the horizon, and why is there a twilight before sunrise and after sunset? Alhazen calculated that atmospheric refraction causes sunlight to scatter in the air when the Sun is 19° below the horizon. This calculation led him to ask questions about the density of the atmosphere and the **altitudes** of clouds that reflected sunlight while the Earth's surface was still dark. The influence of Alhazen's work on subsequent optical research should not be underestimated.

ASTRONOMY AND OTHER SCIENCES IN THE ARABIC WORLD

Because of its value in calendrical calculations, amongst other uses, astronomy was highly cultivated by the Arabs, but this cultivation must be understood within the context of the other sciences. The study and practice of medicine grew to new heights in the Arab world through the use of translated clinical texts by Aristotle, Claudius Galen, Hippocrates and other Greek writers, as well as Indian and Persian sources. Human health was seen as deriving from a balance of the four humours within the body: Yellow Bile (Fire); Black Bile (Earth); Blood (Air); and Phlegm (Water). Because these humours were sensitive to the planets and their movements, Arabic physicians made use of astrology in their diagnoses and prognoses. Avicenna (AD 980–1037) was one of the greatest physicians of the Middle Ages.

Experimental chemistry and alchemy also developed greatly. Gebir devised scientific distillation techniques, which led to the discovery of new substances such as nitric acid and *al-kohol*, from grape juice. Medieval European alchemy was built upon Arabic foundations, as was the supposedly astrologically influenced metallurgical chemistry. It was believed that the Sun, Moon and planets formed gold, silver and other metals in the Earth; alchemists believed they could speed up this process in their laboratories. Algebra, analytical mathematics and the concept of zero were all Arabic innovations, while numerous scientific terms – **zenith**, nadir and many star names – are also of Arabic origin.

UNDERSTANDING THE CELESTIAL MOTIONS

During its first few centuries, Arabic astronomy was strongly influenced by Indian and Persian ideas, though with Al Battani (*c.* AD 850–929) classical Greek ideas came to predominate. While the basic cosmology of Aristotle's Four Elements and nine celestial spheres was absorbed into Arabic thinking, Arabic astronomy came to be grounded in a tradition of practical observation, made in the hope of producing mathematical models that explained the true motions of the heavens. In consequence, astronomers in Baghdad, Cairo and Spain all became exasperated with the **equants** and **epicycles** of the Ptolemaic system, as well as the less precise nature of his values for equinoctial precession and other constants. Al-Sufi produced the first major Islamic star chart in *c.* 964 in Baghdad, while *Zij* tables – astronomical handbooks based on fresh observational data – were compiled by Al-Biruni, Al-Battani, Ibn Yunis and, in Spain, Arzachael. Yet while Ptolemy and other Greeks often worked with relatively small

instruments, and rounded up their numbers to convenient wholes, the Arabs were increasingly concerned with precision, both of observation and of calculation. Observatories were built across their lands, sometimes housing large instruments with finely divided scales. One illustrious centre for astronomical research was established at Maraghah in Iran, while the most famous Islamic observatory was established at Samarkand.

▲ *The peoples of the Middle East* began to formulate ideas about the celestial sphere during medieval times.

▶ *The astrolabe* was perfected by the medieval Arabs. Its star map and astronomical scales enabled it to be used in navigation, time-finding and teaching.

THE ASTROLABE

The most famous and widely used medieval astronomical instrument was the astrolabe. Its earliest form, the planisphaerum, was described by Ptolemy in AD 125, though it was the Arabs who developed and popularized the instrument. In its mature, or planispheric, form it consists of a star map, in brass filigree, rotating upon a series of zonal or 'climate' plates. These plates are engraved with zenith, horizon and hour-angle coordinate lines, each one projected to suit the sky as seen from its particular **latitude**. An astrolabe would usually contain five thin brass climate plates, engraved on both sides, for 10 separate locations. These plates would be placed inside the hollowed-out circular 'mater' of the astrolabe, with the Latin-named *rete*, or network, of the star map above the climate plate suitable for the current working latitude. At the centre is a brass peg, representing the North Pole, and when the *rete* is turned about it, one can simulate the rotation of the stars and relate them to terrestrial positions from the projected lines on the climate plate. Astrolabes were immensely versatile observing and computing instruments that could be used for time-finding (for prayers), direction-finding (for Mecca) and calculations in spherical trigonometry.

Aristotle ▶ p. 54 Newton ▶ p. 42 Ptolemy ▶ p. 30

ARABIC SCIENCE

HISTORY OF ASTRONOMY

33

HISTORY OF ASTRONOMY

The History of Astronomy
MEDIEVAL EUROPE

While many developments in classical astronomy were made by the Greeks, some had been absorbed into the Roman – Latin-speaking – West during the early centuries of the Christian era. Even if the works of the Greek scientists were not known in their complete versions, some of their ideas survived in Latin poetic digests or encyclopedic works, most notably by Macrobius (*c*. AD 400), Cassiodorus (*c*. AD 550) and Boethius (AD 480–524). Despite the apparent cultural retrenchment of first-millennium Europe, astronomy was too useful a subject to ignore. In particular, it was still cultivated in the relatively un-troubled and prosperous Christian Byzantine world which covered much of modern Greece, Turkey and Southern Russia, and an elaborate Byzantine bronze calendar mechanism, with geared dials and moving scales, survives in the Science Museum in

London, from *c*. AD 520. But the most important astronomical writer to emerge out of this period was the Venerable Bede (AD 675– 735) of Northumbria in England; he was also the first known English astronomer. His great concerns were calendrical astronomy, time determination and computation. With the new stability enjoyed by Europe under the Emperor Charlemagne (*c*. AD 800), monastic schools, in which astronomy was taught as a 'liberal art', began to flourish.

EUROPE'S TWELFTH-CENTURY RENAISSANCE
Contrary to popular opinion, scientific knowledge in medieval Europe was anything but moribund, especially after the Crusades brought western Christians into contact with the Arab world. The resulting twelfth-century Renaissance affected physics, philo-sophy, medicine, architecture and astronomy. For if one Creator God (shared by Jews, Christians and Muslims) had built the Universe in accordance with reason and number, it was man's duty to understand it. Twelfth-century Latin trans-lations of **Ptolemy**'s *Almagest* and **Aristotle**'s *Physics* brought classical Greek science into northern Europe for the first time. What mattered to medieval astronomers, however, was not original observation, but the acquisition of astronomical tables from Ptolemy and others (sometimes corrected by the Arabs), while the compilation of the Christian Spanish *Alphonsine Tables* (1274) provided a com-putational basis for subsequent western astronomy.

▲ *Astronomers contemplating* the Universe, taken from a medieval manuscript. In the twelfth century, translations of Ptolemy and Aristotle brought classical science to northern Europe for the first time.

The translated writings of Aristotle, with their implicit assumption that both the natural world and the human intellect partook of the same rational causal processes, were crucial to the establishment of western science. It was in optics, however, that medieval European scientists made the first great breakthroughs. Starting from Alhazen's translated writings, Roger Bacon unravelled the optical geometry of the rainbow and may even have invented spectacles around 1270.

RICHARD OF WALLINGFORD
The most influential of all medieval inventions was the mechanical clock. Though gears had been known in the ancient and Byzantine worlds, it was not until about 1325 that a gear train, a falling weight and an escapement to release the power in exact bursts, were put together. Richard of Wallingford (*c*. 1292– 1336), Abbot of St Albans, was one of the first clock designers, building an extraordinary mechanism which not only told the time, but which also rotated circles to replicate the motions of the Ptolemaic Universe. To reproduce the complex motion of the Moon, for instance, he used elliptical and differential gears some three centuries before Jeremiah Horrocks (1619–41) actually proved the lunar orbit to be elliptical. Once the concept of clockwork had appeared, astronomers and inventors across Europe began to build clocks, to tell the time and to replicate the celestial motions. Then, in 1364, Giovanni de Dondi built an even more complex astronomical clock. Detailed descrip-tions of the Wallingford and Dondi clocks still survive. Richard of Wallingford was an accomplished practical astronomer and, though a powerful church-man, seems to have made instruments with his own hands.

THE PROBLEM OF THE CALENDAR
The calendar used across Christian Europe between 45 BC and 1582 had been established by Julius Caesar based on the calculations of Sosigenes. Yet its 365¼-day year was too long by 10.53 minutes. As a result, the dates on the calendar, such as 21 June and 21 December, 21 March and 21 September, no longer corres-ponded to the astronomical **solstices** and **equinoxes**, which fell earlier and earlier in each year as the centuries passed. This posed a serious problem for Christians who, like Jews and Muslims, needed reliable calendars to regulate religious worship. The date of Easter, central to Christianity, was calculated from the full Moon following the spring equinox. But when was the true equinox – 21 March, as in the calendar, or more correctly 11 March, where the astronomical equinox was falling by 1574? The calculation of the correct date for Easter made astronomy very important at this time and various computational formulae were tried to solve the problem. The Venerable Bede had made major advances, and had established the birth of Christ as the starting-point for the reckoning of centuries. Yet a definitive solution to the calendar problem was not found until Pope Gregory's 'Gregorian Calendar' of 1582. This is the calendar we still use today.

GEOFFREY CHAUCER'S ASTROLABE

Geoffrey Chaucer's fame rests on his poetry, yet he was also the first man to write a scientific book in the English language, rather than in Latin. His *Treatise on the Astrolabe* (*c.* 1391) is a detailed manual for the use of that Arabic astronomical instrument, which had already become commonplace across Europe. The brass astrolabe was used by calendar calculators, university teachers and students studying astronomy, in the same way that a modern person might use a planisphere or a computer. Dozens of hand-written copies of Chaucer's *Treatise*, many with working drawings, still survive in libraries, attesting to its medieval popularity. Indeed, this self-contained guide to coordinate astronomy, angle measuring, time finding, and the **celestial sphere** complemented that long-standing medieval best-seller *De Sphaera Mundi* ('The Sphere of the World') written by the Englishman John of Holywood (Latinized: Johannes de Sacrobosco) in about 1240. In addition to the astrolabe *Treatise*, however, Chaucer is believed to be the author of the much more complex *Equatorie of the Planetis* (*c.* 1392), which describes an elaborate brass instrument, the equatorium, used for sophisticated planetary calculation. All of this astronomy assumed the Earth to be motionless in space.

▼ *By medieval times, observational astronomy was becoming a more defined science, sparking a revolution in astronomical knowledge.*

THE MEDIEVAL COSMOS

By 1350 Europe had a sophisticated cosmology that was based on the perfect circles, spheres and epicycles of Ptolemy's *Almagest*, with refinements derived from the Arabs. Yet medieval cosmologists also began discussing space and time in ways which were surprisingly relativistic. Their starting point was theological, as they considered what an all-powerful Creator could do, but their ideas had scientific implications. Could time exist in Heaven, which lay outside the physical Universe, or was time only relative to the realms of astronomical bodies? By 1327, Jean Buridan in France was arguing that both planets and earthly objects move because of a property of physical 'impetus', which had parallels to gravity.

EUROPEAN OBSERVATIONAL ASTRONOMY

Medieval astronomy had employed astrolabes, quadrants and other instruments, mainly to read off celestial positions against existing – often ancient – tables to check the time or calendrical data. Observational astronomy for research purposes in northern Europe was a product of the early Renaissance, when accurate translations from Ptolemy's Greek had revealed discrepancies between where the planets actually were in the sky and where they had been predicted. Around 1460 Johannes Müller (1436–76), writing under the name Regiomontanus, began to measure celestial angles with a set of Ptolemy's Rulers, or three graduated rods. These rods, around 2 m (6 ft) long, carried delicate scales, and could measure vertical angles with unprecedented accuracy. But it was Müller 's colleague, Bernhard Walther, who compiled the first long run of accurate original positional observations (of the Sun's **zenith** distance) by a north European, between 1475 and 1504. The printed observations of the Nuremberg school, of which both Müller and Walther were members, would soon provide vital new material for **Nicolaus Copernicus** (1473–1543). What mattered to Renaissance astronomers, however, was not the visual appearances of astronomical objects, but the accurate measurement of their positions as a way of refining their celestial geometry.

ASTRONOMY AND EDUCATION

Between 1100 and 1500, astronomical knowledge became widespread across Europe, although in many respects it was still extremely conservative in content. This conservatism derived not from any Church oppression, but from the seemingly self-evident belief that if the Greeks had taken not

But it was the infinite Universe, that lay beyond the visible heavens, which most fascinated medieval cosmologists. Thomas Bradwardine (1290–1349), the Oxford scholar and Archbishop of Canterbury, argued in 1344 that an infinite God could create an infinite Universe. By 1464 the German Cardinal Nicholas of Cusa was further considering an infinite number of rotating stars occupying a universe that looked the same from all positions within itself. Though the product of philosophical deductions, this medieval cosmos posed questions about infinity, time and relativity that sound remarkably modern.

▲ *A medieval scholar glimpses the celestial machinery that governs the heavens. This engraving does not, in fact, come from the Middle Ages, but seems to have been devised by the nineteenth-century French astronomical writer Camille Flammarion.*

only astronomy but also medicine, mathematics and philosophy to the limits of human understanding, then succeeding generations could only refine the details and pass on the tradition. Pre-modern astronomy, therefore, was curatorial rather than investigative in its approach. Following the establishment of Europe's great universities – Bologna, Paris, Oxford, Padua and Montpellier after 1100 – astronomy became part of the curricular *Quadrivium*, or four sciences of proportion, along with arithmetic, geometry and music. Ptolemy's spheres and **epicycles**, Aristotle's physics of motion, and the harmonic ratios that linked divine and human reason, became the staple of every Latin-reading medieval undergraduate. Some of them, such as the fictional Nicholas in Chaucer's *The Miller's Tale*, even owned astrolabes and copies of Ptolemy. By the last quarter of the thirteenth century the scholars of Paris were discussing the possibility of whether other worlds could exist, whilst around 1370, the French scientist-bishop Nicole de Oresme suggested that the Earth itself might be spinning on its axis.

The History of Astronomy
NICOLAUS COPERNICUS

For centuries the idea of a geocentric Universe remained unchallenged. There seemed no logical basis for disputing it. So when the Polish astronomer Nicolaus Copernicus formed his theory of a Sun-centred, or heliocentric, system, it was a turning point in the history of science. Copernicus was naturally hesitant in presenting his idea, as it overturned the ancient thinking of Aristotle and Ptolemy. Nevertheless, it was the revolution that was to put right years of misconception about Earth's place in the Universe.

THE RECOVERY OF GREEK ASTRONOMY

The origins of the Copernican revolution lie less in errors found in practical astronomy than in the growth of Renaissance humanism. The influx of scholars and original Greek manuscripts into Europe following the Turkish occupation of Constantinople in 1453 meant that first Italian, and then other European scholars now had access to

◀ *Ptolemy's Rulers, or Triquetrum*
The hinged rods are: A = 10,000 units long, B = 10,000 units, C = 10,000. Yet when an observation is made of the star, rod A crosses rod C at its division 8,542. From this proportion, 10,000:10,000:8,542, the astronomer can calculate the angle X, which is also the altitude of the star, from a table of geometrical chords.

Ptolemy in the original Greek, rather than through Arabic-to-Latin translations. And in this late medieval world which so revered the ancients, pure texts had a higher truth status than potentially defective modern observations. The Constantinople-educated cardinal and scholar Johannes Bessarion was a major driving force in this movement, and his collection of Greek manuscripts was important in enabling Georg Peurbach (1423–61) and Regiomontanus to produce the *Epitome*, or abridgement, of Ptolemy's Greek *Almagest* (dedicated around 1463, published 1496). It was these accurate Greek interpretations of Ptolemy which truly started the astronomical Renaissance. Regiomontanus's *Kalendarium* (1472) and other published tables made possible the more accurate prediction of eclipses (Columbus used the *Kalendarium* to predict the lunar eclipse of 29 February 1504) and a more sophisticated theory of planetary motion. They gave Copernicus his firm foundation.

COPERNICUS AND THE ITALIAN RENAISSANCE

Italy was at the heart of the new astronomy when Copernicus went to study there in 1496 and, since the middle of the century, a fascination with mathematics had become an inextricable part of Italian high culture. Even more important than Ptolemy were Plato's *Dialogues* (*c.* 427–*c.* 347 BC), which also became available in the

▶ *Nicolaus Copernicus, who turned more than 14 centuries of astronomical thinking on its head when he suggested a heliocentric model of the Universe as opposed to the accepted geocentric model.*

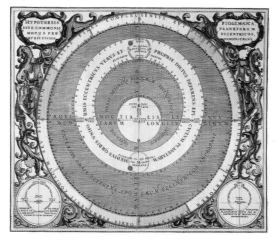

▲ *The Ptolemaic model of the Universe presented a number of problems, the main one involving the retrograde motions of the then-known outer planets, Mars, Jupiter and Saturn.*

original Greek, and which placed geometry at the heart of all truth, for while we can never draw a perfect circle, we can, as Plato said, visualize and describe one in our minds. Surely this perfect mental geometry could not exist if it was not sustained by the perfect originals that were permanently present in the mind of God? Although living 400 years before Christ, Plato's emphasis on the perfection and universality of a Creator God greatly appealed to Christian scholars, who sometimes described themselves as new or 'Neo'-Platonists. Copernicus came to Italy to study law and then medicine, and it is likely that he developed his fascination with astronomy from this broader neo-Platonic culture in which he moved. Yet when he compared the observed motions of the planets with their computed places, theory was far from perfect.

THE PROBLEM OF THE PLANETS

Copernicus returned home to Poland in 1503, and over the next few years was occupied with problems of astronomy, in addition to fulfilling his duties as a canon of Frauenberg Cathedral. One problem which haunted him, as it had haunted Peurbach and Regiomontanus, is why, in Ptolemy's **geocentric** Universe, the planetary periods seemed to be related to the solar period, or terrestrial year. Why, indeed, if everything rotated around the Earth, did the planets 'beyond' the Sun – Mars, Jupiter and Saturn – display annual **retrograde motions**, whereas the Moon, Mercury and Venus did not? These motions could only be explained in Ptolemy's system if one added eccentrically rotating circles, or **epicycles**, the centres of which rolled along the circumferences of greater circles to produce apparent loops in the planet's motion. Then, in 1514, Copernicus composed his

NICOLAUS COPERNICUS

Nikolaj Koppernigk, latinized to Copernicus, was born into a prominent Polish family in 1473. Though his father died when he was a boy, Nicolaus's education was supervised by his powerful uncle, Lucas Waczenrode, Bishop of Ermeland, who no doubt recognized his nephew's abilities and saw their future use to Poland. Nicolaus was first sent to the University of Cracow in 1491, where he received a Latin training in classical literature, law and theology, and then, in 1496, to Italy. He studied in Italy until 1503, gaining doctorates in law and medicine, and acquiring skills that would be important to his perceived

administrative career back in Poland. He also became fascinated with the new Greek learning of the Renaissance humanists, acquiring the familiarity with Greek astronomy and its problems that would one day make him famous. Through his uncle's influence, Copernicus obtained the canonry of Frauenberg Cathedral, which gave him a comfortable income for life. Here, amidst his public and ecclesiastical duties, Copernicus was to spend the next 40 years of his life. And it was here that he would study the pre-Ptolemaic Greek astronomers, make some observations, and quietly develop his heliocentric theory. He died on 24 May 1543, following a stroke.

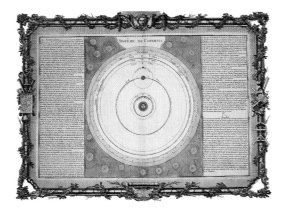

▲ *Copernicus's new* and revolutionary planetary system, showing the Sun at the centre, around which all the planets – including Earth – orbited.

unpublished essay *Commentariolus*, in which he set out his perceived problems in astronomy, noting in particular the increasingly contrived machinery of epicycles and **equants** necessary to make everything turn about the Earth. Was the Universe really so complicated? Or was the true centre of rotation not the Earth, but the Sun, as Aristarchus and other pre-Ptolemaic Greeks had suggested?

THE REVOLUTION OF THE CELESTIAL SPHERES

Between writing the manuscript *Commentariolus*, intended as it was for private circulation only, and 1539, when he was approached by Georg Joachim Rheticus of Wittenberg, Germany, we know virtually nothing about the development of Copernicus's astronomical thought. Over these years, however, he had thought out and written his *De Revolutionibus Orbium Coelestium* ('On the Revolution of the Celestial Spheres'). In 1539 Copernicus allowed Rheticus to read the *Commentariolus*, and then publish a short synopsis of his ideas on Sun-centred astronomy. But Rheticus was too busy to see Copernicus's *De Revolutionibus* through the press, and the task fell upon another young German, Andreas Osiander. Profoundly reverential to the Greeks as he was, Copernicus did not see his radical idea as overturning accepted astronomical thought, but rather as a development of the works of the wiser ancients. *De Revolutionibus* followed a similar structure to Ptolemy's *Almagest*. The Sun-centred system, he argued, explained planetary motion more simply than the geocentric.

If this was the case, though, why was Copernicus so reluctant to publish his theory that he only received a copy of his masterpiece as he lay on his deathbed? And why did Osiander add an almost certainly unofficial preface suggesting that the **heliocentric** theory did not describe the actual heavens, but was simply a calculating device? What were they afraid of?

ADVANTAGES OF THE HELIOCENTRIC THEORY

The idea of a Sun-centred Universe dated back to Philolaus, around 430 BC. Later Greeks such as Aristarchus had also explored its possibilities, yet it was Ptolemy's *Almagest* which anchored subsequent astronomy to the geocentric theory. After all, an Earth-centred universe accorded better with common sense, and fitted in elegantly with the philosophy of Aristotle. Central to this Greek tradition were the concepts of perfectly circular orbits and perfectly uniform planetary motions. Yet the only way of reconciling such perfect theoretical motions with the peculiarities of the lunar orbit and the annual retrograde motion of Mars, Jupiter and Saturn, was to invent epicycles and eccentrics to 'save

THE IMPACT OF DE REVOLUTIONIBUS

Popular legend has it that Copernicus and Osiander feared Church persecution. Yet Copernicus was a Polish cathedral dignitary, and in 1543, the Roman Catholic Church had no established policies regarding science. Osiander, moreover, was a follower of Luther (a founder of the Protestant Reformation). Their concern came not from the Inquisition, but from academic ridicule, for the suggestion that the Earth was hurtling through space affronted not only common sense, but the entire physics of **Aristotle**, which dominated the universities of Europe at the time. According to Aristotle, objects fell to Earth because they were drawn to the centre of the Universe. If the Universe really was heliocentric, a thrown stone should fly to the Sun. And if the Earth was spinning in space, how did anything remain on its surface, and why was the natural order of earth, water, air and fire broken down? Inertial and gravitational theory were as yet unknown, and Copernicus knew that while his heliocentric cosmology was geometrically elegant, and ex-

the phenomena'. As humanist astronomy developed after 1460, the growing number of circles looked increasingly contrived. Copernicus argued that as all planetary periods seemed to relate to the solar year, their motions could be more simply explained by seeing the Sun as the centre of rotation. Because Copernicus's ideas were still rooted in Greek ideas of perfectly uniform circular motions, however, he was still obliged to clutter his system with epicycles. Not until **Johannes Kepler** (1571–1630) abandoned this uniformity principle in favour of elliptical orbits and varying speeds did epicycles fade away.

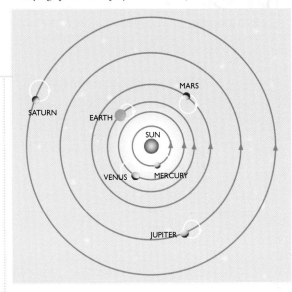

▲ *The Copernican System*
By placing the Sun slightly offset from the centre of the celestial system, Copernicus removed some of the difficulties imposed by Ptolemy's unwieldy system. He retained the idea that the planets' orbits were circular, and his system still called for the planets, including Earth, to move in small circular epicycles.

plained the planetary retrogrades, science seemed to lose far more than it gained by it. Being a naturally conservative and cautious man, Copernicus had no wish to plunge himself into a vortex of academic name-calling. However, *De Revolutionibus* was to give Renaissance astronomy a vital new impetus.

COPERNICUS'S INSTRUMENTS

While many of the forces that motivated Copernicus derived from Greek symmetry and elegance, he was very much aware that observing the heavens was important. Only in this way could one build up tables of the celestial motions alongside which ancient and modern theories could be tested. Observational astronomy in 1500 was about accurate positional measurements made with mathematical instruments. Copernicus made his first recorded observation – of the occultation of Aldebaran by the Moon – in Bologna on 9 March 1497. He was using

the fixed star to check the complex orbit of the Moon. In March 1513, he built a stone tower in Frauenberg from which to observe the heavens. We know that Copernicus owned a set of Ptolemy's Rulers, because **Tycho Brahe** (1546–1601) later left a detailed description of this instrument. Its rods formed a variable triangle and he could use it, with a sine table, to measure the **zenith** angles of the planets. He also had an astrolabe and other instruments. Though Copernicus was not a regular observer, he could use his Rulers to test predictions in Ptolemy's, the Alphonsine and Regiomontanus's tables, as a way of refining his heliocentric theory.

Aristotle ▶ p. 54 Brahe ▶ p. 38 Kepler ▶ p. 38 Ptolemy ▶ p. 30

NICOLAUS COPERNICUS

HISTORY OF ASTRONOMY

The History of Astronomy
TYCHO BRAHE & JOHANNES KEPLER

The next steps in the development of astronomy as a practical science were taken by two men: Tycho Brahe and Johannes Kepler. Tycho, a brilliant and determined young man, relentlessly pursued the continued improvement of instruments that would allow more accurate observations of celestial bodies and their motions. His discoveries led him to develop a theory that combined the Ptolemaic and Copernican universes. Kepler became Tycho's assistant and continued the great man's work after his death, redefining the thinking of the day about planetary motion.

HOW DID THE PLANETS MOVE?
Even by the time the second edition of **Copernicus**'s *De Revolutionibus* was published in 1566, no conclusive proof for the moving Earth existed. It was Tycho Brahe who saw the necessity of a new line of attack in deciding the structure of the Universe. Fresh observations, made with instruments of unprecedented accuracy over a long period of time, were the answer. It was no longer enough to correct the old tables: entirely new ones needed to be built up from superior data. Tycho had come to realize the need for a new instrumental approach by his early twenties, though it was only after establishing Europe's first big research

observatory on the Danish island of Hven in 1576 that things really got going. Regiomontanus, Copernicus and others could only measure their angles to within 10 arc minutes, but Tycho improved this to a single arc minute. In addition to critically accurate planetary tables, Tycho hoped to measure a **parallax**, for he realized that if the Earth moved around the Sun, then as it approached one region of the sky, the **constellations** there should get slightly bigger, while the ones behind correspondingly diminished. Not even Tycho's instruments were advanced enough to prove this, though.

DO THE CRYSTALLINE SPHERES REALLY EXIST?
On the evening of 11 November 1572, Tycho Brahe saw a brilliant star in Cassiopeia, which had not been there previously. He devised an instrument with which he hoped to measure its distance above the Earth. Like everyone else, Tycho accepted **Aristotle**'s doctrine that the heavens beyond the Moon were changeless; the new star, therefore, must be an atmospheric phenomenon. Yet after several months' observation, when the **supernova** (as we now know it to have been) was beginning to fade, he found its position unchanged. As astronomers were already able to measure the lunar distance trigonometrically, Tycho's measurements indicated that the star was actually *in space*. Could Aristotle be wrong?

In 1577 and 1588, two bright **comets** blazed across the sky; these comets, in accordance with Aristotelian theory, should also have been atmospheric bodies – hence their ancient status as portents. Once again, however, Tycho found that the comets were beyond the Moon, in planetary

▲ *The system devised by Tycho Brahe was a compromise that incorporated the accepted logic of both the Ptolemaic and Copernican models of the Universe. Although his plan was incorrect, it paved the way for his assistant Johannes Kepler to further investigate planetary theory.*

space. This raised a question: if the planets were carried around the Earth on perfect crystalline spheres, were the comets actually crashing through them? While Tycho's superior observational data were still insufficient to prove or disprove the **heliocentric** theory, his work on the 1572 supernova and the comets showed more than ever that modern astronomy might actually prove the ancients wrong.

TYCHO'S COMPROMISE COSMOLOGY
Tycho Brahe had been deeply impressed by the mathematical and explanatory elegance of Copernicus's theory, as had many other astronomers across Europe. Indeed, one could use heliocentric theory to calculate planetary places even if one believed the Earth to be fixed. Erasmus Reinhold's *Prutenic Tables* (1551), for instance, were grounded

TYCHO BRAHE AND JOHANNES KEPLER
Tycho Brahe (1546–1601) was not only one of the leading astronomers of the Renaissance, he was also perhaps the most exotic. Born into the ancient Danish aristocracy, he was kidnapped by his childless uncle while still a baby. As a youth, he was sent to study law at Leipzig, Wittenberg and several other European universities, but even by the age of 16, a passion for astronomy was uppermost in his mind. On 24 August 1563 he made his first recorded observation of Jupiter and Saturn in conjunction. He later lost part of his nose in a duel, and collaborated with Paul Hainzel of Augsburg in the construction of a quadrant for accurately observing the planets. Tycho's eccentric brilliance was recognized by King Frederick II of Denmark, who

presented him with the Island of Hven on which he built his famous 'Castle of the Heavens', Uraniborg. Tycho lived and worked at Uraniborg between 1576 and 1598, amidst the research assistants and technicians of Europe's first scientific research academy. It was Tycho's aim to prove astronomical theory by practice and mathematical analysis. Court politics caused him to leave Denmark, however, and he died suddenly in Prague in 1601.

Johannes Kepler (1571–1630) came from the opposite end of the social scale to Tycho. He was the sickly and myopic son of an impoverished mercenary soldier and a woman whom Johannes would one day rescue from a witchcraft charge. A Lutheran Protestant from Weil in Catholic South Germany, his youthful talent won him a place to

train for the ministry at Tübingen, and while never actually ordained, his love of astronomy was imbued with a powerful religious conviction. After working as a lecturer at Gratz in Austria (where the idea of the geometrical solids and the planets came to him) he was taken up by Tycho. Following Tycho's death, however, he was encouraged by Rudolf II, the Holy Roman Emperor, and succeeded Tycho as Imperial Mathematician. He became Mathematics Professor at Linz in 1612, and at Ulm in 1627 he published his *Rudolphine Tables*, based on Tycho's observations. Kepler was the most brilliant mathematician of his age. He predicted the transits of Mercury in 1630 and Venus in 1631. Kepler's optical researches also led him to invent the telescope eyepiece that bears his name.

BRAHE & KEPLER

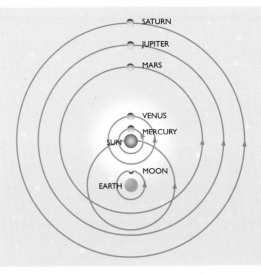

▲ The Tychonic System
Tycho Brahe was not a Copernican: for physical reasons, to explain why heavy objects fell downwards, he held on to the belief that the Earth was the supreme central body. His celestial system was an ingenious compromise in which the Earth was central and yet all other planets orbited the Sun.

in Copernican principles, yet were used by Ptolemaic astronomers. But in spite of his admiration for Copernicus, Tycho could find no physical evidence for the Earth's motion, and he valued the unity of Aristotle's **geocentric** physics. Then, in 1583, Tycho developed a brilliant compromise that seemed like the best of both worlds. In this 'Tychonic System' all the planets rotated around the Sun, but the Sun itself, and all the planets with it, rotated around the Earth. It explained the Sun-related planetary periods, the **retrogrades** of the outer planets, reduced the **epicycles** and kept much of the physics of Aristotle intact. But it no longer seemed plausible to speak of the ancient crystalline spheres, especially as there was now evidence that new stars blazed amongst them and comets crashed through them. Yet if there were no spheres, what sustained and moved the planets in what now seemed to be *empty* space?

KEPLER'S GEOMETRY OF THE VOID

The geometry of the Copernican Universe fascinated the young Johannes Kepler, and in his 1596 book *Mysterium Cosmographicum* he argued that the five regular solids – cube, tetrahedron, dodecahedron, icosohedron and octahedron – could all fit

◀ *Johannes Kepler, who took up Tycho's studies after his death; his investigations into the orbit of Mars revolutionized planetary theory.*

within the orbits of the planets. When so placed, the spaces between three-, four- or eight-sided solids could be made to accommodate the apparent orbital eccentricities previously explained by epicycles. Although it is now known to be wrong, Kepler's theory was brilliant, emphasizing the innate geometry of planetary space. In his mind, what really mattered were the perfect geometrical shapes described in space by the planets, and *not* the imaginary crystalline shells on which they hung. In this respect, his ideas pre-empted the watershed through which subsequent astronomy was to move – a watershed that saw exact mathematical descriptions as more important than contrived mechanical explanations. It was the brilliance of Kepler's *Mysterium* which earned him an assistantship on Tycho's staff in Prague in 1600. Kepler's job was to analyze the complex orbit of Mars, based on 20 years of superlative Tychonic observations, and reconcile them with Tycho's geocentric system. Even after Tycho's sudden death in 1601, Kepler continued with the Mars observations. His solution was to revolutionize planetary theory.

KEPLER'S LAWS OF PLANETARY MOTION

Try as he might, Kepler could not find a combination of perfect circles and epicycles which satisfactorily explained the observed orbit of Mars. There was a discrepancy of eight arc minutes. Now while this small discrepancy was well within the limits of observational error for other astronomers of the period, Kepler knew that Tycho's observations were good to a *single* arc minute. By 1609, he had hit upon a solution: Mars moved not in a circle, but in an ellipse, with the Sun occupying one of the ellipse's foci. This became his first Law of Planetary Motion. In 1619, he further developed elliptical theory to add two new laws. The second law described the exact geometry by which Mars moves fastest close to the Sun and slowest far from it. The third law built upon the second, explaining how orbital velocity decreases as the orbital radius increases. And if these laws applied to Mars, they probably related to the other planets as well. In 1638, the English astronomer Jeremiah Horrocks successfully applied the laws to the lunar orbit. But a question remained: what was the invisible force that locked the planets into such perfect elliptical orbits across the void of space?

Aristotle ▶ p. 54 Copernicus ▶ p. 36

TYCHO'S INSTRUMENTS

Although highly gifted as a mathematician and theorist, it was his recognition of the need to fundamentally improve and broaden the available database of astronomy that made Tycho such a key figure. Theory must be elegant, but it had first and foremost to be judged by observed reality. In 1576, this meant by observations made with instruments of superior accuracy. In his observatory, Uraniborg, Tycho began that approach which has since dominated the whole methodology of western science: the successive perfection of several 'dynasties' or design-types of instruments, each of which eliminated a fault in its predecessor, to produce measurements of unprecedented quality. In particular, Tycho developed a succession of large-radius quadrants to measure the vertical angles of celestial objects, then his sextants and **armillary spheres** to establish their east-west angles. Uraniborg had its own craftsmen and workshops, so that instrument scales, sighting and mounting procedures could be constantly monitored and improved in a relentless pursuit of accuracy. Tycho used his instruments to re-define the fundamental constants of astronomy and draw up new tables of planetary motion. His instruments and observing procedures would change the direction of European astronomy.

▲ Tycho Brahe *at work in his observatory at Uraniborg, where he developed and improved a series of astronomical instruments, including quadrants, sextants and armillary spheres.*

BRAHE & KEPLER

HISTORY OF ASTRONOMY

The History of Astronomy
GALILEO GALILEI

After Tycho Brahe's supernova and cometary studies had undermined belief in the existence of the crystalline spheres carrying the planets, the obvious question was 'What makes the planets move?'. It was to this problem that an Italian named Galileo Galilei turned in 1600. Galileo proved through a series of experiments that Aristotle's ideas of force and weight of a moving body were inaccurate. His telescopic observations also proved beyond doubt that the Earth was not the only centre of rotation in the Universe. Although Galileo never worked out what force did control the planets, his major discoveries in both physics and astronomy have led him to be considered as the father of modern science.

THE PROBLEM OF PLANETARY MOTION

In 1600, the English physician and magnetics researcher William Gilbert suggested that the force that made the planets move could be related to magnetism. Indeed, **Kepler** subsequently considered the idea of the Sun as a rotating magnet, sending out an invisible force that spiralled through space, whipping Mercury, the closest of the planets, into an 88-day orbit, but weakening with

distance, so that Saturn went around in 29½ years. The motions of the planets, without their supporting crystalline spheres, had come to fascinate Galileo by 1600, and he became profoundly sceptical of the explanations offered by vitalist physics. Galileo had already shown **Aristotle** to be wrong in his assumptions by simply dropping a heavy and a light object simultaneously off (legend has it) the Leaning Tower of Pisa. Both objects hit the ground together, though according to Aristotle, the heavier one should have fallen fastest. If Aristotle was wrong about falling objects, what else might he have been wrong about? Although at first he was not completely convinced by the Copernican theories of the Universe, by 1609 Galileo was aware that classical physics and astronomy were seriously flawed.

◀ *Legend has it that Galileo first proved Aristotle's physics to be wrong by dropping objects of different weights off the Leaning Tower of Pisa; according to Aristotelian theory, the heavier object should have hit the ground first, but in fact both objects reached the ground at the same time.*

GALILEO'S TELESCOPIC DISCOVERIES

While Galileo did not invent the telescope (the Dutch spectacle-maker Hans Lippershey achieved this in 1608), he was the first person to recognize its scientific potential. Hearing of the instrument from Kepler, Galileo made a number of telescopes, the best of which could magnify an object by up to 30 times. During the winter of 1609–10, he made a succession of discoveries that transformed astronomy. While studying the Moon, he saw mountains and what he thought were seas (we still use his Latin name, *maria*, for these lunar areas). He also found that the glowing **Milky Way** resolved into countless millions of individual stars when viewed through the telescope. His discoveries did not stop there. He saw that Saturn was a sphere that became strangely elongated at certain times; Jupiter was not only spherical, but it was encircled by four bright moons. This was one of his most significant

▼ *Galileo exhibiting his telescope to a group of noblemen. Although Galileo did not actually invent the telescope, his improvements to its design opened up the skies in a way that would previously have been impossible.*

GALILEO GALILEI

Galileo Galilei (1564–1642) was born into a cultured professional family in Pisa, Italy. He studied at Pisa University and, after abandoning medical studies, took to mathematics, becoming a lecturer at Pisa in 1588. It was in Pisa that Galileo made his first researches into swinging pendulums and falling bodies. He also antagonized the Aristotelian philosophers and displayed a talent for controversy which never left him. In 1591 he became Professor of Mathematics at Padua, which was

◀ *Title page from Galileo's Sidereus Nuncius ('Starry Messenger'), in which he outlined his astronomical theories and observations.*

then the finest scientific university in Europe. It was here, in 1609–10, that he made those telescopic discoveries which, when published in his *Siderius Nuncius* ('Starry Messenger'), revolutionized astronomy and made him an international celebrity. He was taken into the court of Cosimo II, Grand Duke of Tuscany, and basked in his fame. By this time Galileo had become the main defender of Copernicanism, and under Cosimo's protection, enjoyed baiting the conservative Aristotelians. His disputed co-discovery and anti-Aristotelian interpretation of **sunspots** with Christopher Scheiner after 1612, however, made him many enemies amongst the intellectual Jesuit order, which was probably instrumental in engineering his trial in 1633. But the condemnation of such a scientific celebrity by the Inquisition sent shockwaves throughout Europe.

FURTHER TELESCOPIC DISCOVERIES

Galileo's telescopic discoveries changed the whole direction of astronomy by providing new evidence that seemed to back up the Copernican system. Yet for 40 years after 1610, the telescope seemed incapable of further development, because contemporary glass-making techniques limited lens sizes and magnifications. Though some improvements had occurred, it was not until the early 1650s that Christiaan and Constantijn Huygens used a machine of their own devising to form pieces of clear Dutch mirror glass into superb lenses up to 20 cm (8 in) across. The Huygens brothers made lenses up to 37 m (123 ft) focal length, yielding telescopic images of unsurpassed quality. With these long instruments, they discovered the true ring nature of Saturn's 'appendage', along with the Saturnian satellite, Triton, and saw surface detail on Mars and Jupiter. Huygens-type telescopes now made it possible for Gian Domenico Cassini (1625–1712) to discover four additional Saturnian satellites, the Cassini Division, and produce the first accurate tables of Jupiter's moons, timed with Christiaan Huygens' successful adaptation of Galileo's pendulum to clockwork after 1658. Robert Hooke (1635–1703) and Giovanni Baptista Riccioli (1598–1671) started observing the Moon, planets and Orion Nebula with long telescopes, while Johannes Hevelius (1611–87) built one of a record 46 m (150 ft) focal length. And none of these discoveries was compatible with Aristotle.

▶ **Galileo with his telescopes,** *the development of which marked a turning point in astronomical discovery and understanding.*

discoveries – it showed that Jupiter itself was a centre of revolution, proving at a stroke that everything did not turn about the Earth. Over the next few years Galileo found spots on the Sun, which indicated a solar rotation period of 28 days. While none of these discoveries actually proved the Copernican theory, they seriously damaged the credibility of **Ptolemy** and Aristotle.

GALILEO AND THE PENDULUM

In addition to his telescopic discoveries, Galileo was one of the founders of modern physics, replacing the vitalistic physics of Aristotle with mathematical observations of bodies in motion. His first great contribution came in 1583 when he was only 19 years old. Sitting in the great cathedral in Pisa, Galileo noticed that each time the verger lit one of the lamps suspended on long chains, he set it

SCIENCE AND CHRISTIANITY

The culmination of forces which led to Galileo's trial and condemnation for heresy before the Roman Inquisition in 1633 were more about the Catholic Church's political authority than about Christian belief, for Galileo was a devout man. Between 1610 and 1614, indeed, Galileo was lionized by bishops and cardinals, for theologians accepted that while the Earth appeared to be 'fixed' in the Bible, new interpretations of Biblical texts relating to natural phenomena, discovered by God-given human intelligence, were acceptable. The Renaissance Church was not fundamentalist. Galileo's undoing sprang from his disputes with the

swinging. According to Aristotle's physics, the arc and velocity of swing were a combined product of the force of the push and the weight of the swinging object. Lamps, after all, were metallic or 'earthy' objects, and when hanging in the alien element of air, wish to get as close to Mother Earth as possible. After its initial push, therefore, a heavy lamp got 'tired', swinging slower and slower until it stopped. By timing the lamp swings with his pulse, Galileo discovered that the period of the swing stayed the same during all oscillations, from the initial big swings down to the last. Further experiments, conducted with different weights and lengths of pendulums showed that it was oscillation length, *not* the weight or force, that mattered. A mathematical relationship clearly existed between the length of swing and the Earth's pull.

GALILEO'S REVOLUTION IN PHYSICS

It was in the period following his telescopic discoveries in 1610, when he was living in Florence, that Galileo began his great researches into physics. Building upon his earlier work with pendulums and objects falling from towers, Galileo began to analyze motion from the standpoint of number and geometry. He found that not only did *all* objects – be they loaves of bread or cannon balls – fall at the same velocity, but their speed doubled every 9.8 m (32 ft) of fall. Galileo then found that when a polished metal ball was rolled down an inclined groove, it also accelerated in its descent in a way that could be expressed in exact mathematical terms. His researches into buoyancy, impact and resistance all pointed in the same direction: what mattered was

highly Aristotelian Jesuit scientists, and from his partiality for ridicule. When, in 1615, he started drawing theological consequences from Copernicanism, the priests were irritated by the layman amateur. After Galileo was first warned against teaching the still-unproven Copernican doctrine as truth, in 1616, he wisely drew back. Then in 1623, one of his old supporters became Pope Urban VIII, though Galileo was still unable to persuade him to lift the ban of 1616. In 1632, Galileo published his pro-Copernican *Dialogue*, after the Roman censors had authorized it. But Urban believed that Galileo was ridiculing him, and Galileo was found guilty of a relatively minor heresy charge, and required to recant.

not the substance of the objects, but their common mathematical relationship to the Earth when in free motion. In fact, the exact mathematical ratios he produced, underlying physical motion of the Earth's surface, and their similarity to the perfect motions of the heavens, flatly contradicted Aristotle, who had taught that only astronomical motions were perfect: earthly motions were essentially chaotic, because of the warring elements. Galileo had opened a way for a wholly new physics.

GALILEO'S INTERNATIONAL DISCIPLES

It would be difficult to think of any post-1630 physicist who was not, in some way, a disciple of Galileo. Galileo's ideas profoundly influenced the physics of the Frenchman René Descartes (1596–1650), while another Frenchman, Pierre Gassendi (1592–1655), performed a dramatic experiment in 1640 which demonstrated why objects were not flung off the spinning Earth. He dropped a ball from the masthead of a speeding galley, and saw that it fell *not* astern of the mast, but at its foot. If the ball and the galley comprised one inertial system, so did balls, towers and people on the Earth's surface. Despite Galileo's bad relations with the Catholic Jesuits, many of their scientists taught Galilean ideas, at the same time emphasizing that Copernicanism was only a theory, and not a proven truth. Galileo's ideas were warmly taken up in Protestant England and Holland, where the Pope had no power. Jeremiah Horrocks, and his friends William Crabtree and William Gascoigne, were ardent Copernicans by 1636, while John Wilkins not only wrote the first English textbook on Galilean astronomy in 1639, and seriously discussed a lunar journey, but **Christiaan Huygens** (1629–95) was to write about life on other worlds. But there still remained the question: what force made the planets move?

HISTORY OF ASTRONOMY

The History of Astronomy
THE NEWTONIAN REVOLUTION

By 1640, Galileo and Kepler had shown that while the planets appeared to move in a vast and seemingly empty universe, their orbits were governed by exact mathematical laws. The nature and source of this governing force remained a mystery, however – although it seemed 'mechanical' and not 'mysterious' in character. Perhaps there were also astronomical laws, such as Kepler's, related to the law which made terrestrial bodies fall in exact ratios of acceleration, potentially linking earthly and celestial physics in one set of mathematical expressions? While no one could yet prove that the Earth rotated around the Sun, the growing body of physical and telescopic evidence had undermined the geocentric cosmology by 1670.

RENÉ DESCARTES AND ROBERT HOOKE

The first comprehensive system of physics devised in the post-Aristotelian world was that of René Descartes, a profound admirer of **Galileo** and **Kepler**. Descartes developed what became known as the Mechanical Philosophy, which argued that all movement was a product of impacts from corpuscles, or tiny physical bodies. Indeed, corpuscles of different size were the basis of everything, and space was filled with the smallest

▶ *Isaac Newton, whose profound discoveries changed the direction of scientific thinking; they included the laws of motion, theories of gravity and work on the nature of light.*

of these. Indeed, these corpuscles occupied the Universe in the way that a vast warehouse might be packed with tennis balls: wherever you hit a ball, it will send a shockwave throughout the rest. Swirling 'vorticles' in this universe of corpuscles, therefore, could move the planets and **comets** or transmit light, in accordance with exact mathematical laws. The drawback with Descartes' physics, however, was its highly speculative character, as 'thought experiments' ultimately predominated over actual experiments. Robert Hooke, a great admirer of Descartes, and a gifted experimental scientist, was fascinated by the mechanical, or vibratory, nature of physical forces and, amongst other things, he was the first scientist to suggest – from experimental criteria – the wave nature of light. In 1669 he tried, unsuccessfully, to measure a stellar **parallax**, and during the 1670s began to build upon Kepler's three laws, taking the first steps towards a *gravitational* theory of planetary motion.

EXPERIMENTS TO QUANTIFY GRAVITY

Through the establishment of the English Royal Society after 1660 and the French Scientific Académie soon after, it is clear that science and astronomy were rapidly becoming subjects of

international prestige. Under the auspices of the Royal Society, Hooke began a series of experiments that he hoped could quantify the mysterious attractive force of the Earth. In 1662 he set up a set of scales on the roof of Old St Paul's Cathedral in London. Each of the pans contained one of two identical spheres. One sphere was let down on its weighed thread, tied to the balance arm, until it hung 28 m (90 ft) below. Hooke had hoped that the suspended sphere, being closer to the Earth, would now register as slightly heavier than its companion in the balance pan, and that the resulting weight discrepancy could produce a mathematical expression for Earth's pull on the two spheres. Unfortunately, the instruments of the time could not be made delicately enough to give a reliable result.

All over Europe, though, scientists were trying to reach the same expression, and small steps were taken. In 1672, French astronomers found that an accurate pendulum clock lost two minutes per day on the Caribbean island of Cayenne, even though it had previously been regulated to keep perfect time in Paris. Was Cayenne further from the Earth's centre than Paris, and was the pull on a swinging pendulum different in these places?

SIR ISAAC NEWTON AND GRAVITY

By the 1660s, Robert Hooke, **Christiaan Huygens** and many other scientists across Europe were investigating the 'gravitas' or physical properties of bodies, in the hope of discovering the secret force that lay at the heart of Galileo's and Kepler's laws. It was in 1666, however, that the young Isaac Newton had his *annus mirabilis*, or 'wonderful year', which saw the beginning of his own insights into gravitation. According to a story he later told William Stukeley, it was triggered by his realization that the force that pulls an apple from a tree is probably the same as that which makes the Moon revolve around the Earth. This raised the question: why did the Moon not collide with the Earth? From his knowledge of the works of Galileo and Kepler, Newton began to analyze the problem mathematically. He noticed that while the 'falling' Moon always misses the Earth, some

SIR ISAAC NEWTON

Born prematurely on Christmas morning in Woolsthorpe, Lincolnshire, Newton (1642–1727) was not expected to survive until the evening, though he lived for 85 years. His father, also Isaac, had died three months previously, and his mother Hannah soon married a wealthy clergyman who disliked the young Isaac and wanted him out of the house. He was, therefore, brought up by grand-parents, boarded out, and sent to Grantham Grammar School. It is likely that his childhood rejection lay at the heart of his life-long reclusive tendency. It was only as a young man at Trinity College, Cambridge, in the mid-1660s that he first

bloomed. Newton had enormous powers of mental concentration and an ability to wrestle with the most abstruse mathematical problems for days on end without speaking to anyone. He was also an experimental scientist of genius, as his work on optics testifies. Elected to the Royal Society in 1672, his isolated disposition always prevented him from being communicative. But after solving the gravity problem, in *Principia* (1687), he rapidly won international renown, and became the archetype of a scientific genius. Taking the positions of Master of the Mint and President of the Royal Society, he was the most powerful figure in British science, although he remained difficult to deal with throughout his life.

THE NEWTONIAN REVOLUTION

NEWTON'S OPTICAL DISCOVERIES

Newton's *annus mirabilis* of 1666 was not only wonderful because of his first insights into gravitation, but also because of his revolutionary work on light. **Aristotle** had taught that true light was white, coming as it did from the perfect realm of the heavenly bodies, but when it entered the terrestrial realm, it became contaminated and broke down into colours. Newton acquired a pair of prisms in 1666 and, on producing the familiar spectrum from one prism, was surprised to find that he could reconstitute the white light by passing it through the second. Surely one could not reconstitute perfection from the decomposed? He also found that each of the six colours into which the **spectrum** always fell, from red to blue, were physical absolutes, and could be broken down no further. By 1672, Newton had overturned the established theories of classical optics. The spectrum breakdown during **refraction** suddenly made sense of the **chromatic aberration** fringes found in refracting telescope images. This led him to devise a new way of forming optical images in 1668: by using a metallic parabolic mirror which reflected rather than refracted light. His 'Newtonian' Reflecting Telescope was shown to the Royal Society in 1672, and won Newton his first recognition.

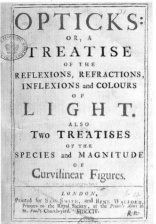

◀ Title page from Newton's great work Opticks, in which he laid out his theories on the properties of light, most notably the breaking up of a beam of white light into its constituent colours through the use of a prism.

◀ Reverend John Flamsteed, the first Astronomer Royal of the Royal Observatory, Greenwich; when he was appointed, Flamsteed was expected to provide all his own instruments.

force always swings the Moon back around the Earth. The calculations he made on the basis of this observation 'answered pretty nearly'. The whole affair, however, was made vastly more complicated by the presence of the Sun and its influence on both bodies. This produced that 'three bodies problem' which Newton confessed made his head 'ake'. It was to be another 12 years before Newton made his breakthrough. Robert Hooke drew Newton's attention to the mathematical potential of Kepler's second, or 'Areas' law, when analyzing the variable motions of orbiting bodies.

NEWTON'S *PRINCIPIA MATHEMATICA*

Early in 1684, Sir Christopher Wren, himself a former Oxford astronomy professor and eminent mathematician, offered a prize to anyone in the Royal Society who could produce a convincing mathematical expression for the gravity force. Despite attempts by some of the eminent English scientists of the day, no one had the answer.

At around the same time, the young Edmund Halley (1656–1742) visited Newton in Cambridge, and discovered that he had already solved the problem of why the Moon moves around the

THE INSTRUMENT REVOLUTION: GREENWICH OBSERVATORY

Between 1655 and 1670, three new instruments revolutionized precision astronomical observation. These were: the telescopic sight, which enabled the critical alignment of measuring instruments; the micrometer, which allowed measurements to be made in the field of view of a telescope; and the pendulum clock, whereby accurate star **transits** could be timed. Suddenly, the accuracy ceiling had risen from Tycho's one-arc minute to 10 **arcseconds**, promising vastly improved data upon which to base mathematical analysis. The Royal Observatory, Greenwich was founded in 1675, in the hope that an astronomical solution to the **longitude**

Earth in an elliptical orbit. Halley urged the secretive Newton to publish his findings, and in 1687 *Principia Mathematica* appeared, paid for by Halley himself. One of the keys to Newton's achievement, however, had been his invention of fluxions, an early form of calculus, and a revolutionary new mathematics, capable of expressing changing quantities in physics. *Principia* did not invent the idea of gravity, what it did was give it clear mathematical definition. The book was truly revolutionary, changing the direction of all subsequent astronomical and physics studies.

In the work, Newton dismissed the idea of saying what gravity was, and concentrated on elucidating the laws it followed. He saw gravity itself as an unknowable property of God; what man could do, however, was trace its exact effect on the masses of bodies between which it acts. By defining such things as **mass**, velocity and attraction, it became possible to express every form of motion, from Galileo's pendulum to planetary rotations, within one Universal Law.

▼ The famous Royal Observatory at Greenwich, established in 1675 in an effort to solve the problem of longitude at sea.

problem at sea using the lunar parallax could be found. The Reverend John Flamsteed became the first Astronomer Royal at Greenwich, though he had to provide the clocks and instruments from his own resources. Between 1675 and his death in 1719, Flamsteed began an entirely new catalogue of the northern heavens. He was a meticulous observer, and devised and established the working procedures for a '**meridian** transit' observatory as they would survive down to the twenty-first century. Without the new standards of accuracy established at Greenwich, the predictive power of Newtonian gravitation would have lacked its physical basis in observed nature.

HISTORY OF ASTRONOMY

The History of Astronomy
EARLY MODERN COSMOLOGY

At last the stage was set for scientists to make real inroads into the understanding of cosmological bodies. Through the increasingly powerful telescopes built by William Herschel, the changing nature of the stars was discovered, the seeds of the universal laws of gravity were sown and cosmology finally gained a firm observational basis. The study of heavenly bodies was vastly opened up and with the predicted return of Halley's Comet in 1758 the era of new modern cosmology was deemed to have arrived.

WILLIAM AND CAROLINE HERSCHEL

Sir William Herschel (1738–1822, knighted 1816), the son of an army bandsman, was born in Hanover, Germany. Coming to England as an army oboist in 1757, he decided to stay, and rose to eminence as a fashionable musician in Bath, where he was joined by his younger sister, Caroline (1750–1848). Caroline recorded in her journal that William took up astronomy as a hobby and soon found that he could make reflecting telescopes of superlative quality. Though entirely self-taught in both optics and astronomy, Herschel became fascinated by the nature of the

THE IMPACT OF NEWTONIAN GRAVITATION

Newton's Gravitation Laws offered a mathematical key to the ancient problems of how astronomical bodies move, and throughout the eighteenth century astronomers refined the details. For example, the English astronomer James Bradley discovered two new motions of the Earth that accorded perfectly with Newtonian physics. When trying to measure the annual **parallax** of the star Gamma [γ] Draconis in 1728, Bradley detected an annual motion for the star that was at 90° to the expected parallax. Bradley chose Gamma Draconis because it passes directly overhead in London, and to discount any errors caused by atmospheric **refraction**, he observed it with a Zenith Sector – a long refracting telescope, hanging like a pendulum in the **meridian**. Bradley realized that the motion must be caused by the Earth approaching and receding from Gamma Draconis every six months. Bradley's discovery of the aberration, as he called this motion, was the first observed proof that the Earth moves in space.

In 1748, using the same Zenith Sector, Bradley discovered a second motion, a slight rocking of the Earth on its axis, caused by the gravitational attractions of the Sun and the Moon. This was called nutation, from *nutare*, the Latin for 'to nod'.

EDMOND HALLEY AND STARS

One of the great questions being asked by astronomers at this time concerned the size and permanence of the stellar Universe. Since antiquity, astronomers had spoken of the stars as fixed, or

◀ *William Herschel and his sister Caroline, observing the stars and planets. William is credited with the discovery of the planet Uranus, while Caroline discovered no fewer than eight comets.*

stellar Universe, and started to sweep the sky through a series of zones, to build up a whole-sky picture of double-star and nebula distribution. His great discovery came in March 1781, when he chanced upon the previously unknown planet Uranus; this shot him to international fame, and won him the friendship of King George III. Abandoning his musical career in 1782, William moved with Caroline to Slough, near Windsor, and began a 40-year career in cosmology. Caroline soon revealed herself to be an astronomer of great talent. In addition to assisting William, Caroline discovered eight comets in her own right, and after his death in 1822 was honoured by the Royal Astronomical Society, receiving its Gold Medal, for completing his work.

▶ *Edmond Halley, whose observations of the changing positions of stars forced astronomers to reassess their belief in the fixed nature of the Universe.*

unchanging, but in 1715 Edmond Halley published a paper describing six stars that no longer looked the same as they had done to earlier observers. **Brahe**'s **supernova** of 1572 had disappeared altogether, whereas other stars, such as Mira, varied in brightness. In 1716 Halley drew attention to the existence of six 'lucid spots' or telescopic **nebulae**. In 1718, moreover, he further discovered that three bright stars – Aldebaran, Sirius and Arcturus – had shifted position in relation to other stars since **Ptolemy**'s time. They had independent proper motions in space. Halley went on to ask some of the most searching questions in cosmology: how could stars, when studied with telescopes, be found to have changed in position and brightness over human history, and how did the stars relate to the nebulae? And why, if the Universe contained uncountable millions of telescopically visible stars in all directions, did the sky ever go dark at night? This problem would later be redefined as Olbers' Paradox. Indeed, instead of being fixed, the stellar Universe seemed both infinite and dynamic.

STAR CLUSTERS AND EIGHTEENTH-CENTURY BLACK HOLES?

By 1720 Halley was arguing that if the Universe was finite, and had a centre, then it would *implode* under the force of its own gravity. The Universe's stability, therefore, meant that it must have no centre and an infinite **mass** distribution.

In 1767 the English scientist John Michell published his work on the Pleiades. Increasingly powerful telescopes showed more and more stars in the Pleiades cluster, which led Michell to ask what forces were drawing the stars together. Michell also argued that it was statistically unlikely that the large number of **double stars** in the sky were simple line-of-sight doubles, but that they were probably connected pairs. Unfortunately, at the time it was impossible to demonstrate that gravity was the agent that caused clusters and pairs of stars; proving this would fall to future astronomers. In 1784, Michell calculated that if a star or cluster of stars attained a particular mass, then its gravitational attraction would be so great

JOHN HARRISON AND LONGITUDE

The Royal Observatory at Greenwich, had been founded in order to catalogue the heavens so that the complex orbit of the Moon could be used to enable navigators to fix their **longitudes** at sea. To find one's longitude, in fact, it was necessary to have an exact timekeeper so that the time (and hence, angular) difference between a ship and its port of departure could be fixed. The Moon's parallax among the stars was one such timekeeper, but it was not until improved tables, instruments and calculating techniques became available, after around 1760, that finding the **latitude** by the Moon, or 'lunars', became feasible in practice. There was another method of fixing the longitude angle, and this depended on carrying an accurate clock on board ship. A challenge was thrown down, offering a prize to the man who could solve the longitude problem and invent an accurate timekeeping mechanism for use at sea. It was John Harrison (1693–1776), a Yorkshire carpenter, who set himself the life-long task of developing this chronometer, or sea clock, capable of keeping time to the one minute of error per month allowed by the Longitude Act of 1714. This was a huge task, for Harrison had to design mechanisms to compensate for temperature changes and the rolling of the ship. By 1761, he had produced four chronometers, each one an improvement on the one before, and while the Admiralty backed the cheaper lunars method, Harrison eventually earned his £20,000 prize in 1773.

▼ *Illustration dating from* about 1750, showing the growing popularity of astronomy and instruments at that time.

THE RETURN OF HALLEY'S COMET

Since remote antiquity, comets had been objects of fear and confusion, coming from nowhere and seeming to follow no fixed laws. Tycho Brahe, however, had pioneered their scientific study, and John Flamsteed suggested from his observations that the bright comet of 1680 passed in its orbit behind the Sun. But it was Sir Isaac Newton's *Principia*, Book III (1687), which showed that comets were astronomical bodies moving in mathematical orbits.

In 1705, Edmond Halley published an analysis of cometary records, and argued that the bright comets of 1682, 1607 and 1531 were apparitions of the same body, moving in an elliptical orbit between the Sun and Jupiter. Halley argued that this comet was a planet-like body moving under the same gravitation laws as everything else, and predicted that it would return in 1758 or 1759. Astronomers across Europe and America kept watch, but it was a German amateur astronomer, Georg Palitzch, who caught the first glimpse of

the returning Halley's Comet on Christmas night, 1758. Cometography became one of the great triumphs of eighteenth-century astronomy. Charles Messier, Caroline Herschel and others discovered new comets, while Leonhard Euler and Heinrich Olbers developed the gravitational mathematics that tamed these once frightening objects.

▼ *Halley had stated* that comets orbited the Sun and were subject to the same laws of gravitation as the planets; the return of Halley's Comet in 1758, just as the astronomer had predicted, proved him right.

that even light would not be able to escape from it. It would be what we now call a **black hole**. The ideas of Halley, **Immanuel Kant** (1724–1804), Thomas Wright (who had suggested that the **Milky Way** was a ring of stars) and Michell opened up wonderful possibilities for scientific cosmology.

COSMOLOGY ACQUIRES AN OBSERVATIONAL FOUNDATION

All these cosmological discoveries were the products of very localized studies. What was needed to progress was a massive increase in observational data upon which broader theories could be built. The next important advance came in 1781, when Charles Messier (1730–1817) in Paris published his catalogue of 103 fuzzy objects that had often confused **comet** hunters. What were these nebulae, and why did they seem most common around the fringes of the Milky Way? In the early 1780s, William Herschel began the first all-sky survey for nebulae and star clusters, and by the end of the decade, Herschel had discovered 2,000 such objects. At first he believed that the nebulae were immensely remote clusters of stars, not all of which could be resolved individually, but which produced a common glow. He suggested that gravity was probably the force causing this 'condensation' of star fields into nebulae. The Pleiades and the Hercules Cluster (Messier numbers 45 and 13) were probably fairly close. By 1791 Herschel had realized that, in addition to stars, space also contained shining

fluid, or glowing gas, as we would call it. Did this fluid power the stars? And was the once 'fixed' Universe really a dynamic place of stellar births and extinctions?

HERSCHEL'S TELESCOPES

One major component in Herschel's success as a cosmologist was his genius as a mirror maker, which enabled him to build telescopes of greater space-penetrating power than ever before. Herschel was one of the first scientists to realize the difference between simple magnification and light grasp. The bigger the **aperture** of a telescope, the more starlight can enter, focus into the observer's eye, and enable him to see deeper into space. Up to this time, only small-diameter glass lenses had been built and so Herschel set about improving the reflecting telescope developed by **Newton** a century before. Instead of glass, reflecting telescopes deploy a mirror made of bright metal, figured into an exact parabolic curve to catch and focus the light. Such mirrors can be cast in much greater sizes than glass lenses. Herschel's first telescopes had mirrors of 16 cm (6½ in) in diameter, with a focal length of 2 m (7 ft). Herschel's work won the support of King George III and from 1781 onwards, he continued to develop mirrors that would allow deeper observation. By enabling him to see farther into space than ever before, Herschel's telescopes transformed the observational basis of cosmology.

Brahe ▶ p. 38 Kant ▶ p. 136 Newton ▶ p. 42 Ptolemy ▶ p. 30

EARLY MODERN COSMOLOGY

HISTORY OF ASTRONOMY

The History of Astronomy
THE NINETEENTH CENTURY

Throughout the nineteenth century, large numbers of scientists and astronomers raced to make the next big discovery about space, the Universe and the rules governing everything in it. Bigger and better telescopes were constructed, revealing new data about the nature of stars and nebulae. The new science of astrophysics took off in a big way; the chemical make-up of stars and the Sun was discovered. But as astronomy became even more costly to research, and governments and universities backed it, the science passed from the wealthy amateurs to paid professionals.

GRAVITY AND BINARY STARS
In 1802 **Herschel** published a list of **double stars** in which the position of one or both component stars had changed since 1782. In short, these were binary (or triple) stars moving around each other as a physical system, just as John Michell had first suggested in 1767. But there still remained the question: what was the force that governed their motion? Gravity was the obvious candidate, but to prove it would need decades more of observation, and greater developments in Newtonian mathematics, especially calculus. Astronomers across Europe were now using powerful telescopes and delicate eyepiece micrometers to monitor the motion of Herschel's binaries, as well as discovering scores of new ones. Wilhelm Bessel, Wilhelm Struve, Sir John Herschel (Sir William's son), Sir James South and many others were competing in the binary race to see if the motions of these stars corresponded

with Newtonian criteria – and in 1830, the French astronomer Felix Savary found that they did.

Also crucial to understanding the binaries were the rapid developments in mathematical analysis pioneered by **Pierre Simon LaPlace** (1749–1827) and his circle, for Paris led the world in higher mathematics at the time. But if the laws of gravitation had been shown to apply to stellar systems, the next question was: did they also apply to the remote **nebulae**?

NEW TELESCOPE TECHNOLOGIES
The international investigation of **binary stars** only became possible in the wake of a new type of refracting telescope. Though John Dollond had patented the first achromatic lenses in 1758, it was the Munich optician **Joseph von Fraunhofer** (1787–1826), who made fundamental advances after 1805. Fraunhofer produced bigger-**aperture**, clearer, better-figured lenses than ever before, using his own secret processes. Realizing that a telescope was only as good as its stand, he also developed the counterpoised, clock-driven German **equatorial mount**. Fraunhofer's refractors yielded exquisite star images that could be automatically guided for hours on end, thereby enabling delicate binaries to be accurately measured.

By the 1830s, Lord Rosse and William Lassell were applying the technology of the Industrial Revolution to the development of the reflecting telescope. Metal casting had improved since Herschel's day, making it possible to cast superior 'speculum' metal mirrors, while both Rosse and Lassell used steam engines and heavy machinery to impart perfect optical curves to mirrors of 2, 3, 4 and 6 feet aperture. Lassell solved the complex engineering problem of mounting a large reflecting telescope in the equatorial plane in 1845. Its sharp, bright, steady images made it the ancestor of every modern big reflector.

◀ *Model of the 6-ft telescope built by the Irish astronomer Lord Rosse; this telescope allowed more detailed observations of distant galaxies.*

THE PROBLEM OF THE NEBULAE
The laws of gravitation had been shown to apply to the stars by 1830, solving one of the great intellectual problems in cosmology. But did gravity also apply to the nebulae? And what were the nebulae made of: individual stars, or Herschel's luminous fluid? It was while in pursuit of answers to these questions that the Anglo-Irish aristocrat Lord Rosse had built his 36-in (90-cm) and 72-in (183-cm) reflectors at Birr Castle, Ireland, in 1840 and 1845. Until this date, all the 5,000 or so recorded nebulae (including those discovered in the

▼ *The nineteenth century saw the rise of a new type of astronomer – the Grand Amateur. These were usually wealthy gentlemen who had the money to fund research and experiments. Professional astronomers at the time were restricted in the work they were able to conduct. Pictured is the interior of the Radcliffe Observatory, Oxford, in 1814.*

⌛
PROFESSIONAL AND GRAND AMATEUR ASTRONOMERS
Nowadays, all front-rank astronomical research is performed by salaried professionals, but 150 years ago, things were very different. While it is true that in autocratic countries such as Prussia or Russia, government money paid for prestigious research, in libertarian Britain and America low taxation meant that scientists often had to pay their own way. Official salaried scientists in Britain, such as the Astronomer

Royal, were expected to keep to routine positional work that was useful for Admiralty tables. If one wanted to investigate binary stars, nebulae or astrophysics, however, the work had to be paid for privately. Self-funded or Grand Amateur scientists dominated British astronomical research. They included Lord Rosse, William Lassell (a brewer), Sir James South (a surgeon), Isaac Roberts (a builder) and privately wealthy gentlemen. In the US there was William Cranch Bond (who founded the Harvard

Observatory), the spectroscopists Henry Draper (a doctor), Lewis Rutherfurd (a lawyer), and others. Others were industrialists, such as the telescope designer James Nasmyth. Their money allowed the development of the equatorially mounted reflecting telescope, the spectroscope and photography. These Grand Amateurs also ruled the Royal Society and Royal Astronomical Society, often making the professionals feel like hired hands. But with the end of Victorian prosperity, their star was setting.

THE DISCOVERY OF NEPTUNE

Following Herschel's discovery of Uranus in 1781, its orbit was computed using Newtonian theory. By the 1820s, however, Uranus was straying from this predicted orbit and astronomers were faced with two possible explanations. Either there was an as-yet unknown body further out in space that was disturbing Uranus, or **Newton's laws** were wrong. In the early 1840s an Englishman and a Frenchman, John Couch Adams (1812–92) and Urbain J. J. Le Verrier (1811–77), began quite independently to investigate this problem. While equally brilliant as mathematicians, Le Verrier was confident and already famous, whereas Adams was shy and had only just completed his Cambridge undergraduate degree. Both men worked on the assumption that there was an unknown body acting on Uranus. Working from the discrepancies between observed and predicted places for Uranus since 1781, they both identified the section of sky where the unknown planet must be. Adams got a preliminary result in September 1845, but his failure to reply to the letters of the Astronomer Royal meant that his work was not followed up. Conversely, Le Verrier persuaded the Berlin Observatory to search in his predicted place, and on 23 September 1846, Neptune was found. Both men were later honoured for their work.

southern hemisphere) had appeared only as vague, glowing patches. But when Lord Rosse looked at the nebula M51 in 1845, the 72-inch-aperture telescope revealed a distinct right-hand spiral, spangled with stars, subsequently named the 'Whirlpool'. Lord Rosse detected structures and star fields in about a dozen other nebulae which suggested (but did not prove) that they were gravitational systems of stars. But what about the thousands of nebulae inside which no structure was visible, such as the Great Nebula in Orion? Were these nebulae island universes so vastly remote from the **Milky Way** that even Rosse's 72-inch could only pick up a dim glow? Or were they made of gas?

THE RISE OF ASTROPHYSICS

The first scientific use of the **spectroscope** was in chemical analysis in the laboratory, when scientists realized that each chemical element, when made incandescent in a beam of light, writes its unique signature in the **spectrum**. In 1859, Robert Bunsen (1811–99) and Gustav Kirchhoff (1824–87) in Heidelberg identified sodium in the sunlight entering their laboratory. By the early 1860s, spectroscopes were being attached to telescopes, and two wonderful facts emerged. Firstly, astronomical

objects contain the same basic chemical elements as we have on Earth. Secondly, the elements seem to be mixed in different ways in individual stars. A new key to cosmology, in addition to bigger telescopes, had suddenly become available. Angelo Secchi (1818–78) in Rome produced a spectral classific-ation of stars in 1863, while in England, Sir William Huggins detected gas in six nebulae in 1864. The difference between stellar and gaseous nebulae became immediately obvious when examining them through the spectroscope. Huggins and his wife Margaret, together with Henry Draper (1837–82) in New York, photographed the spectra of thousands of objects over four decades, while Sir Norman Lockyer and the Parisian Jules Janssen (1824–1907) studied the physics of the Sun. By 1880, the new science of astrophysics was transforming cosmology.

EVIDENCE FOR THE BIG GALAXY UNIVERSE

Bigger telescopes, spectroscopy and astronomical photography changed the whole scope of astronomy after 1860. Yet while it was now possible to define the chemical make-up of the Universe, the big questions remained unanswered: just how big was it? And did gravity rule everything? Of course, astrophysics had shown that stars *seemed* to have life histories, but how far away were the nebulae? Were they remote island universes, or was every-thing contained within one big galaxy? Astronomers realized that if they could detect change in the nebulae, then they were probably relatively local. Then in 1885, a new star flared up in the Andromeda Nebula, M31, and within days accounted for 10%

▲ *Sir William Huggins, who was a pioneer in recording stellar spectra photographically during his 40-year career.*

of the light of the entire nebula. Surely, astronomers argued, if a single star was so bright, then the nebula could not be too remote. Also in the 1880s, Draper, Andrew Common (1841–1903) and Isaac Roberts (1829–1904) were obtaining photographs of galactic structures, from which it was hoped that movement would be detected. In 1900, Roberts first detected the spiral arms of the Andromeda Nebula, showing it to look like Rosse's Whirlpool seen side on. Surely gravity was at work here. All the evidence seemed to argue not for island universes but for one big galaxy.

THE SPECTROSCOPE

In 1666, **Newton** split sunlight into its component colours using a prism and since this time further experiments had been conducted to try and under-stand the nature of light more comprehensively. In 1802 William Wollaston (1766–1828) found that when he substituted a fine slit for Newton's pinhole as his light source, a few inexplicable black lines were seen amongst the colours. Wollaston's black lines were re-discovered by Fraunhofer, who believed that he could use them as neutral, or colourless light, when testing telescope lenses. Fraunhofer then refined the Newtonian apparatus, by putting his prism before the object glass of a theodolite (a surveying instrument for measuring horizontal and vertical angles), and in 1814–15 he mapped the positions of 574 black 'Fraunhofer lines' seen in sunlight. But the precise nature of these lines remained a mystery.

The breakthrough came in 1849 when Leon Foucault (1819–68) found that when he burned sodium in an electric arc in a ray of sunlight passing into a spectroscope, the black 'D' lines in Fraunhofer's solar map suddenly blazed yellow. But it was Bunsen and Kirchhoff who took Foucault's discovery to new heights by realizing that while an element burned in a laboratory flame produces a coloured line, light from the Sun passes through a solar layer which results in its producing a black line. By incandescing laboratory chemicals and painstakingly matching the positions of their colours to the black Fraunhofer lines, one could identify elements in the Sun.

▶ *A nineteenth-century spectroscope, with seven prisms clearly showing, used to view the spectra of prominences outside a solar eclipse.*

Fraunhofer ▶ p. 82   W. Herschel ▶ p. 44   LaPlace ▶ p. 172   Newton ▶ p. 42

HISTORY OF ASTRONOMY

The History of Astronomy
THE TWENTIETH CENTURY

The great discoveries made in the nineteenth century only laid the groundwork for the greater ones to follow in the the twentieth. The Great Debate raged: was there just one big galaxy or was space made up of infinite numbers of island universes? As the century progressed, the idea of an expanding Universe, born out of a massive Big Bang came to be accepted. Cosmologists turned their attention to the questions of just how big the Universe was. By the dawn of the new millennium, astronomy had made immense advances, technically and conceptually, since antiquity. Yet still so many questions remained unanswered.

CHANGING PERSPECTIVES
By 1900 most astronomers were confident that the fundamental discoveries in astronomy and physics had been made. Gravitation had been shown to provide answers for motion of all kinds, from laboratory mechanics to stellar dynamics, while the bulk of deep-space evidence seemed to argue in favour of one big galaxy – based on the **Milky Way** system – rather than scattered island universes. In many ways, however, the titanic changes which were to engulf twentieth-century physics and cosmology were first glimpsed in the laboratory rather than the observatory. The 1887 experiments of Albert Michelson (1852–1931) and Edward Morley (1838–1923), to detect the Earth's motion against that mysterious ether which was so often invoked to explain anomalies in planetary motion, drew a disturbing blank. At the same time, the solid Daltonian theory of the atom was being gradually undermined by new physical evidence. Exactly what were the matter-penetrating **X-rays** discovered by Wilhelm Röntgen (1845–1923) in 1895 and what were the forms of energy being emitted by uranium and radium? After 1908, **Ernest Rutherford** (1871–1937) began to reinterpret the atom in terms of the **electrons** and **protons** that we know today, while **Albert Einstein**'s (1879–1955) $E = mc^2$ was turning accepted concepts of space and time on their heads.

THE BIG AMERICAN TELESCOPES
At the same time as laboratory physicists were developing new and peculiar concepts of matter and energy, another form of instrumentation was about to revolutionize observational astronomy: the big American telescopes. Early twentieth-century America enjoyed a combination of circumstances which proved indispensable to the new astronomy. Wealthy millionaires such as James Lick, Leland Stanford and Charles T. Yerkes were keen to immortalize their names in the endowment of great observatories and scientific universities, creating unprecedented employment opportunities for research scientists, backed up by scholarships which could attract the brightest students. And the rapidly advancing technologies made even larger telescopes possible, invariably mounted in prime dark-sky locations. At first, the race was on for big **aperture** refractors, and by 1898 the Lick and Yerkes Observatories had 36-in (91-cm) and 40-in (1-m) telescopes respectively. But it was the 60-in (1.5-m, 1908), and especially the 100-in (2.5-m, 1917) reflectors on Mount Wilson, California, that were to change astronomical thought in the twentieth century. For it was with the Mount Wilson 100-in (2.5-m) instrument that **Edwin Hubble** (1899– 1953) would settle the question of the 'island universes' after 1920. And when a 200-in (5-m) reflector was built on Mount Palomar in 1948, the vastness of the Universe was confirmed.

▼ *The 100-in (2.5-m) reflector* at the Mount Wilson Observatory in California.

▲ *Henrietta Leavitt* (right), whose studies of the Small Magellanic Cloud led to the discovery of the period–luminosity correlation which was to establish the existence of galaxies beyond our own.

HENRIETTA LEAVITT AND THE CEPHEID STARS
Astronomers needed a yardstick into space, however, if a breakthrough in deciding between the big galaxy and the island universes theories was to be made. Between 1908 and 1912 Henrietta Leavitt, in her analysis of photographic plates of the Small **Magellanic Cloud** taken by the Harvard Observatory's telescope in Peru, discovered a curious thing: all the **variable stars** which showed a light-output curve similar to Delta [δ] Cephei – or **Cepheid variables** – shared a mathematical relationship. The brighter the **absolute magnitude** of the Cepheid, the longer it took for the particular star to go through its light cycle. Because all the stars in the Magellanic Cloud were virtually the same distance from Earth, they provided an ideal laboratory in which to quantify the **period–luminosity** patterns of bright and dim Cepheids.

THE SIZE OF THE UNIVERSE
It was to be the period–luminosity correlation of the Cepheids which finally enabled astronomers to decide between the big galaxy and island universe cosmological models. From a complex statistical analysis of Cepheids in **globular clusters** of stars, Harlow Shapley (1885–1972) at Mount Wilson presented evidence in 1917 which argued that there really was only one big galaxy, some

ALBERT EINSTEIN AND RELATIVITY

Nineteenth-century physics had developed a model of the Universe in which three dimensions of space and an even flow of time explained all motions in terms of gravity. Yet anomalies in Mercury's orbit and several problems in laboratory physics failed to fit the theory. These questions were to be solved by Einstein's theory of relativity. In the Special Theory (1905) Einstein demonstrated that time and motion were not absolute, but relative to the observer. But it was his General Theory of 1915 which opened the doors to an entirely new cosmology, to explain the relationship between matter, energy, space, time and gravity. For one thing, Einstein's theory provided an exact explanation for the advance of the **perihelion** of Mercury's orbit. His 1915 theory also predicted that, due to the relation between electromagnetism and gravity, light could be bent by astronomical bodies. **Sir Arthur Eddington** (1882–1944) provided a spectacular vindication for this prediction when he measured starlight deflected by the Sun during the 1919 eclipse. Einstein's equations also predicted that the Universe was expanding – as Hubble and Milton Humason had observed – and explained the gravitational cause of redshift. Relativity is the key to modern cosmology, bringing laboratory physics, optical and radio astronomy together to explain space and time.

◀ *The greatest theoretical physicist of modern times: Albert Einstein. Einstein's theories of relativity revolutionized ideas about the nature of the Universe and space–time.*

300,000 **light years** across. Yet when Vesto Slipher (1875–1969) and Heber Curtis (1872–1942) quite independently studied the spectra and sky distribution, not of globular clusters but of **spiral galaxies**, a different picture began to emerge.

One of the biggest arguments in favour of the big galaxy was the absence of spirals, clusters and other complex objects from the region of the galactic plane. Curtis realized, however, that when photographing edge-on spirals a band of dark obscuring matter always existed around the galaxies' extremities. If our own Galaxy was but a spiral amongst spirals, then the reason why we see nothing in the galactic plane was probably because the light from the island universes could

not penetrate our own galactic dust bands. In short, the spirals were probably 'inconceivably distant', and not part of our own Galaxy, argued Curtis. The pros and cons for the big or island galaxies formed the subject of the Great Debate that took place between Shapley and Curtis in Washington in 1920, though their arguments were still evenly matched.

Then, in 1923, Hubble, working with the 100-in reflector at Mount Wilson, discovered a Cepheid in the spiral nebula in Andromeda, M31. Using Henrietta Leavitt's period–luminosity ratio, he originally calculated it to be 900,000 light years away, and unequivocally an island universe – what we now term a galaxy. And if the Andromeda spiral was an island universe, it became likely that the other spirals were likewise.

Building on the knowledge that receding objects display spectra that are displaced to the red end – or **redshifts** – astronomers had discovered by 1940 that the dimmest cosmological objects invariably had the biggest redshifts. Was the entire Universe expanding from the site of a formative Big Bang, or was it forever renewing itself in a Steady State? Only better data could decide.

SPACE PROBES

By the mid-twentieth century the Big Bang, expanding, Universe model was generally accepted. Then the information available to astronomers was yet further increased by the new technologies of space probes and computing science. Though planetary probes dated back to 1957, by 1989 the unmanned Voyager mission was relaying images which fundamentally altered previous knowledge of the outer **Solar System**. The landing of the

💡 HUBBLE'S CONSTANT AND THE EXPANSION OF THE UNIVERSE

By 1925, Hubble had shown the Andromeda 'nebula' to be an island universe, or galaxy, in its own right, though his estimates for its size and distance were both much smaller than those accepted today. With his colleague Milton Humason, Hubble began to collect data on the redshifts of distant spirals, especially if he could find Cepheids in them. A redshift is found only in deep-space objects that are receding from our own Galaxy. The faster the recession, the more the object's light is displaced, or shifted, towards the red end of the spectrum. In 1929, Hubble presented evidence indicating that the most distant galactic objects were receding fastest. This was the observational germ of the later Big Bang theory, especially as Hubble's and Humanson's observations corresponded with Einstein's equations, which said that the Universe should be expanding. During the 1930s, Hubble and Humason collected data from a number of remote galactic objects, and found that a clear redshift correlation held: the more pronounced the velocity of recession, the greater the object's spectral redshift. Hubble's Constant for the rate of expansion of the Universe came to lie at the heart of the Big Bang theory.

Sojourner wheeled space laboratory on Mars in 1997, combined with images of the surface from probes in orbit, took geological studies to the planets. But it was the Hubble Space Telescope, launched in 1990, which produced thousands of breathtakingly detailed images which will keep cosmologists occupied for many years to come.

⎍ RADIO ASTRONOMY

Since time immemorial, astronomers had worked in one electromagnetic **wavelength**: visible light. And while in 1859 scientists discovered that outbursts of solar energy could induce electric currents in telegraph wires, it was not until 1932 that Karl Jansky (1905–49) accidentally detected radio emissions from the Sagittarius region of the Galaxy.

In 1937, Grote Reber (b. 1911) built a 9-m (31-ft) antenna to detect astronomical radio sources. It was the enormous advances in electronic engineering during World War II, however, that really made radio astronomy possible. Sir Martin Ryle (1918–84) at Cambridge pioneered long, fixed-baseline radio telescopes, while **Sir Bernard**

Lovell (b. 1913) at Jodrell Bank, Manchester, developed the steerable dish. Radio telescopes operated by collecting those long-wave radio signals that originate in the depths of space, and which can pass through cosmic dust and earthly clouds. This new observational window transformed astronomy after 1950. Radio telescope link-ups or **interferometers** across the world made it possible to identify the exact positions of radio sources, to match them up with visible light sources and to detect **pulsars** and other objects. And in addition to radio, modern astronomers also study X-ray, **gamma ray** and other non-optical sources, thereby transforming our knowledge of the composition of space.

The History of Astronomy
SUMMARY

I t is impossible to think of a subject as old as astronomy in cultural isolation. For dealing as it does with those great celestial cycles that measure out human lives, as well as pre-empting questions about infinity, design and purpose, astronomy has played a profoundly formative role in shaping civilization. Indeed, from the first realization by the ancient Egyptians that the pre-dawn rising of Sirius heralded the flooding of the Nile, to modern-day questions about what could have existed before the Big Bang, astronomy has always challenged us on our deepest levels. Astronomy also inspired some of the world's oldest surviving literature, such as the Egyptian regeneration myths of Nut and Ra, the Greek story of the impudent young Phaethon who almost set the world ablaze with the Sun's chariot, as well as numerous passages in the Bible. Astronomy, moreover, has always required mankind to

▼ *Astronomy inspired some of the earliest myths and legends for the ancient peoples, such as that of the Egyptian Sky Goddess Nut.*

bring together a powerful collection of insights and techniques to help us to reach our conclusions: observation, philosophy, geometry, technology and questions about infinity. And among other things, astronomy formed the archetype for all the sciences.

A MODEL FOR ALL THE SCIENCES

- One of the earliest intellectual puzzles posed by astronomy to ancient cultures was why the celestial cycles were so regular in contrast with the earthly forces of wind, fire, death and regeneration, which seemed so chaotic.
- If everything that exists had been created by one all-powerful divine intelligence, as the Greek Neoplatonists, the Jews, Christians and Muslims all agreed, why could not that same logic clearly present in the heavens also be traced in medicine, biology, physics and chemistry? This approach gave rise to the first rational taxonomic studies in the life sciences, and to applied geometry in terrestrial physics.
- While **Galileo** discovered the mathematical key to physics, to explain the motions of flying arrows and falling weights by 1630, it was not until after 1800 that mathematical predictability came to chemistry with Dalton's atomic theory and Mendeleev's Periodic Table.
- In 1866 Gregor Mendel unlocked the key to the life sciences in the mathematics of heredity. This in turn led Crick and Watson to discover that cell replication itself appeared to be governed by mathematical sequences no less precise than the degrees of a planetary orbit in the double helix of the DNA code.
- Without the example of astronomy to follow, one might argue that the modern sciences would never have developed.

THE IMPORTANCE OF INSTRUMENTS

- The regular motions of the heavens, combined with the 360° circle, enabled astronomers to measure and describe the heavens in remarkably exact terms by 300 BC.
- The **armillary sphere**, quadrants, triquetrum, sundial and astrolabe facilitated the theoretical modelling of the Universe, as **epicycles**, ellipses and other geometrical constructs could be tested against the hard evidence of observed degrees and minutes.
- Such instruments enabled that sophisticated flowering of astronomy that we see in **Ptolemy**, the Arabs, late medieval and Renaissance Europe.

- The designs and accuracy of these instruments remained relatively unchanged for centuries, as techniques of circle division and the physiological resolving power of the naked eye imposed fixed limits beyond which measurements could not be made.
- After 1570, two fundamental changes took place. Firstly, **Brahe** invented new techniques for dividing circles down to single arcminutes, while at the same time initiating what would become an ongoing engineering revolution in instrument design and manufacture. Secondly, Galileo broke through the eye-resolution barrier when he brought into use astronomy's most far-reaching instrument: the telescope, in 1610.
- When improved divided instruments were combined with telescopes, the accuracy of astronomical measurement improved over a hundredfold in less than a century.
- The superior instruments of Flamsteed, Halley and Bradley after 1675 not only clinched the Copernican theory, but also produced the data on which Newtonian gravitation stood.
- Better object glasses, **Herschel**'s great reflectors, **spectroscopes**, photography, big American reflectors, radio telescopes and the Hubble Space Telescope all followed in the same tradition.

▼ *Galileo's development of the telescope opened up the wonders of the Universe as they had never been seen before.*

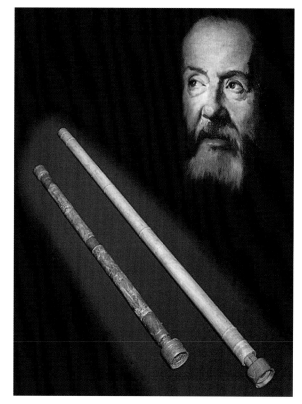

SUMMARY

HISTORY OF ASTRONOMY

WATERSHEDS IN ASTRONOMICAL UNDERSTANDING

- For two millennia, from Hipparchus to **Kepler**, the physical explanation of planetary motion in terms that conformed both to instrumental observations and to mathematical theory was the overwhelming concern of astronomers.

- The abandonment of crystalline spheres and **geocentric** epicycles in favour of Sun-centred ellipses and variable planetary speeds through seemingly empty space by 1620 was perhaps the biggest watershed in the history of astronomical thought. Without it, Newtonian gravitation would not have been possible. And without gravitation, those concepts of **mass**, attraction, and inertia so central to the development of post-1700 mathematical physics could not have existed.

- Though the infinite Universe had been speculated upon in the Middle Ages, it was seventeenth- and eighteenth-century telescopic observations that gave it physical plausibility.

- Each increase in optical power, from Galileo's first views of the **Milky Way** to Herschel's catalogues of dim **nebulae**, suggested that the Universe went on to infinity.

- It was that watershed instrument, the telescope, which made all this possible, augmented by observations at radio and other non-visible **wavelengths** in the twentieth century which revealed a Universe that was infinite and unbelievably strange.

- This eventually revealed a Universe based on Einsteinian relativistic concepts of matter, energy, time and motion, quantum physics, and populated by particles acting in warped space, in an expanding Universe.

- Running through all of this progressive knowledge is a concept of astronomy (and extending from it, the rest of the sciences), which is experimental. From Hipparchus's epicycles to Einsteinian space, theory had to be matched by observed reality, and not just speculation. And as astronomers recognize that by describing and explaining observed reality, they cannot presume to speak upon the un-observed origins, purposes, or ultimate ends of the Universe.

▶ *The Copernican system, the revolutionary new model of the Universe which suggested that the planets revolve around the Sun, rather than the Earth, as had been believed previously.*

THE PUBLIC'S FASCINATION WITH ASTRONOMY

While few people today would claim that modern amateurs are likely to change astronomical thought, the international explosion of amateur astronomy is profoundly important to the science. For not only do serious amateur observers monitor **asteroids**, **supernovae** and other objects of interest to the professionals, but they also form a vital bridge to the general public. The amateur community has been a major driving force in taking astronomy to school children and to ordinary people. And through the extensive media coverage of the spectacular pictures returned from space probes and the Hubble Space Telescope, with the inevitable questions which they pose about origins and infinity, the public's fascination with astronomy is greater than ever before.

PROFESSIONAL AND AMATEUR ASTRONOMERS

- Over the greater part of scientific history, astronomy was done by people who earned their livings elsewhere. Greek astronomers kept libraries or taught students. Medieval astronomy professors were also primarily university teachers, as were their counterparts in the Islamic world.

- Research as we know it today did not really exist prior to the sixteenth century, and before that time astronomy largely consisted of checking the sky against accepted tables, and passing on an accepted tradition.

- It was the Copernican revolution which unleashed modern research astronomy, as scientists tried to find physical evidences for whether the Earth did or did not move.

- This demanded new and more accurate data, and figures such as Brahe and Galileo realized that by pushing human perception to ever-new heights, either by graduated instruments or through telescopic power, harvests of fresh physical evidence could be forthcoming.

- This new emphasis on a progressive technology and the skills to use it effectively demanded new types of funding, and in continental Europe especially, governments and monarchs footed the bill. Figures such as Brahe, Cassini, and Bessel came to do astronomy as full-time paid occupations.

- Though the self-funded Grand Amateur tradition developed in response to the political and economic circumstances prevailing in Victorian Britain, front-rank astronomical research had become thoroughly professionalized by the time of **Hubble**.

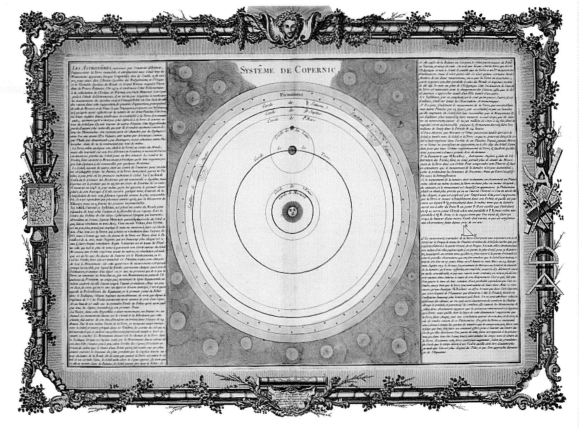

THE LAWS OF PHYSICS

Everything we know of the Universe tells us it is an orderly place, founded upon the same laws of nature that prevail on Earth. Light that has spent billions of years crossing the cosmos is identical to the light from a glowing campfire. The same gravity that plucks an apple from a tree drags torn-up stars into the black holes in the hearts of remote quasars. If beings elsewhere are contemplating their place in the cosmos, then we can be sure they, too, are made of atoms built from the same electrons, protons and neutrons that we find in our own corner of the Universe.

It was not always thought to be so. Until four centuries ago it was taken for granted that the heavens and the Earth occupied different realms and were governed by different laws. With the telescope came the idea that the heavens could be explored as the Earth was being explored; the planets became places and the stars became suns and the Earth was seen to be but a small part of a uniform universe.

Today's physics makes no distinction between the heavens and the Earth. Cosmologists can, and do, write equations that they hope will describe the origin and evolution of the Universe as a whole and perhaps predict its destiny. Those equations are based on a small number of physical laws laboriously uncovered on Earth and found to apply universally. Some of these ideas challenge everyday experience and can be hard to understand, but the fact that we can understand as much as we do about the Universe is remarkable enough.

THE LAWS OF PHYSICS

The Laws of Physics
INTRODUCTION

Physics is built up from a small number of principles that have been repeatedly tested and refined by experiment and observation. Although some of these have been given the status of 'laws', nothing in physics is sacrosanct. Newton's law of gravity, for example, has given way to Einstein's, and that in turn will be replaced by something more

◀ *Albert Einstein's* special and general theories of relativity changed the course of physics in the early years of the twentieth century. Now scientists are working on what is expected to be the next step – a quantum theory of gravity.

comprehensive still when a quantum theory of gravity is developed. This chapter reviews the basic principles of classical physics (including relativity), which still remain essential to understanding astronomy and astrophysics.

WHAT IS PHYSICS?

A dictionary might define physics as the study of matter and energy and the interactions between them. As such, physics is the most fundamental of the disciplines of science, dealing with very basic questions about the nature of the world in which we live. It establishes the foundations upon which other disciplines, including astronomy and astrophysics, are constructed.

Physics is built up from a small number of principles that govern the behaviour of matter and energy. The US physicist **Richard Feynman** (1918–88) once posed a question: if only one sentence of scientific knowledge could be passed to future generations, what should it be? His proposal was, 'All things are made of atoms – little particles that move around in perpetual motion, attracting each other when they are a little distance apart, but repelling upon being squeezed into one another'. Feynman's sentence is the principle of atomic theory.

This chapter will look at Newtonian mechanics, thermal properties of matter, **electromagnetic radiation** and relativity. The large and profound area of quantum physics is dealt with in the following chapter.

FOUR FORCES

All the phenomena known to physics can be understood in terms of no more than four fundamental interactions, or forces, between particles: these are the gravitational and electromagnetic forces, and the strong and weak nuclear interactions.

◔ The gravitational interaction acts between all bodies that possess mass. It is the weakest of the four, but the only one with sufficient range to act over distances on the scale of the Universe as a whole.

◔ The electromagnetic interaction acts between electric charges. It is much stronger than gravity, but is effectively neutralized over long distances by the fact that there are equal numbers of positive and negative charges. It binds atoms together.

◔ The strong interaction acts between protons and neutrons. It is very short-range, but binds the nuclei of atoms together

◔ The weak interaction is responsible for the decay of certain nuclear particles.

Classical physics is concerned only with the gravitational and electromagnetic interactions.

MAKING PROGRESS IN PHYSICS

Physics, like most areas of science, has grown by formulating, testing and rejecting many different *hypotheses* about the Universe. A hypothesis is essentially an educated guess, a plausible explanation for an observational fact.

Until the end of the nineteenth century it was believed that all space was filled with a substance called the ether. There was no evidence for the ether, but it was a long-standing hypothesis formulated to provide light with a medium in which to travel. One prediction from the ether hypothesis was that the speed of light should vary according to the direction of the Earth's motion in space. When the speed of light was found to be constant, the hypothesis had to be either amended or replaced. Both solutions were attempted, but when **Albert Einstein** (1879–1955) showed that an ether was unnecessary the hypothesis was finally abandoned.

A body of related hypotheses that have survived many tests can attain the status of a *theory*. A good theory must be able to account for all the experimental facts within its remit *and*

ARISTOTLE

Aristotle (384-322 BC) was one of the greatest of Greek philosophers, whose systematic and wide-ranging study of science formed the basis of Western thought for centuries. Aristotle considered the laws of physics in the heavens to be quite different from those on Earth. The 'natural' motion of the heavenly bodies was in circles (his universe was a construction of some 55 concentric spheres). The 'natural' motion

of objects in the terrestrial realm was either towards the centre of the Earth (for objects consisting predominantly of the elements of 'earth' and 'water') or away from the centre of the Earth (for objects consisting predominantly of 'air' and 'fire').

Aristotle also maintained that a force was needed to keep a body in motion. An arrow flying through the air, for example, was propelled by air displaced from the front rushing to the back and pushing the

arrow along. Because such motion depended on the presence of the air, Aristotle concluded that a vacuum was impossible, or else the planets would not be moving at all. Aristotle's highly systematic approach to science, flawed though it was, survived for 2,000 years. But today Aristotle's ideas on motion seem eccentric, and very little of Aristotle's 'natural philosophy' survived the rise of physics that began with **Galileo Galilei** (1564-1642) and Newton.

cosmic rays ▶ p. 337 electromagnetic radiation ▶ p. 338 galaxy ▶ p. 339 quasar ▶ p. 342 radiation ▶ p. 343 spectroscopy ▶ p. 344 spectrum ▶ p. 344

◀ **The technique** of spectroscopy is a powerful tool used by physicists and astronomers. Coloured emission lines, such as these from a Wolf-Rayet star, can yield information about the object's composition, temperature and other properties.

predict the outcomes of new experiments and observations in such a way that the validity of the theory itself can be put to the test. A good theory is capable of being proved wrong.

MODELS IN PHYSICS

Physics proceeds through the construction of *models* of the natural world. A model is a kind of analogy that represents a simplified version of reality.

An example of an extremely successful model is the 'ideal gas'. The model portrays the molecules of a gas as simple particles of zero size that bounce off the walls of their container but otherwise have no interactions. This very simple model accounts for much of the behaviour of gases, even though we know that molecules are much more complicated than simple particles. By extending the model, for example by allowing the particles to have a definite size, to attract each other at short range, and to bounce off each other on colliding, it can be made to conform even more closely to reality.

Not all models can be visualized as easily as the ideal gas. In some cases, such as the curved space of Einstein's relativity, the most precise expression of a model is in the language of mathematics. Sometimes more than one model is necessary. **Isaac Newton** (1642–1727) modelled light as a stream of particles. **Christiaan Huygens** (1629–95) modelled it as a wave motion. We now know that no one model adequately represents the nature of light: both the particle and wave models are needed to understand all the subtleties of light.

ASTROPHYSICS

Physics applied to astronomy becomes astrophysics, the science concerned with the physical conditions in the Universe beyond the Earth.

The first successful astrophysical theory was created by Newton, when he discovered that the same force of gravity that caused apples to fall from trees also guided the planets in their courses around the Sun. By extending terrestrial science to the cosmos, he showed that the whole Universe might be understood in terms of the same physical laws. But astrophysics really took off in the mid-nineteenth century with the rise of **spectroscopy**. Forming a **spectrum** of the light received from the stars made it possible to determine their composition, temperature, motions and other properties.

Astrophysicists ask such questions as how stars are born and how they die, how energy is generated inside them, how **radiation** is produced and absorbed, the nature of the gas and dust lying between the stars, how stars move within **galaxies**, and the source of the mysterious **cosmic rays**.

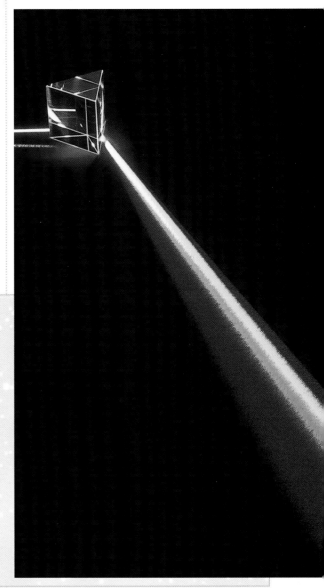

▶ **Isaac Newton** was the first person to show that the coloured bands of light that appeared when white light was passed through a prism were actually components of the light itself and not an effect simply introduced by the prism.

🧪 EXPERIMENT IN PHYSICS

The ultimate test of a hypothesis is *experiment* – does it hold up in the real world? This chapter and the next have many examples of experiments that have changed the course of history by showing that a long-held model no longer applies or that a new kind of model fits the facts better.

Newton's simple experiment with a prism demonstrated that the colours of the spectrum were present in white light rather than being introduced by the prism itself. Young's demonstration that two beams of light could, paradoxically, combine to produce darkness was the turning-point in the success of the wave model of light. **Heinrich Hertz**'s experiment with radio waves was clinching evidence for **Maxwell**'s theory of electromagnetism. The

Michelson–Morley experiment showed that the Earth's supposed motion through the ether was undetectable, opening the way to Einstein's relativity. **Sir Arthur Eddington**'s measurement of the deflection of starlight by the Sun showed that Einstein's theory of gravitation fitted the facts better than that of Newton.

In astrophysics experiments are not possible, but their role is filled by observation. The long struggle to understand the nature of **quasars**, for example, has been guided by testing numerous hypotheses against observation.

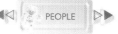

The Laws of Physics
LAWS OF MOTION
NEWTON'S LAWS OF MOTION

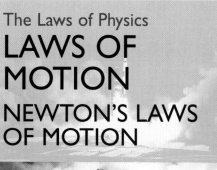

All of physics is underpinned by an understanding of how bodies move in response to forces that act upon them. The principles of classical mechanics were established by Sir Isaac Newton in the seventeenth century and published in 1687 in a landmark book called the *Principia*. Classical mechanics is founded upon Newton's three laws of motion, which together give a complete description of the motion of bodies in space. During the twentieth century classical mechanics had to be modified by the insights of relativity and quantum mechanics, but for all normal purposes the laws of motion set down by Newton remain valid.

PHILOSOPHIÆ
NATURALIS
PRINCIPIA
MATHEMATICA.

Autore *JS. NEWTON,* *Trin. Coll. Cantab. Soc.* Matheseos Professore *Lucasiano,* & Societatis Regalis Sodali.

IMPRIMATUR·
S. PEPYS, *Reg. Soc.* PRÆSES.
Julii 5. 1686.

LONDINI,
Jussu *Societatis Regiæ* ac Typis *Josephi Streater.* Prostant Venales apud *Sam. Smith* ad insignia Principis *Walliæ* in Cœmiterio D. *Pauli,* aliosq; nonnullos Bibliopolas. *Anno* MDCLXXXVII.

SPEED, VELOCITY AND ACCELERATION

Although the words 'speed' and 'velocity' mean the same in everyday life, physicists make a clear distinction between them. The *speed* of an object is the distance it has travelled divided by the time taken to travel that distance:

$$speed = distance/time$$

In SI units, speed is measured in metres per second (m/s). A speed of 1 m/s is equal to approximately 3.3 feet per second (ft/s).

The *velocity* of an object is a measure both of the speed and the direction of motion. Velocity is described as a vector quantity because it involves direction. It can be represented by an arrow, whose length (the magnitude of the vector) represents the speed and whose direction represents the direction of motion. Velocity has the same units as speed.

The *acceleration* of an object is the rate of change of velocity. An object is said to accelerate if either its speed is changing, its direction is changing, or both. Like velocity, acceleration is a vector, with both a magnitude and a direction. The SI unit of acceleration is the metre per second per second (m/s^2).

NEWTON'S FIRST LAW

Before the work of the Italian physicist **Galileo** it was commonly held that that the natural state of any object was to be at rest and that a force, or some other influence, was required to keep a body in motion. Galileo asserted that this was not so, that the natural state of an object was to be in motion and rest was just a special case of motion with zero velocity. The English physicist **Newton** took Galileo's assertion – the principle of inertia – as his first law of motion.

Newton's first law of motion states that an object at rest will remain at rest and an object in uniform motion will continue in uniform motion unless acted upon by unbalanced external forces (that is, forces that do not happen to be of such strengths, and in such directions, that they cancel each other out).

In practice, because of the pervasive influence of gravity, there are hardly any circumstances in which an object can be at rest while acted on by no force at all. A body floating in intergalactic space would be the

◀ *In 1687 Isaac Newton* published his Principia, *the book that was to establish the new science of classical mechanics. It covered the laws of motion, the theory of gravity and an explanation of the tides.*

▲ *Newton's first law of motion refers to objects at rest in space, yet even isolated stars move under the influence of distant gravitational forces.*

closest approximation – but balanced forces are common.

The first law tells us that if the velocity of an object is changing, the object is being acted upon by unbalanced forces. For example, planets move in elliptical orbits around the Sun. Their speeds and directions of motion are constantly changing and the first law asserts that a force must therefore be acting upon them.

NEWTON'S SECOND LAW

Newton's first law enables the physicist to recognize when unbalanced forces are acting but it does not say how big they might be or what their effect is. The second law places mechanics on a quantitative footing. It specifies how to calculate the motion of an object that is acted upon by a given unbalanced force. In particular, it relates the unbalanced force acting on an object to the acceleration of the object.

The two are related through the concept of inertia, the property of a body that causes it to resist attempts to change its state of motion. This rather vague concept is made clearer by identifying inertia with the **mass** of a body, a measure of how much matter it contains. The greater the mass, the greater the inertia. The SI unit of mass is the kilogram (kg), approximately equal to 2.2 lb.

Newton's second law of motion states that the acceleration of an object is directly proportional to the unbalanced force acting upon it, is in the same direction as the force, and is inversely proportional to the object's mass.

acceleration of the object = force acting on the object/mass of the object, or $a = F/m$

The second law also leads to a definition of the SI unit of force, the newton (N), which is the

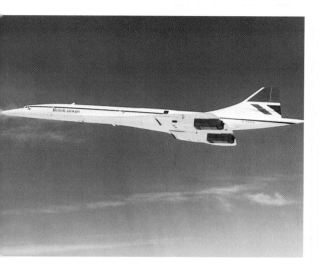

▲ *An aircraft* in level flight at a constant speed is acted on by two pairs of balanced forces. Gravity is balanced by lift from the wings and atmospheric drag is balanced by thrust from the engines.

force required to accelerate a mass of one kilogram by one metre per second per second.

Newton's second law is the most useful of the three laws, because if the force is known, then the subsequent motion of the object can be predicted in all circumstances. It may conversely be used to infer the forces acting on an object by observing its motion. But it can tell us only how a body responds to forces – it says nothing about the origin or nature of the forces themselves.

 ## BALANCED FORCES AND EQUILIBRIUM

An object may be acted upon by more than one force at the same time. The acceleration of the object is then determined by the sum of the acting forces. Sometimes the sum of the forces may be zero – there is no net force acting on the object. In this case the forces are balanced and the object is said to be in mechanical equilibrium. In accordance with Newton's first law, such an object either remains at rest or moves with a constant velocity.

A person standing still is acted on by two forces: the force of gravity acting downwards and the supporting force of the ground acting upwards. The two forces balance and there is no acceleration. The person remains at rest.

An object may still be in equilibrium even if it is moving. An aircraft cruising at a steady speed in level flight is acted on by four forces: gravity acts downwards and is balanced by the lift of the wings; the thrust of the engines acts forwards and is balanced by atmospheric drag. The four forces balance and there is no acceleration. In this case the aircraft continues to move at constant velocity.

FRAMES OF REFERENCE

Newton's laws deal with the motions of objects, but motions are relative. What is the speed of the Earth? To us the Earth appears at rest. Yet we know it moves at 30 km/s (18.6 mi/s) around the Sun, and the Sun itself is moving at 230 km/s (143 mi/s) around the centre of the Galaxy. The Galaxy, in turn, is moving through space at around 600 km/s (375 mi/s). The speed of the Earth, like all speeds and velocities, depends on the frame of reference in which it is measured, and the frame itself may be in motion.

It is always possible to choose a frame of reference in which an object has zero velocity.

NEWTON'S THIRD LAW

The first two laws describe the motion of single objects. The third law concerns the forces that act between two bodies.

Newton's third law states: when two bodies interact, the force exerted by the first body on the second is equal in magnitude and opposite in direction to the force exerted by the second body on the first.

This tells us that forces come in opposed pairs. For example, the gravitational force exerted on an apple by the Earth has its counterpart in the force exerted on the Earth by the apple. The force pulling the apple down is the same strength as the force pulling the Earth up: they have the same magnitude but act in opposite directions.

The law applies only where the two forces act on *different* bodies. Newton's third law is not concerned with pairs of opposed forces acting on the same body, such as the thrust and drag on the aircraft in the illustration above left.

A commonly quoted version of the third law is: to every action there is an equal and opposite reaction. This can be misleading, as it does not make clear that two bodies are involved. It also implies that one of the two forces is a consequence of the other, whereas in fact they are entirely symmetrical.

A falling object, accelerating under gravity, will appear at rest to an observer falling alongside it. In this case the observer's frame of reference is accelerating at the same rate as the object.

To avoid confusion, physicists prefer a special class of reference frames known as inertial frames. An inertial frame is one in which Newton's first law holds true. If a frame is inertial, the other two laws will also hold and all forces and accelerations will be measured correctly. Inertial frames can be in motion relative to each other, but only with constant velocity, not accelerated. Newton's laws apply only in inertial frames.

▲ *In accordance* with Newton's second law, the force exerted on a rocket by its engines is enough to accelerate it to such a speed that it can place a satellite in orbit around the Earth or, in the case of this Saturn V, send a spaceship to the Moon.

NEWTON'S LAWS IN EVERYDAY LIFE

The first law can be seen in operation in a game of ice-hockey. When the puck is hit it moves over the ice in a straight line at almost constant speed. The friction between the puck and the ice is so low that there is effectively no unbalanced force acting on the puck, so it continues in uniform motion.

To demonstrate the inertia of an object at rest, lay a card on top of a cup and place a coin in the centre of the card. If you flick the card away with your finger and thumb, the coin will drop into the cup. Its inertia causes it to remain at rest while the card moves beneath it.

The second law becomes apparent whenever it is necessary to push-start a stalled car. The same force will accelerate a small car to a given speed much faster than a large car.

An ice rink is a good place to demonstrate the third law. If a stationary skater pushes on a stationary companion, both skaters will move apart, showing that opposite forces are acting on both, even though only one is 'doing the pushing'.

THE LAWS OF PHYSICS

ENERGY & MOMENTUM

The concepts of energy and momentum are fundamental not only to physics but science as a whole. The idea of momentum as a measure of the 'amount of motion' of a body was recognized by Newton, but the concept of energy was not accepted until the middle of the nineteenth century. Today the powerful principle of the conservation of energy is universally accepted: as energy is converted from one form to another or transferred from place to place, the total amount of energy within any isolated system (and the Universe as a whole) remains unchanged.

◀ *James Joule, whose nineteenth-century experiments with different types of energy gave support to the law of conservation of energy.*

ENERGY, WORK AND POWER

Newton's second law states that a force acting on a body will cause it to accelerate. The amount of energy transferred to the body in the process is called the work done by the force and is given by:

$$\text{work done by a force} = \text{force} \times \text{distance moved in the direction of the force} = Fd$$

The SI unit of work and energy is the joule (J), which is the work done by a force of one newton acting through one metre.

In physics, work has no direct connection with physical exertion. A person does work in lifting a sack of potatoes from the floor, but does no work on the sack while carrying it horizontally at constant speed from one room to another, since the distance moved is at right angles to the gravitational force acting on the sack.

Power is the rate at which work is done or at which any kind of energy is transferred from place to place or converted from one form to another:

$$\text{power} = \text{energy transferred}/\text{time}$$

The SI unit of power is the watt (W), where one watt is equal to one joule per second.

KINETIC ENERGY

When an object is accelerated from rest, the work done on it by the accelerating force is identical to the kinetic energy gained by the object. Kinetic energy is the energy associated with the motion of an object and is given by:

$$\text{kinetic energy} = \text{1/2 mass} \times \text{velocity}^2 = \tfrac{1}{2}mv^2$$

Because kinetic energy depends on the square of the speed, fast-moving objects can carry a great deal of energy. An asteroid impacting on a planet at 60 km/s, for example, has twice the speed but four times the kinetic energy of a similar asteroid moving at 30 km/s, and can do four times as much damage.

Kinetic energy is also associated with the rotation of an object and is then given by an analogous equation:

$$\text{kinetic energy} = \text{1/2 rotational inertia} \times \text{angular velocity}^2$$

The rotational inertia (also called moment of inertia) is a measure of how mass is distributed within the body and the angular velocity is the angle rotated by the body each second.

Stars, planets and galaxies generally possess kinetic energy due both to their motion through space (translational kinetic energy) and also to their rotation (rotational kinetic energy).

OTHER TYPES OF ENERGY

Many other types of energy have been identified. Important ones include the following:

🌍 Gravitational energy is the energy possessed by a mass by virtue of its position within a gravitational field. The initial stages of star formation involve the release of gravitational energy as the star contracts.

▼ *All bodies in motion have kinetic energy. Fast-moving objects such as this bullet in flight, or an asteroid impacting with a planet, carry high amounts of energy. They can cause significant damage because kinetic energy is dependent on the square of the object's velocity.*

JAMES JOULE

James Prescott Joule (1818–89) was a British physicist whose experimental demonstrations that different forms of energy could be converted into each other led to the acceptance of the principle of conservation of energy.

Largely self-taught, Joule was one of several scientists who struggled to make sense of the concept of energy in the early nineteenth century. One early success was his measurement of the heat produced by an electric current passing through a resistance, a phenomenon now known as Joule heating. In collaboration with **William Thomson**, later to be Lord Kelvin (1824–1907), he also inves-

tigated the cooling effect of expanding gases, which is now known as the Joule–Thomson effect.

He is best known for an experiment in which slowly falling weights turned a system of paddles inside a barrel of water. The paddles churned the water and raised its temperature. Joule measured the heat generated and the gravitational energy lost by the falling weights and transferred to the water. He could then calculate the rate of exchange between the two – the amount of heat equivalent to a unit of energy.

Although many scientists contributed to the development of the concept of energy, Joule's experimental work is regarded as decisive. The SI unit of energy is named the joule in his honour.

- Electromagnetic energy is the energy associated with electric or magnetic fields. Radiant energy is the electromagnetic energy carried by an electromagnetic wave such as light.
- Chemical energy is released in chemical reactions and is ultimately derived from the kinetic and electromagnetic energy of atoms and molecules. The energy released by burning fossil fuels is solar radiant energy that was stored as chemical energy by living organisms many millions of years ago.
- Nuclear energy is the energy released in nuclear reactions. It is the source of power for the Sun and other stable stars.
- Internal energy (loosely called 'thermal' energy) is the energy possessed by an object due to the kinetic energy of the molecules within it and the electromagnetic energy arising from the forces acting between them.

Any form of energy can, in principle, be converted to any other form and all are measured in the same unit, the joule.

MOMENTUM

Another quantity associated with the motion of an object is its momentum, which can be thought of as the 'quantity of motion'. The momentum of a body is given by the product of its mass and its velocity.

$$ \text{momentum} = \text{mass} \times \text{velocity} $$

The SI unit of momentum is the newton-second (N s).

The change in momentum of an object acted on by a force is known as the impulse, and is given by:

$$ \text{impulse} = \text{force on object} \times \text{time} \\ \text{over which force acts} $$

It follows that force can be regarded as the rate of change of momentum.

Although both depend on mass and speed, momentum should not be confused with kinetic energy. Momentum is a vector, whose direction is the direction of the velocity, while kinetic energy has no direction associated with it. The practical importance of momentum is that it is conserved during collisions between objects, while kinetic energy is generally not conserved.

ANGULAR MOMENTUM

An object rotating on its axis, such as a star or a planet, possesses angular momentum, analogous to

CONSERVATION LAWS

Although classical mechanics is essentially concerned with the forces acting between bodies, an alternative (but equivalent) understanding can often be achieved through the use of the conservation laws. These powerful laws allow scientists to analyze complex situations without making detailed calculations of the interactions between bodies. In each case, the law is exact when the system under consideration is isolated – that is, it does not interact with its surroundings in any way. No systems are completely isolated, but that does not diminish the value of the conservation laws in practice.

In classical mechanics there are four conservation laws:

1. *In any isolated system, the total amount of mass remains constant.*
This is the simplest of the laws, and expresses what once seemed to be an experimental fact, that mass can neither be created nor destroyed.

2. *In any isolated system, the amount of energy remains constant.*
This may not be as apparent as the conservation of mass, because energy may appear in many different forms.

Einstein's theory of relativity showed that mass and energy are equivalent and interchangeable. The sum of mass and energy is conserved, but for all normal purposes the separate laws hold good.

3. *In any isolated system, the amount of momentum remains constant.*
It is possible for the total momentum of a system to be zero. For example, a rocket on the launch pad has zero momentum relative to an observer stationary on the ground. When the engine ignites, the momentum carried upwards by the rocket is precisely matched by the momentum carried downwards by the exhaust gases. The total momentum of the rocket and exhaust remains zero.

4. *In any isolated system, the amount of angular momentum remains constant.*
Angular momentum is conserved independently of linear momentum and is traditionally illustrated by a pirouetting ice skater. As the skater pulls in her arms, her rotational inertia decreases and her angular speed increases to conserve angular momentum. For the same reason, a gas cloud contracting to form a star will rotate faster as its radius decreases.

Similar laws relate to the conservation of electric charge and to certain properties of fundamental particles.

The conservation laws can be exemplified by the collision of two galaxies. The physics of such a collision would be extremely complex to analyze, involving clouds of gas, magnetic fields and many billions of stars. Yet the total amounts of mass, energy, momentum and angular momentum would not change during the collision.

the linear momentum of an object moving in a straight line. A body rotating about an axis of symmetry has an angular momentum given by:

$$ \text{angular momentum} = \text{rotational inertia} \\ \text{of body} \times \text{angular velocity} $$

Angular momentum is a vector, with its direction aligned along the axis of rotation. Despite the name, angular momentum is a not a kind of momentum but is a different quantity and is measured in different units. The SI unit of angular momentum is the joule second (J s).

In the case of a planet or a satellite in a circular orbit, the

angular momentum of the orbiting body around an axis passing through the centre of the circle is given by:

$$ \text{angular momentum} = \text{mass of body} \times \\ \text{orbital speed} \times \text{radius of orbit} $$

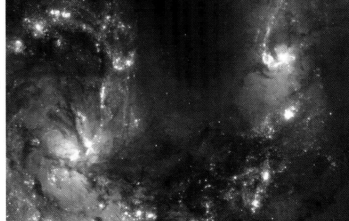

▶ *Although this spectacular collision between two galaxies is immensely complicated, astrophysicists can make sense of it by using the conservation laws.*

LAWS OF MOTION

The Laws of Physics
GRAVITATION & ORBITS
GRAVITY

THE LAWS OF PHYSICS

Newton's laws of motion imply that the planets are kept in their orbits by forces, but what are those forces? Yet another of Newton's discoveries was that the familiar force of gravity extends throughout the Universe, every object attracting every other object. The gravitational force falls off with the inverse square of the objects' separation. The same force that causes apples to fall also keeps the planets in their courses. Newton's description of gravity is incomplete and has given way to the deeper understanding provided by Einstein's general theory of relativity, but the Newtonian theory is good enough for all normal purposes.

HENRY CAVENDISH

Henry Cavendish (1731–1810) was a British physicist who made the first experimental measurement of the gravitational constant, known as G.

Born into an aristocratic family, Cavendish was a wealthy recluse. He was a skilled experimenter who made important contributions to many fields of physics and chemistry. He discovered several of the electrical phenomena later attributed to **Michael Faraday** (1791–1867) and others, but owing to his reclusiveness much of his work remained unknown until many years after his death.

In 1798 Cavendish mounted a small lead sphere at each end of a metal bar suspended at its centre by a wire. He placed two larger lead spheres nearby so that the smaller masses would be attracted towards them. By measuring the tiny angle through which the bar turned, Cavendish was able to calculate the force on the spheres and so work out the value of G.

With G known, he could then calculate the mass and density of the Earth for the first time. While modern methods have improved on Cavendish's technique, the measurements are demanding and G remains the most poorly known of the fundamental constants of nature.

NEWTON'S LAW OF UNIVERSAL GRAVITATION

Newton's insight into the nature of gravity is reputed to have come when he saw an apple fall from a tree and realized that the same force attracting the apple to the Earth also kept the Moon in its orbit.

Newton's law of universal gravitation states: every object in the Universe attracts every other object with a force proportional to the product of their masses and inversely proportional to the square of the distance between them. It is conventionally written in algebraic form as:

$$F = \frac{Gm_1m_2}{r^2}$$

where m_1 and m_2 are the masses of the bodies and r is the distance between them. G is a fundamental constant of nature called the universal gravitational constant. Its value is 6.67×10^{-11} N m² kg⁻².

The importance of the law is that the gravitational influence of a mass, its gravitational field, extends to infinity in every direction, its strength falling as the inverse square of the distance. Every mass attracts every other mass.

Although it is the weakest of the four fundamental forces of nature, gravity is the only

one that extends over interstellar distances, and it largely determines the structure of the Universe. It is of no consequence on the atomic scale, where electric and nuclear forces prevail.

MASS AND WEIGHT

In everyday language 'mass' and 'weight' have similar meanings, but scientifically they are quite distinct. Mass is a measure of the amount of matter in a body and is measured in kilograms. Weight, by contrast, is a measure of the gravitational force acting on a body, and is measured in newtons.

An object in a uniform gravitational field has a weight given by:

$$\text{weight} = \text{mass} \times \text{gravitational field strength} = mg$$

The gravitational field strength, g, varies slightly with position on the Earth's surface. Its mean value is approximately 9.81 N kg⁻¹. Because g may vary from place to place, the weight of an object of fixed mass also varies. An object on the Moon would have one sixth of its weight on Earth, while the same object on Jupiter would weigh 2.5 times as much.

We commonly speak of our 'weight' in pounds or kilograms, when we really mean our mass. A spring balance is affected by the weight of an object on it, but is usually graduated in units of mass because g is for all practical purposes constant wherever the balance is used.

FREE FALL AND WEIGHTLESSNESS

An object moving freely in a gravitational field, with no other forces acting on it, is said to be in free fall. Stars and planets are in free fall, as are moons and artificial satellites.

Galileo showed that all objects in free fall move with the same acceleration, regardless of their mass or composition. Near the surface of the Earth, this acceleration is numerically identical to g, which is sometimes known as the acceleration of free fall or the acceleration due to gravity.

The term 'weightlessness' is often used to describe the condition experienced by astronauts in free fall. But although they do not feel their weight, they are still being acted upon by gravity.

It is sometimes said that astronauts are weightless because they are beyond the reach of the

◀ *The mathematician and scientist* Henry Cavendish, *whose measurement of the gravitational constant (known as G) helped physicists calculate the mass and density of our planet for the first time.*

barycentre ▶ p. 336 binary star ▶ p. 336

▲ **Astronauts experience** *weightlessness because they are in free fall in a gravitational field. In low Earth orbits the strength of gravity is only slightly less than on the Earth's surface.*

Earth's gravity. This is a fallacy. The gravity at the height of the International Space Station, for example, is 88% of that at the Earth's surface. Astronauts are weightless because they are in free fall, not because there is no gravity in orbit.

CENTRES OF MASS AND GRAVITY

As is common in physics, the law of universal gravitation is strictly framed in terms of forces between particles, where a particle is an object with mass but with negligible size. The distance between two particles is defined uniquely. This is not the case for real objects, such as planets, which have a finite size. In many cases, however, a finite body can be treated as an equivalent

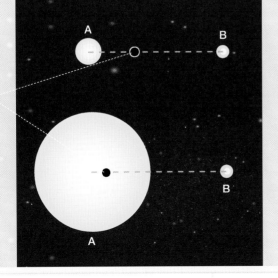

Centre of mass

GRAVITATIONAL ENERGY

An object in a gravitational field possesses gravitational energy by virtue of its position. If it is allowed to move in response to the field, gravitational energy will be converted into kinetic energy.

The gravitational energy possessed by a pair of bodies of masses m_1 and m_2 is given by:

$$\text{gravitational energy} = \frac{Gm_1m_2}{r}$$

where r is the distance between them. In a region of uniform gravitational field, such as near the surface of the Earth, the change in gravitational energy due to a mass, m, moving through a height, h, is given by the simpler formula:

$$\text{gravitational energy} = mgh$$

Stars rely on gravitational energy as an important source of power at several points in their lives. For example, as a gas cloud collapses to form a star, its gravitational energy is converted first to the kinetic energy of the cloud itself and then to the internal energy of its constituent particles, so causing the temperature to rise to the point where nuclear reactions can begin.

particle of identical mass placed at a point called the centre of mass of the body. Newton's laws then apply to the centre of mass.

Where the body is symmetrical, as is the case with most astronomical bodies such as stars or planets, the centre of mass is at the geometric centre of the body. A freely rotating body will always rotate about an axis passing through its centre of mass.

Two or more objects, such as a multiple star system or a star and its circling planets, will have a common centre of mass, known as the **barycentre**.

The centre of gravity of a body is the point at which its weight appears to act. Except where an object is in a highly non-uniform gravitational field, its centre of mass for all practical purposes coincides with its centre of gravity.

◀ *Centre of Mass*
In the upper pair of orbiting stars, the mass of star A is double that of star B. They orbit around a centre of mass one-third of the distance between the centres of the stars. In the lower pair, star A is 10 times more massive than star B. In this case, the centre of mass is much closer to the middle of star A, whose orbital motion is slight compared to star B.

THE NATURE OF MASS

Mass plays two subtly different roles in classical mechanics. It is identified in Newton's second law as inertia, the tendency of a body to resist changes in motion. It also appears in Newton's law of gravitation as the source of the gravitational field. These two kinds of mass, inertial mass and gravitational mass, are quite different properties of a body and the question arises as to whether their apparent identity is anything more than a coincidence.

One consequence of this identity – as Galileo had proposed – is that all objects in a gravitational field fall with the same acceleration, regardless of their mass or composition. Newton himself performed many experiments with pendulums to test the identity of the two kinds of mass and more sophisticated experiments have been carried out to the present day. None has revealed any difference between the two types of mass.

The puzzle was resolved in the twentieth century when **Einstein** showed that the identity of inertial and gravitational mass is a necessary consequence of the nature of time and space as set out in the general theory of relativity.

TIDES

When a body is placed in a non-uniform gravitational field it experiences tidal forces. The Moon, for example, pulls more strongly on the side of the Earth facing it than on the opposite side. The result is a stretching force, which causes the oceans to be pulled out into an ellipsoid, with bulges raised on either side of the Earth. As the Earth rotates beneath the bulges, the oceans rise and fall twice a day. The Sun also raises tides in the oceans, but of lesser height.

Tides can also be raised in the solid body of the Earth. Such body tides are observed both on Earth and on the Moon. Close **binary stars** (stars that orbit each other under their mutual gravitational attraction) suffer extreme tidal distortion and are drawn out into egg-shapes.

Friction caused by the continuous deformation of a body by tidal forces leads to tidal heating. In the case of Io, one of the moons of Jupiter, tidal heating melts rock and drives volcanic activity.

The loss of energy to tidal heating eventually causes the body to stop rotating. This has already happened to the Moon, whose rotation has been braked by tidal heating and now keeps the same face towards the Earth, an example of captured rotation.

THE LAWS OF PHYSICS

Einstein ▶ p. 88 Faraday ▶ p. 70 Galileo ▶ p. 40 Newton ▶ p. 42

GRAVITATION & ORBITS

THE LAWS OF PHYSICS

ASTRONOMICAL ORBITS

An orbit is the path described by any object moving freely under gravity. The path of a falling apple is no less an orbit than the path of the Moon. Earth and the other planets in the Solar System move around the Sun in elliptical orbits, first described in the early seventeenth century by Johannes Kepler. They were later explained by Newton as inevitable consequences of the law of universal gravitation. Newton's approach shows that in reality the orbits are more complex than the simple ellipses of Kepler. Newton's theory allows us to extend our understanding beyond the Solar System to the motions of the stars and galaxies.

KEPLER'S LAWS

A breakthrough in understanding the motions of the planets was made by **Johannes Kepler** (1571–1630). He formulated three laws of planetary motion based on his analysis of the meticulous observations of Mars made by **Tycho Brahe** (1546–1601). The first two were published in 1609 and the third in 1619.

1. *The orbit of a planet around the Sun is in the form of an ellipse, with the Sun at one focus.*
Until the time of Kepler the planets had been assumed, largely for philosophical reasons, to move in circles. He was the first to recognize the true form of their paths.

2. *The line connecting a planet to the Sun sweeps out equal areas in equal times.*
This apparently obscure law accounts accurately for the changing speed of a planet in its orbit. A planet moves fastest when it is at its closest point to the Sun and slowest when it is at its farthest point, thus conserving its angular momentum with respect to the Sun.

3. *The square of the orbital period of a planet is proportional to the cube of the semi-major axis.*
The orbital period of a planet is the time it takes to complete one revolution of the Sun. The major axis of the orbit is its greatest diameter, and the semi-major axis is half that distance. The third law expresses an intimate connection between the period and semi-major axis. Once we know the orbital period, we can immediately calculate the average distance from the Sun. If the period, P, is in years and the semi-major axis, a, is in astronomical units (the average distance of the Earth from the Sun), then:

$$P^2 = a^3$$

NEWTONIAN ORBITS

Kepler's laws describe the observed motions of the planets, but offer no explanation of why the planets move as they do. Isaac Newton, through his laws of motion and the universal law of gravitation, demonstrated that the planets move under the influence of gravitational forces that act between each planet and the much more massive Sun. He showed that each of Kepler's laws followed from this assumption.

The elliptical orbits of Kepler's first law are a direct consequence of the inverse square law of gravitation. So, too, are parabolic and hyperbolic orbits of which Kepler was not aware, but which can sometimes be followed by comets. The third law also follows directly from the law of gravitation. The second law, curiously, has nothing to do with the law of gravitation, but is simply an expression of the conservation of angular momentum.

Edmond Halley (1656–1742) used Newton's analysis to calculate the orbits of comets. He found that what had been thought to be observations of different comets at intervals of about 76 years were actually sightings of a single regularly returning comet. He successfully predicted it would return in 1758 (16 years after Halley's death), and since then it has been known as Halley's comet.

TYPES OF ORBITS

The shape of an orbit is characterized by its eccentricity. When the eccentricity is 0, the orbit is a perfect circle. Eccentricities between 0 and 1 describe elliptical orbits, with the longer ellipses having the greater eccentricities. The Earth's orbit has an

▶ *A nineteenth-century diagram showing the phases of the Moon. In the 1950s astronomers introduced a new system of timekeeping – Ephemeris Time – which was based upon the predicted position of the Moon in its orbit.*

▲ *The type of orbit in which a body moves is determined by its eccentricity. At 0 eccentricity, the orbit is perfectly circular; at precisely 1 the orbit is parabolic. Comets such as Hale–Bopp move in highly elliptical orbits.*

⧖ EPHEMERIS TIME

Time has traditionally been reckoned by the rotation of the Earth, with one complete turn being 24 hours. By the 1930s it was clear that the Earth did not rotate smoothly and that the length of the day was not constant.

In the 1950s astronomers introduced a new, smoother-running time scale which they called Ephemeris Time (ET). It was based on the orbits of bodies in the Solar System. By using Newton's laws, astronomers predicted the positions of the Sun, Moon and planets for any given ET and listed them in a table called an ephemeris. In principle, the time could then be found by observing the positions of these bodies and looking in the ephemeris to see at what ET they were expected to be in that position. In practice, ET was determined by measurements of the position of the Moon. This was very cumbersome.

From 1960 the SI second (the scientific unit of time) was defined as the duration of the second of ET, before it was replaced by the more practical atomic second in 1967. In 1984 ET was itself replaced for astronomical purposes by Terrestrial Time (TT) and Barycentric Dynamical Time (TDB), both of which are defined in terms of International Atomic Time (TAI).

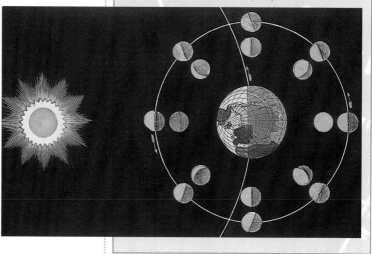

GRAVITATION & ORBITS

THE LAWS OF PHYSICS

IS THE SOLAR SYSTEM STABLE?

Classical mechanics brings a reassuring order to the Solar System. The planets move not at the whim of mystical forces, but in response to the gravitational pull of the Sun and of each other. Provided their positions and velocities are known with sufficient precision, can their paths be predicted into the indefinite future?

Since the 1980s mathematicians have come to realize that the answer is 'no'. New insights from chaos theory have revealed that there are regions of the Solar System where motions are unpredictable, no matter how carefully observations are made.

One region is in the **asteroid belt** between Mars and Jupiter. At certain bands in the belt an asteroid may be suddenly thrown out of its orbit by the influence of Jupiter, but it is not possible to say when or where this may happen. Another example is Hyperion, one of the small satellites of Saturn, which is tumbling unpredictably in its orbit, under the influence of the large moon Titan.

Experiments with digital models of the Solar System show that even the motions of the larger planets cannot be foreseen beyond a few million years. This does not mean the Solar System is going to go haywire – only that we cannot say whether it will or not.

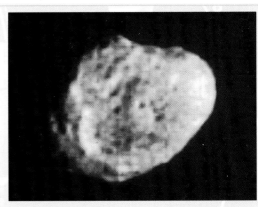

▲ *Hyperion is a 400-km* moon of Saturn whose chaotic axial rotation cannot be predicted even by the most powerful computers.

eccentricity of 0.017. The most eccentric planetary orbit is that of Pluto (0.248), while the orbit of Halley's comet has an eccentricity of 0.97, taking it from within the orbit of Venus to beyond the orbit of Neptune.

If the eccentricity is precisely 1 the orbit is parabolic and the orbit is open – the body will only make one passage of the Sun. Where the eccentricity is greater than 1, the orbit is hyperbolic. Many long-period comets appear to be in parabolic

ANATOMY OF AN ORBIT

The elliptical orbit of a planet around the Sun has two points of special mathematical significance, called the foci, one of which is occupied by the Sun. The line connecting the Sun to the planet is the radius vector.

The long axis of the ellipse is known as the major axis. Half this distance, the semi-major axis, is equal to the average distance of the planet from the Sun. The short axis is the minor axis.

The point of closest approach to the Sun is known as the **perihelion**, and the farthest point is the **aphelion**. Similar terms are used for orbits around the Earth (**perigee**, **apogee**) and around other stars (periastron, apastron).

The amount by which the ellipse is flattened, as compared with a circle, is described by the **eccentricity**, which is defined as the distance between the foci divided by the length of the major axis.

The inclination of the orbit is the angle the plane of the orbit makes to the plane of the Earth's orbit (which is also known as the plane of the ecliptic). The point at which a planet crosses the ecliptic from south to north is the ascending node and the corresponding southgoing point is the descending node. Orbits around other bodies make use of a variety of other reference planes.

or hyperbolic orbits as they pass through the inner **Solar System**, although it is difficult to distinguish their paths from extremely elliptical orbits.

MORE COMPLEX ORBITS

Kepler's laws strictly apply only to the orbits of solitary planets around solitary stars, so-called two-body systems. Where three or more bodies are interacting, Kepler's laws do not hold exactly, although they are good approximations for the Solar System, where the dynamics are dominated by the Sun.

In most cases the full Newtonian treatment is required, and can lead to surprising findings. One such is the **Lagrangian points**, five stable points in a two-body system where smaller bodies can be 'balanced'. The Trojan **asteroids**, for example, are two groups of asteroids held at the Lagrangian points ahead of and behind Jupiter.

With multiple bodies interacting, it becomes possible for an object to be captured into a closed orbit and, conversely, for an object to be ejected from an orbit that appeared to be stable. This commonly happens to comets as they pass through the inner Solar System.

Stars also move in orbits described by the universal law of gravitation. In a binary system, two stars revolve around their common centre of mass, each describing an elliptical orbit. But within the central regions of

galaxies stars may move in complex paths under the influence of billions of similar stars.

ESCAPE SPEED

A comet moving in a parabolic orbit has just sufficient energy to leave the Solar System and never return. At any point in its orbit it is moving at the local escape speed.

The speed required to escape the gravitational pull of a body of mass M at a distance r is given by:

$$\text{escape speed} = \sqrt{\frac{2GM}{r}}$$

For example, the escape speed is 11.2 km/s (7 mi/s) from the surface of the Earth and 2.4 km/s (1.5 mi/s) from the Moon. The escape speed from the Sun at the Earth's orbit is 42 km/s (26 mi/s).

Escape speed only applies to objects in free fall. A powered rocket can leave the Earth without reaching 11.2 km/s, since the escape speed drops the farther the rocket gets from Earth. Providing it is moving in excess of the local escape speed when the engines are shut off, it will not fall back to Earth. It does not matter what its angle of ascent is.

▼ *Orbital Curves and Conic Sections*
This graphic shows how the four kinds of orbital curve – circle, ellipse, parabola and hyperbola – correspond to different conic sections. A conic section is made by slicing through a cone.

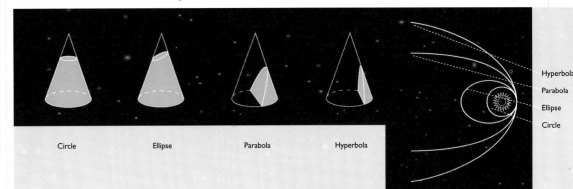

Circle   Ellipse   Parabola   Hyperbola

Hyperbola
Parabola
Ellipse
Circle

ORBITAL MECHANICS

C lassical mechanics allows us not only to understand the motions of the heavenly bodies but also to send satellites and space vehicles beyond the Earth's atmosphere. Newton himself proposed that a sufficiently powerful cannon could put a satellite in orbit around the Earth, and the same principles are used today with modern rockets. But steering a spacecraft is not like flying an aircraft; precise orbits are calculated in advance to ensure that the craft will arrive at its destination. The paths taken to the Moon and planets represent an economical compromise between the demands of gravity and the costs of fuel.

▲ *Once a spacecraft is in Earth orbit, it is moving in free fall and its motion follows the rules laid down in Newton's laws, that is, it will continue in the same orbit unless another force acts upon it. Astronauts can change the orbit of their spacecraft by firing their engines to speed up or slow down the craft.*

GETTING INTO EARTH ORBIT

To place a spacecraft into an orbit around the Earth it must be accelerated from rest to an appropriate speed and height. This requires sufficient expenditure of energy to lift the spacecraft against the Earth's gravitational pull and accelerate it to orbital speed. For example, a typical low Earth orbit is at a height of around 200 km (120 miles), which corresponds to a speed of 7.8 km/s (4.8 mi/s) and an orbital period of about 88 minutes. Such orbits eventually decay, owing to frictional drag from the upper atmosphere. Higher orbits are longer-lived but require a greater expenditure of energy to get there.

An orbital insertion point is the point in space where the launch rocket's engines shut down when the craft has attained the correct height and speed for the intended orbit. For any orbital insertion point there is a speed that will put a satellite into a circular orbit. Giving it a higher speed will push it into an elliptical orbit, with the point of insertion at the lowest point.

The inclination of an Earth orbit is the angle between the plane of the orbit and the plane of the equator. It is also equal to the highest latitude at which the satellite will be seen overhead. In a polar orbit, with an inclination near 90°, the satellite will eventually pass over every part of the globe as the Earth turns beneath it.

MOVING AROUND IN SPACE

Once free of the atmosphere, a spacecraft spends most of its time in free fall, following an orbit prescribed by Newton's laws of motion. Flying

ARTHUR C. CLARKE

Arthur Charles Clarke (b. 1917) is a British science writer and science-fiction author who was the first person to propose geostationary communications satellites and whose novels often reveal a realistic appreciation of orbital mechanics.

After working on wartime radar, Clarke published an article in *Wireless World* in 1945 in which he foresaw a global communications network, including worldwide television, linked by relay stations in geosynchronous orbits. This was 12 years before the first satellite was launched, 20 years before the first geosynchronous communications satellite (Early Bird), and more than 40 years before direct broadcast satellites fulfilled his prediction of beaming television straight into people's homes.

In contrast to many earlier works of science fiction, Clarke's novels of space exploration are informed by a deep understanding of orbital mechanics and the constraints it places on the navigation of space vehicles. He worked with Stanley Kubrick on the acclaimed film *2001: A Space Odyssey*, released in 1968, which was itself based on one of Clarke's early short stories. The film was the first to make an accurate portrayal of interplanetary flight.

Clarke has published more than 60 books and has lived in Sri Lanka since the 1950s. He was knighted in 1998.

▶ *Arthur C. Clarke, best-known for his science-fiction novels; Clarke's mass appeal lies in his real understanding of classical mechanics and the possibilities of space travel.*

Hohmann orbits ▶ p. 339   meridian ▶ p. 341

THE LAWS OF PHYSICS

a spacecraft is nothing like flying an aircraft. The purpose of the engine is not to drive the spacecraft along, but to adjust its total energy so that it can change from one orbit to another.

The energy of a spacecraft in Earth orbit is the sum of its kinetic energy and its gravitational energy; the latter depends only on its distance from the centre of the Earth. One paradoxical consequence is that there is a trade-off between height and speed; a spacecraft moves more slowly in a high-altitude, high-energy orbit, than in a low-altitude, low-energy orbit.

For example, if the commander of a space shuttle wishes to catch up with a space station some way ahead in the same circular orbit, he or she should reduce its speed by briefly firing its engines against the direction of motion. This drops the shuttle into an elliptical orbit of shorter period, whose apogee touches the original orbit. The period is chosen so that by the time the shuttle returns to the apogee it arrives at the same time as the space station. The shuttle then fires its engines again, this time in the direction of motion, to match the orbital speed of the station.

GEOSYNCHRONOUS ORBITS

The period of an orbit lengthens with distance from the Earth. Seen from the ground, satellites in high orbits move more slowly across the sky than satellites in low orbits. At a height of 36,000 km (22,350 miles) the period becomes 24 hours. This means that a satellite placed in such a geosynchronous orbit will complete exactly one

revolution each day and will appear to remain very nearly above the same **meridian** on Earth. If the orbit is inclined to the equator, an observer on the ground will see the satellite describe a figure-of-eight track on the sky every 24 hours. If the orbit is circular and precisely over the Earth's equator, the satellite will appear to be motionless in the sky, or geostationary.

Although expensive to reach, geostationary orbits are used where a constant view of the same hemisphere is needed or where unbroken communication is essential. Applications include communications satellites, meteorological satellites and TV direct broadcast satellites. Because the satellite appears stationary, ground antennas need not be steerable and can be constructed more cheaply.

LUNAR TRAJECTORIES

The complex path followed by an Apollo spacecraft to the Moon consisted of a series of orbits linked by velocity changes at appropriate points. Between 1968 and 1972 the Apollo program sent nine spacecraft to the Moon, six of which landed, and they all used the same procedure.

The first step was to place Apollo into a circular parking orbit around the Earth. At the appointed time the engine was fired to increase the velocity to 10.8 km/s (6.7 mi/s). This put Apollo into a figure-of-eight orbit, with the cross-over occurring at the point where the gravitational attractions of the Earth and Moon are equal and opposite, and with the top of the loop beyond the far side of the Moon.

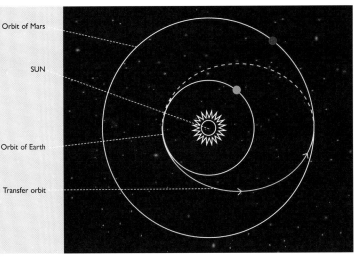

▲ Transfer Orbit
Spacecraft are sent to Mars on a special trajectory called a Hohmann transfer orbit, an ellipse that just touches the orbits of the two planets.

There the rocket was fired again, this time to slow the spacecraft sufficiently to place it into a near-circular orbit around the Moon. To achieve the landing, the lunar module was detached from the main vehicle and its engine was fired to lose energy and put the module into a descending elliptical orbit. At the low point of the orbit, only 15 km (9 miles) above the Moon, the engine was fired continuously to take the module down to the surface.

INTERPLANETARY TRAJECTORIES

Flights to other planets within the Solar System can be even more complex than moonflights. For much of the journey, however, the spacecraft can be considered to be moving in an elliptical orbit around the Sun.

The most economical route for sending a spacecraft to another planet without the aid of another celestial body's gravity is a **Hohmann transfer orbit**, named after Walter Hohmann, a German space-travel theorist of the 1920s. This is an ellipse that just touches the orbits of the Earth and the destination planet. It is commonly used for probes to Mars and Venus, but requires a great deal of fuel to reach more distant destinations.

A more fuel-efficient solution, especially for the outer planets, is to use the gravitational fields of more accessible planets to speed a spacecraft on its way. In the gravity-assist, or gravitational slingshot, technique the spacecraft makes a close approach to a planet in such a way that it picks up energy. For example, in 2004, after the completion of the seven-year flight of the Cassini mission to Saturn, the spacecraft will have been boosted by close approaches to Venus, Earth, Venus again, and Jupiter.

▼ Cassini Trajectory
The trajectory of an interplanetary spacecraft is not always a simple ellipse. The looping path of the Cassini spacecraft allowed it to pick up energy by close passages to Venus, Earth, Venus again and finally Jupiter on its seven-year voyage to Saturn.

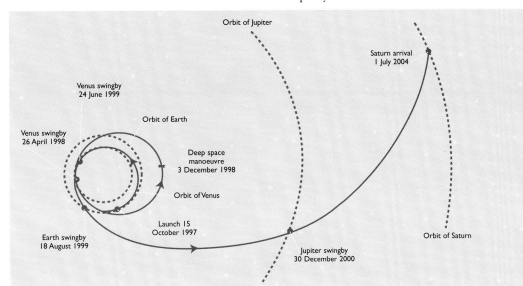

Orbit of Jupiter

Saturn arrival
1 July 2004

Venus swingby
24 June 1999

Orbit of Earth

Venus swingby
26 April 1998

Deep space
manoeuvre
3 December 1998

Orbit of Venus

Launch 15
October 1997

Earth swingby
18 August 1999

Orbit of Saturn

Jupiter swingby
30 December 2000

The Laws of Physics
PROPERTIES OF MATTER
GASES, LIQUIDS & SOLIDS

Everything we see around us, from the smallest grain of sand to the largest galaxy, is made of atoms. These tiny particles, far too small to be visible to the human eye, are the smallest units into which matter can be divided while still retaining its identity. Each of the chemical elements is made of a different kind of atom. Atoms can combine to form larger particles called molecules. The three familiar states of matter – gases, liquids and solids – can be understood in terms of their constituent atoms and molecules and the forces that act between them.

ATOMS AND MOLECULES

All matter is made of small particles called atoms. Atoms are typically 0.1–0.5 nanometres (nm) in diameter (1 nm = 10^{-9} m), with masses of order 10^{-27} to 10^{-25} kg.

Each atom consists of a compact nucleus, which contains most of the **mass**, surrounded by a cloud of **electrons**. Any atom can be characterized by the number of **protons** in its nucleus (the atomic number) and the number of **neutrons** (neutron number). The number of protons and neutrons together is known as the mass number. Atoms of the same element all have the same atomic number but may have different mass numbers. In a normal atom the number of electrons will be the same as the number of protons; if they differ, the atom acquires an electric charge and is called an **ion**. There are known to be at least 110 different elements, of which about 90 occur naturally on Earth.

Atoms may join together to form molecules. A molecule is the smallest unit of a substance that still retains its chemical identity. Hydrogen gas molecules, for example, contain two atoms of hydrogen (H_2). Water contains two atoms of hydrogen and one of oxygen (H_2O) and ammonia contains one atom of nitrogen and three of hydrogen (NH_3).

KINETIC THEORY

A fundamental understanding of how the bulk properties of matter arise out of the behaviour of atoms and molecules is based on the kinetic theory of gases. The theory supposes the existence of an ideal gas that is deemed to have the following properties:

- The gas is made up of large numbers of particles called molecules.
- The size of the molecules is negligible compared to the distances between them.
- The molecules are in constant random motion and obey **Newton's laws**.
- The molecules collide elastically with each other (that is, kinetic energy is conserved).
- No other forces act between the molecules.

In practice, gases are not ideal but the theory has been very successful in explaining the properties

▲ **Hydrogen gas** is widespread in the Universe. This glowing nebula, NGC 6559, consists of ionized hydrogen which is being heated by stars which have recently formed within it.

of real gases and with a few modifications can be extended to an understanding of all three states of matter: gases, liquids and solids.

GASES

In a gas, the molecules are widely separated and most of a gas is empty space. Occasionally a molecule will collide with another, but otherwise there is no interaction between them. Gases tend to expand, and a gas in a container will fill it completely and uniformly.

A volume (V) of gas in a closed container exerts a pressure (P), owing to the bombardment of molecules on the walls of the container. The gas has a temperature (T) related to the average

JOHN DALTON

John Dalton (1766–1844) was a British school-teacher and chemist who built on the ancient Greek notion that all matter was made of atoms and added ideas of his own – notably that they could combine to form molecules.

Dalton's early interest in meteorology led to a study of mixtures of gases, from which he derived his law of partial pressures: that is, that the pressure of a gas mixture could be regarded as the sum of the pressures that each gas would exert if occupying the same volume alone, under the same conditions.

As a chemist, Dalton was interested in the proportions in which elements combined to form compounds and what this revealed about the structure of matter. In 1808 he published his theory that everything consisted of small, indivisible spheres called atoms, and every element was composed of a particular kind of atom of a particular mass. Dalton showed that the existence of atoms implied that different compounds of the same elements would form in ratios of small whole numbers. Some of his proposals were incorrect – he thought water consisted of one hydrogen atom and one oxygen atom – but he laid the foundations of modern chemistry and atomic theory.

In his honour, the atomic mass unit (approximately the mass of the proton or neutron) is sometimes known as the dalton.

▲ **John Dalton** revived the ancient idea that all matter was made of atoms and put it on a firm experimental footing. He showed how atoms could combine to form molecules and laid the foundations of modern chemistry.

electron ▶ p. 338 ion ▶ p. 340 ionization ▶ p. 340 mass ▶ p. 341 molecule ▶ p. 341 neutron ▶ p. 341 Newton's laws ▶ p. 341 proton ▶ p. 342 Solar System ▶ p. 343

THE LAWS OF PHYSICS

kinetic energy of the molecules. If any or all of these three properties change, the changes for an ideal gas are related by the simple equation:

$$\frac{P_1 V_1}{T_1} = \frac{P_2 V_2}{T_2}$$

where the subscripts 1 and 2 refer to the initial and final state of the gas.

At low pressures, such as those within interstellar clouds, the behaviour of most gases is very close to that of an ideal gas. At high pressures, such as those in the interior of stars, the theory has to be modified.

In astrophysics it is very common for gases to be **ionized**, that is, the atoms or molecules have lost or gained one or more electrons. An ionized gas is called a plasma.

LIQUIDS

If a gas is cooled (that is, the kinetic energy of the molecules is reduced), under certain conditions it will condense to form a liquid. Liquids are much more compact than gases, with higher densities. In a liquid the molecules are weakly held together by attractive forces but are still free to move over each other. Because the molecules are touching, most liquids resist compression. A liquid can flow, but although it can conform to the shape of its container, it does not fill it in the same way as a gas.

✳ AVOGADRO'S PRINCIPLE

Amount of substance is the actual number of any specified entity, such as atoms, molecules or ions, present in a sample. It is measured in moles, where one mole (mol) is equal to the number of carbon atoms in exactly 12 grams of carbon. It follows that the mass in grams of one mole of any substance is numerically equal to the relative mass (taking the hydrogen atom as 1.0) of its constituent particles. For example, one mole of hydrogen atoms has a mass of approximately 1 g, while one mole of hydrogen molecules has a mass of approximately 2 g. One mole of water molecules (H_2O) has a mass of approximately 18 g.

The number of atoms or molecules in a mole is known as the Avogadro constant (or Avogadro's number) and is approximately 6.022×10^{23}. Avogadro's principle states that, under similar conditions of temperature and pressure, equal volumes of any gas contain equal numbers of molecules. To be more specific, one mole of any gas at a pressure of one atmosphere and a temperature of 0°C will occupy 22.4 litres.

💡 FORCES BETWEEN MOLECULES

In an ideal gas, no forces act between the molecules except during collisions. While this is a reasonable assumption for gases, it completely fails to account for the properties of liquids and solids. In fact there are both attractive and repulsive forces between molecules.

Except when they are ionized, molecules carry no net electrical charge, since the number of protons in the nuclei of the constituent atoms is matched by an equal number of surrounding electrons. However, some molecules are permanently polarized, with a negative end and a positive end, and certain unpolarized molecules can become polarized when they approach each other. When

polarized molecules come within a few diameters of each other, weak attractive forces come into play between their positive and negative charges. These are called van der Waals forces, after the Dutch physicist Johannes van der Waals (1837–1923), who showed how the kinetic theory could be modified to take them into account. The forces are responsible for gases condensing to form liquids and solids, something not possible for an ideal gas.

When two molecules approach even more closely, a strong repulsive force appears between the electrons of the two molecules. This is the force which causes molecules to bounce off each other. The balance between the attractive and repulsive forces determines the spacing between molecules when they condense to form a solid.

Liquids can exist only over a narrow range of temperature and pressure, and are rare in the Universe. Liquid water is common on Earth but has not been found elsewhere in the **Solar System**, although it may exist on Europa, one of the moons of Jupiter. Hydrogen is believed to exist as a sea of metallic liquid below the atmospheres of Jupiter and Saturn.

SOLIDS

If a liquid is cooled further it eventually becomes a solid. The atoms or molecules in a solid are locked in position and are no longer free to move. This gives solids their qualities of rigidity and resistance to deformation. The atoms can, however, vibrate, and the energy associated with the vibration depends on the temperature of the material.

The properties of a solid are determined not so much by its composition, the types of atoms or molecules that make it up, but by its structure, the way the atoms are linked together. In crystalline solids, the atoms are arranged in a regular pattern. Metals form crystals in which the outer electrons of all the atoms are free to move throughout the crystal. This explains why metals are good conductors of heat and electricity.

Some solids form more than one type of crystal. Carbon, for example, can be found as graphite, which is very soft, or as diamond, which is extremely hard. The only difference is the crystal structure. In amorphous solids there is no regular arrangement of atoms. These include materials like glasses and plastics.

Solids do not feature much in astrophysics but they are, of course, of great importance in planetary science.

▼ *In crystalline solids such as this copper sulphate crystal, the atoms are organized in a fixed pattern, and the properties of different types of solids depend upon this structure.*

THE LAWS OF PHYSICS

HEAT

Thermodynamics is a large and complex branch of physics concerned with the relationship between heat, work and internal energy. It arose from the development of steam engines in the nineteenth century. This section focuses on simple concepts of the transfer of heat and its effects on the bodies emitting and receiving it. Heat can be transferred by conduction, radiation or convection, all three of which are important in astrophysics and planetary science. Heating a body not only raises its temperature but can also change its state from solid to liquid to gas.

TEMPERATURE

The mean kinetic energy of a body's molecules is proportional to its temperature. The SI unit of temperature is the kelvin (K). On the thermodynamic scale of temperature absolute zero – the temperature at which it is at its lowest possible energy – corresponds to 0 K and the triple point of water (where ice, liquid water and water vapour can coexist) is defined as 273.16 K. This Kelvin scale is used extensively in astrophysics.

The Celsius scale, in everyday use in most of the world, originally defined the freezing point of water at atmospheric pressure as 0°C and the boiling point as 100°C. It is now defined in terms of the Kelvin scale, with the size of the degree Celsius being identical to the kelvin. The freezing point of water is 0.01 K below the triple point, from which it follows that absolute zero is -273.15°C.

The two scales can be converted by the following relations:

$$T (K) = T (°C) + 273.15$$
$$T (°C) = T (K) - 273.15$$

where T (K) and T (°C) are the temperatures in kelvins and degrees Celsius, respectively.

INTERNAL ENERGY AND HEAT

The energy of the molecules that make up a body, whether it be gas, liquid or solid, can be considered as having two components. One is the kinetic energy associated with the random motion of the molecules, sometimes known as thermal kinetic energy. This is directly related to the temperature of the body. The other is the potential energy arising from the attractive and repulsive forces between the molecules. The sum of these two kinds of energy over all the molecules is known as the internal energy of the body.

Internal energy is sometimes referred to rather loosely as 'heat energy'. Modern practice prefers to regard heat as the energy transferred from a hot object to a cooler object by virtue of the temperature difference between them. By this definition, a body cannot be said to contain heat, but only to have heat transferred to or from it.

When heat is transferred *from* a body the internal energy of the body falls. When heat is transferred *to* a body, the internal energy of the body rises. Heat can be thought of as the effective transfer of internal energy from one body to another.

▶ *The granulated surface of the Sun, a typical star, is the result of heat transfer by convection cells, with currents of hot gas rising from the depths to the cooler surface, losing heat and sinking again.*

THERMAL EQUILIBRIUM

An object is in thermal equilibrium if its temperature is both unchanging and uniform. Two or more such objects are in thermal equilibrium with each other when their temperatures are the same. It follows that there is no net transfer of heat within or between bodies in thermal equilibrium.

Strict thermal equilibrium is rare and cannot exist where heat is being transferred. A star, for

ENTROPY AND THE ARROW OF TIME

Most laws of physics, such as Newton's laws, would work equally well if time were reversed. The second law of thermodynamics is an exception. One way of stating the second law is in terms of entropy, which is a measure of the amount of disorder in a system: left to itself, the entropy of a system will either increase or remain the same, but can never decrease.

If you open a bottle of perfume in a room, the whole room fills with scent. Molecules of perfume leave the surface of the liquid in the bottle and because of random collisions with molecules of air spread evenly through the room. They are less ordered than they were – being confined in a limited space is one way of being orderly. The entropy of the room has increased. Now we imagine stopping each molecule and reversing its motion. The molecules retrace their steps, repeating their collisions in reverse until they end up back in the bottle. Entropy has now decreased. The molecules are obeying Newton's laws, but the reverse process violates the second law of thermodynamics and indeed common sense. Some events, such as the diffusion of a vapour into a room, are irreversible, not because the reverse processes are impossible in an absolute sense, but because they are so improbable that they will never happen even in a period equal to many times the present age of the Universe.

While Newton's laws are reversible, the second law of thermodynamics seems to imply a preferred direction of time, corresponding to our subjective distinction between past and future. Philosophers have long argued about the significance of the second law to the nature of time.

example, radiates heat from its surface to the cooler surroundings. In the interior of a star, heat is transferred from the hot centre to the cooler surface, but the change of temperature is so gradual that a small volume of the star can be considered to be in local thermal equilibrium. This very good approximation makes it possible to understand the physics of stellar interiors.

The Universe as a whole is certainly not in thermal equilibrium, as shown by the enormous disparity between the centres of stars, which are at tens of millions of kelvins, and intergalactic space, which is at only a few kelvins. However, if the Universe continues to expand indefinitely, it may come into equilibrium in the very distant future.

CONDUCTION, RADIATION, CONVECTION

There are three possible processes for transferring heat between bodies. In *conduction*, heat is transferred by the vibration of atoms or molecules. If one end of an iron bar is placed in a fire, the rapidly vibrating atoms cause similar vibrations in their neighbours, passing the energy along the bar. Conduction is not a significant mechanism for heat transfer in stars, although it is important within the bodies of planets.

In *radiation*, heat is transferred by electromagnetic waves. All bodies emit electromagnetic radiation because of their temperature, and if this radiation is absorbed by another body its internal energy will rise and heat has been transferred. The most familiar example is solar heating, which warms the Earth and makes it possible for life to emerge and survive. Heat energy transfer by radiation is widespread in astrophysics, within stars as well as between stars and planets.

In *convection*, heat is transferred by the bulk motion of hot fluid. When a pan of soup is heated on a stove, heat is transferred from the bottom of the pan to the top by rising and falling convection currents. Convection is an important mechanism for heat transfer within stars and within the interiors of planets.

HEAT CAPACITY AND LATENT HEAT

When heat is transferred to a body its internal energy rises, its molecules move faster and its temperature rises. The *specific heat capacity* of a substance is the amount of heat needed to raise the temperature of one kilogram of the substance by one kelvin. Water, for example, has a specific heat capacity of about 4190 J/kg K.

If a solid is heated its temperature rises until the energy of the molecules eventually becomes sufficient to overcome the attractive forces between them and the solid starts to melt. The temperature remains the same until all the solid is melted and then rises again. Further heating of the liquid to boiling point will separate the molecules completely as the liquid vaporizes and becomes a gas. The *specific latent heat* is the amount of energy needed to change the state of one kilogram of a substance without changing its temperature. For example, to melt ice at 0°C requires 330 kJ/kg, and to boil water at 100°C requires 2,256 kJ/kg. Conversely, the same latent heat is given out when a gas condenses to form a liquid and when a liquid freezes to become a solid.

▶ *Conduction, Radiation and Convection*
Conduction: Even though the blacksmith is using long tongs, he can feel heat being conducted along them from the hot metal.
Convection: Heated fluids move in circular currents called convection cells. These disturb the surface of a liquid.
Radiation: Molten gold emits heat as electromagnetic radiation.

💡 RATES OF COOLING

An object such as a star or planet can lose energy only through its surface. For a given temperature, the internal energy contained within a uniform body is proportional to its volume, which is in turn proportional to the cube of its radius. The area through which energy can be lost is equal to its surface area and this is proportional to the square of its radius. It follows that the rate at which a body loses energy is inversely proportional to its radius – small bodies cool faster.

Consider two planets of identical composition, one twice the diameter of the other. If both formed at roughly the same time and at the same temperature, the larger body will lose energy faster, but as it has eight times as much energy to begin with, the smaller body will have a cooler interior.

Conduction

Convection

Radiation

✦ LAWS OF THERMODYNAMICS

The first and second laws of thermodynamics were formulated in the nineteenth century as a result of the development of the steam engine and concern the relationship between work and heat. The 'zeroth' and third laws were added in the twentieth century.

0. The zeroth law clarifies the definition of thermal equilibrium. If bodies A and B are each in thermal equilibrium with body C, then A and B are in thermal equilibrium with each other.

1. The first law applies the general principle of conservation of energy to thermal physics and shows that the internal energy of a body can be changed only by heating or by doing work. The change in internal energy of a body is equal to the work done on the body plus the heat transferred to the body.

2. The second law puts constraints on the direction in which a change can take place. It can be stated in many ways, but a common one is: heat cannot of itself flow from a cooler body to a hotter body. Another version states: an isolated system tends to move towards a state of greater disorder. The connection between the two formulations is that a system is more *disorderly* (on the atomic level) when its temperature is more *uniform*.

3. The third law is concerned with absolute zero. It states that it is not possible to reach absolute zero in any actual process, although it is possible to approach arbitrarily close to it.

THE LAWS OF PHYSICS

THE LAWS OF PHYSICS

The Laws of Physics
ELECTRO-MAGNETIC RADIATION
ELECTRIC & MAGNETIC FIELDS

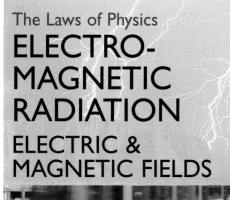

Electric and magnetic fields appear equally mysterious. The first is produced by electric charges at rest, the second by electric currents, which are simply charges in motion. Both kinds of field influence the movement of electrons at a distance and both can be pictured as lines of force in space. Are they really different or are they aspects of the same phenomenon? The first step towards unifying the two fields was taken by Michael Faraday, who found that electric fields could be created by varying magnetic fields. But a complete theory of electromagnetism, including the prediction of electromagnetic waves, had to wait for the insights of James Clerk Maxwell.

ELECTRIC CHARGE AND CURRENT

Electric charge is a fundamental property of matter that has some parallels to **mass**. But unlike mass, charge comes in both positive and negative forms. **Protons** carry positive charge while **electrons** carry an equal amount of negative charge.

The SI unit of electrical charge is the coulomb (C). The magnitude of the charge on the electron or the proton, known as e, is 1.602×10^{-19} C. Larger charges are always made up from multiples of this elementary charge. Bodies containing equal numbers of positive and negative charges are said to be electrically neutral. A neutral body will become charged usually by gaining or losing a number of electrons.

Charge, like mass, energy, momentum and **angular momentum**, is a conserved quantity: in any isolated system the amount of electric charge remains constant.

An electric current is a flow of electric charge. The SI unit of current is the ampere (A), which corresponds to a flow of one coulomb per second. The direction of the current corresponds to the motion of positive charge, which means that current is in the opposite direction to a flow of electrons.

COULOMB'S LAW

The law that governs the forces acting between electrical charges was established by the French physicist Charles Augustin de Coulomb in 1785. It is analogous to Newton's law of universal gravitation and can be stated in a similar way: every charged object in the Universe attracts every other charged object with a force proportional to the product of their charges and inversely proportional to the square of the distance between them.

The law can be stated algebraically as:

$$F = \frac{q_1 q_2}{4\pi\varepsilon_0 r^2}$$

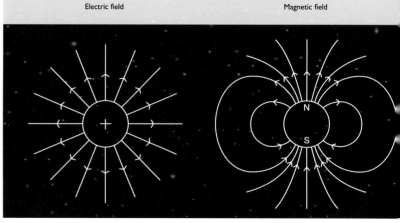

▲ *Electric and Magnetic Fields*
A charged sphere (left) produces a radially symmetric electric field. A magnetized sphere, on the other hand, always has two poles which are connected by loops of magnetic field lines. This is similar to the magnetic field of a star or planet.

where q_1 and q_2 are the charges, r is the distance between them and ε_0 is a fundamental constant called the permittivity of free space.

An important difference between electric forces and gravity is that because the charges can be either positive or negative, the electrical force can be either attractive or repulsive. Like charges repel each other, but unlike charges will attract. Atoms are bound together by attractive forces acting between the positive charges on the protons in the nucleus and the negative charges on the electrons.

ELECTRIC FIELDS

Rather than think about two charges acting on one another at a distance, physicists prefer to use the concept of an electric field. The charge is presumed to alter the space around it in such a way that a second charge will experience a force. The second charge is then responding directly to the local electric field. The strength of the field is defined as the force in

MICHAEL FARADAY

Sir Michael Faraday (1791–1867) was a British chemist whose pioneering discoveries in electricity and magnetism laid the foundations for modern electromagnetic theory and the electrical engineering industry. Trained as a bookbinder, Faraday started his scientific career as assistant to Sir Humphry Davy at London's Royal Institution in 1812. Davy's interest was chemistry, and Faraday soon became skilled at chemical analysis and other experimental techniques.

This was the time when great progress was being made in electrical science and in 1821

he discovered the basic principles of the electric motor, only a year after Hans Christian Oersted had announced that a compass needle could be deflected by a magnetic field. Ten years later he demonstrated the phenomenon of electro-magnetic induction, whereby a current flowing in one coil of wire could, through its changing magnetic field, induce a current to flow in a neighbouring coil. In the same year he built the first electrical generator.

Faraday also developed the laws of electro-chemistry, describing what happens when an electric current flows through a chemical solution

and causes it to decompose. Apart from his many experimental discoveries, Faraday introduced the concept of electric and magnetic fields and how they could be depicted by 'lines of force'.

▲ *Michael Faraday, one of the greatest experimental scientists of all time, uncovered fundamental connections between electricity and magnetism.*

angular momentum ▶ p. 336 electron ▶ p. 338 mass ▶ p. 341 proton ▶ p. 342

THE LAWS OF PHYSICS

OERSTED'S COMPASS NEEDLE

Electric and magnetic forces have been known for centuries. The ancient Greeks knew about static electricity and by the twelfth century many countries were using the magnetic compass for navigation. But it was not until the nineteenth century that an experimental link was discovered between electricity and magnetism.

The turning point came in the winter of 1819. Hans Christian Oersted, a Danish physicist, was giving a lecture demonstration to his students when he noticed that a compass needle was deflected when a current passed through a wire lying over it. It was a small effect, but startling.

In subsequent experiments Oersted mapped out the extent of the magnetic influence and concluded, in modern terms, that the wire was surrounded by a magnetic field whose direction was circular around the wire and always at right angles to it.

Oersted's discovery, that electricity and magnetism were somehow connected, had a mixed reception. Some scientists, led by Coulomb, insisted that such an effect was impossible, but others, such as Faraday and Ampère, immediately saw its significance. Within a few years Ampère had demonstrated that two currents will exert a magnetic force on each other and formulated his theory that *all* magnetic fields were caused by electric currents.

▲ *The discoveries of electromagnetic phenomena in the nineteenth century led to the rapid growth of the electrical engineering industry. This clean and quiet urban train is powered by electric motors supplied by current from an electrified rail.*

newtons that would act upon a charge of one coulomb. The SI unit of electric field strength is the volt per metre (V/m).

Electric field is a vector quantity, with both magnitude and direction. A field can be visualized by lines of force that trace the direction of the field in space. Lines are close together where the field is strong and farther apart where it is weak.

FORCES AND FIELDS

The motion of a charge in electric and magnetic fields can be calculated by combining Newton's second law of motion with a 'force law' describing the force acting upon the charge. In this case, the force law has two parts: one describing the force due to the electric field and the other describing the force due to the magnetic field. Taken together they are called the Lorentz force law.

The electric force on the charge, q, is given by:

$$F_e = qE$$

where E is the electric field. The force on a positive charge is in the same direction as the electric field.

The magnetic force on the charge is given by a rather more complicated expression:

$$F_m = qvB\sin\theta$$

where θ is the angle between the velocity of the charge, v, and the magnetic field, B. This shows that no force will act on the charge if it is stationary or if it is moving along a field line. Only charges moving across the direction of the magnetic field will experience a force and be deflected. In this case the direction of the force is at right angles to both the velocity and the field.

MAGNETIC FIELDS

If a charge is in motion, the space around it is modified not only by the electric field, but also by a magnetic field. While electric fields are produced by electric charges, magnetic fields are produced by electric currents. Even the magnetic field of a bar magnet arises from the motions of electrons within the iron atoms of which it is made.

The SI unit of magnetic field is the tesla (T). An older but still used unit, the gauss, is equal to 10^{-4} T.

Like electric fields, magnetic fields are vector quantities and can be visualized with the aid of field lines. A current passing along a straight wire produces a magnetic field that circles the wire. If the wire is wound into a loop, the magnetic field passes through the loop. If the wire is wound into a cylindrical coil, called a solenoid, the magnetic field passes along the axis of the solenoid and emerges at each end, looking much like the field of a bar magnet. Such fields, with a 'pole' at each end, are called dipole fields.

The opposite poles are conventionally labelled north and south. Like poles repel each other while unlike poles attract. The magnetic fields of planets and stars approximate to that of a bar magnet, and are caused by electric currents circulating inside them.

ELECTROMAGNETIC INDUCTION

As long as they remain static, electric and magnetic fields can be considered as quite separate phenomena. But when either a magnetic field or an electric field is changing, then they become intimately linked.

The principle of electromagnetic induction was first discovered by

Faraday and later extended by **Maxwell**. Faraday demonstrated that a changing magnetic field induces an electric field whose magnitude is proportional to the rate of change of the magnetic field. So if a bar magnet is suddenly pushed into a coil of wire an electric field is briefly induced which drives electrons along the wire, producing a current.

Maxwell found that the reverse is also true: a changing electric field induces a magnetic field whose magnitude is proportional to the rate of change of the electric field. In each case the induced field is at right angles to the field that produces it. Faraday's induction is the principle behind transformers and generators, and Maxwell's induction permits the creation of electromagnetic waves.

▼ *In a thundercloud large differences in electric charge build up between parts of the cloud or between the cloud and the ground until they are neutralized by a sudden flow of current in the form of a lightning flash. The current can be as high as 20,000 amperes.*

Maxwell ▶ p. 72

ELECTROMAGNETIC RADIATION

ELECTROMAGNETIC WAVES

THE LAWS OF PHYSICS

T he concept of a wave has many applications in physics and astrophysics. Electromagnetic waves were predicted by Maxwell, who showed that varying electric and magnetic fields could sustain each other and travel through space at the speed of light. The several different kinds of electromagnetic waves, including gamma rays, X-rays, ultraviolet radiation, infrared radiation and radio waves, differ from visible light only in their wavelength and frequency. Together, this spread of wavelengths is known as the electromagnetic spectrum. Electromagnetic waves are emitted by vibrating electrons and absorbed when they cause other electrons to vibrate in sympathy with them, or they may be emitted and absorbed in changes of the electronic state of atoms and molecules.

▲ *James Clerk Maxwell formulated a theory of electromagnetism that predicted the existence of electromagnetic waves more than 20 years before they were discovered.*

WAVE MOTION

A wave is a regular disturbance in a medium that transports energy from one place to another. A familiar example is a wave on a rope. If one end of a taut rope is shaken, a wave passes along it. Another example is a water wave. If a stone is dropped into a still pond, waves spread out in all directions and are able to transfer energy to objects floating on the surface.

An important characteristic of wave motion is that the disturbed medium remains in place after the wave has passed. A good example is the 'wave' occasionally seen in sports stadiums. Spectators stand up and sit down in a co-ordinated fashion, in such a way that a wave passes through the crowd without the spectators leaving their places.

In some waves, such as waves on a rope, water waves or the 'wave', the medium moves up and down or side to side as the wave goes by. These are known as transverse waves. In other waves, such as sound waves in air and certain seismic waves, the medium moves back and forth along the direction of travel. These are called longitudinal waves.

The direction in which a transverse wave vibrates is known as the **polarization** of the wave.

GENERATION OF ELECTROMAGNETIC WAVES

The entire scope of classical electromagnetism can be represented by four equations known as Maxwell's equations. By combining them in a certain way it is possible to produce another equation which represents the motion of a wave.

Electromagnetic waves may be produced by making electric charges vibrate. The charges are normally electrons. The vibrating electron acts as a changing electric current, which, by the laws of induction, produces a changing magnetic field. The magnetic field in turn produces a changing

electric field, which itself induces another magnetic field. The fields have now broken away from the electric charge that generated them and are moving freely through space.

Energy is exchanged backwards and forwards between the electric and magnetic fields as the wave moves through space, the two fields sustaining each other as they rise and fall. Both the electric and magnetic fields are perpendicular to the direction of motion, so these are transverse waves.

The speed of the wave is set by the need for the fields to change at the correct rate to sustain each other. All electromagnetic waves move at the speed of light, about 300,000 km/s (about 186,000 mi/s).

ELECTROMAGNETIC SPECTRUM

In classical physics, the frequency of an electromagnetic wave is set by the vibration frequency of the electrons that produce it. Rapidly vibrating electrons emit high-frequency radiation, while slowly vibrating electrons emit low-frequency radiation.

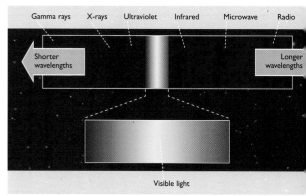

▲ *The Electromagnetic Spectrum*
Visible light occupies only a fraction of the electromagnetic spectrum. The bands of shorter wavelength radiation (ultraviolet, X-ray and gamma ray) and longer wavelength radiation (infrared, microwave and radio wave) are all studied in modern astronomy.

JAMES CLERK MAXWELL

James Clerk Maxwell (1831–79) was an eminent British physicist who developed the mathematical theory of electromagnetism and predicted the existence of electromagnetic waves.

In the 1850s Maxwell tried to understand the discoveries of Faraday and others about electricity and magnetism and devised a mathematical representation of the electric and magnetic fields and how they depended on each other. (Today this representation is written as four simple equations called Maxwell's equations.) From this work, Maxwell showed how all the observed phenomena of electricity and magnetism could be understood.

Most remarkably, by thinking of space as an elastic medium, he discovered that electric and magnetic fields could travel through space in the form of a wave. The speed of the wave could be predicted and came out to be very close to the speed of light. Maxwell proposed in 1864 that light was nothing less than one form of electromagnetic wave, and that an infinite range of 'invisible light' of longer and shorter wavelengths should also exist. It was not until 1888 that **Hertz** demonstrated that electromagnetic waves actually existed.

Maxwell also contributed to thermodynamics and the theory of colour vision and proved that Saturn's rings could not be solid or fluid, but must consist of a large number of independent particles.

ELECTROMAGNETIC RADIATION

blueshift ▶ p. 337 electromagnetic spectrum ▶ p. 338 polarization ▶ p. 342 redshift ▶ p. 343

The range of possible frequencies and wavelengths is known as the **electromagnetic spectrum**. While there is a continuous gradation in these properties through the spectrum, they are conventionally classified into a small number of ranges. The boundaries are not precise and definitions may vary between different scientific disciplines.

- Gamma rays, at the extreme short-wavelength end of the spectrum, have wavelengths shorter than 0.01 nm.
- X-rays are longer, with wavelengths in the range 0.01–10 nm.
- Ultraviolet radiation lies beyond the short-wavelength end of the visible band, between 10 and 400 nm.
- Visible light, the most familiar kind of

electromagnetic radiation, occupies a narrow range of wavelengths between 400 nm (violet) and 700 nm (red).
- Infrared radiation lies beyond the long wavelength end of the visible spectrum and stretches between 700 nm and about 1 mm.
- Radio waves extend beyond wavelengths of about 1 mm.

ABSORPTION OF ELECTROMAGNETIC WAVES

When an EM wave passes through a material, the electric fields in the wave transfer energy to the electrons by making them vibrate. The effect of the material on the wave depends on its frequency and how strongly the electrons are attached to their atoms.

In a transparent material the electrons are bound to their atoms. Little energy is absorbed and the wave passes through except at particular frequencies that cause the electronic state to change. In an opaque material the electrons change state at visible frequencies and the energy is absorbed. Some of the energy is radiated as a reflected wave while the rest is passed on to the molecules as heat. Metals are opaque at visible wavelengths because the outer electrons of the atoms are free to move throughout the metal.

DOPPLER EFFECT

If you stand beside the road while a speeding ambulance goes by you will notice that the pitch of the siren seems to drop suddenly as the vehicle goes past. This effect was investigated by an Austrian physicist called Christian Doppler who published his analysis in 1842.

As the ambulance comes towards you the sound waves from the siren are compressed and arrive at your ear with a shorter wavelength than they would have if the vehicle was stationary. Their frequency – the number striking your ear each second – is also higher, so the pitch sounds

higher. When the ambulance is moving away, the sound waves are stretched out; they arrive with a greater wavelength and lower frequency, and the pitch sounds lower.

The Doppler effect applies to any kind of wave motion and occurs whenever the observer is moving with respect to the source of the waves. For speeds that are small compared with the speed of light, the amount of change (the Doppler shift) of an electromagnetic wave is given by:

$$\lambda_1 = \lambda_0 \left(1 + \frac{v}{c}\right)$$

where λ_0 is the emitted wavelength, λ_1 is the observed wavelength, v is the relative speed of the source away from the observer, and c is the speed of light. When v is positive (the source is moving away) the wavelength is lengthened (a **redshift**), and when v is negative (source moving towards the observer) the wavelength is shortened (a **blueshift**).

▼ *The Doppler Effect*
The Doppler effect changes the wavelengths of light from stars and galaxies in motion. Galaxy A is at a constant distance from Earth so the Doppler effect does not occur. Galaxy B is receding from Earth and the wavelengths of light from it are 'stretched', shifting them towards the red end of the spectrum ('redshifted'). Galaxy C is approaching Earth and the light is squeezed to shorter wavelengths ('blueshifted').

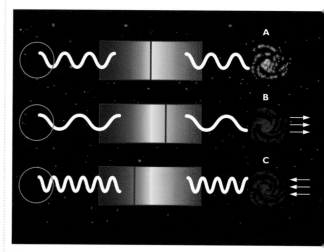

DESCRIBING WAVES

Any kind of wave can be characterized by its wavelength, period or frequency, and speed. The wavelength is the distance between successive crests or troughs. The period is the time between successive crests or troughs. The frequency is the number of crests or troughs passing a fixed point in a given interval of time, and is the inverse of the period. The SI unit of frequency is the hertz (Hz) which is equal to one crest or trough passing each second. Frequencies of radio waves are commonly measured in kHz or MHz.

Period and frequency are related by:

$$\text{frequency} = \frac{1}{\text{period}}$$

The speed of a wave is the speed at which crests or troughs pass a fixed point. Water waves, for example, travel at a few centimetres per second while sound waves in air move at around 330 m/s (about 1,100 ft/s). Electromagnetic waves move at the speed of light, which is approximately 3.0×10^8 m/s (approximately 186,000 mi/s).

Wavelength, speed and frequency are connected by the simple relation:

$$\text{speed} = \text{wavelength} \times \text{frequency}$$

For electromagnetic waves, wavelength and frequency can be converted by:

$$\frac{\text{wavelength}}{\text{in metres}} = \frac{3.0 \times 10^8}{\text{frequency in hertz}}$$

$$\frac{\text{frequency}}{\text{in hertz}} = \frac{3.0 \times 10^8}{\text{wavelength in metres}}$$

MAKING WAVES

Different types of waves can be illustrated with a Slinky, a children's toy that consists of a loosely coiled spring of metal or plastic. If two people stretch one between them on a smooth floor it is possible to send waves along it. One person gently shakes one end of the Slinky from side to side. A transverse wave passes along it. Electromagnetic waves are of this nature. If the end of the Slinky is now given a brisk push to compress it, a longitudinal wave passes along it. Sound waves, and all kinds of pressure waves, are longitudinal.

The emission and absorption of waves can be demonstrated in a bath. Fill a bath and leave the water to settle. Place a floating object, such as a cork, at one end. At the other end dip a finger in the water and waggle it gently to produce waves. The waves travel along the bath and make the cork bob up and down. Energy has been transferred from the finger to the cork. In a similar way, the energy of vibrating electrons may be passed by an electromagnetic wave to other electrons that absorb it.

THE LAWS OF PHYSICS

LIGHT

Light is a form of electromagnetic radiation, and because it is the only such radiation to which our eyes are sensitive it has special significance for us. Much of the behaviour of light can be understood without any knowledge of its wave nature. The science of geometrical optics is based on the laws of reflection and refraction, which have been known for centuries, and is used in the design of optical instruments of all kinds. The human eye can also be understood as an optical instrument, but the perception of colour, while rooted in physics, is largely subjective.

EARLY IDEAS ABOUT LIGHT
The ancient Greeks thought that light was an emanation from the eye, a kind of feeler that went out and touched objects. Around AD 1000 the Arabian mathematician Alhazen saw correctly that light was reflected from objects into the eye and investigated the behaviour of mirrors and lenses.

But as in mechanics, much of our current understanding of light can be traced to Newton, although his contribution was by no means as definitive as it was for the physics of motion. It had been known for centuries that a prism could make a rainbow of colours from white light, but it was Newton who showed that the prism was not introducing the colours itself but separating them out from the white light. He proposed that white light was composed of different types of rays, each of which was bent, or refracted, by a different angle when it passed into glass.

Newton thought that the rays were streams of particles, and to account for refraction he deduced that the particles travelled faster in glass than in air. He was wrong on both counts, but such was Newton's influence that the essential understanding that light is a wave motion had to wait another century. It would take until the early 1900s before physicists realized the complete answer was that light was both a particle and a wave.

REFLECTION OF LIGHT
When a ray of light strikes a glass surface part of it is reflected and part of it passes into the glass. The angle of incidence is the angle between the ray and the normal, a line at right angles to the surface at the point of reflection. The angle of

▲ **Reflection occurs** when light hits a surface and bounces off it. On a largely smooth surface such as this calm water, the light is reflected in a regular way, creating images. On bumpier surfaces, the light is scattered irregularly, a process called diffuse reflection, which does not give rise to images.

reflection is the corresponding angle made by the reflected ray. The law of reflection says that the angle of incidence is equal to the angle of reflection. Reflection may be improved by coating the glass with a thin layer of a metal such as aluminium or silver.

Reflection from shiny surfaces is known as specular reflection, and occurs because the surface is very smooth, any bumpiness being smaller than the wavelength of the light. Where the surface is rough, light is scattered in all directions and the law of reflection does not hold. This is called diffuse reflection. Most objects are visible to our eyes because of diffuse reflection.

REFRACTION OF LIGHT
Refraction is the bending of a ray of light when it passes from one medium into another. Light travels more slowly in a material medium than it does in a vacuum. The difference in speed is measured by the medium's refractive index, n, which is the ratio of the speed of light in a vacuum to the speed of light in the medium. Water, for example, has $n = 1.33$ while glass has $n = 1.5$. The refractive index of air is very close to 1.00.

The behaviour of a refracted light ray is given by Snell's law of refraction, which states that:

$$\frac{\sin\theta_i}{\sin\theta_r} = \frac{n_r}{n_i}$$

where θ_i and θ_r are the angles of incidence and refraction, and n_i and n_r are the refractive indices of the media on the incident and refracted sides of the boundary.

If a ray is incident from a medium with higher refractive index to a medium of lower refractive index (for example, from glass to air) at an angle greater than a certain critical angle the 'refracted' ray will be reflected back from the boundary. This phenomenon of total internal reflection is used in reflecting prisms and to keep light from leaking out of optical fibres. The critical angle for glass is about 42°.

FERMAT'S PRINCIPLE OF LEAST TIME
Around 1650 Pierre de Fermat, a French mathematician, proposed a way of explaining the laws of reflection and refraction. Fermat asserted that the path taken by a ray of light in going between two points was the path that required the least time.

Fermat's principle explains why a ray bends when entering a denser medium. For example, suppose a lifeguard on a beach wants to reach a swimmer in the sea some way along the shore. If the lifeguard can run faster than he can swim, then he should not run straight towards the water, since that would mean swimming the greatest distance. It would also be a mistake to run along the beach until he is level with the swimmer, since that would also take too long. Even the direct straight-line route is not the quickest; it turns out that the lifeguard should run to a point on the water line a little farther along the beach and then swim the rest. The resulting path is precisely analogous to a light ray being bent at a refracting surface.

One consequence of Fermat's principle is that light must travel more slowly in a dense medium

▼ **Reflection and Refraction**
A. Reflection: When a ray of light (the incident ray) is reflected from a flat surface, the angle of incidence from the normal (the line perpendicular to the surface) equals the angle of reflection.
B. Refraction: When an incident ray passes from air into glass, it is refracted. The angle of refraction is determined by three things: the angle of incidence, the refractive index of air and the refractive index of the glass.

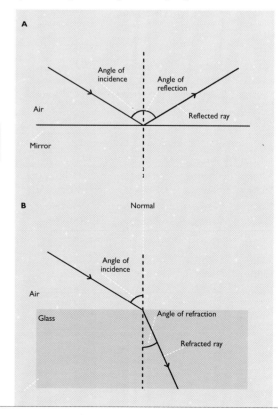

ELECTROMAGNETIC RADIATION

aperture ► p. 336

than in air, contrary to Newton's belief that it must travel faster, based on the idea that light is a stream of particles.

COLOUR

The spectrum of visible light is conventionally divided into seven colours: red, orange, yellow, green, blue, indigo and violet. These have no physical significance, but represent the mind's subjective interpretation of the eye's physiological response to light of different wavelengths.

The eye has three kinds of colour receptors, sensitive to red, green or blue light respectively. Any colour, no matter how subtle, can be represented by a combination of signals from these three receptors. Likewise, an image on a television screen or a computer monitor is made up of glowing red, green and blue dots. These colours are known as the additive primary colours. In printing, colour matching is achieved by mixing a different set of colours: cyan, magenta and yellow. These are complementary to red, green and blue and are known as the subtractive primary colours.

► *Light is refracted when it passes from one medium to another, for example through a lens like this. The light is slowed as it passes through the lens, causing it to change direction.*

HOW THE EYE WORKS

The human eye is essentially a camera. The cornea and the lens make light converge on to the light-sensitive retina. Fine focusing is achieved by altering the shape of the lens. The amount of light admitted to the eye is controlled by the iris, whose central **aperture**, the pupil, can vary in diameter from 1 to 8 mm. The eye is sensitive to light in the range 400–700 nm.

The retina contains light-sensitive cells called cones and rods. The cones work best in good illumination and are responsible for colour vision. Under normal lighting conditions the eye is most sensitive around 550 nm, corresponding to yellow-green light. Cone cells are found all over the retina but predominate in the central region, the fovea, where vision is most acute.

The rod cells are more evenly distributed. They do not sense colours, but are more sensitive to low light levels than the cones. At night, when the eye has become adapted to the dark, vision is almost entirely due to the rods and peak sensitivity is around 500 nm. In these conditions the fovea is relatively insensitive, and a faint star is best seen by letting its image fall to one side of the fovea, a technique known as averted vision.

MIRRORS AND LENSES

The laws of reflection and refraction are the basis of the design of mirrors and lenses. Both can be used in astronomical telescopes to gather and focus light to form images. Any astronomical object is so distant that its rays of light form a parallel beam and this simplifies the design of the instrument.

A converging (or concave) mirror is shaped like a shallow bowl, with the reflective surface on the inside. A parallel beam of light falling on a converging mirror is reflected from the surface and brought to a focus at a point in front of the mirror. The distance from the centre of the mirror to the focal point is the focal length. The ratio of the focal length to the diameter of the mirror is the focal ratio or f-ratio. Converging mirrors are used as the primary mirrors in reflecting telescopes.

The simplest kind of converging mirror has a spherical surface. This has the disadvantage that rays arriving at different distances from the centre of the mirror are brought to a focus at slightly different points, a phenomenon called spherical aberration. It is corrected by shaping the surface to the form of a paraboloid.

A converging (or convex) lens is thicker in the middle than at the edge. A parallel beam of light falling on a converging lens is refracted at both surfaces and brought to a focus at a point behind the lens. The distance from the centre of the lens to the focal point is the focal length, and the focal ratio is defined in the same way as for a mirror. Converging lenses are used as the objective lenses of refracting telescopes (that is, the lenses first encountered by incoming light).

A diverging (or convex) mirror is shaped like a shallow dome, with the reflective surface on the outside. Such a mirror is often used as a secondary mirror in a reflecting telescope. A diverging (or concave) lens is thicker at the edges than in the middle and is commonly used in the eyepieces of telescopes and in spectacles to correct short sight.

Because the speed of light in glass, and so the refractive index, depends on the wavelength of the light, a simple lens focuses different wavelengths at slightly different points. This chromatic aberration results in images having coloured fringes. It can be reduced by using two lenses (converging and diverging) of different types of glass, designed to bring red and blue light to the same focus. Mirrors do not suffer from chromatic aberration.

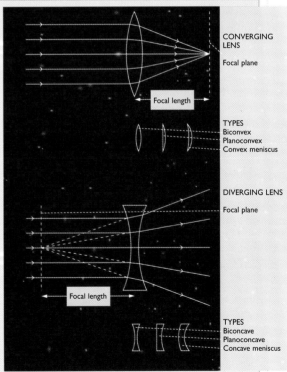

CONVERGING LENS
Focal plane

Focal length

TYPES
Biconvex
Planoconvex
Convex meniscus

DIVERGING LENS
Focal plane

Focal length

TYPES
Biconcave
Planoconcave
Concave meniscus

▲ *Converging and Diverging Lenses*
Parallel rays of light passing through a converging lens are brought to a focus at a point in the focal plane, forming a real image. For a diverging lens, the parallel rays diverge from the lens as if they originated from a point in the focal plane on the incident side, a so-called virtual image.

THE LAWS OF PHYSICS

WAVE MODEL OF LIGHT

T he idea of light being a wave has been accepted only within the last 200 years. Three phenomena in particular can be explained only by a wave model. One of these is diffraction, the spreading of light as it passes through a narrow gap or the bending of light as it grazes the edge of an obstacle. The second is interference, the strange effect in which two beams of light can combine to produce a brighter light in some places and in others complete darkness. Finally, the phenomenon of polarization shows that light must be not only a wave, but a transverse wave.

DEVELOPMENT OF WAVE THEORY

Newton's view that light was made of particles held sway for more than a century, after the publication of his great treatise *Opticks* in 1704. However, some of Newton's contemporaries had proposed that light was a wave.

One of these was Newton's rival, Robert Hooke (1635–1703), who had studied the phenomenon of diffraction and proposed in 1665 that light could be a wave. Christiaan Huygens also devised a wave theory around the same time (though it was not published until 1690) and showed in detail how it could explain diffraction. Both of these ideas were published well before Newton's own theory.

The discovery of polarization by Erasmus Bartholin (1625–98) and by Huygens himself added to the mystery. It implied that if light was made of waves they had to be transverse rather than longitudinal, an idea which seemed inconceivable to the proponents of the wave hypothesis. It was not until Thomas Young demonstrated the phenomenon of interference in 1801 that the road was open to the modern theory of light as a transverse wave, which was comprehensively set out in mathematical form by Augustin Fresnel (1788–1827) in 1818.

DIFFRACTION

Diffraction is the bending or spreading of a beam of light as it passes near an obstacle or through an aperture. The amount of spreading depends on the size of the aperture compared with the wavelength of light. As a rule, a wave passing though an aperture of diameter D metres will spread by a small angle θ radians given by:

$$\theta = \frac{\lambda}{D}$$

where λ is the wavelength of the light in metres. This shows that the spreading is greatest where

▲ **Diffraction provides** *clinching evidence that light travels as a wave. Here the shadow of the point of a needle shows characteristic bright and dark fringes caused by diffraction of light around it.*

▶ **Spikes are** *sometimes seen on photographs of stars taken with reflecting telescopes. They are caused by diffraction of light around the four vanes that support the secondary mirror.*

▼ **Diffraction of Water Waves**
Water waves encountering a narrow gap in an obstacle will diffract through the gap and spread out through almost 180°. The gap then acts as a point source of waves.

CHRISTIAAN HUYGENS

Christiaan Huygens (1629–95) was a Dutch mathematician and astronomer who was one of the first to propose that light was a form of wave motion.

In 1690 he published his theory of light in *Traité de la Lumière*, although he had done most of the work by 1678. Huygens showed that each point on a wavefront, the surface connecting crests of a wave, could be regarded as a source of 'secondary wavelets' that spread out in all directions. The envelope containing these wavelets is a new wavefront. The shape of a wavefront at any moment determines how the waves will move and develop. The theory was successful at explaining diffraction, although he viewed the waves as more like pulses than the regular undulations we understand today by a 'wave'. Huygens also assumed light waves were longitudinal, which made it impossible to explain the phenomenon of polarization, discovered over a century later. Yet Huygens' principle, as it is called, provides a powerful and easily visualized model for the propagation of light and is still used today in teaching the physics of diffraction.

Huygens also contributed to classical mechanics, developing the concepts of momentum and kinetic energy, and he devised the first pendulum clock. He built his own improved telescope, with which he discovered Titan, the largest of the moons of Saturn.

▲ **Christiaan Huygens**, *one of the first scientists to propose that light was a wave.*

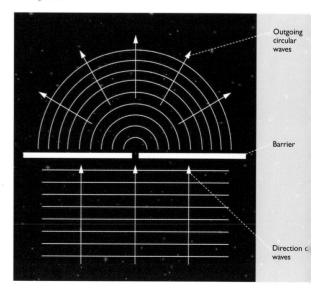

Outgoing circular waves

Barrier

Direction of waves

ELECTROMAGNETIC RADIATION   interferometer ▶ p. 340

THE LAWS OF PHYSICS

YOUNG'S SLITS

Although Huygens had published a wave theory of light in 1690, it was not until 1801 that Thomas Young, a British physician and physicist, provided compelling evidence that light did indeed propagate as a wave.

Young let a beam of sunlight fall upon a screen containing two pinholes. Light passing through the pinholes fell on a second screen. If light were a stream of particles, then a brightening should be seen where the light from the two holes overlapped. In fact, Young saw bright and dark fringes. The only way these fringes could be understood was by interference occurring between light waves arriving from each pinhole. In the bright fringes the waves were reinforcing each other and in the dark fringes the waves were cancelling each other out.

Young's work was initially received with hostility (it conflicted with Newton's teachings that light was a particle phenomenon) but later support from French physicists eventually revealed it to be a turning-point in the understanding of the nature of light. The so-called 'double-slit experiment' is regarded as a classic of physics and very similar experiments were later to prove equally decisive in the acceptance of the quantum theory of light and matter.

apertures are small and wavelengths are long. Diffraction sets the maximum resolution of a telescope (the diffraction limit) because it causes the images of, for example, stars, which should be pinpoints of light, to spread into small disks. Optical telescopes (which use short wavelengths) can produce much sharper images than can individual radio telescopes (which use long wavelengths), even allowing for the radio instruments' much greater size.

Diffraction of radio waves about obstacles explains why in mountainous territory it is much easier to receive distant broadcasts from long-wave radio stations (wavelengths about a kilometre) than from television stations broadcasting with wavelengths of less than a metre.

The pattern of bright and dark fringes sometimes seen in diffracted light is an example of interference, which is discussed in the next section.

INTERFERENCE

When two waves meet they interfere with each other. If the crests of one wave coincide with the crests of the other then the two waves add to produce a wave of greater amplitude, a condition

THE LUMINIFEROUS ETHER

If light is a wave, then what is it a wave in? A sound wave is transmitted by the movements of air molecules, but what is the medium that transmits electromagnetic waves? Until the end of the nineteenth century physicists assumed that all of space was filled by a jelly-like substance known as the luminiferous ether. A light wave was simply a disturbance in the ether.

To match the known properties of light waves, the ether had to possess some unusual properties of its own. It had to be a massless, elastic solid that

known as constructive interference. If the crests of one wave coincide with the troughs of the other then the two waves will cancel each other out, a condition known as destructive interference. For light, patterns of constructive and destructive interference appear as bright and dark bands, which are known as interference fringes.

Interference can be used in sensitive instruments known as **interferometers** to detect whether two waves are in phase or, if they are not, to measure the phase difference between them. In this way, interferometers can be used to measure very small differences in length. In astronomy, interferometers are used to make precise measurements of wavelength, to make very high-resolution images of objects, and to test the optical quality of telescope mirrors.

POLARIZATION

The direction in which a transverse wave vibrates is known as its polarization. For light and other electromagnetic waves the direction of vibration is conventionally taken to be the direction of the electric field (the associated magnetic field is at right angles to this). In ordinary light, the waves vibrate in random directions and the light is said to be unpolarized. In *linear polarization* (or *plane polarization*) the waves all vibrate in the same plane. A beam of unpolarized light can be regarded as a mixture of two linear polarizations at right angles to each other. In circular polarization the direction of vibration changes steadily.

Unpolarized light can become polarized by reflection from a smooth surface, by scattering from molecules or dust particles, or by passing through a polarizing filter.

In astronomy, the state of polarization of light from an object can give clues to the nature of the emission mechanism and of the interstellar medium through which the light has passed.

filled all of space and permeated all matter. It had to be immensely stiff to sustain wave speeds of 300,000 km/s (186,000 mi/s), yet offer negligible resistance to the motion of the planets. It had to be all these things and undetectable, too.

When Maxwell predicted that electromagnetic waves would always move with the same speed in free space it was taken for granted that the speed was to be measured with respect to the ether.

But there is no ether. The clinching evidence that the ether did not exist was given by the experiment of Albert Michelson and Edward Morley in 1887 but the explanation had to wait for Einstein.

METHODS FOR DETECTING LIGHT

In photography light energy is absorbed by crystals of silver halide in a gelatin emulsion. The crystals can vary in size from 0.02 to 2 micrometres, with the smallest crystals giving the highest resolution and the largest crystals giving the most sensitivity (the highest speed). When the emulsion is developed, crystals which have been exposed to light are turned into dark particles of silver, so forming a photographic negative.

In conventional colour photography the emulsion contains three layers sensitive respectively to red, green and blue light. Astronomical colour photography is more usually done by making three exposures through red, green and blue filters, and then combining them to make a colour image.

Other methods of detecting light rely on the photoelectric effect, in which light falling on a metal surface gives up its energy to electrons that are then ejected from the metal, forming an electric current. Many detectors use this principle, including television cameras, image intensifiers and photomultipliers.

The charge-coupled device (CCD) is a wafer of silicon divided up into a large number of light-sensitive cells. When an image falls on the CCD, the light releases electrons and an electric charge builds up in each cell in proportion to the intensity of the light.

▼ *The spectacular colours sometimes seen in a film of oil on water are caused by interference between light reflected from the top and bottom surfaces of the film. Since each colour has a different wavelength, their fringes will occur in different positions, making up a complex pattern.*

THE SPECTRUM BEYOND LIGHT

Light occupies but one narrow band of the electromagnetic spectrum that extends for many orders of magnitude to both longer and shorter wavelengths. Divisions between the types of radiation are arbitrary. There is a continuous spectrum of wavelengths and whether a particular type of wave is called an X-ray or a gamma ray is purely conventional. But the physical characteristics of the waves do vary systematically with wavelength.

RADIO

The longest electromagnetic waves, from about 1 mm upwards, are known as radio waves. They were discovered by the German physicist Heinrich Hertz in 1888 while testing the predictions of Maxwell that electromagnetic waves should exist at longer wavelengths than those of light.

Radio waves occupy the widest stretch of the electromagnetic spectrum and the region is often subdivided into standard bands for convenience (see table). The shortest waves, less than about 10-30 cm, are called microwaves.

STANDARD RADIO FREQUENCY BANDS

Wavelength	Frequency	Band	
1–10 mm	300–30 GHz	EHF	Extremely High Frequency
10–100 mm	30–3 GHz	SHF	Super High Frequency
0.1–1 m	3–0.3 GHz	UHF	Ultra High Frequency
1–10 m	300–30 MHz	VHF	Very High Frequency
10–100 m	30–3 MHz	HF	High Frequency
0.1–1 km	3–0.3 MHz	MF	Medium Frequency
1–10 km	300–30 kHz	LF	Low Frequency
10–100 km	30–3 kHz	VLF	Very Low Frequency

The radio band is of great commercial importance, being used for television and radio broadcasting as well as for other telecommunication purposes.

The relatively long wavelengths and low frequencies allow radio waves to be generated under precisely controlled conditions by causing an electric current to oscillate in a suitable circuit.

INFRARED

The infrared region of the spectrum lies to the long-wavelength side of visible light and was discovered in 1800 by the German-British astronomer **William Herschel**, who placed a thermometer beyond the red end of a spectrum of sunlight formed by a prism. The thermometer rose, indicating that it was being heated by radiation invisible to the eye.

Infrared radiation, which extends from 700 nm to 1 mm, is sometimes known as 'radiant heat' because it is emitted by objects at normal temperatures and can be felt as warmth on the skin. This term can be misleading, since all forms of electromagnetic radiation, including visible light, can warm matter if they are absorbed.

Radiation at infrared wavelengths is readily emitted and absorbed by vibrating or rotating molecules. Absorption by carbon dioxide and water molecules make the atmosphere opaque in most of the infrared band. These molecules are also largely responsible for the greenhouse effect: infrared waves emitted from the warm surface of the Earth are absorbed in the atmosphere, so raising its temperature.

The infrared region of the spectrum is roughly divided into the near, mid and far infrared, but there is no agreement on where the boundaries lie.

▼ *Infrared telescopes*, such as this, the UK Infrared Telescope (UKIRT) on Mauna Kea in Hawaii, are used to observe relatively cool objects in space; these include dust clouds and newly forming stars. Since infrared waves are longer than those of ordinary light, they are not absorbed by the particles in the dusty star-forming regions, revealing objects hidden at visible wavelengths.

HEINRICH HERTZ

Heinrich Rudolf Hertz (1857–94) was a German physicist who discovered radio waves and showed that they travelled at the speed of light. Hertz was educated in engineering, physics and mathematics and held several posts at universities in Germany. In 1884 he put Maxwell's equations into the form in which they are known today and made clearer the symmetry between electric and magnetic fields.

In 1888, in order to test Maxwell's theory, Hertz generated electromagnetic waves by causing an oscillating spark to jump between two electrodes at the focus of a parabolic reflector. He detected them with a similar apparatus that produced a small spark in response to the waves. He showed that the waves were reflected and refracted in a similar way to light and exhibited the phenomena of diffraction, interference and polarization. He even measured their speed, which turned out to be the same as that of light. This new type of electromagnetic radiation, now called radio, was the first to be discovered after light, infrared and ultraviolet radiation.

Hertz not only confirmed that Maxwell's theory was correct, but also laid the foundations of radio communications. The SI unit of frequency is named the hertz (Hz) in his honour.

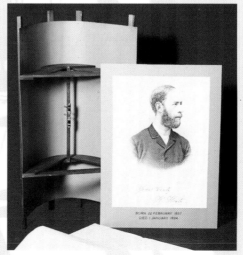

▲ *Heinrich Hertz* and a model of the machine with which he proved the existence of electromagnetic waves.

ELECTROMAGNETIC RADIATION

ionization ▶ p. 340

THE DISCOVERY OF X-RAYS

In 1895 a German physicist, Wilhelm Röntgen, was studying cathode rays at his laboratory in Würzburg. Cathode rays, now known to be streams of electrons, were then still mysterious. They were produced at high voltage inside an evacuated glass tube, and revealed themselves by causing a fluorescent coating inside the tube to glow green.

One day in November Röntgen was working in his darkened laboratory when he noticed a faint sparkling on his work bench. A piece of paper painted with the fluorescent coating was glowing green, even though it was far from the cathode ray tube. Röntgen soon determined that a new kind of ray was being produced where the cathode rays hit the inside of the glass tube.

The new rays travelled in straight lines, cast sharp shadows and affected photographic plates. To his astonishment Röntgen found that the rays could pass through the human body and reveal the shadows of the bones within. He called them 'X-rays' (in Germany they are known as Röntgen rays) but it was several years before they, too, were confirmed to be a type of electromagnetic wave.

In recognition of his discovery, Röntgen was awarded the first Nobel Prize for physics in 1901.

ULTRAVIOLET

No sooner had Herschel discovered infrared radiation beyond the red end of the spectrum than a German pharmacist, Johann Ritter (1776–1810), investigated the region beyond the blue end. In 1801 he discovered an invisible type of radiation that affected photographic plates and became known as ultraviolet radiation (UV). The UV has wavelengths in the range 10–400 nm.

UV waves carry more energy than visible light does, and the higher-frequency radiation is able to knock electrons out of atoms (**ionization**), break chemical bonds and damage molecules. Ultraviolet radiation from the Sun is responsible for sunburn and certain types of skin cancer and is most dangerous in the 290–320 nm region known as UVB. Most of the UVB from the Sun is absorbed by atmospheric ozone.

When UV rays strike certain materials the energy is absorbed and re-emitted in the form of visible light, a phenomenon known as fluorescence if the re-emission is immediate, or as phosphorescence if it persists after the stimulating UV ceases.

▲ *Fluorescence is often seen in nightclubs where the dancers are illuminated by UV light. Clothing washed with detergents that include so-called 'optical brighteners' will appear to glow when UV light falls upon it.*

X-RAYS

X-rays were discovered by the German physicist Wilhelm Röntgen (1845–1923) in 1895. Coming soon after Hertz's discovery of radio waves it was expected that they too would prove to be electromagnetic waves, but they seemed at first to show none of the characteristics of light, such as reflection and refraction, that also characterized radio waves. It took until 1912 for physicists to discover that this was because their wavelength was extremely short, in the range 0.01–10 nm, many times shorter than light waves and comparable to the dimensions of atoms.

The most familiar application of X-rays is in medicine and industry, where their penetrating properties are used to great advantage to make images of otherwise hidden structures. Their short wavelengths also make X-rays useful for studying the structure of complex organic molecules. A beam of X-rays directed on to the molecules in the form of a crystal will produce a diffraction pattern from which the structure of the molecule can be deduced.

X-rays are generated when high-speed electrons collide with a heavy metal target and give up their kinetic energy as electromagnetic radiation. This is the principle of the X-ray tubes used in most industrial and medical applications.

GAMMA RAYS

Gamma rays were identified in 1900 by a French chemist, Paul Villard (1860–1934), but it was not until 1912 that they were recognized as the most energetic form of electromagnetic radiation, with extremely short wavelengths of less than 0.01 nm.

In many ways they behave like high-energy X-rays, and there is no consensus on the wavelength at which X-rays end and gamma rays begin. A useful distinction is their source: gamma rays proper are emitted from the nuclei of atoms either during nuclear reactions or as a result of spontaneous radioactive decay. Gamma rays are detected by counting them one by one. In that sense, they behave more like particles than like waves. Their discovery added to the suspicion around the turn of the century that the very successful wave theory of electromagnetic radiation could not be the whole story.

▼ *The penetrating power of X-rays is routinely used in medicine to examine broken bones.*

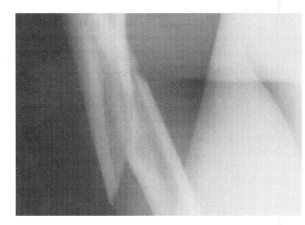

MEASURING THE SPEED OF ELECTROMAGNETIC WAVES

All electromagnetic waves travel at the same speed in free space: the speed of light. Its value is very close to 300 million m/s. The first successful attempt to calculate its value was made in 1676 when a Danish astronomer, Ole Roemer (1644–1710), timed the eclipses of Jupiter's moons. They were delayed by different amounts when the planet was at different distances from the Earth because of the resultant variations in the light's travel time. In 1728 James Bradley (1693–1762) measured the apparent displacement of stars at different times of year, an effect that depends on the speed of light.

In 1849 Armand Fizeau (1819–96) made the first terrestrial measurement of the speed of light by timing the flight of flashes of light to a distant mirror and back. Similar techniques were used by others and by the end of the nineteenth century the speed of light was known to better than two parts in 10,000.

Since then the accuracy has continued to improve to the point when it became possible to measure the speed of light to an accuracy limited not by the apparatus but by the definition of the metre itself. Since 1983 the speed of light has been defined to be 299,792,458 m/s exactly and the metre is now defined as the distance travelled by light in a fraction, 1/299,792,458 of a second.

W. Herschel ▶ p. 44

ELECTROMAGNETIC RADIATION

SPECTRA

The band of colour produced by shining white light through a prism is known as a spectrum. It is caused by different wavelengths of light travelling at different speeds in the glass and being refracted by different angles at the surfaces. A similar effect is achieved by a diffraction grating, a simple device that splits white light into several coloured beams. Either can be used as the basis of instruments to form and study the spectra of astronomical objects. The general appearance of the spectrum depends upon the physical conditions in the source of light and any absorbing matter along the line of sight.

▼ *As a beam* of white light enters a prism, the different wavelengths of light slow down by different amounts, causing the beam to bend and fan out into different colours. It is bent again on leaving the prism, so creating the familiar band of colours known as a spectrum.

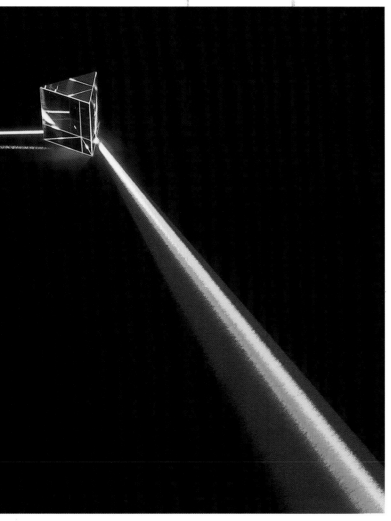

DISPERSION

In free space, light and other electromagnetic radiation always travel at the same speed, very nearly 300,000 km/s. This is not true in other media, where interaction with the electrons in the constituent atoms causes light to travel more slowly.

In fact, different frequencies of light travel at different speeds in a given medium, a phenomenon known as dispersion. In most media the speed of light decreases with frequency (increases with wavelength), so that red light travels faster than blue light. In glass, the difference in speed across the visible spectrum is about 1%.

The speed of a water wave depends in a more complicated way on its wavelength. The slowest speed for a water wave is 23 cm/s (9 in/s), for a wavelength of 17 mm (0.7 in). Longer or shorter water waves move at faster speeds. Ocean seismic waves (tsunamis) have wavelengths of 100–200 km (60–120 miles) and move at speeds of typically 200 m/s (660 ft/s).

PRISMS

As discussed previously, the ratio of the speed of light in free space to its speed in a medium is the refractive index of the medium. Because the speed of light in a medium is directly related to its refractive index, dispersion causes the refractive index to vary with wavelength. So blue light is refracted more than red light and bends more when passing into a denser medium.

A prism is a simple device for using dispersion to separate light into its component wavelengths. It consists of a glass block of triangular cross-section. A beam of white light falling on the prism is refracted once on entering the glass and again on leaving it. Since each wavelength is bent by a different amount, the result is that the different wavelengths emerge at different angles, and the white light is separated into all the colours of the rainbow.

DIFFRACTION GRATINGS

Another device for separating light into its component wavelengths is the diffraction grating. It consists of a sheet of glass ruled with closely spaced opaque lines typically 1–2 micrometres (40–80 millionths of an inch) apart. The clear spaces between the lines act as numerous parallel slits.

A beam of light falling on the grating is diffracted through the slits and light waves spread out from each one. In the straight-through direction all the waves are in phase and interfere constructively with each other, so a beam of light passes in a straight line through the grating. But constructive interference will also occur at angles where the waves differ by a whole number of wavelengths. This means that several other beams will emerge from the grating at different angles given by:

$$\sin\theta = \frac{n\lambda}{d}$$

where θ is the angle through which the light is deviated, n is the number (order) of the beam, with the straight-through beam corresponding to $n = 0$, λ is the wavelength of the light and d is the distance between the slits.

It can be seen that for each order, longer wavelengths are diffracted through bigger angles, so the effect of the grating is to separate a beam of white light into a fan of many colours. A complete spectrum is produced for each value of n.

TYPES OF SPECTRUM

Several different types of spectrum may be distinguished according to the conditions in which the radiation is emitted.

- A continuous spectrum is emitted by a hot, opaque object such as a star. It looks like a smooth band of colours.
- An emission line spectrum is emitted by hot, low-pressure gas. It consists of a number of bright lines (emission lines) of different colours. It is created when atoms and molecules in the object change their states, emitting light of definite frequencies.
- An absorption spectrum is formed when light from a hot body passes through cooler material such as the outer atmosphere of a star or an interstellar cloud. The cooler material absorbs some of the light from the continuous spectrum, leaving a number of dark lines (absorption lines) at the missing wavelengths.
- A reflection spectrum is seen when light from one source is reflected from another object, such as a planet or a dust cloud. Emission or absorption lines in the spectrum of the source may be seen in the reflection spectrum.

ELECTROMAGNETIC RADIATION

absorption line ▶ p. 336 black hole ▶ p. 336 chromatic aberration ▶ p. 337 emission line ▶ p. 338 pulsar ▶ p. 342

THE LAWS OF PHYSICS

NEWTON'S PRISM

In the mid-1600s **Isaac Newton** investigated why telescopes suffered from coloured fringes around their images, a phenomenon known as **chromatic aberration**.

Glass lenses had much in common with glass prisms, and people believed that a prism modified light and introduced colours into it. Newton let a narrow beam of sunlight fall upon a prism and observed the spectrum cast on the wall of his study. He noticed that the band of colours was elongated; the prism was not modifying the light but, he proposed, fanning out the coloured rays of which white light was composed.

To test his hypothesis he passed the spectrum through a second, inverted, prism and found that the colours could be recombined into a beam of white light. When he passed the spectrum through a prism at right angles to the first, each colour was bent by a different amount.

For another test, he isolated colours from the spectrum and passed them into a second prism to see if new colours were introduced. But in every case, although bent by the second prism, red remained red and blue remained blue. Newton had shown conclusively that white light is a mixture of colours which are refracted by different amounts.

More than one type of spectrum may be present at the same time. The spectrum of the Sun, for example, consists of a continuous spectrum together with bright **emission lines** and dark **absorption lines**. The entire solar spectrum may itself be observed as a reflection spectrum in the light reflected from clouds.

RAINBOWS

A rainbow is seen wherever drops of rain on one side of an observer are illuminated by sunlight shining from the opposite side.

Rays of white light from the Sun are refracted on entering the raindrop, are reflected off the back and refracted again as they emerge. Because of dispersion inside the raindrop, the white light is separated into colours each emerging at a slightly different angle.

An observer facing away from the Sun sees a circular bow centred on the point immediately opposite the Sun. The outer edge of the bow is red and the inner edge is violet. Only the parts of the bow occupied by raindrops and illuminated by the Sun are seen. The red part, for example, is made up of drops refracting red light. From an aircraft, a complete circle is often visible.

Usually a second, fainter bow with the order of colours reversed can be seen outside the main one. This is caused by light making two reflections inside the raindrop before emerging.

The inner edge of the rainbow often shows narrower coloured fringes (supernumerary bows) caused by diffraction effects.

◄ *A rainbow is a good example* of the refraction of light. When the Sun shines through a drop of rain, the light is dispersed and refracted and emerges as separate colours, each one at a slightly different angle, causing the arched shape. The radius of the primary bow is 42° and that of the secondary bow is 51°.

▶ *A faint object spectrograph* on the William Herschel Telescope. Such devices are used to create a photographic or electronic image of a spectrum.

SPECTROSCOPY

Spectroscopy is the branch of physics concerned with the formation and interpretation of spectra.

Instruments used for observing spectra are variously known as spectroscopes, spectrometers, or spectrographs. A spectroscope is a simple instrument designed for visual use, a spectrometer is a device for measuring wavelengths and sometimes intensities and a spectrograph is a device which forms a photographic or electronic image of a spectrum (a spectrogram). However, the terms are often used interchangeably.

Spectroscopic instruments use either a prism or a diffraction grating to form a spectrum. A diffraction grating is more common, as there is a simple relationship between the wavelength of the light and the angle through which it is diffracted. It is also effective over a wider range of wavelengths than a prism.

In astronomy, where almost all objects have to be observed remotely, spectroscopy is a powerful tool for studying the physical conditions in planets, stars, galaxies and interstellar space. Among its many applications, astronomers use spectroscopy to measure the chemical composition of stars and planetary atmospheres, the temperatures of gaseous nebulae, the strengths of magnetic fields in space, the rotation speeds of galaxies, the expansion rate of the Universe, and the masses of **black holes**.

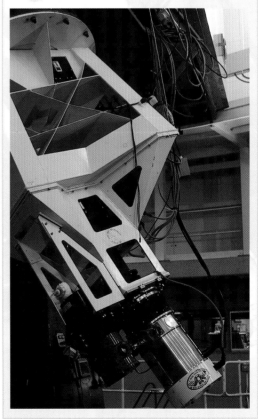

DISPERSION OF RADIO WAVES IN SPACE

An important application of dispersion occurs in radio astronomy. The space between the stars is filled with a very tenuous gas of ionized hydrogen. The gas is dispersive, and while it has negligible effect on visible light, it causes radio waves to be delayed by observable amounts. The lower its frequency, and the longer its wavelength, the more slowly a radio wave travels through interstellar space. This is known as interstellar dispersion.

The phenomenon is useful for the study of **pulsars**, radio sources which emit regular pulses

of radio waves many times a second. If the pulsar is observed simultaneously at two different frequencies, the higher-frequency waves from each pulse will arrive shortly before the lower-frequency waves. By measuring the delay between the pulse arrival times at the two frequencies, astronomers can measure the distance to the pulsar, if they know the density of the ionized gas along the line of sight. Conversely, if they have an independent measure of the pulsar's distance they can estimate the density of the ionized gas. By observing many hundreds of pulsars it has been possible to map the distribution of ionized gas in the Galaxy.

Newton ▶ p. 42

ELECTROMAGNETIC RADIATION

LINE SPECTRA

The bright and dark lines seen crossing the spectra of light sources remained mysterious throughout the nineteenth century. Bright lines appear in the spectra of hot gases and complementary dark lines appear where a continuous spectrum is viewed through an intervening cool gas. By the middle of the century it was clear that the lines formed patterns, and each pattern was characteristic of one of the chemical elements. Astronomers could then do what no one had ever thought possible – determine the chemical composition of the Sun and stars. It was the beginning of astrophysics, but still there was no explanation of why atoms emitted and absorbed light at these discrete wavelengths.

EMISSION LINES

If a gas is heated to a high enough temperature it will begin to glow. But it will not emit light across the spectrum in the same way as a solid object. Provided it is not so dense that its

molecules interact significantly with each other, a hot gas will emit light at a small number of discrete wavelengths. If the light from a hot gas is examined with a spectroscope it appears not as the familiar rainbow of colours but as a number of bright, coloured lines against a dark background. These are called emission lines.

The most familiar example of an emission line spectrum is seen in sodium street lamps, which have a characteristic yellow glow. The lamp contains a hot sodium vapour which emits most of its light in a strong yellow line (actually a close pair of lines – the sodium D-lines) with a wavelength of 589 nm.

Another example is the pink glow often seen in photographs of **nebulae**. This is caused by a red emission line in the spectrum of hydrogen gas, the H-alpha line, which has a wavelength of 656 nm.

Many more lines like these were discovered as spectroscopy developed through the nineteenth century.

▶ *The world's first* operational atomic clock, constructed at the UK National Physical Laboratory in 1955. Like modern atomic clocks, it was based on an absorption line in the spectrum of the caesium atom.

ABSORPTION LINES

If a light source with a continuous spectrum is viewed through a cool gas then dark absorption lines appear in the spectrum. The wavelengths of the lines are identical to the wavelengths of emission lines from the same gas when heated.

The Sun's spectrum has many absorption lines, due to light from the surface (the photosphere) being absorbed in the cooler, overlying atmosphere (the chromosphere). The most prominent lines were first noticed in 1802 by a British chemist, William Wollaston (1766–1828),

JOSEPH VON FRAUNHOFER

Joseph von Fraunhofer (1787–1826) was a German physicist and instrument maker who discovered dark absorption lines in the spectrum of light from the Sun.

The largely self-taught Fraunhofer began his career in 1806 with an optical instrument company in Munich. In 1814 he devised a way of measuring the optical properties of glass by using the bright yellow emission line in the spectrum of sodium as a standard. He devised the modern spectroscope by using a telescope to study the spectrum produced by a prism. He found that the spectrum of the Sun was crossed by numerous dark lines, one of which corresponded in wavelength to the bright emission line found in the spectrum of sodium. Fraunhofer went on to catalogue more than 500 of these absorption lines – now known as Fraunhofer lines – and the system of letters he used to label the more prominent lines is still used.

In 1821 he constructed the first diffraction grating and used it instead of a prism to form a spectrum and make precise measurements of the wavelengths of the solar absorption lines. Although

it was left to others to interpret his discoveries, Fraunhofer is recognized as the founder of astronomical spectroscopy.

▲ *Joseph von Fraunhofer* was the first person to study systematically the dark absorption lines in the Sun's spectrum that now bear his name.

ATOMS AS FREQUENCY STANDARDS

Line spectra are not confined to the visible spectrum. Absorption lines at radio frequencies are used as the basis of atomic clocks, which underpin the world timekeeping system.

The caesium atom has an absorption line at a wavelength of 3.26 cm, in the microwave region of the radio spectrum. In an atomic clock, a beam of caesium atoms is made to fly down an evacuated tube. Radio waves generated by an oscillator are directed across the beam and the frequency of the waves is gradually varied. When the frequency exactly matches that of the absorption line, the atoms absorb the radio energy and change their magnetic state. The changed atoms are deflected by magnets on to a detector, generating a signal that is used to maintain the frequency of the oscillator. This frequency, the caesium frequency, is precisely 9,192,631,770 Hz. The clock counts the cycles from the oscillator.

Since 1967 the SI second has been defined as the duration of 9,192,631,770 cycles of the caesium frequency. The best caesium clocks can keep time to one second in 10 million years.

ELECTROMAGNETIC RADIATION

THE LAWS OF PHYSICS

THE DISCOVERY OF HELIUM

Helium is the second most abundant element in the Universe, but it was discovered on Earth only after it had been discovered in the Sun.

In 1868 a French astronomer, Jules Janssen (1824–1907), travelled to India to observe a total eclipse of the Sun. During a total eclipse the Sun's atmosphere becomes briefly visible, and one object of the expedition was to examine the spectra of the hot clouds of gas (**prominences**) that can be seen at this time. Janssen found a bright orange line which he assumed to be the yellow line of sodium. In the same year Janssen,

along with the British astronomer Sir Norman Lockyer (1836–1920), devised a method of taking spectra of the Sun's prominences without the need to wait for an **eclipse**. The orange line was confirmed, but it was not at the same wavelength as the sodium line. Lockyer proposed that this was the signature of a new element, unknown on Earth, which he called helium, from *helios*, the Greek word for Sun.

Helium was isolated on Earth 27 years later by the British chemist Sir William Ramsay (1852–1916), who noticed that a gas emitted from certain radioactive minerals showed the same orange line seen in the Sun.

but he failed to recognize their significance. They were rediscovered in 1814 by the German physicist Joseph von Fraunhofer, who recorded several hundred of them. More than 30,000 so-called Fraunhofer lines have now been mapped in the solar spectrum, and they are an intricate record of the composition of the outer layers of the Sun.

Absorption lines are also seen in the spectra of other stars. Most of them are stellar Fraunhofer lines but some of them arise in the cold interstellar gas along the line of sight and give clues to the physics and chemistry of the interstellar medium within our Galaxy. Remote quasars also have absorption lines, and these carry information about intergalactic space.

IDENTIFICATION OF ELEMENTS

In 1860, two Germans, Robert Bunsen (1811–99) and Gustav Kirchhoff (1824–87), showed that each element produces its own characteristic set of lines, which may appear either in emission or

absorption. The sodium D-line, for example, appears in emission as a bright, closely spaced double yellow line, and in absorption as a dark double line against the yellow part of the spectrum. The wavelengths of the bright and dark lines are identical. By comparing the Sun's spectrum with the spectra of laboratory samples Bunsen and Kirchhoff made the first analysis of the Sun's composition.

The idea was quickly taken up by the British astronomer Sir William Huggins (1824–1910), who showed in 1863 that the stars are made of the same elements as the Sun, and a year later discovered that several nebulae have emission line spectra and are therefore made of hot gas.

Whether or not the lines of a particular element appear in a spectrum depends in a complex way on the physical conditions in the source, especially the temperature, and are not necessarily an indication of how much of the element is present. The most abundant elements in the Sun are hydrogen and helium, and many lines belonging to these two elements are present in the Sun's spectrum; but also very prominent are calcium, sodium, magnesium and iron. Cool stars often show lines from molecules such as titanium oxide, which are not found in the spectrum of the Sun.

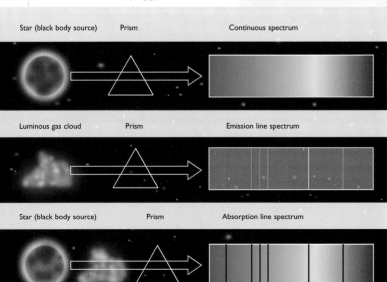

Star (black body source)   Prism   Continuous spectrum

Luminous gas cloud   Prism   Emission line spectrum

Star (black body source)   Prism   Absorption line spectrum

◀ *Emission and Absorption Spectra*
(A) A hot opaque object, such as a star, emits a continuous spectrum. (B) A hot, thin gas emits light at certain wavelengths only, forming an emission line spectrum. (C) Where a cool, thin gas lies between the star and the observer, it absorbs light from the star, forming an absorption line spectrum.

REDSHIFTS AND BLUESHIFTS

The measured wavelengths of emission and absorption lines in astronomical objects often differ from the wavelengths of the same elements in the laboratory. This is because the gas in which the lines arise is moving with respect to the observer, and the wavelengths of the lines are being changed by the **Doppler effect**.

The lines from many stars are either **redshifted** or **blueshifted** depending on their movement away from the Earth or towards it, and the amount of the shift indicates the **radial velocity** of the star.

An emission line from the hydrogen atom, with a wavelength near 21 cm, is the strongest feature in the radio spectrum. By measuring the precise

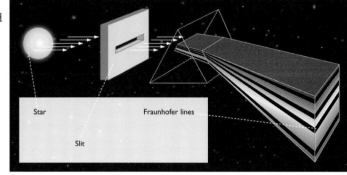

Star   Fraunhofer lines

Slit

▲ *The Spectrum of a Star*
Atoms in the atmosphere of a star absorb light at specific wavelengths, causing dark lines in its spectrum, called Fraunhofer lines. Spectroscopy, the study of such spectra, reveals much about the composition, age and distance of objects in the Universe.

wavelengths of the 21-cm lines coming from the **Milky Way**, astronomers can identify many clouds of gas moving at different speeds and have used them to map the spiral structure of the Galaxy.

LINE BROADENING

The Doppler effect also affects the width of the spectral lines. The atoms in a gas are all moving at different speeds, and in consequence the emissions from each atom will have slightly different wavelengths. The result is that the line is slightly broadened, and the lines from a hot gas are broader than the lines from a cold gas, an effect known as Doppler broadening. The line width tells astronomers the temperature of the gas.

Doppler broadening can also indicate turbulence in the gas or even that the whole source is rotating, expanding or contracting. Any of these conditions would broaden the lines.

Sometimes single lines that appear broadened are actually split into two or more lines. This *Zeeman effect* is apparent in the light from sunspots and indicates that a strong magnetic field is present. The amount of splitting is a measure of the strength of the field.

CONTINUOUS SPECTRA

T he smooth, continuous spectrum of colours emitted by hot bodies, the Sun included, was an in-superable challenge to the physicists of the nineteenth century. Some could explain the shape of the spectrum at short wave-lengths, others could explain it at long wavelengths, but none could account for the whole spectrum – the so-called black-body spectrum – in any way that made sense. It was a fatal blow to the contem-porary understanding of the Universe.

THERMAL RADIATION

All bodies emit radiation by virtue of their temperature. The particles within them are in continual random motion, and the radiation they emit is known as thermal radiation. The hotter the body, the more radiation it emits, and this is one of the ways in which heat can be transferred from one body to another.

The form of the spectrum depends on the nature of the body emitting it. A rarefied gas emits a line spectrum when heated, but dense gases, liquids and solid bodies emit a different kind of spectrum called a black-body spectrum.

Black-body radiation has a characteristic, continuous spectrum that rises from low-intensity levels at short wavelengths, reaches a broad peak, and then falls away towards longer wavelengths. A black-body spectrum has the same shape no matter what the object is made of – there are no emission or absorption lines – and the position of the peak and the brightness of the radiation depend only on the temperature. In reality, no object has a spectrum that conforms exactly to that of a black body, but stars, planets and other astronomical bodies can be good approximations and that is why the concept is useful.

WIEN'S DISPLACEMENT LAW

One of the characteristics of black-body radiation is that the wavelength of peak emission depends on the temperature of the body. For example, as a piece of iron is heated, the peak of its spectrum moves from the infrared into the red part of the visible spectrum and then to successively shorter wavelengths.

▲ *A piece of metal* glows different colours according to its temperature. The coolest part is dull red, with a peak in the infrared part of the spectrum, and the orange, yellow and white parts are progressively hotter.

The German physicist Wilhelm Wien (1864–1928) studied this relationship and, in 1893, came up with an expression which is now known as **Wien's displacement law**. It relates the wavelength of the peak of the spectrum to the temperature of the body:

$$\text{wavelength of peak emission } (\mu m) = \frac{2900}{T(\text{K})}$$

For example, the Sun, at a temperature of 5,770 K, peaks at a wavelength of 0.5 μm, which is in the middle of the visible spectrum. The Earth, which has an average temperature of around 288 K, radiates most strongly around 10 μm, which is well into the infrared. The **cosmic background radiation**, which has a black-body spectrum of 2.7 K, peaks near 1.1 mm.

Cool stars are red, because their spectrum peaks in the red or infrared. Moderately hot stars look white, because the peak comes in the middle of the visible spectrum, while very hot stars look blue, because the peak lies in the blue or ultraviolet.

STEFAN–BOLTZMANN LAW

The amount of energy radiated each second by a **black body** (its power or **luminosity**) is also related to temperature. The relationship was found by two Austrian physicists, Josef Stefan (1835–93) and Ludwig Boltzmann (1844–1906). Stefan derived the law through the study of the radiation from hot bodies while Boltzmann explained it on thermodynamic principles in 1884. The

Stefan–Boltzmann law states that the luminosity of a black body is proportional to the fourth power of its temperature. For a spherical black body of radius R, the luminosity L is given by:

$$L = 4\pi R^2 \sigma T^4$$

where T is the temperature and σ is a fundamental constant known as the Stefan constant. The law means that doubling the temperature of a body raises its luminosity by a factor of 16.

A more practical version of the law can be used for the properties of stars, which are good approximations to black bodies:

$$\frac{L}{L_\odot} = \left(\frac{R}{R_\odot}\right)^2 \left(\frac{T}{T_\odot}\right)^4$$

where $L_\odot$, $R_\odot$ and $T_\odot$ are the luminosity, radius and surface temperature of the Sun.

RADIATION FROM NEBULAE

Another type of thermal radiation is seen from hot gas clouds such as nebulae, which consist essentially of ionized hydrogen. Because the nebula is transparent, the radiation from its atoms escapes from the cloud before the interior can attain thermal equilibrium. The spectrum of a nebula, while still thermal, is not that of a black body but is more complex.

The spectrum from an ionized nebula looks like that of a black body at longer wavelengths where it is opaque or optically thick, but falls below it at shorter wavelengths, where it becomes transparent or optically thin. In the optically thin region the shape of the spectrum will depend in a complicated way on the detailed emission mechanisms at work in the nebula. In an ionized gas there will be a continuous spectrum due to interactions between the electrons and the positive ions, as well as an emission-line spectrum from the atoms themselves.

NON-THERMAL RADIATION

While thermal spectra are very common in astronomy, there are other kinds of continuous spectra that are non-thermal in origin.

The most important is synchrotron radiation, emitted by relativistic electrons (that is, moving near the speed of light) in magnetic fields. A relativistic electron entering a magnetic field is forced to move in a spiral path, and as it does so it radiates away its energy.

Synchrotron radiation, which is named after the type of particle accelerator in which it was

ELECTROMAGNETIC RADIATION

BLACK BODIES

A black body is defined as an object that completely absorbs all radiation that falls upon it, and is therefore 'black' at all wavelengths. An ideal absorber is also an ideal emitter, and a black body is therefore an ideal thermal radiator.

The usual model of a black body is a cavity, such as a hollow sphere, with a small hole in it. A ray of light entering the hole will be scattered around inside the sphere until all its energy is absorbed. A negligible amount will re-emerge from the hole, so the hole itself has the properties of a black body. If the sphere is then heated the spectrum of radiation emerging from the hole (sometimes called cavity radiation) will conform very closely to that of a black body. This is because light is emitted and absorbed many times inside the sphere before it escapes, and is able to attain thermal equilibrium with the walls. Any lines in the spectrum that would indicate the composition of the sphere will be smeared out into a continuous spectrum after multiple scatterings. Indeed, black-body radiation is characteristic of the interior of an object in thermal equilibrium.

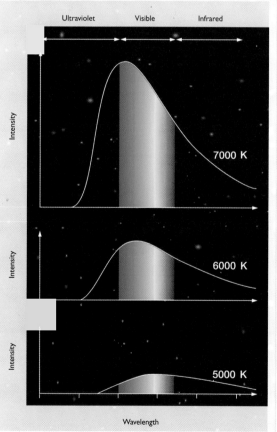

▲ *Black-body Radiation at Different Temperatures*
These graphs show the relative intensity of each colour in the light emitted by a black body at three different temperatures. Note that as the temperature of a black body rises, its peak emission moves towards shorter wavelengths.

THE ULTRAVIOLET CATASTROPHE

Why do black bodies radiate the way they do? Why does the spectrum have a peak whose wavelength depends on temperature? These questions puzzled nineteenth-century physicists who were trying to understand the theory of black-body radiation.

Wilhelm Wien, using an analogy from the speeds of molecules in a gas, in 1896 devised a curve that fitted the short wavelengths and the peak very well. By 1900, as measurement techniques improved, it was clear that Wien's formula did not hold up at the longer wavelengths of the black-body spectrum.

Meanwhile a British physicist, Lord Rayleigh (William Strutt, 1842–1919), used the model of a cavity to work out from first principles the spectrum of radiation that should emerge from it. Everything then known about electromagnetic radiation, based on the very successful theory of Maxwell, indicated that the amount of radiation should rise sharply at shorter wavelengths. Instead of a peak in the spectrum, Rayleigh showed that as the wavelength approached zero the radiation should rise to infinity. This devastating clash of theory with observation became known as the ultraviolet catastrophe. It marked a complete failure of contemporary physics, and heralded a revival of an old idea: perhaps light was made of particles after all.

first observed, has a very broad spectrum and can appear at all wavelengths from radio to X-rays. The spectrum rises at lower frequencies, distinguishing it from thermal radiation which generally falls. The radiation is also strongly polarized and is emitted in a beam, rather than uniformly in all directions.

The electrons themselves must first be accelerated to relativistic speeds, and this may happen far from the magnetic field where they give up their energy. Synchrotron radiation is seen from the magnetic fields in the disk of the Milky Way. The electrons are cosmic-ray particles, which are believed to have come from **supernova** explosions in the distant past. Other important sources of synchrotron radiation include supernova remnants, pulsars and **quasars**.

▼ *The North America Nebula is a vast cloud of ionized hydrogen heated by a young, hot star. The gas emits thermal radiation over a wide spectrum.*

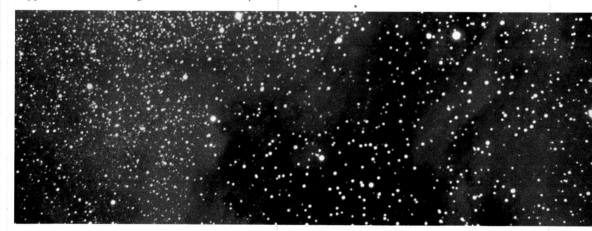

ESTIMATING TEMPERATURES

The concept of a black body is widely used in astrophysics, especially when estimating the temperatures of stars and other objects.

If the radius and luminosity of a star are known then the Stefan–Boltzmann law can be used to estimate its *effective temperature*. This is the temperature of a black body that would have the same luminosity as the star.

If the spectrum of the star can be obtained, then the wavelength of peak emission could be used to get a temperature from Wien's displacement law. But a black-body spectrum can be characterized by measurements of brightness at just two wavelengths, such as the standard photometric bands U (ultraviolet), V (yellow) and R (red). Any two of these can be used to derive the *colour temperature* of the star.

If the distance and radius of a star are known, then it is possible to calculate the temperature of a black body that would match the star's brightness at a chosen wavelength. This *brightness temperature* is widely used in radio astronomy to estimate the temperature of a cloud of gas. If the cloud completely fills the telescope's field of view, then the brightness temperature can be calculated without knowing the size or distance of the cloud.

ELECTROMAGNETIC RADIATION

THE LAWS OF PHYSICS

The Laws of Physics
RELATIVITY
SPECIAL RELATIVITY

By 1900 Newtonian mechanics and Maxwell's formulation of electromagnetism were well established. With a few exceptions, Newton seemed to account for all mechanical phenomena and Maxwell seemed to account for all electromagnetic phenomena. But there remained a few puzzles, and one was the speed of light. What precisely was meant by the speed of the waves that was calculated from Maxwell's equations? In what frame of reference was the speed to be measured? The answer came from a little-known physicist named Albert Einstein, and it would challenge the most deeply held notions about time and space.

PRINCIPLES OF RELATIVITY

A young German physicist called **Albert Einstein** (1879–1955) had puzzled in his youth about light. What would a light wave look like if one could catch up with it? For 10 years he worked his way towards a new view of the nature of time and space, which he published in 1905. His special theory of relativity was based on two simple principles:

1. The laws of physics remain the same in all inertial frames.
2. The speed of light in free space has the same value in all inertial frames.

The first principle is a reassertion of Newton's first law in a more general form. It implies that there is no such thing as absolute velocity; only relative velocities have meaning. The second elevates an experimental finding to the status of a principle. To others the constancy of the speed of light was perplexing and disturbing, but to Einstein's intuitive mind it expressed a fundamental truth: there was no ether.

Einstein showed that these two principles led to a new kind of physics which conformed with Newtonian mechanics for bodies moving at normal speeds, but diverged sharply when speeds became comparable with the speed of light.

LENGTH CONTRACTION

One famous consequence of special relativity was presaged by two physicists, George Fitzgerald (1851–1901), who was Irish, and Hendrik Lorentz (1853–1928), who was Dutch. In the 1890s they independently proposed that the results of the Michelson–Morley experiment could be made consistent with the existence of an ether if objects physically contracted in their direction of motion. This was an *ad hoc* proposition, with no physical justification, but Einstein found that a contraction of similar form, but utterly different in nature, was an inevitable consequence of his theory of relativity.

An object of length l_0 moving with a speed v with respect to an observer would be measured in the observer's frame of reference to be of a length l given by:

$$l = l_0 \sqrt{1 - \frac{v^2}{c^2}}$$

But in the frame of reference of the object, distances in the observer's frame of reference are also reduced.

Although the Lorentz–Fitzgerald contraction is measurable, it is not generally observable in a literal sense. An object moving close to the speed of light would look to the eye not contracted but distorted in a more complex way.

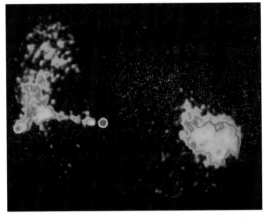

▲ *This **radio image** of the quasar 3C179 (centre) shows a jet of material (left) being ejected at a substantial fraction of the speed of light. Einstein's theory of special relativity is used routinely to understand objects like this.*

TIME DILATION

Of more fundamental importance than length contraction are the implications for the nature of time. The cost of accepting the principles of relativity is that the concept of a universal 'now' has to be abandoned. Observers, in general, will not agree on the time of events, nor even whether events are simultaneous or not. Time becomes relative to the observer, not universal.

Consider a clock which ticks at regular intervals t_0, when at rest. If the clock is carried on a spaceship moving at a speed v with respect to an observer on the Earth, the ticks in the observer's frame of reference would be measured to be of an interval t, given by:

$$t = \frac{t_0}{\sqrt{1 - \frac{v^2}{c^2}}}$$

This phenomenon of time dilation is often summed up by the phrase 'moving clocks run slow', but it should be understood that neither clock is 'slow' in its own frame of reference: each observer judges the other's clock to be running slow, since it is only their relative speed that comes into the equation.

Although time dilation is a difficult concept to grasp, and certainly conflicts with our common-sense view of the world, it has been verified in numerous experiments and is an everyday part of modern physics and astrophysics.

MASS AND ENERGY

A third consequence of special relativity can be summed up in the famous equation $E = mc^2$. It tells us that a mass m is associated with an energy E, where c is the speed of light. In other words, mass and energy are alternative ways of describing the same thing. The consequence of this is that the separate laws of conservation of mass and of energy no longer hold. What is conserved is

THE FOURTH DIMENSION?

It is often said that time is the fourth dimension. There is nothing new or remarkable about this. In Newtonian mechanics there are three dimensions of space and one of time. Any event can be located by specifying where it happened (three coordinates) and when (one coordinate). But whereas for Newton space and time exist independently of each other, for Einstein they do not. In Einstein's universe space and time are entwined in a four-dimensional entity called space–time.

This means that changes in intervals of time for one observer are associated with changes in distance in space for another, but the separation between events in space–time is the same for both. It is in this sense that we live in a four-dimensional universe.

It remains true, though, that space and time are different in kind. The equations of relativity treat space and time slightly differently, but space and time are intimately interwoven.

MICHELSON–MORLEY EXPERIMENT

Once light was established as a wave, physicists assumed it propagated in the 'luminiferous ether', much as sound waves propagated in air. No-one had ever detected the ether, and this disturbed a young American physicist, Albert Michelson (1852–1931), who was developing precise methods of measuring the speed of light.

Maxwell had argued that the motion of the Earth around the Sun should create an 'ether wind' through the Earth. He showed that a beam of light sent to a mirror and back would take slightly longer to complete the journey in the direction of flow than across the flow, but doubted that the difference would ever be measurable. Michelson rose to the challenge. Together with his colleague Edward Morley (1838–1923), he set up an instrument of his own design (since called a Michelson interferometer) to measure this tiny difference.

Despite many measurements with the apparatus oriented at different angles and at different times of year, Michelson and Morley announced in 1887 that they could find no evidence of an ether wind. It seemed that the speed of light measured on Earth remained the same, no matter how the Earth itself was moving. This made no sense, and the only way out, as Einstein saw, was to abandon the ether and replace it with a completely new set of ideas.

mass–energy: kilograms and joules are different units for the same quantity.

The equivalence applies to all kinds of energy, whether kinetic, gravitational, electromagnetic or nuclear, and explains why an object cannot be accelerated to the speed of light. As a body is accelerated, it gains kinetic energy. But, according to Einstein, that kinetic energy itself possesses mass, so to accelerate the body further becomes more difficult. As the speed of light is approached the mass–energy rises towards infinity, and that is why the speed of light cannot be reached.

THE SPEED OF LIGHT

When Maxwell showed that electromagnetic waves would move at a certain fixed speed, it was tacitly assumed that this was the speed through the ether. In relativity there was no ether, so what was the significance of the speed of light?

At first sight Maxwell's theory fails the first principle of relativity. It says that stationary charges cause electric fields and moving charges (electric currents) cause magnetic fields. But if the theory is to apply in all inertial frames, it should

TWIN PARADOX

One of the most famous puzzles arising from special relativity is the so-called twin paradox. Two identical twins, Ann and Betty, are astronauts. Ann is assigned to the first flight to the nearest star, Alpha Centauri, a distance of 4.4 light-years. She sets off at 90% of the speed of light while Betty has the job of ground controller. When Ann gets to Alpha Centauri she turns around and comes home at the same speed.

When Ann and Betty meet up again, 9.8 years have passed on Earth according to Betty. But Ann has aged only 4.3 years. Time dilation ensures that, according to Betty, time on Ann's spaceship passes more slowly than on the ground, so she is not surprised to find that Ann is now 5.5 years younger than she is. This may be strange, but it is not yet a paradox.

But now consider Ann's point of view. From her ship she sees Betty receding at 90% of the speed of light and then approaching at the same speed on the return trip. If only relative speeds matter, time dilation should now operate in the opposite sense, and Ann should find that Betty is now the younger. While they are separated, they can disagree about the passage of time. But when they are together on Earth again, they can't both be the younger!

The resolution is that the situation of the twins is not symmetrical. Betty remains in an inertial frame throughout, while Ann has to change frames at least three times in her journey – when she takes off, when she turns her ship around at Alpha Centauri, and when she lands. It is the changing frames that cause one twin to age more than the other.

apply in frames in which the stationary charges appear moving or the moving charges appear stationary. In that case, what becomes of the electric and magnetic fields? Einstein showed that electric and magnetic fields are simply different faces of the same phenomenon; an electric field to one observer is a magnetic field to another, but both observers will agree on the force acting on a charge and hence its acceleration. In particular, both will agree that electromagnetic radiation moves at the speed of light in their own frame of reference. Maxwell, it turned out, had unwittingly anticipated special relativity by some 40 years.

▼ *Modern accelerators can boost fundamental particles almost to the speed of light so that physicists can study their behaviour.*

GENERAL RELATIVITY

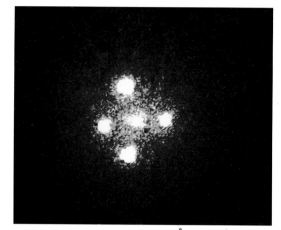

THE LAWS OF PHYSICS

No sooner had Einstein stunned the world with his elegant new theory of space and time than he embarked on an even more ambitious quest to develop a theory of gravity. The general theory of relativity (GR) was even more radical than the special theory in that it sought to abolish Newtonian gravity altogether. Instead of forces acting over a distance, Einstein introduced the notion of curved space–time. Masses distort the space–time around them, he said, and a planet orbiting the Sun is doing no more than following the four-dimensional equivalent of a straight line through curved space–time.

EQUIVALENCE PRINCIPLE

Einstein's starting point for developing his theory of gravity was a thought experiment very like the following one.

Imagine two small laboratories, one stationary on the Earth and the other in a spaceship, far out in interstellar space, but being accelerated at the same rate as the acceleration due to gravity. Observers in both laboratories would feel the same force 'downwards' but without looking out of the window, neither would be able to tell whether they were stationary in a gravity field or accelerating in free space. Einstein's great insight, a leap of faith, was that this was more than coincidence and that the two situations were literally indistinguishable. There is no experiment that would allow the observers to decide which laboratory they were in.

This *equivalence principle* also implies that the two kinds of mass in Newtonian mechanics – inertial and gravitational mass – are necessarily identical. Inertial mass determines how much force is required to make an object accelerate (the laboratory on the spaceship). Gravitational mass determines the force experienced by the object in a gravitational field (the laboratory on Earth). See The Nature of Mass, page 61.

GRAVITATIONAL SHIFT

One prediction that came out of the equivalence principle, later confirmed by the full GR theory, is that time at a point in a gravitational field runs more slowly than time in free space. This means that the wavelength of light from a star is changed by an amount that depends on the difference between the gravitational fields at the source and at the observer.

◀ *A manuscript showing Einstein's early search for his general theory of relativity, the theory that was to change the way scientists thought about space and time.*

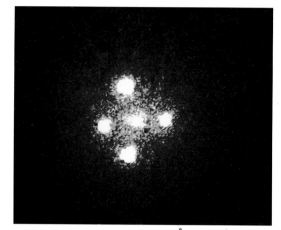
▲ *A gravitational lens photographed by the Hubble Space Telescope. The central blob is a foreground galaxy while the four surrounding blobs are four gravitationally lensed images of a quasar behind the central galaxy. This X-shaped effect is known as an Einstein Cross.*

Although predicted by Einstein in 1907, the gravitational shift was not demonstrated until 1960. It has important applications in astrophysics, since the wavelength of light emitted from the surface of a star will, to an observer in free space, be redshifted by the gravitational field. It is almost as if the light loses energy as it climbs away from the star. Gravitational redshifts of spectral lines have been detected in the spectra of **white dwarf** stars, which have an unusually strong gravitational field at their surfaces.

Modern atomic clocks are so accurate that the gravitational shift has to be taken into account when comparing clocks at different heights above sea level.

ALBERT EINSTEIN

Albert Einstein (1879–1955) was German-born, but took first Swiss and then US citizenship. He devised the two theories of relativity and made important contributions to quantum physics. Einstein's approach to physics was highly intuitive: as a boy he questioned what a beam of light would look like if he could catch up with it. In 1905, while working in the Swiss patent office, Einstein published two landmark papers: one in which he introduced the concept of the **photon** (which won him the Nobel Prize in 1921) and the other in which he set out the theory of special relativity. But it was not until 1909 that he secured his first academic post, at Zurich University, and became an established member of the scientific community.

The general theory of relativity, published in 1915 when he was director of a research institute in Berlin, explained gravity not as a force between particles but as a distortion in the fabric of space–time. The latter part of his life was spent unsuccessfully trying to unify gravity with electromagnetism. In 1933 Einstein left Nazi Germany for Princeton where he was to remain. In 1939 he alerted the US government to the possibility of the Nazis developing an atomic bomb and later campaigned for nuclear disarmament.

▲ *Einstein saw that gravity should be understood not as the force between masses but as the effect of distortions in the four-dimensional fabric of space–time.*

EDDINGTON'S ECLIPSE OBSERVATION

One of the predictions made by general relativity was that light should be bent by passing through the curved space around a massive object like the Sun. Einstein showed that light just grazing the surface of the Sun would be deflected by 1.75 seconds of arc, causing the apparent positions of stars to be displaced by the same amount. But how could such a tiny angle be measured so close to the Sun?

The answer came in May 1919, when astronomers led by the British astrophysicist **Arthur Eddington** travelled to the island of Príncipe, off the coast of equatorial Africa, to observe a total solar eclipse. During the minutes of totality the sky would be dark enough for stars to be visible close to the edge of the Sun.

When Eddington measured the positions of the stars on photographic plates taken during totality, he found that they were slightly shifted away from the Sun, and the displacement was just that predicted by Einstein. The space around the Sun was indeed curved.

When the news arrived back in Europe, Einstein became an instant scientific celebrity, and remained so for the rest of his life.

EINSTEIN'S THEORY OF GRAVITY

Einstein realized that physics could be simplified if the old concept of an inertial frame of reference, in which the laws of physics are always the same, was replaced by a local inertial frame, defined as any frame in free fall. If physics is formulated in local inertial frames, then gravity vanishes from the picture because gravity cannot be detected by an observer in free fall. Unlike Newtonian inertial frames, which extend over all space, local frames apply only over a small region.

In ordinary, three-dimensional space the shortest distance between two points is a straight line and Newton's first law is that free bodies move in straight lines. GR still asserts that free bodies

EARTH

SUN

B ☆   ★ A

Apparent True
position position

◀ *Deflection of Light Passing Around the Sun*
The curvature of space produced by the mass of the Sun causes the light from a star in position A to be deflected. As a result, the star is viewed from Earth as if it were at position B.

follow straight lines, but now the straight line (or rather its analogue) is drawn not in space but in four-dimensional space–time. Free fall, not rest or constant velocity, becomes the natural state of motion.

Gravity, said Einstein, should be considered as a curvature of space–time. The attraction towards a distant mass is then replaced by a purely local phenomenon: free bodies are simply moving in response to the local curvature of space. The planets are kept in orbit around the Sun not by the 'force' of gravity, but because the curvature of space constrains them to move in closed paths.

For most practical purposes the paths predicted by Einstein are those predicted by Newton, but GR starts to deviate from Newtonian mechanics in strong gravitational fields. The first success for GR was Einstein's demonstration that a small anomaly in the motion of the planet Mercury, long known to astronomers but inexplicable by Newtonian mechanics, was simply a consequence of the curvature of space near the Sun.

The theory can be summed up as: matter tells space–time how to curve, space–time tells matter how to move.

LIGHT IN GRAVITATIONAL FIELDS

General relativity makes two important predictions about the behaviour of light in gravitational fields.

The first is that light will be deflected when passing near a massive object like the Sun. The GR view is that the light is following a straight line in curved space–time. The bending was first detected at an eclipse of the Sun in 1919 and has been verified several times since.

The second prediction is that the time taken for light to travel between two points in a gravitational field is greater than for an identical distance in free space. This was demonstrated by measuring the time for radio signals to be returned from interplanetary spacecraft on the far side of the Sun, notably the Viking Mars landers in 1976.

Both the bending of light and the time delays can be observed in **gravitational lensing**, in which a massive galaxy distorts the light from more distant galaxies.

GENERAL RELATIVITY AND THE UNIVERSE

Space–time is locally curved by the presence of mass. Each star curves the space–time around it and each galaxy curves space–time in a way that

CURVED SPACE

General relativity replaces gravity with curved space–time. But exactly what is meant by space being curved?

If a triangle is drawn on a flat sheet of paper, the sum of its angles will always be 180°. No matter what the size and shape of the triangle, this will always be true because it is a necessary consequence of the paper being flat. If a triangle is now drawn on a sphere, the angles will add up to more than 180°; the exact sum will depend on the size and shape of the triangle. This is a consequence of the surface of a sphere being curved.

The analogy carries over into four-dimensional space–time, but it is not of course possible for us to see the curvature from outside, as it is with the surface of a sphere. But we can still use the triangle test. If the angles add up to exactly 180° space is said to be flat; if the sum is greater than 180° space has positive curvature and if less than 180° space has negative curvature.

The curvature of space near the Sun is, in fact, slightly positive, but it is enough to keep the Earth and the other planets in their orbits.

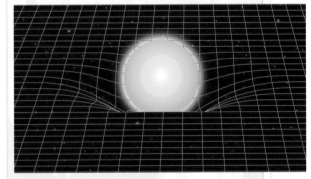

▲ *Curved Space*
The curvature of space–time by a massive body can be likened to the distortion of a flat rubber sheet when a ball is placed upon it.

is the sum of the curvatures produced by its stars. Clusters of galaxies likewise curve space–time on larger scales, and the question arises whether the Universe as a whole has a curvature. Is the whole of space–time somehow curved by the matter within it, and what is the form of this curvature?

It is possible using GR to model the Universe as a whole, by assuming all the mass of the galaxies to be spread smoothly through space. Using that model one can then predict the kinds of universe that are possible and how they evolve with time.

At the present time comparison of the models with observations indicates that the Universe is very nearly flat, with negligible overall curvature.

The Laws of Physics
SUMMARY

T he physics outlined in this chapter is sometimes known as 'classical' physics (though some would exclude the theory of relativity from that title) as opposed to the 'modern' physics of quantum mechanics. Most of it was already well established by the end of the nineteenth century, and indeed there was a sense of satisfaction among physicists at that time that our understanding of the physical world was in some sense nearing completion. That optimism could be excused, since the strangeness of relativity or quantum physics is perceived only in extreme situations. In our everyday world we do not move fast enough for the distorted time of special relativity to be apparent; we do not live in strong enough gravitational fields for the warped space of general relativity to make itself known; and, of course, the scale of our existence smoothes away the quantum graininess that defines the world at the level of the atom. For most of us, most of the time, classical physics will do. But astronomers are not confined to the everyday world. On a cosmic scale relativity supersedes Newton's laws, and on the atomic scale quantum mechanics rules instead of Maxwell's equations. And it is to the physics of the very small that we turn next.

LAWS OF MOTION

◉ Newtonian mechanics is based upon three simple laws which relate the motion of an object to the forces acting upon it.

◉ The laws of physics take the same form in any inertial frame, which is a frame of reference moving at constant velocity.

◉ Energy, the ability of a body to do work, comes in many forms, including kinetic, gravitational, electromagnetic and nuclear.

◉ The principle of conservation of energy states that the energy within an isolated system remains constant, no matter what processes are taking place within it.

◉ Three other mechanical quantities are also found to be conserved in isolated systems: **mass**, momentum and **angular momentum**.

▲ *Some comets,* those moving on parabolic or hyperbolic orbits, have enough energy to escape from the Solar System.

GRAVITATION AND ORBITS

◉ Newton's law of universal gravitation states that every object in the Universe attracts every other object with a force proportional to the product of their masses and inversely proportional to the square of the distance between them.

◉ Mass is the amount of matter in a body, weight is the gravitational force acting on that body.

◉ Kepler's three laws, which describe the orbits of planets around the Sun, are consequences of Newtonian gravitation.

◉ Orbits may be circular, elliptical, parabolic or hyperbolic according to their energy. An object moving on a parabolic or hyperbolic orbit around the Sun has sufficient energy to escape from the **Solar System**.

PROPERTIES OF MATTER

◉ All matter is made of particles called atoms, which combine to form **molecules**. The three states of matter – gases, liquids and solids – can be understood in terms of the interactions between their constituent particles.

◉ The temperature of a body is a measure of the random kinetic energy of its particles, which can be changed by the transfer of heat. Heat can be transported by conduction, convection or radiation.

◉ A body is in thermal equilibrium if its temperature is uniform and unchanging.

ELECTROMAGNETIC RADIATION

◉ Electric fields are produced by stationary electric charges or changing magnetic fields. Magnetic fields are produced by moving electric charges (electric currents) or changing electric fields.

◉ Electric and magnetic fields can travel through space in the form of electromagnetic waves. Such waves are generated by vibrating electrons.

◉ Light is the most familiar form of electromagnetic wave.

◉ The colour of light depends on its **wavelength**. White light is a mixture of wavelengths. The perception of colour, which is subjective, is due to three kinds of receptors in the retina of the eye, sensitive to red, blue and green light respectively.

◉ The laws of reflection and refraction are used to design mirrors and lenses as components of optical systems.

◉ The wavelike nature of light is revealed by the phenomena of **diffraction**, **interference** and **polarization**. Until the twentieth century it was believed that the waves propagated through a medium called the ether that filled all of space.

◉ Other forms of electromagnetic radiation include radio waves, **infrared** radiation, **ultraviolet** radiation, **X-rays** and **gamma rays**. The shorter types behave more like particles than like waves.

SUMMARY

THE LAWS OF PHYSICS

WAVE OR PARTICLE?

Newtonian mechanics treated the Universe as a system of interacting particles and was very successful. **Newton** assumed that light, too, was made of particles, and such was his influence that evidence to the contrary was overlooked for a century.

After the interference experiments of Young, however, the particle theory of light was no longer sustainable and the wave interpretation, proposed initially by **Huygens** and others, was rediscovered and seen to account well for diffraction, interference and polarization. By the mid-nineteenth century the wave nature of light had been accepted.

Around the same time rapid progress was being made in electricity and magnetism, with **Faraday** uncovering intimate links between the two. This was completed by **Maxwell**, whose four equations not only accounted for all electromagnetic phenomena to date, but predicted a new one: that disturbances in the electric and magnetic fields could fly through space in the form of waves. The speed of the waves was calculated to be the speed of light. Towards the end of the century, when **Hertz** discovered Maxwell's predicted radio

waves, it seemed as though physics was approaching completion. But there were several things that did not seem quite right.

- 🌑 It was not at all clear what the speed of light that came out of Maxwell's equations actually meant. If it was the speed through the ether, why did the speed of light seem to have the same value irrespective of the motion of the Earth?
- 🌑 Spectroscopists had discovered patterns of emission and absorption lines that seemed characteristic of each chemical element. Why did atoms only emit and absorb at certain precise wavelengths?
- 🌑 Atoms in bulk emitted the continuous spectrum called the black-body spectrum. It had a peak whose wavelength depended on the temperature of the body, and fell away at longer and shorter wavelengths. But the form of the spectrum defeated the best physicists of the day.

The first problem was cleared up by **Einstein** in his spectacular special theory of relativity. But the other two required yet another radical solution, foreseen by Einstein himself and fellow German

physicist **Max Planck**. They asserted that, yes, light was a wave, but it was also a particle. Depending what experiment you did, sometimes light would look like a wave and sometimes like a particle. The particle of light was the **photon**, and it would usher in a revolution even more startling than relativity.

▲ *At the beginning* of the nineteenth century most physicists assumed that light was made up of particles, but later experiments showing diffraction (pictured) and interference seemed to prove its wavelike properties.

They all travel at the speed of light, which is now defined to be 299,792,458 m/s exactly.
- 🌑 White light can be separated into its component wavelengths by passing it through either a prism or a diffraction grating. The band of colours so produced is known as a **spectrum**, and the branch of physics concerned with making and interpreting spectra is called **spectroscopy**.
- 🌑 Some spectra contain bright or dark lines (**emission** or **absorption lines**), which are characteristic of the chemical elements present in the source or in the intervening medium.
- 🌑 All objects emit thermal radiation by virtue of their temperature. The continuous spectrum of thermal radiation from an opaque body takes the form of **black-body radiation**. Classical physics is unable to explain the form of the black-body spectrum.
- 🌑 Non-thermal radiation from astronomical objects is produced when **electrons** moving close to the speed of light lose their energy in magnetic fields.

RELATIVITY

- 🌑 Einstein's special theory of relativity introduced a new way of looking at space and time as two aspects of a four-dimensional 'space–time.' Time is no longer universal, but different for each observer. Light has the same speed in all

inertial frames and no longer requires an ether.
- 🌑 Mass and energy are two facets of the same phenomenon. They are no longer conserved separately but are conserved jointly as mass-energy.
- 🌑 Einstein's general theory of relativity interprets the force of gravity as the curvature of space–time. 'Matter tells space how to curve, space

tells matter how to move.' The theory can be applied to the study of the Universe as a whole.
- 🌑 Newton's inertial frames become local inertial frames, defined as any frame of reference in free fall.

▼ *White light is made up* of a continuous range of colours, each with its own wavelength. Rainbows are caused by the refraction of white light in drops of rain, splitting it into its spectral colours.

IN SEARCH OF QUANTUM REALITY

This chapter explains the strangest and yet most successful theory in the whole of science: quantum mechanics. The world's top physicists agree that anyone who is completely comfortable with what quantum mechanics tells us about the world has probably not really understood it. Despite this, it was the single most important scientific discovery of the twentieth century. Quantum mechanics not only underpins modern physics, but the fields of chemistry, electronics and astronomy as well. Without it we could not explain the structure of crystals, nor would we have invented the laser or the silicon chip. Without an understanding of the rules of quantum mechanics there would be no television sets, computers, microwaves, CD players, digital watches, mobile phones and much more that we take for granted today.

Quantum mechanics predicts the behaviour of the very building blocks of matter – not just the atoms, but the particles that make up the atoms – with incredible accuracy. It has led us to our current, very precise, understanding of how subatomic particles interact with each other and connect to form the world around us. At the same time, though, it forces upon us a view of the world that goes totally against our common sense. In this chapter we discover how quantum mechanics came about, its incredible success in describing the world of the very small, and the strange reality it forces us to accept.

In Search of Quantum Reality
INTRODUCTION

In 1900 the German physicist Max Planck announced to the world a new mathematical formula that helped explain some strange experimental observations. By the time Albert Einstein had completed his two theories of relativity 15 years later, the quantum revolution was already well under way. But the baton was soon to be passed to a group of young physicists, who would weave together the few existing strands of the subject into a rich mathematical and philosophical tapestry that is known today as quantum mechanics. This was to become the most powerful and accurate description of the nature of matter so far devised.

THE PINNACLE OF CLASSICAL PHYSICS

It is remarkable that so many of the diverse phenomena we see around us in nature can be successfully explained within a scientific framework that has been around much longer than quantum mechanics. Even more remarkable is the fact that it was due almost entirely to the work of one scientist, **Isaac Newton** (1642–1727). The way objects move and collide, the way they behave when they are pushed and pulled, can all be explained by what is known as Newtonian mechanics. This incorporates Newton's equations of motion along with his law of gravitation. Between them they describe all motion in terms of force, momentum and acceleration.

By the second half of the nineteenth century, the work of **Michael Faraday** (1791–1867) and **James Clerk Maxwell** (1831–79) had led to an

▲ *At the heart of quantum reality is the strange paradox that we can never know precisely what a single atom will do in the future. You could say the same about people – and yet, just as insurance companies use actuarial tables to estimate probable life expectancies of the population as a whole, scientists use quantum statistics to predict the probable behaviour of particles in large numbers.*

understanding of the nature of light in terms of electromagnetic waves. This, together with Newtonian mechanics, is known as classical physics. Even when **Albert Einstein** (1879–1955) showed – first in his special theory of relativity and later in the general theory – that Newton's laws were only approximate and had to be modified when considering high speeds and large **masses**, the underlying philosophy remained the same. All phenomena in the Universe could be understood within this framework: in terms of particles and fields, interacting via the two fundamental forces of gravity and electromagnetism. Thus relativity can be considered as the pinnacle of classical physics.

A NEW QUANTUM REALITY

An important aspect of classical physics is that it describes a universe that is, in principle at least, entirely predictable. That is to say, if we were to know the positions and velocities of everything in the Universe, right down to the individual atoms, at an instant in time, then in principle we could work out their positions at any later time. Of course, in practice this is impossible, since we would need a computer of almost infinite power. However, there is an important point to be made: the Universe of Newton and Einstein is a clockwork universe, in which the whole future can, in principle, be entirely determined.

NIELS BOHR

The Danish physicist Niels Henrik Bohr (1885–1962) is widely regarded as the father of quantum mechanics and one of the most influential scientists of the twentieth century. As a young man he was a keen soccer player, like his younger brother Harald, a mathematician, who represented the Danish soccer team in the 1908 Olympics. In 1911, Bohr went to work with J. J. Thomson (1856–1940) in Cambridge and later with **Ernest Rutherford** (1871–1937) in Manchester, where he developed a model of the atom based on **electron** orbits that

obeyed the rules of the then-new quantum theory. He was awarded the Nobel Prize for this work in 1922.

Back in Copenhagen during the 1920s, Bohr headed a new institute, which now bears his name. It attracted the top young theorists in the world and became the centre for work on quantum mechanics. Bohr famously entered into a long-running debate with Einstein over the meaning of the new theory, developing what has become known as the Copenhagen interpretation.

During World War II, after working tirelessly to help Danish Jews escape the German occupation, Bohr himself escaped to the US, where he worked on the atomic bomb project. He later returned to Denmark, devoting his life to promoting peaceful uses for atomic energy.

▶ *Niels Bohr, considered by many to be the father of quantum mechanics; he won the Nobel Prize in 1922 for his model of the atom based on electron orbits.*

cosmology ▶ p. 337 electromagnetic spectrum ▶ p. 338 electron ▶ p. 338 interference ▶ p. 340 mass ▶ p. 341

THE CENTRAL MYSTERY: THE TWO-SLIT EXPERIMENT

Over the next few pages it will become clear that quantum mechanics provides us with a reality unlike anything we encounter in our everyday world of macroscopic objects. One of the strangest phenomena at the very heart of quantum mechanics is known as wave–particle duality. Nowhere does this show up more sharply than in the 'two-slit' experiment. Here is a brief summary of what is involved, without, at this stage, an explanation of why or how it happens.

The wave nature of light is what leads to the **interference** pattern of light and dark fringes that appears when light passes through two narrow slits in a screen. This feature is always seen with waves. Try a similar experiment by dropping sand through two slits in a screen and, unsurprisingly, a pile of sand will form beneath each slit. There will no longer be an interference pattern, since sand is made up of individual particles (grains).

Now try the experiment with electrons. A special apparatus fires a beam of electrons at a screen with two closely spaced narrow slits in it, enabling a few of the electrons to pass through. On the other side is a second screen that stops these electrons and shows a visible flash of light every time one strikes it.

Electrons are, of course, incredibly tiny. They are most certainly not spread-out entities, capable of overlapping both slits like a wave. Yet the electrons do not form two bright patches where many of them accumulate on the second screen. Instead an interference pattern of light and dark fringes appears, exactly equivalent to that produced by waves.

What is even more incredible is that even when the electrons are fired at the slits one at a time, and those that get through show the tell-tale flashes on the second screen, the same interference pattern slowly builds up. Somehow, each electron that leaves the source as a particle, and arrives at the screen as a particle, behaves like a spread-out wave in between, able to propagate through both slits at once by splitting into two waves that can interfere with each other.

▲ **It is the** *wave-like property of light that produces multi-coloured interference patterns on the surfaces of soap bubbles.*

Quantum mechanics has shown that this view is wrong. It became clear in the early part of the twentieth century that the microscopic world did not behave in a predictable way. For instance, it is a fundamental property of the world of the very small that we cannot know for sure exactly what an atom will do next. Measuring its position and energy at one instant does not mean we can predict definitely where or in what state it will be at any later time. We can only assign probabilities to different outcomes.

QUANTUM PHYSICS TODAY

Quantum mechanics describes the way things behave on the very tiniest scales, much smaller than can be seen with the naked eye. It is the foundation of the fields of atomic and nuclear physics, which together describe the properties of all the elements. Without an understanding of how the basic building blocks of matter – the atoms – behave we would not have been able to understand how they bind together to form the diversity of materials we see around us. Quantum mechanics, therefore, underlies the whole of modern chemistry and materials science, too.

Another success of quantum mechanics has been its explanation of how different materials conduct electricity. Metals such as copper are good conductors because electrons are able to flow freely inside them. This is not the case with insulators, such as wood or plastic, because of the different way in which their atoms fit together.

A third type of material, with properties that lie between those of conductors and insulators, is known as a semiconductor. This conducts electricity only because of the special way in which its atoms behave when grouped together. It is thanks to the rules of quantum mechanics that we understand this phenomenon. Semi-conductor materials are used in the integrated circuits that go into so many of today's modern appliances, as well as in computer chips.

Even more intriguing is the possibility of a new type of computer, far more powerful than anything available today. Such 'quantum computers' are still far from being realized, but progress is currently being made at a rapid pace.

Quantum mechanics has also had a major influence on the field of astronomy. Not only does it help to explain the structure of stars and what we can tell about them from their **electromagnetic spectra**, but it is also the basis of many of the current theories in **cosmology**.

▶ **Lasers are just** *one of the many new technologies developed as a result of the success of quantum mechanics in explaining the behaviour of matter and radiation at the subatomic level.*

QUANTUM REALITY

Einstein ▶ p. 88 Faraday ▶ p. 70 Maxwell ▶ p. 72 Newton ▶ p. 42 Rutherford ▶ p. 100

INTRODUCTION

In Search of Quantum Reality
THE BIRTH OF QUANTUM THEORY

In Einstein's special theory of relativity, light must be regarded as a wave. This had been established earlier by Maxwell, who had shown light to be made up of oscillating electric and magnetic fields. Indeed, light behaves like a wave under most circumstances that we are familiar with. However, by the late nineteenth century cracks began to appear in this view. Certain phenomena seemed to suggest that, fundamentally, light had to be composed of particles – something Newton had suggested long before. Could he have been right after all?

MAX PLANCK

The German physicist Max Planck (1858–1947) grew up in Munich, but studied in Berlin under some of the most influential German physicists of the nineteenth century. He received his doctorate at the age of 21 and soon afterwards joined the faculty of the University of Munich. In 1889 he returned to Berlin University, where he remained until his retirement in 1926.

Planck's radical quantum theory took several years to be accepted. In fact, even Planck himself was not convinced it was correct. He had proposed it in an ad hoc manner in order to solve the black-body radiation problem, but he half suspected that it was just a mathematical trick rather than a true description of the nature of radiation. He spent many years attempting to find a way around his own theory – without success.

As an elder statesman of science, Planck resisted the Nazis during the 1930s. He lost his two sons during Word War II: one was killed in action and the other was executed after being accused of plotting to assassinate Hitler. Planck was rescued by the Allies in 1945 and spent the last two years of his life honoured around the world.

◀ *Max Planck, one of the most influential physicists of the twentieth century, could never reconcile himself to the full implications of his radical idea of the 'quantum'.*

THERMAL RADIATION

Quantum theory evolved as a response to the rather complicated phenomenon of **black-body radiation**, which has to do with the way matter absorbs and emits energy as heat and light.

The heat, or thermal radiation, of the Sun has travelled through the vacuum of space and is closely related to the Sun's light, in that they are both electromagnetic waves. In fact, all objects emit **electromagnetic radiation** over the whole frequency range of the **spectrum**. The nature of this radiation depends on an object's temperature, a phenomenon studied by Josef Stefan (1835–93). Any solid heated above about 700°C (1,300°F) will glow visibly. At lower temperatures a solid still emits radiation, but with an intensity in the visible region that is too weak to be seen. Of course, materials also absorb and reflect radiation falling on them, and it is the reflected rather than the emitted radiation that makes most everyday objects visible to us.

Hypothetical bodies that do not reflect any radiation falling on them would look black (provided that they are not so hot that they glow). Such objects are perfect emitters as well as absorbers of radiation, and the spectrum of **frequencies** over which they radiate their heat depends only on their temperature. Physicists call them **black bodies**. The higher the temperature of a black body, the higher the frequency of the most intensely emitted radiation. This relation, known as **Wien's displacement law**, accounts for the change in colour of materials as they are heated, from lower-frequency red through to higher-frequency blue. It has also enabled astronomers to deduce the temperature of deep space (about -270°C/-454°F) by measuring the frequency of the radiation that reaches us from space, known as the **cosmic background radiation**.

X-RAYS

X-rays are a form of highly penetrating electromagnetic radiation. This radiation is like light but with much higher frequency, putting it far beyond the visible spectrum. The discovery of X-rays in 1895 by the German physicist Wilhelm Röntgen (1845–1923) is regarded by many as marking the beginning of the era of modern physics. X-rays are emitted when certain materials are bombarded by high-energy **electrons** inside a special tube containing a low-pressure gas. They show typical wave behaviour, such as **polarization**, **diffraction** and **interference**, but certain of their features could not, at the time of their discovery, be explained by the prevailing classical theory of radiation developed by **Maxwell**. According to this theory, all electrically charged particles emit electromagnetic radiation

▼ *During the early twentieth century, spectrographs such as this model built c. 1919 were used to observe and measure electromagnetic radiation emitted by black bodies.*

THE BIRTH OF QUANTUM THEORY

PLANCK'S REVOLUTIONARY SUGGESTION

Planck's original proposal was not that electro-magnetic radiation itself came in packets, or quanta, but rather that the atoms of the black body could only *emit* the radiation in packets. It was only later that it became clear that all electromagnetic radiation is quantized.

Planck initially assumed that the atoms of a black body vibrate like tiny electric 'oscillators'.

To obtain a satisfactory fit to the data, he was forced to introduce two new concepts that were to form the foundation of the new quantum theory of matter and radiation.

The first of these ideas was that the energy of the atoms (or oscillators) cannot take on any value but must exist as a set of discrete (discontinuous) energy levels that are simple multiples of the vibrating frequency of the atoms.

The second new concept was that the emission and absorption of radiation by the black body are associated with the atoms 'jumping' from one energy level to another. For instance, when an atom drops to a lower energy level, it emits a single quantum of radiation energy. Conversely, by absorbing a quantum of energy the atom will jump up to a higher energy level. Such emitted and absorbed quanta of electromagnetic radiation are today known as **photons**.

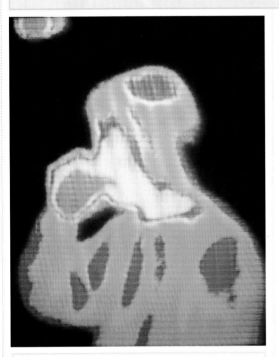

◀ *Human beings,* like all objects, emit thermal radiation. The difference in the temperature radiated by the liquid being drunk is clearly registered by the thermal-imaging camera here.

PLANCK'S CONSTANT

According to Planck's formula, the energy of the smallest bundle of light of a given frequency (a single photon) is given by the frequency multiplied by a constant number. This number is Planck's constant of action (known simply as Planck's constant). It has the symbol h and, like the speed of light, is one of the universal constants of nature.

The relation between energy and frequency, according to Planck, is simple. For instance, the frequency of violet light at one extreme of the visible spectrum is twice that of red light at the other end, and so a quantum of violet light has twice as much energy as a quantum of red light.

Planck's constant is an extremely tiny number and its effects show up only on the subatomic level. Its value depends, of course, on the units in which we choose to measure it. In units of kilograms, metres and seconds, h has a value of 6.63×10^{-34}. This is the reason why we do not see most of the strange quantum phenomena discussed in the forthcoming pages anywhere other than on the subatomic scale.

when accelerated, and the sudden deceleration of the electrons when stopped in the material causes them to emit the X-rays. However, there turned out to be a minimum energy that the electrons could have, below which no X-rays would be produced. This was not what Maxwell's theory predicted and was an indication that something was amiss with the classical theory of light.

THE ULTRAVIOLET CATASTROPHE

Early in 1900, Lord Rayleigh (1842–1919) and James Jeans (1877–1946) came up with a formula that related the amount of thermal energy radiated from a black body to its frequency. They found a serious problem with the prevailing classical theory, which predicted that a black body should emit an infinite amount of radiation at high frequencies, whatever its temperature. This state of affairs became known as the ultraviolet catastrophe, because the term 'ultraviolet' at that time referred to all radiation frequencies higher than those of the visible spectrum. Of course this does not happen in reality, since no body can give off radiation at infinitely high frequencies. On the other hand, the Rayleigh–Jeans formula could not be at fault: it had to be the underlying assumptions of the classical theory of radiation that were flawed.

LUMPS OF ENERGY

In December of 1900, the German physicist Max Planck proposed a simple formula that correctly described the distribution of black-body radiation over all frequencies. His new idea was based on a crucial assumption: that the light emitted by a black body cannot have just any amount of energy. Rather, its energy must be a multiple of tiny bundles that he called 'quanta' (from the Latin for 'how much'). A quantum of

light energy is the smallest unit of energy possible for light of that frequency and cannot be subdivided into smaller units. Thus the radiation emitted and absorbed by bodies came in lumps, albeit extremely minute ones, rather than as a continuous flow.

The energy of a quantum of black-body radiation is proportional to its frequency, according to Planck. Therefore the higher the frequency of the radiation, the higher the energy of a quantum and the less likely that a body would have the energy to radiate a quantum at that frequency. Thus the amount of radiation at high frequencies would be less. Planck's theory of radiation removed the ultraviolet catastrophe and agreed very well with experiment.

This idea that black-body radiation is emitted discontinuously in 'lumps' of energy rather than continuously implies that materials must give up (and absorb) their energy in 'quantum jumps'. This can be likened to the way in which a ball loses potential energy in rolling down a flight of stairs, as contrasted with a smooth slope.

▼ *Ultraviolet Catastrophe*
This graph plots the intensity of radiation coming from a black body as the frequency of the radiation increases. The experimentally observed curve typically rises then falls, but the Rayleigh–Jeans formula predicted that radiation in the ultraviolet range rises sharply to infinity. This serious theoretical discrepancy was called the 'ultraviolet catastrophe'.

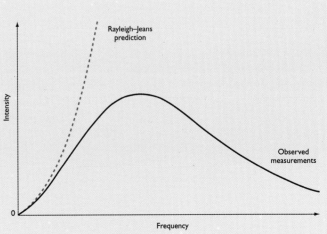

QUANTUM REALITY

In Search of Quantum Reality
LIGHT AS PARTICLES

Even though Max Planck had been able to solve the puzzle of the ultraviolet catastrophe in black-body radiation through his two postulates, physicists were still some way from accepting the particle nature of light. After all, the wave-like behaviour of light had been confirmed by Thomas Young (1773–1829) 100 years earlier. It was Young's research on diffraction and interference in 1803 that had finally laid to rest Newton's corpuscular theory. However, more surprises were to come. Planck's work was just the beginning and was soon to be followed by experimental confirmation that sometimes light does indeed behave as though it is made up of particles.

PHOTOELECTRIC EFFECT

A nineteenth-century discovery that could not be explained by the physics of the time was the photoelectric effect. In 1887 **Heinrich Hertz** (1857–94) performed a series of experiments that confirmed Maxwell's theory of electromagnetism, in the course of which he noticed a strange phenomenon: light that is shone on to a negatively charged metal plate can knock electrons from its surface. This process of ejection of **electrons** from a surface by the action of light is the photoelectric effect. Here, the observed experimental results were in even starker contradiction with the classical wave theory of light than those of **black-body radiation**. This is partly because the effect is more straightforward to explain, and partly because there are in fact three features of this effect that are puzzling.

First, it was found that the energy of the ejected electrons depended not on the intensity of the light that fell on the plate but on its **frequency**. But if light consists of waves, increasing its intensity implies increasing the amplitude of its oscillations, and hence the energy of the light and that of the subsequently ejected electrons. This did not happen.

According to the wave theory, the photoelectric effect should occur at any frequency, provided the light is intense enough to provide the electrons with the necessary energy to escape. What is observed, however, is that there is a cut-off frequency below which no electrons are emitted, no matter how high the intensity.

Electrons should need time to absorb enough energy from the light waves to be ejected. The time needed should be greater if the light is feeble. No time lag was detected, however, with electrons being emitted as soon as light started to impinge on the surface.

It was **Einstein** who successfully explained this effect in 1905 by extending **Planck**'s quantum theory. Each electron is knocked out when it is hit by a single **photon** of light, the energy of which depends on its frequency. It was for this work that Einstein received his Nobel Prize, not for his simultaneous work on relativity.

COMPTON AND EXPERIMENTAL PROOF OF PHOTONS

The particle nature of light received dramatic confirmation in 1923 when the US physicist Arthur Compton (1892–1962) performed an experiment in which a beam of **X-rays** with a specific frequency was scattered from a block of graphite. He found that the scattered radiation had a slightly lower frequency. Again, this effect could only be explained by thinking of the X-rays as made up of photons.

According to the wave theory, the incident X-rays are oscillating electromagnetic waves, which should set electrons in the graphite oscillating with the same frequency. These electrons in turn should give up their energy by radiating electromagnetic waves with the same frequency. This was not what was seen.

Compton explained the drop in frequency of the scattered X-rays by assuming light to be made up of particles. He thought of the X-rays as a collection of photons that collide with the electrons as in a collision between billiard balls. Each individual photon bounces off an electron in the graphite and in the process gives up some of its energy. The rest is explained by Planck's relation: since the scattered photon is left with lower energy it has lower frequency.

THE DUAL NATURE OF LIGHT

We can no longer think of light (or, more accurately, all **electromagnetic radiation**) as purely a wave phenomenon, nor as a stream of particles. Whatever it is, light seems to behave like a wave under some circumstances, as when it undergoes **interference** or **diffraction**, and as a collection of localized particles under other circumstances. The three physical processes of black-body radiation, the photoelectric effect and Compton scattering all show that this 'split personality' of light has to be taken seriously, despite the natural feeling of discomfort when contemplating it for the first time. Today this so-called wave–particle duality is not in doubt. We now know it to be a general characteristic of all physical entities on the quantum scale, not just light.

Light's Dual Nature
▶ ***Light as a Wave: Interference (Near Right)***
In this experiment, circular waves advancing from the barrier with two slits interfere to produce a pattern of dark and light bands on the viewing screen – a clear indication of light behaving as a wave.

▶ ***Light as a Particle: Compton Scattering (Far Right)***
Compton's experiment showed that light could also behave as a particle. When an X-ray photon with frequency f_1 collides with an electron, the photon bounces off at an angle. In so doing, it loses some energy and consequently has a lower frequency, f_2.

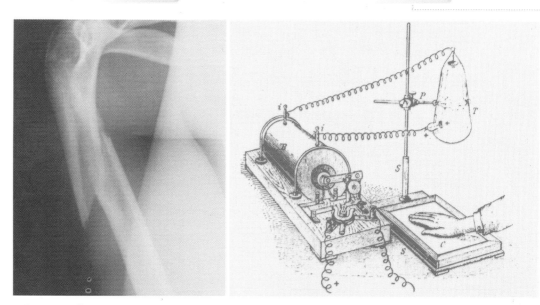

▲ *Compton was able to experiment with X-rays after their discovery by Wilhelm Röntgen in 1895. Modern equivalents of Röntgen's first X-ray machine (right) produce images of the human body such as the leg shown here (left).*

QUANTUM REALITY

EINSTEIN'S NOBEL PRIZE

In 1905, the same year that he published his paper on special relativity, Einstein proposed his new theory of light to explain the photoelectric effect. While Planck is quite rightly credited with being the founding father of quantum theory, the postulates that so successfully explained away the ultraviolet catastrophe in black-body radiation did not go as far as insisting that all radiation is quantized. Instead, Planck went no further than proposing that the atoms in the black body radiate energy in packets. He still believed that free radiation was continuous. Einstein presented quantization as a property of electromagnetic waves themselves, and suggested that all light is ultimately made up of spatially localized bundles, or photons. This was more than Planck was prepared to accept.

Einstein argued that the reason we do not normally see the particle-like nature of light is the large number of photons involved, just as we do not see the water from a running tap as being composed of individual droplets. It was in this way that Einstein was able to explain the three features of the photoelectric effect satisfactorily.

The first feature was the dependence of the ejected electrons' energy on the frequency of the light, not its intensity. This is simply a consequence of Planck's equation.

The second feature of the photoelectric effect is the cut-off frequency. This arises because the threshold for production of the so-called photo-electrons only occurs when a photon's energy is sufficient to release an electron. Increasing the intensity of the light just means more photons. Since photons are so localized in space, the likelihood that any one electron could acquire extra energy through being hit by more than one photon is tiny.

The third feature of the process is that it is instantaneous. This is because the electrons do not have to accumulate their energy from a wave that is spread out in space. Instead, each photon delivers its energy to an electron in a single collision. If this energy is above the necessary threshold, the electron will escape.

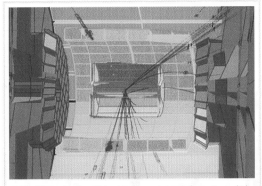

▲ *Particle tracks* of a collision between an electron and a positron, which produces a photon.

PAIR PRODUCTION

In addition to the photoelectric effect and Compton scattering, there is another process whereby photons can give up their energy, called pair production. In this process, a high-energy photon passing near the nucleus of an atom can give up all its energy by being converted into two, almost identical, particles: an electron and a **positron**, the antiparticle of an electron. Positrons are identical to electrons in every way apart from having an opposite electric charge (along with the opposite sign of a certain magnetic property known as magnetic moment). Pair creation is one of the clearest examples of the equivalence of mass and energy. The whole of the photon's energy, as defined by its frequency, is used to create the mass of the two particles, along with their kinetic energy, via Einstein's relation $E=mc^2$.

The opposite process to pair production is pair annihilation. Here a positron, which cannot exist for long in normal matter, interacts with an electron and the two particles are destroyed as their masses combine to produce a burst of pure radiation (a photon). This process takes place whenever any elementary particles and their antiparticles come into contact.

PHOTONS

One of the conclusions of Compton's experimental findings is that the photons must carry momentum since they can scatter the electrons, and momentum is a quantity associated with bodies with **mass**. If the idea of a photon as a localized bundle of energy is literally true, it must have certain rather special properties. To begin with, since photons travel at the speed of light, and yet have finite energy content, they cannot have any mass. This is a lesson learnt from special relativity, according to which no object with mass can ever attain light speed. Photons, therefore, must be massless. This is yet another departure from classical physics, in which particles without mass cannot exist.

Photons are therefore elementary particles with no mass and yet have finite energy and momentum, both of which quantities depend on their frequency and **wavelength**. Indeed it is the relation between momentum and wavelength that shows the clearest connection between light's particle-like nature (momentum) and its wave-like nature (wavelength).

As long as photons do not interact with matter, they have an infinite lifetime. And being electromagnetic radiation, they can never stand still, but must always move at the speed of light. Finally, since photons are pure electromagnetic energy, they can be created out of nothing by being emitted from atoms, and disappear when absorbed by atoms.

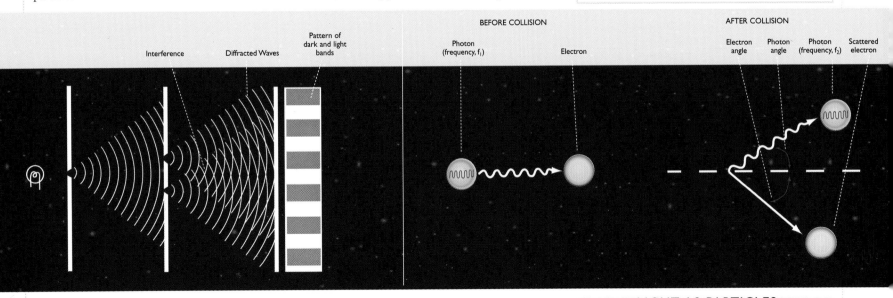

BEFORE COLLISION

Photon (frequency, f_1)   Electron

AFTER COLLISION

Electron angle   Photon angle   Photon (frequency, f_2)   Scattered electron

Interference   Diffracted Waves   Pattern of dark and light bands

QUANTUM REALITY

LIGHT AS PARTICLES

In Search of Quantum Reality
ATOMS

Towards the end of the eighteenth century chemists had begun to distinguish substances that could not be decomposed by known chemical processes. These are now known as the chemical elements and their existence paved the way for the modern concept of atoms, the building blocks of matter. However, the idea that everything is ultimately made up of indivisible units dates back to the ancient Greeks. Indeed the word *atomos* (literally meaning 'indivisible') has its origins in the fifth century BC. But our picture of atoms as the fundamental constituents of the elements has undergone several revolutions, and these gained acceptance only through the careful and painstaking accumulation of experimental evidence.

A BRIEF HISTORY OF THE ATOM

Two figures stand out as having made the greatest contribution to the modern idea of atoms. The first was the Greek philosopher Democritus (*c.* 470–*c.* 400 BC). Although he lived in the shadow of the great Socrates, Democritus came up with some uncannily accurate and modern-sounding scientific notions. For instance he suggested that the **Milky Way** was a vast collection of individual stars, and that the workings of the Universe were as mindless and determinate as a machine. His greatest contribution, though, was the idea that all matter is made up of tiny indivisible particles so

small that nothing smaller is conceivable. However, Democritus was not like modern scientists. His ideas were not the results of experimental observation or mathematical deduction, but were merely intuitive. Furthermore, later philosophers chose to ignore his ideas in favour of those of Plato (*c.* 428–*c.* 347 BC) or **Aristotle** (384–322 BC).

After **Galileo Galilei** (1564–1642) had demonstrated, at the beginning of the seventeenth century, the power of the experimental method, science progressed rapidly. But it was still to be 200 years before the atomic theory became a subject of experiment. In 1803, **John Dalton** (1766–1844) proposed what is now regarded as the modern definition of the atom. His theory asserted that each chemical element is composed of a particular type of atom, which represents the smallest unit of matter in which that element can exist. Atoms retain their separate identities in chemical processes and each type of atom differs in weight from those in other elements.

THE 'PLUM-PUDDING' MODEL

Even before the existence of atoms had been demonstrated experimentally, it had become obvious that they must contain some kind of internal structure. Electrical phenomena required an explanation on the basis that matter contains charged constituents, the flow of which results in electric current. It also became apparent that all charges are simple multiples of an elementary, or unit, charge.

Experimental proof of the existence of 'particles of electricity' was provided by the work of J. J. Thomson (1856–1940) in 1897. His discovery, together with phenomena such as the photoelectric effect, convinced scientists that these negatively charged elementary particles, dubbed

electrons, must reside within atoms. And since atoms are electrically neutral, this implied that they must also contain a positive charge to cancel out the charge of the electrons.

The next question was how the positive and negative charges were distributed in the atom. Thomson proposed a tentative model of the atom in which the electrons were evenly distributed throughout a sphere of positive charge. The size of the sphere could be calculated reliably from quantities known at the time. This became known as the 'plum-pudding' model, but was soon to be superseded by more realistic models.

▼ *This scanning-tunnelling image shows atoms of gold as yellow, red and brown on a background of graphite, shown as green.*

RUTHERFORD 'SEES' THE NUCLEUS

With the discovery of alpha particles in natural **radioactivity**, a method for investigating the structure of atoms became available. In 1911, Johannes Wilhelm Geiger (1882–1945) and Ernest Marsden (1889–1970) carried out a series of experiments in which a narrow beam of these particles was scattered from thin metallic foils. The scattered particles could be detected by the small flashes of light, or scintillations, that were produced when they collided with a zinc sulphide screen. They discovered that the solid foils would allow most of the alpha particles to pass through to the screen without being deflected. However, a few (about 1 in 8,000) would bounce back off the foil.

ERNEST RUTHERFORD

Born in New Zealand, Ernest Rutherford (1871–1937) was to become one of the most influential figures of twentieth-century science. By the time he had demonstrated the existence of the atomic nucleus, through his study of alpha-particle scattering, he had already won a Nobel Prize for chemistry (in 1908 for his work on radioactivity). He was later to go on to make many significant contributions to physics. For instance, through his work on nuclear reactions he was the first scientist to realize the alchemists' dream of transmutation of the elements (changing one element into another).

Rutherford was knighted in 1914 and took over from J. J. Thomson as Director of the Cavendish Laboratory in Cambridge in 1919. He was

President of the Royal Society from 1925 to 1930, and was created Baron Rutherford of Nelson (after his birthplace) in 1931, enabling him to take up a seat in the House of Lords.

His greatest mistake was the belief that nuclear power could never be realized. He famously described as 'moonshine' the notion that humanity could ever harness the energy trapped within atomic nuclei. He died in 1937, two years before the discovery of nuclear fission.

▲ *A Nobel Prize winner in 1908, Ernest Rutherford made numerous contributions to science, notably his demonstration of the existence of the atomic nucleus.*

ATOMS

Rutherford realized the significance of this result and proposed a new model for the atom in which all the positive charge of the atom and essentially all of its **mass** is concentrated in a much smaller region at its centre called the nucleus. The rest of the atom's volume had to be empty space that contained the even tinier negative electrons. His argument was that Thomson's plum-pudding model, in which an atom's mass and charge is evenly distributed throughout its volume, could not possibly account for the few alpha particles that were deflected back through angles of more than 90°. Atoms had to have a hard central core of positive charge, which most of the alpha particles would miss completely. Only those few that came close enough to a nucleus would be deflected back. The atomic electrons, having a much smaller mass than the incoming alpha particles, would have a negligible influence on their trajectories.

From the angles of deflection of the alpha particles, Rutherford was able to derive a remarkably accurate estimate of the size of the atomic nucleus, which proved to be about 10,000 times smaller than the atom.

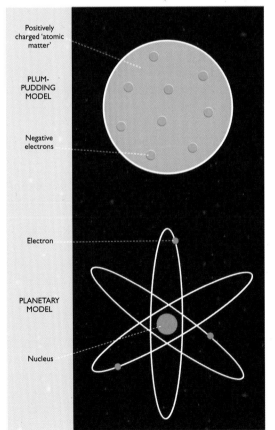

Positively charged 'atomic matter'

PLUM-PUDDING MODEL

Negative electrons

Electron

PLANETARY MODEL

Nucleus

▲ **Outdated Models of the Atom**
J. J. Thomson proposed his plum-pudding model of the atom in which negatively charged electrons were evenly distributed inside a positively charged atomic sphere. Niels Bohr's planetary model of electrons orbiting a nucleus remains a popular, but not really accurate, visualization of the atom.

💡 HOW CAN ATOMS SURVIVE?

Rutherford's model of the atom suggested that electrons exist outside the nucleus at distances thousands of times greater than the radius of the nucleus itself. This picture immediately raises the question of the stability of atoms.

Electrons cannot be at rest within the atom, since the Coulomb force (the electrostatic attraction between the positively charged nucleus and the negative electrons) would pull them into the nucleus. This suggests a planetary model, in which the electrons are in continuous orbit around the nucleus, just as Earth is in orbit around the Sun. However, according to classical electromagnetic theory, an accelerating charge will radiate energy. The orbiting electrons are continuously accelerating towards the nucleus because motion in a curved path involves change of direction, which is acceleration. They should therefore be losing energy and consequently spiralling inwards. This process would happen very quickly (in about 10^{-12} seconds) and atoms should therefore collapse in this time.

It was this problem above all else that prompted Bohr to propose his quantized model of the atom, in which electrons would remain in fixed orbits. Later work by Wolfgang Pauli (1900–58) showed that each electron orbit can only accommodate a certain number of electrons. Electrons can jump to a lower orbit only if it is unfilled.

THE BOHR MODEL

Although it is not really correct, the atomic model proposed by **Bohr** for the hydrogen atom is still taught to schoolchildren today. His theory, which he published in 1913, was based on the two major advances in physics of the early twentieth century: Rutherford's discovery of the nucleus with the electrons in orbit around it and Planck's idea of quantization. Whereas Planck had suggested that atoms emit and absorb energy in packets (quanta), Bohr postulated that a quantity called the **angular momentum** of the orbiting electrons was quantized. This implied that electrons were not free to follow any orbit whatever (as would be possible according to Newtonian mechanics) but only those for which their angular momentum was a multiple of Planck's constant.

This model of a miniature solar system was based on the ideas of the early quantum theory but still contained aspects from classical physics, which we know today to be wrong. The current model of the atom required the full development of quantum mechanics 10 years later.

▶ *Gustav Kirchhoff was the first to observe that elements each produce distinctive patterns of lines, as can be seen in the emission spectra of the gases xenon, krypton and neon (left to right).*

Xenon   Krypton   Neon

🧪 ATOMIC SPECTRA

When an electric current is passed through a gas, the atoms absorb energy through collisions with electrons or **ions** produced by the current. They may then release this energy in the form of **electromagnetic radiation**. This emitted light can be broken up into its component wavelengths using a spectrometer (such as a prism).

When sunlight is passed through a prism it is broken up into a continuous **spectrum** spanning the range of **wavelengths** of visible light. This, as every schoolchild knows, is because the different wavelengths are refracted by different amounts in the prism, causing a narrow beam of 'white' light to be smeared out. However, atomic spectra consist of a set of individual coloured lines, known as a line spectrum. The lines are produced by the instrument: they are images of the slit through which light enters the spectrometer. The line spectrum characterizes the discrete quantum nature of the radiation and can be used to identify the different elements. This technique is known as **spectroscopy** and is a vital analytical tool in astronomy for studying the composition of stars from their light.

The German physicist Gustav Kirchhoff (1824–87) was the first to notice in 1859 that each element produced its own distinctive pattern of lines. The spectrum of hydrogen, the lightest element, was later studied by a Swiss mathematician, Johann Balmer (1825–98), who devised a simple formula that gave the wavelengths producing the different lines. This 'Balmer series' was later used by Bohr to support his atomic model.

QUANTUM REALITY

QUANTUM REALITY

In Search of Quantum Reality
QUANTUM UNCERTAINTY & PROBABILITY

At the beginning of the twentieth century Planck's quantum postulates had made it possible to clear up a number of outstanding problems relating to the nature of light and the structure of atoms. However, the theory that is known today as quantum mechanics was not completed until the mid-1920s, nor was it the work of any one person. A number of new concepts had yet to be put in place to pave the way for the quantum revolution.

THE WAVE NATURE OF MATTER

The work of **Planck**, **Einstein** and Compton that confirmed the particle-like nature of light was, in a sense, only half the story. In 1924 Louis de Broglie (1892–1987) made the bold proposal that this dual nature of **radiation** might also apply to matter. He pointed out that if waves can sometimes behave like a stream of particles, then moving particles might sometimes have wave-like properties. His ideas were soon to be verified experimentally by Clinton Davisson (1881–1958) and Lester Germer (1896–1971), and independently by George Thomson (1892–1975). In both experiments, a beam of **electrons** was shown to undergo **diffraction** – a property of waves.

De Broglie was inspired by Compton's scattering experiments, which were interpreted as involving **photons** colliding with electrons. By treating the photons and electrons on the same footing, de Broglie suggested that the electrons would also have a **wavelength** associated with their momentum.

Although initially applied to free electrons, matter waves can be associated with all bodies, but the wavelength becomes smaller as the **mass** of the body increases. For instance, a tennis ball served at 160 km/h (100 mi/h) has a wavelength of 10^{-34} m, which is many orders of magnitude smaller than the dimensions of atoms and therefore undetectable. That is why we never see any wavelike properties associated with tennis balls. But an electron can have wavelengths of atomic dimensions and will thus readily exhibit its wavelike nature.

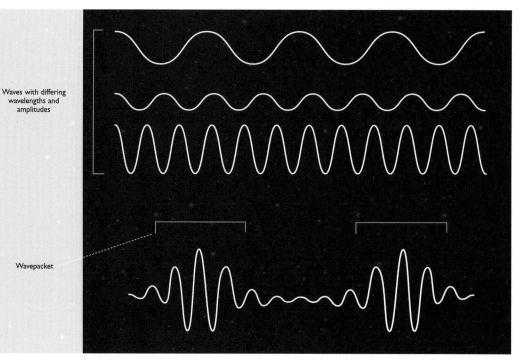

Waves with differing wavelengths and amplitudes

Wavepacket

▲ *Wavepackets*
A superposition of many different waves can give 'wavepackets' (or pulses). De Broglie, whose work inspired Schrödinger, suggested that particles can be considered as localized wavepackets.

HEISENBERG'S UNCERTAINTY PRINCIPLE

A direct consequence of wave–particle duality is one of the best-known and most frequently misunderstood aspects of quantum mechanics. Stated originally as the principle of indeterminacy by the young German prodigy Werner Heisenberg (1901–76) in 1927, it is now known simply as the **Heisenberg Uncertainty Principle**.

According to classical mechanics and our common sense there is no reason why we could

not determine both the precise location *and* the momentum of an object at any instant in time. Indeed, with good enough measuring equipment we should be able to measure these values to as high a degree of accuracy as we wanted. But in the quantum world, where particles sometimes behave like spread-out waves, things are different.

A wave has no definite location in space. However, the de Broglie formula relates the wavelength of a wave to its momentum. Thus a wave with a unique wavelength has a definite momentum but no definite location. Following this line, we can imagine a particle to be just a very squashed-up wave (known as a wavepacket). This can be obtained by combining many waves of different wavelengths in such a way that they interfere and cancel each other out everywhere apart from one confined region where the particle is located. Now we have the opposite situation: an object with a definite location but no definite momentum (since many wavelengths have been combined).

The Uncertainty Principle states that there will always be a small uncertainty in our knowledge of a particle's position and momentum. The more accurately we know one, the less we know about the other. The product of these two uncertainties thus has a lower limit, which turns out to be the **Planck constant**.

◄ *Werner Heisenberg gave his name to the Uncertainty Principle, which states that it is possible to accurately measure either a particle's position or its momentum, but never both at the same time.*

QUANTUM UNCERTAINTY

ELECTRONS ARE PARTICLES *AND* WAVES

A fascinating irony in the history of science is the work of the English father and son physicists Joseph (J. J.) and George Thomson. In 1897, J. J. Thomson showed that electrons are electrically charged particles that are deflected by electric fields inside what is now known as a cathode-ray tube. He was also able to deduce the electron's mass (less than one-thousandth the mass of the lightest atom), showing that it is truly a 'subatomic' particle. While evidence for the existence of electrons had already existed before Thomson's work, he is considered as having provided the final proof. He was awarded the Nobel Prize for physics in 1906 for his discovery of the first elementary particle.

In 1927, Thomson's son George published his work on electron diffraction. His experiments, carried out independently and at about the same time as those of Davisson and Germer in the US, involved passing a beam of electrons through a thin metal foil and observing the diffraction pattern produced on a screen. In this way, he demonstrated the wavelike properties of electrons by measuring the interference effects of their associated waves. He was awarded the Nobel Prize in 1937 (together with Davisson). While the father had won a Nobel Prize for showing that electrons are particles, the son had done the same for showing that electrons are waves.

▲ *The idea of* an abstract wave of probability is more familiar than it might seem. In a crime wave, locations where crimes are more likely to occur are where the wave is at its most intense.

SCHRÖDINGER'S WAVE MECHANICS

The Austrian physicist Erwin Schrödinger (1887–1961) was inspired by de Broglie's ideas on matter waves and set out to incorporate them into **Bohr**'s atomic model. Schrödinger suggested that the electrons could only have orbits into which a whole number of electrons' associated waves fitted. Any orbit in between two successive allowed orbits could only contain a fractional number of wavelengths, and so would not be 'allowed'.

By 1926, Schrödinger's theoretical work culminated in the publication of a famous equation that described not only the behaviour of atomic electrons but of all subatomic systems. Unlike **Newton's laws** of motion, Schrödinger's equation did not describe the motion of particles but rather the propagation of waves. The mathematical principles on which his ideas are based are today known as wave mechanics.

QUANTUM PROBABILITY

How can an electron be both a particle and a wave at the same time? The answer is that it is not both of these. The electron should not be thought of as a spread-out entity that ripples through space. A more accurate statement is that an electron is a point particle whose motion is controlled by wavelike principles, which means that its precise location is not well defined.

Are matter waves real? The accepted view today is that they are abstract waves of probability. We are accustomed to the idea of a crime wave spreading through a city. This is also a probability wave. The locations where it is at its most intense are where crimes are most likely to happen. In the same way, the probability wave associated with an electron specifies the locations at which the electron is most likely to be. It is this probability distribution that one obtains by solving the Schrödinger equation.

▶ *Erwin Schrödinger is* best known for his equation which predicts the properties and dynamics of particles at the quantum level in terms of probabilities.

MEASUREMENT IN QUANTUM MECHANICS

Heisenberg's Uncertainty Principle has so far been phrased as a statement of the wave–particle duality of matter and radiation. The more accurately the position of a quantum entity such as an electron is known (its particle-like nature), the less well its momentum is known (its wavelike nature). But there is no limit to how accurately one of these quantities alone, say the electron's position, can be measured.

The usual way to observe an object is to bounce light off it; that is what we mean by 'seeing' something. Normally, this has no effect on the object. But at the quantum level, a photon can give a significant kick that would knock the observed particle (such as an electron) from its original position. A short-wavelength, high-energy photon will give the electron a stronger kick. However, a gentler photon probe, with a longer wavelength, will provide less resolving power – less accuracy in determining the electron's position. In any kind of microscope, the use of shorter wavelengths permits finer detail to be seen.

By assuming both the photon and the electron to be point particles, it would be possible to calculate the strength of the kick and work backwards to pinpoint the electron's original position. The reason this is not possible even in principle is due to the wave–particle nature of the probing photon.

Thus, quantum uncertainty is not, as is often stated, the result of an unavoidable disturbance due to the clumsiness of the measurement process. Rather it is due to the wave–particle duality of both the light probe and the quantum system that is being measured.

SCHRÖDINGER'S EQUATION AND THE WAVE FUNCTION

Schrödinger's wave equation is one of the best-known and most important equations in physics. It successfully predicts the properties and dynamics of all systems in the quantum world.

Given the momentum of an electron, along with the forces acting on it at any given time, Schrödinger's equation gives information about the state of the electron at any later time. However, unlike Newton's equations of motion, which predict the precise positions and velocities of objects at all times given their initial states, Schrödinger's equation gives only the probabilities of different outcomes.

However, this does not mean that Schrödinger's equation is somehow less accurate than **Newton**'s or **Maxwell**'s equations. It describes the way a quantum system evolves and predicts its properties with high accuracy. Its probabilistic nature reflects the way matter behaves on the quantum level, rather than any shortcoming in the equation itself.

Bohr ▶ p. 94 Einstein ▶ p. 88 Maxwell ▶ p. 72 Newton ▶ p. 42 Planck ▶ p. 96

QUANTUM REALITY

In Search of Quantum Reality
THE NEW QUANTUM MECHANICS

The old quantum theory and the work of de Broglie inspired Schrödinger to develop his wave mechanics in the mid-1920s. At about the same time, Werner Heisenberg developed an alternative formulation, based on more abstract mathematical ideas, that became known as matrix mechanics. Within a year of their work, the English physicist Paul Dirac showed that the two approaches were equivalent, but were just using different mathematical languages. Nowadays, both approaches are used together, although Schrödinger's approach is easier to teach to students in the first instance.

PAUL DIRAC

Paul Adrien Maurice Dirac (1902–84) was one of the twentieth century's greatest theoretical physicists. He was one of the handful of young geniuses, together with the likes of Heisenberg and Pauli, who placed quantum mechanics on a firm mathematical footing. A very quiet and shy man, he was the first to bring together quantum mechanics and special relativity, predicting the existence of antimatter in the process.

He shared the Nobel Prize for physics with Erwin Schrödinger in 1933 for his contribution to the development of quantum mechanics.

▲ *Paul Dirac was one of the greatest theoretical physicists of the twentieth century. His significant mathematical contributions to quantum mechanics earned him a Nobel Prize, shared with Erwin Schrödinger, in 1933.*

THE WAVE FUNCTION

The solution of Schrödinger's equation is a mathematical quantity called the wave function and contains all the information that can be obtained about a quantum system. It depends on both the space and time coordinates of the object it is describing, and is specified by two numbers known as its real and imaginary parts. This property of the wave function is necessary in order for it to exhibit wavelike properties such as **interference**.

Unlike its counterpart in classical wave mechanics, the wave function is not a directly observable quantity. Indeed, it does not have any concrete physical reality and must be regarded simply as an abstract mathematical entity from which we may deduce physically meaningful information. It is only the square of the wave function, called the probability density, that is measurable. For an **electron**, the probability density at a position in space and a moment in time gives the likelihood of finding the electron there and then.

THE VIEW FROM COPENHAGEN

Since quantum mechanics states that we are only able to make predictions about the behaviour of quantum particles such as electrons, we are forced to concede that, at this fundamental level, the future cannot be determined with any certainty. We can only gain statistical information from the collective behaviour of many quantum systems. The commonly accepted philosophical interpretation of this view is embodied in what is known as the Copenhagen interpretation, since that is where most of its ideas were developed (under the guidance of **Bohr**). However, it is wrong to think that quantum mechanics is somehow unable to predict the future with certainty, even given complete knowledge of the present. Rather, it is the fact that we can never have complete knowledge of the present that prevents us from predicting the future with certainty. This is due to the nature of the wave function.

DIRAC AND ANTIMATTER

While Schrödinger's equation replaces Newton's equations of motion when we want to describe the world of the very small, we also know that Newton's equations must be modified in order to describe the world of the very fast, because of relativity. In 1927 Paul Dirac developed a relativistic treatment of quantum mechanics that provided new insight into the nature of matter and led to the discovery of an important concept: the existence of antimatter.

Dirac developed an alternative equation to Schrödinger's which described the behaviour of electrons moving at relativistic speeds. However, the Dirac equation made a strange prediction. It suggested that the electron's energy could be positive negative (a concept that makes no sense). Rather than disregard the negative energy solutions as being of no physical significance, Dirac took them seriously and put them to use.

Stated simply, Dirac suggested not only that electrons with negative energy exist, but also that empty space is full of them. We are normally unable to detect them, just as we are normally unaware of the air that surrounds us. However, a **photon** can interact with one of the negative-energy electrons and, by giving it positive energy, make it observable. The 'hole' this electron leaves behind in the negative energy 'sea' also takes on a reality of its own. It behaves exactly like a particle with positive charge, but is otherwise identical to the electron. This particle is now identified with the antimatter version of the electron, the **positron**. This process of creating an electron and a positron from a photon is known as pair production.

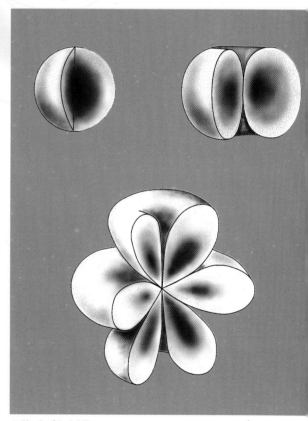

▲ *Clouds of Probability*
This graphic visualizes 'clouds of probability' – in this case, the probability of finding electrons at different locations within a hydrogen atom. The atom is shown in three energy states, and in each state the density of the black shading indicates where electrons are most likely to be found.

QUANTUM REALITY

QUANTUM TUNNELLING

One of the more bizarre consequences of quantum uncertainty and wave–particle duality is the phenomenon of quantum tunnelling, or barrier penetration. Quantum particles are able to 'tunnel' through potential-energy barriers that classical particles are forbidden to penetrate. Consider the following example. To roll up a hill and over the other side, a ball has to be given enough momentum. It will gradually slow down while it climbs the slope as it converts its kinetic energy into potential energy. If it runs out of energy before the top, it will simply roll back

down the way it came. However, if it were to behave quantum-mechanically, there would be a certain probability that it would travel half-way up the slope and then spontaneously disappear and reappear on the other side.

This type of behaviour can be explained using a modified version of the **Heisenberg Uncertainty Principle**. Rather than putting it in terms of uncertainty in a particle's position and momentum, the principle can be stated in terms of uncertainty in the particle's energy and the time for which it has that energy. Thus, a particle may 'borrow' the energy it needs to get over the barrier, provided it

'gives back' the energy within a time specified by the uncertainty relation.

Quantum tunnelling plays an important role in many processes and applications. It is vital for understanding **radioactivity**, and forms the basis for a number of modern electronic devices. An everyday example is in aluminium household wiring. A thin layer of aluminium oxide often builds up on the surface of electrical wiring, forming an insulating layer between two wires twisted together that, according to classical physics, should not allow the passage of electrons. Fortunately, the layer tends to be very thin and electrons can easily tunnel across.

SPIN

In 1894, a Dutch physicist named Pieter Zeeman (1865–1943) conducted an experiment in which he found that atomic line spectra were split into further components when the atoms were subjected to a magnetic field. This Zeeman effect confirmed the idea of Hendrik Lorentz (1853–1928) that atoms consisted of charged particles which could be affected by magnetic fields. Once Bohr developed his orbital model of the atom, the Zeeman effect was interpreted in terms of changes to the electrons' orbits. However, further experiments showed that this could not be the whole story. There seemed to be too many **spectral lines**.

COMPLEX NUMBERS

The numbers we are used to dealing with in everyday life are known as real numbers. They can be positive or negative, fractions or whole numbers. But there is another class of numbers that plays an important role in mathematics. The so-called imaginary numbers have the strange property of involving the square root of a negative number. Here is a short explanation.

The square of a positive number is a positive number, and the square of a negative number is also a positive number (since $- \times - = +$). But mathematicians can also consider a number that when multiplied by itself still gives a negative number. Such a number is known as 'imaginary' and obeys a consistent set of rules. A complex number is a pair of two numbers, one real and one imaginary. The value of the wave function in quantum mechanics is just such a complex number.

Taking the square of the wave function yields a positive real number that is the physically meaningful probability required to make predictions in quantum mechanics. Doing this unavoidably throws away part of the information contained within the complex number, known as 'the phase'. This is the reason for the unpredictability of quantum mechanics.

In 1925 two young physicists, George Uhlenbeck (1900–88) and Sam Goudsmit (1902–78), suggested that along with the electrons' orbital motion there is another source of their **angular momentum**: their spin. Just as Earth orbits around the Sun and spins on its axis, so too do electrons.

An electron's spin, however, is a rather counter-intuitive concept and cannot be thought of in everyday terms. Wolfgang Pauli suggested that spin should be 'quantized', like the orbits of the electrons – that is, they are discrete. In fact, electrons can have only two spin 'states': clockwise (called spin 'up') or anticlockwise (spin 'down').

THE EXCLUSION PRINCIPLE

The reason that the different elements have different chemical properties is due to the way their electrons occupy orbits inside the atoms. Why do they do this? What governs when a particular quantum orbital is full and cannot accommodate any more electrons? The answer was discovered by Pauli, who proposed yet another important principle in quantum mechanics, now known as Pauli's **exclusion principle**.

Pauli pointed out that electrons cannot exist in the same quantum 'state'. The quantum state is defined by a set of quantum numbers specifying (in the case of atomic electrons) the electron's orbital energy and angular momentum, and its spin. Electrons fill up successive 'shells' with each one accommodating a number of

◀ **Wolfgang Pauli** *discovered that the way electrons behave inside atoms determines the different chemical properties of matter. His explanation has become known as Pauli's exclusion principle.*

▲ *Fermions and Bosons*
Fermions obey Pauli's exclusion principle, bosons do not. Imagine fermions as being like members of the audience at a classical music concert (top), and picture the rows of seats being like different states of energy with the front row being the most desirable 'state'. Each audience member can only occupy one of the available seats on each row (or 'energy state'). Bosons, by contrast, behave more like the audience at a pop concert (bottom). As soon as the performers appear, the entire audience rushes to the front of the stage, all trying to achieve the same 'energy state'.

electrons according to well-defined rules. Once a shell is 'full', the next electron has to fill the next one up in energy. Stated another way, the exclusion principle explains why electrons cannot be pushed into the nucleus and matter has rigidity.

Particles that obey the exclusion principle, such as electrons, are known as fermions, after the Italian Enrico Fermi (1901–54). Another class of particles, called bosons, which includes photons, does not obey this principle, and so different bosons are able to occupy the same state.

In Search of Quantum Reality

WHAT DOES IT ALL MEAN?

Quantum mechanics forces on us a view of the world that runs counter to common sense. Particles that can be created out of pure light, travel through two slits at once and tunnel through impenetrable barriers sound like science-fiction rather than the predictions of the most successful theory in science. Physicists have now had three-quarters of a century to come to terms with these counterintuitive concepts, and have proposed a number of different interpretations of quantum reality; but there is still no consensus as to which of them is 'correct'. A few of the better-established interpretations are examined here.

FEYNMAN PATHS

The US physicist **Richard Feynman** (1918–88), renowned for his extraordinary ability as a lecturer and popularizer of physics as well as his role in the development of **quantum electrodynamics**, developed an alternative interpretation of quantum mechanics in 1947. His approach, known as the path-integral, or 'sum over histories', method, at first appears to be even more crazy than the notion of wave–particle duality.

According to the usual picture of the two-slit experiment, a particle must somehow go through both slits at once. This is usually explained as the two parts of the particle's probability wave interfering on the other side of the slits to cause the pattern of bright and dark fringes. Feynman's alternative view was that the particle itself goes through both slits at once; it follows two paths simultaneously. In fact, the **electron** actually follows *all* possible paths between the source and screen, no matter how 'wiggly'. The final position of the particle is thus due to all the possible paths combining, and it is their average we observe. There are no matter waves, no probability waves, and the particle is always point-like.

This view may seem completely crazy: how can a particle follow all possible routes at once? But its conclusions are identical to those of the Copenhagen view involving waves of probability.

BOHMIAN MECHANICS

The standard (Copenhagen) interpretation of quantum mechanics states that, at the quantum level, nothing really exists until it is observed. Until then, an electron, say, is just a wave of probability. This state of affairs was unsatisfactory for physicists like **Einstein**, Schrödinger and de Broglie, who felt that particles must have a real existence, whether or not they were being observed.

Building on earlier ideas of de Broglie, David Bohm (1917–92) published two papers in 1952 in which he provided an alternative view of what is happening to quantum particles such as electrons when we are not looking at them. What Bohm did was reinterpret the meaning of the wave function. He was able to do away with the notion of wave–particle duality for electrons, which he regarded as particles. Thus, in the two-slit experiment, the electron passes through only one of the slits, but its path is guided by a strange kind of energy field known as the quantum potential, which is worked out from the wave function. This quantum potential propagates through both slits and so influences the way the particle moves to form the interference pattern.

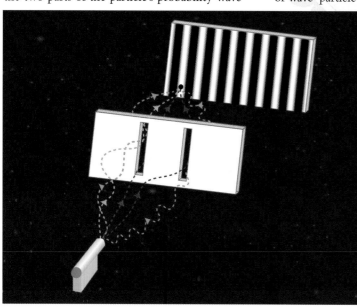

◄ *Sum Over Histories*
Feynman's radical 'sum over histories' theory asserts that an individual particle in a two-slit experiment actually follows every possible path between the source and the screen. What we observe is the average of all the possible paths combined.

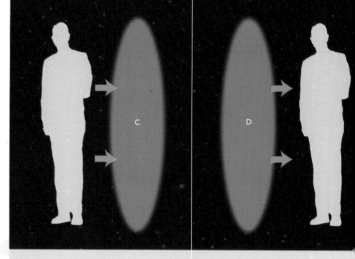

LOCATION A   LOCATION B

▲ *Quantum Teleportation*
Teleportation, science fiction's favourite fantasy, could have a theoretical quantum explanation. Two or more particles can be quantum entangled, and an object comprising such particles could theoretically be moved from place to place. At A, the person is 'encoded' together with auxiliary material C. Material C is quantum entangled with material D. Data is sent non-quantum mechanically from A to B, to enable material D to be reconstituted as the person at B.

While it has a significant number of supporters, Bohmian mechanics is still rejected by many physicists. The usual criticism is that while it removes some of the weird aspects of quantum mechanics associated with the Copenhagen interpretation, it replaces them with a different kind of weirdness to do with the strange way the particles move. However, there is as yet no known way to test experimentally whether Bohmian mechanics is correct.

MANY-WORLDS INTERPRETATION

Originally proposed by Hugh Everett in 1957, this is the interpretation of quantum mechanics favoured by some of the greatest minds in theoretical physics today. Like the Bohmian view, it states that the quantum-mechanical wave function is more than just a mathematical quantity. Now, instead of a measurement of a quantum system forcing just one of a number of alternative outcomes to be realized, the whole Universe splits into a number of parallel realities equal to the number of possible outcomes. So while just one possible outcome of a quantum measurement is observed in our reality, every other possible outcome is observed in alternative parallel universes. Again, while this version of events does away with the most problematic features of the Copenhagen view, many physicists find it hard to accept.

electron ► p. 338   interference ► p. 340   photon ► p. 342   quantum electrodynamics ► p. 342

QUANTUM REALITY

ENTANGLEMENT

When a particle is faced with options, such as decaying or not decaying, or going through slit A or slit B, it can exist in a combination of the two states. This is known as a quantum superposition and is one of the unusual features of the quantum domain that we do not see around us in the classical world. It can also be the case that two quantum particles can be correlated so that the action of one affects the behaviour of the other. If one of the particles is in a quantum superposition of different states, then the other is forced to exist in a superposition also. This type of correlation is known as entanglement.

The most famous example of the entanglement of two particles is due to Einstein who, together with his colleagues Boris Podolsky (1896–1966) and Nathan Rosen (1909–95), devised what became known as the EPR paradox (after their initials). They imagined two particles created from a common source (hence correlated) and sent off in opposite directions. Their intention was to highlight the absurd nature of quantum mechanics. Ironically, such experiments are routinely carried out these days and show that two particles can remain entangled despite being very far apart.

In the past few years, the idea of quantum entanglement has led to a number of fascinating possible applications, such as quantum teleportation and quantum cryptography.

DECOHERENCE

The spread-out nature of the wave function of a quantum system means that something like an electron can only be described in terms of different probabilities of being in certain states. There is even a probability that it can be in two or more states at once. Until recently, physicists did not understand why such quantum superpositions never show up when a measurement is made.

Relatively recently, the idea of decoherence has gained popularity. It states that once a quantum system becomes too entangled with its surroundings, all the quantum weirdness leaks out. For example, the probabilities of an electron being in more than one place at once disappear very quickly as soon as it interacts with many other particles, as happens in any real measurement.

Many physicists claim that this cannot be the whole story since decoherence still does not explain why one particular outcome, rather than any other, is actually realized during a measurement.

COMPLEMENTARITY

According to the uncertainty principle, there is a limit to the precision with which we can simultaneously determine the values of certain pairs of variables, such as the position and momentum of a particle. In 1928, **Bohr** introduced a more general way of putting this in his complementarity principle, in which he stated that the wave and particle aspects of a quantum system such as an electron or photon are complementary, both aspects being necessary for a complete description of its properties. Later in life, Bohr incorporated his ideas on complementarity into other areas of philosophy and even chose the motto 'opposites are complementary' to describe a wide range of phenomena in other areas of science and in human life.

◀ *The idea of complementarity can be found in many philosophical systems. Niels Bohr, proponent of the complementarity principle, adopted the motto 'opposites are complementary' – an idea portrayed in the Chinese symbols for Yin and Yang.*

THE TWO-SLIT EXPERIMENT REVISITED

The two-slit experiment demonstrates that, whatever is going on in the quantum world, it is certainly very strange. The appearance of an **interference** pattern when particles are fired through two slits has been confirmed in the laboratory. Physicists have even obtained an interference effect when they send large carbon molecules, known as buckyballs, through two slits.

An interference pattern is a clear signal of wave behaviour, since it can only be formed when the wave fronts from two sources (the two slits) overlap. If a position on the screen corresponds to a point where the crests of one wave arrive at the same time as the crests of the other then the resulting wave front will be enhanced. On the other hand, where crests of one wave arrive at the same time as troughs of the other, the two waves will cancel each other out. In this way, light and dark fringes appear on the screen.

When this result is found in the quantum world, the explanation is no longer straightforward, since it is not a continuous wave that washes up against the screen, but rather individual flashes of light, corresponding to particles hitting precise locations. An interference pattern appears when the particles, be they **photons**, electrons or atoms, are fired one at a time, each one being fired only when the previous one has hit the screen.

Whatever is going on, there is no simple explanation. Physicists have had three-quarters of a century to come to terms with quantum mechanics, consoling themselves with the fact that at least the mathematics works. How the quantum world is interpreted is for the time being still a matter of philosophical choice.

▼ *The Two-Slit Experiment*
This diagram shows the two-slit experiment with photons. On the left is the experimental set-up in which the photons are counted in detector boxes. On the right are the results of three experiments: A, B and A+B. Photons passing through Slit 1 are red, those that pass through Slit 2 are green. A – Slit 1 only open; the reds directly in line with Slit 1. B – Slit 2 only open; the greens directly in line with Slit 2. A+B – Both slits open; we can no longer say through which slit each photon has passed, as shown by the red/green photons collected in column A+B. The total found in each box is not the sum of those found in experiments A and B.

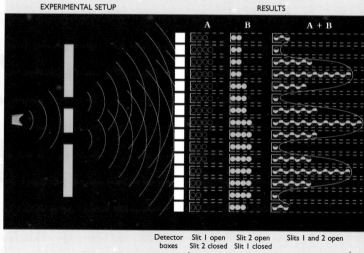

EXPERIMENTAL SETUP	RESULTS
	A B A+B

| Detector boxes | Slit 1 open Slit 2 closed | Slit 2 open Slit 1 closed | Slits 1 and 2 open |

Number of bullets in each box after a fixed time

QUANTUM REALITY

In Search of Quantum Reality
INTO THE NUCLEUS

T he nucleus is a tiny object, positively charged, containing most of the mass of the atom and acting as the focus of the electron orbits. Atomic nuclei also have a rich and complicated internal structure and are of crucial importance to our understanding of the physical world. This is especially true in astronomy: almost everything we see when we look up in the night sky is the result of nuclear processes.

RADIOACTIVITY

The study of **atomic nuclei** began in 1896 with the accidental discovery by Henri Becquerel (1852–1908) that certain uranium salts spontaneously emit **radiation**. This phenomenon, now called **radioactivity**, was soon shown in many naturally occurring atoms. Later work showed that there were three types of radioactive emissions: alpha, beta and gamma radiation. They were distinguished by their ability to penetrate materials and by the ways in which they were affected by electric and magnetic fields. The alpha rays proved to be the nuclei of helium atoms, the beta rays to be fast-moving **electrons**, and the gamma rays to be high-energy **photons**.

All three types of radioactivity were eventually found to originate within atomic nuclei. But even before this became apparent, radioactive particles were found to be ideally suited to studying the structure of other nuclei.

The historical significance of radioactivity in the development of nuclear physics has been profound. In recent years, the use of radioactivity sources has become increasingly important in industrial and medical applications, as well as in scientific research. Indeed, a whole new sub-field of nuclear physics, based on the production of radioactive beams of certain short-lived nuclear species, emerged during the last decade of the twentieth century.

PROTONS AND NEUTRONS

For almost 20 years after the formulation of the **Rutherford** and **Bohr** models of the atom, the basic constituents of the nucleus remained unknown. It was initially thought that the nucleus was made up of electrons as well as of heavy, positively charged particles called **protons**. There would have to be more protons than electrons to make the nucleus positive overall, thus balancing the negative charge of the orbiting electrons. Then, in the early 1930s, a new type of radiation was discovered. James Chadwick (1891–1974) showed in 1932 that this radiation consisted of a new type of particle, approximately equal in mass to the proton, but with no electric charge. This particle was dubbed the **neutron**, a name earlier proposed by Rutherford. Heisenberg then correctly suggested that atomic nuclei consisted of protons and neutrons only, the accepted picture today. Chadwick's discovery was to mark the beginning of modern nuclear physics.

Today there are a number of highly sophisticated theoretical models of the structure of nuclei that describe the way the protons and neutrons can be arranged. It is also known now that protons and neutrons are themselves composed of smaller constituents, called **quarks** and gluons.

NUCLEAR FORCES AND NUCLEAR DECAY

Nuclei exist because there are attractive forces that hold the protons and neutrons together and that counteract the repulsive electrostatic forces between the protons. This short-range strong nuclear force combines with the longer-range and weaker electrostatic force to allow protons and neutrons relative freedom to move around inside nuclei but produces a confining force-field (a 'potential barrier') at the nuclear surface.

When an alpha particle (which consists of two protons and two neutrons) is emitted from a nucleus, it must penetrate the potential barrier at the nuclear surface in order to escape. According to classical physics, this would be impossible. Thanks to quantum tunnelling, however, the alpha particle can get through the barrier. This explanation of alpha radiation was the first successful application of quantum mechanics to nuclear problems.

Beta decay is a process in which an electron is emitted from the nucleus. But since electrons do not normally exist inside the nucleus, the beta-decay process must involve the conversion of a neutron into a proton with the simultaneous creation of an electron. In 1933 Pauli discovered that this process must also involve the creation of a new particle, now known as an antineutrino, in order to interpret the data. To explain the process, beta decay had to be associated with a second, weaker, nuclear force, known rather unimaginatively as the weak force. Protons can also be converted into neutrons within nuclei while emitting a **positron** and a **neutrino**. This is another form of beta decay.

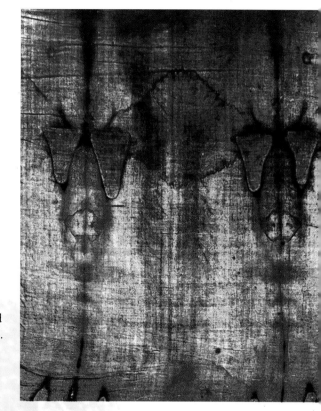

▲ **The phenomenon of** *nuclear decay is used to estimate the age of ancient organic materials such as the Turin Shroud. This is done by measuring the amount of carbon 14 (a radioisotope of carbon) that remains in the tissue.*

Gamma decay is the emission of pure energy, in the form of photons, from nuclei. Before the decay, the nuclei are in highly excited energy states, either because individual protons or neutrons have excess energy or because the whole nucleus is rotating or vibrating. Since the energy of a nucleus must be quantized, the emission of a gamma ray is associated with the transition of the nucleus from a higher energy level to a lower one. The study of nuclear structure and energy levels by means of the gamma rays emitted is known as gamma-ray **spectroscopy**.

NUCLEAR FISSION AND FUSION

The process in which a heavy nucleus, such as uranium, splits into two nuclei of roughly equal size is called fission. Spontaneous fission occurs rarely; to make use of fission, it is necessary to induce it by supplying energy to a nucleus, usually in the form of neutrons (although fission can occur with the absorption of photons). In 1938, Otto Hahn (1879–1968) and Fritz Strassmann (1902–80) conducted the first fission experiment and thus 'split the atom'.

A typical example is the fission of uranium into barium and krypton. This is accompanied by the release of energy and the escape of a few

QUANTUM REALITY

Alpha-particle
(2 neutrons and
2 protons)

Lead nucleus
(spherical)

Deformed
spinning nuclei

*A: Prolate
shape ('cigar')*

*B: Oblate shape
('squashed-sphere')*

Lithium-11,
an example of
a 'halo nucleus'

▲ *Shapes and Sizes of Nuclei*
Nuclei come in a variety of shapes and sizes.

CLASSIFICATION OF NUCLEI

Nuclei are classified according to the number of protons and neutrons they contain. There are around 7,000 or so different possible combinations, most of which are unstable and undergo alpha or beta decay. Each nuclear species is known as a nuclide, of which there are only several hundred stable forms in nature.

Nuclear physicists use the following notation to define nuclides: the nucleus of an element that has chemical symbol X and which contains A protons and neutrons is written as ^{A}X. Thus, the carbon nuclide ^{12}C contains 6 protons and 6 neutrons.

The chemical identity of an atom is determined by the number of electrons orbiting the nucleus, and hence by the number of protons in the nucleus (since the number of negative charges on the electrons must balance the number of positive charges on the protons). The number of protons is known as the atomic number of the nuclide.

Nuclides having the same number of protons but different numbers of neutrons are called isotopes. Every element has several isotopes. Nuclides with the same number of neutrons but different numbers of protons are called isotones. Nuclides with different numbers of protons and neutrons, but the same total number of protons plus neutrons, are called isobars.

neutrons that can then induce fission in neighbouring uranium nuclei. In this way, a self-sustaining chain reaction can take place. If, on average, more than one neutron from each fission process can induce a new fission, the process will quickly escalate, resulting in an explosion a million times more powerful than a chemical explosion involving the same number of atoms. By controlling the number of fission-inducing neutrons, the energy released can be harnessed.

Nuclear fusion is the opposite process, in which two light nuclei are pushed together, causing them to 'fuse' to form a larger, more stable, nucleus. When this happens there is also a great release of energy. It is this process that is responsible for the energy within stars, including the Sun, and is thus the ultimate source of energy that sustains all life on Earth.

▶ *Nucleosynthesis occurs when lighter nuclei fuse to form heavier elements. This process stokes the power of all stars, including the Sun.*

NUCLEOSYNTHESIS

An understanding of the synthesis of the lightest elements in the first few minutes after the Big Bang is considered to be one of the cornerstones of modern **cosmology**. The abundance and distribution of the nuclei of hydrogen, helium and lithium provides a detailed record of conditions in the early Universe. As the Universe cooled down, these light elements began to coalesce under the action of gravity to form stars. The synthesis of nuclei continues within stars as the lighter nuclei fuse to form heavier elements such as carbon and oxygen.

Additional nucleosynthesis occurs in **supernovae**, the violent explosions of the most massive stars.

The products ejected in supernovae spread out into the interstellar medium, where they can form the raw material for a subsequent generation of stars and planets. Indeed, all the heavier elements in our bodies were 'cooked' inside an earlier generation of stars, so that it is true to say that we are all made of stardust.

NUCLEAR POWER

A fission chain reaction can be made to proceed at a controlled rate in a nuclear reactor by ensuring that, on average, only one neutron from each fission process induces a further fission. The remaining neutrons are either absorbed by non-fissionable material or allowed to escape.

Power reactors are designed to extract energy in the form of heat, as the particles and radiation produced in fission are stopped in the reactor material. This heat energy can then be used to generate electricity. Breeder reactors produce additional nuclear fuel as a by-product of the fission, as well as generate electric power. Some countries, such as France, produce more than half their electricity needs from power generated in nuclear reactors.

Issues surrounding the safety of nuclear power have been of understandable concern for the public in many developed and developing nations, in particular those issues involving the processing of radioactive waste products. A number of options are now being considered around the world to solve this problem, including the possibility of transmuting the waste nuclei into harmless products by bombarding them with high-energy neutrons.

Nuclear reactors are also of crucial importance for the production of radioactive nuclear isotopes that are used as tracers in biomedical research. These help to develop diagnostic procedures used in the treatment of diseases such as cancer and AIDS.

▼ *The safety issues surrounding nuclear power stations and the radioactive waste they produce are an ongoing concern for the public; one solution is to use high-energy neutrons to turn the nuclei into harmless waste.*

QUANTUM REALITY

In Search of Quantum Reality
PARTICLE ZOO

By the mid-1930s a handful of elementary particles were known to exist. Apart from the protons, neutrons and electrons that make up atoms, and photons of electromagnetic radiation, physicists had also discovered positrons, the antimatter versions of the electrons. Their theories also predicted the existence of other particles such as the antiproton and the neutrino, both unobserved at that time. Then, in 1935, the Japanese theorist Hideki Yukawa (1907–81) predicted the existence of a new particle, the pion, with a mass somewhere between that of an electron and a proton. Two years later the muon was detected in cosmic rays. Before long the number of elementary particles began to get out of hand. Perhaps they were not elementary after all.

MURRAY GELL-MANN

US physicist Murray Gell-Mann (b. 1929) entered Yale University when only 15 years of age, and was made a full professor at Massachusetts Institute of Technology by the age of 27. Gell-Mann was interested in understanding and classifying the many different elementary particles that were being discovered at the time. He grouped together certain particles that shared common properties and then included in such groups other undiscovered particles which he correctly predicted would be found. Many years earlier, Dmitri Mendeleyev (1834–1907) had predicted the existence of new elements in the same way. The most successful scheme of classification was known as the eightfold way, because it grouped together eight particles that included the proton and neutron. It was this classification that led Gell-Mann and George Zweig to devise the quark model.

Gell-Mann was awarded the Nobel Prize for physics in 1969 for his many contributions to particle physics. More recently, he has made significant contributions to the development of the idea of decoherence in quantum mechanics.

◀ *A Nobel Prize winner for physics in 1969, Murray Gell-Mann has made many contributions crucial to the advancement of quantum mechanics, notably his proposition, with George Zweig, of a new class of particles – quarks.*

COSMIC RAYS

The surface of Earth is illuminated by the light of the Sun and by the far weaker reflected light from the Moon and the light from other stars. But these photons are not the only particles raining down on Earth. **Cosmic rays** also bombard Earth. They consist of many types of particle originating from the Sun and beyond. Many of these collide with air molecules in the upper atmosphere and produce other particles, which arrive at the surface of Earth.

The most abundant particles in cosmic rays are **neutrinos**, but these interact so weakly with normal matter that most of them pass right through Earth without hitting anything at all. Cosmic rays also produce muons, which are like very heavy electrons. Luckily for us, not many of these particles manage to reach Earth's surface.

PARTICLE ACCELERATORS

Scientists study small objects through a microscope by observing the light that bounces off them. However, they are limited to seeing structures that have dimensions larger than a single **wavelength** of the probing light. At the subatomic scale, we need shorter wavelengths (higher frequencies) to resolve the structure. The Planck and de Broglie relations imply a higher-momentum (higher-energy) probe. Since particles such as electrons and protons have waves associated with them, they can be used instead of light to 'see' things. This is the reason particle accelerators (atom smashers) were built. The more deeply an instrument probes the structure of matter and the tinier the objects it is to resolve, the shorter the wavelength and hence the higher the energy of the probe particles that are needed. In order to

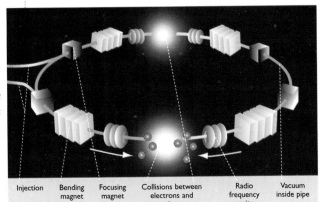

▲ *How a Particle Accelerator Works*
In a particle accelerator ring, beams of particles are guided around the pipe by bending and focusing magnets, and accelerated by passing them through radio-frequency cavities. The particles can be spun around in opposite directions numerous times to increase their speeds before smashing them together in order to study the effects.

| Injection | Bending magnet | Focusing magnet | Collisions between electrons and positrons | Radio frequency cavity | Vacuum inside pipe |

maximize the energy produced, the most powerful accelerators bring two beams of accelerated particles together in a head-on collision.

Another reason for wanting high energy is to enable the creation of new particles from the energy of the colliding particles. This is why the field of particle physics is often referred to as high-energy physics.

The first accelerators were developed in the 1930s to study nuclear reactions. John Cockcroft (1879–1967) and Ernest Walton (1903–95) built a machine that accelerated protons using an electrostatic field. Ernest Lawrence (1901–58) built the first cyclotron, which used a high-frequency alternating electric field to accelerate the protons. Today, the distance scales being probed require huge, and therefore extremely expensive, accelerators. This means that only a limited number of them have been built around the world. Most of the work done in these accelerator laboratories, such as CERN in Geneva, involves large international collaborations.

LIFETIME OF THE MUON

Muons are similar to electrons but are about 200 times heavier. They are unstable and decay into an electron and two neutrinos. In the upper atmosphere many new types of particles, mostly muons, are created by collisions of cosmic rays, and these travel down to the surface of Earth. Physicists have studied the properties of muons and know that they have an extremely short lifetime, just one millionth of a second. This lifetime is only statistical, in that some muons might live for a little longer, some for a little less. But if 1,000 muons are created at once, then after a millionth of a second there will be roughly 500 left.

The muons created in the upper atmosphere are so energetic that they travel towards Earth at an incredible 99% of the speed of light. However, even at this speed it should still take them several muon lifetimes to cover the distance to the surface of Earth (and, more importantly, into the muon detector in the laboratory). Only those few with unusually long lifetimes should be able to complete the journey, but surprisingly, nearly all the muons are able to make it. The reason is special relativity. The muons' time runs more slowly than time in the frame of reference of Earth. So, from the muons' point of view, only a fraction of their lifetime has elapsed by the time they reach the ground.

QUANTUM REALITY

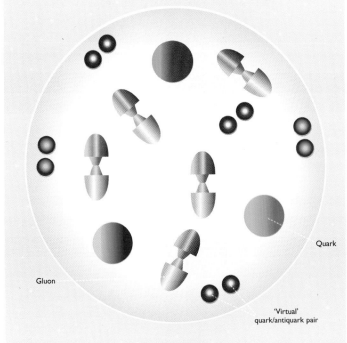

Gluon

Quark

'Virtual' quark/antiquark pair

◄ *Model of a Nucleon with Quarks and Gluons*
A nucleon containing three quarks held together by gluons, which are the carriers of the strong force. This nucleon also contains some 'virtual' quark/antiquark pairs.

CLOUD CHAMBERS AND BUBBLE CHAMBERS

When new particles are produced by a high-energy collision, they must be detected and information about their **mass** and charge must be recorded. Charles Wilson (1869–1959) developed a visual detector by making use of the fact that charged particles can lose some of their energy by **ionizing** atoms in a chamber full of humid air. The **ions** then start the growth of tiny droplets of water and this creates a visible track in the particles' wake. These cloud chambers were later replaced by bubble chambers, in which liquid gas is maintained at a very low temperature. Again, charged particles travelling through the gas will cause ionization and this in turn causes the gas to boil along the particles' paths.

QUARKS

By the second half of the twentieth century, so many new particles had been discovered that physicists began to question whether they were all truly elementary. In the same way that the 92 different kinds of atoms were found to be made up of just three particles (**protons**, **neutrons** and **electrons**), perhaps these particles were in turn built of more fundamental constituents.

In particular, there seemed to be too many different varieties of hadrons, particles that feel the strong nuclear force – these include protons, neutrons and pions. In the early 1960s, in an attempt to restore some economy to the field,

two theorists, Murray Gell-Mann and George Zweig, proposed independently that hadrons are all composed of just a handful of more elementary particles called **quarks**. A few years later experimenters at the Stanford Linear Accelerator Center (SLAC) in California found that high-energy electrons were scattered from protons and neutrons in a way reminiscent of **Rutherford**'s alpha-particle experiments. Now it appeared that tiny compact particles existed inside the hadrons: quarks.

Initially it was thought that there were just three types of quarks, but we now know that there are in fact six different 'flavours', all different in mass. Normal matter (that is, matter composed of atoms containing protons and neutrons) is made up of just two types of quark, known as up and

down. The other flavours are strange, charm, top and bottom. What is more peculiar is the electric charge of quarks. Three have a negative charge equal to one-third of the charge of an electron, and the other three carry a positive charge that is two-thirds that of an electron. Thus two up quarks (charge +2/3) and one down quark (charge -1/3) make up a proton (charge +1), whereas a neutron is composed of two downs and one up.

In addition to electric charge, quarks also carry 'colour charge'. This property is necessary to explain why quarks only occur in threes (inside protons and neutrons), or in pairs (quark/antiquark pairs in pions).

THE GLUONS: STRONG FORCE CARRIERS

Just as photons carry the electromagnetic force and are exchanged when electrically charged particles interact, so particles called gluons carry the strong force and are exchanged among the quarks inside the hadrons. However, owing to the complicated nature of the strong force there are many varieties of gluons, not just one like the photon.

▼ *These **artistically enhanced** particle tracks were captured in the Big European Bubble Chamber at CERN in Geneva.*

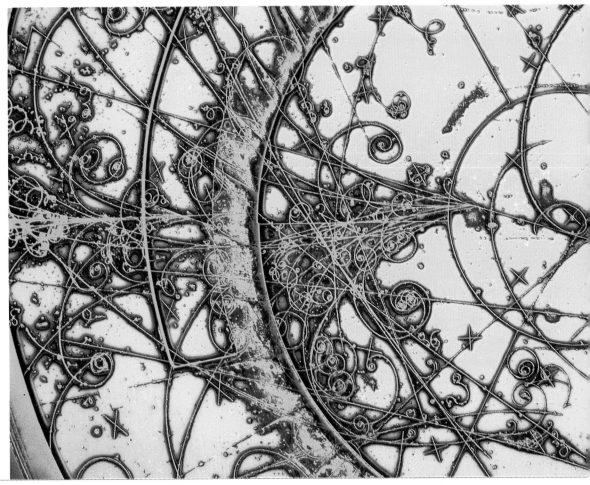

QUANTUM REALITY

QUANTUM REALITY

In Search of Quantum Reality
THE STANDARD MODEL

As well as classifying subatomic particles in terms of their most fundamental building blocks, physicists have tried to understand how these particles interact with each other to make up our physical Universe. To do this, they have needed to understand the origins of the different forces that can act between the particles. There are four such fundamental forces in nature: the gravitational force, the electromagnetic force, and the strong and weak nuclear forces. In their search for simplicity and symmetry in nature, physicists hope to prove that these very different forces actually have a common origin.

QED: THE MOST ACCURATE SCIENTIFIC THEORY

In order to proceed further it is necessary to introduce the idea of a virtual particle. One way of phrasing the **Heisenberg Uncertainty Principle**, which was mentioned in the discussion of quantum tunnelling, is in terms of an energy–time relation. A quantum particle can 'borrow' energy from its surroundings as long as it 'gives it back' within a very short time specified by the uncertainty relation. The more energy the particle borrows, the less time it can keep it. In a similar

way, particles can be created out of nothing, borrowing the necessary energy and quickly disappearing again. These are called virtual particles to distinguish them from real particles that 'own' their energy and **mass**.

In the late 1940s, three physicists independently developed their own versions of a new theory that combined quantum mechanics with **Maxwell**'s theory of electromagnetism. Richard Feynman (1918–88), Julian Schwinger (1918–94) and Sin-itiro Tomonaga (1906–79) developed **quantum electrodynamics** (or QED for short), in which the basic idea is that electrically charged particles such as electrons interact by exchanging a virtual **photon**. In fact, things are very busy in

the quantum world, where a single electron can continually create and re-absorb virtual photons at random. Similarly, a photon can create a virtual **electron–positron** pair, provided they annihilate quickly to form the photon again. QED predicted that empty space was in fact a teeming sea of all sorts of virtual particles appearing and disappearing all the time.

QED has now been experimentally verified as correct to within one part in 10 million, making it the most accurate theory in the whole of science.

GAUGE THEORIES

The mathematical idea of symmetry is very important in physics. A symmetry exists if some property remains the same under a certain transformation. For instance, a sphere looks the same whichever angle it is viewed from, and the age difference between two people remains the same throughout their lives. Physicists speak of a 'global' symmetry when certain relevant properties or laws of physics remain unchanged when a particular transformation is applied equally everywhere.

🪐 THE PARTICLE FAMILIES

This table shows the matter particles, collectively known as fermions. Each row of four particles represents one family or generation. The rows are ordered in increasing mass.

Quarks				Leptons			
charge +2/3		charge -1/3		charge -1		charge 0	
Up	↑	Down	↓	Electron	▼	Electron-neutrino	▽
Charm	✦	Strange	◎	Muon	⬣	Muon-neutrino	☆
Top	⊤	Bottom	⊥	Tau	⬟	Tau-neutrino	○

THE FOUR FORCES AND THEIR FORCE PARTICLES

Each of the four fundamental forces of nature has an associated particle that carries the force. These force-carrying particles are collectively known as bosons.

Force	Force Particle	
Gravity	Graviton	⬇
Electromagnetic	Photon	⟳
Strong nuclear	Gluon	↔
Weak nuclear	Weak gauge bosons	▪▪▪

👨 RICHARD FEYNMAN

The the US theoretical physicist Richard Feynman (1918–88) was one of the most influential and best-known scientists of the twentieth century. He developed the path-integral method of quantum mechanics and invented Feynman diagrams, a schematic and highly powerful method of describing the interactions of subatomic particles. Like many physicists of his generation, he was involved in the development of the atom bomb as part of the Manhattan Project. After World War II he developed, along

◀ **Richard Feynman** was one of the twentieth century's best-known physicists. He shared the 1965 Nobel Prize with Schwinger and Tomonaga for the QED theory.

with Julian Schwinger and Sin-Itiro Tomonaga, the quantum theory of electromagnetism known as QED, for which all three were to share the 1965 Nobel Prize.

Considered by many of his contemporaries as the most original thinker since **Einstein**, he was also well known as a brilliant educator and lecturer. He is famous for his varied and unusual exploits and interests, which included safe-cracking, playing the bongo drums and deciphering hieroglyphics.

In 1986, two years before he died, he sat on a US government commission to look into the cause of the Challenger Space Shuttle disaster earlier that year. During a televised press conference, he dramatically illustrated the flaw in the shuttle design by demonstrating that a ring made of the same material as the seal that failed on the Challenger became brittle when immersed in a glass of icy water.

⧗ THE DIRECTION OF TIME AT A SUBATOMIC LEVEL

One of the many fascinating issues surrounding the nature of time is the fact that we always perceive it to flow in one direction (from past to future) and never in the other direction. However, all equations of physics have the property of not distinguishing between the two directions. Despite this, physicists have found a number of phenomena that happen only one way. The best-known example is described by the concept of entropy, which is roughly equivalent to disorder. For instance, it is overwhelmingly likely that shuffling a pack of cards will make them more disordered and the entropy of the pack increases. If you were to watch a film of a pack of cards being shuffled and ending up with all the suits arranged in order you would know (barring a conjuring trick) that the film was running backwards.

In an experiment carried out at CERN in 1998, a quantum phenomenon was found that also defines a direction to time. It was discovered that it is slightly more likely for antimatter to be converted into matter than the other way round. The experiment suggests that, starting with an equal amount of matter and antimatter in the form of unstable particles known as kaons and antikaons, an imbalance will arise because the antikaons will decay slightly faster. This provides an 'arrow of time' at the subatomic level.

Some theories are more special than this. For instance Maxwell's equations remain the same even when certain transformations are applied 'locally' (that is, differing from one place to another). This means that the electromagnetic force between two particles remains the same even when a change is made to the local electric field strength, since there is an equivalent change in the magnetic field strength. The theory in this case is said to be a gauge theory with local symmetry. When quantum mechanics is introduced we find that QED also has this property, but now it is changes to the wave function that leave the equations unchanged.

ELECTROWEAK THEORY AND THE W AND Z BOSONS

In the 1950s, Chen-Ning Yang (b. 1922) and Robert Mills (b. 1927) further developed the ideas of symmetry by considering what happened to the equations of QED when other symmetries were enforced. Their theory predicted that there would be other 'force-carrying' particles in addition to the photons. Like the photons, these particles did not have any mass, but they did carry electric charge. But did the Yang–Mills theory have anything to do with the real world?

Later on it was found that the weak nuclear force, which is responsible for beta decay, could also be thought of as being due to the exchange of virtual particles. These are known as intermediate vector bosons, and are of three types: W^+, W^- and Z^0. By introducing a theoretical idea called symmetry-breaking, it was discovered that these particles were just the ones predicted by the Yang–Mills theory, but endowed with mass. This meant that under conditions of very high energy the W and Z bosons were closely related to photons. This led to a unified theory of electromagnetic and weak interactions, called the electroweak theory.

QUANTUM CHROMODYNAMICS

The property known as colour charge was introduced to explain why **quarks** arrange themselves in threes inside protons and neutrons. It was linked with colour – which it has nothing to do with in reality – because of an analogy with the way that colours of light combine. The three different colour charges – known as red, blue and green – together form what is known as a colour singlet (something colourless). The theory states that all three-quark systems must be colourless. In the same way, pions, which are made up of a quark and an antiquark, are also colourless, since, for example, red and anti-red will also give something colourless.

In order to put these ideas on a firmer theoretical footing, physicists developed the strong-force equivalent of QED, applied to the strong force rather than electromagnetism, and with colour charge replacing electric charge. This theory is known as QCD, or quantum chromodynamics (from the Greek word *chroma*, meaning 'colour').

One difference between QED and QCD is that in the former there is only one type of electric charge, but in QCD there are eight different ways the quarks can interact via the gluons.

THE SEARCH FOR THE HIGGS PARTICLE

Together, the electroweak theory and quantum chromodynamics explain all known properties of the quantum world and include three of the four forces of nature. This is what particle physicists refer to as the Standard Model. Another particle predicted by the theory is called the Higgs boson, named after Peter Higgs (b. 1929) who introduced the idea. It is thought that this particle is responsible for endowing all others with their mass. In late 2000, the first clear evidence for the existence of this particle was announced at CERN.

▼ *In the particle accelerator at CERN, beams of particles are smashed together at nearly the speed of light to probe the nature of matter and energy. Experiments there in 2000 yielded evidence of the elusive Higgs boson.*

QUANTUM REALITY

In Search of Quantum Reality
THEORIES OF EVERYTHING

While the standard model of particle physics successfully describes all three of the forces that act on the quantum scale, it is far from being the end of the story. It is not a single, unified theory but rather two separate ones: the electroweak theory and QCD. However, their success in unifying the electro-magnetic and weak forces to create electroweak theory spurred physicists on. They are now striving to bring all the forces of nature, including the force of gravity, into a unified superforce described by a single theory.

A GRAND UNIFIED THEORY

Science has come a long way in its search for a single theory that would explain all phenomena in nature. In the nineteenth century, **Faraday** and **Maxwell** showed the forces of electricity and magnetism were really two aspects of the same electromagnetic force. **Einstein** spent the last 30 years of his life searching unsuccessfully for a field theory that would bring together electromag-netism and gravity. When the two nuclear forces were discovered, the job became even harder.

Mathematically, physicists have tried to bring together the electromagnetic, the weak nuclear and the strong nuclear forces into what are known as grand unified theories (or GUTs). The basic idea of unification is to find some underlying symmetry of nature, some property that remains the same while something else is changed. In GUTs, the aim is to put the strongly interacting **quarks** and the weakly

▲ **General relativity** describes the behaviour of matter at very large scales, such as galaxies. Quantum mechanics comes into play at the level of the very small. Together, these two theories hold the key to understanding the Universe.

interacting leptons (the **electrons** and **neutrinos** and their relatives) onto an equal footing. This means that, according to a GUT, a **proton**, composed of quarks, might occasionally decay into a **positron** and a pion, a process involving the transformation of a quark into a lepton. But such a process has yet to be observed experimentally.

Today, GUTs are no longer top of the agenda for theoretical physicists. Most of the effort is now being put into an even more ambitious endeavour: the inclusion of gravity into a theory describing all four forces of nature.

QUANTUM GRAVITY

Despite the shortcomings of the Standard Model, such as the large number of particles it requires, it at least shows that three of the four forces of nature can be described quantum-mechanically. The odd one out is gravity. A good theory of gravity already exists – Einstein's general relativity. But to quantize relativity gravitational attraction needs to be described in terms of the interchange of a quantum of gravity, called a graviton. However, it turns out that this is easier said than done.

EDWARD WITTEN

Ed Witten (b. 1951) is widely regarded as the most brilliant living physicist, and has been called the 'smartest man in the world'. He is one of the pioneers and leading experts in string theory. Another particle theorist, Michio Kaku, has described him as the Picasso of the physics community, setting the trends that others later follow. He works at the Institute for Advanced Study at Princeton, where Einstein spent the last decades of his career.

Witten was a relative latecomer to physics, initially graduating as a history major with an interest in linguistics. But such is his intellect that, once he had chosen to follow a physics career, he became a full professor of physics at Princeton by the age of 28. In 1990, he received the Fields Medal, mathematics' equivalent to the Nobel Prize.

Witten has said that superstring theory and M-theory do not merely combine gravity with the other forces – they *predict* the existence of gravity, in the sense that it falls naturally out of the theory. He believes the progress that is being made in string theory points to a likely revolution in quantum mechanics, suggesting that our current formulation is incomplete.

THEORIES OF EVERYTHING

electron ▶ p. 338   Heisenberg Uncertainty Principle ▶ p. 339   mass ▶ p. 341   neutrino ▶ p. 341   positron ▶ p. 342   proton ▶ p. 342   quark ▶ p. 342

General relativity is normally applied to the world of the very large. It describes the way space–time affects, and is affected by, the presence of **mass** and energy, which causes it to curve. However, this effect is usually appreciable only with very large masses, such as planets, stars, galaxies, even the whole Universe. Quantum mechanics is applied to the realm of the very small, where gravity is usually too weak to be noticed. But on the quantum scale the **Heisenberg Uncertainty Principle** rules. Over very small distances and time intervals, virtual particles of very high masses and energies can jump into existence and then disappear. They can briefly generate intense gravitational fields and strong distortions of space and time. But this view is in conflict with general relativity's assumption of a smooth, geometric space–time. Clearly, a theory of quantum gravity is not going to be easy to find.

SUPERSYMMETRY

In the early 1970s, theorists discovered a new type of mathematical symmetry, called supersymmetry, that has to do with quantum-mechanical spin. Grand unified theories of the strong and electroweak forces imply a connection, or symmetry, between quarks and leptons, the two types of matter particles known collectively as fermions. When the ideas of supersymmetry were applied to the Standard Model, it was found that an even grander connection could be made, this time between fermions and bosons, the force particles. This beautiful unification was seen as a tremendous theoretical success, but was nature really supersymmetric?

Supersymmetry predicts that fermions and bosons should be paired with each other, but the problem is that none of the known particles pair up. This means that all the supersymmetric partners have yet to be discovered.

✳ M-THEORY

A few years after its initial explosion on to the theoretical physics scene, superstring theory seemed to have stagnated. By the early 1990s there were a number of different versions of string theory and no one knew which was the correct one. Then, in 1995, US physicist Ed Witten proposed a solution to this embarrassment of riches. Together with a Cambridge physicist, Paul Townsend, he suggested that the fundamental entities are membranes rather than strings. This is now known as membrane

💡 HIGHER DIMENSIONS

In 1919, a Polish physicist, Theodor Kaluza (1885–1945), proposed a mathematically elegant way of incorporating electromagnetism into the framework of general relativity. The trick was to propose that we lived in five-dimensional space–time, rather than the usual four (three of space and one of time). In Kaluza's theory, electromagnetic waves became vibrations of this hidden fifth dimension. However, there has never been any evidence to suggest that it is a correct description of reality.

▼ *Curled-Up Dimensions*
Two extra dimensions curled up in the shape of a sphere. For clarity they are shown at regularly-spaced intervals, but in reality they would exist at every point in space.

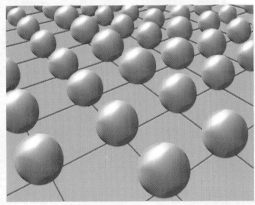

SUPERSTRING THEORY

By the end of the twentieth century, there were a number of candidates for a theory of quantum gravity, an all-encompassing description of reality that would unify all the forces of nature under one mathematical framework. The theory that has received the most attention is known as superstring theory (or string theory for short).

According to string theory, elementary particles are not really point-like objects. Instead, at a scale far more minute than anything we can probe with present technology, these particles are really vibrating strings, like tiny rubber bands.

theory, or M-theory for short. (Many physicists like to think of the M as standing for Magic, Mystery or Mother, since this really would be the mother of all theories.)

Witten was able to unify the different versions of string theory into M-theory by requiring yet another dimension. Thus, M-theory is an 11-dimensional theory. It also has the advantage over the older superstring theories of allowing a more natural merging of gravity with the other three forces.

Theories of higher dimensions made a comeback in the 1970s when physicists began their search in earnest for grand unification. Indeed, the current leading candidate for a 'theory of everything', superstrings, suggests that we live in a 10-dimensional universe, with one dimension of time and nine spatial dimensions. The reason we do not see evidence for this is that the six extra dimensions of space are curled up on themselves so tightly that they cannot be observed. The fundamental strings are not really one-dimensional 'lines', but multi-dimensional objects.

▼ *The Ant and the Garden Hose*
This illustration suggests how curled-up dimensions might be revealed if we were able to zoom in to an incredibly small scale. A garden hose viewed from a distance looks like a one-dimensional line. However, when magnified, it is revealed that the object is in fact two-dimensional. An ant on the hose could not only go along the hose's extended dimension, it could also go around its curled-up dimension.

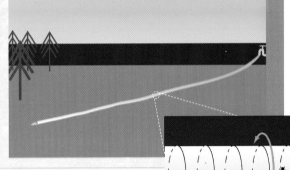

It turns out that with this change, the force of gravity can be unified with the other three forces. In other words, string theory shows a way of resolving the incompatibility problem between quantum mechanics and general relativity.

The original version of string theory was discovered in 1968 by Gabriele Veneziano (b. 1942). However, his theory suffered from a number of problems, such as its prediction of faster-than-light particles (called tachyons), something physicists could not accept. Then, with the discovery of supersymmetry, a new version of string theory emerged in the mid-1980s, which did not suffer from the problems of Veneziano's theory.

Physicists are still a long way from being able to say whether string theory is correct. It is so far removed from what can be tested in the laboratory that it will probably take many more years to confirm or refute. In addition, the mathematics of string theory is so complex that very few people in the world are able to understand its equations. Indeed, much of the mathematics required to unravel its secrets has yet to be developed.

QUANTUM REALITY

In Search of Quantum Reality
SUMMARY

More than any other area of science, quantum mechanics has baffled the greatest minds of the twentieth century. But no one has any real doubt that it is fundamentally correct. It may be lacking some ingredients that would allow it to merge more naturally with general relativity and many physicists also hope that one day we can choose among its varied interpretations. But quantum mechanics has made too many accurate predictions, and led to too many important discoveries and applications, to be very far off the mark.

It might come as a surprise that quantum mechanics should play such an important role in astronomy. After all, its domain is the microscopic world, not the great expanses of space. However, much of the field of astrophysics involves the study of atomic, nuclear and subnuclear processes in stars and the interstellar medium. This is the playground of quantum mechanics. Furthermore, without an understanding of the quantization of the electromagnetic spectrum, astronomers would not have made progress in understanding the composition of stars and galaxies. The light that reaches Earth from space can be analyzed and understood only by understanding the nature of line spectra, which in turn make sense only in terms of quantum mechanics.

CLASSICAL PHYSICS

- Classical physics includes all physics that is not quantum mechanics.
- While sometimes taken to mean the physics of **Newton**, its crowning achievements are actually **Maxwell**'s electromagnetism and **Einstein**'s general theory of relativity.
- An area of current interest is the borderline at which the classical world stops and the quantum domain starts.
- By the late nineteenth century cracks had begun to appear in the classical theory of **radiation**.
- Certain phenomena, such as thermal radiation from **black bodies** and the photoelectric effect, seemed to suggest that light is fundamentally composed of particles rather than waves.

PLANCK'S POSTULATES

- In 1900, **Planck** suggested that the thermal radiation emitted by the atoms of a black body was emitted in packets, or 'quanta'.
- He postulated that the energy of a quantum of radiation was proportional to its frequency.
- His ideas marked the birth of quantum theory.

◀ *At the beginning of the twentieth century, Max Planck suggested that thermal radiation, as emitted by human beings, was actually emitted in 'quanta'. This marked the start of quantum mechanics.*

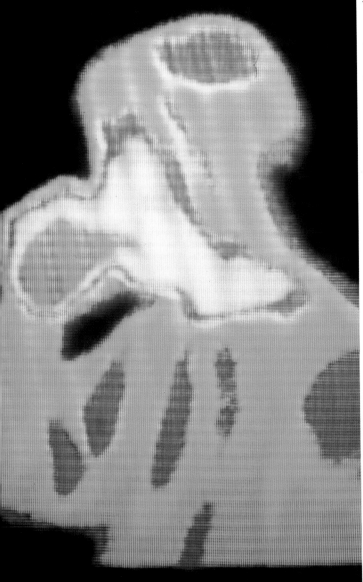

▶ *Werner Heisenberg, whose Uncertainty Principle has become one of the essential theories to understanding quantum physics.*

WAVES AND PARTICLES

- In 1905, Einstein followed the lead of Planck and successfully explained the photoelectric effect in terms of particles of light.
- This was later confirmed by Compton, who needed a particle description of light to explain its scattering from **electrons**.
- In 1924, de Broglie proposed that if electro-magnetic waves can sometimes behave as particles, then moving particles of matter could also sometimes behave as though they were waves, with **wavelengths** that depended on their momentum.

THE BOHR MODEL OF THE ATOM

- **Rutherford**'s alpha-scattering experiments had shown that atoms consisted of tiny dense nuclei surrounded by even tinier electrons.
- **Bohr** provided the first quantum model of the atom in 1913.
- Although not the modern view of the atom, his model successfully explained the stability of the electron orbits.

PROBABILISTIC NATURE OF QUANTUM MECHANICS

- Quantum mechanics states that we are able to make predictions only about the probable behaviour of particles such as electrons.
- We can only ever gain statistical information from the collective behaviour of many quantum systems.
- This is not a shortcoming but rather the way nature behaves at the subatomic level.

THE UNCERTAINTY PRINCIPLE

- The **Uncertainty Principle**, one of the cornerstones of quantum mechanics, was proposed by Werner Heisenberg in 1927 and states that it is impossible to have exact simultaneous determinations of both the position and momentum of a quantum particle. The more accurately one is determined, the less accurately the other can be known.
- The same relationship holds between other pairs of quantities, such as energy and time.

ANTIMATTER

◉ Every elementary particle has a corresponding antiparticle, which has the same **mass** but opposite electric charge, as well as opposite values of other quantum properties.

◉ Antiparticles were initially predicted by **Dirac** as a result of solving an equation and were discovered later, in the 1930s.

◉ When a particle and its corresponding antiparticle meet they are destroyed in a burst of energy, a process known as pair annihilation.

THE COPENHAGEN INTERPRETATION

◉ This is the standard way of explaining the strange results of quantum mechanics.

◉ It was devised and honed by physicists at Bohr's institute in Copenhagen in the mid-1920s and states that although quantum mechanics tells us what to expect when we make measurements, it cannot provide a picture in terms of particles or waves, or what is happening between measurements.

QUANTUM TUNNELLING

◉ Also known as barrier penetration, quantum tunnelling is one of the more bizarre consequences of quantum uncertainty and wave–particle duality.

◉ It involves quantum particles 'tunnelling' through potential-energy barriers that classical particles are forbidden to penetrate.

◉ Without this process, nuclei would not be able to decay by ejecting alpha-particles.

THE TWO-SLIT EXPERIMENT

◉ This is regarded as the classic experiment that highlights the strange features of quantum mechanics.

◉ It shows that individual particles, such as **photons**, electrons and even atoms, have wavelike properties that give an interference pattern on a screen after passing through two narrow slits, even when they are sent through one at a time.

THE ATOMIC NUCLEUS

◉ The positively charged core of an atom, consists of **protons** and **neutrons** packed together very densely.

◉ There are over 7,000 different nuclear species, of which only a few hundred are stable.

◉ The nucleus has a typical diameter of less than 10^{-12} cm.

💡 QUANTUM MECHANICS IN COSMOLOGY

While particle physicists attempt to unify the forces of nature by trying to find a quantum-mechanical theory of gravity, cosmologists now realize that quantum mechanics is also needed to explain some of the ultimate questions about the cosmos.

In the relatively new field of quantum cosmology, physicists such as **Stephen Hawking** (b. 1942) try to provide a quantum-mechanical description of the birth and origin of our Universe. It may seem a contradiction to try and describe the vastness of the Universe with a theory of the microscopic world, but immediately after the Big Bang the Universe was less than the size of an atom and therefore has to be described quantum-mechanically. In fact, the **Uncertainty Principle** would help to remove the problem of singularities (space–time points of zero size), which otherwise occur at the Big Bang itself as well as at the centre of black holes.

◉ The two nuclear processes whereby nuclei can give up their energy are fission and fusion.

◉ In fission a large, unstable nucleus splits into two roughly equal parts and liberates energy, along with several neutrons. These neutrons may initiate further fission.

◉ In fusion two light nuclei are brought together to form a heavier, more stable nucleus, again with the release of energy.

THE PAULI EXCLUSION PRINCIPLE

◉ The rule that states that electrons cannot all occupy the same lowest-energy orbit in atoms but must stack up in successively higher orbits according to their energy and angular momentum.

◉ This phenomenon gives matter its rigidity and explains why a relatively small star such as the Sun will not be able to collapse into a black hole when it runs out of its nuclear fuel.

QUARKS AND GLUONS

◉ Quarks and gluons are are major class of the building blocks of matter, which reside inside atomic nuclei; they are the constituents of hadrons, such as protons and neutrons.

◉ There are six flavours of quarks and each can carry one of three 'colour charges'.

◉ The gluons are the carriers of the strong force and hold the quarks together.

THE STANDARD MODEL

◉ Currently, the Standard Model is the most comprehensive theory of all matter on the quantum scale.

◉ It encompasses both the electroweak theory and quantum chromodynamics, but is not a unified field theory of the strong and electroweak forces, and cannot, therefore, be the last word.

◉ It predicts the existence of the Higgs boson.

SUPERSTRING THEORY

◉ The front-runner for a 'theory of everything'.

◉ It unifies all the forces of nature, including gravity, by suggesting that elementary particles are not point-like but are like tiny vibrating strings.

◉ String theory is highly mathematical and is not yet supported by experimental evidence.

▼ **The next big step** in quantum mechanics is using it to provide a unified theory for the existence of the Universe.

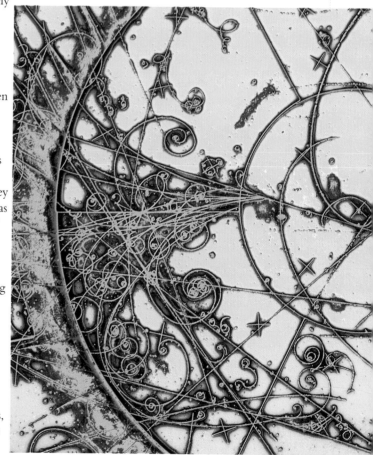

QUANTUM REALITY

THE UNIVERSE:
PAST, PRESENT & FUTURE

This chapter deals with cosmology, the study of the large-scale properties of the Universe. All the early civilizations – those, at least, which had a model of Earth and that part of the Universe which could be observed with the naked eye – believed in an ordered system. Some believed in the control of natural spirits, others by a pantheon of gods. Early cosmologies were geocentric, a system that fitted both observations – the Sun, Moon and stars do appear to circle Earth – and man's understanding of his place in the Universe. This system, usually called the Ptolemaic model after the second-century AD Greek astronomer Ptolemy, was maintained as an article of faith by astronomers for almost 1,500 years. To disagree with the doctrinal view was scientific as much as religious heresy. Ptolemy's geocentric view was finally challenged by the heliocentric model of the Polish astronomer Nicolaus Copernicus, published in 1543. Rapidly adopted by astronomers, the heliocentric model was put on a firm observational and theoretical footing by first Galileo and then Johannes Kepler, while Newton characterized a Universe understandable to man with laws that governed both earthly and heavenly objects – but Einstein showed that the Universe beyond Earth is a much stranger place. Twentieth-century observers such as Harlow Shapley and Edwin Hubble completed the revolution begun by Copernicus by showing that even the Sun had no special place in the Universe, nor does the Universe have any centre at all.

The Universe: Past, Present & Future
INTRODUCTION

 In Isaac Newton's (1642–1727) view the Universe was infinite in extent, the bodies within it moving under the influence of the mutual attraction of gravity. The bodies moved according to Newton's laws of motion – laws which assumed time and space to be absolute. Newtonian cosmology lasted for 250 years until Albert Einstein (1879–1955) changed our understanding of space and time, showing that they are relative, and that gravity is caused by distortions of space–time rather than an attractive force in the conventional sense. Modern cosmologies are built around the framework of general relativity, but incorporate the ideas of quantum mechanics, as an understanding of the very small is needed to explain the very large.

HOW BIG IS THE UNIVERSE?

Nicolaus Copernicus (1473–1543) did not need a telescope to understand that the Sun was the centre of the **Solar System**. The later development of the telescope showed that the Sun was just one of many stars. Earth was not the centre of the Universe, and neither was the Sun, nor even the **Milky Way**. It became apparent that our Galaxy was one of many galaxies: each time a more powerful telescope was built yet more galaxies were seen. The problem for astronomers was measuring the distances to these far-off galaxies. Within the Milky Way, **Cepheid variable** stars are a useful 'standard candle' for measuring

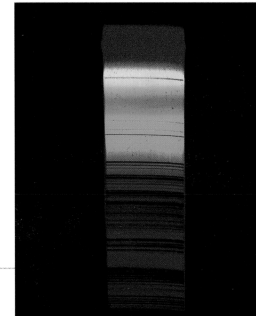

distances, but Cepheids are visible only in the nearby galaxies. It was during investigations of alternative methods of measuring distances to these galaxies that **Edwin Hubble** (1889–1953) proposed that since the fainter galaxies (which he assumed were farther from Earth) had the larger **redshifts** (indicating that they were travelling faster from Earth), then the velocity at which they were receding was proportional to their distance. The implication of this simple statement was profound: the Universe was expanding. Current cosmological models suggest that the Universe is not just big, but that it may be infinite.

HOW OLD IS THE UNIVERSE?

Hubble's discovery of an expanding Universe had a second, equally profound, implication. If the galaxies were receding then they must, at some time, have been concentrated into a small region of space. This means that the Universe must have a finite age. **Hubble's law** can be expressed simply; it relates the distance of a galaxy to its velocity of recession, via a value known as **Hubble's constant**. If the Universe has been expanding at a constant rate, then Hubble's constant is a direct measure of the age of the Universe. The problem is that measuring Hubble's constant is not straightforward, and that cosmological models allow for a variation of the expansion rate of the Universe with time. Nevertheless, it is possible to calculate the age: present best estimates put it at about 13 billion years.

A fascinating effect of the Universe being of finite age is that it solves a problem which had exercised scientists for many years (though it is not a problem which even occurs to most people): why is the sky dark at night? The intuitive answer to this – that the Sun has set – is incorrect, as we shall see later in this chapter.

HOW DID THE UNIVERSE BEGIN?

In 1965 the American physicists Arno Penzias (b. 1933) and Robert Wilson (b. 1936) made the serendipitous discovery that the Universe was not entirely cold, but was suffused with a warmth some three degrees above absolute zero – this is known as the **cosmic background radiation**. Following Hubble's proposal that the Universe

◀ The absorption spectra from objects such as the Sun and galaxies reveal details about their properties. The dark lines which appear on a galaxy's observable light spectrum are the basis for investigating the speed at which that particular galaxy is travelling away from us. The position of the lines changes with velocity. This phenomenon is called redshift, as the lines are shifted towards the redder end of the spectrum. The larger the redshift the faster the galaxy is moving away.

had expanded from a small region of space, it had already been suggested that because this initial phase of the Universe would have been very hot, the **radiation** from it might still be detectable, but redshifted to a much lower temperature. The discovery of the cosmic background radiation was confirmation that this suggestion was true, leading many cosmologists to argue that the observable Universe had begun as a point of infinite density which had exploded, a model which became known as the Big Bang.

COSMOLOGICAL MODELS

Early models of the Universe had assumed it to be static and infinite, although Newton had recognized that such a universe could not be stable in the presence of gravity. Einstein's theory of general relativity changed our understanding of the nature of space and time and of the effects of gravity, but early models based on the new theory also assumed a static, infinite Universe, a representation which presented a similar problem to that encountered by Newton. The discovery of the expansion of the Universe meant that the need for a static model was redundant. Yet despite the evidence for a single point of origin, cosmologists did not immediately adopt the Big Bang as their model. The Steady State theory, a model which involved the continuous creation of matter to account for the assumed large-scale homogeneity

---

⧗ TIMING THE EVOLUTION OF THE UNIVERSE

The study of light from stars and galaxies allows us to look back to a time when the Universe was about 1,000 million years old, i.e. less than 10% of its present age. Although it would be an over-statement to say that the evolution of the Universe from that time to the present is understood, theories can be checked by observation. In the period between the Universe being about 300,000 years old and 1,000 million years old our understanding of its evolution is uncertain. But moving backwards in time from 300,000 years to the dawn, other types of observation become available and it is much more clearly understood. Some epochs of the evolution can be timed to within a few minutes, some even to a few seconds or fractions of a second, our knowledge of the interactions of matter and radiation being well understood from experiments on Earth. It is an irony that although the Universe's age has yet to be determined to better than a few billion years, we can be very precise indeed about the first few minutes following the Big Bang.

PAST, PRESENT & FUTURE

of the Universe, had many virtues – and many proponents. Not until the discovery of the cosmic background radiation was the theory finally abandoned by mainstream cosmologists in favour of the theory of the Big Bang. The evolution of ideas on the nature of the Universe are explored in this chapter.

THE BIG BANG AND THE EVOLUTION OF THE UNIVERSE

Models of the Big Bang marry the two fundamental ideas of the physics of the twentieth century, quantum mechanics and general relativity. This unlikely juxtaposition is required because the huge density of the primordial fireball makes consideration of gravity inevitable, while on the other hand the constituents of the fireball are the fundamental particles of matter and radiation, the temperature of the fireball being too high for atoms to form. The lack of a satisfactory theory combining the two means that the very earliest stages remain poorly understood, but cosmologists have been able to make precise predictions from within a second of the Big Bang onwards.

Present observations of the Universe show it to be littered with galaxies, often in clusters or superclusters. How such structures evolved, and why they are formed of matter rather than an equal mixture of antimatter and matter, are two crucial questions for our understanding of the Universe.

▲ *Galactic encounters such as this one, photographed by the Hubble Space Telescope, can help astronomers understand the history of galaxies. Locked in a gravitational embrace, NGC 2207 (left) and IC 2163 will merge into a single, more massive galaxy billions of years from now. Many present-day galaxies, including our own, are thought to have been assembled from such a process of coalescence.*

HOW WILL THE UNIVERSE END?

Once the question of how the Universe began has been addressed, the other fundamental question of how it will end must also be considered. Will the Universe end and, if so, by what means? These questions form the final part of this chapter.

▼ *The study of subatomic particles in particle accelerators has helped scientists understand the processes which occurred shortly after the Big Bang.*

✦ TYPES OF REDSHIFT

Fundamental to an understanding of the effects of gravity and the expansion of the Universe is an understanding of redshift, the change in the **wavelength** of light as a result of cosmological processes. Redshift comes in different types. The one with which people on Earth are most familiar is the **Doppler effect**, produced when a source moves relative to an observer. This effect causes the change in pitch of a siren as a vehicle passes a stationary listener. The version of interest to astronomers applies to light rather than sound, and the Doppler effect can create both blue- and redshifts depending on whether the source is approaching or receding.

Gravitational redshift is caused by the effect of gravity on light. Gravity reduces the energy of **photons**, a loss of energy being the equivalent of an increase in wavelength and, thus, a redshift. Gravitational redshift can be considered as the slowing of time in a strong gravitational field.

In cosmology, the redshift of light from distant galaxies is caused by the expansion of the Universe itself, which stretches the wavelength of light as it passes through space.

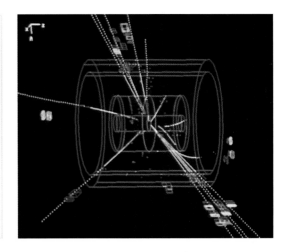

Copernicus ▶ p. 36 Einstein ▶ p. 88 Hubble ▶ p. 122 Newton ▶ p. 42

INTRODUCTION

The Universe: Past, Present & Future

THE EXPANDING UNIVERSE

THE SIZE OF THE UNIVERSE

It is likely that from the first time humans gazed up into the night sky they have wondered just how big and old the heavens were. At first, such questions were almost certainly overshadowed by religious interpretations, but with the Age of Reason came serious consideration of the question of how big the Universe was. These questions are explored below. We start by considering the size of the Universe and the startling discovery that the Universe is expanding.

EDWIN POWELL HUBBLE

The American astronomer Edwin Powell Hubble (1889–1953) was born in Marshfield, Missouri. He studied law at the University of Chicago and was a Rhodes Scholar at the University of Oxford from 1910 to 1913. Back in the United States he decided against a career in law and took up astronomy. After military service during World War I he took a post at the Mount Wilson Observatory, where he remained until his death in 1953. Using Cepheid variable stars Hubble determined the distance to the Andromeda galaxy in 1924. From his study of galaxies he proposed an evolutionary sequence beginning with near-circular **elliptical galaxies** and moving through increasing ellipticity to reach a basic form of **spiral galaxy**. The sequence, he believed, continued as two parallel forms of spiral, forming the famous **tuning fork diagram**. Although the evolutionary part of this classification is no longer believed to be correct, Hubble's basic forms are still used to define galactic types. In 1929 Hubble proposed what is now known as **Hubble's law**, implying that we live in an expanding Universe.

NEBULAE

Astronomers have traditionally used the term **nebula** (Latin for 'mist') to refer to any diffuse patch of light in the night sky. Many nebulae can be seen with the simplest telescopes – and some with the naked eye – and as early as 1781 Charles Messier (1730–1817) had drawn up a catalogue of about 100 nebulous-looking objects. With improved telescopes many of the nebulae proved to be clusters of stars, but others remained enigmatic. For many years a Great Debate raged over the nature of these misty objects. Some astronomers argued that they were clouds of gas between the stars of the **Milky Way**, while others maintained that they were huge star systems far outside our Galaxy. Interestingly, both views are now known to be correct. Some objects, such as the Orion Nebula, are indeed gas clouds within the Milky Way (and in modern usage a 'nebula' is such a cloud), but others are distant galaxies, far removed from our own.

THE ANDROMEDA GALAXY

The first galaxy shown to lie definitely outside the Milky Way was the spiral galaxy in the constellation of Andromeda (M31 in the **Messier catalogue**). In 1924, using the 100-in telescope at Mount Wilson in California, Edwin Hubble took photographs of **Cepheid variable** stars in M31. Measurements of Cepheids within the Milky Way already existed, but the stars Hubble observed were much fainter, indicating that the Cepheids in M31 lay much farther away. About 20 years earlier US astronomer Harlow Shapley (1885–1972) had estimated the size of the Milky Way from observations of globular star clusters. Hubble's images of the Cepheids in the Andromeda galaxy indicated that its distance was 10 times greater than the diameter of our Galaxy – M31 must be extragalactic. Further observations showed that Andromeda's apparent size and brightness meant that, allowing for its distance, it must be a stellar system larger than the Milky Way.

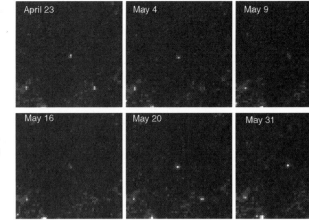

◀ *Edwin Hubble determined that the Universe is composed of a multitude of galaxies of stars, and that they are moving apart from each other at a speed that increases with their separation. This was the first observational evidence that the Universe is expanding. The Hubble Space Telescope is named in honour of his contributions to cosmology.*

April 23   May 4   May 9   May 16   May 20   May 31

▲ *Astronomers use Cepheid variable stars as reference points for measuring distances in the Universe. Here, the rise and fall in brightness of a Cepheid variable over a period of five weeks is recorded by the Hubble Space Telescope. This Cepheid lies in M100, a spiral galaxy in the Virgo cluster, 55 million light years away. Observations such as this have allowed astronomers to determine the value of Hubble's constant.*

REDSHIFTS AND BLUESHIFTS

The nineteenth-century discovery of emission and absorption lines in the spectra of gases revolutionized observational astronomy, allowing astronomers to determine for the first time the chemical composition of the gas emitting the light. But the **spectral lines** also had another use. **Doppler shifts** in the expected **wavelength** of the lines allowed astronomers to measure the velocity of the source along the line of sight (the so-called **radial velocity**). Observations of the spectral lines of stars in the Milky Way showed that they were moving randomly, with radial velocities of up to a few hundred kilometres per second.

Astronomers use the symbol z to designate the change in wavelength, defined as:

$$z = \frac{\lambda_1}{\lambda_0} - 1$$

where λ_0 is the emitted wavelength of the line and λ_1 is the observed wavelength. The shift z is called a **redshift** when the object is moving away and a **blueshift** (or a negative redshift) when the object is moving towards the observer. For small values of z (less than about 0.1) the radial velocity can be found simply by multiplying z by the speed of light (300,000 km/s). At higher values of z the equations of special relativity must be used.

REDSHIFTS OF GALAXIES

In 1913 the American astronomer Vesto Slipher (1875–1969), working at the Lowell Observatory in Arizona, measured the shift in the spectral lines of the Andromeda galaxy. He showed that it was approaching Earth at about 300 km/s (190 mi/s), similar to the measured speeds of nearby stars and so consistent with M31 being within the Milky Way. By the early 1920s he had extended the work to another 40 nebulae. Slipher discovered that 36 of them had redshifts, but only four had blueshifts. He also discovered that most of the redshifts indicated recession velocities of many thousands of kilometres per second. Both results were surprising. If, like the stars of the Milky Way, the galaxies were moving at random, there should be roughly equal numbers of blueshifts and redshifts. Moreover, the velocities were much higher than those of stars in the Milky Way.

▼ *Hubble's Law*
According to Hubble's law, the velocity at which a distant galaxy is receding is proportional to its distance. Galaxy B is twice as far away as galaxy A, so its recessional velocity is proportionally greater.

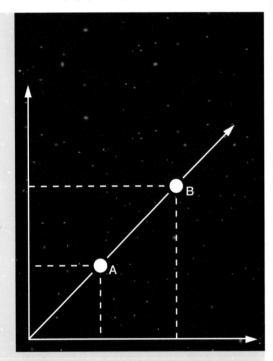

ocity of cession

Distance

CEPHEIDS AS DISTANCE INDICATORS

Galaxies can now be detected at distances ranging from a few million **light years** to many billions of light years, but no single technique exists to measure distances over the entire range. One of the best uses Cepheid variable stars. Henrietta Leavitt (1868–1921), an American astronomer who was head of photographic photometry at the University of Harvard, had been studying Cepheids in the Small Magellanic Cloud (SMC) – a small galaxy which orbits the Milky Way and forms part of the Local Group. In 1912 she found that the greater the average brightness of the Cepheids the longer was their period of variation. Assuming, correctly, that the Cepheids of the

SMC were at roughly the same distance, she concluded that the average luminosity of the stars was related to their period. Shortly after, Harlow Shapley recognized the significance of this **period–luminosity relationship** as a means of calculating distances. Measurement of the period allows the true luminosity to be calculated, which is then compared to the observed brightness to give the distance to the star.

Although it is a powerful tool, the Cepheid method can be used only for relatively nearby galaxies. In galaxies farther than about 150 million light years away the images of individual stars become too indistinct, even with today's telescopes, while at greater distances individual stars cannot be detected at all.

HUBBLE'S LAW

Hubble noticed that although Slipher's data showed that a few bright galaxies had blueshifts, the fainter galaxies had only redshifts. He also noticed that galaxies with the largest redshifts tended to have the smallest images, and so were presumably farther away. Using this limited data, Hubble made the bold proposal that the galaxies were receding from us at a rate that was proportional to their distance. This seemingly straightforward relationship has a profound implication – at some finite time in the past the whole observable Universe must have been concentrated into a very small region, i.e. the Universe had a beginning.

Hubble's law states that the recessional velocity, v_r, of any distant galaxy is proportional to its distance, d, and is normally expressed as a simple equation:

$$v_r = H_0 d$$

The value H_0 is known as **Hubble's constant**.

HUBBLE'S CONSTANT

One of the challenges in observational cosmology is to measure an accurate value for Hubble's constant. When the Hubble Space Telescope was in the planning phase a number of key projects were identified in which the telescope would make major advances in astronomy. One of these was to measure the distance to the nearest large cluster of galaxies – the Virgo cluster – and so provide an accurate value of Hubble's constant. The improvement in resolution over ground-based telescopes would allow Cepheid variable stars in the Virgo galaxies to be measured for the first time and so give a reliable distance. Observations by several groups of astronomers have now resulted in a distance to the Virgo cluster of 55 million light years. This accurate distance, together with the known redshift of the galaxies, gives a value of Hubble's constant of $H_0 = 65$ km/s/Mpc, with an uncertainty of 10%. This means that the recession speed of galaxies increases by 65 km/s (40 mi/s) for every million **parsecs** distance from Earth.

▼ *The Expanding Universe*
In the 'expanding balloon' model of the Universe, each galaxy recedes from every other one and no galaxy can be said to be central. In the time it has taken the balloon to expand from size 1 to size 2, galaxies A, B and C have all receded from one another with speeds proportional to their distances.

THE HUBBLE FLOW

The recession of galaxies due to the expansion of space is sometimes called the Hubble flow. Superimposed on this flow are random motions of the galaxies caused by the gravitational effects of local concentrations of matter – clusters and superclusters of galaxies. These random velocities are referred to as 'peculiar' velocities and are typically 300 km/s (190 mi/s) for galaxies in small clusters but can be 10 times higher for superclusters. Our own Galaxy is known to be moving at some 600 km/s (380 mi/s) with respect to the Hubble flow.

In general it is not possible to tell how much of a galaxy's redshift is due to the Hubble flow and how much is due to its peculiar velocity. To measure the Hubble constant successfully the Hubble flow must be much greater than the likely peculiar velocity of the galaxy. This requires the galaxy to be at a redshift greater than 0.001 (for small clusters) or 0.01 (for superclusters). In practice this means that measurements of Hubble's constant must be made on galaxies at distances of about 100 million light years or more.

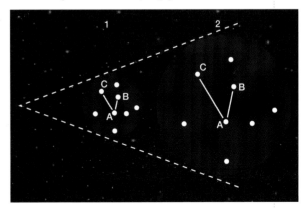

THE EXPANDING UNIVERSE

PAST, PRESENT & FUTURE

123

THE AGE OF THE UNIVERSE

How old is the Universe? Hubble's discovery of the recession of the galaxies suggests a simple way to find out – measure the speeds at which they are flying apart and work out how long ago they must all have been in the same place. The answer, known as the Hubble time, is directly related to Hubble's constant and with some adjustments for the changing speed of the expansion is the cosmologists' best estimate of the age of the Universe. But first we consider an apparently unrelated question – why is the sky dark at night?

OLBERS' PARADOX

On a moonless night the sky is dark. This apparently obvious observation is, in fact, a profound one, the problem of the darkness of the night sky having exercised astronomers for several hundred years. The reason appears to be obvious – the Sun is below the horizon. But what about the light from all the other stars? Although stars are far away (the apparent answer to the problem) there are a lot of them. If we draw a line of sight outwards from Earth then in an infinite Universe every such line

▼ *Olbers' Paradox*
The resolution to Olbers' paradox lies in the fact that light from the most distant stars cannot have travelled for longer than the age of the Universe – between 13 and 15 billion years – and that light from an infinite number of stars could not have reached Earth.

▶ *For centuries* astronomers have been puzzled by the following problem: why does the night sky appear so dark when it is peppered with so many stars – and beyond them, we now know, countless galaxies? This conundrum is referred to as Olbers' paradox.

would eventually reach a star. Since stars are similar to the Sun the whole night sky would, therefore, appear as bright as the surface of the Sun.

The problem goes deeper. Given enough time the Universe would heat up to the surface temperature of the Sun. But the night sky is dark and the Universe is not at the surface temperature of the Sun. This problem is referred to as Olbers' paradox after the German astronomer Heinrich Olbers (1758–1840). Although it is popularly believed that Olbers first enunciated the night sky problem, it actually pre-dates him by several centuries. The English scientists Thomas Dygges (in the sixteenth century) and Edmond Halley (1656–1742), and German astronomer **Johannes Kepler** (1571–1630) all recognized the problem.

WHY THE SKY IS DARK AT NIGHT

Olbers thought that the explanation for the dark night sky was that light from distant stars was absorbed by interstellar material such as dust clouds. But this explanation, while appearing plausible, does not hold up. Given enough time, the absorbing material would itself heat up and become luminous.

To understand the correct solution of Olbers' paradox we need to introduce the concept of the *observable Universe*. Because the speed of light is finite, we can see no farther than light can travel in the age of the Universe. Whether or not the Universe as a whole is infinite, the *observable* Universe is always finite and bounded by a sphere whose radius is approximately the age of the

Universe multiplied by the speed of light. Stars beyond that distance cannot be seen because their light has not had time to reach us. When astronomers speak of the 'edge' of the Universe they mean the edge of the observable Universe – a source of much public confusion.

So most lines of sight from Earth reach the edge of the observable Universe before they

 RADIOACTIVE DATING
Radioactivity is the spontaneous disintegration of an **atomic nucleus**. Certain heavy elements are naturally radioactive and the decay of one of these nuclei (termed the 'parent') creates a nucleus of a different element (the 'daughter'). Often the daughter nucleus is itself radioactive, but occasionally the daughter is stable and serves as a tracer of the now-decayed parent nucleus. Radioactivity provides us with clocks which can be used over timescales ranging from a few thousand years (or even less for certain nuclei) to several million years. Nuclei with very long half-lives are especially useful as radioactive decay is a statistical process. If the half-life of a nucleus is 1,000 years, half the atoms in a sample will decay in that time. The statistical nature of the process also means that some nuclei will decay in much less than 1,000 years, even though the lifetime of a specific nucleus cannot be predicted. As even the smallest sample contains millions of nuclei, decays occur frequently even for long half-life isotopes.

For example, uranium-238 decays to lead-206 with a half-life of 4,500 million years. Thus if the ratio of the two elements in any sample of rock is measured the age of the sample can be determined. The age of the Solar System has been determined by dating meteorites in this way.

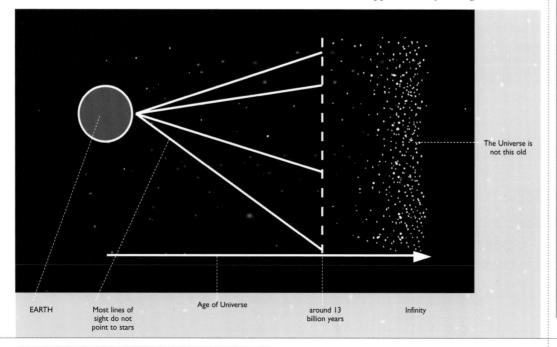

EARTH | Most lines of sight do not point to stars | Age of Universe | around 13 billion years | Infinity

The Universe is not this old

THE EXPANDING UNIVERSE

atomic nucleus ▶ p. 336 galactic halo ▶ p. 339 globular cluster ▶ p. 339 nuclear bulge ▶ p. 341 supernova ▶ p. 344

reach a star. The night sky remains dark because the Universe has a finite lifetime.

It is often said – incorrectly – that the sky is dark because the Universe is expanding. Distant galaxies are receding and their light is redshifted to lower energies, so light from distant stars is not as bright as light from nearer stars. While the light from distant galaxies is indeed redshifted, calculations show that the loss of energy is insufficient to account for the dark sky.

THE HUBBLE TIME

So how old is the Universe? Hubble's law states that due to the expansion of space galaxies are receding at a speed proportional to their distance, the constant of proportionality being Hubble's constant, H_0. If the expansion of the Universe has been steady, then the time which has elapsed since the Big Bang can, in principle, be calculated by selecting any distant galaxy (since the relationship holds for all galaxies) and dividing its distance from Earth by its recession velocity. In fact, because of the way in which Hubble's law is framed, the age of the Universe estimated in this way is simply $1/H_0$ and this is often referred to as the Hubble time. If we know Hubble's constant we also know the Hubble time. The current best estimate of the Hubble time is 15 billion years.

HAS THE EXPANSION
BEEN CONSTANT?

Although the Hubble time is an important step in obtaining an accurate age for the Universe, we must also understand how the expansion varies with time. If the rate of expansion has changed, then the Hubble time will be not be the same as the age of the Universe. The galaxies are moving apart against their mutual gravitational attraction and we might expect that this pull is slowing the rate of expansion. If so, the expansion would have been faster in the past, suggesting that the real age

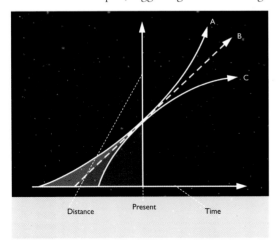

Rates of Expansion

The distances between galaxies in an expanding universe increase with time. Line B shows a constant rate of expansion. Line C shows how the rate of expansion might naturally be expected to slow down due to gravity. However, observations show that the expansion rate might actually be increasing, as shown by line A.

of the Universe is somewhat less than the Hubble time. Surprisingly, however, current observations do not support a slowing of the expansion; if anything, they show the expansion is accelerating, a phenomenon we explore later in this chapter. Incorporating this unexpected discovery into the calculations, the best current estimates of the age of the Universe lie at around 13 billion years.

Obviously the Universe must be older than any object within it, and so there are two checks which can be made on this calculated age, both of which are estimates of the ages of stars.

▲ **Omega Centauri,** *the largest and brightest of the 150 or so globular clusters that surround our Galaxy. Globular clusters formed early in the history of our Galaxy and their stars are mostly old. Measurement of the ages of stars in such clusters therefore places a lower limit on the age of the Galaxy and hence the Universe as a whole.*

THE OLDEST STARS

When a cluster of stars is formed it contains stars with a range of masses. The most massive stars are the most luminous and these hot, blue stars dominate the light from the cluster. However, the massive stars go through their life cycles very quickly compared to smaller stars and so as the cluster ages the blue stars will disappear first and its light will be increasingly dominated by the smaller, redder stars. This process can be used to estimate the age of a cluster. Observations have shown that the oldest clusters in our galaxy are the **globular clusters**, found mostly in the **galactic halo** surrounding the **nuclear bulge**. Their stars appear be up to

Kepler ▶ p. 38

💡

The explanation of the darkness of the night sky is a reminder of just how empty space is. Stars are actually very small in comparison to the space between them, the emptiness of space being very difficult to convey in diagrams or models. So empty is space that if a line is drawn from Earth right through the observable Universe it is very unlikely to pass through a star, a fact which means that even with the most powerful telescopes the images of stars will never overlap. The same does not hold true for galaxies. With the long exposures of the Hubble Space Telescope it is possible to record about 100,000 galaxies per square degree of the sky, an area which corresponds to about four times the size of the Moon's image as seen from Earth. In these exposures the galaxies cover about 1% of the field of view, so that it is likely that in the foreseeable future telescopes will record views of the sky in which galaxy images overlap. This overlap of galaxy images, but not of stars, seems to imply a contradiction, but in reality it does not as stars occupy just a tiny fraction of the space within a galaxy. Galaxies are, therefore, themselves almost empty space.

12 billion years old. Since theories of the evolution of the Universe suggest it was about one billion years old when galaxies first formed, the globular clusters offer good supporting evidence for the age of the Universe being about 13 billion years.

THE AGE OF CHEMICAL ELEMENTS

The second method of checking the age of the Universe involves another assessment of the age of the oldest stars. Studies of nucleosynthesis, the creation of the elements, show that the heaviest elements found on Earth were created in **supernova** explosions. Some of these elements, such as uranium and thorium, are radioactive with very long half-lives. By comparing the abundances of these elements found in stars to those of the products which result from their radioactive decay, astronomers can estimate the time which has elapsed since the elements were created, a technique which is also used to estimate the age of certain rocks on Earth. To use the technique we have to estimate how much of the radioactive elements were created in the explosion, a rather uncertain calculation. However, the method suggests that the oldest stars in our Galaxy are between 10 and 20 billion years old, consistent with the age of the Universe obtained from Hubble's constant.

THE COSMIC BACKGROUND RADIATION

The expansion of the Universe seems to point to all the matter we can see in the Universe having expanded from a tiny volume many billions of years ago. We can even estimate when that occurred. But the clinching evidence that the Universe began with the Big Bang did not come until the 1960s, with the discovery of a faint microwave glow covering the whole sky. This glow, the cosmic background radiation, proved to be the fading light of the Big Bang itself. Not only was there a Big Bang, but the Big Bang had been hot.

DISCOVERY OF THE COSMIC BACKGROUND RADIATION

Like many of the most important discoveries in science, the discovery in 1965 of the cosmic background radiation was an accident. Arno Allan Penzias (born in Munich in 1933, but raised in the United States after his family escaped Germany as refugees) and Robert Woodrow Wilson (b. 1936) were working at the Bell Telephone Laboratories in New Jersey on the development of receivers for communication satellites. They were investigating the problem of radio noise from the sky using a 6-m (20-ft) horn antenna. Penzias and Wilson had expected that the radio noise – which was causing

▲ *Dr Robert Wilson (left) and Dr Arno Penzias (right), joint 1978 winners of the Nobel Prize for physics, pose in front of the radio telescope which allowed them to detect background radiation. Their investigation proved that radiation from the early Universe could still be detected and provided evidence for the Big Bang.*

problems for satellite communications – would be less severe at shorter wavelengths, but found instead that the sky was unexpectedly bright at a wavelength of 7.3 cm, in the microwave region of the radio spectrum.

The pair assumed at first that the problem was in their antenna (they found pigeons nesting in it), but calculations showed that it could not have generated all the noise they detected. Noise in electronic equipment arises from the random motion of **electrons**: it can be greatly reduced by cooling the equipment with liquid nitrogen, but not eliminated, the residual noise being expressible as an equivalent temperature of its source. Penzias and Wilson found that the effective temperature of the sky background radiation was about 25% higher than their calculated contribution from the

antenna and concluded that the noise from the sky corresponded to the glow of a **black body** at a temperature of about 3 K (later to be more accurately measured). They also discovered that the radiation was independent of position in the sky – wherever they looked they saw the same amount of radiation.

THE ORIGIN OF THE RADIATION

Penzias and Wilson's investigation of the radiation they had discovered indicated that it did not come from specific stars or galaxies – for instance it is not concentrated along the Milky Way – but came from all over the sky. The spectrum of the radiation matched that of a cool black body at a temperature of about 3 K.

By coincidence, a few miles away at Princeton University, Robert Dicke (1916–97), an American astrophysicist, and Jim Peebles (b. 1935), a Canadian cosmologist, were developing their own theory of the first moments after the Big Bang. They realized that radiation from the hot gas that filled the Universe should still be visible, but redshifted by the Hubble expansion from visible light to very weak radio waves. It would have a black-body spectrum but now at a very low temperature: they estimated about 10 K. Dicke and Peebles had begun an independent search for this 'remnant' radiation in the microwave part of the radio spectrum when they heard of the discovery of Penzias and Wilson.

What neither group knew at that time was that the radiation they had found had been predicted 20 years earlier by the Russian-born, but naturalized American, George Gamow. In the 1940s Gamow had argued that radiation from the early, hot, dense Universe might still be detectable. He and his colleagues calculated that because of the cosmological redshift, the radiation would now be at a much longer wavelength than when it was emitted and its black-body spectrum would now correspond to a temperature of only 5 K.

Both groups published their findings together. The cosmic background radiation is now known to have a temperature of 2.725 K, redshifted by a factor of about 1,000 from the time when it last interacted with matter. It is regarded as compelling confirmation that the Universe did begin in a hot Big Bang. For their discovery Penzias and Wilson shared the 1978 Nobel Prize for physics.

GEORGE GAMOW

George Gamow (1904–68) was a Russian physicist who predicted the existence of the cosmic background radiation 20 years before it was discovered.

Trained in nuclear physics in Leningrad (now St Petersburg), Gamow spent several years on visits to Germany, Denmark and Britain, working on radioactivity and the quantum theory of the atomic nucleus. This led him to study the nuclear reactions that supply the energy for stars. In 1933 he finally left the Soviet Union for the United States, settling at George Washington University in Washington, DC. During World War II he worked on the Manhattan Project to develop the atomic bomb and later contributed to research on the hydrogen bomb.

In the late 1940s, while researching the physics of the Big Bang, Gamow and his students realized that the radiation emitted when the Universe was a hot, uniform gas should still be detectable, but redshifted to a temperature only a few degrees above absolute zero. But it was not until two decades later that Penzias and Wilson fulfilled his prediction, and then only by accident.

Gamow later applied his skills to biochemistry, making important contributions to the study of the DNA molecule and the genetic code, and to writing a series of popular science books in which his hero, Mr Tompkins, explores the worlds of particle physics, relativity and cosmology.

THE EXPANDING UNIVERSE   black body ▶ p. 336 electron ▶ p. 338 photon ▶ p. 342

COBE'S SEARCH FOR ANISOTROPY

The initial measurements of Penzias and Wilson indicated that the cosmic background radiation was isotropic – that is, it was uniform across the whole sky, appearing equally bright in all directions. This was a very important observation as it implied that the radiation originated before galaxies had had time to form, in a Universe consisting only of a primordial gas of uniform density. However, theories of the evolution of the Universe required that the density of the prim-ordial gas could not be completely uniform. In order to form clusters of galaxies small variations in density must have existed, and these would show up as very small variations in the brightness of the radiation.

The Cosmic Background Explorer (COBE) satellite was launched in 1989 to search for such variations. The results from the satellite confirmed that the spectrum of radiation had the shape of a black-body spectrum at a temperature of 2.725 K with a deviation of just one part in 10,000, but also that there were changes in intensity across the sky. Although tiny, corresponding to differences of just a few parts in 100,000, the changes are consistent with the expected density differences within the primordial gas. When the discovery was announced in 1992 the press had a field day about the discovery of these 'ripples in time', but cosmologists were as much relieved as excited: if COBE had not found the fluctuations then the Big Bang theory would have been in serious trouble.

▼ *NASA's Cosmic Background Explorer* (COBE) satellite produced a microwave map of the whole sky, eventually collecting four years of data. Cosmology theory indicates that the density fluctuations observed (witnessed by the different colour concentrations) could ultimately be responsible for the formation of galaxies.

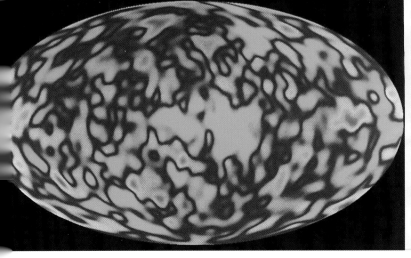

WHERE WAS THE BIG BANG?

When astronomers talk about the Big Bang, it is easy to imagine that it was a gigantic explosion into a void, something like a bomb going off with debris flying away in all directions. Hubble's discovery of the receding galaxies even seems to put us at the very centre of the explosion. Why else would all the galaxies be rushing away from us? Does this mean we are at the centre of the Universe?

This picture, though very common, is completely wrong. The idea of the Big Bang happening at a particular location in space has no meaning – it happened everywhere. It occurred at every point in space, including the space that is now within your body. It is better to think of the Big Bang as an extremely rapid expansion of space rather than an explosion, more like a balloon than a bomb.

In fact, the surface of an expanding balloon is quite a good two-dimensional analogy to the expansion of three-dimensional space. If a number of dots are marked on a balloon and the balloon is then inflated the dots all move away from each other. Any one dot will see all the others moving away at a speed proportional to their distance (just like Hubble's law), yet no dot can be said to be at the 'centre' of the spherical surface. Likewise the expansion of space ensures that an observer anywhere in the Universe will see galaxies rushing away in accordance with Hubble's law.

For similar reasons the Universe has no edge. Just as none of the dots on the balloon can be said to be at the 'edge' of the spherical surface, no galaxy can be said to be at the 'edge' of the Universe. The Universe looks much the same from all points in space, an expression of the cosmological principle we shall meet later in this chapter.

When we look out at the cosmic background radiation we are looking back in time and seeing **photons** coming from a region of space where the Universe was less than half a million years old. An observer looking back towards us from that distance would see exactly the same thing – the radiation from our part of space long before our Galaxy formed. The fact that the radiation is all around us indicates that the Big Bang did indeed happen through-out all space.

◀ *Spiral galaxy NGC 4603* is receding from us at at a speed of nearly 2,500 km/s (1,500 mi/s). Lying over 100 million light years away in a cluster of galaxies in Centaurus, it is the most distant galaxy in which Cepheid variable stars have been detected.

▼ *An artist's concept* depicting crucial periods in the development of the Universe, from a tiny fraction of a second after the Big Bang up to the Universe as we see it today, at an estimate of 13 billion years later.

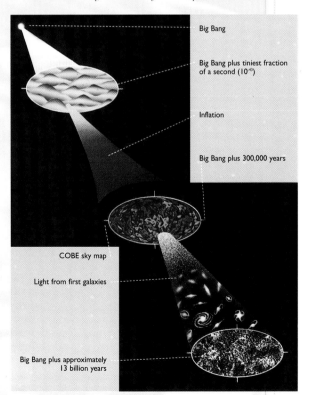

Big Bang

Big Bang plus tiniest fraction of a second (10^{-43})

Inflation

Big Bang plus 300,000 years

COBE sky map

Light from first galaxies

Big Bang plus approximately 13 billion years

PAST, PRESENT & FUTURE

The Universe: Past, Present & Future

THE BIG BANG

EVIDENCE FOR THE BIG BANG

Measurements of the motions of distant galaxies have led cosmologists to the view that the Universe was formed from a single event which has become known as the Big Bang. From the Big Bang emerged not only the matter to form all that we see now in the Universe, but time and space as well. Since time started at the Big Bang it is meaningless, in science, to ask what happened before. In this section we look at both the theoretical and experimental evidence for the Big Bang, gradually moving backwards in time to the dawn of the Universe.

MATTER AND RADIATION

We live in a Universe dominated by matter. Almost all of astronomy is concerned with the forms that this matter takes. **Radiation**, as a form of energy, plays only a secondary role. But that has not always been the case, for the relative importance of matter and radiation has changed as the Universe has expanded. Because

of the equivalence of **mass** and energy expressed in **Einstein**'s equation, $E = mc^2$ (frequently referred to as the 'newspaper equation' as it is the only one editors feel brave enough to print), both matter and radiation contribute to the mass density of the Universe. As the Universe expands the number of particles of matter per unit volume falls. The same is true for radiation, the number of **photons** per unit volume falling in the same way, but another effect also comes into play. Since the energy of a photon depends on its **wavelength**, the higher the energy the shorter the wavelength. As the Universe expands the wavelength increases (by the cosmological **redshift**) and so the photon

---

THE EVOLUTION OF THE UNIVERSE

The reason that the early history of the Universe can be delineated with any confidence is that until recombination the entire Universe was very nearly in thermal equilibrium at every point and every instant. Bodies in thermal equilibrium are relatively simple to understand. Since thermal equilibrium is characterized by a single quantity – temperature – the nature of the Universe at any time can be deduced from a knowledge of its temperature, which in turn is related to its size and hence its age.

Event	Age	Temperature
End of Planck era	10^{-43} s	10^{31} K
End of inflationary era	10^{-34} s	3×10^{28} K
Formation of nucleons	10^{-6} s	3×10^{12} K
Helium production	1 s	10^{10} K
End of radiation dominance	10,000 yr	10^{4} K
Recombination	3×10^{5} yr	3,000 K
Galaxy formation	~10^{9} yr	30 K
Present day	~10^{10} yr	3 K

---

energy falls. The contribution to the mass density from radiation therefore falls more rapidly than that from matter leading to an increasingly matter-dominated Universe.

RADIATION IN THE EARLY UNIVERSE

The energy density of radiation, most of which resides in the **cosmic background radiation**, is now quite feeble in comparison to the energy density of matter, but this situation was reversed in the early history of the Universe. When the Universe was less than 10,000 years old most of its energy was in the form of radiation. A radiation-dominated Universe is simpler to understand theoretically than a matter-dominated one. As the Universe evolved, it passed through a series of critical times – known as epochs – each identified by a temperature. One of the interesting aspects of the study of this early phase of the Universe is the astonishingly precise timing of the various epochs, a precision which allows times soon after the Big Bang to be calculated to fractions of a second, despite the present age of the Universe being uncertain by at least a billion years.

THE TEMPERATURE OF RADIATION

All hot bodies emit radiation. Although this is clearly true for something as hot as the Sun, it is also true for any body whose

---

10^{-43} seconds 10^{-34} seconds   1 second

Planck era   Inflationary era   Helium production

THE BIG BANG

temperature is higher than absolute zero (0 K). Those bodies which absorb all the energy which falls on them – i.e. they reflect no radiation – are known as **black bodies**. A black body is also a perfect radiator of energy and it is possible to measure its temperature from the wavelength of the characteristic peak in its spectrum. As the temperature of the black body rises, the wavelength of peak emission shortens. The early Universe behaved as a black body and so it is possible to define a temperature from the spectrum of the cosmic background radiation. As the Universe expands, the wavelength of this radiation lengthens and so the temperature of the Universe decreases. Since wavelength and temperature are related, if the Universe expands by a factor of 10, wavelengths increase by a factor of 10 and the temperature of the Universe falls by a factor of 10.

THE FIRST ATOMS

It is known from experiments that when an **electron** inside an atom is given an energy of just a few electron volts it can escape from the atom. The energy can come from collisions bet-ween atoms, and at temperatures of about 3,000 K or more the collisions are energetic enough for electrons to escape. It follows that atoms could not exist in the Universe until its temperature had fallen below 3,000 K, since any atom that did form would be broken up again by collisions. Before this time the Universe was a sea of **atomic nuclei** and electrons – a plasma rather than a gas. This plasma was opaque, radiation being continually emitted and re-absorbed so that matter and radiation were at the same

temperature. When the Universe cooled below this critical temperature stable atoms could form and so the link between matter and radiation was broken – they were 'decoupled' – and the two evolved independently of each other. Space became transparent for the first time and radiation could pass more-or-less freely through the Universe. For this reason the period during which atoms formed is sometimes known as the decoupling era, although it is more often referred to as the recombination era (a slightly inappropriate name since nuclei and electrons were forming for the first time rather than recombining).

OUR EARLIEST VIEW OF THE UNIVERSE

Since the recombination era, the cosmic background radiation has passed through the Universe without interacting with matter. Before recombination the Universe was opaque, so the background radiation provides us with our earliest view of it. The temperature of the radiation is now 3 K, and since it was 3,000 K at recombination we know the Universe has expanded by a factor of 1,000 since then. With current cosmological models this places the start of recombination at a Universe age of about 300,000 years. The most distant galaxies that can be detected with present-day telescopes are seen at a time when the Universe was about 1,000 million years old.

We have no observations of the Universe when its age lay between one million years and 1,000 million years, and the evolution of the Universe in this period, including the era of galaxy formation, remains very uncertain.

The cosmological background radiation can be detected all over the sky, so the answer to the frequently asked question 'where do we look to see the Big Bang?' is straightforward. The Big Bang did not happen in a single place; the Universe was created in the Big Bang and its after-effects can be seen everywhere.

🧪 PLASMAS

At very high temperatures gases become plasmas in which the atoms are broken down into electrically charged electrons and nuclei. The attraction between these charged particles gives a plasma properties which are different from those of normal gases. For example, various forms of wave motion can arise in a plasma and these can be excited by the absorption of radiation. Consequently plasmas are opaque over a wide range of frequencies, unlike normal gases which absorb at specific frequencies, the so-called **spectral lines**.

Plasmas are now being created on Earth in an effort to develop controlled nuclear fusion which, it is hoped, will form the basis of a future generation of power reactors. One problem which has to be overcome is how to contain the plasma, since it is destroyed when it comes into contact with the walls of the reactor. Devices such as the large 'tokamaks' (named for an early Russian experimental contain-ment) use magnetic fields to confine the plasma within a doughnut-shaped vessel.

▼ *The Big Bang*
This stylized diagram shows the principal evolutionary stages of the cosmos from the earliest formation of matter from energy just after the Big Bang to the Universe of galaxies of the present day.

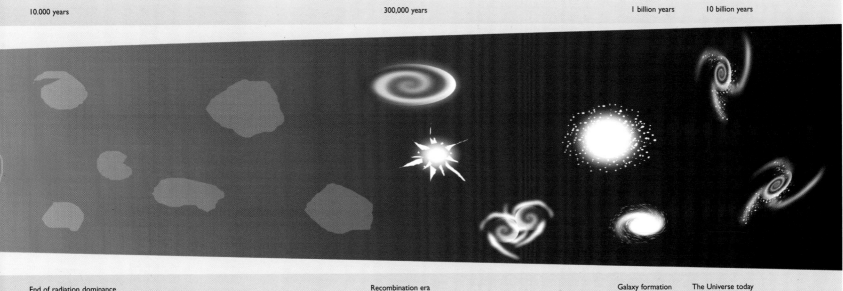

| 10,000 years | 300,000 years | 1 billion years | 10 billion years |

| End of radiation dominance | Recombination era | Galaxy formation | The Universe today |

PAST, PRESENT & FUTURE

THE FIRST PARTICLES

We can see no further back than 300,000 years after the Big Bang, so any deeper understanding of the early Universe must come from theoretical models. The primordial gas that emerged from the Big Bang consisted almost entirely of hydrogen and helium with small amounts of other species. The observed abundances of these elements are in good agreement with theoretical predictions. One mystery concerns the 'dark matter' which seems to make up the bulk of the mass of the Universe. Much of it may be made of unknown exotic particles, and these, too, would have been created in the Big Bang.

THE PRIMORDIAL GAS

Stars condense out of the interstellar gas in a galaxy, but what was the composition of the primordial gas from which the first stars condensed? It cannot be the same as the present interstellar gas which has been enriched with new elements synthesized in earlier generations of short-lived massive stars and ejected back into space. To learn about the make-up of the primordial gas we must look for members of the first generation of stars to be formed.

Such first-generation stars are found in the **globular clusters** which lie in the spherical halo which surrounds the disk of our Galaxy. These halo stars are composed of 75% hydrogen and 25% helium, with small traces of lithium. If these elements existed before the first stars were formed they must have been created in the very high temperatures of the early Universe.

▲ *A Hubble Space Telescope view of part of the Cygnus Loop, the remains of a star that exploded as a supernova some 30,000 years ago. Supernova explosions such as this produce the heavy elements we find in nature today, from which planets and life are formed. The primordial gas from which galaxies formed consisted of only the lightest elements, hydrogen and helium, plus a small amount of lithium.*

MAKING THE FIRST NUCLEI

The simplest nucleus after hydrogen is deuterium (also known as heavy hydrogen) which is just a **neutron** joined to a **proton**. Both particles would have formed in equal numbers in the first moments following the Big Bang, but because of the slight difference in mass, more protons than neutrons survived the initial reactions to be available to form nuclei. At 1 s, when the temperature had dropped to 10^{10} K, the Universe consisted of 87% protons and 13% neutrons.

Formation of deuterium could not get underway until the Universe had cooled to around 10^9 K – at around 100 s – since at higher temperatures the nuclei would break up as quickly as they formed. On the other hand, free neutrons decay into protons and electrons with a mean lifetime of about 15 minutes, so the deuterium formation must essentially have been finished by then.

Most of the deuterium made in this short period was used up in making bigger nuclei, especially helium, but a small amount was left over, the amount being very sensitive to the density of the Universe at the time of formation.

▼ *Evolutionary Eras of the Universe*
The temperature and key evolutionary eras of the Universe are shown at different times after the Big Bang. This graph is not to scale.

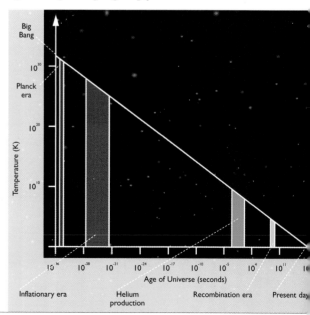

THE BIG BANG

Since there is no significant production of deuterium in stars, essentially all the deuterium we see today in interstellar gas – about 0.002% – is primordial. This figure is consistent with the present density of the Universe and is a good test of the Big Bang model. In fact, such is the faith in the Big Bang theory that the abundance of deuterium is considered a more reliable indicator of the present density of the Universe than direct observation of matter in stars and galaxies.

Once there was a significant number of deuterium nuclei, then helium – with two protons and two neutrons – could begin to form from them. Helium formation was also essentially complete after 15 minutes. Allowing for the decay of neutrons, the amount of helium produced in this time is predicted to be 24–25%, and this figure varies very little with the initial conditions. Cosmologists were both delighted and relieved to find that, within measurement uncertainties, this is precisely the amount of primordial helium present in the Universe – any figure that was very different would simply be incompatible with a Big Bang.

MAKING FUNDAMENTAL PARTICLES

How did electrons, protons and neutrons come into existence? These particles, which are the building blocks of all matter, would have been created even closer to the instant of the Big Bang, when the Universe consisted almost entirely of radiation. The collision of two photons can produce particles – because of the equivalence of mass and energy – but very high energies are required: the photons must be **gamma rays**. To produce an electron in gamma-ray collisions requires a temperature of 10^{10} K, which existed when the Universe was much less than one second old. To produce a proton or a neutron, which are about 2,000 times more massive than an electron, the temperature must be even higher, 10^{13} K. Such temperatures existed only during the first millionth of a second (one microsecond) after the Big Bang.

All particles would have been created along with equal numbers of their antiparticles, and the absence of antiparticles in today's Universe is a problem we will return to later in this chapter.

THE MYSTERIOUS DARK MATTER

Most of the mass of the Universe seems to be in the form of invisible material known as **dark matter**. If that is true then the dark matter must also have been created in the early Universe.

💡
EVIDENCE FOR DARK MATTER

One of the most interesting discoveries of the past decades is that the visible matter of the Universe, which we see as stars and gas, is a relatively insignificant portion of the whole. It seems that by far the bulk of the Universe is made up of unseen 'dark matter'. There are two main pieces of evidence which point towards its existence.

The first comes from studies of the rotation rate of galaxies, including our own Milky Way, which show that most of a galaxy's mass lies in a halo perhaps 10 times the diameter of the visible galaxy.

Although there are some stars in the halo they cannot account for all the mass.

The second comes from observations of clusters of galaxies. It is possible to measure the speeds of the individual galaxies from their **Doppler shifts**, and these show that the galaxies are moving much faster than one would expect from the total mass of the visible galaxies. Again, there must be a large quantity of dark matter in the space between the galaxies which cannot be seen but which is making its presence felt by its gravitational pull.

Current estimates of the amount of dark matter in the Universe range from 10 times the amount of visible matter to as much as 300.

Some of the dark matter may be gas, dust and remains of burned-out stars, but there is simply not enough of this material to account for the dark matter. And there is even a strong theoretical reason why most of the dark matter cannot be made of the ordinary **baryonic** matter that makes up the visible Universe. As we have seen, the present abundance of deuterium places strict limits on the density of matter in the early Universe and hence on the density today. It turns out that there is simply too much dark matter for it to be baryonic – at least some of it has to be made of something else. But what?

Very little is known about the particles of dark matter because they have never been detected. But it is possible to predict some of their properties. For example, dark-matter particles must be stable or they would not have survived into the present Universe; aside from the deuterium argument they cannot be baryons or they would be bound to the protons and neutrons of atomic nuclei; and they cannot carry electric charge or they would be found, like electrons, trapped around atomic nuclei. The fact that dark-matter particles have never been detected in experiments at particle accelerators suggests that they might be so

massive that present accelerators are not powerful enough to produce them. Perhaps they are created very rarely or they could be so elusive that they have been created but never detected and so never recognized.

▲ *Carl David Anderson (1905–91) observed events in a cloud chamber which confirmed Paul Dirac's prediction that a high-energy photon could create an electron and its antiparticle (which is often referred to as a positron).*

which can be achieved are about 1,000 GeV which corresponds to a temperature of about 10^{15} K. Temperatures as high as this existed during the first ten thousandth of a microsecond after the Big Bang. Almost all the particles produced are unstable and decay in very short times, leaving only the stable particles – the proton, electron and **neutrino** – together with photons.

❙
PARTICLE PHYSICS IN THE LABORATORY

Our knowledge of fundamental particles comes from the study of **cosmic rays** – particles from space which can be collected high in Earth's atmosphere – and from experiments with particle accelerators. Within particle accelerators the energy of the particles in the beams is used to create new particles. From the energy required to create the particles we can calculate the temperature of the early Universe at which these particles were produced. At present the highest particle energies

TOWARDS THE INSTANT OF CREATION

H ow far back towards the beginning can we push our understanding of the Universe? Cosmologists are confident that they can tackle times as early as 10^{-32} s, the so-called 'inflationary era' when the Universe ballooned dramatically before settling down to its present, more leisurely expansion. Inflation solves several cosmological puzzles, including the uniformity of the cosmic background radiation and the value of the current density of the Universe. Ultimately our physics takes us back to 10^{-43} s but no further – here the laws of physics themselves break down and further speculations are left to the creative minds of theoretical physicists.

THE HORIZON PROBLEM

When regions of the Universe have the same temperature the natural assumption is that this has come about because they have been in physical contact, probably through the exchange of radiation, allowing the temperature to equalize. An example is the cosmic background radiation, which is uniform to one part in 10,000. Unfortunately, the Big Bang model does not allow such interaction. The radiation was emitted in the recombination era when the Universe was only

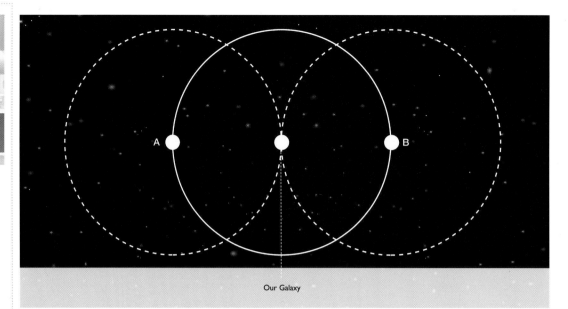

Our Galaxy

▲ *The 'Horizon Problem'*
The solid circle drawn around our Galaxy represents our cosmic horizon. This is the limit to the distance we can see, set by how far light can have travelled during the lifetime of the Universe. It follows that no radiation can have passed between points A and B, on opposite sides of the horizon.

300,000 years old. When we measure the radiation from any direction we are looking at gas which is so far away that the radiation from it has taken almost the age of the Universe to reach us. But then, when we look in the opposite direction, we find that the same is true. Obviously no radiation can have passed between these two regions to equalize their temperatures and so physical contact between the two appears to violate the principle of causality. This is the 'horizon problem'.

THE INFLATIONARY ERA

In 1981, the American physicist Alan Guth (b. 1947) suggested a solution to the horizon problem. He proposed that during its earliest stages – between roughly 10^{-34} s and 10^{-32} s – the Universe underwent a period of extremely rapid expansion, during which the expansion rate was actually accelerating. This period is known as the inflationary era (often shortened to 'inflation'), and during it the Universe's size increased dramatically. If we consider a small region of the early Universe, one so small that physical contact has established thermal equilibrium, we see that inflation can expand this small region to a size so great that our entire observable Universe can lie within it. In order to solve the horizon problem the expansion must be by a factor of at least 10^{30}, and many cosmologists believe it might be vastly more than that.

STEPHEN HAWKING

The English theoretical physicist Stephen Hawking (b. 1942) was born in Oxford (on the 300th anniversary of Galileo's death). Having graduated from Oxford University he moved to Cambridge to study for his PhD. He was made Lucasian Professor of Mathematics at Cambridge in 1980. Two of the eminent theoretical physicists who had previously occupied this chair were **Newton** and **Paul Dirac** (1902–84). Hawking's work has been an attempt to synthesize the interests of these two giants, gravity and quantum mechanics. After early work on relativity, Hawking concentrated on the problems of gravitational singularities, publishing important work on black holes and the Big Bang. In particular he has shown that black holes are not an irreversible energy sink, but by the processes of quantum mechanics slowly evaporate by the emission of thermal radiation. The rate of evaporation is inversely proportional to the mass of the black hole, so as the black hole diminishes in size the rate of mass loss increases.

In 1988 Hawking published *A Brief History of Time*, a popular account of the current thinking in cosmology. The book was an outstanding success, reprinted 10 times in the first year of publication and several times since then. It was another major achievement for a man afflicted by a crippling motor neurone disease, which has confined him to a wheelchair and requires him to talk by means of a voice synthesizer.

▲ *The bulk of Stephen Hawking's theoretical physics has been devoted to the study of black holes and their properties as well as the Big Bang. He demonstrated that thermal radiation could be emitted from black holes, thus turning previous theory about matter's inability to escape from these very dense bodies on its head.*

THE BIG BANG

Heisenberg Uncertainty Principle ▶ p. 339   quantum mechanics ▶ p. 342   Schwarzschild radius ▶ p. 343

Although this solution may appear contrived, inflation simultaneously explains several features of our observed Universe, which is one of the credentials required of a powerful scientific idea. One of these is that inflation forces the average density of the Universe to be extremely close to what is called the 'critical' density. Another is that inflation provides a possible mechanism by which galaxies may form. We shall return to both these points later in this chapter.

THE LIMITS OF PRESENT PHYSICS

Present theoretical physics is built around the cornerstones of **quantum mechanics** and general relativity, the former dealing with processes on the small scale, including the interactions of fundamental particles, the latter explaining phenomena on the large scale, particularly the effects of gravity. Quantum mechanics makes predictions which can be checked experimentally using particle accelerators. At present, the particle energies available from accelerators are limited to about 1,000 GeV which, in cosmological terms, equates to a time when the Universe was about 10^{-12} seconds old and its temperature was about 10^{15} K. Theoretical physicists attempt to understand earlier eras through extrapolation of known physics, but they know that there will come a point at which their present theories will prove inadequate. At very high particle energies, around 10^{19} GeV, both quantum mechanics and general relativity are needed simultaneously and at present no theory linking the two exists – there is no quantum theory of gravity. This very early phase of the Universe is called the Planck era and covers the time to when the Universe was 10^{-43} seconds old and its temperature was about 10^{31} K.

THE PLANCK ERA

The problem which arises in the Planck era can be understood by considering two scale lengths associated with fundamental particles.

Quantum theory postulates that all particles have a wave nature and so behave differently to classical particles; in particular they have an inherent 'fuzziness' which means their positions cannot be pinned down precisely. A measure of this fuzziness is the Compton wavelength, which depends on the particle's mass and for a proton or neutron is around 10^{-15} m. This can be regarded as one measure of the size of the particle.

The second scale length derives from general relativity – it is the **Schwarzschild radius** of the particle and is proportional to the particle's

mass. If an object's size is less than its Schwarzschild radius then it is a black hole. Now we can ask, what is the mass of a particle for which the Compton wavelength is equal to the Schwarzschild radius? The answer is about 10^{-8} kg, a quantity called the Planck mass, and the associated size is 10^{-35} m, known as the Planck length.

The significance of these quantities is that if we project the evolution of the Universe backwards towards zero time there will come a point when the density was such that everywhere in space Planck masses would be contained in regions of the Planck size. At this point – at the Planck time, around 10^{-43} s – general relativity and quantum mechanics come into conflict, and our present laws of physics break down. There is, as yet, no satisfactory quantum theory of gravity that would guide us in understanding the development of the Universe up to the Planck time.

HARTLE AND HAWKING'S THEORY OF CREATION

As we have seen, present theoretical physics cannot be used to study the very earliest Universe. Neither can it deal with the actual origin as the Universe is thought to have derived from a singularity, with zero size and infinite density and temperature. Stephen Hawking, the Cambridge-based theoretical physicist, and Jim Hartle have proposed a method of combining quantum mechanics and general relativity to allow some study of the origin. They point out that **Heisenberg's Uncertainty Principle** allows the conservation of energy to be violated for short periods of time and that the shorter the period of time involved the greater the violation (in energy terms) can be. There is, therefore, a short, but finite, time interval after the singularity in which the energy required to create the whole Universe is available. Hawking also notes that it may be possible to avoid the problem of considering zero time by modifying the concept of four-dimensional space–time. In Hartle and Hawking's theory the four are identical, but their roles change close to the singularity. The origin of the Universe might then not be the start of time, but the point at which the time dimension first begins to behave as such.

CAUSALITY

Causality is the name given in science to the principle that an effect can follow a cause, but cannot precede it. In our world this seems straightforward and obvious, but as a scientific concept it has far-reaching consequences, not least because ideas which violate causality are always rejected by scientists. The theory of special relativity predicts that the measured time interval between two events – a cause and an effect – will be different for two observers if they are in relative motion. The difference they perceive will depend upon their relative speeds – the faster they are travelling relative to each other the greater the difference will be. But despite this change there will be no change in the order of the events. If there were such a change then causality would be violated for one of the observers. Special relativity states that this could only occur if the relative velocity exceeded the speed of light, but as such speeds are forbidden then causality cannot be violated.

Two immediate consequences of this are that 'warp factor' speeds are not possible, and that time travel which involves interference in history rather than mere observation is impossible. They belong solely in the realms of science fiction.

PAST, PRESENT & FUTURE

▲ *An artist's impression* of the Very Small Array (VSA) in Tenerife, Spain – a prime example of the type of observational equipment which is currently allowing astronomers to test theories of cosmology by detecting relic radiation from the early stages of the Big Bang.

Dirac ▶ p. 104 Newton ▶ p. 42

THE BIG BANG

WHERE IS THE ANTIMATTER?

We have seen that the particles which make up the Universe were created from radiation in the earliest moments following the Big Bang. Half of those would have been antiparticles, but if so, where are they now? All the evidence is that the observable Universe is made of matter. Although isolated enclaves of antimatter are not ruled out, physicists are now inclined to the view that there must have been an imbalance of matter and antimatter in the early Universe. The matter we now see around us is then the tiny surplus remaining after the mutual annihilation of nearly equal quantities of matter and antimatter.

▲ *The Athena experiment at the CERN laboratory in Switzerland has created atoms of antihydrogen, formed by a positron orbiting an antiproton.*

MATTER AND ANTIMATTER

The production and decay of all particles is governed by a number of conservation laws which insist that certain properties must not change during the decay. One of these laws governs electric charge: if a particle with a positive charge is created, one with a negative charge must also be created so that the net change of charge is zero.

Conservation of electric charge is not the only law which limits the type of particles which can be created from radiation. Protons and neutrons are part of a family called baryons, and the total number of baryons must also be conserved. So every time a proton is created from photons (which are not baryons), an antiproton must be created as well to keep the change in baryon number to zero. Similarly, electrons belong to a family known as leptons and the number of leptons must also be conserved. The creation of antiprotons, antineutrons and antielectrons (often called **positrons**) would have created a Universe consisting of half matter and half antimatter. Which we call 'matter' and which 'antimatter' is arbitrary, just as is our definition of positive and negative charge. But the conservation laws of particle physics leave scientists with a real problem – where is all the antimatter?

LOOKING FOR ANTIMATTER

Antiprotons, antineutrons and antielectrons (positrons) should form antiatoms (atoms of antihydrogen have been produced in the laboratory). Particles and antiparticles should have been produced in equal numbers in the early Universe so the Universe should now consist of equal amounts of matter and antimatter. To decide whether the symmetry exists it is necessary to look for antimatter, but how can this be done? Astronomers identify atoms in distant objects by the spectral lines they produce, but antiatoms would emit the same radiation. The photons which would be produced from antimatter are identical to those produced from matter because the photon is one of the few particles which is identical to its antiparticle.

Neutrino astronomy offers a possibility of distinguishing matter from antimatter because the antineutrino is a distinct particle, but neutrinos are elusive and so far have only been detected from the Sun and from a supernova in the **Large Magellanic Cloud**. This proves that both the Sun and the stars in the Large Magellanic Cloud are made of matter, but we have no evidence about more distant galaxies.

ANNIHILATION

If antimatter does exist in the Universe then it must be very well separated from matter or the two would annihilate. The annihilation process is the reverse of the creation process which brought the matter of the Universe into existence and is subject to the same conservation laws. This means that a particle can annihilate only with its own antiparticle, i.e. a proton with an antiproton, but not a proton with an antielectron. When an electron and antielectron annihilate their masses are converted into gamma rays. When a proton annihilates with an antiproton their mass is converted into pions, which rapidly decay into gamma rays and neutrinos. It is known, therefore, that annihilation of matter and antimatter would create large fluxes of gamma rays, and it is these gamma rays which might indicate that the Universe was symmetric in the two forms of matter.

ANDREI SAKHAROV

Andrei Dimitrievich Sakharov (1921–89) was born in Moscow and gained his doctorate from work on cosmic rays. During the 1960s his call for a nuclear test-ban treaty and improved civil rights for Soviet citizens brought him into conflict with the authorities. He was awarded the 1975 Nobel Peace Prize, but in 1980 was exiled as a dissident. He was pardoned by President Mikhail Gorbachev in 1986.

Though best known in the West as a dissident, Sakharov was a brilliant physicist. In 1967 he argued that a matter Universe could have emerged from the initial creation process if that process had produced a very small excess of matter over antimatter. Matter and antimatter would later have annihilated, and at the end of the annihilation process the slight excess of matter would have remained to evolve into the Universe we see now. The annihilation produces photons, and recognizing that the microwave background contains around one thousand million photons for every proton in the Universe Sakharov concluded that the initial asymmetry needed only to be one part in one thousand million. He also pointed out that such an asymmetry must arise from the decay or interaction of fundamental particles and should, therefore, be detectable by experiments in particle accelerators.

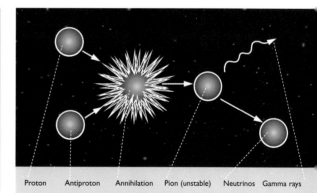

| Proton | Antiproton | Annihilation | Pion (unstable) | Neutrinos | Gamma rays |

▲ *Proton/Antiproton Annihilation*
A particle can only annihilate with its own antiparticle. In this example, the annihilation of a proton and antiproton produces an unstable pion which decays into gamma rays and neutrinos.

PAST, PRESENT & FUTURE

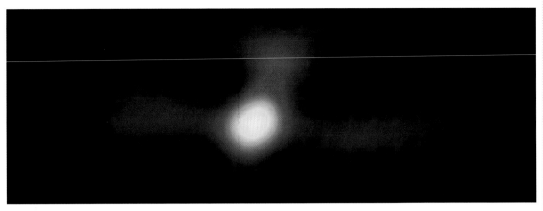

▲ *Small-scale annihilation* of matter and antimatter occurs in our Galaxy between electrons and positrons, producing a faint glow of gamma rays mapped here by the Compton Gamma Ray Observatory. At centre is the nucleus of the Galaxy with the galactic plane either side and a diffuse cloud of positrons above it.

ANTIMATTER IN THE UNIVERSE

To date, gamma-ray telescopes, notably the Compton Gamma Ray Observatory, have failed to find any evidence for large-scale annihilation taking place in the Universe. This negative result allows two important conclusions to be drawn. First, there can be no stars constructed of anti-matter within our own Galaxy. If there were then annihilation would be taking place at the star's surface between stellar antimatter and interstellar gas which is known to be matter. Similarly, clusters of galaxies must either be made entirely of matter or of antimatter or there would be annihilation between a galaxy and the cluster gas.

There is also evidence from cosmic rays, the high-speed particles which continuously bombard the top of Earth's atmosphere. The most energetic of these particles are believed to emanate from other galaxies, but all experiments on the make-up of the particles imply that they are matter rather than an equal mix of matter and antimatter.

▼ *The Compton Gamma Ray Observatory* has failed to find convincing evidence for any large-scale annihilation in the Universe. This means either that no large amounts of antimatter exist or, if they do, they must be well-separated from matter.

Although the evidence strongly suggests that we live in a matter Universe, it does not preclude the possibility of well-separated structures of each form. However since gravity acts symmetrically on matter and antimatter there is no known mechanism which could create such a separation. The conclusion is, therefore, that we do indeed live in a matter Universe. Since the initial creation processes of the Big Bang would have produced equal numbers of particles and antiparticles, physicists have been searching for mechanisms which would create the imbalance we see today.

POSSIBLE ASYMMETRIES IN PARTICLE PHYSICS

Although protons, neutrons and electrons can be created (together with their antiparticles) directly from radiation, many of them were produced by the decay of heavier, unstable particles. These would have been created earlier when the Universe was much hotter and, con-sequently, particles were much more energetic. Russian physicist Andrei Sakharov argued that it was likely to be these decay processes that yielded an excess of matter over antimatter. If so then it would be necessary to violate some of the

If human beings are ever to explore beyond the Solar System a new form of fuel is required to propel the rockets. There are two priorities for any such fuel. One is that it offers high energy for a given mass: a major fraction of the mass of today's rockets is the fuel. Secondly the fuel must offer a high velocity for exhaust gases because rockets are accelerated by momentum transfer between the vehicle and its exhaust gases. On both these criteria the ultimate rocket fuel is antimatter since in annihilation with matter 100% of the total mass is converted into energy, and the products of the annihilation, gamma rays, travel at the speed of light. The problems with antimatter are the difficulties of producing and storing it. At present physicists have succeeded only in making a few antiatoms in particle accelerators. If, in the future, methods of producing and storing large quantities can be devised, antimatter would be ideal for space flights.

conservation laws which normally govern the interactions and decays of fundamental particles.

Sakharov's mechanism involves the violation of two laws – conservation of baryons and leptons and 'charge conjugation', the principle that matter and antimatter particles should behave symmetrically. Violation of the first would mean that a heavy baryon, such as might have fleetingly existed in the early Universe, could sometimes decay not into a proton but into a positron. Experiments are currently underway to search for such a decay. Violation of charge conjugation would allow the non-conservation of antibaryons to occur more frequently than that for baryons. Examples of such a violation are already known but not at the level required to explain the asymmetry according to Sakharov's mechanism.

COMPTON GAMMA RAY OBSERVATORY

The Compton Gamma Ray Observatory (GRO) was one of NASA's four Earth-orbiting 'Great Telescopes' – the others being the Hubble Space Telescope launched in 1990, the Chandra X-ray satellite launched in 1999 and the Space Infrared Telescope Facility (SIRTF) scheduled for launch in 2001. GRO was named after US physicist Arthur Compton (1892–1962), one of the pioneers of quantum mechanics whose work concerned the nature of gamma rays and the way in which they interact with matter. It was launched in 1991 and reentered Earth's

atmosphere in 2000. GRO had four independent telescopes. The Burst and Transient Source Experiment (BATSE) monitored the whole sky for short bursts of gamma rays with energies between 50 keV and 600 keV. The Oriented Scintillation Spectrometer Experiment (OSSE) looked for the gamma-ray spectral lines expected from radioactive decay and electron-positron annihilation. The Imaging Compton Telescope (COMPTEL) surveyed the sky at gamma-ray energies from 1 MeV to 30 MeV, while the Energetic Gamma Ray Experiment Telescope (EGRET) was a spark-chamber detector sensitive to gamma rays from 20 MeV to 30 GeV.

THE BIG BANG

The Universe: Past, Present & Future
MODELLING THE UNIVERSE
SPACE & TIME

From the time of the ancient Greeks – and perhaps even before – humanity's insatiable desire to understand the world has led us to consider the origin and form of the Universe. In this section we consider cosmological models; that is, models which deal with the Universe as a whole rather than considering in detail the objects it contains. We begin with ideas on space and time, then consider other ideas before looking at observations which can be made to check the theories.

NEWTONIAN COSMOLOGY

Although the Greeks had engaged in 'thought experiments' on the nature of the Universe, the first attempt at a physical model was made by the English scientist **Isaac Newton** based on his concepts of absolute space and time. Newton recognized that the force of gravity would dominate on the largest scale, but this led to a problem when considering whether the Universe was finite or infinite. If it were finite Newton realized that this would necessarily mean it had a centre, and that would require everything in it 'to fall down into the middle of the whole space and there compose one great spherical mass'. The fact that this had not happened – and that there was no evidence that it was happening – led Newton to the view that the Universe was infinite. In such a Universe there was no 'middle' and so no preferred direction for gravity. The objects within the Universe would then be spread evenly through it so that gravity effects would cancel out.

GRAVITY IN AN INFINITE UNIVERSE

However, an infinite, static Universe in Newtonian cosmology leads immediately to two other major difficulties. One, as we have seen, is Olbers' paradox, the darkness of the night sky, while the other is the difficulty of obtaining a state of equilibrium in an infinite Universe. Newton reasoned that if stars were spread evenly through space, then the gravitational pull on the Sun due to distant stars in any given direction should be exactly balanced by the pull of stars in the opposite direction. The net force of the whole Universe on the Sun should add up to zero. But what if the Universe were infinite? The opposing forces would then be infinite also, but can two infinite forces really balance each other? Newton showed that the slightest unevenness in the distribution of stars would cause a net pull on the Sun, and on all other stars, which would quickly lead to the Universe collapsing in on itself. As a consequence, Newton found that in an infinite Universe he had no satisfactory way of treating gravity.

OBSERVATIONS OF THE COSMOS

Until the early twentieth century, advances in cosmology were limited by the lack of observational evidence from beyond the Milky Way. The English philosopher Thomas Wright (1711–86), noting the high density of stars in the direction of the Milky Way in comparison to the few in other directions, suggested that the Universe took the form of an infinite slab, his work leading to Immanuel Kant's more modern view of our Galaxy and the system of galaxies.

William Herschel (1738–1822), in his investigations of the structure of the Milky Way, introduced some of the observing techniques used by modern cosmologists. Having failed to measure the distances to individual stars by the **parallax** method he realized that he could gain some measure of the distances from the apparent brightnesses of stars. This does not work well for individual stars because they may have very different luminosities, but it works better for groups of stars where it may be assumed that their average properties are the same. Using this 'star-gauging' technique he was able to confirm Thomas Wright's argument on a quantitative basis. Following **Hubble**'s discovery that the spiral 'nebulae' were external galaxies, cosmologists adapted Herschel's technique of star-gauging to their studies of distant galaxies as we shall see later.

IMMANUEL KANT

Immanuel Kant (1724–1804) was born in Königsberg, Prussia (now called Kaliningrad and part of Russia), just three years before the death of Newton. He studied at the town's university, and later taught there, and appears to have lived a life of amazing orderliness. Now seen as one of the greatest figures in the history of western thought, Kant made significant contributions to science as well as philosophy.

In his *Theory of the Heavens*, published in 1755, Kant argued that the stars of the Milky Way were a rotating disk maintained by gravity. He also argued that some of the **nebulae**, the diffuse patches of light in the night sky, were elliptical, and 'systems of the same order as our own'. He conjectured that those other milky ways might form clusters and that such clusters might be found throughout the Universe. In addition to making these very early speculations on the nature of our own and other galaxies, Kant considered the Universe as a whole, concluding that 'there is here no end but an abyss of a real immensity, in the presence of which all the capacity of human conception sinks exhausted', a thought that has struck a chord with many since.

▶ *The German philosopher* Immanuel Kant argued that stars in the Milky Way were arranged in a rotating disk held together by gravity, and went on to speculate that some elliptical nebulae were separate systems of stars like the Milky Way.

nebula ▶ p. 341   parallax ▶ p. 342   redshift ▶ p. 343

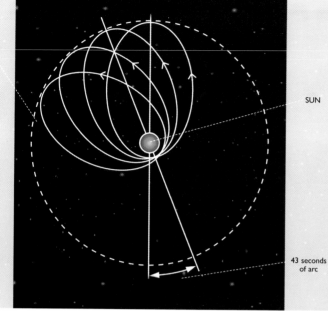

Orbit of Mercury

SUN

43 seconds of arc

▲ *Mercury's Advance of Perihelion*
Einstein's theory of general relativity predicts that the perihelion of Mercury's orbit around the Sun is advancing at 43 seconds of arc per century (discounting other gravitational influences). The slowly shifting orbit will trace out a full circle every three million years. For clarity, the advance has been greatly exaggerated on the diagram.

RELATIVISTIC COSMOLOGIES

The real breakthrough in our understanding of the Universe came with **Einstein**'s publication of his theory of general relativity in 1916. This provided not only a new interpretation of gravity but also introduced new concepts of space and time. Whereas in Newtonian physics space was just an emptiness through which bodies moved, in general relativity space has a character of its own, possessing curvature and with the possibility of it being either expanding or contracting. In general relativity cosmology is the study of space and time as well as the bodies within it.

When Einstein started to consider cosmological models based on general relativity he found the same problem that Newton had faced 250 years before – namely, how to treat gravity in an infinite Universe. Einstein could only find models in which space was either contracting or expanding. Today this is not seen as a problem as it is known that the Universe is indeed expanding, but Einstein was working a decade before Hubble's work on the recession of distant galaxies. Einstein, like other cosmologists at that time, was looking for a static Universe and to achieve this he introduced an extra term in his equation, which he referred to as the cosmological constant, to prevent his models expanding or contracting. The need for a model of a static Universe disappeared with Hubble's work and Einstein immediately realized that his 1917 models which predicted the expansion had been on the right track after all.

EARLY COSMOLOGICAL MODELS

General relativity has now become the basis for all cosmological models. Fundamental to these models is the work of two cosmologists whose characters could hardly be more different. The Russian mathematician Aleksandr Friedmann (1868-1925), working in St Petersburg, the city of his birth, published his model in 1922. Friedmann saw cosmology as an exercise in mathematics and he made little attempt to relate it to the physical Universe: he did not, for instance, link his model with Hubble's work despite it clearly showing the possibility of an expanding Universe. Indeed, it is said that Friedmann only became interested in cosmological models after noticing an error in Einstein's 1916 work, which Einstein first disputed and then later accepted. Today, cosmological models with a cosmological constant of zero are termed Friedmann universes.

Independently of Friedmann the Belgian Georges Lemaître (1894–1966) produced a similar model, but one more closely linked to the physical world. Lemaître studied engineering at the University of Louvain and was decorated with the Belgian Croix de Guerre during World War I. After the war he changed his field of study to mathematics and physics, then in 1923 he was ordained as a Catholic priest, a calling which many would see as being at odds with work on models of the Universe. Lemaître discussed the role of radiation in the Universe and also the relationship between expansion and **redshift**. He was also the first cosmologist to point out that an expanding Universe implied both a finite age and an origin in a point of extreme density. The latter he called the 'primeval atom'. For his work Lemaître is often called 'the father of the Big Bang'. His role in suggesting a physical origin to the Universe, a question which had previously been the domain of religion, is somewhat ironic in view of his vocation.

▼ *An early cosmological model* attributed to Nicolaus Copernicus. Astronomy has come a long way from its early beginnings with Copernicus and his heliocentric Solar System, which correctly envisaged Earth and all the other planets revolving around the Sun.

THE COSMOLOGICAL CONSTANT

In order to overcome the problem of unstable static models Einstein introduced an extra term into his equations. He called it the cosmological constant (although it is, in fact, a force rather than a constant). In essence the constant allowed for the existence of a very weak repulsive force, detectable only on the very large scales associated with cosmology, which would balance gravity and so allow a static Universe. Hubble's subsequent discovery of the recession of the galaxies removed the need for a static Universe and Einstein retracted his proposal for a cosmological constant, calling it 'the biggest blunder' of his life.

Other cosmologists were less willing to abandon the idea. For example Lemaître proposed models of the Universe using several different values for the constant. He also showed that by introducing the constant it was possible to remove the conflict which existed at the time between different estimates of the age of the Universe. Moreover the assumption of space being filled with a hitherto unknown form of energy now seems less unreasonable because of developments in quantum mechanics. According to quantum mechanics the vacuum is not empty, but filled with a sea of virtual particles. Many cosmologists now consider that observations alone can demonstrate the existence, or otherwise, of a cosmological constant.

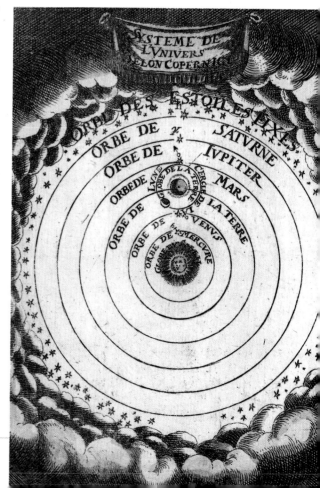

SYSTEME DE L'VNIVERS SELON COPERNIC

ORBE DES ESTOILES FIXES

ORBE DE SATVRNE
ORBE DE IVPITER
ORBE DE MARS
ORBE DE LA TERRE
ORBE DE VENVS
ORBE DE MERCVRE

PAST, PRESENT & FUTURE

Einstein ▶ p. 88 W. Herschel ▶ p. 44 Hubble ▶ p. 122 Newton ▶ p. 42

PAST, PRESENT & FUTURE

THE EXPANSION OF SPACE

In developing mathematical models of the Universe, cosmologists have been guided by the 'cosmological principle', the notion that the Universe must look much the same from any point in space. Although not yet fully supported by observation, the principle underpins most of the progress in theoretical cosmology. In particular, the principle is consistent with models from general relativity in which the expansion of the Universe is a consequence of the uniform expansion of space itself. An extension of the principle holds that the Universe must look the same at all points in time as well as in space, and this provided the philosophical basis of the Steady State theory, the main competitor to the Big Bang model until the late 1960s.

THE COSMOLOGICAL PRINCIPLE

Ever since the sixteenth century, when **Copernicus** showed that the Sun, and not Earth, lay at the centre of the **Solar System**, astronomers have been increasingly reluctant to accept that Earth is in a privileged position in the Universe. The strength of this view was reinforced in the early twentieth century when Harlow Shapley showed that the Sun did not lie at the centre of the Galaxy and later when our Galaxy was shown to be but one of innumerable similar galaxies. This view that there was nothing special about Earth, or the Sun, or even our own Galaxy has become enshrined in what is termed the 'cosmological principle'. This principle states that on the large scale the Universe is isotropic (it looks the same in all directions) and homogenous (it looks the same from all points). It follows directly from the principle of homogeneity that the Universe can have no edge.

The problem with the cosmological principle is in defining what is meant by the 'large scale', a definition which is far from trivial. The Solar System clearly does not satisfy the principle. Neither does the Galaxy nor the **Local Group**. It is only beyond the scale of **superclusters** of galaxies that the principle may hold. Despite the fact that the principle cannot, as yet, be justified by observation it has nevertheless played an important role in theoretical cosmology, as any cosmological model which does not conform to the principle is considered unsatisfactory.

THE STEADY STATE THEORY

In the 1950s some cosmologists argued that the cosmological principle ought to be extended so that the Universe should not only be homogenous in space, but also that its large-scale properties should not change with time. To differentiate such an extension from the basic principle the extension became known as the 'perfect cosmological principle'. The perfect principle led to the development of the 'Steady State' theory of the Universe, primarily by Austrian-born English physicist Hermann Bondi (b. 1919), the Austrian-American Thomas Gold (b. 1920) and English astronomer Fred Hoyle.

Since the expansion of space is moving galaxies apart and, therefore, reducing the overall density of the Universe, the perfect cosmological principle requires that new matter is continuously created in space, this matter eventually condensing to form galaxies and so maintaining a constant density of the Universe. The required rate of creation is surprisingly low, about one hydrogen atom per

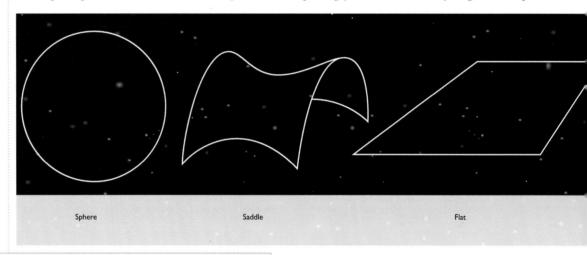

Sphere   Saddle   Flat

▲ *Cosmological Models*
Three two-dimensional models are commonly used to visualize the curvature of the Universe. The sphere has a closed, finite surface with no edge. The saddle, whose shape would extend geometrically, represents an open, infinite universe. The intermediate model is the open universe with a flat surface.

cubic metre of space every 10^{10} years. The proponents of the Steady State theory were also influenced by Mach's principle, which implies that the laws of physics would change in an evolving Universe. However, despite its attraction most cosmologists abandoned the Steady State theory after the discovery of the **cosmic background radiation** in 1965. The radiation was a natural consequence of the Big Bang, but could not be explained by the Steady State theory.

FRED HOYLE

The English astrophysicist and cosmologist Fred Hoyle (b. 1915) was born in Bingley, Yorkshire and after being educated at the University of Cambridge he taught there, becoming Plumian Professor of Astronomy and Experimental Philosophy in 1958. In 1972 he left Cambridge to become Professor-at-Large at Cornell University. In 1948, together with Hermann Bondi and Thomas Gold, Hoyle proposed the Steady State theory of the Universe. Hoyle was also responsible, together with Margaret and Geoffrey Burbidge (b. 1925) and William Fowler (1911–95), for the theory of nucleosynthesis, the production of the heavy chemical elements in nuclear reactions in stars. The theory – which became known by the acronym B²FH after the initials of the four – applied nuclear physics to astrophysics, and argued that the synthesis of the elements found on Earth was the result of a series of nuclear reactions which occur in supernova explosions and the final stages of evolution of red giant stars.

In 1967 Hoyle founded the Institute of Theoretical Astronomy in Cambridge of which he was the first director. More recently he has popularized the idea, with Chandra Wickramasinghe, that life did not originate on Earth but evolved in space and was brought to our planet by **comets**. Hoyle is also a successful author, both of popular books on science and of science-fiction. He was knighted in 1972.

comet ▶ p. 337 cosmic background radiation ▶ p. 337 Doppler shift ▶ p. 337 Local Group ▶ p. 340 quasar ▶ p. 342 Solar System ▶ p. 343 supercluster ▶ p. 344 wavelength ▶ p. 344

MACH'S PRINCIPLE

Ernst Mach (1838–1916) was an Austrian physicist and philosopher whose ideas influenced many contemporary physicists. Although his name is now chiefly associated with the ratio of the speed of an object to the speed of sound in the same medium – the Mach number – he made important contributions in many fields, including cosmology. He argued that the laws of physics and the physical

▲ *The Austrian philosopher and physicist Ernst Mach formulated Mach's principle, which argued that a body's inertial mass (its reluctance to be accelerated), such as that of a jumbo jet in flight, is not a property of the aircraft itself but a product of the interaction occurring between that object and the surrounding matter in the Universe.*

constants, such as the speed of light, were determined by the matter in the Universe. Therefore, if the overall properties of the Universe, such as its density, changed then the laws and the constants should also change. In particular he argued that the inertial mass of a body, i.e. its reluctance to be accelerated, is not a property of the body itself, but is an interaction between the body and the rest of the matter in the Universe. This is known as Mach's principle. For this inertial effect to be true, i.e. so that it is not solely the matter in the Solar System which affects local objects, but distant matter as well, then the force must be very long range. Einstein was greatly influenced by Mach's principle when developing his theory of general relativity. Ironically Mach was not convinced by relativity and, indeed, opposed it.

Mach's principle has influenced several cosmological models. Those proposed by **Paul Dirac** in 1937 and Pascual Jordan in 1944 involved a change in fundamental constants in an evolving Universe, while a belief in the constancy of the laws and constants of physics led to the formulation of the Steady State theory.

THE NATURE OF THE COSMOLOGICAL REDSHIFT

When the redshifts of distant galaxies were first discovered it was initially assumed that these represented a **Doppler effect** similar to that produced by the motions of stars in the Galaxy. Indeed, many popular accounts of cosmology still give this interpretation. When the first cosmological models based on general relativity were proposed it was recognized that there was another explanation of the redshift, based on the expansion of space–time. General relativity predicts that as radiation travels through space its **wavelength** increases to keep up with the expansion. On this view the shift is a progressive stretching, occurring throughout the journey rather than due to relative motion between source and observer. As a consequence, light from a more distant galaxy, which has been travelling through space for a longer time, has had its wavelength stretched by a larger factor.

By this interpretation the redshift is not only a measure of distance but also a measure of the relative size of the Universe when the light was emitted. It can be shown that radiation of redshift z was emitted when the Universe was a fraction $1/(1 + z)$ of its present size. So light from a distant **quasar** with a redshift of $z = 4$ started on its journey when the Universe was only one fifth

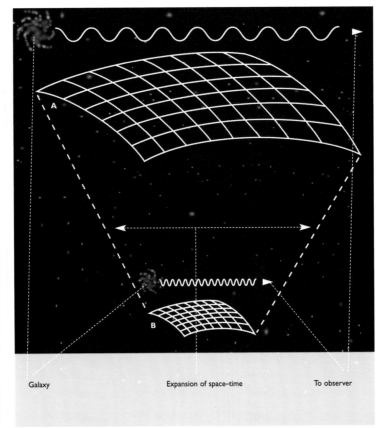

Galaxy   Expansion of space–time   To observer

the present size and its wavelength is now five times as long as when the light was emitted. Since volume is proportional to the cube of distance, it follows that any given volume of space has expanded by a factor of $5 \times 5 \times 5 = 125$ since then. At $z = 4$ the same number of galaxies were contained in a volume 125 times smaller than today, implying that the density of the Universe was 125 times greater than it is now.

LOOK-BACK TIME

The cosmological redshift from a distant galaxy is a measure of how long the radiation has been travelling through expanding space. It is, therefore, a measure not only of the distance to the galaxy, but also of the age of the Universe when the radiation was emitted. The difference between the present age of the Universe and its age at the time of emission is known as the look-back time.

Unfortunately the relationship between the redshift and the look-back time depends on how rapidly the Universe has expanded in the past which, as we have seen, is still very uncertain. But in all models of the Universe the greater the redshift the greater the look-back time, making redshifts both a convenient and a reliable way of putting observations of the Universe in chronological order. However, the exact relationship between redshift and look-back time differs from model to model. As an example, in a matter-dominated universe with a density equal to the critical density (i.e. similar to our Universe), the Universe is seen at about one third its present age at a redshift of $z = 1$, and at about 9% of its present age at a redshift of $z = 4$, the largest galaxy redshift commonly seen with large telescopes.

In general cosmologists prefer to use redshift instead of distance and look-back time instead of age, when talking about remote galaxies.

◀ *Cosmological Redshift*
According to general relativity, as radiation travels through space its wavelength increases to match the expansion of space–time. The wavelength of light from a more distant galaxy (A), which has been travelling through space for a longer time, is consequently longer than light from the nearer galaxy (B).

PAST, PRESENT & FUTURE

THE GEOMETRY OF THE UNIVERSE

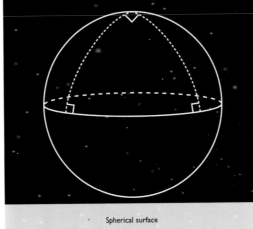

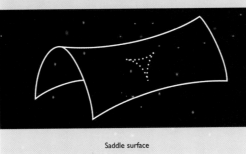

General relativity tells us that the Universe as a whole may be curved. The only way to find out is by making sensitive measurements. One method is to see how the numbers of galaxies change with distance into space, but this presents problems because the galaxies themselves have been changing with time. Another is to measure the apparent sizes of distant galaxies, but that too is inconclusive. The best method is to measure the size of the fluctuations in the cosmic background radiation, and such observations seem to imply that the Universe is not curved but flat. If correct, that has profound implications for the future of the Universe.

THE CURVATURE OF SPACE

Modern cosmological models, based on general relativity, predict that there will be an overall curvature of space–time in the Universe, super-imposed on the small-scale distortions due to the gravitational fields around local concentrations of **mass**.

Visualizing curvature in four dimensions is difficult and it is easier to consider a two-dimensional analogy. Consider first a sphere. A sphere has a closed surface which is finite in extent, but has no edge. Such surfaces are classified as having positive curvature. A universe with positive curvature would be a closed universe. Surfaces with negative curvature are infinite in extent, an example being a saddle. The saddle here is not the conventional one which clearly has finite

Spherical surface

Saddle surface

▲ *Geometry on Curved Surfaces*
On a spherical surface the angles of a triangle always add up to more than 180°, whereas on a saddle surface the angles always add up to less than 180°. These properties could be used to calculate the curvature of space-time if it were possible to accurately measure sufficiently large triangles in space.

dimensions, but the mathematical equation of a saddle which extends the surfaces to infinity. A universe with negative curvature would therefore be infinite – an open universe. An intermediate state between these two extremes would be a flat universe. The flat surface is open, with neither positive nor negative curvature.

CAN WE OBSERVE THE CURVATURE?

The curvature of space-time might be expected to produce bizarre effects. For example, in a closed universe light could travel right around the Universe and create multiple images of galaxies. If observed, such an effect would confirm that the Universe was closed, but unfortunately it cannot be used as an observational tool because we can see no farther than the boundary of the observable Universe.

Yet observations should still allow us to detect and measure the curvature of the Universe by measuring the geometrical properties of space-time. The method of constructing large triangles and measuring the angles is not open to cosmologists, but there are two other geometrical properties which can be used. One is the relationship between the radius of a sphere and its volume, the other the relationship between the radius of a sphere and its surface area. Both of these depend on the geometry of space and are used when calculating the number of objects within a given distance from Earth, and in calculating the apparent brightness of sources at different distances.

GALAXY COUNTS

As we have seen, William Herschel realized that he could investigate the distribution of stars in the Milky Way by counting the numbers of stars with different apparent brightnesses. The apparent brightness of a star falls off with distance because the star's radiation is being spread out over a larger and larger spherical surface. The relationship between apparent brightness and distance therefore depends on the relationship between the surface area of a sphere and its radius. In flat (Euclidean) space this effect leads to the well-known inverse-square law, but the relationship would be different in curved space–time.

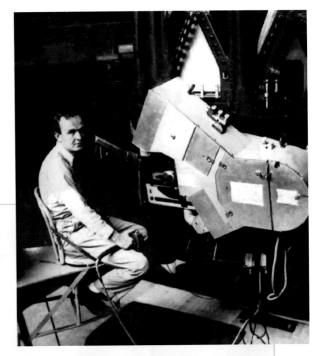

▲ *Allan Sandage, Edwin Hubble's former assistant and now one of the world's leading cosmologists, has been at the forefront of attempts to measure the precise value of the Hubble constant.*

ALLAN SANDAGE

The American astronomer Allan Rex Sandage (b. 1926) was born in Iowa City and studied at Illinois University and Caltech before joining the Hale Observatories as an assistant to Edwin Hubble. Sandage was a major figure in observational astronomy in the second half of the twentieth century and has played a leading role in many of the controversies which have been a feature of the subject. His early work was a continuation of Hubble's attempt to measure the

geometry of the Universe using optical observations of distant galaxies. In 1960 he made the first optical identification of a quasar. With Thomas Matthews (1824–1905) he found a faint optical object coincident with the compact radio source 3C48. Sandage noted the object's unusual spectrum, which was soon shown to be due to a large redshift. More recently he has engaged in a vigorous debate over the value of **Hubble's constant** and hence the age of the Universe, arguing for a low value of the constant and a large age for the Universe.

Hubble constant ▶ p. 340 mass ▶ p. 341 radio galaxy ▶ p. 343

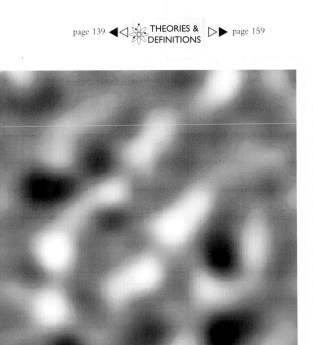

▲ **Fluctuations in** the cosmic background radiation, as observed in the constellation Ursa Major by the Cosmic Anisotropy Telescope (CAT), a radio telescope at Cambridge, England. CAT looks for fluctuations on a smaller scale than the COBE satellite. Each 'blob' represents a patch of sky which is about 0.0001 degrees hotter (white) or cooler (black) than average. These hot and cold blobs, seen here as they were only 100,000 years after the Big Bang, will by now have grown into large superclusters of galaxies.

✳ THE 3C CATALOGUE OF RADIO SOURCES

Astronomers over the centuries have compiled catalogues of certain classes of objects containing their positions and other information. A catalogue involves a great deal of painstaking observation and provides a framework in which astronomers can do their research. Some catalogues remain in use for centuries – the numbers in Messier's catalogue of nebulae drawn up in the eighteenth century are still used to identify the brighter nebulae.

A widely used catalogue of radio sources is the Third Cambridge (3C) Catalogue which contains information on 471 objects measured by Cambridge radio astronomers in their attempt to use radio sources as a test of cosmological models. The data came from a survey of the sky visible from Cambridge made with a multi-element interferometer developed by **Martin Ryle** (1918–84) and his colleagues. The survey had unexpected results, leading to the discovery of compact radio sources which, after optical identification, proved to be quasars. 3C48 was the first quasar to be identified.

Many well-known sources are still identified by their number in the 3C catalogue, perhaps the best example being 3C273, the brightest quasar. Of course many objects occur in different catalogues: one such is the Crab nebula which is M1 in the Messier catalogue and 3C144 in the 3C Catalogue.

Herschel assumed that all stars were emitting the same amount of light and that space was flat. He then used his observations to determine the distribution of stars in space. In modern cosmology we assume that, on average, galaxies have a standard luminosity and are distributed uniformly in space, consistent with the cosmological principle. We can then use galaxy counts to determine the curvature of space–time, an idea which was first used by Hubble and continued by his assistant Allan Sandage. However we need measurements over a very large scale to detect the curvature and optical telescopes do not, at present, detect galaxies at large enough distances. A related but more sensitive test, which has been exploited by Sandage and others, is to plot the apparent brightness of a galaxy against its redshift, but this has also so far failed to distinguish between different models.

RADIO GALAXY COUNTS

When **radio galaxies** (galaxies which are unusually powerful radio emitters) were discovered in the 1950s it was clear that they could be detected at much greater distances than could normal galaxies using optical telescopes. Although very little was known about these galaxies, they were thought to be suitable objects with which to probe the structure of the Universe. Two groups of radio astronomers, in Cambridge in the UK and at Sydney in Australia, surveyed the sky and counted the number of galaxies at different apparent brightnesses. Unfortunately the data showed that, on average, radio galaxies at greater distances (and seen at an earlier epoch in the Universe) were more powerful than the nearer ones. Such an evolution in the population of radio galaxies meant that they could not be used as probes of the geometry of space. When optical astronomers tried to use quasars to measure the curvature of space they found the same problem. However the radio surveys did provide the first evidence against the Steady State theory which, while permitting individual galaxies to evolve, required the properties of the whole population of radio galaxies to remain constant.

APPARENT SIZES

In flat space the apparent (i.e. angular) size of an object varies inversely as its distance. So, if the distance is increased by a factor of two, the apparent size of the object is halved. In curved spaces the relationship between apparent size and distance is different and if this relationship is measured it can be used to determine the curvature of space–time. An interesting effect which occurs in some cosmological models is that the apparent size of an object does not continue decreasing with increasing distance – at a certain distance the apparent size reaches a minimum and beyond that distance it increases again.

Again this technique, as with galaxy counts, is limited by our ability to detect galaxies at sufficiently large distances. However, even if that obstacle were overcome, there is evidence that galaxies have evolved, with many of the large elliptical galaxies in the present Universe having formed from the merging of small spiral galaxies. It is possible, therefore, that observations of the apparent size of clusters of galaxies would offer a better means of detecting the effects of curvature in the future.

STRUCTURE IN THE BACKGROUND RADIATION

The most convincing evidence about the curvature of space comes from an analysis of the small variations across the sky in the intensity of the cosmic background radiation, emitted at the recombination epoch when the Universe was only 300,000 years old. These variations correspond to small fluctuations in the density of the primordial gas, which would subsequently lead to the formation of galaxies. By comparing the observed size ('scale size') of these variations with the expected size, we can determine the geometry of the Universe. The most recent measurements of the scale size were obtained in 1998 by balloon-borne detectors flying high over Antarctica – the BOOMERanG experiment. To a very good approximation, space turns out to be flat. In other words the mass density of the Universe is at the so-called critical value.

▼ **The Hubble Space Telescope** has dramatically increased the number of known galaxies. But even this number has not yet led to to a definitive conclusion on the geometry of space. The yellow arrow indicates a highly redshifted galaxy, one of the most distant ever seen.

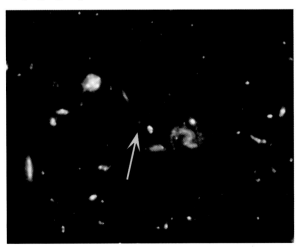

PAST, PRESENT & FUTURE

The Universe: Past, Present & Future
LARGE-SCALE STRUCTURES

O ne of the key problems in cosmology is the formation of large-scale structures such as galaxies and clusters of galaxies. We have no observations of galaxies being formed, but they are expected to have grown by gravitational condensation around slight fluctuations in the initial density of the primordial gas. Since most of the mass in the Universe is locked up in the mysterious dark matter, this will certainly have played a central role in the process and perhaps a decisive role. As for the fluctuations themselves, they may have arisen from random quantum events in the first instant of the Universe.

WHEN DID GALAXIES FORM?
Galaxies are the building blocks of the Universe and so one of the key problems of astronomy is how galaxies formed, why they have the sizes they do and why there are distinct types. A major obstacle to our understanding is that we have no observations of galaxies being formed. The **cosmic background radiation** gives us a picture of the Universe when it was 300,000 years old, a time when there was very little structure and no galaxies. By contrast the farthest back in time we can currently detect individual galaxies is at **redshifts** of $z \sim 5$ which corresponds to when the Universe was

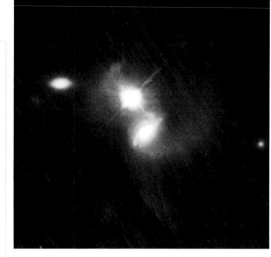

▲ **This Hubble Space Telescope** image of a distant quasar reveals that it consists of at least two galaxies in the process of merging. Such images support the idea that present galaxies were formed by the accumulation of smaller ones.

about one billion years old. Although the galaxies we see from that era look different from those in the present Universe, the major part of galaxy formation had clearly taken place by that time. We return to this topic in Contents of the Cosmos.

CLUSTERING OF GALAXIES
Galaxies are not uniformly distributed across the sky, but are found in clusters, a feature which has been recognized since **Hubble**'s work early in the twentieth century. Structure on scales larger than clusters is difficult to discern from two-dimensional images of the sky and so cosmologists use measurements of redshifts as an indication of distance and prepare maps of slices of the sky with redshift distances drawn radially from the centre. Such maps show that the clusters of galaxies are themselves arranged in **superclusters**. On even larger scales these superclusters link to form filamentary structures with vast voids separating them. On the largest scale the Universe therefore has a sponge-like structure.

Galaxy clusters are probably the largest bound structures in the Universe, that is the largest structures in which the individual components are held together by gravity and are unable to escape. Larger structures are yet to form, though were one able to study the Universe in the far distant future we would expect that much larger structures would have had time to form.

GRAVITATIONAL CONDENSATION
A key process in the formation of large-scale structures must be condensation under gravity. Once any region of the primordial gas becomes slightly denser than its surroundings it will continue to accumulate material by gravitational attraction. The English physicist Sir James Jeans studied the formation of stars by this process. He showed that for a gas cloud with a certain temperature and density, condensation would occur only if the cloud was above a critical size. The same ideas can be applied to galaxy formation with the added factor of the expansion of the Universe.

Two very different processes could explain the very large range of structures found in the Universe. One is the 'top-down' approach, in which the largest structures formed first but then fragmented to create the smaller structures: this is what happens when stars form. The second possibility is 'bottom-up', with the smallest structures, the galaxies, forming first and larger structures being created by galaxies merging. Evidence from the Hubble Deep Field Survey in 1995 shows that there were many more small galaxies in the early Universe than there are now. It is thought that the giant **elliptical galaxies** found at the centres of clusters formed from the merging of these smaller galaxies, and so the bottom-up picture is currently favoured.

The application of Jeans's theory to galaxy formation suggests that the first condensations in the Universe were on the scale of globular star clusters (less than one million solar masses) and as many as a million of these condensations merged to create each of the galaxies which occupy the present-day Universe.

THE ROLE OF DARK MATTER
Since **dark matter** is thought to make up around 90% of the matter of the Universe, it must have played a key role in any gravitational condensation processes. Our knowledge of the properties of dark matter is limited because we have not, as yet, detected in the laboratory the particles which make it up. Without this knowledge, the role played by dark matter remains uncertain, and different scenarios have been

JAMES JEANS
Sir James Jeans (1877–1946) was a British mathematician and physicist who made several contributions to astrophysics and cosmology. Much of his early career was spent teaching at the University of Cambridge, in England, and at Princeton University in the US, but in 1923 he joined the staff of the Mount Wilson Observatory in California where he remained for 20 years.

In 1902 Jeans propounded his theory of star formation in which he showed that a cloud could only condense under gravity if it met certain initial conditions of density, size and temperature. His findings remain the starting point for all modern work on the formation of stars and galaxies. Rather oddly, perhaps, he rejected the idea that the **Solar System** had condensed in a similar fashion, preferring the view that the material which formed the planets had been drawn out of the Sun by the gravitational pull of a passing star.

He later worked on cosmology and in 1928 was the original proposer of the Steady State theory of the Universe, in which matter was continuously created to maintain the average density of the expanding Universe.

Jeans was a distinguished popularizer of science and wrote several books on physics and astronomy as well as being a well-known lecturer and broadcaster. He was knighted in 1928.

THE HUBBLE DEEP FIELD SURVEY

The deep field survey carried out by the Hubble Space Telescope in 1995 revealed galaxies much farther away than any previously studied. The images of the sky were taken in the direction of the north galactic pole, away from the disk of the Milky Way with its obscuring dust and gas. Over 300 exposures were made, each with an exposure time of 15-40 minutes and each in four colours to cover the spectrum from infrared to blue. The faintest galaxies captured by the survey are four billion times fainter than can be detected by the human eye. Although the survey covered an area of sky with diameter only 3% of the that of the Moon, more than 1,500 separate galaxies were identified. One of the most interesting features is the large number of faint blue galaxies which were present in the early Universe but are not present now. One explanation is that the present elliptical galaxies were formed from mergers of these early, smaller galaxies.

◀ **Three portions** of the sky seen in the Hubble Deep Field (HDF) observation, the deepest-ever view of the distant Universe, like a cosmological core sample. These views capture an assortment of galaxies at various stages of evolution, some dating back to within a billion years of the Big Bang.

studied. Cosmologists distinguish between 'cold dark matter', which would consist of slow-moving massive particles, and 'hot dark matter' which would consist of fast-moving particles much lighter than an **electron**. Structure formation from hot dark matter is considerably less efficient, because the fast-moving particles are not easily condensed, and the most successful models are based on cold dark matter. Direct detection of dark matter particles would revolutionize these studies.

STRUCTURE FROM THE INFLATIONARY ERA

Gravitational instability explains how condensations grow once they are formed, but some initial irregularity is needed to start the process. It is now widely believed that such irregularities probably arose in the inflationary era when the Universe was extremely young. The **Heisenberg**

▲ **Inflation in the early Universe** is thought to be a phase change, rather like that which takes place when water freezes, the ice crystals corresponding to structures which could have formed at that stage.

Uncertainty Principle ensures that even apparently empty space is a seething turmoil of particles popping in and out of existence. The rapid expansion during inflation – by a factor of at least 10^{30} – stretches these quantum fluctuations until their size is astronomical rather than microscopic. One can say that while inflation is trying its best to make the Universe perfectly smooth, **quantum mechanics** defeats it and ensures that residual irregularities are left over. Which is just as well, because from these irregularities galaxies, stars and eventually we ourselves were formed. According to the inflationary model, all structures in the Universe owe their existence to the quantum nature of reality.

SURVEYS OF LARGE-SCALE STRUCTURES

One of the early problems with studying the large-scale structure of the Universe was that while mapping positions of galaxies on the sky is straightforward, measuring redshifts to distant galaxies was a very slow process. This problem has now been overcome by two surveys which use optical fibres to increase the rate at which redshifts to distant galaxies can be measured.

The first survey, begun in 1998, was the '2dF' (Two Degree Field) using the 3.9-m (12.8-ft) Anglo-Australian telescope in New South Wales. During each 45-minute exposure optical fibres pipe the light from up to 400 galaxies at a time

into one of two spectrographs. As many as 2,000 redshifts can be measured each night, and the survey aims to measure redshifts of 250,000 galaxies in an area of the sky close to the north and south galactic poles.

The second survey, the SDSS (Sloan Digital Sky Survey – the Alfred Sloan Foundation provided the initial funding), is a joint American-Japanese experiment based at the Apache Point Observatory in New Mexico. SDSS is a five-year project covering about one quarter of the sky close to the north galactic pole, but also scanning a section of the southern sky for much fainter objects. SDSS is actually two surveys, since both photometric and redshift information is being collected.

▲ **The Sloan Digital Sky Survey (SDSS)** is a major project to map the large-scale distribution of galaxies in the Universe, allowing cosmologists to refine their theories of how the Universe evolved.

Hubble ▶ p. 122

LARGE-SCALE STRUCTURES

PAST, PRESENT & FUTURE

The Universe: Past, Present & Future

THE FUTURE OF THE UNIVERSE

THE BIG CRUNCH

T he evidence suggests that the Universe was created in a single cataclysmic event – the Big Bang – and we have looked at the models that cosmologists have proposed to explain the present form of the Universe and its contents. In this section we consider how the Universe will end. First we look at the density of the Universe and how it determines whether the Universe will continue to expand or else stop expanding and contract to a Big Crunch. At the moment cosmologists are leaning towards the view that the Universe will expand forever, and to make sense of the observations they are dusting off an old idea that Einstein abandoned long ago – the 'cosmological constant'.

THE FUTURE OF THE EXPANSION

Since gravity is expected to slow the expansion of the Universe down, one factor which affects the future rate of expansion is the average density of the Universe. To determine the future of the expansion, we need to know how the density compares with the so-called critical density, which is the density for which the expansion will come to a stop in the infinite future. The density is often expressed in terms of a 'density parameter' which is given the symbol Ω (Greek 'omega'). When expressed this way, $\Omega = 1$ means that the density of the Universe is equal to the critical density. If Ω is greater than 1 the expansion will eventually stop and the Universe will start to contract. Such a Universe will return to a very hot, dense state which is usually called the Big Crunch. A Universe with Ω less than 1 will continue to expand forever.

The density of the Universe also determines the geometry of space–time. If $\Omega = 1$ then space–time is Euclidean, i.e. the Universe is flat. If Ω is greater than 1, space–time is positively curved, like a spherical surface, while if Ω is less than 1, space–time is negatively curved, like a saddle surface.

The critical density is directly related to **Hubble's constant**, and for the current best value of the constant the critical density is about 1×10^{-26} kg/m³, equivalent to about six hydrogen atoms per cubic metre.

This all assumes that **Einstein**'s cosmological constant, the hypothetical force he introduced to balance gravitational attraction, is zero. If the constant turns out not to be zero, then the critical density alone will not be enough to determine the fate of the Universe, but if Ω is less than 1 the expansion will continue indefinitely.

THE PRESENT DENSITY OF THE UNIVERSE

So what is the average density of the Universe? Several components contribute towards it. One, of course, is the density of normal matter, comprising **protons**, **neutrons** and **electrons** and now mostly existing in the form of stars. The most reliable way of measuring this component is not by adding up the amount of visible matter in the present Universe, but from observations of the amount of deuterium and lithium produced in the early Universe. A second

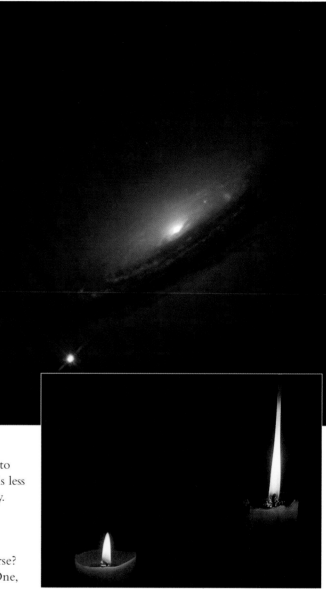

▼ *A Standard Candle*
The calculation of Hubble's constant involves measuring the distances to remote galaxies. This can be done using the common Type Ia supernovae as 'standard candles' – a standard radiation source whose luminosity acts as a reference from which the distance of the host galaxy can be deduced. The main picture shows a Type Ia supernova (the bright spot at lower left) in the outskirts of galaxy NGC 4526.

component is the mass equivalent of the **cosmic background radiation**, once the dominant component but now negligible. The largest but most uncertain component is the **dark matter**, which may exceed the density of ordinary matter by a factor of 10 or more. Taking all these into account the best estimate of the average density is about 2×10^{-27} kg/m³. This is about five times smaller than the critical density. From density measurements alone, it appears that the Universe will continue to expand.

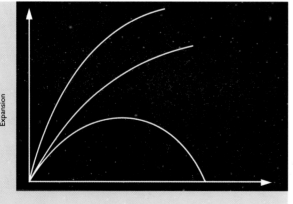

◀ *Types of Expansion*
The scale of the Universe changes differently over time, depending on which of three types of universe is predicted: open, flat or closed. (A) An open universe expands forever. (B) A flat universe also expands forever but at a rate that diminishes ever-closer to zero. (C) A closed universe expands to a finite size then collapses.

Expansion

Time

PAST, PRESENT & FUTURE

THE FUTURE OF THE UNIVERSE

VARIATION OF THE HUBBLE CONSTANT WITH TIME

Are there any other ways of predicting the fate of the Universe? Hubble's constant is a measure of the expansion rate of the Universe, so if we could measure the constant at an earlier era and compare it with the present value we should be able to tell whether the expansion is slowing, quite independently of measurements of the density. This is possible because the finite speed of light means that we see very distant **galaxies** in the Universe at an earlier era. However, in order to measure Hubble's constant in the past we need to measure the distances to these very remote galaxies which is very difficult as they are too far away to detect **Cepheid variable** stars in them. An alternative, more powerful 'standard candle' is the Type Ia **supernova**. Unlike other forms of supernovae these are all thought to release the same amount of energy and to reach the same peak **luminosity**. Unfortunately they are rare, occurring in a typical galaxy at a rate of a few per century, so many galaxies must be monitored regularly to find enough supernovae to make useful measurements of distance.

A large-scale collaboration of astronomers, known as the Supernova Cosmology Project and led by Saul Perlmutter at Berkeley, USA, has investigated the expansion of the Universe using distant supernovae. By monitoring large numbers of remote galaxies the team has discovered 43 supernovae in galaxies with **redshifts** between 0.2 and 1.0. Surprisingly they found that Hubble's constant was *smaller* in the past, indicating that the Universe was expanding more slowly than it is now, quite the opposite of what was expected. In other words the expansion of the Universe is accelerating.

RETURN OF THE COSMOLOGICAL CONSTANT

Such an acceleration requires a repulsive force to overcome gravity, a force similar to the cosmological force (called the cosmological constant) first introduced by Einstein but then abandoned. Further evidence for the existence of a cosmological force comes from the BOOMERanG measurements of the variations in the intensity of the cosmic background radiation across the sky. As we have seen, these measurements showed that the geometry of space is flat, which is just what we would expect if the very early Universe had an inflationary phase. Flat space goes hand in hand with the density of the Universe being equal to the critical density, but the measurements show that the density falls far short of this, even when dark matter is included.

MISSING GALAXIES

Bright, nearby galaxies are easily recognized by their characteristic **spiral** or **elliptical** shape. More distant galaxies are fainter and smaller but, up to certain distances, can be distinguished from stars by their fuzzy images.

In the last 10 years astronomers have come to realize that they may have overlooked some classes of galaxies, even in our own neighbourhood. One obvious case is that of small galaxies containing relatively few stars which are therefore less luminous. Less obvious cases are galaxies containing large numbers of stars, but where these stars are either more widely spaced, or more tightly packed, than usual. The galaxies with very widely spaced stars will have a low surface brightness and consequently will be very difficult to detect against the background light of the sky, a background which is present from all observing sites. By contrast, the images of galaxies with tightly packed stars may not be easily distinguished from those of stars in our own Galaxy. The images of stars should be much smaller than those of even the most distant galaxies, but they are often blurred, either by atmospheric turbulence or by diffraction effects in the telescope. To locate these compact galaxies it is usually necessary to obtain a spectrum of the source.

Present estimates of the number of galaxies which had previously been overlooked indicate that they could double the amount of normal matter in the Universe. However this still does not remove the need for dark matter, especially as dark matter is required within individual galaxies to account for the motions of their stars.

This discrepancy would be removed if a cosmological constant exists, because the energy associated with the force would itself contribute to the density of the Universe. If the Universe is indeed at the critical density, then the BOOMERanG result is consistent with two-thirds of the density coming from the cosmological constant and one-third coming from matter. In the future, as the Universe expands, the contribution from the matter will decrease, so the importance of the cosmological constant will increase and the expansion will continue to accelerate.

▶ *The BOOMERanG measurements of the cosmic background radiation have contributed to the growing belief in the existence of the cosmological constant: the factor that Einstein originally included in his theory, but later dismissed as his 'biggest blunder'.*

THE LABORATORY SEARCH FOR DARK MATTER

One of the biggest unknowns in measuring the density of the Universe is the nature and amount of dark matter. Theories indicate several possible candidates for the particles of dark matter, one of the most promising being WIMPs (weakly interacting massive particles). These are expected to be neutral particles with masses more than 100 times that of the proton. If dark matter is present in our Galaxy then WIMPs should be all around us and detectable in laboratories on Earth.

Several searches are underway, mostly using detectors set deep underground to avoid them being swamped by the effects of **cosmic rays**. One such experiment, deep in a potash mine at Boulby in Yorkshire, looks for **photons** created when a WIMP interacts with an **atomic nucleus** in a crystal of sodium iodide. But even though the mine blocks the cosmic rays, naturally occurring radioactivity still produces a detection rate a hundred times greater than that expected from WIMP interactions, making identification of real WIMPs very challenging.

Other experiments include one at the south pole which looks downwards through the ice cap and another being planned for a new powerful particle accelerator at CERN, the European Laboratory for Particle Physics in Switzerland. To date no unambiguous indication of a WIMP has been seen, although an Italian-Chinese group working beneath the Apennines has found more light flashes in summer than in winter. This would be expected since in summer Earth's velocity is in the same direction as the Sun's and so Earth sweeps faster through the Galaxy and through the WIMP cloud.

THE DEATH OF THE UNIVERSE

The evidence so far indicates that there is not enough matter in the Universe to halt the expansion – the Universe will expand forever, although not necessarily at the same rate. That means that the Universe itself will never end, though the same cannot be said of the objects within it. Stars will burn out, galaxies will collapse and even black holes will evaporate. And if current theories are correct, even the protons that make up matter may spontaneously decay at some distant point in the future.

HEAT DEATH

In the late nineteenth century Lord Kelvin (William Thomson) used thermodynamics to show that the Universe must be running down. In the present Universe energy exists in many different forms, such as gravitational, nuclear, thermal, etc. The conservation of energy tells us that we can convert one form to another, but that we cannot create or destroy energy. The second law of thermodynamics tells us that although we can convert other forms of energy into thermal energy – energy associated with the random motion of particles – the reverse process is never completely efficient, i.e. it is always easier to go one way rather than the other.

Kelvin therefore predicted that after a sufficiently long time all energy will be converted into thermal energy. Kelvin did not need to know

anything about the form of the Universe or of its structures to predict this: he was merely applying the fundamental laws of thermodynamics. He went on to note that since heat always flows from hot to cold everything in the Universe will eventually reach the same temperature. Once that has happened the thermal energy cannot be reconverted into other forms of energy: the Universe will have suffered 'heat death'; although the energy content of the Universe will still be high it will be randomized and useless, and life will be unable to exist.

Kelvin may have predicted the death of the Universe, but astronomers are still interested in the actual methods in which its structures will end. Here we explore those endpoints.

THE DEATH OF STARS

The way stars end their life cycles depends upon their **masses**. Stars with masses similar to the Sun end as **white dwarfs**, while those which are much more massive die catastrophically in supernova explosions. During the explosion the core of the star is compressed to extremely high densities, the final form the core takes being dependent on its mass. If the core has a mass of less than a few solar masses it becomes a **neutron star**, if the core is more massive than this it becomes a **black hole**.

New stars are continuously being created from interstellar gas, and some gas is recycled in supernova explosions, but eventually all the gas will be used and star formation will cease.

The rate at which stars go through their life cycle depends upon their masses, the lowest mass stars having lifetimes which are greater than the current age of the Universe. But eventually – present estimates are that it will take 10^{14} years, that is about 10,000 times the present age of the

Universe – the final generation of stars will have reached the end of their life cycles. The Universe will then consist of galaxies populated with white dwarfs, neutron stars and black holes.

THE DEATH OF GALAXIES

Galaxies are vast arrays of stars moving in orbits controlled by gravity. Because these orbits are not affected by the evolution of individual stars galaxies will eventually become systems of dead stars. But over a long period of time there are processes which alter the stellar orbits and so lead to an evolution of galaxies. In one process two stars approach each other close enough to exert

▼ *Death of a Star*
The Sun, a typical star midway through its probable 10-billion-year life, will end up as a white dwarf. But before that it will expand to become a swollen red giant, looming over the parched and deserted surface of Earth.

WILLIAM THOMSON, LORD KELVIN

William Thomson was born in Belfast in 1824, but moved to Glasgow at the age of eight when his father was appointed Professor of Mathematics at the university there. William was precociously talented and after graduating from Cambridge was himself appointed a Professor of Natural Philosophy in Glasgow when he was just 22. He held the post until 1899.

Thomson's early work was in electromagnetism where he gave a mathematical basis to the discoveries of Michael Faraday, paving the way for James Clerk Maxwell's great synthesis. He was also a practical man and was knighted for his

consultative work on the transatlantic telegraph cable which began operation in 1866.

He also helped develop the theory of thermodynamics, and took part in a spirited debate on the age of Earth, which he estimated from its rate of cooling to be as low as 20 million years – no-one then knew that the temperature of Earth is maintained by radioactive decay.

Thomson became Lord Kelvin in 1892, and is best remembered for the introduction of the absolute scale of temperature which bears his name. His scale is independent of any physical substance, in line with his desire to create an international system of standards. Kelvin died in 1907: he is buried in Westminster Abbey beside **Isaac Newton**.

THE FUTURE OF THE UNIVERSE

black hole ▶ p. 336 Hawking radiation ▶ p. 339 mass ▶ p. 341 neutrino ▶ p. 341 neutron star ▶ p. 341 positron ▶ p. 342 red giant ▶ p. 343 white dwarf ▶ p. 344

▲ *The Laser Interferometer* Gravitational Observatory (LIGO) at Hanford, *in Washington State is hoping to detect gravitational radiation from binary or collapsing stars. Such radiation, if it exists, may contribute to the death of galaxies.*

strong gravitational pulls on each other. In such approaches one star can occasionally be flung out of the galaxy. As a consequence the galaxy will gradually evaporate its stars. The second star loses energy and moves towards the galactic centre. Stars may also migrate towards the centre of the galaxy as they radiate gravitational radiation. As a result a black hole forms at the centre (if one does not already exist) and increases in mass until it eventually contains all the mass of the galaxy.

THE DEATH OF BLACK HOLES

Stephen Hawking has shown that black holes radiate energy by the loss of so-called **Hawking radiation**. This means that the black hole also loses mass, the rate of loss being inversely proportional to the mass of the hole. As a black hole loses mass this rate of evaporation increases, the black hole finally disappearing in a single burst of gamma rays. Calculations show that black holes with the mass of a star will survive for about 10^{65} years, while a black hole with the mass of a galaxy will have a lifetime of 10^{98} to 10^{100} years.

THE DEATH OF MATTER

Matter which is not captured by black holes will consist of protons, neutrons and electrons, together with the particles of dark matter. These fundamental particles are considered to be stable, but stability in this context means only that they have survived to the present age of the Universe, i.e. the particles have lifetimes greater than about 10^{10} years. Grand Unification Theories of the forces of Nature suggest that these particles may decay, allowing such reactions as a proton decaying into a **positron** and a gamma ray. This decay would mean the violation of the conservation of baryon number but, as we have seen, such non-conservation has already been invoked to explain the preponderance of matter in the Universe. The decay of the fundamental particles brings our interest in the Universe to an end: it has become a dark, empty place filled with photons and **neutrinos**, too low in density and energy ever to create matter again.

THE END OF LIFE?

It is not impossible that human life can outlive the death of the Sun. Present estimates are that the Sun will heat up sufficiently to render life on Earth untenable in about 1 billion years, long

💡 THE PROTON AND THE UNIVERSE

Some of the Grand Unification Theories (GUTs) predict that protons are not really stable particles. Several experiments have attempted to measure, or place a limit on, the lifetime of the proton. The most sensitive of these experiments is Super-Kamiokande, located outside Tokyo, Japan. The detector consists of 50,000 tonnes of water surrounded by 11,000 photomultiplier tubes, housed in a laboratory deep underground to shield it, as far as possible, from sources of background radiation. If a proton decays in the detector the positron emitted produces a weak flash of light which would be picked up by the photomultiplier tubes. So far no proton decay events have been witnessed.

Proton decay, if it occurs, will eventually bring about the disintegration of the material Universe, so the life story of the Universe is intricately tied up with the life story of the proton. Several arguments can be used to set limits on the decay rate, i.e. the mean lifetime of the proton. One limit is set by the fact that the Universe has already survived for 13 billion years. This means that the decay rate must be less than 10^6 decays per kilotonne of material per day. A more stringent limit comes from the consideration of the radiation dose which the decay process would deliver to living tissue. For this dose not to have been large enough to have prevented the evolution of life on Earth the rate must be less than 1 decay per kilotonne per day which equates to a proton lifetime of 10^{16} years. The fact that no events have been detected so far in SuperKamiokande means that the proton lifetime must be greater than 10^{32} years. It is intriguing that the proton lifetime not only sets a limit to the age of the Universe, but to the very existence of ourselves within it.

before the Sun's swelling to a **red giant** engulfs the planet. The Sun's red giant phase is due to begin in a few billion years. A billion years seems long enough, on the present evolutionary speed of mankind, for interstellar travel to have become a reality. If we have not already annihilated ourselves or our planet it is likely that humanity will have colonized the planets of stars similar to the Sun. Will our race have evolved to such an extent that life will be possible in a dead galaxy? And if so, will it evolve to allow intelligence and consciousness to exist when matter itself ceases to exist? Cosmology, indeed science as a whole, can currently offer no answers to these questions.

Hawking ▶ p. 132 Newton ▶ p. 42

THE FUTURE OF THE UNIVERSE

PAST, PRESENT & FUTURE

The Universe: Past, Present & Future

SUMMARY

Most cosmologists now believe that the Universe started with the Big Bang, the origin, perhaps, being a quantum fluctuation which created a singularity of very small size and very high density. The Big Bang created everything, not only the radiation which, eventually, formed the matter which condensed to form all the structures of the Universe, but space and time itself. To ask what existed before the Big Bang is meaningless – there was no before. To ask what lies beyond the Universe is meaningless – there is no beyond, the Universe has no edge.

Quantum mechanics and general relativity have helped us to explore the evolution of the Universe, with observations from the latest telescopes aiding that exploration, but the answers to some fundamental questions await a unified theory, a quantum theory of gravity. Even then the earliest phase of the Big Bang will probably remain a mystery, the laws of physics breaking down when densities rise to infinity. Equally uncertain is the end of the Universe. Observations suggest that it will continue to expand for ever rather than collapsing back to a single point. If this is true, stars and galaxies, and matter itself, may cease to exist.

HOW OLD IS THE UNIVERSE?

◉ When Messier published his catalogue of **nebulae** in 1781 a debate was still raging over their nature. Were they clouds of gas between the stars of the Milky Way, or were they beyond the Milky Way and star systems in their own right? Although the answer to the questions was 'both', it was a nebula from Messier's list (M31, the Andromeda galaxy) which was first shown, by **Edwin Hubble**, categorically to lie beyond the **Milky Way**.

◉ Vesto Slipher measured the shift in **spectral lines** within the spectrum of light from the Andromeda galaxy and showed it was moving towards Earth. Later he measured shifts to other objects which also proved to be extragalactic, discovering that the majority were redshifted, a surprising result since it was assumed that there would be an equal spread of **redshifts** and **blueshifts**.

◉ Using Slipher's measurements and assuming that galaxies had much the same luminosities, Hubble was able to show not only that the more distant galaxies were all receding from Earth, but that their recession velocity was proportional to their distance. This meant that the Universe was expanding and that at some point in the past it had been compressed into a small region. The present best estimate of when it began to expand is 13 billion years ago.

THE HOT BIG BANG

◉ It has been said that great scientists can be identified at an early age because they look at the night sky and ask why it is dark, a question which does not occur to most people. The reason for the dark night sky puzzled scientists for many years, but is explained by the fact that the Universe has a finite age.

◉ The Big Bang theory argues for a Universe which began in a state of infinite density and infinite temperature. The discovery of the 3 K **cosmic background radiation** supports the theory. The radiation is isotropic, supporting the cosmological principle that all parts of the Universe have similar histories.

◉ **Einstein**'s most famous equation, $E = mc^2$, tells us that radiation and matter are two forms of energy and are interchangeable. At temperatures of about 10^{15} degrees the collision of two **photons** can create a **proton**. Such temperatures existed during the first millionth of a second after the Big Bang. At a lower temperature, 10^{10} degrees, which existed when the Universe was still much less than one second old, such collisions could produce electrons.

◉ Within about one second the temperature of the primordial fireball had fallen to about one billion degrees, allowing protons and **neutrons** to interact to form helium nuclei.

◉ Despite the existence of protons and helium nuclei, not until the temperature of the Universe had fallen to 3,000 K could stable

▼ *Two spiral galaxies are caught in the process of a leisurely merger. Strong tidal forces from the larger and more massive galaxy, NGC 2207 (left), have distorted IC 2163, flinging out stars and gas into long streamers stretching for 100,000 light years. Dust lanes in the spiral arms of NGC 2207 can be seen silhouetted against IC 2163. The large concentrations of gas and dust in both galaxies may well erupt into regions of active star formation in the near future.*

▲ *Omega Centauri, the largest and brightest of the 150 or so globular clusters that surround our galaxy. Globular clusters formed early in the history of our Galaxy and their stars are mostly old. Measurement of the ages of stars in such clusters therefore places a lower limit on the age of the Galaxy and hence the Universe as a whole.*

atoms of hydrogen and helium form. At that time the Universe was about 300,000 years old. The fact that the cosmic background radiation is now 3 K tells us that the Universe has expanded by a factor of 1,000 since that time.

COSMOLOGICAL PROBLEMS

◯ One of the problems with the creation process of matter is that it should produce equal amounts of matter and antimatter. If that happened we would expect the two to annihilate and we would not be here to discuss the issue. But observations suggest that the majority of the Universe is predominantly made of matter. The investigation of this asymmetry has led to profound implications for the laws governing the creation and decay of fundamental particles which are still being investigated.

◯ Recent observations of the cosmic background radiation show that it is not completely isotropic, but has variations in intensity consistent with the density variations required in the early Universe for galaxy formation to start. Galaxies formed and fragmented to form stars, and it is known that the chemical elements heavier than helium were formed by nuclear reactions within those stars and then distributed across the Universe by stellar explosions.

◯ Although much is known about the early Universe and about the Universe we now observe there are still fundamental questions to be answered. How did the Big Bang take place? Was it just a **quantum mechanical** fluctuation? What happened during its earliest existence (a period known as the Planck era, which covers the time up until the Universe

THE ANTHROPIC PRINCIPLE

To understand the workings of the Universe, physicists look for fundamental laws which have wide applications. They then use these laws to show relationships between phenomena which, at first sight, seem very disparate. Examples are Newton's law of gravity which explains not only how apples fall on Earth, but also how the planets orbit the Sun, and Maxwell's laws of electromagnetism which show the connections between electricity, magnetism and radiation.

One unsatisfactory aspect of this description of the Universe is that it contains a number of independent fundamental constants, such as the gravitational constant, the speed of light and Hubble's constant, whose values have to be determined by observation and which cannot be predicted. The 'anthropic principle' attempts to explain the values of these constants by arguing that they are the only ones consistent with the evolution of intelligent life in the Universe – since we are asking the questions, these are the only answers we should expect to get!

Two examples will show how the principle is used. One requirement for the presence of intelligent life is that there has been sufficient time – of the order of a billion years – for the slow process of evolution to work. So the lifetime of stars, which depends directly on the value of the gravitational constant, must be sufficiently long. Likewise, the very existence of stars and galaxies requires that the early Universe was

was 10^{-43} seconds old, its temperature about 10^{31} K) where laws of physics break down. What is needed to explain this is a quantum theory of gravity which, at present, has not been developed.

◯ A second fundamental question addresses the ultimate fate of the Universe. To answer this requires a knowledge of the geometry of the Universe and the amount of **mass** contained within it. The geometry is difficult to measure, as is the mass density because of the probable existence of **dark matter**. If dark matter exists, how much of it is there? However, estimations of both the geometry and the mass density have been made. These show that the Universe is flat, but that there is insufficient matter to stop it expanding. The expansion will go on for ever, perhaps even accelerated by the pressure of a cosmological constant, with the Universe ultimately becoming a place devoid of matter and, presumably, of life.

expanding sufficiently slowly for condensations to take place and this places a restraint on the value of Hubble's constant.

The anthropic principle also provides an explanation of several coincidences which are found in Nature. One of these occurs in the nuclear reactions in which the element carbon – a vital building block of living organisms – is synthesized in stars. A carbon nucleus is formed when three helium nuclei collide and fuse together. The probability of this happening would be vanishingly small if it were not for a 'resonance' – an effect that makes the nuclei appear very much bigger than normal and the collision more likely – at precisely the right energy. It seems very lucky that such a resonance exists, but if it did not we would not be here worrying about it.

The anthropic principle also provides answers to fundamental questions which might normally be considered outside the realm of physics. For example, planetary orbits would not be stable if space had anything but three dimensions. The lack of stable orbits would mean there would be no sites for intelligent life to develop, so we could not expect to find ourselves in a Universe in which space did not have three dimensions.

Not all physicists are happy with the use of the anthropic principle to answer such questions. Although the principle does not argue that the Universe was made for the benefit of mankind, it does give human beings, the observers, a special role in the Universe and, to some physicists, this is counter to the objectivity of physics.

▼ *Spiral galaxy NGC 4603 is receding from us at a speed of nearly 2,500 km/s (1,550 mi/s). Lying over 100 million light years away in a cluster of galaxies in Centaurus, it is the most distant galaxy in which Cepheid variable stars have been detected.*

SUMMARY

149

CONTENTS OF THE COSMOS

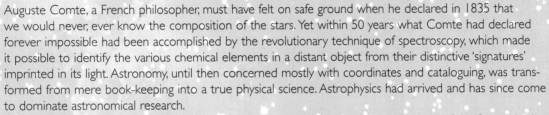

Auguste Comte, a French philosopher, must have felt on safe ground when he declared in 1835 that we would never, ever know the composition of the stars. Yet within 50 years what Comte had declared forever impossible had been accomplished by the revolutionary technique of spectroscopy, which made it possible to identify the various chemical elements in a distant object from their distinctive 'signatures' imprinted in its light. Astronomy, until then concerned mostly with coordinates and cataloguing, was transformed from mere book-keeping into a true physical science. Astrophysics had arrived and has since come to dominate astronomical research.

The new breed of astrophysicists were able not only to determine the composition of the Sun and other stars but also, in league with practitioners of another novel discipline, that of nuclear physics, to understand why stars glow – a simple fact that had until the middle of the twentieth century been inexplicable. Step by step, astrophysicists have now pieced together the entire life story of a star from cloudy birth via bountiful middle age into swollen seniority and to its ultimate demise, either as a fading relic or, in extreme cases, a monumental explosion which can punch a black hole into the fabric of space and time.

In the wider Universe, twentieth-century astronomers have found that stars inhabit huge conurbations termed galaxies. The most beautiful specimens assume a swirling spiral shape, while others are rounded or irregular. A prominent minority exhibit violent activity at their centres, most notably the intensely luminous quasars. The origin of all these objects can be traced back to shortly after the origin of the Universe.

Contents of the Cosmos
INTRODUCTION

F or most of its extent, the cosmos is an ocean of empty space, but its darkness is relieved from place to place by brilliant galaxies of stars. We live within such a galaxy, our Sun being merely an average star among billions of others. The Sun's powerful radiance in our daytime sky is due solely to our front-row view of it from within the Solar System, not to any inherent supremacy in its size or luminosity. Nearby stars are scattered over the night sky, the most prominent of them making up the familiar constellation patterns, while their more distant relatives crowd upon each other into the faint band of the Milky Way, marking the plane of our Galaxy.

SEEING STARS

During the seventeenth century, astronomers came to realize that stars must be other Suns, but unimaginably distant – even when observed with the newly invented telescope they stubbornly remained as mere points of light. Until the middle of the nineteenth century, stellar astronomy was mostly concerned with measuring the positions of these points with ever-greater precision, an exercise known as astrometry. Initially the main purpose of these observations was practical and commercial, notably for timekeeping and navigation. But, gradually, results of pure scientific import began to be extracted from this painstakingly mined ore. Most important were **parallaxes**, from which the distances of the stars could at last be directly calculated. It turned out that even the closest were so far away that their light had taken many years to reach us – in other words,

▲ **Stars are collected** into huge assemblies called galaxies such as this one, called the Whirlpool. Galaxies formed early in the history of the Universe, probably from the accumulation of smaller parts. We can study galaxy formation by looking far off into space and hence back in time.

stars were separated by distances best expressed not in the puny everyday units of measurement on Earth, but in **light years**.

Knowing the distance of the stars, astronomers could gauge how bright they really were in comparison with the Sun. Additional information could be squeezed from the trickle of starlight by the **spectroscope**, a device borrowed from physicists in the second half of the nineteenth century. Examined through the spectroscope,

the stars were confirmed to be balls of incandescent gas at enormously high temperatures. Astronomers could estimate a given star's surface temperature and **luminosity** to work out how big it must be. Such calculations of the sizes and brightnesses of stars emphasized that the Sun was only an average specimen in all regards. Seen from another star, our Sun would itself appear as an unremarkable point of light.

LIVES OF THE STARS

Magnificent **nebulae** are the birthplaces of stars. The same process that gave rise to our Sun and its retinue of planets 4.6 billion years ago can be seen replayed elsewhere today. Among the most enchanting of these present-day stellar nurseries is the Orion Nebula, visible with the naked eye and binoculars as a faintly luminous area beneath the belt of Orion and displayed in its full glory on long-exposure photographs. Large nebulae produce hives of stars, known as **open clusters**, which sparkle with the blue light of their hot, young

J. L. E. DREYER AND THE NGC

John Louis Emil Dreyer (1852–1926) was a Danish-born astronomer who spent most of his working life in Ireland, first as an assistant at the observatory of Lord Rosse, where he observed faint nebulae and star clusters with Rosse's huge telescope, and then at the Armagh Observatory, where he was director from 1882 to 1916. At Armagh, Dreyer continued his interest in what are now termed deep-sky objects, publishing in 1888

the *New General Catalogue of Nebulae and Clusters of Stars* (popularly known as the NGC). This was a revision and expansion of a catalogue published earlier by John Herschel and contained 7,840 objects. Many of these turned out to be galaxies, the existence of which was not recognized at the time. In 1895 and 1908 Dreyer published two supplements, called the Index Catalogues, which added over 5,000 newly discovered objects. Objects in these catalogues are still widely referred to by their NGC and IC numbers.

INTRODUCTION

THE COSMOS

HOW FAR THE STARS?

Early attempts to measure the distances of the stars proved unsuccessful, because none of them showed any detectable shift in position (termed parallax) as Earth orbited the Sun. In 1685, **Isaac Newton** (1642–1727) used an indirect method to gauge the distance to Sirius, the brightest star in the night sky, based on the brightness difference between it and the Sun. He concluded that Sirius was nearly a million times more distant than the Sun – actually something of an overestimate, but a powerful pointer to the true enormity of the cosmos around us.

As the precision of telescopes improved so did hopes of measuring the microscopic parallax shifts that stars would show if they were at distances comparable to that which Newton had estimated for Sirius. Results finally came simultaneously from both hemispheres. In 1838 the German astronomer F. W. Bessel (1784–1846) announced that the star 61 Cygni had a parallax that placed it 10.3 light years away, only one light year closer than the currently accepted distance. Shortly thereafter, Thomas Henderson at the Cape of Good Hope in South Africa announced that Alpha Centauri showed an even greater parallax – now known to be the closest naked-eye star to the Sun.

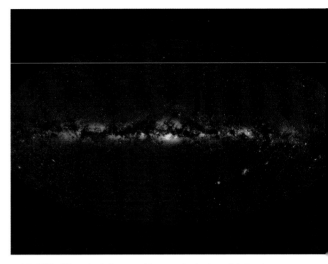

▲ **The Milky Way**, a band of stars stretching across the sky, is our disk-shaped Galaxy seen from within. This composite view shows a 360° panorama of the Milky Way, with the Galactic bulge at the centre.

offspring – it is a seeming contradiction that blue light is described as 'cold' yet the hottest stars are actually the bluest, while the coolest ones are red.

Stars power themselves by nuclear reactions, converting hydrogen into helium at their centres. Once the hydrogen is used up, the star dies, but how soon this happens – and exactly how it happens – depends on the star's **mass**. Perhaps surprisingly, the most massive stars have the

▼ **Giant clouds** of gas and dust in space give rise to open clusters such as this pair in the constellation Perseus, known as the Double Cluster. The stars are powered by nuclear reactions at their centres. The most massive stars are the hottest and brightest and burn out the quickest.

shortest lives, greedily consuming their hydrogen fuel in only a few million years before exploding as brilliant **supernovae**. Less massive stars, the Sun included, live for many billions of years. They gently divest themselves of their outer layers with their dying breaths before fading into obscurity as diminutive **white dwarfs**.

STELLAR REMNANTS

Something truly remarkable happens to the core of a star during a supernova explosion. The very stuff of which the core is composed is crushed together with such force that its **electrons** and **protons** combine to form **neutrons**, creating a super-dense object known as a **neutron star**, no larger than about 20 km (12 miles) across. As the debris of the supernova explosion clears, the neutron star becomes detectable as a **pulsar**, beaming out radio waves like a radar scanner as it spins every few seconds or faster. But in the most extreme cases the core of the star is crushed even further, disappearing into the oblivion of a **black hole**, a region in space with a gravitational pull so intense that nothing can escape, not even its own light. Black holes can grow without limit by swallowing gas and even other stars, particularly in areas where there is plenty to feed their appetite such as at the centres of **galaxies**.

Originally purely theoretical concepts, black holes have now emerged as the key components in some of the most energetic objects in the Universe, such as active galaxies and **quasars**.

THE MILKY WAY – AND BEYOND

Immanuel Kant (1724–1804) proposed in 1755 that the band of the **Milky Way** could be understood as an enormous edge-on disk of stars. This was the first successful description of the shape of what we now term our Galaxy. But whether there was anything other than empty space beyond the boundaries of the Galaxy remained undecided until the 1920s, when the American astronomer **Edwin Hubble** (1889–1953) established that certain spiral-shaped nebulae such as that in Andromeda were in fact distant accumulations of stars – separate galaxies. As Hubble and others looked ever-farther into the Universe it became clear that galaxies extend in all directions with no end in sight.

Unlike stars, galaxies are not forming today so it is more difficult to study their origin and evolution. However, when we look far out into space we are also looking back in time, to an era when galaxies were still young. There are many signs of disturbance in these younger galaxies, evidently due to collisions and mergers. It seems that many galaxies have been built up over billions of years from the gradual accumulation of smaller parts. During these encounters, gas and stars may fall towards a supermassive black hole at the centre of the galaxies, creating a small but intensely bright disk of hot gas which we see as an **active galactic nucleus (AGN)** – or, in the most extreme cases, a quasar. Once considered the most baffling objects in the cosmos, quasars are now recognized as a normal stage in the evolution of many galaxies.

THE COSMOS

spectroscope ▶ p. 344 supernova ▶ p. 344 white dwarf ▶ p. 344 Hubble ▶ p. 122 Kant ▶ p. 136 Newton ▶ p. 42

INTRODUCTION   155

Contents of the Cosmos
STARS
PROPERTIES OF STARS

Looking into the sky on a clear, dark night we see thousands of sparkling stars of different brightness. At the heart of each of them is a nuclear reactor producing energy that makes the star shine. Our impression of a star's brightness is affected both by how much light it gives out (its luminosity) and by how far away it is. Disentangling the effects of luminosity and distance has enabled astronomers to compile a gallery of star types. They range from small, cool glow-worms with a fraction of the Sun's output up to dazzling supergiant beacons pumping out the power of 100,000 Suns or more. It turns out that our own Sun is a mid-range star in almost every way.

APPARENT BRIGHTNESS AND STAR NUMBERS

A star's brightness to the eye is known as its **apparent visual magnitude**. Ancient astronomers divided stars into six brightness classes, the faintest being sixth **magnitude** and the brightest first magnitude. Modern astronomers have refined this scale so that a star of magnitude one is exactly 100 times brighter than a star of magnitude six. Objects that are brighter still are assigned negative magnitudes, whereas fainter objects (below naked-eye visibility) have progressively larger positive magnitudes. There are over 8,000 stars of sixth magnitude and brighter, although only half this number will be above the horizon at any one time and you will need exceptionally clear skies, as well as first-rate eyesight, to see them all.

STELLAR DISTANCES AND ABSOLUTE MAGNITUDE

Since the apparent magnitudes of stars depend on their distances, to compare stars' brightnesses objectively we need to know how far off they are. For the nearest stars, distances can be measured by simple trigonometry, like that used by surveyors (see Parallax, page 155), while for those farther afield comparisons can be made with similar stars nearby to deduce approximate values. Once a star's distance from our Sun is known, its apparent magnitude can be corrected to give its **absolute magnitude**, corresponding to the brightness it would have at a standard distance of 10 **parsecs** (32.6 **light years**).

THE MOTION OF STARS

Although stars appear to be fixed in their **constellation** patterns, precise measurements reveal that they do slowly move relative to each other. This transverse **proper motion** across the sky is larger for nearby stars than more remote ones. Barnard's Star, a **red dwarf** only six light years away, has the largest proper motion of all, covering the apparent width of the Moon in 180 years. Stars also move towards or away from us in the line of sight, the speed being known as their **radial velocity**. These two components of motion arise from the stars' individual orbital movements around the centre of our Galaxy and, as a result, the appearance of the constellations gradually changes over long periods of time.

▲ *The temperature of a star can be gauged by its colour; the dramatic variety of colours can be seen here in the Sagittarius star cloud.*

COLOURS AND TEMPERATURES OF STARS

In much the same way that an iron poker in a fire glows first dull red, then orange, yellow and finally white as it gets progressively hotter, so the temperature of a star's visible surface is indicated by its colour. Yellow stars, of which the Sun is one, have surface temperatures in the region of 6,000°C (10,800°F). A cooler star, such as Betelgeuse, the bright red star in one corner of

▼ *Cecilia Payne-Gaposchkin was a pioneer for female astronomers and astrophysicists. She made a great contribution to the calculation of stellar chemical compositions.*

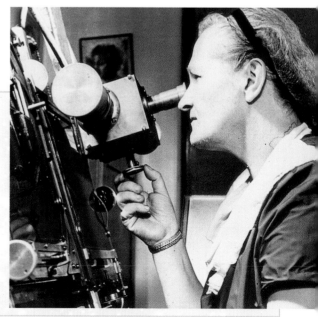

CECILIA PAYNE-GAPOSCHKIN

Cecilia Payne (1900–79) was born in Buckinghamshire, England, and gained a scholarship to Cambridge University in 1919, taking a science BA in 1923. By that time she had discovered a passion for astronomy but to develop her interest further she had to leave male-dominated Cambridge and pursue her doctorate at the Harvard College Observatory in Massachusetts, USA. There she gained her PhD in just two years for a brilliant thesis entitled *Stellar Atmospheres*, in which she used the recent calculations of the Indian astrophysicist Meghnad Saha (1893–1956) to interpret the spectra of stars according to temperature and to deduce their chemical compositions. She went on to demonstrate that hydrogen was the main constituent of stars, which had not been appreciated until then. She made many fundamental contributions to astrophysics, especially on the subject of variable stars in which she shared an interest with Sergei Gaposchkin, whom she married in 1934. Of equal importance was her demonstration that a woman could compete effectively in a discipline hitherto dominated by men, and in which women had seldom been more than assistants.

STARS

THE COSMOS

PARALLAX

The key to measuring distances to the nearest stars is provided by the technique of parallax, which was first successfully applied in 1838 by the German astronomer Friedrich Wilhelm Bessel when he determined the parallax of the star 61 Cygni, approximately 11 light years away. Parallax is the shift in the position of a nearby star with respect to more remote background stars caused by our change in vantage point as Earth moves around the Sun. The diameter of Earth's orbit provides a baseline, and the size of the parallax shift is related directly to the distance of the star. The distance corresponding to a shift in position of 1 **arcsecond** is called a parsec and is equal to 3.2616 light years. In practice, all parallaxes are smaller than this; the nearest star, Proxima Centauri, has a parallax of 0.77 arcseconds. Distance measurement was revolutionized by the European Space Agency's Hipparcos satellite, which measured parallaxes of over 100,000 stars with an accuracy far greater than was possible from the ground.

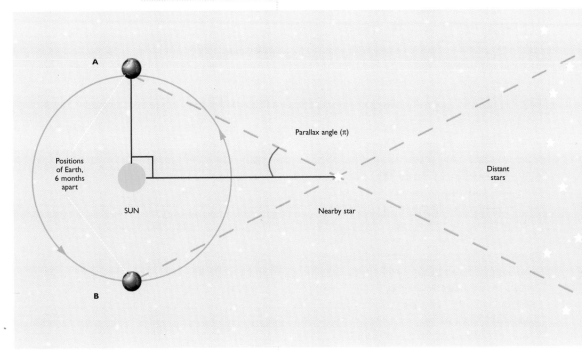

▲ *Stellar Parallax*
As Earth orbits the Sun, the line of sight to a nearby star varies from position A to position B. This causes the star regularly to change its apparent position in relation to more distant stars. The parallax angle π is defined as one-half of the total angular motion. If the parallax angle is 1 arcsecond then the distance to the star is 1 parsec.

Orion, with a temperature of only 3,650°C (6,600°F), actually emits most of its energy as **infrared** radiation. In the opposite corner of Orion, Rigel shines with a blue-white colour, since its surface temperature is over 11,000°C (20,000°F). Even hotter stars exist, with surface temperatures up to 50,000°C (90,000°F) or more, which pour much of their energy out as **ultraviolet** radiation which is blocked by Earth's atmosphere and has to be studied by spacecraft. The colour of a star can be measured to give a **colour index**, which is closely related to the surface temperature, provided that the star is not artificially reddened by interstellar dust in the line of sight.

STELLAR SPECTRA AND THE COMPOSITION OF STARS

The rainbow-like **spectrum** of a star is interrupted at many points by dark **absorption lines** produced by atoms and **ions** (and, for the cooler stars, **molecules**) in the star's atmosphere. The precise pattern is determined by many factors but the dominant one is the temperature: the hotter the gas, the more violently the atoms, ions and **electrons** collide with one another. This influences how many electrons the **atomic nuclei** can retain (their state of **ionization**) and also the level of **excitation** of those electrons in

their orbits. Density is another factor: the more rarefied the stellar atmosphere, the easier it is for electrons to avoid being captured by the positively charged ions. The composition of the gas in the star's surface layers can be determined from the signature of absorption lines in its spectrum once the temperature is known (from the star's colour) and its density has been estimated (from the degree of ionization of elements such as iron).

SPECTRAL TYPES

A shorthand description of a star's basic properties is provided by its **spectral type**, designated by one of seven capital letters: O, B, A, F, G, K, M, running from the hottest stars to the coolest. Such spectral types are allocated purely from the appearance of the star's spectrum; this is compared with a set of standard stellar spectra that are arranged so that the pattern of features (primarily absorption lines) changes progressively with surface temperature. For example, in stars of B type, which are among the hottest, lines in the visible spectrum of neutral (i.e. un-ionized) helium reach their greatest strength; in cooler stars, these lines become less prominent because the electrons in the helium atoms are not excited enough to cause significant absorption in these

lines, whereas in still-hotter stars (O-type) the helium becomes ionized and produces an entirely different set of lines.

The odd sequence of letters used in spectral classification is a legacy of early attempts to find a pattern among stellar spectra, before the effects of temperature were recognized. Subclasses 0 to 9 were introduced to afford a finer distinction. On this scheme, the Sun is classified as a G2 star.

PHOTOMETRY AND STAR COLOURS

Measurement of the light of stars is known as photometry and in its simplest form can be done with the naked eye, by estimating a star's brightness in comparison with others of known magnitude. More accurate measurements can be made photographically, but the most precise results are obtained electronically with a photoelectric photometer or a **charge-coupled device** (**CCD**) camera at the focus of a telescope. If special coloured filters are interposed between telescope and detector then, for example, ultraviolet (U), blue (B) or visual (V) apparent magnitudes can be obtained – the V filter approximates the eye's response to light. By comparing the light obtained through such different filters a star's colour index is obtained, from which certain properties of the star can be deduced. For example, by subtracting the visual (V) from the blue (B) magnitudes the star's temperature can be estimated – this is possible because hotter stars emit more light at blue **wavelengths**, and vice versa.

THE COSMOS

THE HERTZSPRUNG–RUSSELL DIAGRAM

T he Hertzsprung–Russell (HR) diagram is no more than a graph displaying the brightness and temperature characteristics of stars, yet it has provided essential clues to the evolution of stars and has been a valuable tool in determining the distances of those stars beyond the reach of parallax measurements. The diagram was first plotted by Henry Norris Russell in 1913, but it was later appreciated that similar work had been done independently around the same time by Ejnar Hertzsprung.

DEFINITION OF THE HR DIAGRAM

In an HR diagram, the intrinsic brightnesses of a sample of stars are plotted against some measure of their surface temperatures. Various parameters can be used for such a plot: the vertical axis could be the star's apparent magnitude (for stars of the same distance, as in a cluster), absolute magnitude (for any star whose distance is known) or **luminosity** (the total energy output across the whole spectrum). The horizontal axis might be spectral type, surface temperature or, most likely, colour index (when the diagram is referred to as a colour–magnitude diagram). As the figure shows, the outcome is not a random scattering of points. Stars populate very specific regions of the diagram, and this gives clues to their physical nature and their stage of evolution.

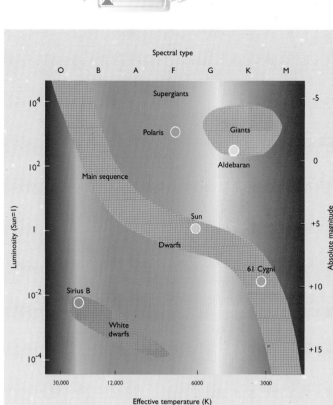

◀ *The Hertzsprung-Russell Diagram*
The HR diagram is a graphical way of comparing the different types of stars. It plots their surface temperatures (or spectral types) against their luminosities (or absolute magnitudes). Most stars lie along the band that runs from top left to bottom right, called the main sequence. Red giants and supergiants are at upper right, and white dwarfs at lower left.

GIANTS AND DWARFS

The well-populated band running across the HR diagram from top left to lower right is known as the **main sequence**; our Sun, of spectral type G2, falls about halfway along it.

Stars on the main sequence are termed **dwarfs** even though the largest of them, the O3 stars at the top, are 15 times bigger than the Sun. In turn, the Sun is 10 times larger than an M8 dwarf at the bottom end.

One significant fact revealed by the diagram is that stars of similar surface temperature can have very different luminosities. To take this into account, stars are assigned a luminosity class which

is appended to their spectral type. Main-sequence stars are said to be of luminosity class V, so the full description of the Sun becomes G2V. Above and to the right of the main sequence are stars of greater luminosity than dwarfs of similar temperature. These stars are **giants** (luminosity class III) or, for the more extreme cases in the top and upper right of the diagram, **supergiants** (class I, with sub-divisions denoted by lower-case letters a, ab and b). Rigel, a blue supergiant, is described as a B8Ia star. At the other extreme, in the lower left part of the diagram are to be found the **white dwarfs**, the fading cores of stars that have run out of fuel.

Dwarfs, giants and supergiants differ in both size and luminosity. For example, a giant star of the same spectral type as the Sun (G2) is eight times larger, while a G2 supergiant is 130 times larger than the Sun. Having a greater visible surface than the Sun, these stars are also much more luminous, around 35 times in the case of a G2 giant and some 30,000 times for a G2 supergiant. The larger stars have more tenuous atmospheres, and this is imprinted in the spectrum in the form of more ionization and sharper lines than dwarfs, allowing luminosity classification from their spectra.

👨 HENRY NORRIS RUSSELL AND EJNAR HERTZSPRUNG

Henry Norris Russell (1877–1957) was born in Oyster Bay, New York, and spent much of his academic life at Princeton University. He was one of the most influential scientists of the first half of the twentieth century, making major contributions to laboratory and theoretical **spectroscopy** and to stellar astrophysics, interests he combined to make the first reliable determination of the abundances of the elements in the Universe. His pioneering work on stellar evolution, starting with the HR diagram, was followed by equally

seminal work on eclipsing **binary stars**, which provide one of the most fertile testing grounds for theories of stellar structure.

Ejnar Hertzsprung (1873–1967), born in Roskilde, Denmark, came to astronomy only at the age of 29, combining an attention to detail and accuracy with dedication as an enthusiastic observer to produce a wealth of valuable results on magnitudes, colours, parallaxes and proper motions of stars, including visual binaries and variables. He was the first to realize that for stars of a particular colour there was a wide range in luminosity – in other words, he discovered giants and dwarfs.

◀ ▶ *E. Hertzsprung (left) and H. N. Russell (right) revolutionized the study of stellar evolution through the diagram that now bears their names.*

▲ *Globular clusters* are dense balls of old stars in which stars down to about the mass of the Sun have already evolved into red giants. The age of such clusters can be deduced from the point at which stars have turned off the main sequence.

CLUSTERS AND DISTANCES

Amongst the most splendid binocular sights in the sky is the Pleiades, a young open cluster of stars in Taurus. Important characteristics of stars in such clusters are that they formed at about the same time, from the same large cloud of gas and, most especially, that they are all at about the same distance from us. Hence their apparent brightness is proportional to their true brightness so, when their apparent magnitudes are plotted against their spectral types, the standard pattern of an HR diagram emerges. If a second cluster is plotted on the same diagram its pattern of dwarfs, giants and supergiants will be the same but vertically displaced from the first. This vertical difference, which is best measured using the well-defined band of the main sequence, is because of the difference in distance of the two clusters, and the amount of displacement tells us the difference in distance. If the distance of one cluster is known, the distance to the other can be found by this process, known as main-sequence fitting.

THE HR DIAGRAM AND STELLAR EVOLUTION

It was once thought that stars moved along the main sequence on the HR diagram as they evolved. That is now known not to be the case – a star's position on the main sequence is fixed by its **mass** (the quantity of material it contains). The most massive stars (up to about 100 times the mass of the Sun) populate the upper end of the main sequence as O-type stars, while more modest ones (say 2.5 solar masses) are to be found among the A stars lower down. G-type stars have similar masses to the Sun, while stars with only half the mass of the Sun and less inhabit the bottom end of the main sequence at M0 and beyond.

▶ *Open star clusters* such as this loose grouping called NGC 2264 in the constellation Monoceros are so young that even their most massive stars have barely had time to evolve into red giants.

INTERSTELLAR ABSORPTION

Absorption of light by gas and dust in interstellar space affects measurements of the colour and luminosity of stars. Interstellar material is usually too tenuous to be noticeable to the eye, but the gas within it produces an absorption-line spectrum which is superimposed on a star's own spectrum while, more seriously, the dust not only dims starlight but also reddens it by preferentially absorbing blue light, altering a star's colour index.

PRINCIPAL SPECTRAL TYPES

Spectral Type	Surface Temperature (K)	Characteristic Features
O	50,000–28,000	Ionized helium; often emission lines
B	28,000–10,000	Neutral helium
A	10,000–7,400	Hydrogen
F	7,400–6,000	Metals, hydrogen
G	6,000–4,900	Ionized calcium (H and K lines), metals
K	4,900–3,500	Calcium, metals, molecules (CH, CN etc.)
M	3,500–2,000	Molecules, esp. metal oxides, carbon

LUMINOSITY CLASSES

I	Supergiants
II	Bright giants
III	Giants
IV	Subgiants
V	Dwarfs (main sequence)

Typically, along a path in the plane of the Galaxy where dust and gas are concentrated, the visual (V) magnitude of a star is dimmed (i.e. increased) by 1.8 magnitudes for every 1,000 parsecs, while the blue (B) magnitude is increased by three times the change in V. The effect of reddening on an HR diagram where apparent magnitude is plotted against colour index is to slide the points downwards and to the right. However, the spectral type is not changed and so the amount of reddening can be estimated and corrected for.

As time passes, the most massive stars, which are disproportionately profligate with their fuel, show the first signs of age. They swell up to become supergiants, cooling at the surface so that their spectral type changes but remaining fairly constant in luminosity because of their increased size. These changes are reflected in a move to the right in the HR diagram. Eventually they die and disappear, while stars of somewhat lesser mass start to evolve to become giants, also moving to the right on the diagram. And so it continues over time, with stars of progressively lower mass leaving (or turning off) the main sequence. It is possible to estimate the age of a star cluster from the turn-off point that stars have reached on the main sequence. However, the lowest-mass stars hardly move on the HR diagram because their lifetimes are so long – longer even than the age of the Universe.

THE BIRTH OF STARS

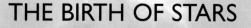

Like humans, stars are born, live their lives, changing their appearance as they do so, and eventually die. Only now is the full picture of the first stage in that stellar life cycle starting to emerge. In particular, the recent use of infrared telescopes on satellites, and instruments observing at submillimetre wavelengths on high, dry mountain sites, has allowed astronomers to probe the dense, dusty clouds where stars are born. Star birth continues throughout the Galaxy today, and presents us with an action replay of the processes that led to the formation of our own Sun 4.6 billion years ago.

COLLAPSING CLOUDS

In the space between the stars lie cold, dark clouds of gas and dust that can be seen only where they are silhouetted against a brighter background. Hydrogen is the main constituent of these clouds, much of it in the form of molecules (H_2). Many other molecules, some of considerable complexity, are also to be found within them; hence they are known as **molecular clouds**. The largest examples, the giant molecular clouds (GMCs), contain a million solar masses or more of gas and are hundreds of light years across.

A GMC can become unstable, perhaps as the result of a shockwave from a nearby supernova, as is believed to have prompted the formation of our Sun. Gravity will start to pull the particles towards the centre of the cloud, gently at first. As the cloud collapses its density increases, the pull gets stronger and the process accelerates. The GMC will fragment, each portion undergoing its own continued collapse. Along with the increased density, the temperature of the gas rises as particles jostle each other, and the fragment begins to rotate.

YOUNG STELLAR OBJECTS

Perhaps 100,000 years after the GMC started to collapse the individual fragments turned into **protostars**. These glow dull red from the heat released by the compression of their gas but have not yet started to produce their own energy by nuclear fusion. If plotted on the **Hertzsprung–Russell (HR) diagram** a protostar would lie to the upper right, in the realm of large and luminous but cool bodies. As the collapse into true stardom proceeds, the protostar progresses downwards on the HR diagram and to the left, joining the main sequence once the core is hot and dense enough to ignite hydrogen-burning nuclear reactions. This marks its transition into a true star.

New stars make their first appearance as **T Tauri stars**, named after the prototype in Taurus. Such objects are still surrounded by their birth clouds which partially or totally obscure them at optical wavelengths, but they also exhibit surface activity of their own which causes erratic brightness changes. Some, particularly the more massive young stars, are associated with bright blobs of gas known as Herbig–Haro objects. It seems that these bright blobs are produced where high-speed gas flowing outwards from the poles of T Tauri stars collides with nearby dense clouds, which pushes the clouds away from the stellar nursery.

STELLAR NURSERIES

Depending on its size, a GMC can spawn anything from a few dozen to several thousand stars, producing a star cluster or a much more extended group known as an association. A large association will have a core of hot stars of spectral types O and B and hence is known as an **OB association**, although this is not to say that a whole range of lower-mass stars is not also present; they are simply drowned out by the light of the hotter stars. When such massive stars are not

◄ A Bok globule – a cloud of gas which may be condensing into a new star – seen silhouetted against a brighter nebula.

present, a collection of lower-mass T Tauri stars forms a **T association**. The members of an association or cluster usually drift apart over millions of years, leaving the stars isolated, although the denser clusters may remain united by gravity.

At first, the newly born stars will still be shrouded in the remains of the gas and dust from their formation, but this will eventually be driven off by the radiation pressure and winds of atomic particles that blow from their surfaces. Now uncloaked, the stars will shine forth and the ultraviolet radiation from the hottest of them, of spectral types O and B, will ionize the gas nearby, producing a glowing **emission nebula**. One famous example is the Orion Nebula. A burst of star formation took place here just a few million years ago and is still continuing in the dark clouds behind the visible nebula, unseen to the eye but detectable at infrared wavelengths.

STELLAR FAMILIES

On the smaller scale, the gravitational attraction between neighbouring protostars is sufficient for them to become linked into multiple star systems, orbiting around their common centre (or centres) of mass. In general, multiple star systems are not long-lived, being disrupted by encounters with other stars in the cluster, particularly in dense clusters. However, the simplest kind of multiple, the twin or binary stars, do form in large numbers: something like

▼ A spectacular image of a Herbig-Haro object, a jet of gas half a light year long ejected by a young star at the bottom left of this image.

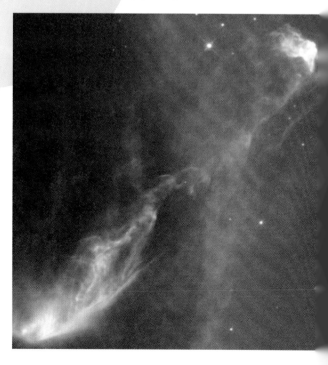

▲ *A fantastically sculptured* column of gas in the Eagle nebula, illuminated *by hot young stars off the top of the picture. Finger-like protrusions contain denser knots of gas where stars are forming.*

ANGULAR MOMENTUM IN STARS

A clump of gas trying to contract to become a star is faced with a serious difficulty: how to avoid spinning so fast that the coalescing material is not flung off again into space. This is the problem of **angular momentum**, and it can be illustrated by the example of an ice skater who, spinning on one blade, can rotate faster by drawing her arms in close to her body. Thus if a collapsing cloud is rotating, even very slowly at the start, it will spin faster and faster as it shrinks because angular momentum must be conserved.

Most stars, particularly the lower-mass ones like the Sun, actually seem to be spinning rather slowly, evidently having managed to lose most of their angular momentum. How has this trick been achieved? For these cooler stars, magnetic fields play an important role: they connect the body of the star to material spun off the surface, enabling that material to carry off the star's angular momentum into space and hence braking its spin. More massive stars do not have such magnetic assistance and most continue to spin quite rapidly.

half of all stars have a companion in stable orbit. Most of these probably formed when a single, rapidly rotating protostar split, although some undoubtedly result from the capture of one star by another after birth.

BROWN DWARFS – STARS THAT FAILED

If the mass of a protostar is less than about 8% that of the Sun (80 times the mass of Jupiter), the gas at its centre will never get hot or dense enough to trigger nuclear burning. Such failed stars are called **brown dwarfs**. Although not burning hydrogen,

THE INFRARED ASTRONOMICAL SATELLITE

The Infrared Astronomical Satellite (IRAS), a collaboration between the Netherlands, the UK and the USA, was launched by NASA in January 1983 to conduct the first survey of the sky at far-infrared wavelengths. The significance of this region of the spectrum for star formation is that during the early stages of the process, the interstellar clouds and collapsing cores emit most of their energy in the infrared. Thus the IRAS survey could highlight the location and structure of stellar nurseries. However, the 60-cm (24-in)

MOLECULES IN SPACE

Amazingly, a considerable amount of chemistry is going on in the space between the stars. How can this happen in such a near-vacuum? The key to the process is the stars themselves. All stars, except the least massive, will end their lives having synthesized heavy chemical nuclei in their cores, and in their death throes will spill these pollutants into space, either during their **planetary nebula** stage in the case of lower-mass stars or in supernova explosions for the most massive stars.

Some of this material, particularly carbon, will cool and condense into grains before it gets very far from the dying star. The surfaces of these grains act as catalysts for chemical reactions, producing a wide range of molecules of surprising complexity (including alcohol, C_2H_5OH). These have a part to play in the early stages of further star formation because molecules can soak up heat energy and radiate it away from the infalling cloud, cooling it and allowing the collapse to continue. Molecules are also valuable tools for the astronomer: the lines in their spectra allow temperatures and densities to be determined accurately, and they can also be detected by radio and submillimetre telescopes in regions obscured to optical telescopes. This permits the structure of star-forming regions to be mapped in detail.

brown dwarfs are still quite warm, with surface temperatures of 1,000– 2,000°C (1,800–3,600°F) from their collapse and perhaps from their failed attempts to ignite as stars. Hence they can readily be detected at infrared wavelengths. Brown dwarfs can be recognized by the presence of the light element lithium in their optical spectra: in any 'real' low-mass stars, convection would carry lithium into the interior where it would be destroyed by nuclear reactions, but this does not happen in brown dwarfs. If the gas ball were even smaller, below a mass of about 10 Jupiters, it would be classified as a planet.

telescope and detectors had to be cooled to just three degrees above absolute zero by liquid helium to prevent its own radiation swamping the weaker signals from space. When the last of the cooling helium had evaporated, in November 1983, the highly successful mission came to an end. In addition to its maps of the interstellar medium, IRAS recorded more than a quarter of a million individual sources and discovered five **comets** and numerous **asteroids**. The Infrared Space Observatory (ISO) was launched by the European Space Agency (ESA) in 1995 to follow up IRAS's discoveries.

THE COSMOS

STARS

159

LIFE ON THE MAIN SEQUENCE

A new-born star emerging from its cocoon of dust and gas settles on to the main sequence, where it will spend most of its life. Here it 'burns' its initially plentiful hydrogen fuel by nuclear reactions in its core which produce the energy that makes it shine. During this time it will be generally well-behaved, maintaining a steady surface temperature and luminosity. Just where a star will be positioned on the main sequence and how long it will spend there depends critically on its mass. Two cases will illustrate the diversity of stellar lifestyles.

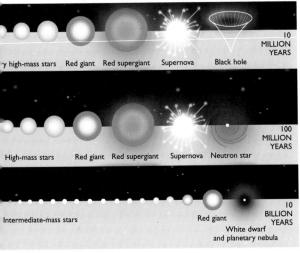

▲ Lives of the Stars
Stars lead very different lives, according to their mass. Very high-mass stars (up to 100 times the mass of the Sun) are profligate with their nuclear fuel; their eventual gravitational collapse results in a black hole. High-mass stars (up to 20 times the Sun's mass) also burn fuel rapidly; they explode as supernovae and collapse to form neutron stars. Intermediate-mass stars, such as the Sun, fuse hydrogen for billions of years before swelling to become red giants and then white dwarfs.

▶ Proton–Proton Reactions
These are the three primary nuclear reactions of the proton-proton chain. Such reactions produce most of the energy within stars of similar mass to the Sun while they are on the main sequence.

KEY
- Proton
- Neutron
- Positron
- Neutrino
- Gamma ray

1. Hydrogen + Hydrogen → Deuterium
2. Deuterium + Hydrogen → Helium-3
3. Helium-3 + Helium-3 → Helium-4 + Hydrogen + Hydrogen

LOW-MASS STARS

A low-mass star such as the Sun can expect to spend more than 10 billion years on the main sequence, undergoing only a small decrease in surface temperature and a small increase in size. This long-term stability makes low-mass stars the most likely environment for finding planets on which life might have developed. The leisurely lifestyle is a consequence of the modest rate at which these stars consume their hydrogen fuel. At the centres of such stars, where temperatures are 15 million°C (27 million°F) and densities 150,000 kg/cubic metre (9,400 lb/cubic ft), the main nuclear reactions are the **proton–proton chain**. In this, the nuclei of hydrogen atoms are progressively combined to form a helium nucleus, with the loss of a small amount of mass and the associated release of a large amount of energy. This energy is carried most of the way to the star's surface by **photons** (i.e. in the form of **radiation**), but over the last 15% or so (more for the smaller stars) convective motion of gas does the job. As the central store of hydrogen is used up, the core becomes richer in helium, leading eventually to a crisis in the star's life (see Old Age and Death of Stars, page 162).

HIGH-MASS STARS

Massive stars are altogether more profligate with their energy reserves and have correspondingly shorter lifetimes. They burn their hydrogen in a set of reactions known as the **CNO cycle**, which involves using the nuclei of carbon (C), nitrogen (N) and oxygen (O) as catalysts. This process runs at high core temperatures and produces enormous amounts of energy. For example, a star of 15 solar masses (corresponding to spectral type B0) has a central temperature of about 35 million°C (63 million°F); it pours out nearly 40,000 times as much energy as the Sun, using up its hydrogen fuel in only 12 million years or so.

The high luminosity of the most massive stars, of spectral type O, drives an outflow of gas from their surfaces called a stellar wind, a more extreme version of the **solar wind**. Stellar winds carry off considerable amounts of matter during the star's main-sequence lifetime. For example, a star of spectral type O5, with a mass of 60 Suns and a surface temperature of 42,000°C (76,000°F), will blow away more than a quarter of its mass during its 3.5-million-year lifetime on the main sequence. In fact it is the ability of intense radiation to accelerate gas in the atmosphere of a star that limits the mass a star can have. If the energy generation rate is so high that the radiation pressure is greater than the surface gravity, then the star will literally blow itself apart. This maximum luminosity is known as the Eddington limit. It puts a ceiling on stellar masses in the region of 150 solar masses.

THE ROTATION OF STARS

Since star-forming clouds have such difficulty in losing angular momentum, it is no surprise that stars arriving on the main sequence are rotating rapidly. This shows up through the existence of

ARTHUR EDDINGTON

Arthur Stanley Eddington (1882–1944) was born into a Quaker family in Kendal, Cumbria, England. He had a brilliant passage through school, college and university, becoming Plumian Professor at Cambridge in 1913, where he spent the remainder of his career. He was one of the early and most authoritative proponents of **Albert Einstein**'s (1879–1955) theory of relativity and was closely involved in one of the expeditions to photograph the solar eclipse of 1919 which demonstrated the bending of light by a gravitational field, in this case the Sun's. Eddington worked in many areas of astronomy, such as stellar dynamics and the nature of the interstellar medium, but one in which he played a major role was in the application of physical and mathematical principles to describe the structure of a star; this was encapsulated in his benchmark book *The Internal Constitution of the Stars* (1926). Towards the end of his life he devoted his considerable intellectual powers to a search for connections between the fundamental constants of nature.

▲ *Our Sun might be the life source of everything on Earth, but within the Universe it is just a typical yellow dwarf.*

sequence lives. Rotation speeds tend to climb with increasing mass, reaching in excess of 300 km/s (190 mi/s), or approximately once a day, for some O and B stars, sufficient to fling off material which may form a ring of hot gas around the star.

STELLAR ATMOSPHERES AND BEYOND

Whatever may be going on deep inside a star, almost all the information arriving at Earth comes solely from the outermost layers, known as the atmosphere. The appearance (colour, variability and especially spectral features) of this relatively thin region reveals the nature of the star within: its temperature, luminosity, composition and a host of other characteristics. However, the atmosphere is not a featureless blanket around the star but is in a state of turbulence. Many stars cooler than about spectral type F are now known to have cycles of activity just like the Sun's, typically lasting a few years, while others have their activity provoked by their companions in close binary systems. The most noticeable effects, including huge flares, are seen on K and M type dwarfs, known appropriately as **flare stars**. Curiously, the more massive stars are rather better-behaved in this respect, at least while growing up on the main sequence.

▲ **The members of** the open cluster M6 are on the main sequence except for the brightest star, which has already evolved to become an orange-coloured giant.

✳ COMPUTER MODELLING OF STARS

We cannot cut open a star to see its interior, so the principal way of discovering how one works is to model it on a computer. Firstly it is necessary to have mathematical formulae describing the relationships between the atomic particles within the star, which are governed by the internal pressure, temperature and composition. The way in which the rate of energy generation is affected by temperature and density needs to be understood, especially in the central regions, as do the mechanisms which transport that energy to the surface, i.e. radiation and convection. We also need to know the conditions at the star's surface and other details such as its rotation speed. Finally, the equations describing the balance of pressure and gravity at each depth need to be known. If all goes well with the calculations, the result is a table of all the parameters from core to surface of the star, and predictions of the amount of energy that emerges. This can be compared with what is actually observed. The final step is to 'evolve' the star in the computer, allowing the model to change according to the use of nuclear fuel in the interior and mass loss from the surface. This art has now been brought to a stage of some sophistication and whole evolutionary tracks across the HR diagram are routinely calculated for a range of initial masses. Such work allows the construction of lines on the HR diagram showing the location of stars of different masses at specified times; fitting these predictions to observational data allows the ages of star clusters to be found.

broad absorption lines in their spectra: one side of the star is approaching the observer on Earth while the other side is receding, causing the contributions to the spectrum from different parts of the star's disk to be **Doppler shifted**, some to the red and some to the blue. The more rapid the rotation, the greater the spread of Doppler shifts and the broader the line becomes.

Over time, the lower-mass stars slow down as their magnetic fields link the stars with their stellar winds which carry off angular momentum. Such stars, including our Sun, have equatorial speeds of just a few kilometres (or miles) per second, taking a few weeks to spin once. The same is not true of stars with more than a couple of solar masses. These lack strong magnetic fields, because their outer layers are not in convective motion, and they continue to rotate rapidly throughout their main-

⎕ MEASURING MAGNETISM IN STARS

In the presence of a magnetic field, some lines in a spectrum are split into two or more components with slightly different wavelengths; this is known as the Zeeman effect, after the Dutch physicist Pieter Zeeman (1865–1943) who first detected it. The amount of splitting depends on the strength of the magnetic field. Weak fields cause a broadening of the line rather than a splitting, and even with strong fields the difference in wavelength between the components is small. However, the different components have the

valuable attribute that their light is polarized so they can be separated and measured by a suitably equipped spectrograph. Using this technique, magnetic fields have been measured on the Sun and on a number of other stars where the absorption lines in the spectrum are sharp enough to yield good results. A group of stars known as peculiar A stars (or Ap stars) have magnetic fields thousands of times stronger than the Sun's. As a result, certain elements such as silicon, chromium and europium are concentrated in spots on their surfaces, so that the stars' spectral features vary as the star rotates.

THE COSMOS

OLD AGE & DEATH OF LOW-MASS STARS

Three-quarters of a new-born star's mass is hydrogen, which fuels the star's central furnace for most of its life. However, there comes a time when every star needs to seek alternative energy resources. The arrival of this stage is accompanied by significant changes in the star's outward appearance, notably a vast increase in size and luminosity. Eventually its fuel resources become depleted entirely, the star's distended outer layers slip away and it fades into oblivion. Here is a glimpse into the future of stars with masses about twice that of the Sun and less, corresponding to main-sequence spectral types of A5 and cooler.

RUNNING OUT OF HYDROGEN

Over a star's main-sequence lifetime the energy-generating nuclear reactions gradually convert all the hydrogen at its core into helium 'ash'. As fuel stocks run out at the centre, hydrogen burning moves out into a shell around the core where supplies are still plentiful. In response, the star's outer layers expand and cool, signalling the end

of its days on the main sequence. No energy is now being generated in the helium-filled core, which begins to contract and heat up in what will become a crucial restructuring. Stars of two solar masses reach this turning point in their evolution less than two billion years after birth, while for the smallest of all, the lightweight dwarfs of spectral types K and M, the timescale is longer than the present age of the Universe, so none of those have yet left the main sequence.

BECOMING A RED GIANT

The swelling star sets off on a major journey across the HR diagram, moving to the right as its surface temperature drops and then upwards to reflect its increasing luminosity, into the realm of the **red giants**. The star's outer layers balloon into a tenuous envelope while its surface cools to perhaps 3,600°C (6,500°F); such a star is classified as an M0 giant. In the case of the Sun, it is uncertain whether this expansion will engulf Earth or stop just short.

Inside the star, gravitational contraction increases the core's temperature to an astounding 200 million°C (360 million°F) and its density to 10 million kg/cubic metre (620,000 lb/cubic ft), enough to bring helium into play as a source of fuel. The reaction, which begins suddenly in what

▼ *Astrophysicist Subrahmanyan Chandrasekhar (right).*

is termed the **helium flash**, requires three helium nuclei to be combined almost simultaneously to produce a carbon nucleus. This is known as the **triple-alpha process** because a helium nucleus is termed an alpha particle.

The onset of helium burning in the core leads to further structural readjustment. The star falls back near-vertically down the **giant branch** in the HR diagram, where it remains until it starts to squeeze out the last of its energy reserves by the initiation of helium shell burning. The end is nigh.

SUBRAHMANYAN CHANDRASEKHAR

Subrahmanyan Chandrasekhar (1910–95), born into a cultured and gifted Indian family (his uncle was Nobel Laureate C.V. Raman), became one of the leading theoretical astrophysicists of the twentieth century. He graduated in Madras in 1928 and went to Cambridge, England, in 1930 to work on the application to stellar structure of a new theory of gases. He showed that the matter in the interiors of white dwarfs was degenerate and that, provided the mass was less than 1.4 Suns (now known as the **Chandrasekhar limit**), such a body could cool without further gravitational collapse. Chandrasekhar's interests were wide-ranging and he published mathematically elegant monographs and papers on many topics, including stellar dynamics, radiative transfer of energy, hydrodynamics, stellar structure and black holes. He was managing editor of *The Astrophysical Journal* for almost 20 years and recipient of numerous honours, notably a share of the Nobel Prize for physics in 1983.

Hydrogen-burning shell
Inert helium
Helium-burning core

Envelope

▲ Last Gasp of a Giant
A – As a red giant finally expires, its outermost layers are ejected into space. B – Millions of years later, high-energy radiation propels a second wave of material outwards. C – Driven by the stellar wind, particles sweep through the slower-moving outer layers like a snowplough, pressing the gases into a ring. D – The fast inner gases find weak spots in the ring and most of the material leaks out into the interstellar medium. Planetary nebulae (gas clouds expelled by old stars) at three corresponding stages are shown: X – Radiation forces the bubble of gas outwards in this young planetary nebula, NGC 7027. Y –This middle-aged planetary nebula, IC 3568, has a well-defined sphere of gas. Z – Gas escapes through the outer layer of NGC 3132, distorting its shape.

▲ A Red Giant
By the time a star becomes a red giant, the nuclear fusion of hydrogen over billions of years has already made a core of burning helium, contained within a helium shell. Surrounding this is a further shell of burning hydrogen, beyond which is the tenuous envelope.

THE COSMOS

STARS

THE LAST GASP – A MIRA VARIABLE

With helium now burning in a shell, and possibly also some residual hydrogen burning in a shell surrounding that, the star is becoming unstable. It ascends the HR diagram again on a track known as the asymptotic giant branch, taking it to even greater extremes of size and luminosity. This time it becomes an even cooler M giant than before and, more interestingly, a **long-period variable** or **Mira star** (they are named after their prototype, Mira or Omicron [o] Ceti). Such stars pulsate in size with a period of a year or so, being about two or three times larger when at their biggest than at their smallest. During the pulsation cycle their surface temperatures range from about 1,600°C (2,900°F) to 2,300°C (4,200°F) – that is, if their tenuous outer layers can be thought of as a surface, since they have densities lower than a vacuum produced in an Earth-bound laboratory. It is mainly the change in temperature rather than size that is responsible for the variations in the visible light from Miras; at minimum, much of their energy is emitted in the invisible infrared. Another feature of this phase of evolution is extensive mass loss in a 'wind' of particles from the star's surface; half the star's mass, or more, can seep into space in this fashion. Equally significantly, huge convection cells in the star's outer layers can dredge carbon from the star's core to the surface. Carbon is blown off by the stellar wind and condenses into solid dust grains in interstellar space.

PLANETARY NEBULAE – STRIPPING TO THE CORE

Attempts to ignite any remaining fuel via a series of helium shell flashes produce bursts of energy that push away the star's outer layers, denuding the hot core. Any remaining hydrogen and helium is burning at the core's surface (not, like a normal

💡 **EXTINCTION OF LIFE ON EARTH**

As the Sun evolves, the benign conditions that currently sustain life on Earth will irrevocably change. Evidence already exists that relatively small variations in the Sun's output can have serious climatic consequences (see The Maunder Minimum, page 198). During the next billion years, models predict that the surface temperature of the Sun will hardly change and the luminosity will increase by only 1%, so life on Earth is secure from major solar changes on that time-scale. Five billion years presents a more worrying scenario. By then, the Sun's luminosity will be 50% greater than now. Eight billion years from now, the Sun will fill the daytime sky as a red giant and its energy output will have soared to 10 times what it is today. Life of even the simplest kind will have been extinguished on our parched and roasted planet, while our descendants will long ago have fled to a younger or more long-lived star.

EVOLUTIONARY TIMESCALES

Spectral Type	Main-sequence Mass (Suns)	Main-sequence Life (10^6 y)	Main Sequence to Red Giant Timescale (10^6 y)	Ascent of Giant Branch (10^6 y)
O5	60	3.4	Becomes an LBV★	
B5	6.9	60	4	20
A5	2.0	1650	42	100
F5	1.4	6200	155	200
G5	0.92	11,000	5,000(?)	1,000(?)
K5	0.67	45,000	Not known	?
M5	0.21	190,000	Not known	?

★LBV = luminous blue variable (star never becomes a red giant or supergiant)

star, at its centre), producing temperatures as high as 150,000°C (270,000°F), far in excess of the surface temperature of even the hottest main-sequence star. Ultraviolet light pouring forth from the exposed, intensely hot core illuminates the thinning envelope of ejected gas around it. A planetary nebula is the result. This misleading name comes from the fact that some of them resemble the small disks of planets when viewed telescopically; they have nothing to do with planets.

The central star zooms to the left in the HR diagram as its envelope clears and the last traces of hydrogen and helium burn up. The high-energy photons pouring from its surface cause the ions in the expanding, low-density nebula around it to glow with wonderful colours. Planetaries show great diversity of structure. One reason is the differing angles at which we view them, but another is that the mass is often not ejected symmetrically, sometimes because of the influence of a binary companion. The light show lasts a few tens of thousands of years before the planetary nebula shell disperses.

◀ *The Egg Nebula, a cloud of dust and gas ejected from a former red giant, the core of which is is hidden by a cocoon of dust (the dark band at the centre). The arcs are denser shells which arise from variations in the rate of mass ejection from the central star on time scales of 100 to 500 years.*

▶ *The Hourglass Nebula, a young planetary nebula, showing different colours due to ionized nitrogen (shown red), hydrogen (green) and doubly ionized oxygen (blue).*

WHITE DWARF – STELLAR CINDER

Although it shines brilliantly at first, the central star of a planetary nebula is bereft of fuel. Soon it enters its twilight years, cooling and sliding down the HR diagram to become a white dwarf. Such stars are burnt-out cinders with diameters not much bigger than Earth yet with a mass similar to that of the Sun. Physics dictates that the maximum mass a star in the white dwarf state can have without collapsing further is 1.4 solar masses, known as the Chandrasekhar limit. Even stars of intermediate mass, up to about eight solar masses, can end up as white dwarfs, but to do so they must lose their excess mass in the red giant and planetary nebula stages.

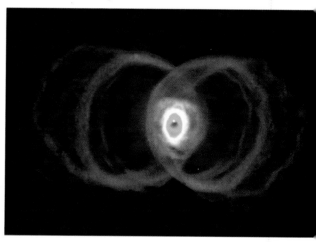

THE COSMOS

OLD AGE & DEATH OF HIGH-MASS STARS

Massive stars are rare, in part because fewer are born than their lower-mass brethren, but also because their evolution is so rapid that, astronomically speaking, they are gone in a flash. If their stay on the main sequence brief but brilliant, their evolution afterwards is spectacular, giving rise to some of the brightest objects in the Universe and making crucial contributions to the chemical mix of the Galaxy. Indeed, if it were not for massive stars that lived and died early in the history of the Galaxy, the raw materials of which Earth and life are composed would not exist.

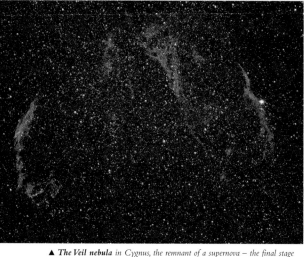

▲ *The Veil nebula in Cygnus, the remnant of a supernova – the final stage in the death of a massive star.*

DASHING ACROSS THE HR DIAGRAM

Once the central hydrogen supplies of a high-mass star have been turned into helium and the core contracts to seek new sources of energy, the outer layers expand and the star takes off near-horizontally to the right on the HR diagram. How far it gets depends critically on its mass: if the initial mass is much above 45 Suns, what was already serious mass loss from the surface becomes even more extreme, turning it into a violently unstable luminous blue variable, or LBV. In this case, the star's spectral type never gets cooler than about A0. Less massive stars can proceed fully to the right of the HR diagram to become red

supergiants. Those stars close to the 45-solar-mass threshold evolve not into supergiants but the even more remarkable hypergiants, in a mad dash lasting just 100,000 years.

HYPERGIANTS

The hypergiants are among the most luminous stars in the Galaxy, perhaps 100,000 times brighter than the Sun in the visual part of the spectrum. They have masses above about 30 Suns but below the 45-solar-mass threshold of the luminous blue variables. They have spectral types ranging from B to M and have been allocated the luminosity class Ia+ or even 0, setting them above the brightest normal supergiants. Hypergiants are often to be found moving around the HR diagram, even within a human lifetime, these excursions being accompanied by sporadic events of mass-loss which generate shells of material around the star. A well-known example, Rho Cassiopeiae, is to be found near the familiar W-shape of stars in Cassiopeia (see Star Chart 2, page 260).

WOLF–RAYET STARS – STRIPTEASE ARTISTES

Whether a star evolves into a luminous blue variable or remains a hypergiant, mass loss from its surface is a dominant feature of its later life. After such stars leave the main sequence, a fierce wind with speeds up to 2,000 km/s (1,200 mi/s) strips off their outer hydrogen layers at a rate of 10 million million tonnes per second. With the outer covering removed, the products of the nuclear burning in the star's interior come into view, giving rise to strong, broad **emission lines** in its spectrum, which originate in a dense expanding envelope.

Such specimens are known as **Wolf–Rayet stars** after their French discoverers, Charles Wolf (1827–1918) and Georges Rayet (1839–1906). Those showing prominent nitrogen lines are designated WN stars. Later in the star's evolution, when the helium produced in the CNO reactions is utilized to produce carbon (and the mass loss continues), the nitrogen lines disappear to be replaced by strong carbon and oxygen lines; these are classified as WC stars.

DESPERATE FOR ENERGY

Thus far, the massive stars have only achieved what their much smaller siblings accomplished, turning hydrogen into helium and carbon, albeit on a much shorter timescale. However, at this point, despite losing considerable mass through stellar winds, they still have sufficient mass to

crush their cores, tapping into new sources of energy, and this they proceed to do. Firstly they turn their carbon into neon and magnesium, then they burn their neon to make oxygen and more magnesium. Not content with that, they change magnesium into silicon, and oxygen into sulphur. Then, in an alchemist's last fling, they transmute silicon into iron. All these processes are run in turn, building up an onion-like structure inside the star. With each desperate new attempt to squeeze out energy, stars with initial masses between about 5 and 20 Suns meander repeatedly from left to right and back again on the HR diagram. On each trip the star passes through what is termed the **instability strip**, where they pulsate as a **Cepheid variable**. Ultimately, the star is left with a heart of iron. But, because of its atomic structure, iron will not fuse with anything to yield more energy. The star has reached the end of the line.

BANKRUPT – GOING SUPERNOVA

In a doomed attempt to extract yet more energy, the iron core tries to shrink further. However, as the core's mass is above the Chandrasekhar limit, gravity takes over and the star collapses. The energy created by this infall, instead of going to support the core against further collapse as previously, is soaked up by the iron nuclei which disintegrate into alpha particles (helium nuclei). The implosion proceeds at incredible speed, with considerable amounts of energy being carried off on a tide of

▼ *A **Wolf–Rayet star**, WR124 in Sagittarius, surrounded by a cloud of hot gas being ejected in a furious stellar wind, as seen by the Hubble Space Telescope. Wolf–Rayet stars are short-lived and extremely hot – in this case 50,000 Kelvin.*

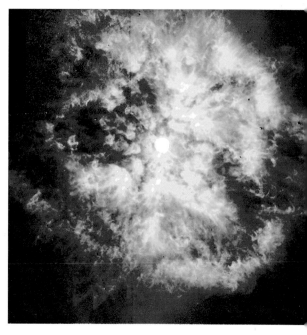

SUPERNOVA 1987A

On 23 February 1987 astronomers had no inkling that possibly the most dramatic celestial event of the twentieth century was about to unfold. That morning, automatic detectors working deep underground at Kamioka in Japan and Lake Erie in the USA registered an unexpected burst of neutrinos, elusive particles that are produced in vast numbers at the death of a massive star. Not until the following morning did Ian Shelton, a Canadian astronomer at Las Campanas Observatory in Chile, notice that his just-developed photographic plate of the Large Magellanic Cloud showed a bright star that had not been there the night before. Thus was discovered the first naked-eye supernova since Kepler's Star of 1604.

The significance of this event cannot be overestimated. For the first time, a bright supernova could be studied with all the tools of modern astronomy, right across the **electromagnetic**

spectrum. Indeed, it was the International Ultraviolet Explorer (IUE) satellite that pinpointed which star had actually exploded. On previous charts of the area there were three stars very close together in the sky; after the event, two were recorded by IUE but the third, known by the catalogue name Sanduleak -69°202, had vanished. What was surprising was that the exploded star had earlier been classified as a blue supergiant, not a red one as theory predicted. However, some quick theoretical footwork on what turned out to have been an unusual supernova solved the mystery, indicating that the star had been towards the lower level of the mass range for supernovae, perhaps initially about 20 solar masses. Possibly it was on its final excursion to the left of the HR diagram when it exploded.

▶ *Supernova 1987A photographed by the Hubble Space Telescope seven years after it exploded, showing rings of gas lost from the star during its red supergiant stage about 20,000 years ago.*

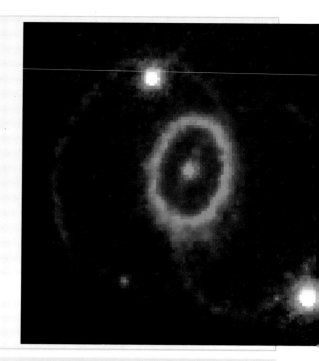

neutrinos in just a fraction of a second. At the star's centre, protons and electrons merge to become **neutrons**. When a ball of such nuclear material a few kilometres (or miles) across has formed, the collapse halts. The impact of further infall is reflected in a huge shock wave which races back through the outer layers of partially processed material. It is then that many exotic heavy chemical elements are produced by the process known as rapid neutron capture (the **r-process**). The shock wave blasts the outer layers into space at speeds up to 20,000 km/s (12,500 mi/s). A **supernova** of Type II appears, shining with the luminosity of 10 billion Suns.

▶ *A Supernova is Born*
With no more hydrogen left to fuse into helium, the big squeeze starts. The gravitational collapse makes the core heat up, and a supernova is conceived. With increasing temperatures, new elements are created in a sequence that ends with the formation of a compressed ball of iron. At this point, no more reactions can occur. The iron core crushes in on itself in the 'maximum scrunch', triggering an explosion that sends shock waves outwards. Seconds later, trapped neutrinos stream out. In a matter of hours, the shock waves burst through the surface, hurling matter into space. All that remains is a city-sized neutron star.

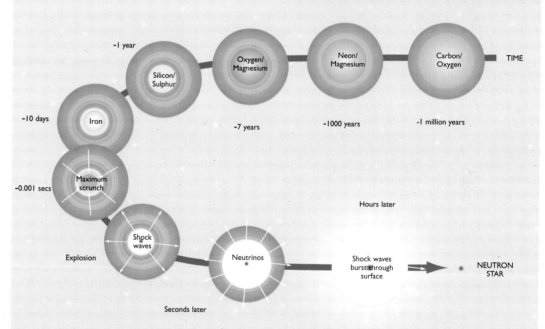

TIME

-1 year · Silicon/Sulphur · Oxygen/Magnesium · Neon/Magnesium · Carbon/Oxygen

-10 days · Iron · -7 years · -1000 years · -1 million years

-0.001 secs · Maximum scrunch

Explosion · Shock waves · Neutrinos · Shock waves burst through surface · NEUTRON STAR

Seconds later · Hours later

A FAMILY OF SUPERNOVAE

On a photograph, one supernova looks much like another – a bright star shining more brilliantly than any other in its host galaxy for weeks on end. Yet careful study has established that there are two principal types of supernovae with distinctly different origins. Type II, as mentioned above, are the explosion of high-mass stars at the ends of their lives, flinging off several solar masses of gas into space; famous examples include Supernova 1987A and the supernova of 1054 that gave rise to the Crab Nebula.

Type I supernovae are broken into further subclasses. Type Ia is thought to arise from the detonation of a white dwarf in a close binary system. A white dwarf is a low-mass star that normally has no right to explode. What causes its misbehaviour is gas from the companion, which spills on to the white dwarf as the companion expands towards the end of its own life. This extra mass compresses and heats the white dwarf, igniting the carbon and oxygen in its core in a nuclear explosion. Since white dwarfs are of low mass, only about one solar mass is ejected as the star disrupts itself.

Astronomers distinguish a supernova's type on the basis of features in its spectrum: those of Type II show the existence of hydrogen, from the ejected outer layers of the star, whereas those of Type I lack hydrogen, since the star had no hydrogen left when it exploded. Confusingly, despite their spectral similarities, the subtypes known as Ib and Ic seem to have different origins from the Ia variety – they apparently result from the explosion of single massive stars that have lost their outer layers, probably Wolf–Rayet stars. Hence they may actually have more in common with Type II supernovae than Type Ia.

THE COSMOS

STELLAR REMNANTS

Throughout a star's life the dominant factor controlling its lifestyle is its mass. Exactly the same is true in death: there are three possible end points for a star's existence, and mass determines which one should be its destiny. For massive stars, a neutron star or even a black hole is their fate, while for stars of intermediate and low mass, white-dwarf status lies in store. In all these objects, the physical conditions are far more extreme than anything encountered on Earth.

WHITE DWARFS

White dwarfs are the endpoint in evolution for the majority of stars, the core left behind after a star of low to moderate mass has shed its outer layers as a planetary nebula. Even though they are common, their dimness makes them difficult

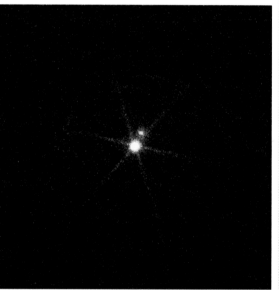

◀ *Sirius,* the brightest star in the night sky, and its white-dwarf companion, seen here at X-ray wavelengths.

to spot. The first was not seen until 1862, when Alvan Clark (1804–87) chanced to turn a large new telescope on Sirius, the brightest star in the night sky, and noted that it had a close companion some 10,000 times fainter. Despite the name 'white' dwarfs they actually range in colour from blue to red, depending on the surface temperature. Over billions of years they cool and fade, from a starting point of more than 100,000°C (180,000°F) to around 4,000°C (7,200°F). The usual classification of spectral types does not apply; their surface composition, a legacy of the parent star's earlier nuclear burning, is more important than surface temperature.

No energy is being generated in the cores of white dwarfs but they are supported against collapse by the ultra-high-density state of matter within them, known as **electron degeneracy**. In this exotic state, the electrons stripped from atoms form a 'sea' which cannot be further compressed by gravity, as long as the total mass of the star is no more than 1.4 Suns (the Chandrasekhar limit). So high are the resulting densities, some billion kg/cubic metre (60 million lb/cubic ft), that a teaspoon of material from a white dwarf would weigh several tonnes. A typical white dwarf has a diameter only slightly greater than that of Earth and a mass of 0.6 Suns. Being small and dense they have enormously strong gravitational fields, some 100,000 times that at the surface of Earth.

NEUTRON STARS AND PULSARS

If a dying star retains too much mass to qualify as a white dwarf it will collapse further, into a **neutron star** only about 20 km (12 miles) across. Under extreme gravitational pressure, electrons combine with protons to form the chargeless particles called neutrons. The bulk of the star becomes a 'sea' of free neutrons with a density over a million times greater than that of a white

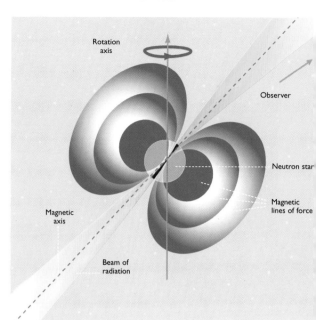

▲ *A Pulsar Lighthouse*
The lighthouse effect of a pulsar is caused by a rapidly rotating neutron star that emits radiation in two beams, directed along the opposite poles of its magnetic field. The beams swing round with the rotation of the star, giving the effect to a distant observer of a pulsing signal.

dwarf, topped by a solid crust consisting of a latticework of atomic nuclei.

Neutron stars are probably left behind by many supernovae, although if their weight-loss routine is insufficient to take them below about three solar masses then further – and final – collapse awaits, into the oblivion of a **black hole**. Some neutron stars may arise from the addition of gas to a white dwarf from a companion star in a binary system, tipping it over the Chandrasekhar limit.

Neutron stars were predicted theoretically in the 1930s but the first was not discovered until 1967, when radio astronomers detected rapidly pulsating sources that were dubbed **pulsars**. Neutron stars have since been found in binary systems where gas from a companion falls on to the neutron star, heating up and emitting **X-rays**. Two neutron stars have been observed flashing optically, both in the centres of supernova remnants: the Crab Nebula, left over from the supernova in Taurus seen by Chinese astronomers in 1054, and the Vela supernova remnant.

Neutron stars have strong magnetic fields, because the magnetic field of their parent star became concentrated when it collapsed, and this is the key to why they emit pulses of energy. Charged atomic particles race along the magnetic field lines, beaming high-energy radiation in their direction of travel. Since the field is particularly strong at the magnetic poles, the beam shines out

JOCELYN BELL

Jocelyn Bell (b. 1943) was a student working on her doctoral thesis at the Mullard Radio Astronomy Observatory, Cambridge, England in 1967. The topic was interstellar scintillation, the 'twinkling' of radio waves from distant sources by intervening interstellar material, as a method of searching for quasars. Towards the end of the year she noticed some strange squiggles on the chart depicting readings from a particular part of the sky. When examined in more detail, they turned out to be pulses of radio energy repeating precisely

every 1.3373 seconds. Soon other sources were found. The original designation of these sources as LGM (Little Green Men, from their suspiciously artificial regularity) was soon dropped, and she and her supervisor, Professor Antony Hewish (b. 1924), were able to announce the discovery of pulsars in 1968. Their true nature was still unknown, but other astronomers soon established that they were rapidly rotating neutron stars. Hewish was later awarded a Nobel Prize for physics for his work in radio astronomy; she is now Professor Jocelyn Bell-Burnell.

STARS

THE COSMOS

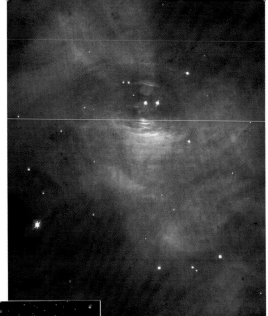

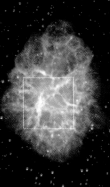

▲ The Crab Nebula, *M1, the remains of a supernova explosion. At its heart is a pulsar, whose energy illuminates the entire nebula.*

most intensely from there. And because the magnetic axis is usually not aligned with the rotation axis, the beam sweeps around like a lighthouse as the star spins, from once every several seconds to hundreds of times a second depending on the pulsar concerned. The powerful radiation from a pulsar spans the entire electromagnetic spectrum but it has to be paid for somehow. It is taken from the rotational energy of the star – so, over thousands of years, the pulsar gradually slows down. However, this spin-down is not always smooth; occasional sharp increases, called **glitches**, are recorded, apparently due to 'starquakes' in the pulsar's crust.

BLACK HOLES

The fate of a stellar core with a mass greater than about three Suns is a black hole. Nothing can support a core with such a mass against the

💡 **SS433 – A STAR THAT WON'T FIT THE PIGEONHOLE**

Close to the plane of the Milky Way, not far from the bright star Altair, lies an old supernova remnant, W50. Near its centre is a faint star known as SS433, from its number in a catalogue of stars with unusual spectra compiled by Bruce Stephenson and Nicholas Sanduleak. This curious object was discovered to be a strong radio source and, later, a variable source of X-rays. Most amazing of all was the discovery in its optical spectrum of emission lines of hydrogen with three components: one was quite strong and moved slightly in a 13-day cycle, while the two weaker components showed enormous shifts, corresponding to velocities

inexorable squeeze of its own gravity and it undergoes the ultimate collapse to become a **singularity**, a point in space with a gravitational pull so intense that even light cannot escape. Because we cannot see a black hole directly, its presence must be inferred from the effect it has on its surroundings. In practice that means looking at **binary stars** in which the black hole component orbits a visible companion. One of the best examples is Cygnus X-1, in which a blue supergiant is partnered with what seems to be a black hole – at least, something fairly massive and invisible is tearing material from the visible star, emitting X-rays as it attempts to swallow the gas it has ensnared (see also Black Holes, pages 172–75).

SUPERNOVA REMNANTS

Neutron stars and black holes are not all that remains after a supernova has exploded. Equally important is the gas that is shot back out into space, highly contaminated with heavy chemical elements forged in the explosion. Once it has dispersed into the interstellar medium, this gas can later be swept up into new generations of stars. These new stars will have an enriched chemical composition and will evolve slightly differently from their predecessors. More to the point, the heavy elements made in the supernova explosion are essential for the creation of rocky planets – and of life.

A number of supernova remnants have been discovered in the Milky Way and nearby

◀ IC 443 *in Gemini is another supernova remnant with a pulsar near its centre.*

of 40,000 km/s (25,000 mi/s), which changed from positive to negative and back again every 164 days. What's going on here? It seems that a relatively normal star is shedding material into an **accretion** disk around a neutron star or black hole. This binary has a period of 13 days and the strong hydrogen emission comes from the disk. Material approaching too close to the compact object is either sucked into it or ejected in jets perpendicular to the disk. The whole system is precessing like a top in a period of 164 days so that the jets end up spraying their material into space like a high-speed garden sprinkler. It is these jets that produce the weak but dramatically moving hydrogen lines.

galaxies, generally through their strong radio emissions, although some also shine brightly at other wavelengths. Among the most striking is the Crab Nebula, M1 in Taurus (see Star Chart 3, page 262), the remains of a supernova seen by Chinese astronomers in 1054. The total energy output of the Crab exactly balances the energy output of the pulsar found blinking 30 times a second at its heart, so it is clear what is powering the remnant. The Crab is a young remnant, a mere 10 light years across and expanding at about 1,500 km/s (900 mi/s), but much older examples exist. An example is the beautiful Cygnus Loop, 150 light years across; its expansion has been slowed by the interstellar medium to just 130 km/s (80 mi/s) over the course of perhaps 40,000 years.

HIGH-ENERGY ASTRONOMY FROM SPACE

The highest-energy processes in astronomy, usually involving exceptionally hot gas, naturally create the highest-energy radiation, so to study these it is necessary to look at gamma-ray, X-ray and extreme ultraviolet (EUV) wavelengths. However, Earth's atmosphere blocks out such radiation, so to carry out studies of celestial sources in these regions of the spectrum, telescopes and detectors have to be carried into space. Initially rockets were used to give brief glimpses of this exciting Universe, but satellites are now used to permit extended studies. Instruments aboard such satellites have used many techniques developed by particle physicists in their Earth-bound laboratories, but increasingly space scientists are creating new devices specifically for astronomical applications. Simultaneous observation of objects by several space-borne and ground-based instruments operating across the spectrum is now routine.

THE COSMOS

THE STELLAR ZOO – SINGLE STARS

Stars, whether single or double, are not always well-behaved. Most will exhibit variability in their brightness or spectrum at some time during their lives, or display some other peculiarity. The most noticeable changes are in a star's light output. Such intrinsic variability can be a natural part a star's evolution as it crosses certain areas of the Hertzsprung–Russell diagram, particularly the so-called instability strip where stars pulsate in size; this strip stretches from the realm of the yellow supergiants down to main-sequence stars of types A and F, although it is by no means the only danger zone.

PULSATING VARIABLES

Pulsation in stars – a puffing out and subsequent collapse of their outermost layers – is a common cause of variability. To find out why it happens, we must delve beneath the star's surface to a zone where temperatures are hot enough for hydrogen and helium to become ionized or, deeper down, for helium to be doubly ionized. Compression of this ionized zone makes it less transparent to light (its **opacity** is said to increase), damming up the flow of energy from below. Pent-up energy pushes the ionized layer outwards like a piston and the star expands until the opacity drops. The radiation can escape, the pressure eases and the gas settles back, applying renewed pressure so the process repeats. If conditions are right, such pulsations can be maintained with great regularity over long periods of time.

Probably the most important category of pulsating stars are the Cepheids, named after their prototype Delta Cephei. These are yellow supergiants of type F and G which oscillate on anything from a daily to monthly basis. As they do so their visible light changes as much as sixfold (two magnitudes), their temperature by over a thousand degrees and their diameter by about 10%. Cepheids are highly luminous and so can be seen at great distances – even in other galaxies – which makes them valuable as distance indicators despite their relative scarcity.

Fainter than Cepheids, but almost as valuable as distance indicators, are the **RR Lyrae stars**, found in great numbers in **globular clusters**. These stars have a similar range of brightness variation to the Cepheids but much shorter periods, no more than about a day. In the HR diagram they are located on the horizontal branch between the main sequence and the red giants at spectral types A and F, and are stars of low mass burning helium in their cores, following the helium flash.

At the bottom of the instability strip, joining it to the main sequence, we find the **Delta Scuti stars**, which oscillate with periods of a few hours or even less, changing in brightness by only a few per cent, undetectable except with sensitive instruments. Other varieties are listed in the table of variables opposite.

RED VARIABLES

Away from the instability strip, many red giants and supergiants pulsate to some extent because of their bloated size. The most numerous of all known variables are the Mira stars, whose large range of brightness makes them easy to spot. They change by up to 11 magnitudes with periods from months to years, hence their alternative name 'long-period variables'. Less predictable in behaviour are the semiregular variables, which have a smaller range of variability and periods that are not as well defined as the Mira stars; the most prominent example is Betelgeuse. Those whose fluctuations show no detectable periodicity at all are termed irregular variables.

R CORONAE BOREALIS STARS

Among the stranger types of variable, the R Coronae Borealis stars stand out. These are supergiants of spectral types F or G, rich

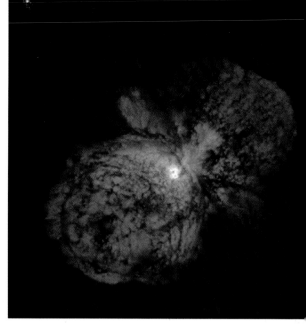

▲ *Eta Carinae, a supermassive star that flared up in the 1840s to become the second-brightest star in the sky. It threw off two lobes of gas and a large thin equatorial disk, shown in this Hubble Space Telescope photograph.*

in carbon. Atmospheric instabilities that occur every few years cause them to eject great clouds of carbon into space, where it immediately turns to soot and blocks the light from the star. Amateur astronomers should keep an eye on R Coronae Borealis itself: it is normally near 6th magnitude but can plummet unpredictably by as much as nine magnitudes and may take months to recover (see Star Chart 5, page 266).

LUMINOUS BLUE VARIABLES AND ETA CARINAE

The most massive stars, of spectral type O, are forever losing huge quantities of gas from their surfaces, but when they leave the main sequence matters become far worse and they become luminous blue variables (LBVs). Violent instabilities set in, resulting in episodes of huge mass loss, probably every few hundred or thousand years, which cause the star to brighten by several magnitudes and for its spectral type to change as its outer envelope of gas lifts off into space: a blue (B-type) supergiant will, for a while, look like an even more super-supergiant of cooler (A) type. The expanding shell will produce striking emission lines in the spectrum, with adjacent

◄ *Mira, a pulsating red giant star, seen at the bottom of these pictures near maximum brightness (left) and fading. Stars like Mira form the largest-known group of variables.*

STARS

THE COSMOS

PROPERTIES OF SOME INTRINSIC VARIABLE STARS

Variable Name	Type of Star	Typical Period	Typical Range
Supernova II	Massive	Once only	12 mag.
Supernova Ia	White dwarf	Once only	20 mag.
Classical nova	White dwarf	1,000 yr(?)	10 mag.
Dwarf nova	White dwarf disk	Weeks	Few mag.
Cepheid	F, G supergiant	1–100 days	Up to 2 mag.
RR Lyrae	A, F giant	0.2–1.0 days	1 mag.
Mira	M giant	100–1,000 days	10 mag.
RV Tauri	F–K supergiant	30–150 days	4 mag.
Delta Scuti	A dwarf	Hours	0.1 mag.
Beta Cephei	B subgiant	Hours	0.1 mag.
T Tauri	Young A–M	Irregular	1 mag.
LBV	O, B supergiant	Decades	Few mag.
R Coronae Borealis	F–K supergiant	Years	10 mag.
Flare stars	K, M dwarfs	Days	1–6 mag.

absorption lines where the shell absorbs light from the star below. Such dual structures are known as **P Cygni lines**, since they were first identified in the spectrum of one of the most famous LBVs, the high-luminosity blue supergiant P Cygni.

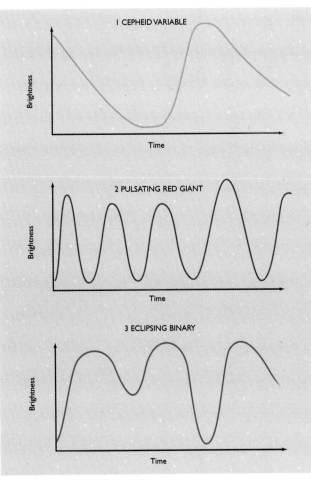

Light Curves of Variable Stars
The rise and fall in brightness of three kinds of variable star. 1. Cepheid variables such as l Carinae go through a regularly repeating cycle as they pulsate in size. 2. Mira, a pulsating red giant, exhibits more erratic variations – here, four cycles are shown. 3. In the case of Beta Lyrae, an eclipsing binary, a close pair of stars regularly orbit each other, causing two separate dips in brightness as first one star and then the other is obscured.

Perhaps the most remarkable LBV of all is to be found in the southern sky: Eta [η] Carinae, which has an estimated mass of around 120 Suns, near the upper limit physically possible for a star. Early in the nineteenth century it began to brighten, attaining a peak of magnitude -1, second only to Sirius, in 1843. It subsequently declined, by seven magnitudes over 14 years, sinking below naked-eye visibility. Since the 1940s it has slowly been on the rise again, reaching 5th magnitude in 1998. Eta Carinae itself is difficult to see, being embedded in thick nebulosity expelled during its outbursts. It is also surrounded by a much larger nebulosity containing a stellar nursery with a cluster of young hot stars (see Star Chart 16, page 288).

CHEMICALLY WEIRD STARS

While most stars have spectra that suggest similar chemical composition to that of the Sun, certain oddities are to be found. Cool **subdwarfs** are deficient in heavy elements compared with ordinary main-sequence stars because they were formed early in the life of the Galaxy, before chemical enrichment by massive stars had really got going. Barium stars are giants that had the misfortune to be too closely partnered to a star that completed its evolution and shot out the products of its nuclear burning all over its companion, especially barium. We have already met the magnetic Ap stars (see Measuring Magnetism in Stars, page 161), but they have non-magnetic relatives, the metallic-line A stars (Am), which are every bit as strange. These show an excess of most heavy elements but are strangely deficient in some, notably calcium and scandium; evidently, some elements are floated up by radiation pressure from within the star while

DISTANCE AND VARIABLES

Variable stars provide important rungs on the ladder of the cosmic distance scale. The first rung is the determination of parallax, followed by the process of main-sequence fitting on the HR diagram. Beyond that, Cepheid variables are extremely valuable because their luminosity is related to their period of oscillation – so, if a star can be identified as one of these, its absolute magnitude can be determined by measuring its period, the brightest ones having the longest periods. In the case of the RR Lyrae stars the matter is even simpler because they all have about the same absolute magnitude, +0.5. If a star can be recognized as an RR Lyrae, its apparent magnitude can be used to calculate the distance. For much greater distances, out to nearby clusters of galaxies, Type Ia supernovae can be used since they all reach about the same absolute magnitude at their peak.

others sink under gravity. Such stars must rotate slowly to avoid stirring up the atmosphere, and most seem to have achieved this through the braking effect of tidal forces supplied by a companion star. Hotter relatives of the Am stars show even stranger chemical anomalies: the mercury–manganese stars exhibit the elements mercury and manganese to excess, often with other rarities such as platinum.

OBSERVING VARIABLE STARS

One area where amateur astronomers can make useful contributions is in the observation of variable stars. At its simplest, this can mean keeping an eye out for exploding stars (novae) with the naked eye or binoculars, or repeatedly photographing selected areas of sky. For those wanting to monitor a known variable, a chart is required showing a selection of nearby stars of constant brightness covering the magnitude range of the star; these can be obtained from an astronomical society, or can be drawn up from an atlas and star catalogue. At intervals the magnitude of the variable is estimated by eye by comparing it with the nearby constant stars. When sufficient estimates have been made a graph called a **light curve**, showing the star's variations over a period of time, can be drawn. Particularly suitable targets for amateurs are the red giant variables such as Mira, which exhibit large magnitude changes and require long-term monitoring.

THE COSMOS

THE STELLAR ZOO – DOUBLE STARS

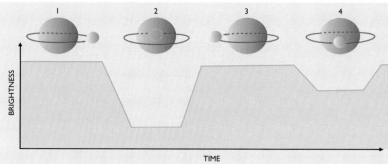

Roughly half of all the stars in the sky are found to be in binary systems, in which the two companions orbit their common centre of mass with periods ranging from hours to many thousands of years. In keeping with Kepler's laws, the wider the separation between the stars, the longer their orbital period. Binary stars may be born together from a rapidly rotating protostar that split, or one may capture the other in a close encounter after birth. Close pairs may affect each other's evolution by transferring mass. Some may eclipse each other, causing variability in brightness.

THE STRUVE DYNASTY

Over the course of 150 years the Struve family has made major contributions to the understanding of binary stars. The first to take an interest was Friedrich Georg Wilhelm Struve (1793–1864) who began to compile a double-star catalogue at the Dorpat (now Tartu) Observatory, Estonia, in 1814. His major catalogue of visual double stars was published in 1837; he had discovered 2,343 of those listed. In 1839 the Pulkova Observatory was established in Russia with the then-largest refractor in the world (15 in, 38 cm); F. G. W. Struve became its first director, although the reins of the double-star program were taken over by his son Otto Wilhelm Struve (1819–1905) who subsequently became director in 1861. Otto added more than 500 doubles to a new catalogue published in 1850. Both of Otto's sons became astronomers: Karl Hermann Struve (1854–1920) and Gustav Wilhelm Ludwig Struve (1858–1920), although they were more interested in motions in the Solar System than double stars. G. W. L. Struve's son, another Otto Struve (1897–1963), continued the family tradition with research into the astrophysics of binary stars. He emigrated to the US in 1921, becoming the director of Yerkes Observatory, Wisconsin, and was closely involved in founding the McDonald Observatory in Texas. He concluded his career as director of the National Radio Astronomy Observatory in West Virginia.

◀ *An Eclipsing Binary*
When the orbital plane is aligned (or nearly) with the observer, the light curve regularly alters according to the relative positions of the two stars. The combined brightness of the two stars gives the peak levels at positions 1 and 3. At position 2, the large, cooler star eclipses the small, hotter one, producing the large dip in the light curve. At position 4, the small star obscures only part of the large star, causing less of a drop in brightness.

VISUAL BINARIES

If the components of a double star are far enough apart as seen from Earth then they may be seen separately, or resolved, through binoculars or a telescope. The larger the telescope's **aperture**, the closer the stars that can be distinguished, although the ease of doing this will also depend on the relative brightness of the pair: a bright star may drown out the light of a feebler companion. Increasingly sophisticated instruments such as **interferometers** are now used by professional astronomers to separate ever-closer binaries.

Some apparent doubles are not connected and are simply viewed along a similar line of sight; these are called visual doubles, but they are less common than true binaries. If a true binary is observed for long enough, signs of orbital motion will be detected, as the stars gradually change their separation and orientation relative to each other; careful measurement of these quantities can be used to determine the masses of the stars (see Orbits and Masses, opposite).

SPECTROSCOPIC BINARIES

Many stars are too close together to be resolved directly through a telescope, but there is another way of detecting their duplicity and that is by **spectroscopy**. If the lines in a star's spectrum are moving cyclically back and forth in wavelength, there is a good chance that this is due to the Doppler shift as the star moves in orbit around its centre of mass with a companion (although it could be a single star that is pulsating in size). If the spectrum of only one component can be seen it is termed a single-line binary, but when the two are of comparable brightness a double-line

▶ *A Spectroscopic Binary*
The sequence of changing spectral lines reveals the presence of a spectroscopic binary. When both stars move across the observer's line of sight, the lines merge (1 and 3). When one star is approaching and the other is receding, the lines become separated (2 and 4).

binary may be observed. By monitoring such a spectroscopic binary throughout its orbital cycle, astronomers can calculate the orbital elements, quantities which describe the characteristics of the orbit: its period, shape, size and orientation in space, and the speed of the stars. This is one of the techniques used to discover many otherwise invisible planetary companions of stars.

ECLIPSING BINARIES

By chance, the orbits of some spectroscopic binaries are orientated side-on to us, so that one star periodically passes in front of the other, causing an eclipse and reducing the total light we receive from the binary for a while. Such a system is termed an eclipsing binary. For example, in the eclipsing binary Algol in Perseus (Star Chart 2, page 260) a bright, hot main-sequence star is paired with a cool giant which is difficult to see against the glare from its companion, except during eclipse when the hot star is obscured. A light curve can be plotted of the brightness variations caused by the eclipses which gives valuable information about the binary, including the shape of the stars, for it turns out that they are not always spherical:

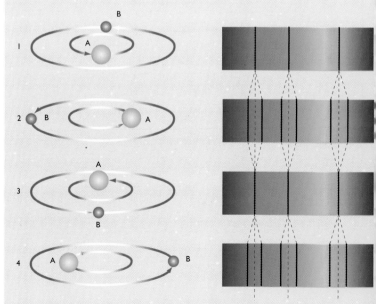

aperture ▶ p. 336   interferometer ▶ p. 340   orbital period ▶ p. 341   spectroscopy ▶ p. 344

THE COSMOS

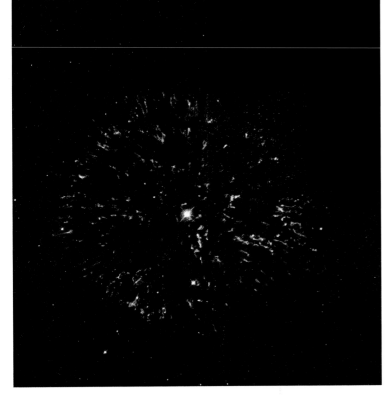

▲ *Ejected gas* surrounding Nova Persei, which erupted in 1901. At maximum the nova was around 200,000 times as luminous as the Sun. This photograph was taken by the Hubble Space Telescope nearly 100 years after the eruption.

when the components are close together, they can deform each other into egg shapes which leads to additional variability as the observer sees the star either side-on or end-on. In the closest pairings the stars seem to be touching: these are termed contact binaries.

ORBITS AND MASSES

The only direct method to 'weigh' a star is through its gravitational interplay with another star and the application of Newton's laws. For a visual binary (i.e. one in which the two stars can be seen separately through a telescope), the calculation requires accurate observations of the relative position of the stars over an entire orbit, or at least the majority of it; allied to this, the star's distance is required to give the orbit a physical size. For spectroscopic binaries, only the eclipsing type are useful, since only in this way can we tell the inclination of the orbit. So far, most spectroscopic binaries discovered have been short-period systems, where the velocity changes are fairly large; modern techniques are now enabling accurate velocities to be measured for those with longer periods, in the 100-day region, and correspondingly smaller velocity changes; this promises reliable masses for many more stars.

CATACLYSMIC BINARIES AND NOVAE

If the components in a binary are sufficiently close they can seriously interfere with each other, sometimes with spectacular results as in the cataclysmic variables. In such systems, a cool dwarf star (spectral type K or M) is partnered by a white dwarf. The two are less than one million km (600,000 miles) apart – only about twice the distance of the Moon from Earth – and orbit with periods less than half a day. Gas is drawn off the red dwarf by the superior gravitational strength of the white dwarf, spiralling on to an accretion disk around the compact star. The point where the infalling stream of gas hits this disk is a bright hot spot. Instabilities in the disk every few days or weeks cause surges in brightness of as much as 250 times (six magnitudes). The result is termed a dwarf nova.

Gas eventually spirals down through the accretion disk on to the surface of the white dwarf. When sufficient gas has built up, a 'hydrogen flash' can occur, a nuclear explosion which blows off a thin outer layer and the star's brightness flares up 25,000 times (11 magnitudes) or so in a nova explosion. The name 'nova', Latin for 'new', was given in ancient times when astronomers saw stars appear when none had been seen before and concluded that they were indeed new arrivals in the firmament. Dozens of novae are thought to occur in our Galaxy each year, although only occasionally does one become bright enough to be visible to the naked eye. Nova eruptions may repeat after hundreds or thousands of years, and several such recurrent novae are known. In extreme cases, where the composition of the white dwarf is just right, the nuclear detonation may be so severe as to trigger a supernova of Type Ia.

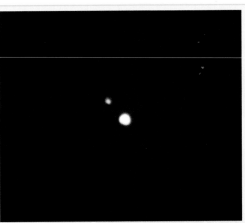

▲ *The yellow and blue* components of the double star Albireo.

SEEING DOUBLES

Double stars provide one of the most fascinating observing targets for amateur astronomers. With even a small telescope it can be instructive to examine a sample of doubles in which the components are of markedly different temperature and thus different colour. For example, the nicely separated stars of Gamma [γ] Andromedae have been described as orange and 'sea green', while those of Beta [β] Cygni (Albireo) are yellow and 'sapphire blue'. A selection of the best double stars is given in the Watching the Sky chapter. Many amateurs pride themselves on just how close a pair they can make out in their telescopes; it requires good optics, steady air and is, ultimately, limited by the size of the main lens or mirror. William Rutter Dawes (1799–1868), an English double-star observer, estimated that two stars can be resolved under good conditions if their separation is greater than $116/D$ arcseconds, where D is the telescope aperture in millimetres. This is now known as Dawes' limit.

▼ *An artist's impression* of the astounding view that might be had from the surface of a hypothetical planet orbiting a double star.

THE COSMOS

Contents of the Cosmos
BLACK HOLES

Of all the objects that make up the Universe probably the most astounding are black holes. Stars reveal themselves by the radiation they generate, but black holes are different: they may be unseen by telescopes, yet are capable of swallowing whole stars. They are the science-fiction writer's dream – to those writers, black holes offer doorways to parallel universes and the possibility of time travel. This section looks at the theory of black holes and then moves on to explore the evidence for their existence.

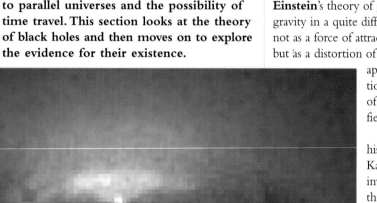

◀ *A disk of hot gas* 3,700 light years in diameter, swirling around a black hole with a mass of 300 million Suns at the centre of the elliptical galaxy NGC 7052. Our Universe is filled with objects which will eventually lose their struggle with gravity and collapse to form a black hole. Black holes have such strong gravitational fields that nothing can escape, not even light.

BLACK HOLES BEFORE EINSTEIN

Using **Newton**'s theory of gravity, the French mathematician Laplace introduced the concept of a black hole around 200 years ago, although he did not use the name or suggest the more bizarre effects later attributed to them. At that time scientists thought light was a stream of particles and Laplace's view was that the pull of gravity would slow down light that was escaping from a star. If the light's speed was less than the escape velocity for the star the light would be unable to escape and would fall back into the star.

THE SCHWARZSCHILD RADIUS

Einstein's theory of general relativity explains gravity in a quite different way from Newton's – not as a force of attraction between two objects but as a distortion of space–time. From this new approach followed two predictions: light follows the curvature of space–time, and gravitational fields affect the rate of clocks.

Soon after Einstein published his theory, the German physicist Karl Schwarzschild (1873–1916) investigated the behaviour of the gravitational field around a single compact object.

According to general relativity, if the body's gravity is sufficiently strong, space–time becomes so highly curved around the body that light is unable to escape from it. Schwarzschild showed that this occurs when the radius of the body is less than a certain critical value, which increases with the body's **mass**. This critical radius is now known as the **Schwarzschild radius**. In Newtonian theory it equates to the distance at which the **escape velocity** from the body becomes equal to the speed of light, equivalent to the Laplacian view that light has been slowed and fallen back to the object.

PIERRE SIMON, MARQUIS DE LAPLACE

One of the greatest of French mathematicians and astronomers, Pierre Simon (1749–1827) was born in Normandy, the son of a farmer. After studying at Caen he went to Paris, rapidly rising to become Professor of Mathematics at the École Militaire, working on the motions of Saturn and Jupiter, on Saturn's rings and Jupiter's moons. In 1785 he became a member of the French Academy of Science and was recognized as France's leading mathematician. Having survived the Revolution Pierre Simon was made Marquis de Laplace by Louis XVIII in 1817. Between 1799 and 1825 Laplace published the five huge volumes of his *Mécanique Céleste* ('Celestial Mechanics'), thought to be the finest work on the subject since Newton's *Principia Mathematica* (1687). It is claimed that after Napoleon had read the work – the Emperor's own interest in mathematics had led an upsurge in French science – he asked Laplace why he had omitted any reference to the Creator. Laplace is said to have replied that there was no need for such a reference, believing that though God had created the Universe, it was now an ordered, self-running system, a belief which heralded the Age of Reason.

In an earlier book, *Système du Monde*, published in 1796, Laplace, perhaps aware of an earlier suggestion by the Englishman John Michell (1724–93) who had made a similar observation in the 1780s, noted that if a star was big enough or heavy enough, the force of gravity acting on light particles would prevent them from escaping. As Laplace put it, 'the attractive force of a heavenly body could be so large that light could not flow from it'. Although his reasoning was wrong – Laplace assumed light consisted of particles which would be influenced by gravity – and he did not use the term, Laplace had defined the black hole.

CLOCKS NEAR A BLACK HOLE

A prediction of the theory of general relativity is that gravity affects time as well as space. If an astronaut carrying a clock approaches a black hole a distant observer would see that clock slowing down, the effect becoming more noticeable as the astronaut approached ever closer to the event horizon. To the observer the astronaut would move more and more slowly and would, in fact, take an infinite time to reach the **event horizon**. From the astronaut's point of view the reverse would be true. He would see the observer's clock running fast and, as he approached the event horizon, he would see the entire future of the Universe flash by. The effect of a gravitational field on the rate of clocks is demonstrated by **spectral lines**, since their precise frequencies can be regarded as accurate clocks. A spectral line emitted from gas near a black hole is seen at a lower **frequency** (or longer **wavelength**) by a distant observer. This is referred to as a gravitational **redshift**.

In another consequence of the same principle, the speed of light as measured by a distant observer is reduced by a gravitational field. This effect can be seen in a **gravitational lens**, where multiple images of a distant **quasar** are produced by light following different paths through the gravitational field of an intervening galaxy. As seen from Earth, brightness variations in the various images of the quasar are delayed by different amounts because of the paths followed by the light. The effect of a gravitational field on the speed of light is at its most extreme at the event horizon of a black hole, where the speed as measured by a distant observer would be reduced to zero. Again, this is an effect which would not be noticed by an astronaut falling freely into a black hole. He would experience no gravitational field and would measure the speed of light as its normal value.

BLACK HOLES

EVENT HORIZON

The sphere around a black hole at the Schwarzschild radius is termed the event horizon. The event horizon is often thought of as the surface of the black hole, but it is not a material surface. Rather, it marks the location where the distortion of space–time by gravity becomes extreme. It is called a horizon because no information from within it can reach the outside world. Imagine, for example, trying to signal to the outside world with light rays (or radio waves, which would behave in the same way). The rays follow the curvature of space–time around the black hole, but within the event horizon the curvature is so great that all light rays return to the black hole. However, there is nothing to stop things, for example an unfortunate astronaut, from falling inside the event horizon – they may not even notice until it is too late and escape is impossible.

▲ *From the accretion disk* around a black hole with the mass of two billion Suns at the core of the elliptical galaxy M87, a bluish jet of atomic particles emerges at nearly the speed of light. This photograph was taken with the Hubble Space Telescope.

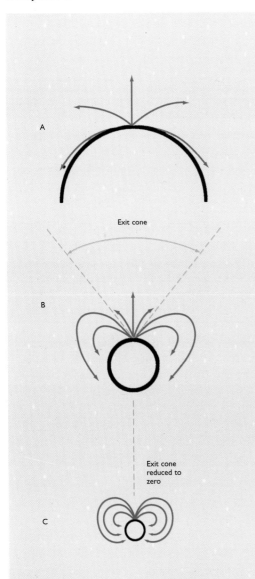

A

Exit cone

B

C

Exit cone
reduced to
zero

◄ *Light Rays Around a Black Hole*
In a weak gravitational field (A), the light rays follow curved paths, but they all escape from the source. In a stronger gravitational field (B), only light rays within a region called the 'exit cone' can escape. When a black hole is at the Schwarzschild radius (C), its gravitational field is so strong that the exit cone has shrunk to zero and light cannot escape.

SINGULARITIES

What happens to matter that falls into a black hole, through the event horizon? In the Schwarzschild model (which assumes the black hole is non-rotating), at the centre of a black hole is a **singularity**, a point of infinite density and zero radius where the laws of physics inevitably break down. Anything falling into such a hole would in effect be crushed out of existence at the singularity. However, in the real world black holes are likely to be rotating rapidly. In the early 1960s the New Zealand mathematician Roy Kerr (b. 1934) pointed out that a rotating black hole would have a ring-like structure – a very thin (almost zero thickness) ring of matter with high, but not infinite, density surrounding empty space. Such structures sound like something out of science fiction, for a ring of sufficiently large diameter could allow a spaceship to pass through, emerging into who knows what.

Rotating black holes (i.e. ring singularities) might have two separate event horizons – and, stranger still, if the black hole is rotating fast enough these might coincide and cancel each other out leaving a so-called 'naked' singularity, not hidden from view. Such naked singularities seem to defy the laws of physics, but some physicists have postulated that they might be produced not only by spinning black holes but may also be left behind when a black hole has completely evaporated by emission of **Hawking radiation**.

GENERATING RADIATION FROM BLACK HOLES

Matter falling towards a black hole can provide the energy for a powerful radiation source, but the gravitational energy released must be converted into heat outside the Schwarzschild radius. If matter fell directly into the hole, all its kinetic energy would be lost within the event horizon. But, like water disappearing down the plughole of a bath, matter approaching a black hole usually has **angular momentum** and so does not fall straight in – instead it creates a disk swirling around the black hole, known as an accretion disk. As in any orbiting system, the matter in the accretion disk moves faster the closer it approaches the event horizon, heating up through friction to temperatures as high as a billion degrees in its fastest-moving inner part. This heat is converted into high-energy **radiation** in the form of **X-rays** and **gamma rays**. As much as 30% of the kinetic energy of the infalling matter can be converted into radiation in this way, about 10 times better than the efficiency of the nuclear processes which power the Sun. In addition, the difference in speed between various parts of the accretion disk generates huge magnetic fields which are thought to power the jets often associated with black holes.

spectral line ▶ p. 343 wavelength ▶ p. 344 X-ray ▶ p. 344 Einstein ▶ p. 88 Newton ▶ p. 42

THE COSMOS

BLACK HOLES

CREATING A BLACK HOLE

The Sun, or any similar star, would become a black hole if it could be compressed to just 6 km (3.7 miles) across, with a corresponding enormous increase in its density. Gravity, if unresisted, would do this in about 15 minutes. So why are stars not collapsing as a result of their own gravitational forces, and if they are not collapsing now, will they do so at some future time? The answer to the first question is the most straightforward – normal stars are supported by thermal gas pressure. The interior of a star is so hot that the rapid movement of the gas atoms is sufficient to resist gravity. But when a star runs out of fuel the thermal gas pressure falls and can no longer counteract gravity. The star collapses, but the end point of the collapse is dependent upon **quantum mechanics**.

STELLAR BLACK HOLES

Most stars avoid collapsing into black holes even at the end of their lives. Stars with the mass of the Sun, or less, end up as **white dwarfs**, where gravity is balanced by the resistance of the tightly packed **electrons**. Heavier stars end in **supernova** explosions in which the star's core is compressed to even greater densities than in a white dwarf. If the core's mass is less than about three solar masses it becomes a **neutron star** in which gravity is balanced by the densely packed neutrons. It is only when the core has a mass greater than three solar masses that gravity wins outright and a black hole is formed.

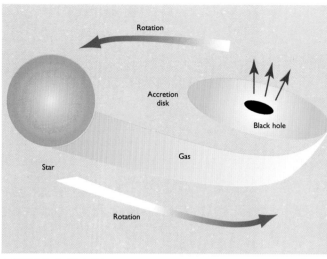

▲ *Tidal Pull of a Black Hole*
Where a black hole and a star are in a close binary system, the tremendous tidal pull of the black hole drags gas from its companion star into the black hole's rotating accretion disk, causing X-ray radiation to be emitted.

Black holes are difficult to detect unless they are in binary systems accompanied by a normal star. In such cases, an accretion disk of hot gas from the normal star accumulates around the black hole, creating a powerful X-ray source (see Generating Radiation from Black Holes, page 173; Mass Transfer in Binary Stars, page 175). However, accretion disks can also form around neutron stars in binaries, so we cannot assume that every X-ray-emitting binary contains a black hole. In favourable circumstances we can measure the mass of the compact object, which provides the most reliable way of distinguishing a black hole from a neutron star. In most known binary X-ray sources the compact object has a mass about that of the Sun, and so is probably a neutron star, but in a few cases the mass is much greater than three solar masses, beyond the theoretical limit for a neutron star. These must, it seems, be black holes.

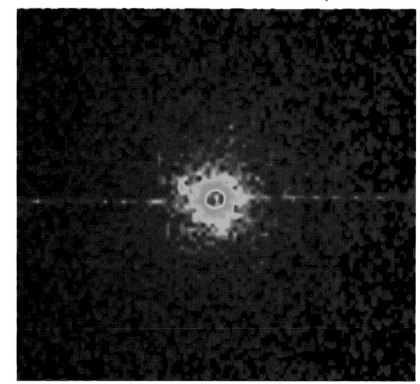

◀ *An X-ray image* of a stellar black hole. Most stars do not collapse to form black holes, as gravity is usually balanced by the degeneracy pressure of electrons or neutrons, causing them to become white dwarfs or neutron stars. Only when gravity wins do the centres of stars cave in to form black holes.

GALACTIC BLACK HOLES

Many galaxies are known to have objects at their centres with masses ranging from a few million solar masses, in the case of our own Galaxy, to several billion solar masses in other galaxies. Yet these enormous masses are crammed into remarkably small volumes, in some cases less than the size of the Solar System. It is widely believed, although there is as yet no definitive proof, that these objects are gargantuan black holes, and they are thought to provide the power sources for **active galactic nuclei**, including quasars. How could such supermassive black holes have come about? One possibility is that they were produced in the early Universe, in which case they could have acted as seeds around which the primordial gas condensed to form galaxies. Another possibility is that the black holes built up inside galaxies as a result of gravitational interactions between stars. Computer studies of the close approach of two stars show that occasionally one star will lose enough energy to allow it to sink towards the centre of the galaxy. An accumulation of such stars at the centre could lead to the formation of a black hole. With the subsequent addition of more stars and gas the black hole would continue to grow.

HAWKING RADIATION

In 1974 **Stephen Hawking** (b. 1942) of Cambridge University, England, showed that a black hole, even without an accretion disk, will emit **thermal radiation**. This represents a loss of energy, and therefore mass, so the black hole will slowly evaporate, eventually disappearing in a burst of radiation. How can this happen, since nothing is supposed to be able to escape from a black hole? The answer lies in the space around the black hole. According to quantum mechanics, space is not an empty vacuum in the conventional sense but is a 'soup' of so-called virtual particles which are continually popping in and out of existence. The virtual particles are created as pairs – one a particle and the other an antiparticle – and if such a pair is created just outside the event horizon of a black hole, one may be captured by the black hole while the other escapes. The separation process extracts energy from the gravitational field and this may be sufficient to convert the virtual particles into real particles. The result is that there is always a cloud of particles and antiparticles around the event

THE COSMOS

▲ *Stephen Hawking, who demonstrated that naked black holes (those without accretion disks) can also emit thermal radiation. The smaller the black hole, the greater the gravitational tidal force it produces. This explains why a smaller black hole is capable of greater radiation than a larger black hole.*

horizon of a black hole; these pairs are continuously annihilating and producing Hawking radiation. A small black hole has stronger tidal forces and therefore produces more intense radiation, so the evaporation process speeds up as the black hole loses mass. A black hole with the mass of Earth has a temperature a few hundredths of a degree above absolute zero, but one with the mass of a small **asteroid** would glow intensely at many billions of degrees.

MASS TRANSFER IN BINARY STARS

In interstellar space there is very little gas to be captured by an isolated black hole. To produce enough radiation to be detectable from Earth a black hole needs to be part of a binary system, the second body being a star which supplies the material to form an accretion disk. In binary systems where one of the stars is a compact object – a white dwarf, a neutron star or a black hole – the gravitational field of the compact object distorts the companion star. If the stars are close enough the distortion not only pulls the companion star into a pear-shape, but can disrupt it altogether. The dis-

ruption mechanism is identical to that by which the Moon raises tides on Earth – gravity is stronger on the near side of Earth and weaker on the far side. Similar tidal forces distort any star near a black hole and, if the star is close enough, may cause it to disrupt, providing gas for an accretion disk. The tidal distortion of the star becomes greater as the star nears the black hole's event horizon but, paradoxically, the tidal forces are greatest around less massive black holes. The difference in gravitational force across the star depends on the ratio of the distances of its near and far sides from the black hole. For a given star, the bigger the black hole the smaller the ratio and the smaller the tidal forces.

PRIMORDIAL BLACK HOLES

In the exceptionally hot and dense conditions of the early Universe, immediately after the Big Bang, it is possible that black holes were produced with an almost unlimited range of masses, from that of an atom to that of a cluster of galaxies. The very smallest ones, if they ever existed, would already have vanished due to evaporation by Hawking radiation, while those with an original mass of around a billion tonnes, similar to that of a small asteroid, would be in

their final stages of evaporation (and hence at their most luminous) about now. A black hole of this mass would have a diameter about the same as that of an atomic nucleus (for comparison, a black hole with the mass of Earth would have a diameter of a few centimetres – just over an inch). Such mini black holes would appear to be so hot that the Hawking radiation from them would be in the X-ray region of the spectrum. But, so far, X-ray telescopes have failed to find evidence for black holes of this type.

▼ *Cygnus X-1, a strong X-ray source in the constellation Cygnus, was the first object recognized to contain a black hole. The black hole has an estimated mass at least eight times that of the Sun and orbits a visible blue supergiant star (arrowed) 8,000 light years away.*

CYGNUS X-1

One of the first cosmic sources of X-rays was discovered in the 1960s in the constellation Cygnus. Observations of the spectrum of the X-rays from Cyg X-1 showed that they were probably produced from gas at an extremely high temperature, while later observations revealed variations in brightness on timescales down to a few thousandths of a second. Such rapid fluc-tuations imply a very compact source, and the very high temperature implies the existence of an accretion disk. Cyg X-1 was therefore a good candidate for a binary system consisting of a 'normal' star paired with a neutron star or black hole. Although the compact source was invisible at optical wavelengths, astronomers were able to identify the companion star: it was a hot, blue **supergiant** of spectral type O, of 9th magnitude as seen from Earth. The orbital period of the binary, revealed by the **Doppler shift** in the spectrum of the visible star, is 5.6 days. Analysis of the system shows that the compact object has a mass of more than eight times that of the Sun, too great to be a neutron star and hence a strong candidate for a black hole.

Sh2-101

Cygnus X-1

Contents of the Cosmos
OUR GALAXY
GALAXY STRUCTURE

An arc of faint light spans the sky from horizon to horizon on a clear, moonless night. This band is the Milky Way, composed of many billions of distant stars arranged in a flattened disk. The richest part of the Milky Way is the great cloud of stars towards Sagittarius, where lies the hub of this huge, slowly spinning disk. Enveloping the disk in a faint halo is an entourage of old stars, some of them grouped into globular clusters, as well as a still more massive halo composed of invisible dark matter. Together, these constituents comprise the Galaxy – the name is written with a capital G to distinguish it from other galaxies.

JAN HENDRIK OORT

Jan Oort (1900–92), a Dutch astronomer, made many significant contributions to Galactic astronomy. His early work included one of the first attempts to understand the mechanism of spiral structure and the earliest determination of the density of matter near the Sun. Oort also explained that stars which are observed to be moving at high speeds relative to the Sun are actually members of the galactic halo and so do not share the Sun's rotation around the Galactic centre. Oort's greatest claim to fame is that he was the first to realize the crucial importance of radio astronomy for studying the structure of our Galaxy. He wondered whether there is a spectral line at radio frequencies that we can detect. The answer turned out to be yes: his student Hendrik van de Hulst (1918–2000) calculated that it is emitted by atomic hydrogen at a wavelength of 21 cm (8 in). This is valuable because atomic hydrogen is common in the disks of galaxies and because absorption is negligible at radio wavelengths. In the 1950s Oort and collaborators used 21-cm (8-in) hydrogen observations to make the first map of the spiral arms of our Galaxy. Nowadays, such observations are routinely used to map the distribution of gas in distant galaxies and to measure their rotation. Oort is also famous for his proposal in 1950 that there is a cloud of comets around our Solar System, now termed the **Oort Cloud**.

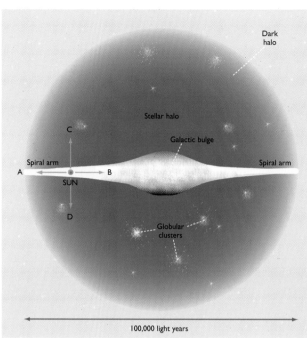

◀ The Milky Way: Side View
Seen from the edge, our Galaxy has a central bulge with spiral arms stretching out to form a flattened disk. From our viewpoint near the Sun, the Milky Way looks like a band of starlight because when looking in the plane of the disk (A or B) we see many more stars than when looking at right-angles (C or D).

Labels on diagram: Dark halo; Stellar halo; Galactic bulge; Spiral arm; Spiral arm; C; A; B; SUN; D; Globular clusters; 100,000 light years

COMPOSITION

Our Galaxy is built on a massive scale. Its disk contains about 100 billion stars fashioned into a spiral shape like a Catherine wheel, extending out to about 75,000 **light years** from the Galactic centre. Along with the stars in the disk are atoms and **ions** of gas (mostly hydrogen, the simplest and most abundant substance), curtains of obscuring dust and giant gas clouds containing complex molecules. Star formation continues in these gas clouds even today.

The **Solar System** lies roughly 25,000 light years from the centre in one of the Galaxy's spiral arms. The Sun follows a roughly circular orbit around the centre, once every 220 million years, at a speed of about 220 km/s (140 mi/s). Our spiral arm is known as the local, or Orion, arm – the arm closer to the centre is called the Sagittarius arm, while the one farther out is the Perseus arm.

The Galaxy's overall **mass** is dominated by its surrounding halo of **dark matter** which probably extends out to about 750,000 light years. The mass of this halo is 10 times greater than the mass in visible stars yet its composition remains a mystery. Possible components of the dark-matter halo include massive **black holes**, dim stars, or elementary particles that are relics of the Big Bang.

THE GALACTIC BULGE

The dead centre of the Galaxy is marked by an unusual radio source called Sagittarius A★, whose diameter is about the same as that of Earth's orbit around the Sun. The motions of stars very close to the centre allow us to estimate the mass of Sagittarius A★, and the answer is astoundingly large: a few million times that of the Sun. Such a large mass squeezed into such a small volume means that Sagittarius A★ is probably the site of a supermassive black hole, such as seem to exist in the **nuclei** of other galaxies. Enveloping the Galactic centre is a bulge of some 10 billion stars with a profile like a flattened ball (a shape known as a spheroid). Dust obscures the Galactic bulge from view at optical **wavelengths**, except in a few places, most notably the area termed Baade's Window in Sagittarius. Baade's window apart, the bulge is best studied at infrared and radio wavelengths, which can penetrate the obscuring dust. One significant discovery in recent years has been that the Galactic bulge is actually a rotating bar, seen nearly end-on from Earth.

▼ Star fields of the Milky Way, *overlain by dark clouds of dust that lie in the spiral arms of the Galaxy. The star fields become denser and brighter towards the centre of our Galaxy in Sagittarius, at the bottom.*

OUR GALAXY

THE COSMOS

OUR CONCEPTION OF OUR GALAXY

In 1906 Jacobus Kapteyn (1851–1922), a Dutch astronomer, set up one of the first international scientific collaborations, whose purpose was to map out the structure of the Galaxy. Astronomers counted the numbers of stars of different brightness in some 200 selected areas of sky and measured their motions. After analyzing the mountains of resulting data, Kapteyn concluded that the Sun lay close to the centre of the Galaxy – an awkward result, as there is no reason why the Sun should occupy such a privileged position. A better idea of the scale of the Galaxy can be obtained by looking away from the disk, at the system of globular clusters which surrounds us. This was done by Harlow Shapley (1885–1972), a US astronomer, who demonstrated around 1918 that the Galaxy was far larger than Kapteyn had estimated and that the Sun lay well away from the centre. Robert Trumpler (1886–1956), another American, showed where Kapteyn had gone wrong. Kapteyn's results were badly affected by interstellar dust, which dims the light from distant stars and gave the erroneous impression that their numbers fell off equally in all directions around us.

THE GALACTIC DISK

Outside the central bulge, most of the stars and gas in the Galaxy lie in a flat disk about 2,000 light years thick, concentrated into a hotch-potch of arms coiled around the hub. The spiral pattern is traced out most vividly by young, hot stars, **open clusters** and glowing **nebulae** like the one in Orion. The disk also contains abundant atoms of hydrogen in its so-called neutral state, i.e. not ionized. Hydrogen atoms have the useful property for astronomers of giving out radio waves at a wavelength of 21 cm (8 in), which can be easily detected by radio telescopes. Radio observations of hydrogen have enabled astronomers to trace the spiral arms of the Galaxy and to measure the Galaxy's speed of rotation at various points between the Sun and the centre. Atoms of hydrogen

in the Galactic disk move on circular orbits with a speed of about 220 km/s (140 mi/s) irrespective of their radial distance. Analyses of the velocities of the satellite galaxies, including the **Large Magellanic Cloud**, suggest that the rotation speed remains almost constant out to distances of 150,000 light years from the centre at least. This is surprising: if most of the mass in the Galaxy were confined to the visible stars, then the rotation speed would be expected to fall off with increasing distance in accordance with **Kepler's laws**. It is this behaviour of the Galaxy's rotation that provides evidence for the dark-matter halo.

THE STELLAR HALO

Enshrouding the disk of the Galaxy is the stellar halo, much less massive than the flat disk and different in composition from the dark-matter halo. Visually, the most obvious features of the stellar halo are the 150 or so **globular clusters**, but there are also many individual stars. Stars in the halo have a low content of heavy elements and are therefore thought to be among the oldest objects in the Galaxy, formed before the Galaxy's interstellar hydrogen became enriched with the products of **supernova** eruptions. Whereas the stars in the disk are moving on roughly circular orbits around the Galactic centre, those in the halo (including globular clusters) have highly elliptical paths that plunge through the disk at all angles.

THE DARK HALO

The most massive – and most mysterious – component of the Galaxy is the dark halo. What can provide so much mass yet remain invisible? One possibility is hitherto unidentified, weakly interacting particles. Experiments searching for these elusive particles are being carried out, although so far

▲ *Star-forming clouds* of hydrogen gas in the spiral arms of our Galaxy show up as pink patches against the starry background of the Milky Way. The most prominent clouds are the Lagoon Nebula (M8), below centre, and the Omega Nebula (M17) and the Eagle Nebula (M16), above centre.

THE BAR IN THE MILKY WAY

We live at an awkward vantage point, rather far out in the disk of our Galaxy. The Galactic bulge is hard to study because it is almost completely obscured by dust at optical wavelengths. Consequently, only recently has it come to be realized that there is a bar at the centre of our Galaxy. Bars are rapidly rotating, elongated structures composed of stars and gas which occur in nearly half of all

spiral galaxies. The Galactic bulge is really a bar, viewed almost end-on down its long axis so it looks almost spherical in projection. Confirming evidence for a bar comes from small but un-mistakable asymmetries in the distribution of stars around the Galactic centre, as well as distortions in the motions of gas detected at 21-cm (8-in) wavelengths. The size of the bar is uncertain but it probably extends out to about 10,000 light years from the centre.

the results are not clear-cut. Another possibility is that the dark halo is composed of dim stars, in particular **white dwarfs**, whose presence can be detected from the small-scale **gravitational lensing** effect they have on the light from background stars. In the 1990s, such so-called microlensing events were detected, suggesting that perhaps as much as a quarter of the dark halo of our Galaxy is accounted for by faint, low-mass stars. Black holes also remain a possible component of the dark halo.

supernova ▶ p. 344 wavelength ▶ p. 344 white dwarf ▶ p. 344

OUR GALAXY

Contents of the Cosmos
GALAXY CONTENTS

Galaxies contain two distinct populations of stars. Those in the spiral arms of our Galaxy are classified as Population I, the Sun being a typical example. Population I stars exhibit a variety of ages, and include the very youngest ones, but a common factor is that they are all comparatively rich in elements heavier than hydrogen and helium. Those stars belonging to the Galactic halo and central bulge are termed Population II – the constituents of globular clusters are typical examples. Population II stars are very old, often nearly as old as the Universe itself, and contain only a very small proportion of heavy elements – about 1% the amount present in the Sun, consistent with them having been formed within the first billion years or so of the Galaxy's history.

THE SOLAR NEIGHBOURHOOD
Stars in the Sun's neighbourhood are overwhelmingly members of Population I. The inventory within 30 light years of the Sun is believed to be reasonably complete, although this is a very small volume of the whole Galaxy. Within this distance there are roughly 250 **main-sequence** stars and six white dwarfs

▼ *The Coma star cluster, an open cluster in Coma Berenices. Clusters such as this consist of relatively young stars, termed Population I, and lie in the spiral arms of galaxies, whereas older stars (Population II) are located in the Galactic halo and central bulge.*

(the closest stars of all are listed in the Reference section). Moving out from the solar neighbourhood our knowledge becomes progressively less complete. Even within 150 light years of the Sun, perhaps as many as half the stars remain undetected.

ASSOCIATIONS AND OPEN CLUSTERS
Many of the brightest stars in our night sky lie in a broad band inclined at about 16° to the Galactic plane, called Gould's Belt after the US astronomer who studied it in the late nineteenth century. Young stars in the Galactic disk often belong to associations or open clusters, having formed together from a giant molecular cloud (see The Birth of Stars, p. 158–59). Gould's Belt seems to be a conglomeration of several associations, including ones in Orion, Scorpius and Centaurus – signs of a recent burst of star formation in our region of the Galaxy. Open clusters are smaller and denser than associations, containing up to a few thousand stars in a volume no more than 50 light years across. They are among the most conspicuous tracers of spiral arms, in our Galaxy and others. Prominent nearby examples are the Hyades and Pleiades. Young open clusters are often still embedded in remains of the gaseous nebulae from which they formed.

GLOBULAR CLUSTERS
Globular clusters are the largest individual components of our Galaxy – huge spheres 100 light years or so across typically containing

▲ *The Pleiades, a nearby young open cluster easily visible to the naked eye in the constellation Taurus. Its brightest members are blue giants. Clusters of hot, young stars like this mark out the spiral arms of galaxies. Note the diffuse nebulosity which reflects light from the stars.*

hundreds of thousands of old stars (but little or no gas or dust). There are about 150 globular clusters arranged in a roughly spherical halo around our Galaxy, the most distant of them over 300,000 light years from the Galactic centre. Globular cluster ages are estimated by calibrating the turn-off point of their stars from the main sequence on the **Hertzsprung-Russell diagram**, from which we deduce that their ages are comparable with the age of the Universe itself – 10 billion years or more. Globular clusters swarm around other galaxies, too, particularly giant **ellipticals**.

⌛ **STELLAR POPULATIONS AND THE AGE OF THE UNIVERSE**

Walter Baade (1893–1960), a German-born US astronomer, is responsible for the idea of stellar populations, which led to the solution of a major riddle regarding the age of the Universe. Before the 1950s, geologists argued (correctly) that Earth was about 4.5 billion years old, whereas astronomers argued (incorrectly) that the age of the Universe was about 2 billion years. The astronomers' argument was based on the distances to other galaxies, from which the **Hubble constant** and age of the Universe could be calculated. Baade resolved the conflict in ages by recognizing that one class of pulsating variable stars on which the distance scale was based, the **Cepheids**, are young stars belonging to Population I, while another class, the **RR Lyraes**, are much older and belong to Population II. As a result of their differences in composition, each of the two populations of variables has a distinct relationship between their periodicity of pulsation and luminosity – previously, it had been assumed that there was a universal period-luminosity relation which led to the incorrect distances. Baade's revision of the astronomical distance scale doubled the known size (and hence age) of the Universe, and subsequent observations have increased these figures still further.

OUR GALAXY

Cepheid variable ▶ p. 337 cosmic rays ▶ p. 337 elliptical galaxy ▶ p. 338 Hertzsprung-Russell diagram ▶ p. 339 Hubble constant ▶ p. 340 main sequence ▶ p. 341 parallax ▶ p. 342 RR Lyrae variables ▶ p. 343

THE COSMOS

HIPPARCOS, FAME AND GAIA

Hipparcos, a European Space Agency (ESA) satellite, measured accurate positions, **parallaxes** and proper motions for over 100,000 stars in our Galaxy between 1989 and 1993. It revolutionized our knowledge of the Sun's neighbourhood by providing – for the first time – large samples of stars with accurate distances and velocities, building up a three-dimensional picture of the Galaxy around us. Following on from the pioneering success of Hipparcos, NASA will launch FAME (Full-sky Astrometric Mapping Explorer) in 2004 to extend the survey to 40 million stars. Beyond that, ESA is planning the GAIA (Global Astrometric Interferometer for Astrophysics) mission for launch by 2012. This will provide accurate distances and radial velocities on every object in our Galaxy brighter than 20th magnitude – roughly a billion stars – revealing the three-dimensional structure of our Galaxy all the way to its central bulge.

▲ *The Hipparcos Satellite was launched by the European Space Agency (ESA) to measure the positions, parallax and movement of more than 100,000 stars in our Galaxy.*

THE INTERSTELLAR MEDIUM

The space between the stars is not as empty as it may first appear – it contains gas, dust, magnetic fields and **cosmic rays**. Collectively, this is termed the interstellar medium, or ISM. The composition of the gas in the ISM ranges from simple hydrogen atoms to complex carbon-containing molecules; it also has a wide range of temperature and density. Some of it is very cold and relatively dense, with a temperature of about 20 K (-425°F) and a concentration of about a thousand molecules per cubic centimetre, as exemplified by the giant molecular clouds from which stars form. Less cold and dense are clouds of neutral (i.e. non-ionized) hydrogen,

termed H I regions, which have a temperature of 100 K (-280°F) and a concentration of about 20 atoms per cubic centimetre – these clouds give out the 21-cm radiation that is detected by radio astronomers. Most impressive visually are the clouds of hot, ionized gas known as H II regions, at a temperature of a few thousand degrees K (or F), familiar to observers as the bright nebulae surrounding young stars, such as in Orion. Finally, there is a very hot (million-degree) gas of low density, heated by the blast waves from supernovae, which fills two-thirds of the volume of interstellar space.

Dust always occurs with gas in the ISM. Interstellar dust is akin to soot and very fine sand, consisting of tiny grains of carbon, silicon and even small ice crystals. These dust grains absorb and scatter starlight, not merely dimming the light from stars but also reddening it. When dense, dusty clouds are illuminated by nearby stars, light scattered by the dust can be directly observed as a bluish reflection nebula, such as the one surrounding the Pleiades.

THE SAGITTARIUS DWARF AND THE MAGELLANIC CLOUDS

Surprisingly, the nearest galaxy to ours was not discovered until 1995 because it is obscured from view almost directly behind the Galactic centre. Called the Sagittarius Dwarf, it lies about 80,000 light years from the Sun and is presently being broken up by the gravitational forces of the Milky Way. Next closest are the Large and Small Magellanic Clouds, about 170,000 and 200,000 light years from the Sun respectively, both visible to the naked eye in the southern sky (see Star Chart 20, page 296). The Large Magellanic Cloud has about a hundredth the mass of our Galaxy and is about 64,000 light years wide, while the Small

▲ *The Large Magellanic Cloud (LMC), one of the Milky Way's closest neighbours. Both the Large and the Small Magellanic Clouds are visible to the naked eye from the southern hemisphere. Astronomers have predicted that both clouds will spiral to destruction towards the Galactic centre in another 10 billion years.*

Magellanic Cloud has about a tenth the mass of the Large Cloud. These two irregular galaxies are enveloped in a common cloud of hydrogen gas, invisible to the eye but traceable by radio astronomers, from which emerges a swathe of gas called the Magellanic Stream that extends over more than 100° of the southern sky. The gas of the Magellanic Stream was probably torn from the Clouds by our Galaxy's gravitational forces during an encounter – close approaches between the Magellanic Clouds and the Galaxy are thought to have occurred about 300 million years and five billion years ago. In fact, the Small Magellanic Cloud itself was probably broken off the Large Cloud during one of those approaches. Both Magellanic Clouds are expected to suffer the same fate as the Sagittarius Dwarf, spiralling in to the Galactic centre in another 10 billion years or so. This story of the accretion of small galaxies has probably happened many times in the history of our Galaxy, as it has in other galaxies.

▼ *The Local Group*
The largest members of the Local Group are two spiral galaxies of similar size and mass: the Milky Way Galaxy and the Andromeda Galaxy (M31). Both of these galaxies have clusters of dwarf galaxies associated with them.

THE LOCAL GROUP

The Local Group is the name given to the cluster of about three dozen galaxies, all bound together by gravity, to which our home Galaxy belongs. All galaxies within about three million light years of our own are generally considered members. Two spiral galaxies dominate the Local Group: one is our own, the other is the Andromeda Galaxy (M31). The Andromeda Galaxy is similar in size and mass to our Galaxy, but its central bulge and disk are larger and it has less gas. The most common galaxies in the Local

Group are dwarfs, which cluster around the two big spirals – our Galaxy has an entourage of 12 dwarfs, ranging from the nearby Sagittarius dwarf to the distant Leo I and Leo II. The Andromeda Galaxy has a similar number of dwarf companions, although here the sample is probably still incomplete. For a list of known galaxies in the Local Group see the Reference section.

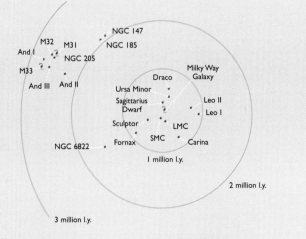

THE COSMOS

Contents of the Cosmos
GALAXIES
GALAXY TYPES

Dotted like islands throughout space, galaxies are collections of many billions of stars bound together by gravity, usually with a good deal of gas, dust and dark matter as well. Although enormous and highly luminous, galaxies are so far off that almost all of them are invisible to the naked eye, and even the brightest appear as little more than faint smudges through a small telescope. Initial studies were restricted to a taxonomic classification according to their appearance, but over the past half-century we have come to understand a great deal about the content, origin and development of these impressive structures.

SPIRAL GALAXIES

Most visually striking of all galaxies are the spirals. They can take a wide variety of forms, ranging from impressive swirling 'grand design' systems in which two spiral arms unwind from their central hubs, such as M51, M81 and M101, to so-called 'flocculent' (or 'fleecy') systems that contain many short, disjointed segments of arms, an example being M94. Spirals have **masses** ranging from about a billion to a million million solar masses and diameters from 10,000 to over 200,000 **light years**.

We can examine spirals from all angles since some are presented face-on to us while others are sideways-on. Seen face-on, spirals display a whirlpool-like appearance that immediately conveys the impression of orderly circulation, like a cup of coffee that has been stirred. This impression of rotation is confirmed by measurements of the **Doppler shifts** in the light from the stars and radio emissions from the **interstellar gas**. Rotation regiments the stars into a fairly flat plane, in the same way that the orbits of the planets in our Solar

▼ *The Whirlpool galaxy, M51, was the first galaxy in which spiral structure was detected. Modern studies have shown that spiral arms are due to density waves which cause stars to bunch up and also compress the galaxy's gas to make new stars.*

▲ *The barred spiral galaxy NCG 1365. The defining feature of the barred spiral galaxy is the long structure – the bar – across the centre of the galaxy. In these cases, the spiral arms of the galaxy usually protrude from the ends of the bar, rather than from the central bulge.*

System are confined to the plane of the **ecliptic**. Indeed, the side-on view of spirals clearly illustrates that their stars are arranged in a disk that is very thin by comparison with the galaxy's diameter, as in the case of our own Galaxy.

In addition to the eye-catching arms, most spirals possess a central bulge, shaped like a flattened sphere, in which the stars follow more random orbits about the galaxy's **nucleus**. The stars of a spiral galaxy's bulge are mostly old – Population II stars as in our own Galaxy – while the encircling disk contains stars with a mixture of ages, including highly luminous young **supergiants**, star clusters and glowing clouds of star-forming gas, all highlighting the structure of the arms. A spiral galaxy is categorized according to how tightly its arms are wound and the size of its central bulge (see Galaxy Classification, opposite).

BARRED SPIRAL GALAXIES

Almost half of all spirals have a cigar-shaped structure across their centres known as a bar, consisting of stars, gas and dust, like a straight inner extension of the spiral arms. Bars can be up to five times as long as they are broad; sometimes, several bar-like structures of different sizes and with different orientations are found in a single galaxy. The bar can act like a giant mixer within the host galaxy, stirring together stars with different properties from different radii. Dust is often concentrated in tight lanes along the leading edges of the bar, while regions of star formation and gas clouds are often found near the ends of bars. Spiral arms originate from the ends of bars, not from the central bulge, and rings sometimes surround bars – a ring can be thought of as a tightly wrapped spiral arm. Our own Galaxy has a bar. The overall dimensions and masses of barred spirals are the same as those of ordinary spirals.

THE GREAT DEBATE

Well into the twentieth century, astronomers were deeply divided over the nature of galaxies, or 'spiral nebulae' as they were then known. Some thought that these were separate 'island universes' comparable to our Milky Way, while others argued that they were smaller, nearby structures, perhaps other solar systems in the making. In an attempt to resolve the issue, a public debate was staged in Washington, DC in 1920 between two proponents of the opposing theories: Harlow Shapley and Heber D. Curtis (1872–1942). This showdown is now known as the Great Debate.

Shapley had recently established the scale of the Milky Way from a study of the distribution of globular clusters, and found it inconceivable that anything could lie yet farther away. He therefore marshalled a series of arguments to show that the spiral **nebulae** had to be nearby. For example, the spiral nebulae seem to avoid a region on either side of the plane of the Milky Way, known as the zone of avoidance, suggesting that they are somehow influenced by the Galaxy. Similarly, the fact that almost all the spiral nebulae are receding, as measured by their Doppler shifts, also implied a direct influence by the Milky Way.

Curtis, an acknowledged expert on spiral nebulae, countered by pointing out that many edge-on nebulae display a dark band running across their centres, which he explained as arising from a belt of obscuring matter around each nebula – dust, as we now know it to be. If the Milky Way were similar to these systems, he reasoned, then it too should contain such an obscuring band which would hide any spiral nebulae behind it. But he had to admit that he had no explanation as to why the spiral nebulae are moving away from us.

At the time, the Great Debate was inconclusive. But the issue was resolved in the following decade, when **Edwin Hubble** (1889–1953) demonstrated that spiral nebulae do indeed lie far outside our Milky Way and that their recession can be explained by the overall expansion of the Universe.

ELLIPTICAL GALAXIES

Less flamboyant than their spiral cousins are the elliptical galaxies. These are almost featureless conglomerations of old stars arranged into spherical or elliptical shapes, bright at their centres and fading to obscurity in their outskirts. The only obvious distinguishing feature of an elliptical galaxy is its overall shape: some appear almost completely round, while others are squashed and elongated. Care must be taken in interpreting these shapes, as they reflect the angle of view as well as the intrinsic morphology – for example, a galaxy shaped like a rugby ball (or an American football) would appear round if viewed end-on, but flattened if viewed from the side. For a long time it was assumed that the flattening of an elliptical, as in a spiral galaxy,

▶ *Elliptical galaxies have no spiral arms, nor any gas to make new stars. Their outlines range from near-perfect spheres to more flattened or elongated shapes. Elliptical galaxies consist mainly of old stars.*

was due to its rotation. However, observations have shown that elliptical galaxies rotate little, if at all, and their flattening must be attributed simply to the fact that their stars move faster, and hence travel farther, in one direction than the other.

Elliptical galaxies display a wider range of **luminosities** than spirals: at the lower end are dwarfs containing a few million stars, like oversized **globular clusters**, while the largest are supergiants (known as cD galaxies) that can be 20 times as luminous as our own Galaxy. The stars in ellipticals are mostly old and there are usually no clouds of cold gas from which new stars can form.

GALAXY CLASSIFICATION

One of the first systematic attempts to classify galaxies according to their appearance was made by Hubble in 1926 with the introduction of his tuning fork diagram. This scheme is still widely used. In **Hubble's classification**, elliptical galaxies are designated by an E followed by a number that indicates how flat the galaxy appears, from E0 (perfectly round) to E7 (the most elongated). Spiral galaxies are designated S and barred spirals SB. The letters a, b or c impart further information about the appearance of the spiral, progressing from galaxies with large central bulges and tightly wound spiral arms to those with small bulges and well-defined, open spiral structure. Lenticular galaxies are termed S0, the '0' indicating the absence of spiral structure. Finally, Hubble introduced the abbreviation Irr for the irregular galaxies that do not fit within this scheme. Our own Milky Way Galaxy lies somewhere between b and c in the spiral sequence; since it is almost certainly barred, its classification would be SBbc.

Other, more sophisticated, schemes have since been introduced. For example, unbarred and barred galaxies are sometimes termed SA and SB, with those containing only weak central bars designated SAB. A spiral galaxy may have 'r' added if it is seen to be surrounded by a faint ring of light or 'pec' if it is in some way peculiar in appearance. However, the simplicity of Hubble's original scheme means that it is still widely used to give a shorthand description of a galaxy's appearance.

SPIRAL DENSITY WAVES

The origin of spiral structure in a galaxy is not simple. The spiral arms cannot just be persistent bright features. Like athletes forced to stay in lanes on a running track, stars orbiting closer to the centre will complete circuits around the galaxy more quickly than those on the outskirts. Thus any features like spiral arms would be rapidly eradicated as stars near the centre lap those further out. The true nature of spiral structure was uncovered in the 1960s, when it was realized that spiral structure could be explained by what is termed a **density wave**. Just as a traffic jam on a motorway can persist long after the accident that caused it has been cleared, so an enhancement of stars in a spiral arm can be a long-lived feature in a spiral galaxy. Like the cars in a traffic jam, stars are slowed and close up as they pass through the arms, creating the enhanced density; interstellar gas is compressed by the density wave, too, leading to the formation of bright, young stars that light up the spiral arms, making them visually prominent. Although this theory elegantly explains spiral structure in galaxies, how a density wave is created in the first place and how long it persists remain unsolved issues.

THE COSMOS

Hubble ▶ p. 122

▲ *Lenticular galaxies such as NGC 3115 exhibit characteristics midway between those of spirals and ellipticals. Their stars are arranged in disks with central bulges as in spirals, but they have no spiral arms and their smooth appearance is reminiscent of the configuration of elliptical galaxies.*

LENTICULAR GALAXIES

Lenticular galaxies appear to form a transition between spiral galaxies and ellipticals; they are so named because they are lens-shaped. Like spirals, these galaxies have disks and (often very prominent) central bulges; they can also contain central bar features. However, they lack any spiral structure and so have the smooth, featureless appearance of ellipticals. Indeed, if a lenticular galaxy happened to be oriented face-on to us it would be difficult to distinguish from a round elliptical – a number of lenticulars have probably been misclassified as ellipticals. Lenticular galaxies contain mostly old stars and little gas or dust, and are more common in rich clusters of galaxies. Whether lenticulars are former spirals that have somehow lost their gas or whether they formed as they are today is uncertain.

DWARF SPHEROIDALS

Dwarf spheroidals are the most common galaxies in the Universe – half the members of the **Local Group** belong to this type – but, being small and faint, they are not easy to see. A typical dwarf spheroidal has a diameter of a few thousand light years and a mass of few million Suns. The first examples were not discovered until after Hubble had drawn up his classification scheme, so they do not fit on to the tuning fork diagram. They are similar in shape to ellipticals but are much less centrally condensed – in fact, they appear as little more than a loose scattering of stars. Dwarf spheroidals contain little or no gas or dust so there is no present-day star formation.

STARBURST GALAXIES

A small proportion of galaxies are undergoing an intense surge of star formation. In many cases, the activity in these so-called starburst galaxies seems to result from a collision or near-miss with another galaxy, which has compressed the gas clouds to form new stars. Starburst galaxies are particularly prominent in the **infrared**; this strong infrared emission comes from dust within them that has been heated by the young, luminous stars. An example of a starburst galaxy is M82 in Ursa Major, which appears to have been disturbed by an encounter with the nearby spiral M81 and other members of the small cluster to which they both belong.

IRREGULAR GALAXIES

Although most galaxies fit into the above categories, a stubborn minority do not. It is therefore necessary to add the catch-all category of irregular galaxies. These lack the symmetry of spirals and ellipticals, giving them a ragged appearance. Their lack of aesthetic beauty, combined with the fact that they tend to be smaller and fainter than other galaxy types, means that irregulars do not feature prominently in most popular books on astronomy. However, they are very common, accounting for more than a third of all galaxies. Two of our closest galactic neighbours, the **Magellanic Clouds**, are both irregulars. The Large Magellanic Cloud has a noticeable bar-shape and so is classified as a barred irregular.

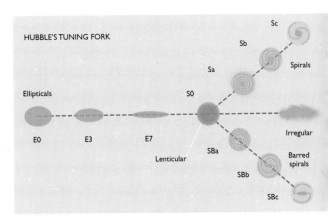

THE TUNING FORK DIAGRAM

Edwin Hubble set out the various classes of galaxy in a diagram resembling the shape of a musician's tuning fork. Hubble's arrangement started with elliptical galaxies, running from the roundest to the flattest, followed by the lenticular systems (S0). The diagram then divided into two branches, one for ordinary spirals (S) and one for barred spirals (SB). Along each branch the galaxies were arranged in increasing order of complexity, starting with those that contained the weakest spiral structure and working towards the complex 'grand design' galaxies with their open two-armed structure. The size of the central bulge of stars also decreases along this sequence of spiral galaxies; since a bulge looks much like an elliptical galaxy, this arrangement places the ellipticals very naturally next to the most bulge-dominated spirals. Irregular galaxies are added at the ends of the spiral galaxy sequences.

▼ *The starburst galaxy M82. Starburst galaxies are undergoing unusually high levels of star formation. In the case of M82, this is due to a near-collision with its much larger neighbour galaxy M81 which is estimated to have occurred about 200 million years ago.*

GALAXIES

active galactic nucleus (AGN) ▶ p. 336 black hole ▶ p. 336 infrared ▶ p. 340 Local Group ▶ p. 340 Magellanic Clouds ▶ p. 341 quasar ▶ p. 342 X-ray ▶ p. 344

MORPHOLOGY AND ENVIRONMENT

Different types of galaxy generally inhabit very different environments. Most of the galaxies in rich clusters such as the Coma Cluster are elliptical or lenticular, and very few are spirals. Conversely, lower-density regions are dominated by spirals, with few large ellipticals to be found. For example, in our own low-density Local Group, the only three bright galaxies are spiral, with the remaining 30-odd galaxies all either dwarf spheroidals, dwarf ellipticals or irregulars. Any theory of galaxy evolution must be able to explain the distribution of galaxy types.

✳ ROTATION CURVES AND THE TULLY–FISHER RELATION

By measuring a galaxy's rotation curve – how fast material orbits at different distances from the galaxy's centre – it is possible to determine the gravitational pull and hence the distribution of mass within the galaxy. In the early 1970s, radio telescopes started observing 21-cm emission from the hydrogen gas at large radii in galaxies other than our own. From the Doppler shift in this emission, the rotational velocity of the gas could be inferred. Instead of declining, these velocities were found to stay almost constant out to hundreds of thousands of light years from the centres, as happens in our own Galaxy. The only way that the speed of rotation can be maintained at large radii is if there is extra gravitating mass out there, well beyond the edge of the galaxy's visible stars – in other words, a massive halo of invisible matter.

Intriguingly, there are close links between the mass of a galaxy's dark halo at large radii and the luminous components in its inner part. In 1977, Brent Tully and Richard Fisher showed that the total luminosity of a spiral galaxy (contributed mainly by the stars in its disk and central bulge) is closely correlated with the rotation speed at large radii (which is dictated by the mass of the dark halo). The physical reason for this Tully–Fisher relation is unknown, but is presumably tied up with the complete history of the galaxy, including its creation and the subsequent formation of stars within it. The relation provides a useful method of estimating the true luminosity of a galaxy from its speed of rotation. An analogous phenomenon is observed in ellipticals, where the luminosity of a galaxy can be deduced from its size, surface brightness and the random motions of its stars (whose velocities are dictated by its mass); this correlation is termed the Faber–Jackson relation, or the Fundamental Plane.

CONTENTS OF GALAXIES

Galaxies are built up from a few major components – stars, gas, dust, dark matter and a central **black hole** – but the mix of these ingredients, and their individual characteristics, varies between galaxies. Here is a summary of the contents of each of the main types of galaxy.

Stars

Stars are the most obvious component of all galaxies. In spiral galaxies, the oldest stars are found in an extended halo of globular clusters and the central bulge, while the disk contains stars with a wide range of ages. The youngest stars of all lie in the spiral arms. In elliptical galaxies, the stars are all older – these galaxies do not possess the cold clumps of gas from which new stars can form. Conversely, irregular galaxies often exhibit very vigorous star formation, probably because collisions between gas clouds are common in these misshapen structures.

Gas

The density of gas in the space between stars is incredibly low, typically only about one atom per cubic centimetre. However, galaxies are so large that this can add up to a great deal: in a large spiral like the Milky Way, the total gas mass is several billion times the mass of the Sun, about 5% the mass of its stars. In spiral galaxies the gas is cool enough to be detected either by the 21-cm radiation emitted by atomic hydrogen (often termed HI) or by the light emitted by ionized hydrogen (HII regions) as it recombines into the atomic form. However, in elliptical galaxies, the gas is so hot – tens of millions of degrees – it forms an ionized plasma so there is no 21-cm radiation or visible light. Instead, the intensely hot gas emits **X-rays**, which can only be detected by satellites. The total mass of hot gas in large ellipticals is around a billion times the mass of the Sun, similar to the amount of cold gas in large spiral galaxies.

Dust

Dust is found in the disks of spiral galaxies, producing the dark lanes running across spiral galaxies seen edge-on and also causing the dark patches seen in our own Milky Way, such as the Cygnus rift and the Coalsack

nebula. Dust is also plentiful in irregulars but is not found in elliptical galaxies because the very high temperature gas in these systems bombards the dust grains, rapidly destroying any that form.

Dark Matter

Stars, gas and dust comprise only about 10% the mass of a typical galaxy. The remaining 90% must be contributed by some other unidentified material, referred to as 'dark matter'. The motions of stars and gas, which are dictated by the gravitational pull of the

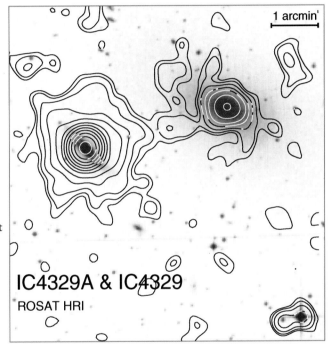

▲ **X-ray emission** *from extremely hot gas in the huge elliptical galaxy IC 4329 (right), overlain with contours on an optical image. To the left is IC 4329A, an edge-on spiral with an active nucleus.*

total mass, indicate that dark matter is distributed in an extended halo stretching well beyond the luminous limit of all galaxies. The nature of this dark matter remains one of the major unanswered questions in astronomy.

Central Black Holes

Most, possibly all, galaxies have black holes at their centre, with masses ranging from 100,000 to a billion Suns. Larger galaxies contain the more massive black holes. Central black holes are thought to provide the energy sources of **quasars** and active galaxies (see pages 184–85). Most galaxies do not behave like quasars because they lack the infall of gas and stars that powers **active galactic nuclei**.

THE COSMOS

QUASARS & ACTIVE GALAXIES

I n the early 1960s radio observations of the sky led to the identification of a baffling new class of object. The objects had large redshifts, much larger than had been seen in galaxies. These redshifts indicated huge distances to the objects, distances which implied correspondingly vast energy outputs. The optical images of the objects, which were star-like, coupled with other properties which were distinctly non-stellar, led the objects to be termed quasi-stellar sources, a name which was rapidly shortened to quasar. Today it is thought that a quasar is just one form of active galactic nucleus (AGN), a term which is used to describe a variety of violent phenomena observed at the centres of some galaxies.

THE DISCOVERY OF QUASARS

The first quasars were discovered as radio sources during the sky survey for the third Cambridge (3C) catalogue carried out in the 1950s by the University of Cambridge, England, although they were not immediately recognized as remarkable. In 1963 **Allan Sandage** (b. 1926) at Palomar Observatory photographed a faint star-like object at the position of one of the radio sources, 3C 48, but was unable to explain its unusual **spectrum**. In the same year the British astronomer Cyril Hazard and others published the accurate position of 3C 273, which allowed Maarten Schmidt (b. 1929) at Palomar to pinpoint a star-like image with characteristics not found in any stars. Most particularly the **emission lines** in the source's spectrum were **redshifted** by an amount greater than that of any galaxy known at that time. Other examples of these curious objects were soon found.

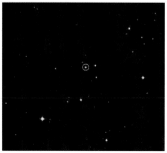

◀ 3C 273 (circled), an apparently starlike object in Virgo that is the brightest of the energetic objects known as quasars. These are now believed to be the active centres of young, very distant galaxies. 3C 273 itself lies about 3,000 million light years away.

OPTICAL PROPERTIES

The most obvious property of quasars is their very large luminosities. This property is not confined to the optical part of the spectrum – they also produce large amounts of **ultraviolet** radiation and X-rays. They also exhibit rapid variations in brightness, especially at X-ray **wavelengths**, which implies that the emission source is small (see The Size of Quasars, opposite). In some cases, changes in brightness of a few per cent occur in a matter of minutes while others fluctuate by larger amounts over weeks or months. One quasar, 3C 446, increased in luminosity by a factor of 20 in a year. The variability of a source gives an indication of its size, demonstrating that many quasars are about one light-day in diameter, similar in size to the Solar System, while many others are no more than one light-month across. Although the variations in luminosity need not be associated with the entire source, changes by a factor of two or more imply that significant sections of the source must be involved.

RADIO PROPERTIES

Maps of quasars at radio wavelengths can display three types of features. At the centre is a compact source which coincides with the star-like object seen by optical astronomers. On either side are two large radio-emitting lobes, arranged symmetrically about the central source but separated from it by many galactic diameters. The third feature, not always seen, is a single jet emerging from the central source and pointing to one of the lobes. The geometry of these features suggests that the energy of the quasar comes from the compact source at the centre, but the relative strengths of the features varies from quasar to quasar and even between frequencies in the same quasar.

ACTIVE GALAXIES

Similar activity to that in quasars is found at the centres of some galaxies. It is not known whether such galaxies are truly different from the majority or whether all galaxies undergo

▲ *Centaurus A, also known as NGC 5128, is the nearest and brightest radio galaxy. It is thought to result from a merger between an elliptical galaxy and a spiral.*

this behaviour for a few per cent of the time. Active galaxies can be conveniently grouped into a few main types. **Radio galaxies**, as their name implies, are unusually strong sources of radio emission. At radio wavelengths they show a similar structure to that of quasars, with symmetric lobes and often a jet, but their centres are not as bright optically. **Seyfert galaxies** (named after the US astronomer Carl Seyfert, 1911–60, who identified the first ones in 1943) have optically bright centres, but are much less energetic than quasars and much more numerous. The great majority of Seyfert galaxies are spirals or barred spirals. Another type of active galaxy is the **blazar**, named after the source **BL Lacertae**. This object was once thought to be a variable star in our Galaxy, but is now recognized as an unusual form of galaxy. Blazars are both more luminous and more variable than standard quasars.

THE COSMOS

DISTANCES TO QUASARS

The redshift of quasars, if due to the expanding Universe, implies huge distances and luminosities up to a hundred times that of a normal galaxy. When these extraordinary properties were first recognized, many attempts were made to find an alternative explanation for the observed redshift, one that would allow the quasars to be closer and, therefore, less luminous. No plausible alternative was found, although it was not until the 1990s when observations from the Hubble Space Telescope showed that many quasars are embedded within very faint (and hence very distant) galaxies that the debate finally ended. It is now generally agreed that the redshift of quasar light is indeed indicative of distance, those with the largest redshifts being among the most distant objects yet detected in the Universe. Hence we see them as they appeared when the Universe was only about 10% of its present age.

FASTER-THAN-LIGHT EXPANSION?

Many quasars eject fragments which seem to be moving at a speed greater than that of light, an apparent violation of one of the basic principles of special relativity. One of the most-studied quasars, 3C 273, shows this phenomenon, a fragment in its jet apparently moving at about six times the speed of light. It is believed that the fragments are actually travelling at just below light speed, but in a direction close to our line of sight. A proper analysis must allow for changes in the transit time of light from quasar to Earth. This can be seen in the following example:

Observers on Earth receive light first from the source at position A and 2 years later they receive light from the same source at position B. If, however, they assume that the source is at position C, then the source will appear to have travelled three light years in two calendar years, i.e. its speed has been 3/2 times light speed (faster-than-light). In reality, the source has moved from A to B, a distance of five light years. Since B is four light years closer to Earth than A, light has taken four years longer to reach the observers from position A than position B, and this difference must be ac-counted for when calculating the expansion velocity. Hence the actual time taken for the source to travel from A to B is 4 + 2 = 6 years, and the true speed of the source is therefore 5/6 times light speed.

THE SIZE OF QUASARS

We can estimate an upper limit for the size of an object such as a quasar from the time it takes to vary in brightness. For example, light takes about five seconds to cross the diameter of the Sun, a distance of 1.4 million km (870,000 miles). Suppose the whole surface of the Sun were suddenly to increase in brightness. An observer on Earth would see the brightening first at the centre of the Sun because this is the point nearest to Earth. Over the next 2.5 seconds the brightness would spread across the Sun's disk, that being the time taken for light to cross the Sun. If the Sun were 10 times larger then the increase in brightness would take 25 seconds. This simple calculation holds for all spherical objects, but it must be applied with caution. Firstly, the variable source might not be spherical, and secondly if the change in brightness is slow then the calculated size will be larger than the true size. Variability therefore sets only an upper limit to the object's size. However, it is always true to say that fast variations in luminosity imply a small source.

A MODEL FOR ACTIVE GALACTIC NUCLEI

All the various forms of active galactic nuclei (AGN) may stem from a single type of source whose appearance depends on its orientation to the observer. In the model shown opposite, a compact source is surrounded by a torus (a ring doughnut) of gas and dust; narrow jets of gas are emitted at velocities approaching the speed of light along the axis of rotation, which is at right angles to the ring. If the line of sight were through the torus, the central source would be obscured and the AGN would be seen as a radio galaxy. But if the torus were tilted to our line of sight, the central source would be visible; this would correspond to viewing a quasar. Blazars would be seen when the observer was looking directly along, or close to, the jet. This would also explain why apparent faster-than-light velocities (see below) are most closely associated with blazars.

THE ENERGY SOURCES OF AGNS

Many galaxies are thought to possess black holes at their centres with masses of millions or even billions of Suns. A supermassive black hole could be transformed into an AGN when sufficient gas collects into an accretion disk around it. The gas comes from the tidal break-up of stars which drift, or are attracted, too close to the black hole. If this happens intermittently, it would explain why only a small percentage of galaxies appear to have AGNs: all galaxies could have black holes, but they would be supplied with gas only in relatively short bursts. In some cases, encounters between galaxies could send more stars and gas tumbling into the black hole, triggering an outburst of activity. The axis of the accretion disk provides a preferred direction along which the observed lobes and jets of quasars are aligned. Magnetic fields generated by the differential rotation of the disk could provide the mechanism for the ejection of the jets.

▼ *Types of Active Galactic Nuclei*
The model for an AGN says that it can be categorized according to the angle at which its radiation emissions are seen. Where the high-energy radiation jet points towards the line of sight (A), the observer sees a blazar. Where the observer looks toward the central black hole over the edge of the torus (B), a quasar is seen. Where the central black hole is obscured by the torus (C), the observer sees a radio galaxy.

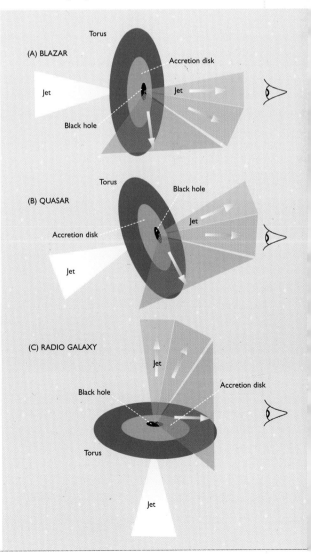

GALAXY FORMATION & EVOLUTION

▲ *NGC 6745, a spiral galaxy that has suffered a collision which compressed its gas to form a trail of hot, blue stars at the top and a 'beak' at the right. The passing galaxy is just out of view lower right.*

A stronomers face a major challenge in explaining how the diverse range of galaxies that exist today can have formed. They have collected a wealth of observational information on the overall structure of the various types of galaxy and their contents, but this still needs to be pieced together into a consistent picture that explains the whole story of galaxy formation and evolution. To do so involves combining these observations with computer simulations and mathematical calculations to uncover the underlying physical processes that are responsible for the observed properties of galaxies.

GALAXY FORMATION: TWO CONTRASTING VIEWS

In the 1960s the view was established that galaxies were born when a large cloud of primordial gas collapsed under the inward pull of its own gravity, like a scaled-up version of the birth of stars. As the cloud contracted it left behind smaller lumps of gas which formed a halo of globular clusters. The key to whether the cloud became an elliptical or spiral galaxy was presumed to be its speed of rotation: if the shrinking gas cloud was rotating fast enough, its outer regions would spread into a disk which would develop spiral arms.

More recently, an alternative scenario has arisen. In this view, the large galaxies seen today grew from the mergers of smaller galaxies which, in turn, had been the products of earlier mergers between still-smaller units. This hierarchical system would first produce an elliptical galaxy. If that elliptical galaxy were to gravitationally attract a new disk of gas it would develop arms and hence turn into a spiral. A mounting array of evidence now supports this hierarchical picture.

MERGING GALAXIES

A collision between galaxies is not as destructive as it might sound, since galaxies consist mostly of empty space – if a galaxy were scaled down until stars were the size of golf balls, then the ball-sized stars would still typically be separated by some hundreds of kilometres (or miles). Their collections of stars can pass freely through each other, but at such close quarters the stars from the two galaxies feel each others' gravitational pulls. Disrupted from their well-ordered orbits, the stars are drawn away from their parent galaxies into huge tidal tails that can stretch for hundreds of thousands of light years.

Unlike stars, the gas in galaxies will collide directly in an encounter. Such collisions compress the gas, sparking the formation of large numbers of stars – so in colliding galaxies it is quite common to see localized bursts of star formation. In extreme cases star formation can spread across the entire galaxy simultaneously, producing a highly luminous starburst galaxy. Another consequence of collisions is that stars and gas can

▼ *Two Theories of Galaxy Formation*
In the monolithic model of galaxy formation, a slowly rotating cloud of primordial gas collapses to form a faster-spinning spiral disk with a surrounding residue of globular clusters. There is growing support for a newer hierarchical theory in which a sequence of small units merging into ever-larger groupings forms a galactic bulge, as gravitationally attracted gas is drawn in to form the disk.

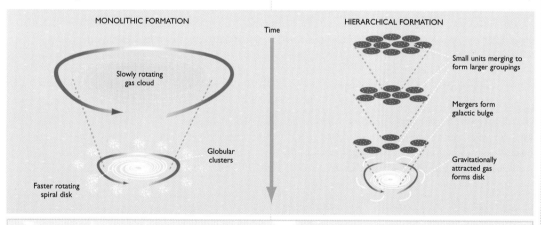

MONOLITHIC FORMATION — Slowly rotating gas cloud — Globular clusters — Faster rotating spiral disk — Time — HIERARCHICAL FORMATION — Small units merging to form larger groupings — Mergers form galactic bulge — Gravitationally attracted gas forms disk

ERIK HOLMBERG

Erik Holmberg (1908–2000) was one of the founding fathers of the study of galaxies. He worked for much of his career in his native Sweden, developing a number of techniques for measuring basic properties of galaxies such as their brightnesses and sizes. One of the standard definitions of the size of a galaxy is named the Holmberg radius in recognition of his work. Perhaps his most innovative contribution was to undertake the first *N*-body simulation; this he did in 1941, before the invention of the computer. Since the strength of a gravitational force decreases with distance in exactly the same way as the apparent brightness of a light source, Holmberg used an array of 74 light bulbs on a table to represent a scale model of two galaxies. The amount of light arriving at the location of each bulb represented the gravitational pull on that part of the system. With this simple apparatus, Holmberg was able to simulate the motions of the galaxies and thus demonstrate for the first time that colliding systems produce tidal tails and eventually merge.

THE EPOCH OF GALAXY FORMATION

When did galaxies form? There seems to have been a 'golden age' of galaxy formation some 10 billion years ago. Hubble Space Telescope views of the remote Universe have shown that objects that we would recognize as galaxies first became common around then. This timescale also fits in with investigations of the star formation history of the Universe, which indicate that star formation peaked around 10 billion years ago, with lower rates of formation before and since.

Unfortunately it is extremely difficult to push back further in time to detect the very first generation of star-forming galaxies. There is some evidence that such **protogalaxies** are enshrouded in dust resulting from an early burst of star formation, absorbing light and re-radiating it in the infrared. The expansion of the Universe redshifts the light into the millimetre part of the spectrum, which is currently almost unexplored. Astronomers are developing a new generation of telescopes to study such wavelengths. One such instrument is the Atacama Large Millimetre Array (ALMA), due to be built on a high plateau in the Andes mountains of Chile.

protogalaxy ▶ p. 342

THE COSMOS

end up being dumped on to the massive black holes at the centres of the original galaxies, so mergers are also associated with activity in the nuclei of galaxies – for example, quasar host galaxies often appear to be merging systems. Ultimately, after hundreds of millions of years, the original stars from the progenitor galaxies, along with new stars created in the collision, will settle down to form a single elliptical galaxy.

GALACTIC ARCHAEOLOGY

Like archaeologists, astronomers use the fragmentary evidence that can be gleaned from nearby galaxies to try to disentangle their complex histories. For example, by examining the spectrum of individual parts of a galaxy (or, for more distant specimens, the entire galaxy) it is possible to determine the mix of stars of various types that generate its light. We can then estimate such fundamental properties as the age of the stellar populations and the abundance of heavy elements in the stars. Since any heavy elements will have been created by previous

▼ *The faintest* and most distant galaxies ever seen are captured in this image from the Hubble Space Telescope called the Hubble Deep Field. As well as the familiar spirals and ellipticals, a range of other galaxy types can be seen at various stages of evolution. The youngest of these may have formed less than a billion years after the Big Bang.

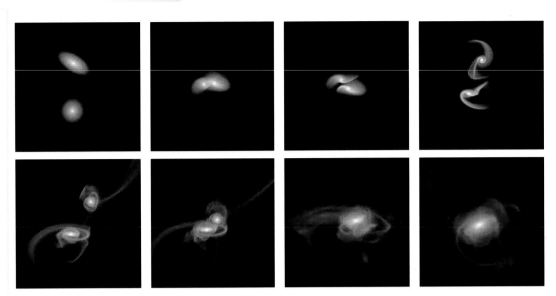

▲ *A computer simulation* showing a future encounter between our own galaxy, the Milky Way, and our nearest neighbour, the Andromeda Galaxy (M31); such simulations have opened up great possibilities in understanding galactic evolution.

generations of stars, these data provide insights into the star-formation history of the galaxy.

Another clue to the galaxy's evolutionary past can be gleaned from Doppler shifts in its spectrum, which reveal the motions of stars and gas in the galaxy. For example, it has been found that around a quarter of all elliptical galaxies contain groups of stars that orbit quite rapidly near the centre of the galaxy in a direction unrelated to the majority of stars in the galaxy. These features appear to represent the partially digested remains of a small galaxy in the final stages of being absorbed by the elliptical galaxy.

OBSERVING DISTANT GALAXIES

Another way of studying galaxy evolution involves comparing nearby galaxies with those far away – because of the time light takes to reach us, we see distant galaxies as they appeared long ago. This technique was first successfully applied in 1978 by the American astronomers Harvey Butcher and

Augustus Oemler, who found that distant clusters of galaxies seemed to contain a greater proportion of spirals than do nearby clusters. So where have the spirals gone in the nearer (and older) clusters? They must, it seems, have collided and merged to create elliptical galaxies, explaining the prevalence of the latter in nearby clusters. The Hubble Space Telescope has since confirmed Butcher and Oemler's findings and also revealed other evolutionary effects. For example, in the past the Universe contained many more irregular galaxies than now; these were presumably the building blocks that merged to form the galaxies we see today.

COMPUTER SIMULATIONS

Along with developments in telescope technology, advances in computing have played a major role in helping astronomers to understand how galaxies evolve. Computer programs can be used to simulate the motions of a collection of massive particles under the influence of their gravitational pull, thereby modelling the behaviour of stars in galaxies. Initially these programs, termed N-body codes, were able to follow the motions of only a few hundred objects, but the development of more sophisticated algorithms and advances in computing power mean that it is now possible to perform simulations with millions of particles. Astronomers owe much of their current understanding of galaxies, including the nature of spiral structure and what happens when galaxies collide, to such simulations. Even larger simulations, which follow the evolution of large-scale structure in the Universe, are now beginning to incorporate the formation of individual galaxies, allowing their complete life history to be investigated.

SEMI-ANALYTIC MODELLING

The ultimate goal in the study of galaxies is to follow their evolution all the way from formation to the present. This quest represents a formidable challenge as it draws on the whole of astrophysics: to understand how a galaxy evolves, one needs to understand everything from the cosmological large-scale structure in which the galaxy forms, right down to the small-scale physics of star formation. Rather than trying to solve the whole problem at once, some astrophysicists have adopted a simpler approach. In this technique, termed semi-analytic

modelling, the poorly understood pieces of physics are reduced to a number of simplified parameters. For example, star formation can be crudely represented by the rate at which gas is turned into stars and the numbers of stars of different types that are created. Since the true values of these parameters are unknown, different estimates must be tried until the predictions of the model tally with all known facts about galaxy evolution, such as the numbers and luminosities of galaxies that we see today and in the distant Universe. In this way, astrophysicists have obtained insights into the physical processes that underlie galaxy formation and evolution.

THE COSMOS

Contents of the Cosmos
PLANET FORMATION

Understanding how Earth and other planets formed has been hampered by the fact that, until recently, ours was the only planetary system known. However, samples of material from early in the Solar System's history still exist in the form of rocky meteorites and comets. Study of these, combined with theoretical models and observations of the environments of new-born stars, has allowed astronomers to build up a picture of planetary origins. In the most widely accepted view, the so-called nebular theory, the planets formed from a disk of gas and dust that surrounded the infant Sun. This implies that planetary formation is a natural accompaniment of a star's birth, so planetary systems should exist around many other stars.

OBSERVATIONAL EVIDENCE: DISKS AND GAPS

Infrared observations have revealed that most very young stars, those less than about 10 million years old, are surrounded by disks of gas and dust, called **protoplanetary disks**. Such disks become less common around progressively older stars, because they dissipate or consolidate into planets. The dust in these disks is a mixture of silicates (rocky particles) and carbon-containing material, as it was in our own **Solar System** at the time of its formation.

About one in six of nearby older stars is also surrounded by a disk, but these disks contain proportionally higher amounts of dust and less gas than those encircling young stars. One of the best-known examples is that around the star Beta [β] Pictoris (see Star Chart 14, page 284). Since dust is expected to fall into the star relatively quickly, the disks around old stars are probably replenished with new dust from **asteroids** or **comets**. Some of the disks are doughnut-shaped, with a hole between the disk and the central star. Planets could orbit in this hole, but if they do we cannot see them with our existing telescopes. In fact our own Sun, if viewed from a neighbouring star, would appear to be surrounded by a dusty ring (the **Edgeworth–Kuiper belt**), with empty space where the planets lie.

FORMATION OF TERRESTRIAL PLANETS

The four inner rocky planets of our Solar System, known as the terrestrial planets, grew from the gradual accumulation of microscopic dust grains in the protoplanetary nebula. Analysis of **meteorites** suggests that heating events – the origin of which is still controversial – melted the various accumulations of dust particles and they stuck together to make blobs up to 1 mm (0.04 in) across called **chondrules**. The chondrules were then cemented together, perhaps with the help of sticky tar-like molecules, to produce boulders called **planetesimals**. Once these planetesimals attained diameters of a kilometre or more (0.6 miles or more) their gravitational field became significant, accelerating their growth as they attracted and merged with other planetesimals. Dating of meteorites (originating from asteroids and Mars) and terrestrial rocks suggests that this planet-forming process took between 10 million and a few hundred million years to complete.

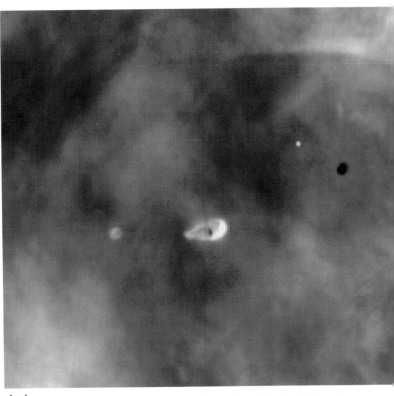

▲ *This Hubble Space Telescope* close-up of part of the Orion Nebula reveals newborn stars surrounded by disks of gas and dust. These are thought to be protoplanetary disks, or 'proplyds', from which planets will subsequently form. At least half the stars in the Orion Nebula are surrounded by such disks, suggesting that planet formation is common in the Galaxy.

FORMATION OF GIANT PLANETS

The outer planets – the gas giants Jupiter, Saturn, Uranus and Neptune – are thought to have formed in a different way, and there are two competing theories. One is that a large rocky core first accreted, much like a more massive version of a terrestrial planet, which then attracted a deep covering of gas. The other proposes that the outer planets may have formed when denser swirls in the orbiting disk collapsed under their own gravitational pull, like a scaled-down version of the birth of the Sun. In the former case we expect the formation of giant planets to have taken a long time – several hundred million years – while the latter scenario would have occurred almost simultaneously with the formation of the Sun. In either mechanism the existence of Uranus and Neptune presents a problem, as the density of gas in the outer regions of the solar nebula where they now lie is thought to have been too low to form such big planets. Possibly Uranus and Neptune initially formed closer to the Sun and later moved outwards. Whatever the case, once full-sized planets have formed the remaining debris is either swept up by collisions or flung out of the planetary system by gravitational interactions or an intense **solar wind**.

▼ *A disk of dust and gas* surrounding the young star Beta Pictoris, photographed by the Hubble Space Telescope. Disks such as this are thought to be the birthplaces of planets. The inner section of the disk is slightly tilted, suggesting the existence of a large planet in a clear zone closer to the central star, which is deliberately blocked out to prevent glare. The colours in the disk have been added by computer to accentuate detail.

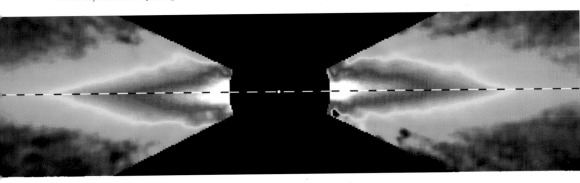

THE COSMOS

IS EARTH UNIQUE?

There is an element of unpredictability in planet formation, particularly during the final stages when collisions occur between large bodies. These impacts can cause chaotic events, changing the face of a planetary system and making each unique. Earth experienced a particularly dramatic event when a Mars-sized body hit it, forming the Moon.

Of course, the most remarkable thing about our planet is that it is teeming with life. Several reasons can be identified. Firstly, Earth lies comfortably within the habitable zone around our parent Sun, where temperatures are suitable for liquid water to exist. The relative circularity of Earth's orbit within this zone may have been another contributory factor.

Life is particularly abundant, and diverse, in tidal areas on Earth, which would not exist but for the presence of a large Moon. Surprisingly, the giant planet Jupiter may also have had a beneficial effect, by diverting comets careening towards the inner Solar System and hence reducing the threat of large impacts on the inner planets.

Given all these random factors, absolute clones of our Solar System containing a planet identical to Earth are unlikely. But, somewhere among the planetary systems now being discovered around other stars, there are almost certainly close matches to our home. Whether these are capable of supporting life is yet to be discovered.

EXTRASOLAR PLANET DETECTION

In recent years a new branch of astronomy has opened up: the detection of planets around other stars. Three techniques have detected such so-called extrasolar planets or **exoplanets**. The most successful has been the **radial velocity** technique, measuring the slight motion of the star towards and away from us as it orbits around its common centre of **mass** with the planet or planets. The star's motion can be detected by the **Doppler shift** in its light. From the period and amplitude of the shift, the mass of the planet and details of its orbit can be determined.

Two alternative techniques are **gravitational lensing** and **transits**. Gravitational lensing relies on the chance passage of a planet in front of a background star. The gravity of the planet will bend the light from the background star like a lens, causing a temporary change in the apparent brightness of the star. The third technique, transits, can yield the most information of all about the planet, but it relies on the planetary system being almost edge-on to us so that the planet passes directly in front of its parent star. The amount of light blocked out by the planet during its transit reveals the planet's size. Effects on the star's light as it passes through the planet's atmosphere can also give clues to the nature of the planet itself.

CHARACTERISTICS OF EXOPLANETS

The planets that are easiest to spot with the radial velocity technique are large and close to their star. Hence, most of the first exoplanets to be discovered were the size of Jupiter or larger and orbited closer to their parent star than Mercury does to the Sun. Such planets, known as 'hot Jupiters', may have formed farther out and then moved inwards as they were slowed by drag from the gas in the surrounding disk.

Many of the exoplanets that lie farther away from their parent stars have orbits that are highly elliptical, unlike the near-circular orbits of the planets in our Solar System. Not until many more exoplanets are known can we tell whether our own Solar System is unusual. However, we can already draw important conclusions, notably that planets are common in the Galaxy and that perhaps the majority of stars have planets orbiting them. Several cases of multiple planets orbiting a single star have been found, confirming that entire planetary systems can arise. Planets tend to be found only around stars that, like our own Sun, are particularly rich in heavy elements.

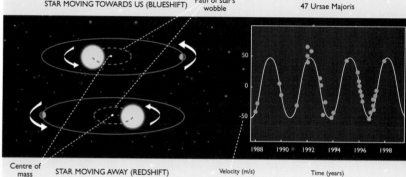

STAR MOVING TOWARDS US (BLUESHIFT)   Path of star's wobble   47 Ursae Majoris

Centre of mass   STAR MOVING AWAY (REDSHIFT)   Velocity (m/s)   Time (years)

1988 1990 1992 1994 1996 1998

Most of the planets known have masses no more than a few times that of Jupiter, and as searches continue the number of low-mass planets is increasing. An object with a mass above about 10 Jupiters is regarded not as a planet but a **brown dwarf**. Surprisingly, brown dwarfs are turning out to be rarer than true planets around other stars.

▲ *Radial Velocity*
Measuring the radial velocity of a star can reveal the presence of an orbiting planet. Both the star and the orbiting planet have a gravitational effect on each other: in effect they are both orbiting around their centre of mass. Since the star is much more massive than the planet its 'orbit' is a mere wobble. The speed and direction of the star's wobble is called its 'radial velocity'. It can be detected by measuring the blueshift (as it moves towards us) or redshift (as it moves away from us) of the star's light. On the graph, the dots show measured data of 47 Ursae Majoris; the line is the sinewave that best fits the data. Calculations from this graph have revealed the existence of a planet two-and-a-half times the size of Jupiter that takes about three years to orbit its star.

PLANET FORMATION THEORIES: A BRIEF HISTORY

A popular theory for the origin of the Solar System in the first half of the twentieth century, developed by the English astronomer **James Jeans** (1877–1946), was that the planets were ejected from the Sun following a collision with a giant comet or another star. But this was rejected when it was realized that such hot gas would disperse, not collect into planets. A rival theory suggested that the planets grew from passing material that was gravitationally captured by the young Sun. Although never disproved, this capture theory now seems less likely than the nebular hypothesis, which is also the oldest suggestion of the three, first proposed by the German philosopher **Immanuel Kant** (1724–1804) in 1755 and independently developed in 1796 by the French mathematician **Pierre de Laplace**. Most modern ideas of planetary formation are based on the nebular hypothesis.

▶ **The Copernican Universe,** *showing the Sun at the centre encircled by Earth and other planets (Jupiter is shown here with its satellites).*

Solar System ▶ p. 343 solar wind ▶ p. 343 transit ▶ p. 344 Jeans ▶ p. 142 Kant ▶ p. 136 Laplace ▶ p. 172   **PLANET FORMATION**   189

THE COSMOS

Contents of the Cosmos
SUMMARY

O ver the past century and a half, astronomers have established a fact that previous generations could only guess at: stars are more than mere points of light, they are balls of incandescent gas similar to the Sun. But stars are not as identical as peas in a pod. They range in size from giants, hundreds of times larger than the Sun with densities less than air, to white dwarfs the size of Earth with densities far greater than any metal. Mounting evidence now suggests that a percentage of stars have planets – and since each planetary system is likely to contain more than one planet, throughout our Galaxy there could be almost as many planets as there are stars. The question of life on them remains unresolved, though.

In the past half century, other objects far more exotic than red giants and white dwarfs have been discovered in the cosmic zoo. They include neutron stars and black holes, left behind after the deaths of massive stars, which are thought to populate our Galaxy in unseen numbers. Way beyond our own Galaxy, astronomers have detected radio-loud galaxies which seem to be ejecting jets of gas at almost the speed of light while, most distant of all, quasars blaze brilliantly as supermassive black holes at their cores devour stars and gas. Entirely unanticipated denizens of the deep Universe almost certainly remain to be discovered. Perhaps there can never be a full census of the contents of the cosmos.

STAR CHARACTERISTICS

- Stars are self-luminous balls of gas. They are born from cold, dense clouds of gas, usually in tightly knit clusters or more widespread starfields termed 'associations'.
- Clusters drift apart over time, but in many cases two or more stars may remain linked by gravity to form a double or multiple system. The orbital parameters of twin stars allow astronomers to calculate accurate stellar **masses**.

- Throughout most of their lives, stars are powered by nuclear fusion reactions that convert hydrogen to helium in their cores.
- A star's brightness as it appears to us on Earth is called its **apparent magnitude**, but this is affected by its distance. The actual **luminosity** of stars can be judged only by comparing them as they would appear at a standard distance, which is chosen to be 10 parsecs. The intrinsic brightness so computed is called the **absolute magnitude**.
- Stars appear different colours, depending on their surface temperatures – the hottest ones being the bluest and the coolest ones the reddest. Stars can be arranged into a temperature sequence, known as **spectral type**, based on features in their spectrum.
- There is a link between the absolute magnitude and spectral type of most stars. This can be shown on a diagram, called the **Hertzsprung–Russell diagram**, which plots absolute magnitude against spectral type. The link is demonstrated by the fact that most stars are found to lie in a band called the **main sequence**.
- The main sequence is also a sequence of stellar mass. The most massive stars are the brightest and hottest and are found at the top of the main sequence. Running down the main sequence, stars become progressively less massive and hence cooler and less luminous.
- Most young stars are surrounded by an orbiting disk of gas and dust left over from their formation. Such disks can consolidate into planets.

STAR DEATH

- The most massive stars burn out the quickest, living for only a few million years. The least massive stars can continue to shine for tens of billions of years before burning out.
- Towards the end of its life, a star begins to run out of hydrogen fuel at its centre. First, hydrogen-burning spreads outwards into the surrounding layers. Then, helium-burning begins in the core. These changes release more energy that causes the star to swell up into a **red giant** or, if it is particularly massive, a **supergiant**.
- In the case of stars with similar masses to the Sun, the outer layers of the red giant drift off into space, forming a glowing planetary

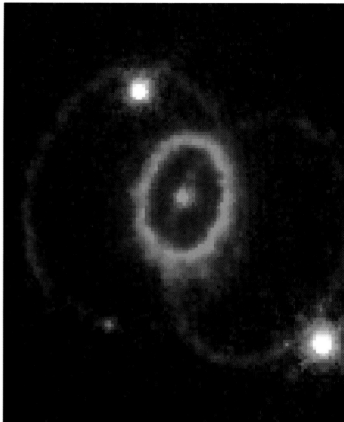

▲ *Supernova 1987A. Supernovae are the final stage in the life-cycle of a massive star, when the central region of the star collapses and the outer layers are flung off into space in an immensely powerful explosion.*

nebula. At its centre, the star's core is exposed as a **white dwarf**, which slowly cools into invisibility.
- The most massive stars undergo successive cycles of nuclear reactions which build up chemical elements of increasing complexity, ending at iron which cannot fuse to create energy. The central core of the star collapses, triggering an explosion which turns the star into a brilliant **supernova**.
- A supernova flings off the star's outer layers into space, producing glowing clouds of stellar debris such as the Crab Nebula in Taurus. The nuclear reactions that occur during supernova explosions create the heavy elements from which planets and life can form around later generations of stars.
- At the centre of a supernova, the **protons** and **electrons** of the star's core are crushed together to form **neutrons**. The resulting object, only about 20 km (12 miles) across, is known as a **neutron star**.
- Atomic particles are focused into beams along the neutron star's magnetic poles. These beams are observed from Earth as a flashing **pulsar** as the neutron spins.

THE COSMOS

DISCOVERY OF THE FIRST EXTRASOLAR PLANETS

Finding planets around other stars has long been a goal for astronomers, but it has been difficult to achieve because any such planets would be too faint to be seen directly from Earth. Astronomers have therefore had to use indirect methods, notably searching for a tell-tale sway in the positions of stars as the planets orbit around them. A report in the 1960s of a wavering of the nearby red dwarf Barnard's Star turned out to be a false alarm due to erroneous data. The first planet of another star detected with certainty was found by astronomers at Geneva Observatory, Switzerland, around 51 Pegasi, a star similar to the Sun, 50 light years away (see Star Chart 13, page 282). The Jupiter-sized planet orbits every 4.2 days at a distance of only 8 million km (5 million miles) from the star, so close it would be roasted. Its presence was detected from a slight periodic Doppler shift in the star's spectrum, using a spectroscopic 'speed gun' so sensitive it can detect changes in a star's motion similar to the jogging speed of a human. Many other planets of similar size to Jupiter are now known. Some stars are orbited by more than one planet – and smaller planets, similar to Earth, could also exist undetected.

◌ If the collapsing core of the star has a mass more than about three times that of the Sun it creates a **black hole**, a region of space with such an intense gravitational pull that nothing can escape from it, not even light.
◌ Black holes can grow to enormous sizes by swallowing gas and stars. Super-sized black holes, with masses of millions of Suns, are thought to reside at the centres of most galaxies.

▼ *Our Galaxy, the Milky Way, is a spiral galaxy approximately 150,000 light years across. Our Solar System lies in one of the Galaxy's spiral arms.*

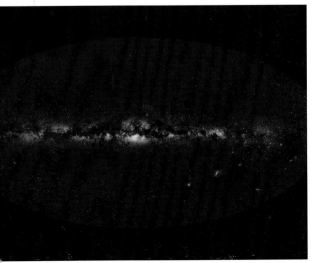

▲ *NGC 6745, an example of a spiral galaxy which has suffered a collision with another galaxy.*

THE MILKY WAY

◌ All the stars in the night sky are arranged in a vast swirling structure called the Galaxy. The hazy band of light known as the Milky Way consists of distant members of our Galaxy.
◌ Our Galaxy is spiral in shape, about 150,000 light years wide. The Sun lies in a spiral arm about a third of the way from the centre to the rim.
◌ The centre of the Galaxy consists of a large bulge of stars about 25,000 light years away in the direction of the **constellation** Sagittarius.
◌ Stars in the Galaxy fall into two main populations on the basis of their age: old stars, of Population II, are found in the central bulge and the Galaxy's surrounding halo; and young stars of Population I, which are found in the spiral arms.
◌ Spherical groups of old stars called **globular clusters** are dotted in a halo around the Galaxy. By observing the distribution of these clusters, astronomers first realized the true size of the Galaxy and the Sun's position within it.
◌ Our Galaxy is accompanied by two small, irregularly shaped satellite galaxies called the **Magellanic Clouds**, visible in the southern sky.

INTO THE COSMOS

◌ Beyond our Galaxy are many other galaxies. Some are spiral-shaped, others are elliptical while still others have no obvious structure at all and are classified as irregular.
◌ In nearly half the spiral-shaped galaxies, the curving arms emerge from the ends of a central bar of stars – these are termed barred spirals. Our own Galaxy may be shaped like this.
◌ Galaxies are the largest individual structures in the cosmos, but they also herd into clusters of various sizes. Our own Galaxy and the Andromeda Galaxy are the main members of a small cluster called the **Local Group**.
◌ Galaxies differ not only in shape but also in content. Whereas spirals contain a mix of old and new stars, with plenty of gas clouds from which more stars can form, ellipticals consist predominantly of old stars with no gas to make new ones.
◌ Nearly all galaxies possess invisible haloes of **dark matter**, which make up the majority of the galaxy's mass. The nature of the dark matter in these haloes remains unknown, but it may consist of faint stars, exotic atomic particles or black holes.
◌ The largest galaxies are supergiant ellipticals, resulting from the merging of other galaxies. Most galaxies are thought to have been built up from mergers of smaller components.
◌ Certain galaxies have unusually bright centres. The most extreme examples of these bright-eyed galaxies are quasars, which release as much energy as an entire galaxy from a region the size of our Solar System. A disk of hot gas circling a supermassive black hole is thought to provide the energy source at the centres of active galaxies and **quasars**.

THE COSMOS

SUMMARY

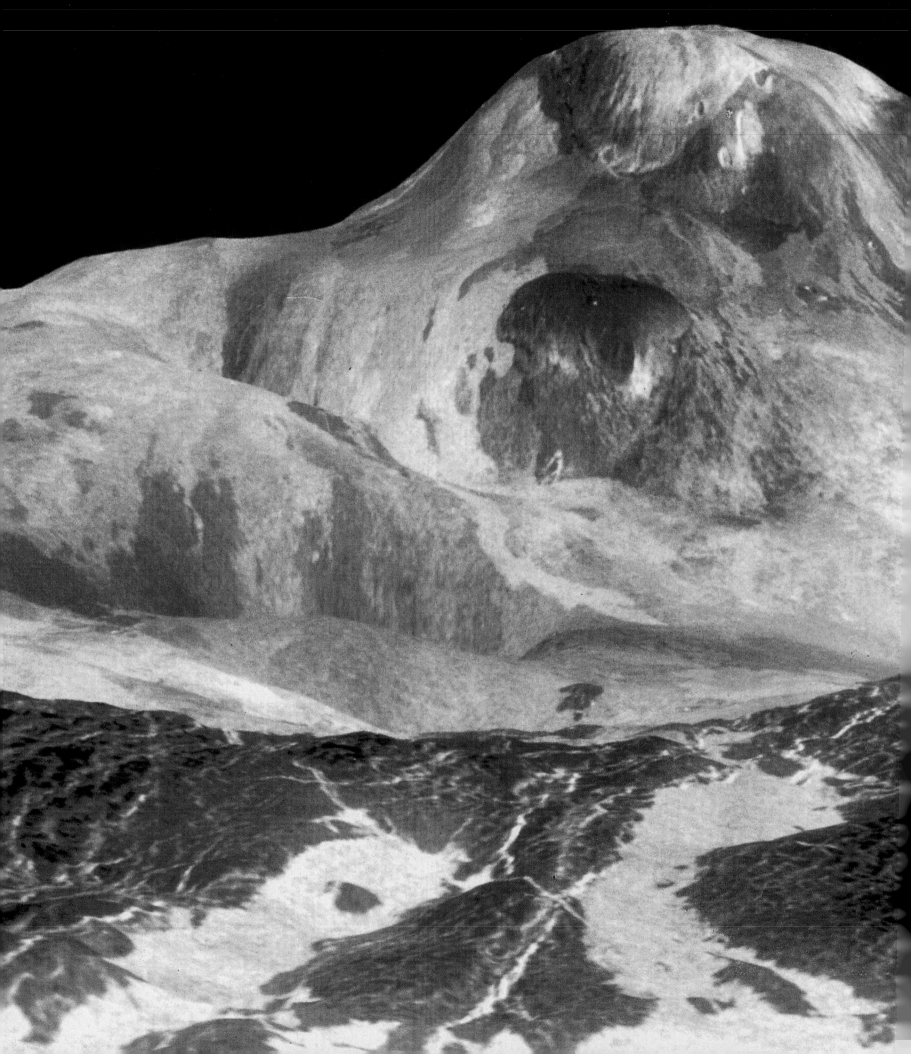

OUR SOLAR SYSTEM

Since antiquity those who have watched the night sky noticed some 'stars' were not fixed: a few appeared to wander slowly across the heavens from night to night; others arrived unexpectedly and sprouted long ghostly tails, blazing across the heavens before fading again. Some appeared as brilliant flashes streaking across the sky in seconds and seemed particularly numerous during certain nights of the year. All these celestial nomads belong to our Solar System, representing a family of planets (meaning 'wandering star'), comets and countless other smaller rocky and icy bodies (asteroids and meteoroids) which circle the Sun.

It was during the Renaissance, when Italian astronomer Galileo Galilei first turned the recently invented telescope towards the sky, that an understanding of the scale and motion of bodies within the Solar System became apparent. Galileo's observations gave credence to the published, but commonly refuted, work of an earlier Renaissance astronomer, Nicolaus Copernicus. Copernicus dared to suggest the Sun, and not Earth, lay at the centre of the Universe, with all the planets revolving around it. The observation that Jupiter had four orbiting satellites presented Galileo with what he considered to be a Copernican system in miniature. Furthermore, the discovery that Venus shows phases clearly demonstrated that it shone by reflected sunlight and must revolve around the Sun, not Earth.

Since the invention of the telescope, the Solar System has shown itself to be an incredibly rich place. During the eighteenth, nineteenth and twentieth centuries the planets Uranus, Neptune and Pluto were discovered respectively, greatly increasing the size of the Solar System. Many thousands of smaller worlds have also been discovered within the asteroid belt, the Edgeworth–Kuiper belt and the Oort cloud. These myriads of minor bodies adrift in the Solar System are the rubble left over from its formation 4.55 billion years ago.

The advent of space exploration has generated an unimaginable wealth of information about the worlds that inhabit our Solar System. By understanding the evolution of other planets we are able to learn more about the evolution of our own.

Our Solar System
INTRODUCTION

At the heart of our Solar System is the Sun. The Sun is just one of a hundred billion other stars which belong to our home Galaxy – the Milky Way Galaxy – a huge stellar system bound together by the force of gravity. The Sun shines by nuclear reactions, converting hydrogen into helium under immense pressures and temperatures within its core. As hydrogen atoms fuse to form helium, a small amount of energy is released. The immeasurable number of reactions occurring within the Sun at any given moment results in its huge energy output.

WHAT'S IN THE SOLAR SYSTEM?

Orbiting around the Sun are nine major planets and a suite of smaller bodies made of rock and ice and ranging in size from a thousand kilometres to particles smaller than house dust. The bodies within the **Solar System** have a profound effect on our home planet. The Sun provides heat and light, sustaining life and fuelling the dynamic processes within the atmosphere and oceans. The Moon raises tides, and is to date the only other body on which humans have set foot. **Asteroids** and **comets** are sources of potential calamities, occasionally striking Earth and changing the evolution of life. The other planets and their satellites provide us with an insight into the evolution of our own.

THE BIRTH OF THE SOLAR SYSTEM

Only recently have we been able to answer the fundamental questions about the evolution and age of our Solar System with any certainty. Surprisingly, the answers come from beyond, as well as within, the Solar System. Using today's most powerful astronomical tools, observations of giant stellar nurseries such as the Orion Nebula reveal newly born stars embedded within dark doughnut-shaped disks of cooler gas and dust. These disks are known as **proplyds** – short for **protoplanetary disks** and are thought to be the birthplace of planets.

Around four and half billion years ago the Sun was also enveloped in a disk of gas and dust born out of a collapsing **nebula**. Under the in-fluence of gravity, material not swept up by our young Sun gradually **accreted** to form small rocky chunks called **planetesimals**. The inevitable collision of planetesimals gave rise to larger bodies and within approximately 10 million years, a relatively short period of time astronomically speaking, the planets had formed.

DATING THE SOLAR SYSTEM

Astronomers are able to date the birth of our Solar System by using **radiometric dating**. Lurking within the Solar System are chunks of rock left over from planetary formation, some of which have survived the passage through Earth's atmosphere to be recovered as **meteorites**. Contained within them are radioactive elements which act as tiny atomic clocks and, when measured, reveal the meteorites' ages. Meteorites are unaltered since their formation, and extracting radiometric dates from them has provided a time frame for planetary formation.

What process triggered the initial collapse of our solar nebula is more of a mystery. One suggestion is that a nearby massive star ended its life in a huge explosion, producing a **supernova**. Supernovae send shock waves through interstellar space in the same way as throwing a brick into a pond sends out large ripples across the water's surface. Slamming into a region of gas and dust, the shock waves of a dying star may have triggered the nebula's collapse, ultimately resulting in the formation of the Sun, planets, and us.

EARTH AND THE MOON

The Earth's Moon is large in comparison to the planet it orbits and scientists often refer to the Earth and Moon as a double-planet system. Earth is the only world we know of that teems with life. Although the initiation of life on Earth is by no means fully understood, we do know that carbon, hydrogen and oxygen, along with energy in the form of heat, are essential ingredients. These elements are found in abundance on other planets and satellites but it seems the balance was not quite right for the proliferation of life to arise there.

The Moon is Earth's only natural satellite. The most likely explanation for its formation was by a cataclysmic glancing blow from an object the size of Mars. This ancient impact sprayed debris into orbit around Earth which

▲ *The Sun and the nine known planets of our Solar System, arranged in an artist's impression (not to scale). Each of the planets has its own characteristic features: Saturn its impressive ring system, Jupiter its raging storm the Great Red Spot, Mars its red, barren landscape, and Earth, our home, the most important feature of all – life.*

eventually coalesced to create the Moon. The Moon is a stark reminder of how, by altering just a few physical characteristics such as size, a world may become dead and lifeless. Owing to its smaller size, gravity on the Moon is only one-sixth of that on Earth. Any atmosphere the Moon may once have had has almost completely leaked away into space.

THE OTHER PLANETS

A casual glance at the planets in our Solar System reveals a distinct dichotomy. The inner, or terrestrial, planets – Mercury, Venus, Earth and Mars – are relatively small and dense with solid rocky crusts. The outer planets – Jupiter, Saturn, Uranus and Neptune (excluding the oddball world of Pluto) – are far larger, lighter, and are composed mostly of hydrogen and helium. They are referred to as the gaseous planets or gas giants. How did such a division occur?

The distinction came about very early during the formation of the Solar System – perhaps during the first 10 million years. Originally, the gas and dust contained within the proplyd from which the planets formed was evenly mixed. However, as the Sun grew larger and radiated more energy, the lighter gases and ices closer in were vaporized and blown outwards leaving behind denser material. The inner planets formed from the dense leftovers. The lighter gases moved outward until they condensed in regions of the outer Solar System where temperatures were cooler. Small rocky planets, already existing in the outer Solar System, became swamped with the lighter gases which they readily swept up, forming the giant gaseous outer planets.

| SUN | Mercury 46.0 to 69.8 million km | Venus 107.5 to 108.9 million km | Earth 147.1 to 152.1 million km | Mars 206.6 to 249.2 million km | ASTEROID BELT | Jupiter 740.5 to 816.6 million km | Saturn 1,352.6 to 1,514.5 million km |

INTRODUCTION

OUR SOLAR SYSTEM

COMPOSITIONS AND ATMOSPHERES

The ancient processes that produced the two distinct planetary categories explain the physical properties of the planets today. The terrestrial worlds have solid crusts with molten cores. Their surfaces record, in varying degrees, the latter stages of the cataclysmic bombardments which formed them. Whilst they are likely to have been relatively similar shortly after their formation, they have followed different evolutionary paths, reflected today by the amount of atmospheric and geological activity on each of them.

Mercury and Mars have tenuous atmospheres whereas Venus's and Earth's are much thicker and more dynamic. Earth is still geologically active with erupting volcanoes, earthquakes and plate tectonics. Mercury's surface is covered with craters, evidence that its crust is much older and that geological activity ceased there billions of years ago. The surfaces of Venus and Mars reveal powerful geological forces have been at play. Mars is home to the Solar System's largest volcanoes and has a huge rift valley. Venus has an abundance of volcanoes and volcanic plains. While present-day volcanism on Venus is open to debate, Martian volcanoes are believed to be extinct.

The outer planets have no solid surface; they are much larger than the terrestrial planets and are composed mainly of hydrogen and helium. Jupiter is the largest planet in the Solar System. Farther out is Saturn with its remarkable ring and satellite system, then Uranus with its 17 moons, and finally Neptune, coldest and most distant of the giants. Each gas giant has a system of satellites and is girdled by a series of rings made of rock and ice.

The ninth planet, Pluto, is the smallest in the Solar System and is composed mainly of rock and ice as hard as steel. Recently astronomers have questioned whether Pluto represents a true planet, since it is likely to belong to a large swarm of icy planetesimals which populate the outer Solar System, known as the **Edgeworth-Kuiper belt**. Pluto is accompanied by its large moon Charon.

◀ *A dense cloud* covering swirls around Venus, obscuring the planet's surface from view. The Venusian atmosphere is highly toxic: the clouds are composed of carbon dioxide and sulphuric acid.

THE DYNAMIC SOLAR SYSTEM

During the sixteenth century the Solar System was considered to be a very orderly place, ruled by the precision of static geometric shapes. In reality, it is far more dynamic. The interactions of gravity between the myriad bodies which populate the Solar System ensure that even today cataclysmic collisions can occur. Planetesimals which change course in the outer Solar System can come hurtling inward toward the Sun before being flung back out to deep space never to return. If such a visitor strays too close to a planet, the gravitational pull of the planet can lead to a direct strike, such as the impact of comet Shoemaker-Levy 9 on Jupiter in 1994 (see page 229). Such a strike here on Earth would have dire consequences; one such may have happened 65 million years ago.

Some scientists have suggested that other planets have suffered large impacts in the past. The fact that Venus rotates anti-clockwise and Uranus rotates on its side like a barrel may indicate these worlds have taken direct hits – perhaps early on in their formation. Other planets may have circled the Sun but have been bumped into extra-solar space by the push and pull of gravity, or a collision with another object. Recently the notion has arisen that planets may live up to their ancient name as 'wanderers'. Observations of the gas giants suggest they may have migrated from the region in which they formed.

COMETS, ASTEROIDS AND PLANETESIMALS

Apart from the planets, the Solar System is littered with the debris left over from its formation. This includes asteroids, comets and planetesimals. The most prominent asteroid belt lies between the orbits of Mars and Jupiter. More than 10,000 asteroids within the belt have been catalogued and many thousands more are likely to exist. The asteroids have a range of compositions and are believed to be the building blocks of a planet that was unable to accrete.

Farther out a plethora of icy bodies are found in the Edgeworth-Kuiper belt, beyond the orbit of Neptune, and the **Oort cloud**, a vast reservoir of small icy worlds which forms a sphere extending out halfway to the nearest star. These bodies represent debris which never became part of the planets and was subequently flung outward by the gravity of these newly formed worlds. The Oort cloud is considered the source of many comets that occasionally sweep in from the cold to the warmer climes of the inner Solar System, vaporizing as they move inward and forming their familiar tails of **ionized** gas and dust.

OBSERVING THE PLANETS FROM YOUR BACK YARD

Even a small telescope, when mounted on a tripod, can reveal the planets in some detail. A 75-mm (3-in) refracting telescope will easily better the view **Galileo Galilei** (1564–1642) had in 1609–10.

Through a modest telescope Mercury and Venus show phases in much the same way that the Moon does as it revolves around Earth. Both are inferior planets – they lie closer to the Sun than we do.

Mars, the first of the superior planets (one which orbits the Sun at a greater distance than the Earth) can also be a fine sight through a telescope. When suitably placed, the icy polar caps and dark surface markings of Mars are clearly visible. The dark markings are often seen to change over a period of several months, owing to vast dust storms that occasionally blow up and rage across the planet.

Perhaps the most striking planets seen through a telescope are Jupiter and Saturn. A casual glance at Jupiter will reveal a number of faint bands of cloud which girdle the planet. The clouds are stretched into belts by Jupiter's rapid rotation. The Great Red Spot, a large storm which has persisted on the planet for at least 300 years, can also be seen. Even binoculars reveal that Jupiter is accompanied by four bright points of light: the satellites Io, Europa, Ganymede and Callisto. Observing over a period of weeks will reveal how they revolve around Jupiter.

Through his telescope Galileo saw two distorted blobs at each side of Saturn that he called 'ears'. These are actually its stunning ring system, clearly visible through a modern telescope. As Saturn slowly revolves around the Sun, it passes slightly below and above Earth's orbit. In doing so it gives an opportunity to look downward and upward at the rings, hence they appear to tilt open and closed and then open again once during the planet's 29-year orbit.

OUR SOLAR SYSTEM

Uranus
2,741.3 to 3,003.6
million km

Neptune
4,444.5 to 4,545.7
million km

Pluto
4,435 to 7,304
million km

Galileo ▶ p. 40

INTRODUCTION

Our Solar System
THE SUN

The Sun is an ordinary star, one of countless millions of similar stars, situated two-thirds of the way out from the centre of a fairly ordinary spiral galaxy, which is itself just one among billions in the observable Universe. The Sun is special only because it is so close to us, a modest 150 million km (93 million miles) away. Its nearest competitor, Alpha [α] Centauri, is 275,000 times farther away – and 40 billion times fainter in our sky. The Sun is single and middle-aged, two attributes that are essential for the existence of life on Earth. If the Sun had been in a binary, which is quite common among stars, Earth's orbit would probably have been disrupted long ago. And if the Sun's energy output were significantly variable, life would not be able to adjust to the fluctuating conditions on Earth's surface.

EVOLUTIONARY STATUS AND FUTURE DEVELOPMENT

The Sun formed almost 4.6 billion years ago from a molecular cloud (see The Birth of Stars, page 158) and settled into its present configuration as a **main-sequence** star, producing energy by converting hydrogen into helium in its core. The central supply of hydrogen is expected to last for a few billion years more, but when it runs out the Sun will undergo a restructuring: the core will shrink and heat to 100 million degrees, hot enough for helium to be converted into carbon. Meanwhile, the outer layers will expand and the Sun will become a **red giant**, extinguishing all life on Earth. Ultimately, when the helium fuel is also depleted, the core will collapse further, the outer layers will be discarded as a **planetary nebula** and the remnant will become a **white dwarf**, a star bereft of energy sources, which slowly cools and fades.

SUN	
Diameter:	1,392,000 km (864,950 miles)
Oblateness:	0.00005
Axial rotation period (sidereal, at 16°):	25.38 days
Equatorial inclination (to Earths orbit):	7.25°
Mass:	$1,989,100 \times 10^{24}$ kg ($4,385,200 \times 10^{24}$ lb)
Mean density:	1.41×10^3 kg/m³ (88.0 lb/ft³)
Surface gravity (equatorial):	274.0 m/s² (898.9 ft/s²)
Escape velocity:	617.7 km/s (383.8 mi/s)
Surface temperature:	5,778 K (9,941°F)
Apparent magnitude:	+4.83
Absolute magnitude:	-26.74

Source: National Space Science Data Center, NASA

THE SUN'S CORE, ENERGY SOURCES

The composition of the Sun is typical of stars of similar age in the **Milky Way** – almost 71% hydrogen, 27% helium and the remainder mainly carbon, nitrogen, oxygen and neon, with a smattering of other chemical elements. This means that the Sun has a rich supply of hydrogen fuel in its core, the region spanning the innermost 20% of the Sun. Here, at a temperature of some 15 million°C (27 million°F) and a density of 150,000 kg/cubic metre (9400 lb/cubic ft), the hydrogen atoms are stripped of their **electrons** and the nuclei collide with sufficient force and frequency to combine in a series of nuclear-fusion reactions to produce helium. The conversion of four hydrogen nuclei into one helium nucleus destroys a small amount of **mass**, m, which is transformed into a large amount of energy, E, via **Albert Einstein**'s (1879–1955) famous equation $E = mc^2$, where c is the velocity of light. In fact, it takes the conversion of something like 4.3 million tonnes of matter per second to power the Sun.

SOLAR STRUCTURE AND ENERGY TRANSPORT

The high temperature and density at the centre of the Sun is maintained by the balance between energy generation heating the gas, so creating pressure to expand, and the forces of gravity trying to crush the outer layers in towards the centre. Meanwhile at the surface, the temperature is only 5,500°C (9,900°F) and the density less than 0.001 kg/m³ (0.00006 lb/ft³). Between the two extremes, the conditions change dramatically (although not very smoothly) but at every level throughout the Sun there exists a balance between pressure and gravity, which keeps the Sun fairly stable. The mechanisms of energy transport change throughout the Sun: for the interior 85%, energy is moved outwards by **radiation** in the form of **photons**, although these are repeatedly absorbed and re-emitted in a 'drunkard's walk' as they progress outwards – a hypothetical photon taking tens or even hundreds of thousands of years to reach the surface from the centre. For the final 15% of the way, energy is transported by convection, where currents of hot gas rise to the surface to be replaced by cooler gas descending.

▼ *Dark sunspots are the most notable features of the Sun's surface. They are constantly changing, some small ones lasting for only a few hours, others up to several weeks. Spots can appear singly or in groups which can stretch for 100,000 km (60,000 miles) or more.*

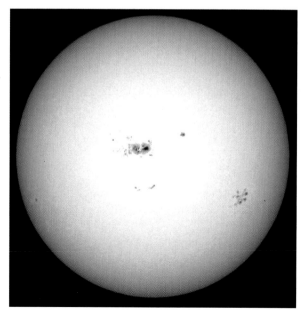

GEORGE ELLERY HALE

George Ellery Hale (1868–1938), born into a wealthy American family, developed an interest in astronomy at an early age and had his own observatory in the prosperous Kenwood suburb of Chicago by 1891. He had a particular interest in the Sun and developed the first spectrohelioscope for studying the solar surface in the light of calcium and hydrogen. In 1892 he became first director of the University of Chicago's Yerkes Observatory, when he persuaded streetcar magnate Charles Yerkes to pay for what is still the world's largest refracting telescope, with a 40-in (1-m) lens. Hale used this telescope to make observations of the Sun's **chromosphere**, but atmospheric conditions at Yerkes were not ideal for solar work and early in the new century Hale found a much better site for his studies, on Mount Wilson near Los Angeles. There he built two major solar-tower observatories and began important programs of photographic observation of the Sun. More significant for astronomy in general, he had a flair for gaining funds to build large optical telescopes and his legacy is the 100-in (2.5-m) Mount Wilson and the 200-in (5-m) Palomar telescopes. He was also a founder, in 1895, of the prestigious *Astrophysical Journal*.

THE SUN

absorption line ▶ p. 336 chromosphere ▶ p. 337 corona ▶ p. 337 cosmic rays ▶ p. 337 dark matter ▶ p. 337 electron ▶ p. 338 granulation ▶ p. 339 ion ▶ p. 340
ionization ▶ p. 340 main sequence ▶ p. 341 mass ▶ p. 341 Milky Way ▶ p. 341 neutrino ▶ p. 341 penumbra ▶ p. 342 photon ▶ p. 342

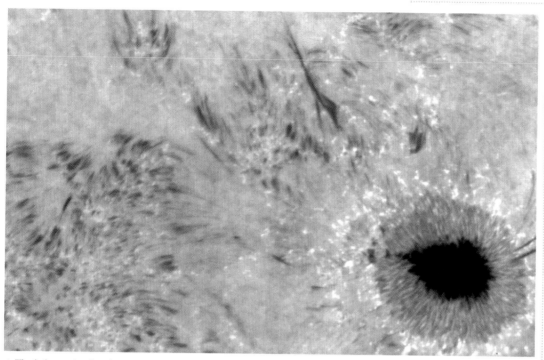

▲ *The dark central region of a sunspot that looks like a hole is known as the umbra. This area is the coolest part of the sunspot — much cooler than the rest of the photosphere — and is surrounded by a hotter outer region called the penumbra.*

THE VISIBLE SURFACE: THE PHOTOSPHERE

The visible surface of the Sun, the **photosphere**, has a temperature of about 5,500°C (9,900°F). Here the most common constituents, hydrogen and helium, are, in the main, not **ionized** although most of the heavier chemical species are at least partially so. This mixture of atoms and **ions**, plus some resilient, simple molecules, absorbs light emerging from the interior to produce a rich **absorption-line** spectrum in which the signatures of at least 65 chemical elements have been found.

High-resolution photographs of the surface reveal a mottled appearance known as **granulation**, which changes on a timescale of 10 minutes or so, caused by rising and falling bubbles of gas about 1,000 km (600 miles) across. On a larger scale supergranulation can be observed, in which

vast convective cells, perhaps 30,000 km (20,000 miles) across, can be traced. Gas wells gently up from the deep interior at the centre of the supergranule and moves almost horizontally towards the edge before descending again. The boundaries of supergranules are regions where strong magnetic fields rise up to the Sun's surface.

SUNSPOTS

Through an ordinary telescope, the most obvious features seen on the Sun's surface are dark **sunspots**, areas with a lower temperature than the rest of the photosphere, down to 3,700°C (6,700°F). Small spots, just a few hundred kilometres across, may last only a few hours, but some grow quite large, perhaps 10 times the size of Earth, and may last for months. A large spot will have a dark central region called the **umbra**, where strong magnetic fields emerge through the Sun's surface; the umbra is generally surrounded by a less cool and filamentary **penumbra**. Cool gas flows horizontally out from the umbra to the pen-

THE SOLAR NEUTRINO PROBLEM

Although the majority of the Sun's energy is produced by a particular set of reactions in which four hydrogen nuclei combine to form helium-4, there are additional but less-common reactions. In one of these, an unstable nuclide of boron, boron-8, is produced which decays to beryllium-8 with the emission of a **positron** and a high-energy **neutrino**. Neutrinos are uncharged particles, of which some 4×10^{34} of this high-energy variety are created every second. They are, however, notoriously elusive but can be detected on Earth using large enough instruments: the first one contained 400 m³ (14,000 ft³) of dry-cleaning fluid and was located underground in the Homestake gold mine, South Dakota, to avoid **cosmic-ray** contamination. Unfortunately the rate of detection is only about a quarter of that predicted, indicating a flaw either in our understanding of the Sun – for example, the exact temperature at the centre which critically affects the nuclear-reaction rates – or our knowledge of the physics of neutrinos, such as their ability to mutate into other kinds of neutrinos. If the latter is the case, then the neutrinos must have mass and they can be counted as **dark matter**, supplying at least part of the missing mass in the Universe.

umbra, to be balanced by hot gas returning to the umbra from the **corona** above. Spots are sometimes single, but there is a marked tendency for them to occur in pairs (or even straggly groups) in which the two major components are magnetically connected (see The Sunspot Cycle, page 198). By watching the movement of sunspots across the solar disk it has been shown that the Sun rotates with respect to the stars in about 25.5 days, although the rate changes with solar latitude, demonstrating that it is not a solid body.

▼ *Observing the Sun at any time, even during an eclipse, is extremely dangerous. Projecting the image of the Sun on to a piece of paper or card is the safest way of viewing.*

OBSERVING THE SUN

The Sun is so brilliant that it can damage the eye within seconds. *Never* look at the Sun directly through a telescope or binoculars, nor stare at it with the naked eye. By far the safest way to observe the Sun is simply to project its image on to a white card or sheet. When this is done, dark sunspots may be seen. Regular observation shows that they change in size and shape from day to day

and move across the Sun's face as the Sun rotates. Advanced observers may use specially made filters, consisting of a thin plastic sheet or glass with a metal coating, which reduce the Sun's light and heat to safe levels for viewing. These filters are placed over the front of the telescope tube, not at the eyepiece (where they may burn or crack). Nonetheless one must be certain that the filters are of proper specification: better safe than sorry.

THE SUN

OUR SOLAR SYSTEM

◄ Solar Eclipse
A total eclipse can only be viewed within the area covered by the umbra. The eclipse is partial in the surrounding penumbra area. An annular eclipse occurs when the Moon is near its apogee and its shadow cone does not reach Earth's surface. When this happens, a thin ring of light remains visible around the silhouette of the Moon.

SUN

MOON

UMBRA

PENUMBRA

EARTH

Total solar eclipse

Annular solar eclipse

ECLIPSES

It is a remarkable coincidence that the angular diameters of the Sun and the Moon in the sky are almost exactly the same. Hence when Earth, Moon and Sun are in perfect alignment, one of Nature's most spectacular events takes place: a total **eclipse** of the Sun, when its brilliant disk is covered by the New Moon. However, because the orbit of the Moon is slightly inclined (by 5° 9′) to that of Earth, eclipses cannot take place every New Moon, only when the Moon is close to the point where the orbits intersect (a **node**).

The eclipse begins at first contact, when the Moon first touches the Sun's western edge (or limb). Progressively, the Moon covers the Sun and the light level dims until second contact, when the eastern edge of the Moon touches the Sun's eastern limb, just before which the last bit of the photosphere shines out like a diamond ring. Briefly, small sections of the photosphere may still shine through gaps in the mountains at the lunar limb, an effect known as **Baily's beads**. Once the photosphere has been completely cut off, there is the brief opportunity to see the red chromosphere, before that too is covered and only the ghostly corona can be seen streaming out into space: totality has begun. All too soon third contact arrives and the reverse process begins, up until fourth contact when the Sun shines again in all its glory.

ABOVE THE SUN'S SURFACE

The chromosphere is a tenuous layer 2,000 km (1,200 miles) thick above the photosphere in which the minimum temperature of the Sun's atmosphere is reached, about 4,100°C (7,400°F). It comes into view during an eclipse, glowing red in the light emitted by recombining hydrogen ions. Above this minimum, the temperature starts to rise again, although the gas density continues to decline. Initially this rise is gradual, but at 2,200 km (1,400 miles) above the surface it increases rapidly through a thin 'transition region' before reaching temperatures of up to 2 million°C (3.6 million°F) in the corona. It is still something of a mystery how this high temperature is maintained. One suggestion is that energy is released when powerful magnetic fields get too close and 'short circuit'. Another possibility is that so-called magneto-acoustic waves can carry energy up from far below the photosphere by tapping the power of the convective motion of the gas; this gas is ionized and therefore strongly bound to the magnetic field. Any wave motion in the gas causes the field to 'twang' like a guitar string, and this 'twang' is felt out in the corona where it vibrates and heats the rarefied gas.

At total solar eclipses the pearly white streamers of the corona can be seen extending several solar radii out into space, although their pattern depends upon the stage of the Sun's 11-year cycle. Close to the maximum of the sunspot cycle, the streamers tend to flow out from the equatorial regions, following complex magnetic loops, while where the lines of magnetic force extend straight out into interplanetary space (known as coronal holes), material flows away from the Sun at a few hundred kilometres per second to produce the **solar wind**; this is estimated to carry off 50 million million tonnes of material per year. At sunspot minimum, the corona is more evenly distributed around the limb.

THE SUNSPOT CYCLE

The systematic study of sunspots, with measurements of their size, number and position, was begun in 1853 by the English astronomer Richard Carrington (1826–75) and continued by the Royal Observatory at Greenwich. It was already known that the number of spots waxed and waned over an 11-year period. A cycle starts with a few spots at mid **latitudes** above and below the Sun's equator. As it progresses, so the number and size of the spots increases and they appear nearer the equator (although the region within two or three degrees of the equator is generally free of spots). At maximum, several groups can often be seen together and the larger ones may persist for a few complete rotations. Eventually the number declines to just a few above and below the equator before the new cycle begins with high-latitude spots.

Spots behave as though they are magnets, with one pole descending into the Sun and the other protruding from the surface. They generally occur in pairs with, for example, those in the northern hemisphere during one cycle having the leading spot showing a north magnetic polarity and the trailing spot a south polarity. In the southern hemisphere, the situation would be reversed with the south-polarity spots leading.

E. W. MAUNDER

Edward Walter Maunder (1851–1928) began his astronomical career as an assistant at the Royal Observatory, Greenwich, in 1873, where he started work on solar photography, recording the positions and sizes of sunspots. He was involved in several eclipse expeditions around the turn of the century, but also made time to found the British Astronomical Association in 1890, play a major role in the Royal Astronomical Society and to edit *The Observatory* magazine. However, he is mainly remembered for his researches on sunspots and in particular for two important results. He created the 'Butterfly Diagram', which depicts how the latitude and size of sunspots changes during the sunspot cycle (a phenomenon recorded by the German astronomer Gustav Spörer, 1822–95). This led to an interest in comparing different sunspot cycles and to the realization, first noted by Spörer, that historical records indicated long periods when there seemed to be essentially no sunspots at all. In their honour we now recognize the **Maunder minimum** (1645–1715) and the Spörer minimum (1450–1540), confirmed by radiocarbon dating of tree rings as being times of low solar activity. The Maunder and Spörer minima also coincide with prolonged cold spells, hinting at the Sun's influence on Earth's climate.

OUR SOLAR SYSTEM

THE SUN

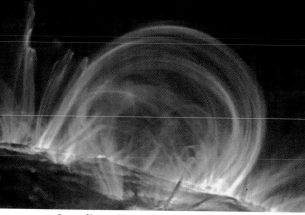

▲ **Loops of hot gas** *follow the Sun's magnetic field lines in the inner corona, as seen at extreme ultraviolet wavelengths by the TRACE spacecraft. These loops, which originate in active regions on the photosphere, are over 10 times larger than Earth.*

In fact, the true solar cycle is more like 22 years because in the next 11-year period everything is turned around, and in the northern hemisphere the south-polarity spots would be leading. The sunspot cycle is thought to be caused by a winding-up and strengthening of the Sun's sub-photospheric magnetic field to the point where it breaks out at the surface to produce active regions and spots; the field can remove energy from the gas in the spots and thus keep them cooler than the surrounding photosphere.

THE ACTIVE SUN

The sunspot cycle controls not only the appearance of spots on the photosphere, but also a host of other activity on and above the Sun's surface. Bright regions called faculae frequently appear before sunspots emerge; these are hotter and denser than surrounding material. When seen in hydrogen or calcium light, they are termed plages. Observations at hydrogen and calcium **wavelengths** can also reveal filaments of gas extending for 100,000 km (60,000 miles), hovering some 40,000 km (25,000 miles) above the photosphere; when these long sheets of material appear at the limb of the Sun, they emerge as **prominences**, showing a range of structures, especially loops and arches. Some quiescent prominences may persist for months, while active ones last only a short time and can sometimes be seen rising away from the Sun's surface. Among the more violent events associated with active regions, where the magnetic field is twisted and complex, are flares, in which a huge amount of energy is released – the equivalent of 10 billion one-megaton bombs in a large one – probably by reconnection of magnetic field lines, in a short time (typically minutes to hours). Very occasionally these

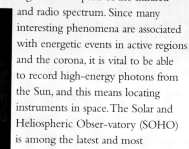

▶ **The Sun** *seen at extreme ultraviolet wavelengths, showing features in the solar corona at a temperature of about one million degrees, recorded by the Solar and Heliospheric Observatory (SOHO).*

can be seen in white light but they also give rise to **X-rays** and **ultraviolet** emission, together with the ejection of energetic particles. Passing through the corona, these create shock waves and produce strong bursts of radio emission.

Yet another manifestation of solar activity is a coronal mass ejection, where perhaps 10 billion tonnes of hot gas are ejected into interplanetary space, pulling with it the coronal magnetic field. Their origin is still uncertain. Earth is not isolated from these violent events – the storms of energetic particles shot out by the Sun often wash across the near-Earth environment, affecting power lines and communications, generating auroral displays in the ionosphere around the north and south magnetic poles and damaging sensitive equipment on spacecraft.

▲ **One of the most** *awesome celestial events we can witness from Earth is a total solar eclipse, when the Moon passes in front of the Sun. The period of totality, when the Moon completely obscures the Sun, lasts only a few minutes.*

FORTHCOMING ECLIPSES OF THE SUN (2001–2010)

Year	Date	Type	Duration	Area of Visibility
2001	21 June	Total	4:57	E South America, South Atlantic Ocean, Africa except N, Madagascar
2001	14 Dec	Annular	3:53	Hawaii, extreme S Alaska, SW Canada, W USA, Mexico, Central America, Caribbean Sea
2002	10–11 June	Annular	0:23	SE and extreme NE Asia, Philippines, N Pacific Ocean, North America except N and E
2002	4 Dec	Total	2:04	S Africa, Madagascar, Australia except N and E
2003	31 May	Annular	3:37	Iceland, N Greenland, Arctic regions, N Europe, N Asia, Alaska, N Canada
2003	23–24 Nov	Total	1:57	Antarctica, S Australasia, extreme S South America
2005	8 Apr	Annular/Total	0:42	Southern Pacific Ocean, part of Antarctica, S United States, Central America, extreme west S America
2005	3 Oct	Annular	4:31	Africa except S tip, Europe, W Asia, Arabia, W India
2006	29 Mar	Total	4:07	Africa except SE, Europe, W Asia, W India
2006	22 Sep	Annular	7:09	E South America, Atlantic Ocean, part of Antarctica, W and S Africa, S Madagascar
2008	7 Feb	Annular	2:08	Antarctica, SE Australia, New Zealand
2008	1 Aug	Total	2:27	Greenland, Iceland, Arctic Ocean, N Europe, Asia except extreme E and SE, Arabia, India
2009	26 Jan	Annular	7:54	S Africa, Madagascar, part of Antarctica, Indian Ocean, S India, SE Asia, W Indonesia, S and W Australia
2009	22 Jul	Total	6:39	China, Mongolia, SE Asia, Japan, Indonesia, W Pacific Ocean
2010	15 Jan	Annular	11:08	E Africa, Madagascar, Arabia, Indian Ocean, S central Asia, India, SE Asia
2010	11 Jul	Total	5:20	Pacific Ocean, extreme W central South America

Source: H. M. Nautical Almanac Office.

VIEWING THE SUN FROM SPACE

From the surface of Earth, observations of the Sun are restricted by the filtering effect of Earth's atmosphere, which confines observations to the visible region together with parts of the infrared and radio spectrum. Since many interesting phenomena are associated with energetic events in active regions and the corona, it is vital to be able to record high-energy photons from the Sun, and this means locating instruments in space. The Solar and Heliospheric Obser-vatory (SOHO) is among the latest and most sophisticated satellites to be involved in Sun watching. Launched in 1995 as a joint ESA-NASA project to take up a carefully balanced position between the Sun and Earth (one of the so-called **Lagrangian points**), it carries a large number of instruments to monitor many aspects of solar behaviour: among them, spectrometers image the Sun at selected wavelengths to reveal coronal holes and bright spots; an **interferometer** makes accurate observations of the motion of the solar surface, highlighting oscillations which can be interpreted like seismic measurements on Earth to shed light on the interior structure of the Sun; and particle detectors record the flow of the solar wind and the more dramatic impact of coronal mass ejections.

OUR SOLAR SYSTEM

THE SUN

Our Solar System
TERRESTRIAL PLANETS
INTRODUCTION

The four planets that orbit closest to the Sun are Mercury, Venus, Earth and Mars. They are all small, rocky worlds and are grouped together as the terrestrial planets, or inner planets. Their location near the Sun is responsible for their bulk similarities which contrast with those of the much larger gaseous worlds beyond the asteroid belt. However, closer inspection by both ground-based and spacecraft instruments reveals that each of the terrestrial worlds has its own persona.

▲ *The four terrestrial planets: Mercury, Venus, Earth (shown with its only natural satellite, the Moon) and Mars. These are also known as the 'inner planets' due to their proximity to the Sun.*

THE EVOLUTION OF WORLDS

Our **Solar System** started life as a large irregular cloud of gas and dust called a **nebula**. It was from this large cloud of gas and dust that the Sun and the planets formed. Under the influence of gravity, material migrated inward adding to the mass of the embryonic Sun. Eventually the Sun grew massive enough for nuclear fusion to start within its core. At this point the Sun began to shine and gave rise to the **solar wind**, a stream of charged particles which blasted lighter gas and dust out of the inner Solar System. Through untold collisions, the denser material left behind finally amalgamated to form the inner planets.

While the inner terrestrial planets were still hot, their interiors differentiated, or separated into layers of different density. The densest material such as iron sank inwards, while lighter elements migrated upwards to form lighter silicate rocks. Differentiation explains why the terrestrial planets

have denser iron-rich cores. A huge amount of heat energy was generated during differentiation, resulting in the continued global melting of most of the planets. With time the surfaces of the terrestrial planets cooled and solidified forming a crust. The newly formed crust was subjected to continuous bombardment from the remaining **planetesimals**. These impactors frequently cracked the crust and gouged out huge basins, allowing hot molten magma to escape in violent volcanic eruptions, creating large volcanic plains. Evidence for these volcanic plains exist today as the maria seen on the Moon, and the heavily cratered volcanic plains on Mercury and Mars.

THE LATE HEAVY BOMBARDMENT

The oldest surfaces in the inner Solar System, those of Mars, Mercury and the Moon, provide a record of a cataclysmic period called the Late Heavy Bombardment. The Late Heavy Bombardment occurred until approximately 3.9 billion years ago. Up until then debris left over from the formation of the planets was still abundant and frequently collided with the young terrestrial worlds, forming the countless craters preserved today. Between 3.9 and 3.4 billion years ago the number of impacts declined greatly. It is between these dates that the earliest known fossils are found on Earth, indicating that as soon as the period of heavy bombardment ceased life was able to gain a foothold.

ROCKY INTERIORS

Although we are unable to see directly into the interiors of the terrestrial planets, a number of observations give clues as to what lies beneath their crusts. These clues come from the degree and style of surface volcanism, the gravitational tug on spacecraft which pass close to them, and, in the case of Earth and the Moon, measuring shock waves that travel through the interior using seismometers. Information about Earth's interior is most complete because we have better access to study it. An iron core, a putty-like mantle and a thin solid crust are the basic elements that compose Earth's interior. The same layers are also believed to exist, with varying thickness, within the other terrestrial planets.

The Mariner 10 spacecraft, which made three close flybys of Mercury in 1974–75, measured a considerable gravitational tug from this tiny world. The inference was that Mercury has a large dense

▶ *The Martian crater known as the 'Happy Face'. The surface of Mars is characterized by such features: craters, mountains, volcanoes and channels. Although there is no evidence of liquid water at the moment, the channels suggest that it may once have existed on the Martian surface.*

core made of iron. The radius of the core is approximately 75% of the total radius of the planet. Venus's interior is thought to be similar to Earth's because of its comparable diameter, density and **mass**. Mars, half as large as Earth and only 11% its mass, has a smaller core. Only when seismometers can be placed on the surfaces of the terrestrial planets will their true interior structures be understood.

TERRESTRIAL TOPOGRAPHY

The topography, or surface features, of the terrestrial planets varies considerably. The surface of Mercury is similar to that of the Moon. Without any substantial atmosphere, both worlds have been pounded by space debris and are covered with craters. Mercury lacks large mountain ranges and is dominated by cliffs, scarps and large impact basins.

▲ *Because of its constant cloud cover the surface of Venus has proved difficult to investigate. Images taken by the Russian Venera landers revealed a harsh, rocky landscape with volcanic plains made of basaltic rocks, similar to those found under the oceans on Earth.*

The surface of Venus can only be imaged by radar, as a permanent cloud deck shrouds its surface. Radar images of Venus returned by Russian and American spacecraft reveal that forces of volcanism and crustal warping have sculpted its surface. Volcanic plains and large shield-style volcanoes hundreds of kilometres across are ubiquitous. Large plateaux and mountain belts are also common, along with coronae, volcanic features characterized

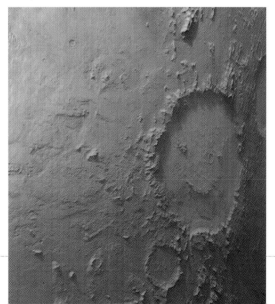

TERRESTRIAL PLANETS

gamma ray ▶ p. 339   mass ▶ p. 341   nebula ▶ p. 341   planetesimal ▶ p. 342
Solar System ▶ p. 343   solar wind ▶ p. 343   ultraviolet ▶ p. 344   X-ray ▶ p. 344

OUR SOLAR SYSTEM

by circular ridges and troughs. Smaller volcanoes tens of kilometres across pepper the surface.

The surface of Mars can be divided into two distinct regions: the northern plains and southern highlands. The southern highlands are more heavily cratered and hence older than the northern plains. A third terrain is the Tharsis volcanic rise which contains four large volcanoes. Mars, along with Earth, has polar ice caps that shrink and grow according to seasonal changes. The caps are composed of frozen carbon dioxide and water.

Earth has the most dynamic surface and hence the youngest, with oceans of water, active volcanoes and the process of plate tectonics – the constant recycling of its surface. The process of plate tectonics explains why its surface has very few preserved craters. Natural and man-made processes are continually reworking Earth's surface environment.

THE ATMOSPHERES OF THE TERRESTRIAL PLANETS

It is in the atmospheres of the terrestrial planets where the greatest divergences appear. Owing to its small size and hence weak gravitational attraction, Mercury has been unable to hold on to any primordial atmosphere it may have once had. Being geologically dead, no active volcanoes exist to replenish it. Today Mercury has a negligible atmosphere. Traces of hydrogen, helium, oxygen, sodium, potassium and argon, probably released from **ultraviolet** radiation hitting the rocky surface, are present. The lack of a substantial

atmosphere gives rise to extreme temperature differences between day and night.

Venus has the thickest and deadliest atmosphere of the terrestrial planets. The atmosphere is composed mainly of carbon dioxide, a gas considerably heavier than nitrogen and oxygen, resulting in the pressure at the surface being 90 times greater than the pressure at sea level on Earth. The thick carbon dioxide atmosphere is also responsible for a strong greenhouse effect, resulting in the stifling global surface temperature of above 400°C (750°F).

Mars also has an atmosphere composed mainly of carbon dioxide, but on the Red Planet the atmosphere is much thinner. At around one-and-a-half times further from the Sun than Earth, Mars has a cold environment that allows the formation of water-ice clouds which float through the atmosphere, transporting water between the North and South Poles. Occasionally global storms develop, whipping up fine surface dust and obscuring the surface for months.

Earth's unique atmosphere consists of mainly nitrogen and oxygen. Different layers of the atmosphere protect life from deadly radiation from the Sun, such as ultraviolet, **X-rays** and **gamma rays**. On a global scale, a moderate greenhouse effect helps warm the Earth's surface. Atmospheric circulation generates climatic patterns, whilst day-to-day weather is a product of the movement of warm and cold air masses and the precipitation (rain, hail, snow and mist) these parcels of air bring.

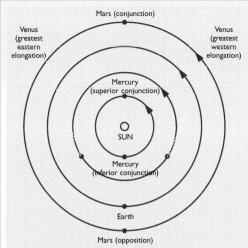

▲ *Elongation, Conjunction and Opposition*
The ease with which the terrestrial planets can be seen depends on their positions in relation to the Sun and Earth. The angle between the Sun and a planet, as seen from Earth, is known as elongation. At conjunction, this angle is around zero and the planet is lost in the Sun's glare. A planet is in opposition when it is directly opposite the Sun.

⬗ OPPOSITION, CONJUNCTION AND ELONGATION

As the planets travel in their orbits, they reach points that are more advantageous than others for earthly observers. Mercury and Venus circle the Sun inside Earth's orbit and are sometimes referred to as morning or evening 'stars' because they never stray very far from the Sun in the sky.

The best time to view Mercury and Venus is when they reach their maximum angular separation from the Sun as seen from Earth, called greatest elongation. At greatest eastern elongation, the planets move as far east in their orbit as seen from Earth and are usually well placed for viewing in the evening sky. At greatest western elongation, Venus and Mercury appear in the morning sky.

Mars, the only terrestrial planet that lies farther from the Sun than Earth, can appear in the sky all night long. Mars and the other outer planets are best seen when they reach opposition. During opposition, Mars sits opposite the Sun in our sky and shines brightest because it lies closest to us. Mars rises as the Sun sets and then sets at dawn, allowing for optimum viewing.

When two celestial objects appear close together in the sky they are in conjunction. A planet in conjunction with the Sun makes for poor viewing. The two inner planets are said to be at superior conjunction when they are on the far side of the Sun, and hence at their most distant from us; in the case of the outer planets this position is simply termed conjunction. The two inner planets can also reach inferior conjunction, when they lie closest to us but are between the Sun and us.

✳ QUESTIONING THE DYNAMO MODEL

The longstanding theory to explain how planets create and sustain magnetic fields is known as the dynamo model. The generation of a magnetic field requires rotation, interior convection, and a fluid, conducting core. To create a self-sustaining magnetic field, the molten, fluid portion of a planet must rotate differentially, with the interior spinning fastest. A magnetic field penetrating this fluid will coil up and strengthen. Rotation and convection carry the field loops upwards, twisting them more and continuously regenerating the field.

For Earth and most of the outer planets, this dynamo model works sufficiently. But upon analyzing the other terrestrial planets, complications with the dynamo model arise.

Mercury is a small planet and therefore loses heat relatively rapidly. Its core should be relatively cold and have undergone a greater degree of

solidification internally. Mercury's slow spin rate coupled with an inactive interior should generate almost no magnetic field. Yet Mercury has a magnetic field, although weak.

Venus's slow rotation rate is likely to be responsible for its deviance from the dynamo model. However, according to the model, the planet should still have a magnetic field (although extremely weak), and no magnetic field at all has been detected on Venus. It is possible that we are viewing Venus in the middle of a magnetic field reversal, when the strength would drop essentially to zero.

Mars is known to currently have a very weak magnetic field. However, in September 1997, Mars Global Surveyor entered orbit around Mars and found isolated fingerprints of an older, stronger magnetic field. These magnetic pockets are preserved in rock which solidified under the influence of a Martian magnetic field much stronger than today's.

MERCURY

Closest of all the planets to the Sun, Mercury swings around its orbit faster than any other, once every 88 days. This diminutive world, under half the diameter of Earth, is the second-smallest planet in the Solar System. It is an airless, waterless ball, with a rocky, cratered surface resembling the Moon in appearance. Being so close to the Sun, Mercury is difficult to see from Earth. Only part of its surface has so far been surveyed by space probes, so our knowledge of it remains sketchy.

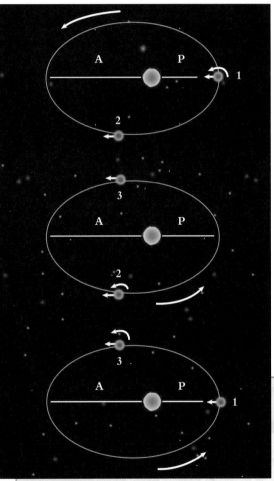

Mercury's Rotation
A = Apehelion, P = Perihelion
At position 1, an imaginary arrow on Mercury's surface points to the Sun; for an observer at this location it would be noon. At position 2, 58.6 days later, Mercury has rotated once on its axis but since it has gone two-thirds of the way around its orbit, the arrow does not point at the Sun. At 3, 58.6 days on, Mercury has spun on its axis for a second time, traversing another two-thirds of an orbit. When Mercury passed through perihelion this time, the arrow pointed directly away from the Sun and it was midnight. Finally, after a third complete spin on its axis and another two-thirds of an orbit, Mercury returns to position 1 and the arrow is back facing the Sun. In all, 176 days have passed from one noon to the next.

SEEING MERCURY

Elusive Mercury can be spotted in the evening or morning sky on only a handful of occasions each year, shining shyly near the horizon for no more than two to three weeks at a time before vanishing again into the twilight. The best times to see Mercury are in the evening sky in the early spring, when it is setting after the Sun, and in the morning sky in the early autumn, when it rises before the Sun. Even then, a pair of binoculars will probably be necessary to pick it out against the twilight glow, looking like a bright, slightly orange star.

So difficult is Mercury to observe through even the largest telescopes that no reliable maps of its surface features existed until the American space probe Mariner 10, which made three flybys in 1974–75. Not surprisingly for a body merely 40% larger than our own Moon, Mercury turned out to look very lunar-like, its surface gouged by craters of all sizes. These craters are the legacy of an intense bombardment by interplanetary debris early in the history of the Solar System – around four billion years ago – which also left its mark on the other planets and their moons. Mercury was open to such strikes because it has no atmosphere to speak of, just a tenuous haze of gas, some of it temporarily captured from the Sun and the rest exuded from the surface rocks, with a surface pressure around a thousand million-millionth that of Earth's atmosphere.

MERCURY'S SURFACE

Mariner 10, on a carefully planned trajectory around the Sun, encountered Mercury three times at six-monthly intervals between March 1974 and March 1975, photographing about half the planet's surface and revealing details down to 100 m (330 ft) across. Much of Mercury's surface in the Mariner 10 photographs could easily be

▲ *The Caloris Basin is the largest formation on the surface of Mercury. This massive lowland plain, partly visible at lower left of the picture, was created by an asteroid collision, and measures 1,300 km (800 miles) across.*

mistaken for the cratered highlands of the Moon. However, there are subtle differences, a result of Mercury's stronger gravity: secondary craters, formed by debris from the main impacts, are found closer to the parent crater than they are on the Moon and crater walls are not as high.

Between the large craters lie smoother areas called intercrater plains, the origin of which remains uncertain. They may be sheets of debris thrown out by vast impacts early in Mercury's history, or they may result from widespread volcanism while the crater-forming bombardment was underway. One type of surface feature unique to Mercury is meandering faults called lobate scarps, several hundred kilometres long and a kilometre or so high,

MERCURY'S ODD ROTATION

A planet's rotation rate is usually regarded as its 'day', but this is misleading in cases such as Mercury where the rotation is a large fraction of the **orbital period**. By the time Mercury has rotated once on its axis, it has moved so far round its orbit that the Sun is now in a completely different direction. As the diagrams show, for the Sun to go once round Mercury's sky (say, from one noon to the next) the planet must rotate three times on its axis, during which time it orbits the Sun twice – a total of 176 Earth days. This arrangement, technically termed **spin-orbit**

coupling, has come about as a result of the Sun's tidal forces which gradually slowed Mercury's rate of spin. Mercury has no moon to lighten the long, dark nights.

Unable to see any definite surface markings on Mercury from which to deduce its rotation period, astronomers had to wait until 1965 before they learned how long Mercury takes to turn on its axis. Radio pulses, bounced off the planet from radio telescopes on Earth, were slightly 'smeared' in frequency by the planet's rotation. The amount of smear in the returning echoes revealed that the planet's rotation period was nearly 59 days, two-thirds of the time it takes to orbit the Sun.

EINSTEIN AND MERCURY

Astronomers in the nineteenth century were faced with a puzzle: the long axis of Mercury's orbit was turning in space slightly more than could be accounted for by the gravitational tugs of the other planets. The amount was only 43 seconds of arc per century, but it was enough to suggest that **Isaac Newton**'s (1642–1727) theory of gravity could be wrong. At first the gravitational effect of an unknown inner planet was suspected. This planet was even granted a name, Vulcan, and some astronomers mistakenly claimed to have seen it. The real answer was provided in 1915 by **Einstein**'s general theory of relativity, evidently resulting from a slight contraction in the planet's size as it cooled.

Mercury's largest formation is a circular lowland 1,300 km (800 miles) wide, fully one-quarter the planet's diameter. This is named the Caloris Basin (from the Latin *calor*, meaning 'heat') because it faces the Sun at Mercury's perihelion. Excavated by an immense meteorite smash, the basin then filled with lava which wrinkled in places as it solidified. On the opposite side of Mercury from the Caloris Basin, shock waves from the hefty impact converged and broke up the surface into blocky, hilly regions that geologists have informally termed 'weird terrain'.

Overall, Mercury's surface rocks are dark grey, reflecting a mere 11% of the sunlight hitting them but – as on the Moon – young craters are surrounded by brighter splashes of pulverized and ejected rock. Mercury's craters are named after artists, musicians and writers. Individuals as diverse as Beethoven, Dickens, van Gogh and Mark Twain are commemorated there.

HOT AND COLD

Mercury has a highly elliptical **orbit** that takes it between 46 and 70 million km (29 and 43 million miles) from the Sun. Seen from Mercury's surface, the Sun appears from two to three times larger than it does from Earth, depending whether Mercury is at its farthest (**aphelion**) or closest (**perihelion**). At perihelion Mercury's daytime surface temperature can exceed 400°C (752°F) on the equator – hot enough to melt tin and lead, were any to be found there.

At successive perihelia, first one side of the planet and then the other is presented to the Sun. The two points on the equator that face the Sun at perihelion, receiving the most intense solar heating, are sometimes termed the 'hot poles'. Yet, without an atmosphere to distribute heat around

which describes the gravitational field around a body as being a curvature of space, rather like a funnel. Travelling on its elliptical path, Mercury passes through different curvatures of space as it alternately dips further in and out of the Sun's gravitational funnel. This causes its orbit to turn by the small additional amount that is observed. The successful explanation of the turning of Mercury's orbit (technically termed the advance of perihelion) was an early confirmation of general relativity. Since then, similar effects have been found for the orbits of Venus and Earth, but the advance is much smaller because their orbits are less elliptical and further from the Sun.

MERCURY

Diameter:	4,879 km (3,032 miles)
Oblateness:	0
Axial rotation period	
(sidereal):	58.65 days
Equatorial inclination	
(to orbit):	0.01°
Mass:	0.33×10^{24} kg (0.73×10^{24} lb)
Mean density:	5.43×10^3 kg/m³ (339.0 lb/ft³)
Surface gravity	
(equatorial):	3.70 m/s² (12.14 ft/s²)
Escape velocity:	4.30 km/s (2.76 mi/s)
Average distance from	
Sun:	57.91×10^6 km (35.98×10^6 miles)
Maximum distance	
from Sun:	46.00×10^6 km (28.58×10^6 miles)
Minimum distance	
from Sun:	69.82×10^6 km (43.38×10^6 miles)
Orbital eccentricity:	0.206
Orbital period (sidereal):	87.969 days
Orbital inclination	
(to ecliptic):	7.00°
Number of moons:	0

Source: National Space Science Data Center, NASA.

▼ *Occasionally, Mercury crosses in front of the Sun, where it can be seen in silhouette as a black dot, an event termed a transit. This series of images shows Mercury near the edge of the Sun during the transit of 15 November 1999, as seen by a Sun-watching spacecraft called TRACE. Other transits of Mercury will take place on 7 May 2003, 8 November 2006 and 9 May 2016.*

the planet, temperatures on Mercury's night side drop below -180°C (-292°F). Near the north and south geographical poles there may be permanently shaded regions where temperatures would remain sub-zero so, surprisingly, Sun-drenched Mercury may have polar caps composed of ice accumulated from cometary impacts.

PAST AND FUTURE

Clues to the nature of Mercury's interior, and also its evolution, come from its unusually high density, second only to that of Earth, due to a large iron core that accounts for two-thirds of its mass. Motions of liquid iron in this core give rise to a weak magnetic field, about 1% the strength of Earth's. Why should Mercury, a dwarf among planets, possess such a disproportionately large iron core? At birth, Mercury may have been twice its present size, but suffered a hit-and-run accident with a stray body of similar size to our Moon which blasted off most of its less dense, outer rocky layers. This collision, if indeed it happened, could also have been responsible for knocking Mercury's orbit into its current elliptical shape.

There is much for future space probes to find out about Mercury. The US space agency NASA intends to launch a probe called Messenger (short for Mercury Surface, Space Environment, Geochemistry and Ranging) in 2004, flying past Mercury twice before going into orbit around it in 2009. Another orbiting probe, which would map the entire planet in greater detail and analyze its surface composition, as well as dropping a lander near one of the poles, is being planned by ESA for launch in 2009 or later.

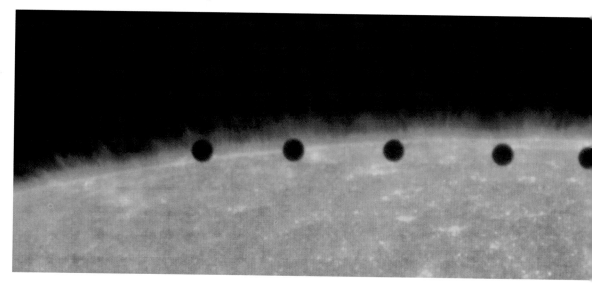

Einstein ▶ p. 88 Newton ▶ p. 42

TERRESTRIAL PLANETS

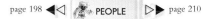
VENUS

The brightest wanderer in the skies is named Venus, after the Roman goddess of love. The second planet from the Sun and closest to Earth, Venus has a mass and diameter similar to our own planet, as well as a substantial cloudy atmosphere, earning it the reputation of Earth's twin. These thick clouds constantly hide its surface and only recently have spacecraft shown that it differs markedly from our own world. Venus differs from the other planets in that it exhibits retrograde rotation, meaning it spins from east to west. It rotates once every 243 Earth days. This slow, back-to-front spin may be the result of a large impact early in Venus's history.

SEEING VENUS

Venus lies closer to the Sun than Earth, and so never strays far from the Sun in the sky. Best seen shortly before sunrise or after sunset it is commonly called the Morning or Evening 'star'. Its brilliance is a result of highly reflective clouds (65% of sunlight which strikes the clouds is reflected away) and its close proximity to Earth. During extremely favourable conditions, Venus has been known to cast shadows and air traffic controllers have even tried to clear the planet for landings after seeing it appear near the horizon.

Because it is located inside Earth's orbit (an inferior planet), Venus has phases like the Moon. These phases are apparent through binoculars or a small telescope. Venus's brightness varies depending on how far it is from Earth and what phase it shows. It is closest to Earth and is at its brightest when it presents a crescent phase. We are unable to see Venus during its fully lit phase, however, because this occurs when the planet is positioned on the far side of the Sun.

VENUS	
Diameter:	12,104 km (7,521 miles)
Oblateness:	0
Axial rotation period (sidereal):	243.02 days
Equatorial inclination (to orbit):	177.36°
Mass:	4.87×10^{24} kg (10.7×10^{24} lb)
Mean density:	5.24×10^{3} kg/m³ (327.1 lb/ft³)
Surface gravity (equatorial):	8.87 m/s² (29.10 ft/s²)
Escape velocity:	10.36 km/s (6.44 mi/s)
Average distance from Sun:	108.21×10^{6} km (67.24×10^{6} miles)
Maximum distance from Sun:	108.94×10^{6} km (67.69×10^{6} miles)
Minimum distance from Sun:	107.48×10^{6} km (66.78×10^{6} miles)
Orbital eccentricity:	0.007
Orbital period (sidereal):	224.701 days
Orbital inclination (to ecliptic):	3.39°
Number of moons:	0

Source: National Space Science Data Center, NASA.

EARTH'S TWIN

Physically Venus is very much Earth's twin. With a diameter only 5% smaller than Earth's, and Venus's similarities in density and mass, it is thought the compositions of the planets' interiors are alike. Both planets have rocky crusts and probably contain metallic cores and thick mantles. Lastly, Venus was formed in the same region of the Solar System as Earth and is its nearest planetary neighbour. But these are where the similarities end. Owing to its dense carbon-dioxide atmosphere, the surface environment is

▲ *Venus, also known as the Evening or Morning Star, is one of the most easily visible objects to the naked eye in the sky. Its brightness is due to the highly reflective clouds that enshroud the planet.*

anything but Earth-like. Venus has no oceans, it never rains and the atmosphere is very dry. The great greenhouse effect means surface temperatures are much hotter than Earth.

THE SURFACE OF VENUS

We owe a great deal of our understanding of the Venusian surface to the flotilla of American and Russian spacecraft which used radar to cut through the cloud decks, revealing the surface below, and others which have landed on the surface. In particular the Magellan spacecraft orbited Venus and mapped 98% of the surface between 1990 and 1994, revolutionizing our understanding of the planet.

Underneath the clouds lies a tortured landscape, preserving the scars from geological processes which have run rife for hundreds of millions of years. Peppered with volcanoes of all sizes, cut by a maze of faults and ridges, and plastered with sheet-like lava flows hundreds of kilometres wide, the geological history of Venus is extremely complex.

Globally, Venus differs from Earth in that the majority of its surface is fairly uniform in height. Earth can be divided into ocean basins (low land) and continents (higher land). In contrast, 80% of the Venusian surface has a range in altitude of no more than 1,000 m (330 ft), making it much smoother than Earth. Venus also lacks plate tectonics – the constant movement and recycling of large crustal regions common on Earth.

Although heights are less extreme, mountains and valleys do exist on Venus,

▶ *Volcanoes such as Maat Mons are ubiquitous on the surface of Venus and are an important way in which the planet loses internal heat. This image was computer-generated from radar data returned by the Magellan orbiter probe and its vertical scale has been exaggerated 10 times.*

CARL SAGAN

Carl Sagan (1934–96) was a legend in his own time. At the time of his death, aged 62, he had written more than 10 well-received astronomy books. Half a billion people worldwide saw his television show *Cosmos*, the largest audience ever for a public television series. The companion book of the same name was the highest-selling English-language science book ever.

Sagan's scientific career began in the 1950s when he was a graduate student at the University of Chicago. His doctoral thesis included a section on 'The Radiation Balance on Venus'. The great minds of the time, such as Harold Urey (1893–1981), Fred Whipple (b. 1906), and **Fred Hoyle** (b. 1915), all disagreed about what existed under Venus's thick cloak of clouds. The confusion only increased when observations told of a blisteringly high temperature. Sagan's thesis, building on the work of those before him, pointed to a greenhouse effect as the logical culprit.

In 1962 Mariner 2 flew past Venus and observed limb darkening, indicating that the high temperature was a result of a thick atmosphere and hence a hot surface, which vindicated young Sagan's thesis. With these results, his credibility as a scientist was earned and his climb to popularity began.

OUR SOLAR SYSTEM

testament to huge forces resulting from the shifting of molten material beneath the crust. The most complex deformed terrain on Venus are tesserae (Gk: 'tiles'), high-standing plateaux characterized by a lattice of faults and fractures. Tesserae record many phases of warping and cracking. The exact formation of tesserae is unknown. Some believe it is where Venusian crust is dragged and rucked up like a carpet, caused by molten rock sinking below.

Over 900 impact craters exist on Venus, compared with fewer than 200 on Earth. Almost all the craters appear fresh; very few are cut by faults or flooded by lava. The number and preservation of Venusian craters gives an indication to the average age of the surface, which is estimated to fall between 200 and 600 million years, much younger than the Moon, but slightly older than Earth's.

VOLCANISM

The surface of Venus is dominated by volcanic landforms. More than 80% of the planet is covered with undulating lava plains. Tens of thousands of small volcanoes a few kilometres across litter the surface. Larger shield-style volcanoes, similar to the Hawaiian Island volcanoes on Earth, have produced aprons of complex lava flows hundreds of kilometres wide. In general, Venusian volcanoes are lower and much wider than those on Earth, a result of the crushing atmosphere and weaker crust.

Coronae (Latin: 'crown') are large circular volcanic structures typically hundreds of kilometres across with collapsed centres. They appear unique to Venus. Each corona marks a blister on the crust where heat and molten rock have welled up from below. Since Venus lacks plate tectonics, coronae, along with the hundreds of large volcanoes, are thought to have been important in allowing the planet's internal heat to escape.

✳ THE GREENHOUSE EFFECT

On no planet is the greenhouse effect more evident than on Venus. Venus's constantly hot temperature results from its atmospheric composition. The thick clouds on Venus let about 3% of sunlight penetrate to the surface. Approximately two-thirds reflects off the clouds and the remainder is absorbed. So how does Venus sustain its sweltering temperature of more than 460°C (870°F)?

The answer is a strong greenhouse effect. Sunlight which does reach the surface of Venus is absorbed and then re-emitted as **radiation** with a

A GUIDE TO THE INSIDE

Venus and Earth formed in the same general location of the Solar System and display similar physical characteristics. Because of this, and because the surface of Venus has landforms formed by similar processes to those on Earth, most planetary scientists believe the two worlds have comparable materials in their interiors. Venus has a rocky crust made of basalts similar to those found under the oceans on Earth, as sampled by the Russian Venera landers. Venus also probably has a thick mantle and a molten metallic core.

But if Venus's interior is similar to Earth's, it should have a magnetic field about as strong as Earth's. Missions to Venus have found only an extremely weak magnetic field. Part of the reason for this lack may be the planet's slow rotation.

Examining the surface features should reveal more about the contents and motions beneath. The lower incidence of craters would suggest that Venus may periodically experience large-scale resurfacing, which would indicate interior processes different from those on Earth. The gravity field on Venus corresponds well with surface topography, indicating that the method by which the crust is supported is unlike Earth's.

THE VENUSIAN ATMOSPHERE

The same clouds responsible for Venus's brilliance totally obscure its surface. When **Galileo** first turned a telescope towards Venus in 1610, he saw a featureless globe. Today, even modern telescopes show only faint irregular cloud patterns and can never glimpse the surface below.

The Venusian atmosphere is composed of 96.5% carbon dioxide. The Russian Venera and US Pioneer probes discovered three principal cloud decks lying between approximately 48 and 60 km (30 and 37 miles) above the surface, much higher than clouds on Earth. The clouds revolve

longer **wavelength**. The thick carbon-dioxide atmosphere is more effective at trapping the longer wavelength radiation, which is absorbed by the ground again. Over time, the heat has built up on Venus resulting in the stifling temperatures recorded today. If Venus's atmosphere were much thinner, the surface would be substantially cooler.

Earth, too, experiences a degree of greenhouse warming, resulting in a global surface temperature about 33°C (91°F) greater than if it had no atmosphere at all. If Earth's atmosphere were filled with more greenhouse gases such as carbon dioxide, our planet's atmosphere could evolve into a runaway greenhouse like on Venus.

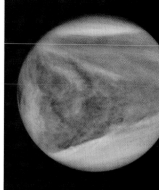

▶ *This view of Venus, taken from orbit by the Pioneer Venus orbiter, shows the swirling mass of thick cloud that obscures its surface.*

around Venus at breakneck speeds, carried by winds up to 350 km/h (220 mi/h), driven by energy from the Sun. The clouds are made from tiny droplets of sulphuric acid making them as corrosive as an acid bath.

Although the temperature at the top of the atmosphere measures a chilly -45°C (-49°F), carbon dioxide is a potent greenhouse gas and the temperature rises steadily toward the surface. On the ground, the true nature of our sister world is revealed. Temperatures soar to over 460°C (870°F), nearly twice as hot as a conventional oven and hot enough to melt lead. Furthermore, the atmospheric pressure at the surface of the planet reaches a crushing 90 times that experienced at sea level on Earth. Far from being Earth-like, the surface of Venus is one of the most hostile known.

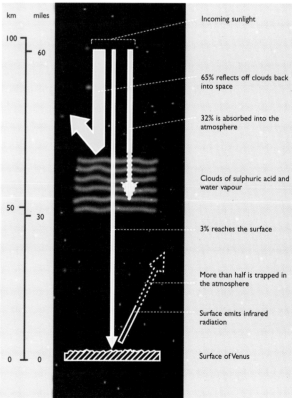

km	miles	
100	60	Incoming sunlight
		65% reflects off clouds back into space
		32% is absorbed into the atmosphere
		Clouds of sulphuric acid and water vapour
50	30	
		3% reaches the surface
		More than half is trapped in the atmosphere
		Surface emits infrared radiation
0	0	Surface of Venus

▲ *The Greenhouse Effect on Venus*
Although only 3% of the Sun's visible light reaches the surface of Venus, this is enough to cause the surface to emit heat as infrared radiation. More than half of this radiation is trapped, causing a greenhouse effect. As most of the atmosphere on Venus is carbon dioxide, this effect is pronounced and results in an average temperature of 464°C.

EARTH

Earth, our home planet, is unique because it is the only place in the Solar System where life is known to exist. It is the largest of the terrestrial planets and lies third from the Sun at a distance of 150 million km (93 million miles). Earth is a dynamic place with active volcanoes and areas of mountain building. The process of plate tectonics constantly recycles the crust of the Earth. Beneath the thin crust lie the mantle, outer core and inner core. Earth's location in the inner Solar System means it has the ideal temperature for large oceans of liquid water which cover two thirds of its surface, and a moist atmosphere which supports and protects its delicate ecosystems.

▲ *Our home planet* Earth with its only natural satellite the Moon. *Although Earth shares some characteristics with the other terrestrial or inner planets, it is truly unique in that it is the only place that we know of that supports life.*

THE SURFACE OF EARTH – PLATE TECTONICS

Earth's crust is divided into large slabs called plates, of which there are two principal types: oceanic and continental. The plates are supported by the denser mantle beneath. The point at which two or more plates meet is called a plate boundary. Here, they may undergo two principal processes. Plates can slide past each other, creating a fault zone such as the San Andreas fault line, where the Pacific

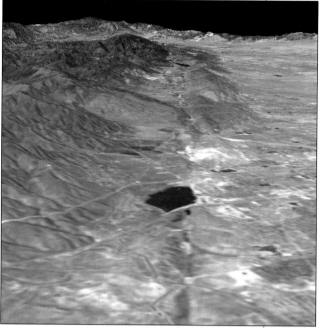

▲ *The crust of the Earth* is divided into plates. This radar image, taken from *the Space Shuttle, shows the San Andreas fault, the meeting point between the Pacific and North American plates – an area of high seismic activity.*

plate slides past the North American plate. Alternatively one plate may descend under another, creating a subduction zone.

A subduction zone is also known as a destructive plate margin, since the plate which is sinking is destroyed in the process. A classic example of a subduction zone is where the small oceanic Juan de Fuca plate (off the coast of western America) is subducting under the larger continental North American plate. Plate boundaries are typically marked by strings of active volcanoes, mountains and earthquake activity. The volcanoes are fed by magma formed when crust melts as it descends into the hotter mantle.

When two plates converge with little or no subduction, significant mountain building often occurs, as one plate buckles against the other. The greatest mountain belt on Earth, the Himalayas, is the result of India crunching into Asia. The release of pent-up energy as two plates grind past each other gives rise to earthquakes.

The opposite of a destructive plate margin is a constructive plate margin. Here crust is formed rather than destroyed. The crust of the oceans is formed by the eruption of lava along mid-oceanic ridges, such as the Mid-Atlantic Oceanic ridge. Molten rock (magma) ascends from the mantle below and is erupted on to the sea floor along a line of submarine volcanoes. The new crust is carried away from the mid-oceanic ridges towards the continents where it will eventually be subducted and melted. The continual recycling of

oceanic crust means no part of the ocean floor is older than approximately 200 million years.

The process of plate tectonics is driven by heat, supplied by convection in the mantle. This great crustal recycling process allows heat to escape from Earth's interior through volcanism, and more importantly, the heat is used to melt subducting crust. It also explains why the majority of the crust on Earth is very young. Other bodies in the Solar System which are covered in craters, such as the Moon and Mercury, have crusts which are much older and have not been affected by plate tectonics for much of their geological history, if at all. These bodies have been geologically dead for a long time.

ROTATION AND REVOLUTION

Rotation denotes the spin of a planet on its axis, creating day and night, and revolution is its orbital motion around the Sun. Earth revolves around the Sun once in roughly 365 days, a period known as a year, moving at a speed of about 108,000 km/h (67,100 mi/h). Earth turns once on its axis, at a speed of 1,600 km/h (994 mi/h) at the **equator**, completing one rotation relative to the Sun every 24 hours.

Earth's rotational axis is not perpendicular to the plane of its orbit but tipped by about 23.5°. As with other planets in the Solar System, it is the tilt of Earth's axis that is responsible for the seasons. When the Earth reaches the point in its orbit where the north axis points towards the Sun, the northern hemisphere experiences summer, and has the greatest number of daylight hours. In the northern hemisphere the longest day of the year occurs around 21 June. This represents the position in Earth's orbit where the northern hemisphere is oriented directly towards the Sun and is known as the summer **solstice**.

At the opposite point in Earth's orbit, six months later, the winter solstice occurs in the northern hemisphere, when the North Pole leans away from the Sun. During the winter solstice the northern hemisphere experiences the fewest daylight hours. Halfway in Earth's orbit between the summer solstice and the winter solstice, Earth reaches the autumnal **equinox**. Between the winter and summer solstice is the vernal or spring equinox. On the equinoxes, the length of day and night is roughly equal at every point on Earth.

OUR SOLAR SYSTEM

CONTINENTAL DRIFT

The seven continents seen today on Earth (Asia, Africa, Antarctica, Australia, Europe, and North and South America) have not always existed as separate entities. The landmasses once formed two large continents: Gondwanaland in the southern hemisphere, and Laurasia in the northern hemisphere. The idea of the continents as having been unified at one point and then separated later in history is called continental drift. The process of plate tectonics is responsible for continental drift.

A number of clues were crucial in formulating the theory of continental drift. First, geologists had long known that similar rock formations and fossils exist in Africa and South America. The puzzle was how could this be when such a large ocean separates the two landmasses. A quick look at a map also reveals how the coastlines of the African and South American continents seem to fit together, as if they were once joined but broke apart in the past. Could this have been possible?

The second line of evidence came from magnetic studies of lavas erupted at mid-oceanic ridges under the oceans. Newly formed lava erupted on to the seafloor contains iron-rich minerals which are susceptible to Earth's magnetic field. While still partially molten, the iron-rich minerals act like iron filings and become preferentially aligned in the direction of the field. Once the rocks have solidified, the direction of the magnetic field is frozen into them.

By studying Earth's magnetic field preserved within sea floor rocks – a technique called palaeomagnetism – scientists uncovered a remarkable pattern. Distinctive stripes on either side of oceanic ridges were discovered which had the same magnetic fingerprint; their palaeomagnetism was the same. Furthermore, stripes either side of the oceanic ridges with the same palaeomagnetism were also the same age. The inference was that crust preserving the same magnetic fingerprint was formed at the oceanic ridges at the same time and carried away by plate movements. This was irrefutable evidence of continental drift and led to the conclusion that Africa and South America were indeed once joined.

EARTH

Diameter (equatorial): .	12,756 km (7,926 miles)
Oblateness:	0.0034
Axial rotation period (sidereal):	23.934 hrs
Equatorial inclination (to orbit):	23.44°
Mass:	5.9736×10^{24} kg (13.170×10^{24} lb)
Mean density:	5.515×10^3 kg/m^3 (344.3 lb/ft^3)
Surface gravity (equatorial):	9.78 m/s^2 (32.09 ft/s^2)
Escape velocity:	11.19 km/s (6.95 mi/s)
Average distance from Sun:	149.60×10^6 km (92.96×10^6 miles)
Maximum distance from Sun:	152.10×10^6 km (94.51×10^6 miles)
Minimum distance from Sun:	147.09×10^6 km (91.40×10^6 miles)
Orbital eccentricity: . . .	0.0167
Orbital period (sidereal):	365.256 days
Orbital inclination (to ecliptic):	0.00°
Number of moons:	1

Source: National Space Science Data Center, NASA.

JOURNEY TO THE CENTRE OF EARTH

Perhaps one of the least-known regions of Earth is its interior. Understanding what lies beneath our feet and the processes acting there is important since they are responsible for shaping the surface of Earth. By studying shock waves (seismic waves) from earthquakes which propagate through Earth, scientists can create a picture of the interior of our planet. Earthquake waves travel through materials of different density at different velocities. Recording the velocities of earthquake waves, and noting the position where

◀ *An image* taken from space of the Kliuchevskaya volcano in Russia, situated on the Kamchatka peninsula bordering the Pacific Ocean. This volcano is a result of ocean crust which is subducting and melting.

▲ *The enchanting sight* of Earth from space. Its isolation in the vast dark expanse forced us to reassess our importance in the Universe and made us realize how unique and precious our planet is.

their velocity changes, has revealed distinct layers and boundaries in Earth's interior. These layers formed early on in Earth's history when the planet underwent differentiation. While still molten, materials of different densities naturally separated into layers. As the interior formed, the least dense materials rose to the surface forming the crust, while the densest materials sank towards the centre.

Seismic wave studies have identified a number of different layers within Earth, including the mantle, outer core and inner core. The densest material is found in the core. Earth's core can be divided into two parts: a solid inner core of iron and an outer core of molten iron. The radius of the entire core is about 3,500 km (2,200 miles). Above the core sits the mantle, 2,900 km (1,800 miles) thick, and consisting of iron and magnesium rocks. The mantle, which made up the bulk of Earth's interior, is not totally molten; it acts more like putty in that it can easily be deformed if a pressure is applied.

On top of the mantle is the crust, made of less dense granitic and basaltic rocks. The crust can be divided into several large plates which ride on the denser mantle. The thickness of the crust varies from 15 km (9 miles) under the oceans to 40 km (25 miles) under the continents.

OUR SOLAR SYSTEM

HOW OLD IS EARTH?

The age of Earth has been dated using techniques which measure the amount by which radioactive minerals have decayed with time. Measurements of radioactive decay put the oldest rocks on Earth at just under four billion years. The dates have been obtained from metamorphic rocks. Metamorphic rocks are those formed from precursor rocks which are baked under very high temperatures and pressures, deep within Earth. In finding the Earth's age, the precursor of the metamorphic rocks must be taken into consideration. However, since these no longer exist in their unaltered form, we must look elsewhere for clues.

The dating of other Solar System material, such as **meteorites** and the Moon, provides an age of 4.55 billion years. Hence the modern consensus is that the Solar System as a whole formed about 4.55 billion years ago. Earth is likely to have started forming at this time as well. Current theories suggest that it continued to grow through the bombardment of planetesimals for some 120 to 150 million years. At that time, 4.44 to 4.41 billion years ago, Earth began to retain its atmosphere and form its core. The rocks that formed at this stage are likely to be those which were later baked to form the oldest metamorphic rocks preserved today.

EARTH'S ATMOSPHERE

Life on Earth would not be possible without its blanket of air that provides warmth, protection, and the oxygen we need to breathe. The atmosphere consists of 77% nitrogen, 21% oxygen, and 2% other gases, including water vapour. Air pressure and density decrease with height. This is apparent in the thin air on mountain tops, where it is more difficult to catch one's breath. Very high up, the atmosphere becomes so rarefied that gas particles collide less often and are able to escape into space.

The atmosphere has several distinct layers. From Earth's surface to a height of approximately 12 km (7 miles) is the troposphere where 75% of atmospheric gases exist. Weather is contained here and the temperature drops with increasing altitude. The stratosphere continues to 50 km (31 miles) and contains the ozone layer, which absorbs most of the harmful ultraviolet radiation from the Sun. The mesosphere, from 50 to 80 km (31 to 50 miles), is the coldest layer. Up to about 1,000 km (600 miles) is the thermosphere. This layer includes both the ionosphere, used to reflect radio waves, and the exosphere. Beyond this lies the magnetosphere, a magnetic bubble around the Earth that traps solar particles, creating the **Van Allen belts**.

ATMOSPHERIC CIRCULATION

Heat from sunlight and Earth's rotation combine forces to affect atmospheric circulation. Through the process of convection, heated air rises while cooler, denser air sinks to replace it. Locally, convection in the atmosphere produces fluffy cumulus clouds and stormy thunderheads. On a global scale, the more direct sunlight that falls on equatorial regions creates hot air, while farther north and south, polar air undergoes less heating and so is colder and denser. Therefore, an atmospheric flow exists between the equator and the poles. The otherwise simple flow of air masses is complicated by Earth's rotation. Spinning Earth creates spiral motions in the atmosphere, anticlockwise in the northern hemisphere and clockwise in the southern hemisphere. Large convection cells, known as Hadley cells, are created by this airflow.

OCEANS, CLIMATE AND WEATHER

Earth's oceans are very efficient at storing energy received from the Sun. They may be thought of as huge storage heaters which regulate Earth's climate and day-to-day weather. The oceans act as a buffer when storing solar energy, that is, they take time to warm up and cool off. Those who live near the coast will know that the sea has a warming effect during the winter, often delaying the onset of winter frosts. Likewise during the summer a cool sea breeze keeps temperatures lower than those experienced inland.

The same process occurs on a larger scale. Large continental landmasses experience a greater diversity of temperatures than a small landmass surrounded by the sea. For example, the British Isles has a much smaller temperature range over the course of a year than central Australia or Canada. The British Isles is said to have a maritime climate rather than a continental climate.

THE OZONE LAYER AND HOLE

About 30 km (19 miles) above the surface of Earth, potentially deadly ultraviolet radiation strikes oxygen in the stratosphere and creates O_3, or ozone. The ozone layer absorbs the harmful ultraviolet rays protecting life from ailments ranging from sunburn to skin cancer. Ironically, the same ozone that is beneficial to life higher up in the atmosphere is detrimental when found down in the troposphere. Near the surface, ozone, or smog, is formed when sunlight breaks down gases emitted from petrol-driven vehicles.

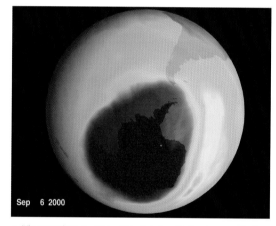

Sep 6 2000

▲ *The ozone layer* lies 30 km (19 miles) above Earth and prevents deadly ultraviolet rays entering our atmosphere. This image shows the continent of Antarctica and the South Pole; the blue area shows the extent of the hole in the layer caused by human activity – a cause of rising concern.

Concern is growing over human activities which alter the amount of ozone in the stratosphere. Since the 1950s, measurements over Antarctica have shown a dramatic decrease in ozone. Chlorofluorocarbons (CFCs) particles found in aerosols, refrigerants and other materials, are the leading pollutants resulting in the depletion of ozone. CFCs can remain in the stratosphere for over 100 years. Ultraviolet rays break up CFCs releasing chlorine. Chlorine breaks down ozone creating oxygen and chlorine monoxide, which then combines with free atomic oxygen to create oxygen and chlorine, thus leaving the chlorine free to repeat the process. One chlorine atom destroys about 100,000 ozone molecules before it combines with nitrogen dioxide.

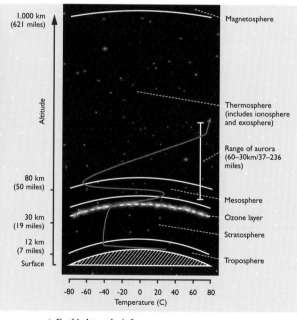

▲ *Earth's Atmospheric Layers*
The temperature of Earth's atmosphere varies greatly with increasing altitude. These changes are shown by the curved orange line as it passes through the principal layers of the atmosphere.

▲ *When the Sun sits on the horizon, its light passes through more of the Earth's atmosphere and the shorter-wavelength light – blue – is scattered. Red light, of longer wavelength, is less affected by scattering so the Sun appears reddened.*

BLUE SKIES, RED SUNSETS

Sunlight entering Earth's atmosphere is scattered by air molecules. Since the shorter-wavelength blue light is scattered more than the longer-wavelength red light, the sky appears blue. At sunset and sunrise, when the Sun is near the horizon, the light from the Sun must pass through more atmosphere. In doing so, blue light undergoes far more scattering so that less reaches the ground. However, red light is relatively unaffected and passes through the atmosphere unhindered. This results in more red light reaching the ground than blue and so creating crimson sunsets and sunrises.

Not all skies are blue. The sky on Mars is coloured salmon-pink owing to the great amount of fine dust suspended in the atmosphere. On the surface of Venus the sky appears a dark yellow-orange colour, as this is the only wavelength of light which can penetrate its thick atmosphere.

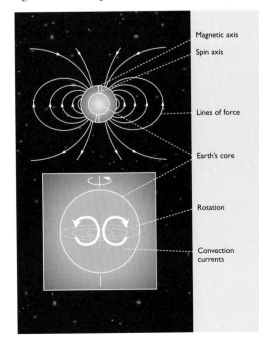

Magnetic axis

Spin axis

Lines of force

Earth's core

Rotation

Convection currents

◀ *Earth's Magnetic Field*
Earth's rotation combines with convection currents to drive fluid motions in the molten core. These motions generate the Earth's magnetic field, whose lines of force extend beyond the surface out into space. The magnetic axis is tilted around 12° from the spin axis of the planet.

MAGNETIC FIELD

Earth's magnetic field can be compared to an immense bar magnet. The north and south magnetic poles protrude from Earth 12° askew from the corresponding north and south spin axes. This configuration is known as a dipole field. Therefore, Earth's magnetic field lines, or lines of force, flow out from the south magnetic pole and loop inwards at the north magnetic pole. Earth's magnetic field will exert a force on any electrically charged particle moving through it. Particles are re-directed from their original paths and spiral along the field lines. The extent of the Earth's magnetic field is called the magnetosphere.

Earth's magnetic field is not constant in strength or direction. Over geological time, the magnetic field occasionally reverses polarity, so the North and South Poles swap over. During a reversal, the magnetic field's strength drops. Measurements in rock show that Earth's poles have changed polarity 171 times in the last 71 million years. Earth's magnetic field is thought to arise by the motion of fluids at the core-mantle boundary.

THE MAGNETOSPHERE AND AURORA DISPLAYS

The highest layer of Earth's atmosphere, the magnetosphere, buffers Earth against the solar wind, a stream of charged particles which radiate away from the Sun. When the solar wind encounters the magnetosphere, the magnetosphere deforms into a shape similar to a teardrop. The pressure of the solar wind compresses the sunward side of the magnetosphere, while the opposite side is drawn outwards, forming the 'magnetotail'.

Sometimes the Sun releases a violent outburst of solar wind, sending a greater number of particles streaming towards the magnetosphere. The increased number of charged particles can give rise to spectacular **aurora** displays. Aurorae occur when **ionized** particles become trapped within the Van Allen radiation belts, two distinct regions within the magnetosphere. The trapped particles spiral along the magnetic field lines within the Van Allen belts, which emerge from Earth at the north and south geomagnetic poles. The particles cascade down into Earth's upper atmosphere, exciting gases and causing them to emit visible radiation in the same way **electrons** strike a television screen and cause it to glow.

The excited gases in the atmosphere glow to produce beautiful green and red aurora displays. Most activity appears in ovals around the poles. In the north, the display is called the Aurora Borealis, and, in the south, the Aurora Australis. The northern and southern lights are sometimes visible at lower latitudes when a particularly strong gust of solar wind encounters the magnetosphere. This phenomenon is more common around the times of peak activity in the solar cycle, every 11 years.

▼ *Aurorae are caused by charged particles ejected from the Sun which become trapped in the polar regions of Earth's magnetic field. The charged particles energize gases in the Earth's upper atmosphere, causing them to glow like a neon sign.*

OUR SOLAR SYSTEM

THE MOON

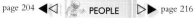

T he Moon is our only natural satellite and has inspired a sense of wonder through the ages, with untold numbers of legends associated with it spanning most cultures. A casual glance at the full Moon reveals distinct surface markings: the light-coloured highlands and the dark lowlands. Through binoculars or a telescope the lunar surface is a fascinating sight, littered with ancient impact craters and mountain ranges. With a diameter of about one-quarter of Earth, the Moon is large compared to the planet it orbits – the Earth-Moon system is often referred to as a double planet. The Moon has a number of important effects for us on Earth, including eclipses and tides.

PHASES OF THE MOON

Everyone is familiar with the phases of the Moon. The Moon shines by reflected sunlight and as it orbits Earth once every four weeks, it progresses through a cycle of phases. The new Moon phase occurs when the Moon lies between the Sun and Earth and hence its unlit side faces us.

▲ **Like Earth,** one half of the Moon is illuminated by sunlight while the other half remains in darkness. The different shapes or 'phases' of the Moon result simply from the fraction of the sunlit half that we see as the Moon orbits Earth.

During the next seven days, as the Moon moves in its orbit, an increasing amount of the illuminated side of the Moon becomes visible to us on Earth. The Moon is said to be waxing or growing larger, and the phase is called a waxing crescent, seen hanging low in the evening sky shortly after sunset. A half Moon occurs seven days after new Moon. This phase is called first quarter Moon, and occurs when the angle between the Sun, Moon and Earth is 90°.

During the ensuing week, 7 to 14 days after new Moon, more of the illuminated side becomes visible. A three-quarters illuminated Moon is called a **gibbous** Moon, and, because it is still growing, this phase is called waxing gibbous. Fourteen days after new Moon, the Moon has moved half way around its orbit and now lies on the opposite side of Earth from the Sun. The phase seen is the full Moon, and the Moon rises as the Sun sets.

Over the next two weeks the amount illuminated as seen from Earth diminishes: the Moon is said to be waning. Gradually, the Moon cycles through a waning gibbous, waning first quarter, waning crescent and back to new Moon, having completed a full orbit of Earth.

SEEING THE MOON

The Moon appears large and bright in our sky only because it is close to us. With the unaided eye, two principal landforms can be seen; dark patches which are the maria, and the brighter highlands called terra. The familiar 'Man in the Moon' is a result of the pattern of dark maria regions on the lunar surface. Other cultures see different creatures in the Moon – for example, a rabbit is common with Asian societies.

Through a pair of binoculars or a small telescope, the Moon is a fascinating object to study, revealing myriads of craters and mountain ranges, as well as smooth lava flows covering the lowland plains. The best time to pick out craters and mountain ranges is when they lie close to the terminator – the line which divides day and night. Close to the terminator the Sun is either

▲ **A close-up of the Moon** in its crescent phase. Being so close to us, the Moon is ideal as a first target for the amateur astronomer – a pair of binoculars will reveal its most prominent features.

rising or setting, and casts long shadows. The low-angled illumination helps throw the topography of the lunar surface into stark relief.

Occasionally, the Moon will pass in front of a bright star or planet. This is called an **occultation**. Because the Moon has a negligible atmosphere, stars passing behind it will wink out instantaneously when they encounter the limb of the Moon, while planets take longer to disappear. Saturn is particularly interesting to watch undergo an occultation. Its rings will first glide behind the lunar limb, followed by the orb of the planet and then the rings on the other side. The whole event can be watched again when the planet reappears on the other side.

▼ **Samples of rock and soil** collected by the astronauts who landed on the lunar surface yielded information not only about the Moon's composition but also about its evolution and history.

MEN WHO WALKED ON THE MOON

Mission	Landing Date	Astronauts who Walked on the Moon
Apollo 11	21 July 1969	Neil Armstrong, Edwin (Buzz) Aldrin
Apollo 12	24 November 1969	Charles (Pete) Conrad, Jr., Alan L. Bean
Apollo 14	9 February 1971	Alan B. Shepard, Jr., Edgar D. Mitchell
Apollo 15	7 August 1971	David R. Scott, James B. Irwin
Apollo 16	27 April 1972	John W. Young, Charles M. Duke, Jr.
Apollo 17	19 December 1972	Eugene A. Cernan, Harrison H. Schmitt

THE SURFACE OF THE MOON

The Moon's surface consists of mountainous highlands (called terrae) and smoother lowlands (called maria). Since the maria reflect a low percentage of light which strikes them (they are said to have a low **albedo**) they appear dark grey from Earth and were mistaken by early astronomers as seas, hence their Latin name *maria* or 'sea'. The maria actually represent lowlands flooded with basaltic lava, and many of them formed when the lunar surface was fractured by large **asteroid** impacts relatively early in the Moon's history.

Craters are most common on the lunar highlands, resulting from the relentless

---

✳ THE GIANT IMPACT MODEL OF FORMATION

A number of models have been proposed to explain the Moon's origin. These include the fission, capture and binary **accretion** models. The fission model claims that when Earth was young and molten it span fast enough to throw a mass of material into space to form the Moon. The capture model claims that the Moon did not form in the same vicinity as Earth but wandered close enough to Earth to be caught in its gravity field. The binary accretion model proposes that the Moon formed after Earth in a leftover **nebula** of debris. Yet none of these account for all the characteristics known about the Moon.

The giant impact theory, developed in the 1960s, does however manage to explain most of the Moon's present characteristics, including the chemical differences and similarities between Earth and the Moon. The theory suggests a Mars-sized body made a sideways strike on the young Earth, after Earth had already formed its iron core. Outer layers of Earth and fragments from the impactor escaped into space and accreted to form the Moon. Evidence for the impact has come from the fact that the Moon and Earth contain similar materials, which suggests a similar origin.

Two lunar missions during the 1990s, the Clementine and Lunar Prospector orbiters, have greatly helped our understanding of the chemical and physical make-up of the Moon. Lunar Prospector added more support to the giant impact model by revealing previously unknown information about the Moon's core. It found that the lunar core was small, making up less than 4% of the Moon's total mass (Earth's core is 30% of its total mass). The radius of the lunar core is somewhere between 225 and 450 km (140 and 280 miles), in agreement with the model's prediction.

---

bombardment of space debris billions of years ago, and astronomers have catalogued a bewildering number of crater shapes and sizes. From ancient ghostly remnant craters, almost obliterated from the surface, to youthful-looking ray craters which retain beautiful bright streaks radiating outwards hundreds of kilometres, each impact has a unique story to tell.

Craters are termed Complex craters if they have flat floors, terraced walls and central peaks – huge mountains of material thrown upward during the last stages of the crater's formation. They are typically tens to a few hundred kilometres in diameter. Two beautiful examples of Complex craters are Tycho and Copernicus, each about 90 km (55 miles) in diameter (located at 43.4°S, 11.1°W and 09.7°N, 20.1°W respectively). Tycho has a distinctive pattern of bright rays composed of debris sprayed thousands of miles across the lunar surface. Rays are typical of young craters. The bright ejecta has not had time to darken by the constant milling action of microscopic dust and **cosmic rays** which rain down onto the lunar surface.

Craters larger than about 320 km (200 miles) across are called impact basins. These lack central mountainous peaks and form multiple rings as they increase in size. Multi-ringed basins are the largest impact structures on the Moon. Most are over 800 km (500 miles) in diameter and result from truly cataclysmic collisions. They are typically flooded by smooth basaltic lava which either welled up through fractures in the lunar crust, or formed from melting of the crust when a vast amount of heat was released as the asteroid-sized impactor struck. The rings themselves are thought to be similar to ripples formed on a pond, but frozen into the lunar crust.

▲ *A total lunar eclipse can occur only when the Moon is full and Earth is lying directly between the Moon and the Sun. Although of limited scientific importance, lunar eclipses can make fascinating viewing.*

NEAR SIDE VERSUS FAR SIDE

Not until 1959, when the Russian spacecraft Luna 3 looped behind the Moon, were we able to see the 'dark' side of the Moon. Because of synchronous rotation, the Moon always keeps its same face towards Earth. Of course there is no real 'dark' side of the Moon – apart from deep polar craters, all areas of the Moon experience two weeks of day followed by a two-week night as the Moon orbits Earth. The side of the Moon not seen from Earth is correctly referred to as the far side.

The near and far side of the Moon differ considerably in appearance. The large, dark maria which are common on the near side are lacking on the far side. This is because the crust here is thicker and so is harder to fracture and it is more difficult for magma to escape to the surface. The far side is a nearly continuous stretch of craters of all sizes. Maria which are found on the far side include Mare Orientale, which peeks over on to the near side on the southwest limb of the Moon as seen from Earth, and Mare Smythii and Mare Marginis, both on the eastern limb.

🪐 LUNAR ECLIPSES 2001–10

Date	Type	Mideclipse	Duration	Location
5 July 2001	Partial	14:55	160 min	E Africa, Asia, Aus, Pacific
16 May 2003	Total	03:40	194 min	C Pacific, Americas, Europe, Africa
9 Nov 2003	Total	01:18	212 min	Americas, Europe, Africa, C Asia
4 May 2004	Total	20:30	204 min	S America, Europe, Africa, Asia, Aus
28 Oct 2004	Total	03:04	220 min	Americas, Europe, Africa, C Asia
17 Oct 2005	Partial	12:03	58 min	Asia, Aus, Pacific, N America
7 Sept 2006	Partial	18:51	92 min	Europe, Africa, Asia, Aus
3 March 2007	Total	23:21	222 min	Americas, Europe, Africa, Asia
28 Aug 2007	Total	10:37	212 min	E Asia, Aus, Pacific, Americas
21 Feb 2008	Total	03:26	206 min	C Pacific, Americas, Europe, Africa
16 Aug 2008	Partial	21:10	188 min	S America, Europe, Africa, Asia, Aus
31 Dec 2009	Partial	19:22	62 min	Europe, Africa, Asia, Aus
26 June 2010	Partial	11:38	164 min	E Asia, Aus, Pacific, . Americas
21 Dec 2010	Total	08:17	210 min	E Asia, Aus, Pacific, Americas, Europe

Eclipse Predictions by Fred Espenak, NASA/GSFC.

OUR SOLAR SYSTEM

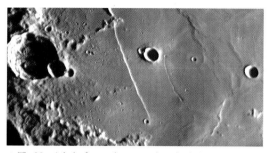

▲ *The Moon's lack of atmosphere* means that it has had no protection from impacts over the aeons. The result is a craggy, scarred surface featuring craters of all sizes and rugged mountain ranges, thrown into stark relief by low sunlight.

SURFACE COMPOSITION

The surface of the Moon is covered with an ancient soil called the lunar regolith. The regolith consists of rock fragments and finer particles which have been ground down through billions of years of bombardment by micrometeorites. Round, glassy spherules are also common within the regolith. These drops are the remains of impacts that melted the surface rocks, and materials flung from large volcanic fire fountains when the Moon was still volcanically active.

Moon rocks can be divided into three main categories: anorthosites, basalts and breccias. The anorthosites are common in the hilly regions and are more numerous than the other types; they are pale, igneous rocks containing aluminium and calcium. Basalts are dark, dense rocks containing iron, titanium and magnesium. The basalts formed when lavas erupted on to the lunar surface and cooled between 3.9 and 3.0 billion years ago. These rocks form the maria regions, the dark grey patches on the Moon seen from Earth. The last type, called breccias, are rocks reconstituted from fragments of anorthosites and basalts,

▼ *Although the* Moon has no water on its surface, tentative evidence suggests that water ice may exist in the deep, permanently shadowed craters at the lunar poles (shown here in blue).

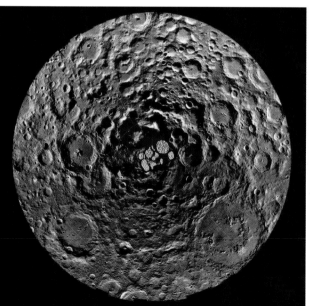

shattered during violent impacts and cemented together by later impacts.

Unlike terrestrial rocks, rocks on the Moon contain no water. However, Lunar Prospector found tentative evidence for water-ice mixed in with the lunar regolith at edges of steep crater walls at the poles. The water is likely to have come from comet impacts. At the end of its mission Lunar Prospector was de-orbited into one of these crater walls in the hope that a plume would arise that could be analyzed for evidence of water, but no plume was detected.

EVOLUTION OF THE MOON

Lunar soil and rock samples returned to Earth by Apollo astronauts and Russian landers have contributed to our understanding of the Moon's evolution. The most likely explanation is that the Moon formed from the debris created when a Mars-sized object impacted Earth, about 50 million years after the formation of Earth. Accretion added material to the Moon starting around 4.5 billion years ago. After a surface formed, leftover debris continued to impact the Moon, heating and melting the crust.

Slowly, the impacts ebbed and the crust had a chance to cool and solidify approximately 4.3 billion years ago. Debris that impacted the surface after this time formed the many craters seen today on the surface. The interior was still hot and molten, and some of the larger impacts fractured the lunar crust and allowed magma to flow outwards, creating the dark maria and lowland basins. The volcanism which formed the maria lasted between 3.9 and 3.0 billion years ago and ended when Moon's interior cooled to such a degree that the crust became thick and im-penetrable, trapping the magma permanently below the surface.

Today, the Moon is considered a dead world. The surface is modified only by the countless micrometeorites and cosmic rays that rain down upon it. Since the Moon has practically no atmosphere, the constant spatter continues to erode the surface and builds up a powdery surface. The footprints left by the Apollo astronauts will eventually fade owing to these minute impacts – but only after millions of years.

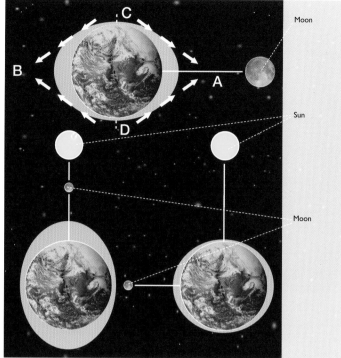

▲ *Tidal Effects on Earth*
1. The gravitational pull of the Moon causes the oceans to bulge at points A and B, and to be reduced at points C and D. 2. When the Moon and the Sun are in line, their gravitational effects combine to produce greater tidal ranges ('spring tides'). 3. When the Moon and the Sun are at right angles, their influences partly cancel each other out, producing smaller tidal ranges ('neap tides').

HOW EARTH AFFECTS THE MOON

Earth and the Moon affect each other due to their similarities in size and physical proximity. As with many satellites in the Solar System, the Moon is locked into a rotation pattern in which its orbital period is the same as the time it takes to turn once on its axis. This relation is called 'synchronous' or 'captured rotation'. Synchronous rotation explains why we always see the same side of the Moon from Earth.

The same gravitational tug-of-war is also slowing the rotation of Earth. Eventually the length of the day on Earth will match the length of day on the Moon. When this happens the Moon will appear to hang permanently in the same part of the sky. The effect has already occurred with Pluto and Charon, the only other double planet in the Solar System. Another result of this orbital dance is that the Moon is very slowly moving away from the Earth. A small amount of gravitational energy is transferred from Earth to the Moon, causing it to recede from Earth by almost 4 cm (1.5 in) a year. As the Moon moves further away so it gets smaller in our sky. An unfortunate consequence of this is that future inhabitants will be unable to see a total solar eclipse because the size of the Moon will no longer appear large enough to cover the Sun.

ecliptic ▶ p. 338 penumbra ▶ p. 342 umbra ▶ p. 344

▲ *Craters such as this*, known as Ptolemaeus, are a defining feature of the lunar surface, formed by the numerous impacts of other bodies on the Moon over billions of years. The size and shape of such craters are defined by the speed and size of the impacting object.

HOW THE MOON AFFECTS EARTH

As people who live by the seashore know, two high and two low tides occur each day. The gravitational forces of the Moon and the Sun are responsible for creating tides within Earth's oceans. The gravitational attraction of the Moon is stronger on the side of Earth closest to the Moon. On this side of Earth, the oceans are pulled in the direction of the Moon, creating a high tide.

However, a high tide also occurs on the opposite side of Earth, the side which faces away from the Moon. This bulge of water, on the opposite side of Earth, forms because Earth is very slightly pulled toward the Moon, leaving an accumulation of water behind. As Earth rotates, so its land masses pass through the two bulges of water, and experience two high and two low tides a day.

But why, in practice, do the tides occur slightly ahead of the Moon and not directly in a line with it? Earth rotates faster than the Moon revolves and friction carries the ocean bulges along with Earth ahead of the Moon. The net effect of the friction slows Earth's rotation slightly.

The alignment of the Moon, the Sun and Earth affects the strength of the tides. During full or new Moon, the Sun, Earth and the Moon are in alignment, resulting in a gravitational pull stronger than average, and creating higher and lower tides, called spring tides. At the Moon's first and third quarter phases, the three bodies form a right angle. With the Moon pulling in one direction and the Sun in the other, the effect on the oceans is less dramatic, resulting in less of a tidal range. These are called neap tides.

LUNAR ECLIPSES

A lunar eclipse occurs when the Moon passes into the shadow of Earth. Since the Sun and Moon must lie on opposite sides of Earth for a lunar eclipse to occur, lunar eclipses can happen only at full Moon. However, a lunar eclipse doesn't occur every time the Moon is full. This is

THE MOON

Diameter:	3,475 km (2,159 miles)
Oblateness:	0
Axial rotation period (sidereal):	27.322 days
Equatorial inclination (to orbit):	6.68°
Mass:	0.07×10^{24} kg (0.15×10^{24} lb)
Mean density:	3.34×10^3 kg/m^3 (208.5 lb/ft^3)
Surface gravity (equatorial):	1.62 m/s^2 (5.31 ft/s^2)
Escape velocity:	2.38 km/s (1.48 mi/s)
Average distance from Earth:	0.384×10^6 km (0.239×10^6 miles)
Maximum distance from Earth:	0.405×10^6 km (0.252×10^6 miles)
Minimum distance from Earth:	0.363×10^6 km (0.226×10^6 miles)
Orbital eccentricity:	0.055
Orbital period (sidereal):	27.322 days
Orbital inclination (to ecliptic):	5.15°

Source: National Space Science Data Center, NASA.

because the Moon does not orbit in the plane of the **ecliptic**; it dips above and below it by about five degrees. So the two conditions of full phase and the Moon crossing the ecliptic must be met before a lunar eclipse can occur.

Unlike a solar eclipse, which has a very narrow path from which totality may be observed, lunar eclipses are visible from anywhere the Moon is above the horizon. Because Earth is much larger

than the Moon, it can take up to an hour or more for the Moon to pass through Earth's shadow. As our satellite first enters the shadow, its eastern edge darkens in stages. First the Moon enters the **penumbra** portion of Earth's shadow, the less intense outer region which begins to dim the lunar surface. The Moon then passes into the darker umbra region of the Earth's shadow. If the Moon is totally immersed in the **umbra**, a total eclipse is seen.

A lunar eclipse does not result in a totally dark Moon; the Moon often appears to glow orange-red. This is due to light passing through Earth's atmosphere and bending towards the Moon. Since light with a longer wavelength is less easily scattered by Earth's atmosphere, far more red light reaches the Moon than any other colour. Particularly dark lunar eclipses are often seen after volcanic eruptions on Earth, when large amounts of ash and dust swamp the upper atmosphere. The ash and dust block light passing through the atmosphere and so less light strikes the Moon and it appears darker.

During a total lunar eclipse, anyone standing on the Moon would see Earth block the Sun and experience a total solar eclipse. Earth would be dark but its surrounding atmosphere would glow brightly due to scattered sunlight.

THE APOLLO MISSIONS

The Apollo missions (1968-1972) were the first, and to this date the only, human spaceflights to orbit another body in space, land on its surface and return. The feat of engineering required to accomplish these goals are among the pinnacles of human achievement. The missions greatly increased our knowledge about the Moon. A large sample of lunar materials collected for analysis, as well as first-person accounts of what it was actually like to experience an extraterrestrial environment, have gone a long way in furthering our knowledge of the Solar System.

CRATER FORMATION

Craters form when a projectile – a piece of rock, ice, or a mixture of both – slams into a solid surface at speeds of many tens of kilometres per second. The impact causes a violent shock wave which ripples through the projectile and the surface it strikes, creating a pressure millions of times greater than Earth's atmosphere. In the ensuing calamity, the projectile either melts and flows like a liquid, or is vaporized by the enormous amount of energy released.

A crater starts to form when rapid decompression of the surface rocks occurs after the initial impact. The decompression 'frees' the recently struck compressed surface allowing a cone of rebounding debris to be flung sideways away from the impact site, forming an ejecta blanket. After the impact, all that is left is a fresh hole in the ground surrounded by upturned material forming the crater rim, caused by the flipping over of rock layers as the cone of ejecta moved outwards.

Beyond the rim lies the ejecta blanket made of excavated material thrown further. Very little, if any, of the projectile remains.

The shape and size of a crater varies depending on the size and speed of the impactor, what the impactor is made of (e.g. ice, rock, or iron) and the type of surface it strikes (hard or soft). Icy comets are less dense than an iron asteroid and, all things being equal, inflict less damage on a planetary surface.

OUR SOLAR SYSTEM

MARS

The fourth planet from the Sun, Mars has been known since ancient times and has always been a special target of scrutiny. Its striking red colour led the ancients to name it after their god of war. One-and-a-half times farther from the Sun than Earth, Mars takes 687 days to complete one orbit. Like Earth, Mars is a rocky planet but is only about half its diameter. Its dusty red surface is also colder than Earth's, insulated from space by a much thinner atmosphere, often stained pink by suspended fine dust. Several decades of unmanned spacecraft missions have revealed the surface of Mars to be very varied, showing features both familiar and alien to our eyes. Now, space probes have shown that there are enormous volcanoes and canyons, and there is strong evidence that liquid water once flowed across its now barren surface. There is also the tantalizing possibility that there may be life on this cold, red world.

SEEING MARS

Whenever Mars is at its brightest in the evening sky, its red colour is readily apparent. As it travels through the zodiacal **constellations** over the course of its orbit, it appears briefly to reverse its path as it is overtaken by Earth in its smaller, faster orbit. At its closest approach to Earth, Mars can be only 56 million km (35 million miles) away, and brighter than any star. It was the Italian astronomer Galileo who first observed Mars using his modest refracting telescope. Although he could detect no surface details, he was able to recognize a distinct phase. A small modern telescope will easily show the phase of Mars, and how it gradually changes from gibbous to full as Earth, Mars and the Sun line up once every 780 days on average. Through a moderate amateur telescope, it is possible to see darker patches on the surface. It was by observing the movement of these patches that, in 1659, the Dutch astronomer **Christiaan Huygens** (1629–95) measured the length of the Martian day to be very similar to our own. Huygens also noticed that Mars had bright polar caps, but it was not until the late eighteenth century that **William Herschel** (1738–1822) suggested they might be made of ice.

Like Earth, Mars has seasons. Its axis is tilted with respect to its orbit at almost the same angle as Earth's but, due to the length of the Martian year, its seasons last twice as long. As Mars's poles alternately point towards to the Sun, the bright polar caps can be seen to wax and wane.

SURFACE

The first successful probe to encounter the red planet was the American Mariner 4 in 1965. During its rapid flyby, its camera snapped just 21 black-and-white images of the surface along a narrow strip. It took three weeks for Mariner 4's primitive systems to transmit the images back to Earth. They revealed a surface pockmarked by impact craters, just like our own Moon. This implied that Mars was geologically dead, with an ancient surface, where little had happened for more than three billion years. The age of the surface could be deduced from the large number of craters: a planetary surface requires billions of years to accumulate craters, as asteroids and meteorites bombard it. On active worlds like Earth, volcanic and tectonic activity, as well as erosion, gradually erase the craters.

The next two flyby missions, Mariners 6 and 7 in 1969, also showed nothing but ancient craters. But the breakthrough came in 1971 with the Mariner 9 mission. Rather than simply fly past Mars, Mariner 9 became the first probe to enter orbit around the planet. When it arrived, the surface was hidden from view by a dust storm, which had engulfed the whole planet. Several weeks later the dust began to settle and Mariner 9 began to see the first features through the haze. Four giant spots appeared through the dust on a part of the planet not witnessed by previous missions: the spacecraft had discovered four huge volcanoes, far larger than anything on Earth. There was also an enormous canyon, 4,500 km (2,800 miles) long, that would stretch from coast to coast of the USA. Named Valles Marineris after the Mariner probe itself, this canyon is up to 7 km (4.3 miles) deep and, in places, is 600 km (370 miles) wide.

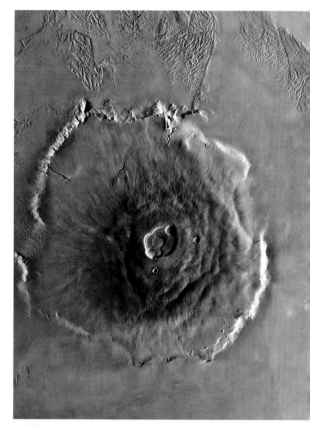

▲ *Orbiting space probes* have revealed some fascinating discoveries about the surface of Mars: volcanoes such as this, Olympus Mons, may have been active as recently as 10 million years ago, a relatively short time in the history of the Solar System's evolution.

THE GREAT MARTIAN VOLCANOES

The four large volcanoes of Mars are the largest in the Solar System. They are located in an area called Tharsis, a great bulge in the Martian crust which is 10 km (6 miles) high at its centre, and about 4,000 km (2,500 miles) across. Along the flank of this bulge are the great volcanoes of Arsia Mons, Pavonis Mons and Ascraeus Mons, all of which dwarf any volcano on Earth. Just off the northwest edge of Tharsis is Olympus Mons, the largest of them all. The summit of Olympus reaches up more than 27 km (17 miles) above mean surface level, and its base is 550–600 km (340–370 miles) across. In

▼ *This close-up image* shows part of the vast Valles Marineris. In total, Valles Marineris stretches for 4,000 km (2,500 miles). It results from faulting in the crust of Mars.

comparison the largest volcano on Earth, Hawaii's Mauna Loa, measures 120 km (75 miles) across its base and 9 km (6 miles) in height. Olympus Mons and the Tharsis volcanoes are similar in nature to the shield volcanoes of the Hawaiian Island chain, formed by magma welling up from a 'hotspot' beneath the crust. On Earth, the movements of the oceanic plates cause the Hawaiian volcanoes to drift away gradually from their heat source – the activity of volcanoes which pass beyond the hotspot dwindles, while new volcanoes begin to grow on the crust directly above the hotspot. On Mars, it seems that plate tectonics never happened, so a volcano above a hotspot would stay there, and keep on growing to the staggering proportions we see today. These large volcanoes have very few craters on their slopes, indicating relatively recent (in geological terms) activity. The sheer size of these edifices suggests that they probably erupted throughout most of Martian history and so are actually among the oldest features on the surface. Observations by NASA's Mars Global Surveyor spacecraft have suggested that Olympus Mons may have last erupted as recently as 10–20 million years ago. Possibly Olympus and the Tharsis volcanoes could erupt again, and a future spacecraft (or even human explorers) might witness the magnificent sight of fresh lava filling the calderas of these sleeping giants.

OTHER ACTIVITY

Not all Martian volcanoes are the same. Alba Patera, north of the Tharsis bulge, is some 1,500 km (930 miles) across but shows very little vertical relief. The large number of craters peppering it suggest that its activity was at a maximum around 1.7 billion years ago. Its low profile means that it erupted very runny lava, the results of which can be seen around the volcano as solidified lava flows

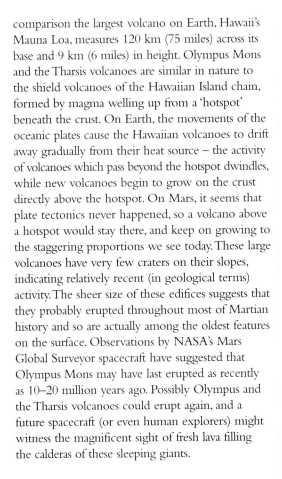

MARTIAN POLAR CAPS

The bright polar caps of Mars can be seen with telescopes from Earth. Made from frozen carbon dioxide and water, they shrink and grow with the Martian seasons. When they are at their smallest, the northern cap is larger, around 600 km (373 miles) across, and represents a residual cap of water ice, left behind after a carbon dioxide seasonal cap has sublimed into the atmosphere. The southern residual cap is only around 400 km (250 miles) across and is thought to be composed mainly of carbon dioxide. Both caps are layered and cut by canyons. These are gradually built up over millions of years, as ice and dust is deposited and then stripped away over the changing seasons.

A WARM, WET MARS?

Martian features such as the outflow channels and river valley networks provide ample evidence that water once flowed on Mars, around three billion years ago. But the current atmosphere is too thin for liquid water to be stable on the surface. It seems reasonable to conclude that the Martian atmosphere was once thicker and thus warmer than it is now. When volcanic activity was widespread, the gases released from volcanoes would have contributed significantly to the atmosphere, allowing water to flow as rivers and perhaps seas and oceans. As the northern hemisphere is considerably lower than the southern, it is possible that most of it was submerged by a great ocean – the Oceanus Borealis. Evidence of much more recent water erosion came from the Mars Global Surveyor spacecraft, which photographed what appears to be small gullies eroded into the walls of Martian craters. These features appear so fresh and unaltered by wind erosion that they must have been formed recently – at least within a million years or so. If this is the case then small eruptions of water on to the surface must still be occurring today, remaining liquid just long enough to carve out these gullies, before evaporating into the thin atmosphere.

and collapsed lava tubes. Other volcanoes, such as Tyrrhena Patera, show evidence that their eruptions were explosive in nature. It is possible that explosive volcanic activity on Mars was driven by rising magma interacting with water or ice in the crust, creating an extremely volatile mixture that erupted furiously at the surface.

Although Mars shows little signs of Earth-like plate tectonics, geological faulting has played a major part in shaping the planet's surface. Rifts and faults can be seen around many volcanoes, but the single-most imposing feature on Mars is the Valles Marineris. This huge canyon begins just to the east of the Tharsis bulge, in a complicated network of fractures called Noctis Labyrinthus, and extends 4,500 km (2,800 miles) east across the planet.

▼ *Evidence that water* may once have flowed on Mars comes from networks of channels on the planet's surface that resemble dried-up river beds. The channels suggest Mars may have been warmer and wetter in its early history.

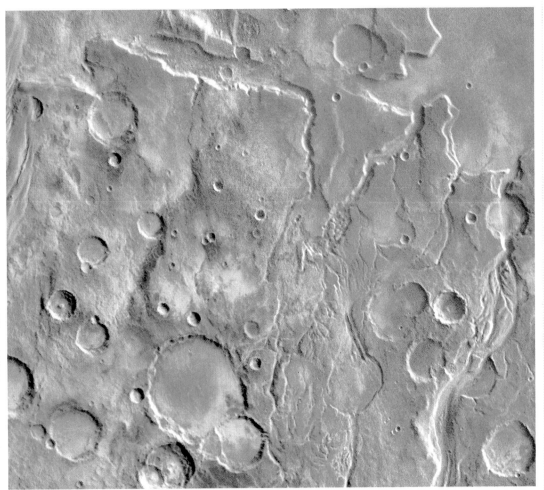

OUR SOLAR SYSTEM

SIGNS OF WATER

Craters, volcanoes and canyons were not the only surface features recorded by Mariner 9 and the later orbiters. Some images showed what appeared to be dried-up riverbeds, distributed across the planet. Many of these channels occur in the ancient, heavily cratered terrain in the southern highlands, and are very similar to river valley networks on Earth. They show tributaries and increase in size downstream. But, in keeping with Mars's large-scale geology, there are other features that dwarf any terrestrial counterparts. To the north-east of the eastern end of Valles Marineris is a low basin called Chryse Planitia. Converging into this lowland are the Martian outflow channels, closely resembling landscapes formed by huge floods on Earth, only much larger. Parts of the upper Kasei Valleys flowing into Chryse Planitia are over 200 km (124 miles) across and some channels can be traced for more than 2,000 km (1,240 miles). In Washington, USA are the Channelled Scablands. These were formed 10,000 years ago when a natural dam of ice gave way and catastrophically released vast quantities of water. This flood reached a discharge rate of 10 million cu m (353 million cu ft) of water

▶ *Percival Lowell, known for his theories about the mythical Martian 'canals'.*

PERCIVAL LOWELL

Percival Lowell (1855–1916) was an American astronomer who founded the Lowell Observatory in Flagstaff, Arizona. He became famous for his observations of the planet Mars. He observed the 'canals' seen by Giovanni Schiaparelli and believed that they were immense artificial structures, built by a native Martian civilization in order to irrigate their vast deserts with water from the polar caps. We now know that these canals were an optical illusion, and few of them corresponded to real features on the Martian surface. But Lowell should not be remembered purely for his canal theories: he also initiated the search for the planet Pluto. Although his calculations were spurious, they did lead the way to Clyde Tombaugh's discovery of the ninth planet in 1930, 14 years after Lowell's death.

per second, which carved out the surrounding landscape. On Mars, the floods are estimated to have reached one hundred times this discharge rate. The presence of chaotic, jumbled terrain at the upper reaches of these Martian channels strongly suggests that the water erupted from under the ground. In the porous rocks of the crust, large amounts of water could be trapped under great pressure by a layer of permafrost. A nearby 'Marsquake' or meteorite strike could release the water, creating the floods. Without its water the original landscape would collapse, forming the chaotic terrain.

INTERIOR

Mars has a distinct core, mantle and overlying crust. Geological modelling coupled with spacecraft data suggests that Mars's dense iron-rich core has a diameter of around 2,900 km (1,800 miles). This is surrounded by a mantle 3,500 km (2,200 miles) thick, and in turn by a thin, light rocky crust whose average thickness is roughly 100 km (60 miles). Orbiting spacecraft have detected a very weak magnetic field, around 1/800th the strength of Earth's. This is most likely the remains of a strong field which Mars once had, before its liquid interior cooled enough to solidify. This would have ended the field-generating dynamo effect of rotating, metallic liquids.

ATMOSPHERE AND WEATHER

The atmosphere on Mars is unlike that of Earth. It is much thinner, with the pressure at the surface less than 1% of sea-level pressure on Earth. It is composed of more than 95% carbon dioxide, with the rest consisting mainly of molecular nitrogen and argon. Traces of water vapour, oxygen and carbon monoxide are also present.

Space probes have photographed thin clouds of water-ice crystals in the atmosphere, and the summit of Olympus Mons is often capped by such clouds which form at high altitude. In fact, astronomers on Earth had reported these clouds, particularly one which regularly formed in the same place, called Nix Olympica (snows of Olympus), now known to be the giant volcano called Olympus Mons.

Dust devils frequently criss-cross the surface and the shadows cast by their tall funnels have

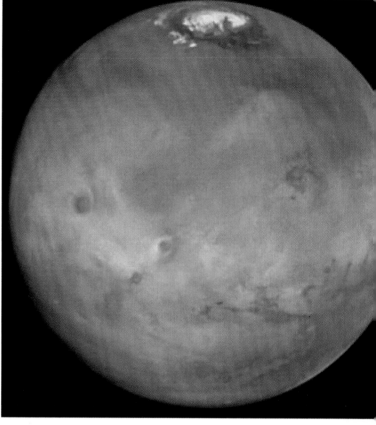

▲ *The Hubble Space Telescope has regularly monitored Mars and its atmosphere. Dust storms, cyclones and weather fronts are common throughout the Martian year.*

been clearly seen from orbiting spacecraft. But the most spectacular phenomena which will confront any future Martian meteorologist are the great dust storms. A combination of very fine dust and winds, coupled with the low gravity, allows dust storms to develop frequently which can engulf the surface of the whole planet.

THE VIEW FROM THE SURFACE

The first probe to land on the red planet was the Soviet Union's Mars 2, on 27 November 1971, but it failed to return any information from the surface. The first successful landings came nearly five years later, when NASA's twin Viking landers safely touched down on 20 July and 3 September 1976. Mission scientists, assuming the sky would be blue, incorrectly assembled the first colour image until the mistake was spotted and the world saw its first view of the bleak landscape, under a red Martian sky. At both landing sites the area was littered with rocks, strewn by volcanic activity and meteorite impacts. It was not until July 1997 that a rover was deployed on the surface. The highly successful Mars Pathfinder, with its rover, Sojourner, landed in an area called Ares Vallis,

LIFE ON MARS?

Mars has long been a focus of particular interest in the search for life on other worlds. When Percival Lowell observed 'canals' on the surface, many people took it for granted that Mars was a haven for life. But spacecraft exploration revealed a barren surface under a thin, poisonous atmosphere. However, many scientists retained belief that there could be primitive life, perhaps microscopic, clinging to life in the harsh environment. When the Viking landers arrived on Mars in 1976, they were armed with equipment to test the soil for biological organisms. In July 1976 Viking 1 reached the surface and carried a mechanical scoop, enabling it to collect some Martian soil which would be analyzed via a number of on-board experiments. The initial

results showed oxygen was produced by the soil, as if microbes were digesting the nutrient liquid. But this excitement was a false alarm, triggered by a simple chemical reaction.

Since the Viking missions, the search for life on Mars has focused very much on Earth, and the extreme environments in which only the most resilient life exists. Bacteria have been studied which are extremely resistant to heat and cold, and toxic acidic conditions. In Antarctica life has even been discovered inside rocks, forming a green layer just below a rock's surface. If life could adapt to these conditions, then why not on Mars? In 1996 NASA scientists announced they had discovered microscopic structures that looked similar to bacteria on Earth in a meteorite from Mars. However, many scientists now believe that these potential fossil

microbes contaminated the rock after it landed on Earth, 13,000 years ago. More space missions are planned to Mars with the aim of searching in places that were inaccessible to the Viking landers. Tools will burrow under the surface and into rocks, exposing areas protected from the harsh radiation from the Sun. It is likely to be many years before we know for sure whether or not there is life on Mars. But many studies of the planet from orbit are providing evidence that Mars was not always so forbidding a place for life to evolve.

▶ **Martian Meteorite ALH84001** *caused a stir in 1996 when scientists announced it contained a number of minute structures thought to be evidence for ancient Martian microbial life. Later, other researchers disputed these claims.*

---

a huge floodplain created billions of years ago by a vast torrent of liquid water. Sojourner was able to reach and analyze different rock types that had been deposited by that great flood. Indeed, many of the rocks around the landing site did appear to be similarly oriented, as if they had been laid down by water.

PHOBOS AND DEIMOS

Mars has two tiny moons, named Phobos and Deimos ('fear' and 'panic') after the two attendants of the god of war. They were discovered by Asaph Hall in 1877 from Washington, DC, USA. These small worlds are irregular in shape, their weak gravity not sufficient to have pulled their mass

MARS	
Diameter (equatorial): .	6,794 km (4,333 miles)
Oblateness:	0.0065
Axial rotation period (sidereal):	24.62 hrs
Equatorial inclination (to orbit):	25.19°
Mass:	0.64×10^{24} kg (1.41×10^{24} lb)
Mean density:	3.93×10^3 kg/m³ (245.3 lb/ft³)
Surface gravity (equatorial):	3.69 m/s² (12.11 ft/s²)
Escape velocity:	5.03 km/s (3.13 mi/s)
Average distance from Sun:	227.92×10^6 km (141.62×10^6 miles)
Maximum distance from Sun:	249.23×10^6 km (154.86×10^6 miles)
Minimum distance from Sun:	206.62×10^6 km (128.39×10^6 miles)
Orbital eccentricity: . . .	0.093
Orbital period (sidereal):	686.980 days
Orbital inclination (to ecliptic):	1.85°
Number of moons:	2

Source: National Space Science Data Center, NASA.

recorded dark-grey, dusty surfaces covered with impact craters. Phobos also possesses a series of almost parallel grooves across its surface. The most likely origin of these giant fractures is from the impact that produced the crater called Stickney, 5 km (3 miles) across, the largest on Phobos. Phobos orbits less than 6,000 km (3,700 miles) from the surface of Mars, circling the planet three times in a Martian day. An astronaut on Mars would see Phobos rise above the western horizon, appearing less than half the size of our Moon, and cross the sky in just 4.5 hours before setting in the east. But, as Phobos's orbit is so tight, our astronaut would never see Phobos from above latitudes of 69°, as it would always be below the horizon.

▼ **Space probes** *have told us a great deal about Mars. In 1997, Sojourner became the first roving vehicle on the planet's rocky surface. Sojourner, seen here examining a rock named Yogi, carried instruments to analyze both Martian soil and rocks.*

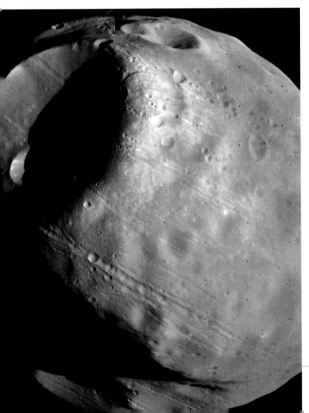

into a sphere. Phobos is the larger of these potato-shaped moons, 27 km (17 miles) across its longest axis, while Deimos measures 15 km (9 miles) long. These small worlds almost certainly originate in the main asteroid belt, between Mars and Jupiter. In the distant past, Phobos and Deimos strayed too close to the red planet and fell under its gravitational influence, becoming satellites of Mars. They were first seen in detail by the Viking orbiters in 1976, which

◀ **Mars has two satellites,** *both thought to be wayward asteroids captured by the planet. Phobos (left) has an ancient surface scarred from billions of years of impacts. The large crater is called Stickney.*

TERRESTRIAL PLANETS

OUR SOLAR SYSTEM

Our Solar System
OUTER PLANETS
INTRODUCTION

Beyond the inner rocky planets and the asteroid belt lies the outer Solar System. Here the distances between the planets increase dramatically. Mercury, Venus, Earth and Mars all lie within 250 million km (155 million miles) of the Sun. But Jupiter, the next planet outwards from Mars, orbits the Sun over three times farther out than the red planet. Then comes the magnificent ringed planet Saturn, at nearly twice the distance of Jupiter. Both of these planets appear as bright stars in the night sky and have been known since prehistory. Jupiter moves slowly against the constellations of the zodiac, completing one orbit of the Sun every 12 years. More distant Saturn takes 29 years to circle the Sun. Uranus was discovered in 1781 by the German-born astronomer William Herschel. The 'new star' he spotted is invisible to the naked eye except under excellent atmospheric conditions. Nineteen times Earth's distance from the Sun, one orbit of this planet takes 84 years. The next planet, Neptune, is half as far again as Uranus is from the Sun, and was not discovered until 1846.

GAS GIANTS

These four planets are not at all like Earth. They are much larger and more massive. Jupiter is the greatest of them all, with a diameter 11 times that of Earth – in fact it could contain every other planet and moon in the Sun's family. The **Solar System** has been described as the Sun and Jupiter, plus debris, with one of the larger pieces of debris being Earth. Saturn is smaller, three quarters the size of Jupiter. Uranus and Neptune are very similar to each other, about four times the diameter of Earth. Seen through telescopes, these four giant worlds reveal no solid surface, but instead are shrouded with thick atmospheres. Rather than being rocky worlds, like those of the inner Solar System, these giants are composed mostly of gas. Beneath the cloud decks of Jupiter, the mainly hydrogen atmosphere is eventually compressed into liquid by the immense pressure. Further towards the centre of Jupiter, the **electrons** are stripped away from the hydrogen atoms and this thick liquid behaves like metal, powering a stupendous magnetic field around the planet. At the very centres of Jupiter and Saturn are thought to exist rocky cores, perhaps the size of Earth.

Uranus and Neptune are composed mainly of water, ice, methane and ammonia, with maybe 10–15% less hydrogen and helium than their larger siblings.

◄ *Sizing Up the Outer Planets*
It is immediately apparent that the outer planets vary greatly in size. Jupiter and Saturn are the true giants, being 11 and 9 times wider than Earth respectively. Uranus and Neptune are similar in size, at around four times that of Earth. The rocky ice-world of Pluto is tiny, less than one-fifth the size of our planet.

▲ *A computer-generated representation of the view between two layers of cloud at Jupiter's equator. We are looking across patchy white clouds towards a hole in the lower clouds (coloured dark blue) through which heat is escaping from Jupiter's interior. At the top of the picture is an upper haze layer.*

BUILDING GAS GIANTS

When the planets were forming, over 4.5 billion years ago, the inner Solar System was too close to the Sun for ices such as water and methane to condense. Consequently, the inner planets are rocky, formed from dust grains which could withstand the radiation from the young Sun. Farther out, there was a boundary beyond which it became cold enough for ices to condense. The outer planets formed by accumulating these ices as well as the rock, and thus grew to the epic proportions we see today. Also, the hydrogen gas

▼ *A false-colour image of Saturn, taken by Voyager 2 in August 1981 from a range of 14.7 million km (9.1 million miles). Seen hovering by the planet are two of its satellites, Dione (above) and Enceladus (below).*

Jupiter
Saturn
Uranus
Neptune
Pluto

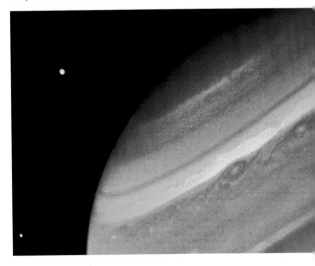

DISCOVERING THE OUTER SOLAR SYSTEM

▼ *Jupiter's satellite Europa is characterized by a fractured and icy crust. Evidence from the Galileo spacecraft suggests that underneath the icy surface there may be an ocean of liquid water.*

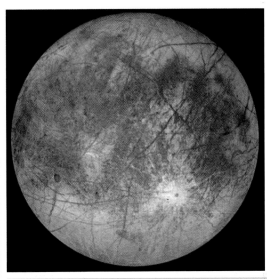

in the solar **nebula** was swept up by these growing gas giants, swelling them further. The relatively weak gravity of the inner planets was unable to retain this light hydrogen gas, but the gravitational force of the gas giants could easily hold on to it.

Because the gas giants formed essentially from the same ingredients as the Sun, the chemical compositions reflect that of our local star, particularly in the case of Jupiter. Had Jupiter grown to be around 80 times its current mass, the temperatures in its core would have been high enough to allow nuclear reactions to occur, with hydrogen fusing into helium and releasing energy. Jupiter would have become a star.

MOONS

Each of the giant planets of the outer Solar System is attended by a large family of moons, over 60 in all, and it is likely more await discovery. The first to be discovered were the four largest moons of Jupiter by the Italian astronomer **Galileo** in 1610. He observed four tiny points of light close to the planet, which changed their locations from night to night. Many of the moons of the outer planets are small, irregularly shaped objects, probably captured **asteroids**. But the larger moons, including the four discovered by Galileo, display great variety. Most are covered with impact craters, dating back billions of years, but others show signs of geological activity. Io, which orbits close to Jupiter, is constantly resurfaced by explosive volcanism, triggered by tidal interactions with Jupiter and another large moon, Europa. And Neptune's icy moon Triton experiences huge erupting geysers of nitrogen gas, driven by the heat of the distant Sun. These worlds formed far from the Sun, in an environment rich in ices, and their relatively low densities suggest that ices are a key part of their compositions.

THE ODD ONE OUT?

During an intensive and systematic search, Clyde Tombaugh (pictured) discovered a new planet in 1930. Using photographs he had taken at the Lowell Observatory, Arizona, he compared portions of the sky taken on different evenings and found that a faint 'star' had moved. The new planet was named Pluto after the god of the underworld. But this planet did not seem to fit in with the rest of the outer Solar System. Instead of a gas giant, Tombaugh had found a tiny, icy world. Pluto has an unusual orbit: it is so elliptical that, at times, it is closer to the Sun than Neptune. Its **orbit** is also inclined to the plane of the Solar System, so there is no chance it will ever actually hit Neptune.

Pluto's place in the Solar System has only begun to make sense fairly recently. In 1992 astronomers David Hewitt and Jane Luu discovered the first **Edgeworth–Kuiper Belt** object, and since then many more have been found. These distant objects are icy relics from the formation of the Solar System, which never gathered together to form a planet like Uranus or Neptune.

EXPLORING THE OUTER SOLAR SYSTEM

Most of our knowledge of the outer planets comes from unmanned robotic spacecraft. Scientists at NASA's Jet Propulsion Laboratory had realized that a chance alignment of the gas giants provided a rare opportunity for a spacecraft to visit each of the four in turn. In this 'grand tour of the Solar System', two identical spacecraft would fly past each planet, using their gravitational pulls to direct them onwards to the next world. The Pioneer 10 and 11 missions provided an initial reconnaissance of Jupiter and Saturn in

1973 and 1979, preparing the way for the Voyager 1 and 2 missions. Their historic encounters with Jupiter and Saturn revolutionized our view of these planets. Never before had such details been seen in the swirling banded clouds, and the variety discovered on their moons had never been anticipated. At Saturn, Voyager 1's trajectory was directed upwards, out of the plane of the Solar System. But Voyager 2 continued on to both Uranus and Neptune. All these worlds are like planetary systems in miniature, with their large entourages of moons and ring systems. After the magnificent views of the outer planets by Voyager, the next

stage of their exploration is well underway. The Galileo spacecraft has been in orbit around Jupiter, performing an in-depth survey of the planet and its moons. The Cassini-Huygens spacecraft is due to arrive at Saturn in 2004, where it will spend several years monitoring changes in the clouds and rings, and charting the surfaces of the moons at high resolution.

Plans to send spacecraft to Pluto and the Edgeworth–Kuiper Belt have yet to be finalized but, sometime in this century, astronomers will have their first close up views of these frigid worlds at the very edge of the Solar System.

JUPITER

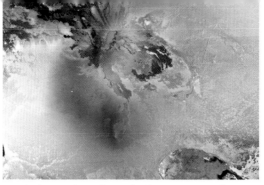

▲ *Each of Jupiter's four Galilean satellites has its own distinctive features. Io (pictured) has a surface sculpted by volcanic activity; Io experiences tidal distortions which release heat, helping to shape its unique surface.*

The fifth planet from the Sun, Jupiter, is by far the largest in the Solar System. It is twice as massive as all the other planets combined, and so voluminous that 1,300 Earths would fit inside it. Made mainly of hydrogen and helium, Jupiter has no solid surface. Its turbulent atmosphere extends downwards until unimaginable temperatures and pressures form a liquid core of metallic hydrogen. Jupiter carries with it an entourage of satellites, including Io, the most volcanically active body in the Solar System, and Europa whose icy surface may conceal an ocean of liquid water.

▲ *Despite its massive size (1,300 times the volume of Earth), Jupiter has no solid surface; it is simply an enormous globe of gas, mainly hydrogen and helium. It is one of the brightest objects in the night sky.*

SEEING JUPITER AND ITS SATELLITES

Jupiter is one of the brightest objects in the sky, outshone only by the Sun, the Moon and Venus. Since Jupiter orbits at a greater distance from the Sun than Earth it is often visible all night long. Jupiter orbits the Sun once every 11.9 years. Hence, as Earth circles the Sun, Jupiter moves approximately one-twelfth of the way around its orbit. One month later, Earth again overtakes Jupiter.

Through a telescope Jupiter is a fine sight. Its complex cloud systems are clearly visible as light and dark bands, running parallel with the planet's equator. These are called zones and belts

respectively. The ribbon-like appearance of the zones and belts results from strong east-west winds that ravage Jupiter. Perhaps the most distinguishable feature on the disk of Jupiter is the huge storm system known as the Great Red Spot. This large storm wells up high above the surrounding cloud decks; it is currently unknown what causes the red coloration, but this may be due to either sulphur or phosphine.

Accompanying Jupiter are four easily observed satellites: Io, Europa, Ganymede and Callisto. These are known as the four Galilean satellites, named in honour of Galileo, who discovered them in 1610.

JUPITER'S ATMOSPHERE

Jupiter's banded structure and spots have been observed telescopically for more than 300 years. However it is only relatively recently that spacecraft have inspected its dazzling cloud formations. In 1979 and 1980, Jupiter was briefly visited by the Voyager 1 and 2 spacecraft, which flew past it in a matter of days after making the three-year journey. In 1995, the Galileo spacecraft became the first man-made object to orbit Jupiter and has returned incredibly detailed images of the planet's turbulent atmosphere.

Jupiter's rapid rotation, 45,000 km/h (28,000 mi/h) at the equator, flattens the planet at the poles and creates a bulging equator, giving Jupiter an oblate appearance. It also organizes the outer atmosphere into regions of alternating wind jets, decreasing towards the poles and creating the banded cloud structure. Eastward winds reach 400 km/h (250 mi/h) in the equatorial region, while westward winds at 17° **latitude** blow at 100 km/h (60 mi/h). The jets constrain the north-south motion of developing clouds and trim them into organized weather systems that remain at fixed latitudes and rotate in the prevailing winds.

AN ENERGETIC ATMOSPHERE

Jupiter radiates 1.7 times more energy than it receives from the Sun. The energy is produced deep within the heart of the planet and rises upward

by convection. It is this which drives Jupiter's atmospheric storms, such as the Great Red Spot, smaller white ovals, and, along with its rotational energy, the bright zones and darker belts. The white zones are areas of cold upwelling clouds, while the darker belts represent areas where the gases descend.

Jupiter has the largest magnetosphere of all the planets. It stretches for some 650 million km (400 million miles) in the opposite direction of the Sun. Within the magnetosphere are radiation belts similar to Earth's **Van Allen belts**. Owing to the stronger magnetic field, the Jovian radiation belts trap far more energetic solar particles, creating lethal doses of radiation, which poses a considerable threat to any future space explorers. As with the other gas giants, the origin of Jupiter's mighty magnetosphere lies within its liquid interior, where a continuous churning driven by internal heat produces a dynamo effect, creating the magnetic field.

JUPITER'S INTERIOR

For many years, what lay beneath Jupiter's upper cloud decks could not be measured. However, in December 1995 the Galileo orbiter released a small atmospheric probe which made the perilous trip into the clouds, sending back a wealth of data for 57 minutes before being crushed and then vaporized by the ever-increasing temperature and pressure. Although descending to just 320 km (200 miles), 0.002 of Jupiter's radius, before contact was lost, the Galileo probe has helped establish Jupiter's atmospheric environment.

The Jovian atmosphere is comprised of three principal cloud decks. The uppermost is composed of cold, wispy cirrus clouds of ammonia ice. In an atmosphere with a composition similar to our Sun, hydrogen combines with nitrogen and carbon to form ammonia and methane. At -150°C (-238°F), the ammonia condenses and forms Jupiter's opaque, upper cloud deck. The middle cloud deck is made of ammonium hydrosulphide. Water vapour below the upper ammonia cloud deck combines with sulphur, forming ammonium hydrosulphide at about 1 bar – the atmospheric pressure at sea level on Earth. The lower deck of cloud is composed of water ice or possibly water droplets where temperatures hover close to the freezing point of water.

Farther down the temperature, pressure and wind speed increase rapidly as Jupiter's immense gravity crushes the atmospheric gases. The Galileo atmospheric probe's dying signal recorded a temperature of 300°C (572°F) and pressure of 22 bars at 150 km (93 miles) below where the probe began taking data. Wind speeds had increased

from 360 km/h (576 mi/h) near the top of the atmosphere to 540 km/h (864 mi/h) and powerful lightning strikes were noted.

Only physical models can be used to appreciate what lies deeper within Jupiter. The models suggest a huge hydrogen- and helium-rich envelope exists for thousands of kilometres, becoming increasingly compressed with depth and capable of conducting electricity, generating a strong magnetic field. Eventually, the atmosphere gives way to a boiling global sea of hydrogen, crushed to such a degree that the gas has become a liquid even at temperatures of many thousands of degrees. The sluggish motion of convective cells in this exotic material is considered responsible for the

---

JUPITER

Diameter (equatorial)*:	142,984 km (88,846 miles)
Oblateness:	0.065
Axial rotation period (sidereal)**:	9.93 hrs
Equatorial inclination (to orbit):	3.13°
Mass:	1898.6×10^{24} kg (4185.7×10^{24} lb)
Mean density:	1.33×10^3 kg/m³ (83.0 lb/ft³)
Surface gravity (equatorial)*:	23.12 m/s² (75.85 ft/s²)
Escape velocity:	59.5 km/s (37.0 mi/s)
Average distance from Sun:	778.57×10^6 km (483.78×10^6 miles)
Maximum distance from Sun:	816.62×10^6 km (507.42×10^6 miles)
Minimum distance from Sun:	740.52×10^6 km (460.14×10^6 miles)
Orbital eccentricity:	0.049
Orbital period (sidereal):	11.86 years
Orbital inclination (to ecliptic):	1.30°
Number of moons:	28

* At the 1-bar pressure level in the atmosphere.
** The rotation period of the planet's magnetic field, assumed to apply to the solid interior.
Source: National Space Science Data Center, NASA.

✳ VARYING GRAVITATIONAL OR TIDAL FORCES

The rate of change of a gravitational field across a body will tend to flex it. If a satellite's orbit were perfectly circular, the satellite would deform into a constant elongated shape. However, this is not the case with the orbits of the Galilean satellites. Gravitational interactions impose **perturbations** that sustain slightly noncircular orbits. As the satellites orbit Jupiter, their internal structures are forced to adjust to a variable gravitational field that imposes tidal stresses causing the satellites to warp physically. Because the satellites orbit Jupiter every 1.8 to 16.7 days, they must adjust rapidly.

---

gigantic dynamo which drives Jupiter's powerful magnetic field, 50 times stronger than Earth's. The models also account for the fact that Jupiter radiates more energy than it receives from the Sun. It is estimated that an incredible central core temperature of 20,000°C (36,000°F), and a pressure of 100 million bars must exist.

THE GALILEAN SATELLITES

Of the inner three Galilean satellites, Io orbits twice as fast as Europa and four times as fast as Ganymede, an arrangement which ensures the three are never aligned. This configuration, and the fact that they rotate and revolve at the same rate (synchronous rotation), means the moons always keep one face towards the planet. However, the mutual forces continually impose stresses, generating tidal distortions which stretch and contract each world. The constant contortion of the satellites heats up their interiors. Strongest at Io and decreasing outwards towards Callisto, these gravitational forces have powered the geological processes throughout their history.

Callisto, the farthest Galilean satellite from Jupiter, experiences the least tidal action and hence has the lowest internal heating. It has a dark, cratered surface, indicating little geological activity has occurred here for billions of years. Ganymede, the next moon inward, has two distinct surface components: areas which have abundant craters, and others which are far less cratered where melting and refreezing have modified the surface. Europa has a cracked, icy surface with very few impact craters, evidence of resurfacing over the entire planet. Recent data from the Galileo spacecraft suggests a water ocean may exist below the frozen surface.

Io is the closest to Jupiter and so experiences the largest tidal forces. Jupiter squeezes Io to

Resistance is generated by the strength of the materials and frictional heating occurs. For the satellites to reach equilibrium, the rate of heat generation must be balanced by the rate of heat loss. Thus, the nearer the satellite is to the planet, the more stress it has to endure. Tidal stresses are a strong function of distance, decreasing as the cube of the distance. This results in a 2.5% change across the diameter of Io. The tidal forces heat the interior, melting it and triggering volcanic activity. Inspection of Io's volcanically altered surface and total lack of water reveals that these processes can drastically modify satellites over the age of the Solar System.

💡 WHY DO THE RED SPOT AND OTHER STORMS LIVE SO LONG?

Evidence of the existence of the Red Spot spans the history of telescopic observations. The intriguing colouration possibly comes from trace molecules (possibly sulphur) dredged up from below. The chemicals at the top of the Great Red Spot strongly absorb violet and blue light, and reflect the remaining red light, hence its colour. The spot extends 23,000 km (14,300 miles) east-west, roughly twice the diameter of Earth, and 12,400 km (7,700 miles) north-south. Prevailing winds blow westwards north of the spot and eastwards south of the spot, creating wind differences of 350 km/h (220 mi/h) across the north-south dimension of the spot. The clouds are trapped and driven into a vast rotating hurricane. The rotation of the Red Spot indicates that the air here is ascending. This replenishment, and the fact that Jupiter has no solid surface to disrupt the flow, allows the Red Spot and other oval weather systems to exist for centuries or longer.

---

such a degree that its interior is completely molten. On the surface, this molten material spews forth in the most violent volcanic eruptions in the Solar System. Ground-based telescopes regularly monitor Io's eruptive activity and have discovered lava erupting on to its surface with temperatures of 1,600°C (2,900°F).

OTHER SATELLITES AND RINGS

After Galileo's 1610 discovery of the Galilean satellites, it was not until 1872 that a fifth moon, Amalthea, was discovered. Eight more were detected in the following 80 years and an additional 12 outer satellites were recently announced. Detection of faint inner satellites is limited by the planet's brightness. Three small, inner satellites and Jupiter's rings were discovered from images obtained by the Voyager spacecraft in 1979.

Jupiter's collection of moons can be subdivided into of four sets: four small irregular ones that orbit inside Io's path (Metis, Adrastea, Amalthea and Thebe), the Galilean satellites and two outer sets. Metis and Adrastea, the two innermost satellites, with diameters less than 50 km (30 miles), are closely associated with Jupiter's tenuous rings. Among the outer satellites, orbital similarities suggest that the satellites are fragments of larger primordial parent bodies.

SATURN

S aturn is the outermost planet easily visible with the naked eye. The second largest in the Solar System, it is in many ways a smaller version of Jupiter, composed principally of hydrogen and helium and with a similar internal structure. While not unique to Saturn, its magnificent ring system is by far the most extensive, complex and brightest of all. Saturn boasts a large family of moons including Titan, the only satellite in the Solar System to possess a thick atmosphere.

▲ *Saturn, the other true giant of the Solar System along with Jupiter, is famous for its complex and dramatic ring system and large number of moons.*

SEEING SATURN

Like Jupiter, Saturn is readily visible in the night sky and has played a major role in the mythology of many cultures. When viewed with a small telescope, Saturn is a spectacular object. Faint bands of colour similar to those of Jupiter, but more subdued, can be seen crossing its yellow disk. The most attractive feature has to be the planet's ring system, which, with a moderate-size telescope, may be resolved into several distinct parts. Titan, Saturn's largest moon, can also be seen as a bright point of light.

Images of Saturn show the planet looks 'flattened'. This is because Saturn rotates on its axis in about 10½ hours, which distorts the planet into an oblate spheroid, resulting in flattening of the poles and an equatorial bulge. As a consequence, Saturn's equatorial diameter is nearly 10% larger than the polar diameter.

SATURN'S ATMOSPHERE

Like Jupiter, Saturn has three main decks of cloud. However, Saturn's frigid environment causes water and ammonia clouds to form lower down than on Jupiter. Convective motion, fuelled by the outward heat transport, does not drive towering cumulus clouds up into the stratosphere. Individual clouds are far less common than on Jupiter and are generally short-lived. Most of them change rapidly in response to prevailing winds.

Measurements of the motion of the clouds reveal a broad equatorial region with eastward winds as high as 1,600 km/h (1,000 mi/h). These winds form alternating east-west jets, with wind speeds decreasing towards the poles. An oval near 70° north latitude and several near 45° to 55° latitude form cloud systems similar to those on Jupiter but are smaller, with dimensions less than 5,000 km (3,000 miles).

SATURN'S INTERIOR

Although Saturn's mass is 95 times that of Earth, the average density is less than that of water. If it were possible to place Saturn on an ocean large enough it would float. The low density and variation in gravity sensed by passing spacecraft indicate that Saturn, like Jupiter, has a small dense core surrounded by a compressed gaseous envelope rich in hydrogen and helium.

Saturn radiates 1.8 times more energy than it absorbs from the Sun. Decay of radioactive isotopes or slow overall contraction of the planet under its own gravity could account for the excess energy. In addition, there may be a region deep in the interior where convective mixing is so small that the heavier helium sinks toward the core, releasing energy. At shallower depths, where the magnetic field is generated, there is evidence of organized convection. Unlike Earth, Saturn's magnetic field is aligned along its rotational axis.

TITAN

Saturn's largest satellite has a diameter two-thirds larger than the Moon. Owing to its greater distance the sunlight at Saturn is a quarter that received by Jupiter, leading to lower surface temperatures than on Jupiter's satellites. Thus, although hydrogen and helium have escaped, Titan has retained a thick nitrogen-rich atmos-

phere. Unlike oxygen and carbon that have reacted with the surface, the nitrogen has formed a frigid atmosphere that contains trace amounts of other molecules and exerts a surface pressure approximately 50% greater than Earth's atmosphere.

Titan revolves around Saturn at a distance of over a million km (625,000 miles). It is tidally locked and rotates on its axis at the same rate at which it revolves. Thus, with a rotational period of 16 days, the distorting forces are small and do not play a major role in atmospheric circulation. Solar **ultraviolet** radiation has interacted with gases in the upper atmosphere, forming thick smog in the upper atmosphere. Large smog particles are likely to sink towards the surface.

The atmosphere of Titan is not only important because it is unique among the moons in the Solar System, but because scientists believe its composition is similar to that of Earth's shortly after its formation. Perhaps locked within the Titanian atmosphere are the clues to how life started on Earth. The Huygens probe and a radar instrument carried aboard the Cassini spacecraft, due to arrive at Saturn in 2004, will help answer this and other questions.

RINGS AND OTHER SATELLITES

A major challenge for planetary scientists has been to account for the plethora of puzzling features

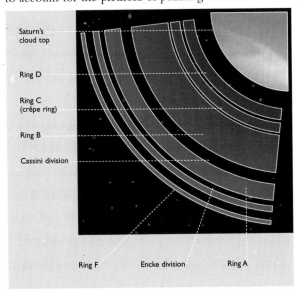

▲ *Saturn's Inner Rings*
Ring A is the outermost part of Saturn's ring system visible from Earth. The Cassini division, about 4,500 km (2,800 miles) wide, separates this from the brightest and widest part, ring B. Going inwards, the next ring is C, or crêpe ring. The fainter D and F rings lie inside and outside the visible rings. Beyond ring F are the even fainter rings G and E, not shown here.

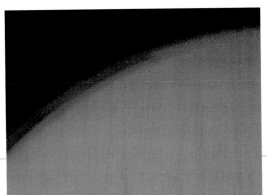

◀ *The surface of Saturn's largest satellite, Titan, is obscured by a haze in the atmosphere; if this could be penetrated and more information learned about the moon, it is possible that answers to the origins of life on Earth may be discovered. Scientists believe that Titan shares many compositional properties with our planet.*

spectrograph ▶ p. 343   ultraviolet ▶ p. 344

OUR SOLAR SYSTEM

SATURN STORMS AND HST

In 1990, a white cloud in Saturn's equatorial region was discovered and its growth noted by amateur observers. Cameras on board the newly launched Hubble Space Telescope were used to study the development of the storm. Although ground-based observers had reported occasional storms, this was the largest storm in 57 years.

Most storms on Saturn are short-lived and disperse in a manner that suggests they are generated by convection which transports heat from the interior. The convective activity carries material to high altitudes where the temperature is low enough that ammonia and water immediately freeze to form white ice. As the rising mass encounters the prevailing winds, the ice clouds serve as markers to reveal the progress of the storm. As the 1990 storm developed, it became apparent that it was similar to two previous equatorial disturbances which had been observed within the previous 125 years. These three storms were spaced at intervals of 57 years, nearly two Saturnian years, implying that these storms may be cyclic in nature.

associated with Saturn's magnificent rings. The rings extend from about 7,000 km (4,000 miles) above the cloud tops out to 74,000 km (46,250 miles). They are made of fragments of dark ice and rock which range in size from a few kilometres to dust as fine as cigarette smoke – most chunks being a few tens of metres across. The rings are the result of larger chunks of rock which have been ripped apart by Saturn's gravitational forces. The plane of the rings is tilted by nearly 27° compared with Saturn's orbital plane.

For convention, the rings have been assigned letters of the alphabet. Three distinct rings are visible from Earth, the A, B and C rings, labelled in order from the outermost to the innermost. A fainter inner ring, the D ring, was discovered in 1969. Farther out beyond the A ring are rings E and F, later discovered by space probes. Earth-based observations show a number of gaps within the rings, notably the Cassini division between rings A and B, named in honour of G. D. Cassini (1625–1712), who discovered the gap in 1675, and the Encke gap, near the outer edge of the A ring. However, much to the surprise of scientists, the gaps are actually filled with thousands of small individual ringlets, as discovered by the Voyager spacecraft.

Voyager made other startling discoveries. Unexpected structures such as braided and kinked rings, and strange ephemeral spoke patterns were found. The F ring has regions which appear

RINGS VERSUS SOLID SHEET

In 1610 Galileo observed Saturn with his 32-power telescope and saw two protrusions at either side of the planet. In 1612 when the protrusions had vanished, Galileo noted, 'I do not know what to say in a case so surprising, so unlooked for, and so novel'. What he had actually observed were Saturn's rings. Continuing observations revealed that the apparent thickness of the rings waxed and waned over a period of about 15 years. Finally, in 1659, **Christiaan Huygens**, a Dutch astronomer, realized that the observations were related to a thin ring around Saturn's equatorial plane.

SATURN

Diameter (equatorial)*:	120,536 km (74,898 miles)
Oblateness:	0.098
Axial rotation period (sidereal)**:	10.66 hrs
Equatorial inclination (to orbit):	26.73°
Mass:	568.5×10^{24} kg (1253.3×10^{24} lb)
Mean density:	0.69×10^3 kg/m^3 (43.1 lb/ft^3)
Surface gravity (equatorial)*:	8.96 m/s^2 (29.40 ft/s^2)
Escape velocity:	35.5 km/s (22.1 mi/s)
Average distance from Sun:	$1,433.5 \times 10^6$ km (890.7×10^6 miles)
Maximum distance from Sun:	$1,514.5 \times 10^6$ km (941.1×10^6 miles)
Minimum distance from Sun:	$1,352.6 \times 10^6$ km (840.5×10^6 miles)
Orbital eccentricity:	0.057
Orbital period (sidereal):	29.46 years
Orbital inclination (to ecliptic):	2.48°
Number of moons:	30

★ At the 1-bar pressure level in the atmosphere.
★★ The rotation period of the planet's magnetic field, assumed to apply to the solid interior.

Source: National Space Science Data Center, NASA.

braided or twisted. Elsewhere kinks and gaps appear within the F ring. Closer inspection of high-resolution Voyager images reveals the culprits responsible for distorting this ring. Tiny asteroid-like moons, called shepherd moons, orbit just inside and outside of the F ring – one moving faster and hence slightly ahead of the other. Each shepherd moon distorts the F ring, tugging it slightly in one direction. The overall effect is that the F ring is constantly 'zipped' and 'unzipped' in a never-ending cycle, resulting in kinks. Further undetected shepherd moons may contribute to the braiding.

Dark radial spoke-line patterns were also seen overlying Saturn's rings. Since the debris in each ring orbits Saturn at a different speed – the closer in the faster it moves in its orbit – spokes should be

Coinciding with these discoveries, **Johannes Kepler** (1571–1630) had formulated his laws of gravity. Although the dimensions of Saturn's rings were not known, it was apparent that the distance around the outer perimeter of the main ring system was 1.5 times greater than that around the inner edge. Kepler's laws required the inner particles to move faster than the outer ones. Thus, the rings could not be a thin, rigid sheet but must be composed of a swarm of particles revolving about the planet. Observational proof of this was not obtained until 1895 when James Keeler (1857– 1900) used a **spectrograph** to show that the orbital speed of the particles decreased outwards across the rings.

▲ *Saturn's rings are made from chunks of ice and rock, some as large as cities and some the size of smoke particles. The rings are transient features caused by the break-up of satellites which strayed too close to Saturn and were pulled apart by the planet's gravity.*

impossible to form since their radial nature would rapidly appear 'sheared.' Scientists have suggested electrostatic charging may be responsible for creating the spokes by levitating fine dust above the ring's orbit. However, the spokes still remain enigmatic.

Beyond the rings, but also interacting with them, lie the intermediate satellites (Mimas, Enceladus, Tethys, Dione, Rhea, Hyperion, Iapetus and Phoebe), which orbit at distances from 3 to 25 Saturn radii. Huge craters visible on their surfaces and the presence of two small moons in the same orbit as Tethys and one in the same orbit as Dione suggest that even larger impacts may have fragmented some of Saturn's satellites. Collisions of this sort may have been the source of the ring material and inner satellites. The outer satellites, including a dozen recently discovered ones, have irregular orbits and are probably captured bodies.

▶ **Looking rather like** the Death Star spaceship from the film Star Wars, Mimas was struck by an asteroid which produced a large crater called Herschel. Had the asteroid been any larger, the impact would have shattered Mimas.

OUR SOLAR SYSTEM

URANUS & NEPTUNE

T he seventh and eighth planets from the Sun, Uranus and Neptune, lurk in the cold depths of the outer Solar System. Their blue–green disks were too dim to be recognized until the eighteenth and nineteenth centuries. Both are considerably smaller than Jupiter and Saturn, but still approximately four times larger than Earth. Although receiving less than 1% of the Sun's energy compared to Earth, Uranus and Neptune have incredibly dynamic atmospheres, with violent storms and some of the fastest winds in the Solar System. Both planets have a system of rings and a suite of moons as exotic as those orbiting Jupiter and Saturn.

SEEING URANUS AND NEPTUNE

Sir William Herschel discovered Uranus by accident in 1781 while surveying the sky. At first Herschel thought he had discovered a faint **comet**, but he soon realized it was a planet. As they followed the motion, scientists noticed the planet was not keeping to its predicted path. This led to a hunt to find an even more remote planet whose gravity was presumably tugging Uranus out of position. In 1845, an English mathematician, John Couch Adams (1819–92), began to analyze the motion of Uranus in the hope of predicting the unseen planet's position. But he was beaten to the punch by French astronomer Urbain Le Verrier (1811–77) who published his own conclusions in 1846. However it was the German astronomers Johann Galle (1812–1910) and Heinrich D'Arrest (1822–75) who actually found Neptune, using LeVerrier's calculations, on 23 September 1846. Today both LeVerrier and Adams are credited with the discovery of Neptune.

Uranus is within the range of the naked eye on a moonless night, away from areas of light pollution, when you know where to look. Neptune is too dim to be seen without the aid of a telescope or good pair of binoculars. Even the largest telescopes only reveal small blue–green disks and occasional vague markings on these distant worlds.

▶ *The Voyager 2 spacecraft flew past Neptune in 1989, revealing a number of prominent storms including a Great Dark Spot rimmed with white cirrus clouds.*

THE ATMOSPHERE OF URANUS AND NEPTUNE

For many years after their discovery, little was known about Uranus and Neptune. Spectroscopic observations had revealed both planets are composed mainly of hydrogen and helium, with their distinctive colour arising from small amounts of methane which absorbs red light. It was long thought that the atmospheres of Uranus and Neptune would be calm in comparison to Jupiter and Saturn, owing to their great distance from the Sun and hence little energy to power them. Indeed, very few atmospheric features had ever been observed on them from Earth.

However, Voyager 2, the only space probe to have encountered Uranus and Neptune, discovered that the two worlds have tumultuous atmospheres. In January 1986, Voyager 2 flew past Uranus. At first sight, scientists were greeted by a featureless disk and presumed their assumptions about the planet having a calm atmosphere were correct. But when the Voyager images were returned to Earth and enhanced, a number of cloud structures, normally hidden by an upper layer of methane haze, became apparent.

Owing to the incredibly low temperatures, most clouds on Uranus (and Neptune) form at significantly lower altitudes than those on Jupiter and Saturn, where temperatures and pressures are greater. The lower altitude means the cloud decks are harder to spot through the upper atmospheric methane haze. Once discovered, scientists tracked the rotation of the faint clouds, whose motion indicate wind speeds of up to 580 km/h (360 mi/h).

Uranus is unique among the Jovian planets in that its axis of rotation is tipped by more than 90°. It is possible that Uranus was struck by a large body which 'flipped' the planet over on its side, resulting in it rolling like a barrel. Owing to this extreme axial orientation of Uranus, the Sun shines directly down on its North and South Poles in turn, as the planet makes its 84-year orbit. The result is each pole has repeated cycles of 42 years daylight and 42 years darkness. Even so, Voyager measured the temperatures at both poles to be similar, suggesting internal heating is more important in warming the planet's atmosphere than solar heating. As with the other gaseous planets, it is this internal heat which drives Uranus's atmosphere.

Voyager 2 sped on and encountered Neptune in August 1989. Surprisingly, Neptune's atmosphere was found to be even more dynamic than Uranus's. Voyager again found a high haze of frozen methane as well as high white cirrus clouds. But the most startling discovery was a series of high-pressure storm systems in the southern hemisphere. The largest was the Great Dark Spot, a large 'hurricane' which whipped around the planet in a little over 18 hours. Further south was the 'Scooter', another smaller storm travelling even faster, revolving around Neptune once every 16 hours. Some of the strongest winds in the Solar System were recorded close to the Great Dark Spot, a staggering 2,000 km/h (1,250 mi/h). Although the Great Dark Spot was similar to Jupiter's Red Spot, ground-based infrared observations indicate that, like storm systems on earth, it has dissolved. Storms on Neptune are again driven by convection, bringing heat upwards up from the warmer interior, powering the violent storms.

THE INTERIORS OF URANUS AND NEPTUNE

By measuring carefully how Uranus and Neptune tugged on Voyager 2 as it flew by, calculations were made to measure the density of these planets, giving a clue as to what lies beneath the clouds. Uranus and Neptune have a greater density than Jupiter and Saturn. Although hydrogen is dominant in their atmospheres, Uranus and Neptune contain less hydrogen than Jupiter or Saturn, and more of their mass consists of heavier gases. Owing to their greater density scientists have speculated that beneath their hydrogen atmospheres an ocean of water with ammonia and methane may exist, with a central rocky core.

▼ *Uranus, its faint rings and some of its satellites, photographed by the Hubble Space Telescope.*

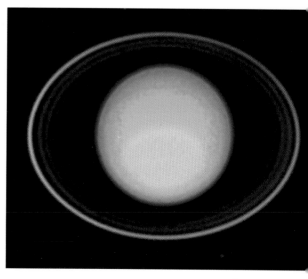

OUR SOLAR SYSTEM

DISCOVERY OF URANUS'S RINGS

In 1977 astronomers organized an effort to observe Uranus as it passed in front of a star – known as an **occultation**, from which they hoped to measure Uranus's diameter more precisely and learn more about its upper atmosphere. The occultation was observed over the Indian Ocean from the Kuiper Airborne Observatory, a telescope mounted in a high-flying jet aircraft. Surprisingly, the intensity of the star dropped abruptly nine times *before* the star was occulted. When it emerged from behind Uranus, a similar pattern of fluctuations recurred,

indicating the existence of nine thin rings around Uranus. Additional analysis of the data revealed four more rings that were called Eta, 4, 5, and 6. In 1989 Voyager instruments revealed two additional, faint rings.

The amount of light blocked by each ring revealed their differing thicknesses. To avoid confusion with Saturn's rings, the densest rings were named Alpha, Beta, Gamma, Delta and Epsilon. The width of Epsilon, the densest ring, is about 100 km (60 miles) while the fainter rings are 12 km (8 miles) wide or less. In contrast, the intervening gaps are hundreds of kilometres wide.

URANUS

Diameter (equatorial)*:	51,118 km (31,763 miles)
Oblateness:	0.023
Axial rotation period (sidereal)**:	17.24 hrs
Equatorial inclination (to orbit):	97.77°
Mass:	86.83×10^{24} kg (191.4×10^{24} lb)
Mean density:	1.27×10^{3} kg/m³ (79.3 lb/ft³)
Surface gravity (equatorial)*:	8.69 m/s² (28.51 ft/s²)
Escape velocity:	21.3 km/s (13.2 mi/s)
Average distance from Sun:	2872.5×10^{6} km (1784.9×10^{6} miles)
Maximum distance from Sun:	3003.6×10^{6} km (1866.4×10^{6} miles)
Minimum distance from Sun:	2741.3×10^{6} km (1703.4×10^{6} miles)
Orbital eccentricity:	0.046
Orbital period (sidereal):	84.01 years
Orbital inclination (to ecliptic):	0.77°
Number of moons:	21

* At the 1-bar pressure level in the atmosphere.
** The rotation period of the planet's magnetic field, assumed to apply to the solid interior.

Source: National Space Science Data Center, NASA.

MAGNETIC FIELDS

As Voyager 2 flew past both Uranus and Neptune it listened for radio emissions surrounding each planet. The hiss of random radio noise would be a sure sign these planets each had a magnetosphere. Not only did Voyager discover that Uranus and Neptune have magnetic fields, the spacecraft also measured that both magnetic fields were tilted in comparison to the rotational axis of the planets. The magnetic fields of Uranus and Neptune are tilted by 60° and 47° respectively, much more than any other planet in the Solar System. As with Earth, Jupiter and Saturn, the **solar wind** distorts the magnetospheres of Uranus and Neptune, creating magnetotails which stretch away from each planet for many tens of millions of kilometres in the opposite direction to the Sun.

The presence of a magnetic field at Uranus and Neptune is important since it gives an insight as to what lies within their interiors. For a planet to produce a magnetic field it must fulfil a number of simple requirements: there must be a region of the planet which is liquid, there must be an internal energy source which sets the liquid in motion and keeps it churning, and finally, the region must be electrically conducting. While Voyager was unable to see beneath the smog and

clouds of Uranus and Neptune, it is likely that an ocean, perhaps of liquid water under very high pressures, exists below.

URANUS'S MOONS

Before Voyager 2 arrived at Uranus, the planet was known to have five moons: Oberon and Titania (discovered in 1787 by William Herschel), Umbriel and Ariel (discovered by William Lassell, 1799–1880, in 1851) and Miranda, the innermost satellite (discovered by Gerard Kuiper in 1948). The satellites revolve around Uranus in the same plane as its equator. Voyager 2 discovered many smaller moon orbiting Uranus.

Surface characteristics of the five largest satellites show increased modification the closer the satellite is to Uranus, as with the Galilean satellites of Jupiter. Oberon and Titania, the larger outermost satellites, are covered with impact craters and stress fractures extending across their surfaces. Umbriel and Ariel, a slightly smaller pair, have distinctly different characters. Umbriel's surface appears dark and old while Ariel displays a maze of fault lines and signs of melting and resurfacing – evidence

that these worlds undergo tidal heating which helps keep their surfaces geologically young. Miranda, the smallest, innermost satellite, has a spectacular surface. Composed of seemingly unrelated structures, it appears to have been shattered by a major collision and then reconsolidated, resulting in the highly complex and disordered surface seen today.

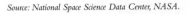

◀ *The surface of Uranus's innermost satellite Miranda is composed of seemingly unrelated structures; it seems to have been collisionally shattered and reconsolidated.*

PLANETARY RINGS: HOW ARE THEY FORMED?

All the gas giants have ring systems. But how did they form? The formation of planetary rings is still not fully understood. For many years it was considered that the rings formed when a moon or stray asteroid entered inside what is called the **Roche limit**, a zone around a planet where the gravitational pull of the planet tears a body apart. French mathematician Edouard Roche first proposed the idea over a century ago. For Saturn, the edge of this zone, the Roche limit, lies close to the outer edge of the main ring system.

However, a number of moons have now been found within the Roche limit of the outer planets. Furthermore, the ring systems for some planets extend beyond their Roche limit. While the Roche limit is likely to play a significant role in the formation of planetary rings, by breaking up small moons and asteroids, ring formation is a more complex process and probably involves the 'shepherd moons' which orbit either side of the rings. The rings themselves are likely to be relatively young, geologically speaking, and are thought to be replenished by dust and fragments from the surfaces of small moons, as well as cometary material.

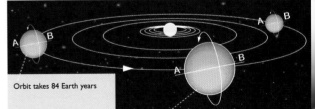

Orbit takes 84 Earth years

Uranus spins on its axis in 17.24 hours

▲ *The Extreme Tilt of Uranus*
Rather than spinning like a top as it orbits the Sun, Uranus has a distinctive feature: its axis of spin is tilted almost parallel to its orbital plane. This means that the polar regions (points A and B) alternately point towards and away from the Sun as Uranus orbits. During an 84-Earth-year orbit, each pole experiences one 42-year day and one 42-year night.

OUR SOLAR SYSTEM

NEPTUNE

Diameter (equatorial)*:	49,528 km (30,775 miles)
Oblateness:	0.017
Axial rotation period (sidereal)**:	16.11 hrs
Equatorial inclination (to orbit):	28.32°
Mass:	102.4×10^{24} kg (225.8×10^{24} lb)
Mean density:	1.64×10^{3} kg/m³ (102.4 lb/ft³)
Surface gravity (equatorial)*:	11.00 m/s² (36.09 ft/s²)
Escape velocity:	23.5 km/s (14.6 mi/s)
Average distance from Sun:	$4,495.1 \times 10^{6}$ km ($2,793.1 \times 10^{6}$ miles)
Maximum distance from Sun:	$4,545.7 \times 10^{6}$ km ($2,824.6 \times 10^{6}$ miles)
Minimum distance from Sun:	$4,444.5 \times 10^{6}$ km ($2,761.7 \times 10^{6}$ miles)
Orbital eccentricity:	0.011
Orbital period (sidereal):	164.79 years
Orbital inclination (to ecliptic):	1.77°
Number of moons:	8

* At the 1-bar pressure level in the atmosphere.
** The rotation period of the planet's magnetic field, assumed to apply
   to the solid interior.

Source: National Space Science Data Center, NASA.

TRITON

In 1846 the English astronomer William Lassell discovered Neptune's only large moon, Triton. This exotic world is about three-quarters the size of our Moon. Triton revolves around Neptune in a **retrograde** or backwards fashion, i.e. in the opposite direction to the rotation of the planet. Its backward motion suggests it may have been captured by Neptune in the past. Even though it currently holds the record as the coldest place in the Solar System, -235°C (-391°F), Triton has an extremely tenuous atmosphere and some geological activity at its surface.

Of particular interest are its active geysers, discovered by Voyager 2, which spray dark gaseous material up to 8 km (5 miles) into the atmosphere before encountering an airflow which sweeps the plume sideways. Fallout from the plumes can be seen as dark diffuse streaks on Triton's frozen surface. A proposed mechanism for the formation of the geysers involves transparent surface ices. Sunlight penetrates the transparent ice causing the surface beneath to sublime as the temperature increases slightly. The gas builds up under the ice and eventually the pressure ruptures the overlying ice, allowing the trapped gas to rush out, carrying fine dark dust with it. The surface reflection of Triton is similar to that of Pluto and its moon Charon, indicating that these bodies may have a similar origin.

NEPTUNE'S RINGS AND OTHER MOONS

Before Voyager 2 arrived at Neptune, Triton and Nereid were its only known satellites. Nereid is a small satellite which orbits the planet in 360 days in a large, highly elliptical, orbit. However, Voyager 2 images identified a further six moons.

The accidental discovery of Uranus's rings in 1977 led to the realization that Saturn's rings were not unique in the Solar System. Ring systems could be a common characteristic of the outer planets. Thus, a concentrated effort was made to determine all possible cases where Neptune would occult a star, in order to search for any rings.

▲ *Triton, Neptune's largest moon, is nearly as large as our Moon, larger than the planet Pluto, and has a surface covered with frozen nitrogen and methane with a temperature of -235°C (-391°F).*

Nearly 50 separate observations had been made before August 1989 when Voyager 2 encountered Neptune. The dimming of starlight was observed on only five occasions, and there was no symmetry about the planet, indicating that Neptune has immature rings with varying particle density around their circumferences. Voyager 2 confirmed these observations. Three thin rings encircle the planet at 1.7, 2.1 and 2.5 planetary radii. These rings contain complex lumps and braided patterns. In addition to the thin rings, there is a broad dusty belt from 2.1 to 2.4 planetary radii.

⧗ THE SHORT LIFE OF NEPTUNE'S GREAT DARK SPOT AND OTHER STORM SYSTEMS

Preceding the Voyager 2 encounter with Neptune, planetary scientists had monitored bright features as they moved across the planet's disk using **infrared** telescopes. The Voyager images revealed that these features were associated with individual weather systems. The largest, the Great Dark Spot, appeared to be a region where gases are up-welling. White clouds rimming the storm system suggested that methane molecules were forced upwards as they encountered the storm, freezing to create white ice cirrus clouds. Other sites displayed similar characteristics and throughout the Voyager encounter, the storm centres drifted in **latitude**. Observational sequences of the Great Dark Spot revealed a rolling motion about the centre; however, unlike Jupiter's Great Red Spot, the latitude of the storm varied as well as its **longitude**. Recent images from the Hubble Space Telescope show no trace of this feature but reveal new dark spots. An explanation of the difference in duration of storms on Jupiter and Neptune may be linked to the latitudinal variations of the prevailing winds on each planet. On Jupiter, jets constrain the north-south motion of storm centres, while on Neptune the strength of the westward winds increases towards the equator. Storms on Neptune which do not remain trapped within longitudinal zones may dissipate rapidly, since they move away from upwelling areas.

OUTER PLANETS

infrared ▶ p. 340 latitude ▶ p. 340 longitude ▶ p. 340 retrograde motion ▶ p. 343

OUR SOLAR SYSTEM

PLUTO

Discovered in 1930 by Clyde Tombaugh, Pluto is the last of the nine known planets in the Solar System. This icy world orbits the Sun once every 248 years and has the most inclined and eccentric orbit of all the planets. At its closest to the Sun, Pluto strays inside the orbit of Neptune. Being the smallest planet in the Solar System, its size and peculiar orbital characteristics suggest it may belong to a group of icy planetesimals known as the Edgeworth–Kuiper Belt. In 1978, Pluto was found to have a large moon, Charon.

▲ *Discovery plates showing the movement of Pluto (arrowed), the ninth planet in the Solar System, over a period of six days in 1930.*

✳ OCCULTATIONS YIELD CHARACTERISTICS OF PLUTO AND CHARON

In 1978 Pluto was discovered to have a large moon, Charon. The plane of revolution of the Pluto-Charon system is inclined 99° relative to its orbit so every 124 years the plane lies along our line of sight. When this occurs, the two bodies mutually eclipse each other. Mutual occultations most recently occurred from 1985 to 1991. Observing the variations in the total reflected light during the eclipses can be used to give insights into the two worlds.

For example, detailed information about the diameters, separation and masses of Pluto and Charon were obtained. The diameters of Pluto and Charon are 18% and 11% of the Earth's respectively, and their centres are separated by about 1.5 Earth diameters. The twin system is tidally locked with both Pluto and Charon having synchronous rotations. This unique feature results in each world keeping the same side facing towards the other. From Pluto, Charon always appears in the same part of the sky.

SEEING PLUTO

Pluto is far too small and remote to be seen with the naked eye. Even large telescopes reveal Pluto as nothing more that a faint point of light. Clyde Tombaugh discovered Pluto at the Lowell Observatory in Arizona by taking photographic plates of the night sky on different evenings and then carefully comparing them for any objects which moved. Finally, in February 1930 he discovered what he was looking for: a tiny point of light which changed position on two plates taken a week apart. The little point of light was Pluto.

THE SURFACE AND INTERIOR OF PLUTO AND CHARON

Pluto and Charon are likely to be similar worlds in terms of surface and internal properties. The Hubble Space Telescope has been able to distinguish surface markings on Pluto, which are likely to be a combination of frost deposits and the result of collisions with smaller bodies which have modified the surface of Pluto over time. An indication of what Pluto may be composed of comes from the planet's density (approximately 2 g/cm³). The density indicates that Pluto is probably a mixture of 80% rock and 20% ice, much like Neptune's largest moon Triton. The brighter areas distinguished by the Hubble Space Telescope are likely to be areas of frozen methane ice. Below an icy crust, Pluto and Charon may have large rocky cores.

THE ATMOSPHERE OF PLUTO

Pluto has a transient atmosphere. In the early 1980s methane gas was observed on Pluto while the planet was nearing its closest approach to the Sun, the period of maximum heating. Because the heating power of the Sun varies by more than 60% throughout Pluto's 248-year orbit, temperatures will drop considerably as Pluto moves away from the Sun and the methane gas should then condense and form fresh white ice on the surface.

▼ *Pluto and its satellite Charon. Charon, discovered in 1978, is believed to be formed from ice thrown off Pluto when another object collided with the planet. The surface of the satellite itself is covered in water-ice.*

PLUTO	
Diameter (equatorial): . . .	2,390 km (1,485 miles)
Oblateness:	0
Axial rotation period (sidereal):	6.39 days
Equatorial inclination (to orbit):	122.53°
Mass:	0.013×10^{24} kg (0.029×10^{24} lb)
Mean density:	1.75×10^3 kg/m³ (109.2 lb/ft³)
Surface gravity (equatorial):	0.58 m/s² (1.90 ft/s²)
Escape velocity:	1.1 km/s (0.7 mi/s)
Average distance from Sun:	$5,869.7 \times 10^6$ km ($3,647.3 \times 10^6$ miles)
Maximum distance from Sun:	$7,304.3 \times 10^6$ km ($4,538.7 \times 10^6$ miles)
Minimum distance from Sun:	$4,435.0 \times 10^6$ km ($2,755.8 \times 10^6$ miles)
Orbital eccentricity:	0.244
Orbital period (sidereal): .	247.7 years
Orbital inclination (to ecliptic):	17.16°
Number of moons:	1

Source: National Space Science Data Center, NASA.

▲ *The best image we have of Pluto's surface, taken by the Hubble Space Telescope. As the planet moves away from the Sun during its 248-year orbit, temperatures drop dramatically and fresh white ice will form on the surface.*

Occultation measurements and the Hubble Space Telescope images indicate that Pluto is over seven times as massive as Charon, and has a brighter surface. Observations have also revealed that Charon loses methane from its surface by evaporation. A torus of methane is thought to surround Charon, some of which is attracted toward Pluto, forming a frost on its surface. Nitrogen and carbon monoxide may also be present within the atmospheres of these two worlds.

IS PLUTO A PLANET?

Some astronomers have proposed that Pluto be reclassified as an asteroid. Beyond Neptune, there may be as many as 35,000 small objects with radii from 50 to 200 km (30 to 125 miles) orbiting within the Edgeworth-Kuiper Belt. Pluto's orbit is so similar to these objects that the planet is likely to be the largest member of this belt (although it the only one with a moon) and is therefore not a true planet. It has, however, yielded a wealth of information concerning the nature of these small, icy worlds.

Our Solar System
ASTEROIDS, COMETS & METEORITES
COMETS & ASTEROIDS

Small bodies swarm throughout the Solar System. They range in size from tiny specks of dust called meteoroids to flying mountains known as asteroids and dirty snowballs that exhale gas and dust to produce glorious comets. All these objects are remnants from the time when the Sun and planets formed, nearly 4.6 billion years ago. They escaped being swept up into the planets because they were either on orbits that were too eccentric, sped too fast or were in parts of the disk where the density was too low for a planet to form. Asteroids and comets are among the most primitive material remaining in the Solar System – the initial building blocks of planets.

WHAT IS AN ASTEROID?

Through a telescope, asteroids appear starlike – which gave rise to the name – but they are in fact irregularly shaped and cratered chunks of rock and metal. Over 10,000 of them are currently known,

▼ *The asteroid Gaspra, photographed by the Galileo space probe, is just one of a myriad of asteroids which orbit the Sun. Asteroids are considered to be the ancient leftovers from the formation of our Solar System.*

and countless more remain to be observed. Most orbit in the **asteroid belt** between Mars and Jupiter, but there are significant exceptions, mentioned below. Jupiter's gravitational pull is thought to have prevented the asteroids from accreting into a planet. Early in the **Solar System**'s history there were probably fewer asteroids than today but of larger size. Collisions have since fragmented them and many of the fragments have been swept up by the planets. The largest asteroid, and also the first to be discovered (by Giuseppe Piazzi on 1 January 1801), is Ceres, 930 km (578 miles) across. Next in order of size are Vesta, 580 km (360 miles) and Pallas, 540 km (335 miles). Numbers increase rapidly as the size decreases.

Asteroids are not smoothly distributed throughout the main belt. At certain distances from the Sun, in regions termed the **Kirkwood gaps**, numbers are few. This is because an object in these regions would have an orbital period that is a simple fraction of Jupiter's **orbital period** – for example, one-third, one-quarter or two-fifths – and hence would quickly be forced into a different **orbit**.

ASTEROID COMPOSITION

We can get some idea what asteroids are made of by analyzing the **spectrum** of sunlight reflected from their surfaces. Their compositions turn out to be diverse, but there are several families that show certain similarities. For example, the C-type asteroids contain carbon and may be similar to carbonaceous chondrites; S-type are mixtures of rock and metal; while M-type consist largely or wholly of metal (iron and nickel). Our most detailed information about asteroid compositions comes from pieces knocked off in collisions that have fallen to Earth as meteorites. In some cases the meteorites can be linked to particular asteroids. Vesta is thought to be the source of some meteorites of achondritic composition, while Eros (a member of the Amor group) appears to have a composition similar to chondritic meteorites.

ASTEROID HABITATS

Outside the main belt, a group of asteroids called the Trojans orbits at the same distance as Jupiter but 60° ahead of or behind it in its orbit. Between the outer planets beyond Jupiter orbits a small group of icy asteroids called the Centaurs, which may have more in common with cometary nuclei (particularly those in the **Edgeworth–Kuiper belt**) than the solid inner asteroids. The first of these to be discovered was Chiron, in 1977, which has shown signs of comet-like activity.

From our point of view on Earth, the most interesting asteroids are the runaways from the main belt, thrown out by the gravitational influences of Jupiter and Mars on to orbits that now bring them our way. There are three such groups of near-Earth asteroids (NEAs): the Amor asteroids cross the orbit of Mars but not that of Earth, while the Apollo and Aten groups cross Earth's orbit – the Atens have

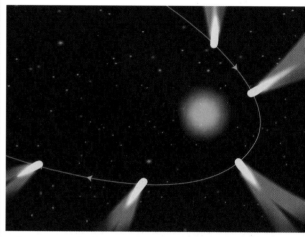

▲ *How a Comet's Tail Develops*
When a comet approaches the Sun on its highly elongated orbit, its frozen nucleus heats up. This releases dust and gas which, in some comets, can form into two tails. Both tails always point away from the Sun, but their characteristics are different: the dust trail is curved by the comet's motion, whereas the gas tail is pushed straight by charged solar particles.

⧗ DISCOVERING THE AGE OF THE SOLAR SYSTEM

We can measure the ages of asteroids, and hence date the origin of the Solar System, by analyzing pieces of them that fall to Earth in the form of meteorites. The technique used, called radioactive dating, relies on the fact that some isotopes are radioactive and decay into daughter isotopes at a known rate. The time taken for half the atoms to decay is called the half-life. The isotopes that are most commonly used for age-dating ancient rocks are rubidium-87 (which decays to strontium-87 with a half-life of 49 billion years) and various uranium isotopes that decay to lead. To age-date a rock, we measure how much of the daughter isotope has built up within it. All asteroid fragments yield similar ages of around 4.5 billion years, older than any rocks on Earth. Thus, the age of the Solar System is believed to be just over 4.5 billion years. The greatest age ever measured for a rock is 4.569 billion years, recorded for parts of the Murchison meteorite which fell in Australia in 1969. These are the oldest-known solids in existence, which probably formed in the dusty disk around the Sun in the early stages of the **accretion** process that built up the planets.

OUR SOLAR SYSTEM

COMET SHOEMAKER–LEVY 9 HITS JUPITER

In 1994 astronomers watched expectantly as fragments of a comet plunged towards the giant planet Jupiter. Carolyn and **Eugene Shoemaker** (1928–97) and their colleague David Levy had discovered the comet in 1993. Calculations showed that it had been in orbit around Jupiter for over 60 years and after a close approach in 1992 had broken into a chain of more than 20 fragments described as a 'string of pearls'. One by one these fragments hit the planet over a momentous week in July 1994.

All the impacts occurred on the far side of Jupiter from Earth, but as the planet rotated the damage they caused came into view: dark spots were created in Jupiter's clouds up to 3,000 km (2,000 miles) across, so large that they were easily visible using amateur telescopes. These spots gradually merged into a dusky belt that took over a year to disperse. If the fragments of Comet Shoemaker–Levy 9 had fallen on Earth they would have blasted out craters 60 km (37 miles) across and caused significant climatic changes. The Jupiter impacts demonstrated the reality of the threat posed to planets by comets and asteroids.

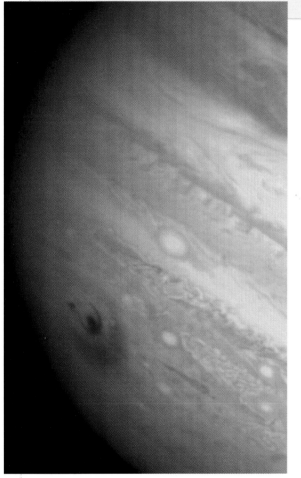

◀ *The 'bruise' caused by the impact of one of the fragments of comet Shoemaker-Levy 9 on the giant planet Jupiter. Evidence of such collisions has demonstrated that our continued existence on our own small world is extremely fragile.*

nuclei spend most of their lives unseen in the far outer reaches of the Solar System, but the gravitational effects of occasional passing stars may send some of them on highly elliptical orbits towards the inner Solar System. As it approaches the Sun, the dirty snowball warms up and releases gas and dust to form a **coma**, perhaps 10 times the diameter of Earth, yet so tenuous that it is transparent. Gas and dust streams away from the coma to produce a tail that, in extreme cases, could stretch from here to the Sun. In fact, comets have two tails, one of dust and one of gas. Gas tails consist of **ionized** molecules which give them a bluish colour; they are almost straight and are carried directly away from the Sun by the solar wind. Dust tails are curved because the dust particles lag behind the comet's motion; they appear yellowish because the particles reflect sunlight. Cometary dust disperses into space where it is eventually swept up by the planets or falls into the Sun. Dust particles from comets produce the bright streaks known as meteors when they burn up in Earth's atmosphere.

COMET CLASSIFICATION

Comets are divided into two broad categories, according to their orbital periods. Short-period comets, which include the famous Halley's Comet, are defined as those with orbital periods less than 200 years. These come from a region

▶ *In ancient times, comets were alleged to be harbingers of doom: the appearance of Halley's Comet in 1066 was taken as a bad sign by the English during the Norman Conquest. Today, bright comets such as Hale–Bopp, shown here, are recognized to be insubstantial 'dirty snowballs' of ice and dust orbiting the Sun.*

the added distinction of having average distances from the Sun that are less than Earth's distance. Astronomers are on the lookout for NEAs because of the collision threat they pose to Earth.

COMETS

Swooping majestically between the planets, a bright comet cuts an impressive figure in the night sky. But in reality comets are highly insubstantial. Their only solid part is the **nucleus**, typically no more than a few kilometres across – a 'dirty snowball' of ice with a dusty crust. Comet

POTENTIALLY HAZARDOUS ASTEROIDS

How likely is it that a giant comet or asteroid might hurtle into Earth, wiping out life as we know it? There are currently believed to be roughly 1,000 near-Earth asteroids of diameter greater than 1 km (0.6 miles) – large enough to cause global problems if they hit us. Strikes of something this size are predicted to occur every 100,000 years or so. For objects with diameters greater than 100 m (330 ft) – big enough to wipe out a city – the number is more like 100,000 and the predicted frequency of strikes is once every few thousand years. The numbers of comets that pose a danger is much fewer. However, such objects are small and difficult to observe so their exact numbers are not well known. Ground-based telescopes are being used to draw up a census of near-Earth objects. Once an object has been observed, its orbit can be calculated accurately to find out if and when it poses a threat. If a collision appears likely, we could send a space probe to deflect it on to a new course, or it could be blown apart by explosives.

around and beyond Neptune called the Edgeworth–Kuiper Belt (or sometimes simply the Kuiper Belt). Long-period comets have orbital periods of thousands or even millions of years, and are thought to originate in a region named the **Oort cloud**. Although the Oort cloud has yet to be directly observed, it is believed to consist of a swarm of cometary nuclei surrounding the Solar System out to 100,000 times Earth's distance from the Sun – roughly halfway to the nearest star.

OUR SOLAR SYSTEM

METEORITES & METEORS

E arth sweeps up interplanetary debris at the rate of around 40,000 tonnes a year. Most of this is in the form of fine dust particles termed meteoroids which either burn up high in the atmosphere to produce a 'shooting star' (properly termed a meteor) or, in the case of microscopic particles, are simply slowed down and settle gently through the atmosphere to Earth's surface. Larger meteoroids produce bright fireballs as they burn up. An object heavier than about 1 g (c. 0.04 oz) can survive its fiery passage and reach the surface of the Earth, where it is termed a meteorite. Meteorites provide us with free samples from the asteroid belt, the Moon and Mars.

▲ *The best-preserved impact crater on Earth is Meteor Crater, near Flagstaff, Arizona, USA. The crater was formed around 50,000 years ago when an iron meteorite struck the ground at 11 km/s (7 mi/s) creating a hole 1.2 km (three-quarters of a mile) in diameter.*

ORIGINS OF METEORITES

Until the nineteenth century the existence of rocks falling from the sky was widely disbelieved. Only after several falls were reliably witnessed and the rocks gathered were meteorites accepted as being extraterrestrial. Orbits of certain incoming meteorites have been calculated from eyewitness descriptions and photographs of their passage through the atmosphere. All those whose orbits have been determined in this way turn out to have come from the asteroid belt – in other words, they are fragments of asteroids.

Age-dating of meteorites confirms that almost all are 4.5 billion years old and hence date back to the formation of the Solar System. But a handful are younger than this. Some, which

▼ *The Leonid meteor shower of 1999, shown on a composite image. The meteors (or shooting stars) appear to stream away from a point in the sky, known as the radiant, in the Sickle of Leo.*

contain the same minerals as lunar samples collected by the Apollo astronauts, have ages in the range of between 4.0 and 2.8 billion years; evidently these were ejected from the Moon by impacts. Stranger still are a group of meteorites of volcanic composition, all but one of which have ages of 1.3 billion years or less. The clue to their origin is that they contain bubbles of gas which exactly match the composition of the Martian atmosphere as measured by space probes. Presumably these gases became trapped within the rocks when they were blasted off the surface of Mars by an impact; the rocks subsequently orbited the Sun for a few million years before finally encountering Earth. As the only Martian samples currently to hand (until future space missions are sent to retrieve some more) these meteorites provide unique insights into the geology of the red planet.

CLASSIFICATION OF METEORITES

Meteorites are divided into three main types on the basis of their composition: stones, irons and stony-irons. Stony meteorites are themselves subdivided into chondrites and achondrites. Chondrites, the most common meteorites of all, are so named because most of them contain rounded objects 1 mm (¹⁄₂₅ in) or so in size known as **chondrules** which were once molten; these

droplets were suddenly heated and then rapidly cooled, which could have happened either in the dust cloud surrounding the young Sun or in impacts on the surface of **planetesimals**. Around 4% of chondrites are of a type called carbonaceous chondrites. These are thought to have the most pristine, unaltered chemical composition of all meteorites, most accurately reflecting the composition of the disk from which the planets formed.

Achondrites are stony meteorites that have been melted. Their parent bodies must therefore have been large enough to retain heat. In most cases this means a big asteroid, but some rare types evidently come from the Moon or Mars.

Iron and stony-iron meteorites are also thought to have come from large asteroids that became sufficiently hot to separate into an iron-rich core and a rocky outer layer, like a small planet. When such bodies were broken apart by collisions, fragments of their cores became iron meteorites. Some stony irons called pallasites are composed of minerals embedded in iron-nickel metal and are thought to come from the boundary between the iron core and rocky mantle of such a body. Other stony-irons, called mesosiderites, apparently formed during a collision between two asteroids of different compositions.

chondrule ▶ p. 337 constellation ▶ p. 337 planetesimal ▶ p. 342 radiant ▶ p. 343

OUR SOLAR SYSTEM

IMPACTS AND CRATERS

Every year, many millions of meteorites land on our planet, most of them falling unseen in the ocean or remote areas. A meteorite weighing more than a few hundred tonnes will form a crater when it hits Earth. It is moving so fast – typically between 15 and 30 km/s (10 and 20 mi/s) – that all its kinetic energy is converted into heat when it impacts Earth, causing an explosion. The explosion will vaporize both the meteorite and part of the rock beneath it, forming a crater which is usually at least 10 times the size of the impactor. If the impactor lands in the sea, a 'crater' will be produced in the water that rapidly infills, forming a tsunami or tidal wave.

A crater-forming impact happens on Earth every 5,000 years or so, causing localized destruction. The very largest impacts have global consequences, chiefly because they throw up a dust cloud into the upper atmosphere that can envelop Earth, blocking out sunlight for years. Events such as this are – fortunately – extremely rare, but they do appear to have happened a few times during Earth's lifetime. One huge strike may have caused the death of the dinosaurs.

INTERPLANETARY DUST AND METEOR SHOWERS

Most of the interplanetary matter encountered by Earth is in the form of particles no more than about 1 cm (0.4 in) in diameter. Entering Earth's atmosphere at high speeds, these particles burn up by friction at around 100 km (60 miles) altitude, creating a column of ionized air that glows briefly to create the illusion of a 'shooting stars'. Some of this dust comes from asteroids and some from comets.

Comets release trails of dust as they pass through the inner Solar System and this dust spreads out along their orbit. As Earth moves around the Sun each year it passes close to the orbits of several comets. Each time it does so it encounters a swarm of dust particles and we see a meteor shower, which can last days or even weeks. All the members of a shower appear to originate from a small area of sky, termed the **radiant**, and the meteor shower is named after the **constellation** in which the radiant lies: for example, the Perseids of August appear to radiate from Perseus and the Geminids of December diverge from Gemini. A 'shower' usually means no more than one or two meteors per minute, although occasionally much higher rates can occur if Earth encounters a particularly dense stream of dust, as happens every 33 years or so with the Leonids, seen in November.

💡 HOW DID THE DINOSAURS DIE?

Sixty five million years ago Earth was subjected to a terrible event. The majority of species alive at that time, including the dinosaurs and many types of marine creature such as ammonites, died out over a few million years – remarkably sudden in geological terms. The period immediately before this event is

▲ *The death of the dinosaurs is thought to have resulted from the impact of an asteroid 65 million years ago. In the ensuing calamity, vast quantities of pulverized rock thrown into the atmosphere blocked out sunlight, drastically altering Earth's climate.*

called the Cretaceous, and the period after the event is called the Tertiary; the point in time between the two is known in geological shorthand as the K/T boundary. In 1978 a father-and-son team, Luis and

Walter Alvarez, discovered a major clue to the event that overtook the dinosaurs: exactly at the K/T boundary is a thin layer of brown clay, rich in the element iridium. Iridium is rare on Earth's surface but more abundant in asteroids, comets and Earth's interior. Its presence points to a catastrophic event at that time – a huge volcanic eruption or a major impact, either of which would have temporarily changed Earth's climate. Further geological investigations have shown that many unusual events took place during a brief period of time. Firstly, a vast area of India called the Deccan Traps is covered in a volcanic outpouring, called a flood basalt, that has a similar age to the K/T boundary. Secondly, some minerals that form only at very high pressures, such as diamonds, have been found at the K/T boundary, indicative of a huge impact. A buried crater of the right age 180 km (112 miles) in diameter has been found in the Yucatán Peninsula, Mexico. Called Chicxulub, it is estimated to have been formed by a body some 10 km (6 miles) in size. So it seems that both a volcanic eruption and an asteroidal (or comet) impact happened around this time. But there's more. Closer inspection of the fossils at that time shows that the dinosaurs had already started to die out, before either the volcanoes or the asteroid hastened their demise.

🧪 SEARCHING FOR METEORITES: WHERE AND HOW

Over the past 30 years the number of known meteorites has increased by about tenfold to more than 22,000 as a result of organized searches. Hot and cold deserts are the best places to look, because in such regions the rocks can survive for long periods without weathering away. Antarctica has been the most fruitful location of all – most of the known lunar meteorites and many of the Martian meteorites have been found there, preserved in a natural deep-freeze. Several features distinguish meteorites from terrestrial rocks. One is the fusion crust, a brown or black skin formed as the outer layers of the meteorite melted during its high-speed passage through the atmosphere. A meteorite's interior is a different colour from the fusion crust, usually grey. Most meteorites contain iron, and so are magnetic to some extent.

▲ *The Nomad explorer vehicle, designed to search for meteorites in the frozen wastelands of the Antarctic, a favourite place for meteorite searches.*

Our Solar System
SUMMARY

W ith the advent of the telescope and interplanetary spacecraft, the true scale and complexity of our Solar System has been brought into sharp focus, and the pace of discovery is accelerating. The Solar System is much larger and contains many more bodies than our ancestors knew about and the gradual accumulation of knowledge, through observation and, later, advanced technology, has allowed us to investigate the motions and composition of the planets, their satellites and other bodies, such as asteroids and comets. Since the 1960s, space probes have been dispatched to study the Sun, and to take close-up views of every planet except Pluto, revealing complex surfaces and atmospheres, and a plethora of exotic minor worlds: satellites, comets and asteroids. Thanks to new observations, we now know the Solar System to be a very different place from the one we knew 40 years ago.

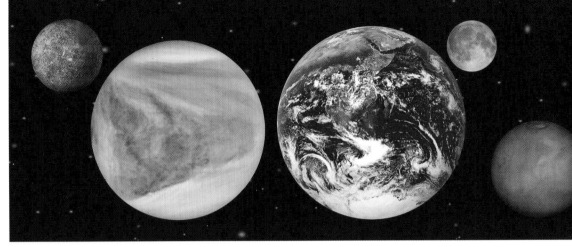

▲ **The four terrestrial planets:** Mercury, Venus, Earth and Mars (left to right). These share similar characteristics, including an iron core and a solid surface. In contrast, the outer planets are largely composed of gas.

THE SUN

◉ The Sun is a medium-sized star, much like countless millions of others in the Universe, situated in an arm of an ordinary **spiral galaxy** called the **Milky Way**.

◉ The Sun condensed from a large cloud of gas and dust. As it shrank under the force of gravity, the newly formed Sun developed a hot, dense core where nuclear reactions began.

Outward pressure from the energy of nuclear reactions pushes against the inward pressure of gravity and keeps the Sun in a balanced state.

◉ The Sun formed about 4.6 billion years ago and is approximately halfway through its 10 billion year life. The Sun is nearer to Earth than any other star and provides a stable source of energy to sustain life on Earth.

◉ At the Sun's core the temperature is 15 million°C (27 million°F), while at the surface it is only 5,500°C (9,000°F).

EARTH

◉ Earth is one of nine planets in our **Solar System**. Large oceans of liquid water covering two-thirds of Earth's surface and an atmosphere that sustains life makes our home planet unique.

◉ Earth is about 4.6 billion years old. As with the other terrestrial planets, it formed from debris within the **nebula** from which the Sun was born. Small dust particles collided and lumped together into larger chunks called **planetesimals**. The planetesimals collected to form the planets.

◉ Unlike the other terrestrial planets, Earth is a dynamic world with active volcanoes, earthquakes and a crust which constantly recycles itself by the process of plate tectonics.

THE MOON

◉ Earth's only natural satellite, the Moon, is about one-quarter its size.

◉ The Moon is thought to be the result of a high-impact collision between Earth and a Mars-sized object early on in the evolution of the Solar System.

◄ **Loops of hot gas**, which originate in active regions on the photosphere, follow the Sun's magnetic field lines in the inner corona.

THE DISCOVERY OF JUPITER'S GALILEAN SATELLITES

The Italian astronomer Galileo Galilei is credited with the discovery of Jupiter's four largest satellites. He published a sketch dated 7 January 1610 and recognized these small bodies as companions orbiting Jupiter. In 1611 Simon Marius (also spelled Mayr and Mayer) claimed he had observed the satellites in November 1609. However, his earliest sketch was dated 29 December 1609. Galileo aggressively pointed out that Marius was using the Julian calendar while he used the Gregorian one. This placed Marius's observations as being one day later than those of Galileo. Galileo also accused Marius of copying his own published works and a bitter argument ensued.

Marius suggested that the satellites be named after Jupiter's mythological lovers. The International Astronomical Union (IAU) did not officially accept the names Io, Europa, Ganymede and Callisto until 1975.

The significance of Galileo's discovery was that he used the observation to argue that large bodies dominated, controlling the motion of smaller worlds and used this system as additional evidence for a Sun-centred Solar System.

- The Moon's size and relative closeness to Earth raise strong tides in the oceans. The Moon also appears virtually the same size in the sky as the Sun, and can cause total **eclipses** when it passes directly in front of the Sun.
- Between 1969 and 1972, the Apollo missions placed 12 men on the Moon. To date these are the only humans who have walked on another world. Along with unmanned Russian landers, the lunar samples returned by the Apollo astronauts have taught us a great deal about how the Moon has evolved.

TERRESTRIAL PLANETS

- Mercury, Venus, Earth, and Mars are known as the terrestrial or inner planets.
- The terrestrial planets share a number of similar physical properties – an iron core, a partially molten mantle and a solid surface crust – which arise from their evolution within a similar area of the Solar System.
- Although the inner planets share a number of bulk characteristics, their surfaces have been sculpted to differing degrees by geological processes such as volcanism, tectonic activity and impact craters.

JOVIAN PLANETS

- The Jovian planets, Jupiter, Saturn, Uranus and Neptune are composed mainly of hydrogen and helium, and are often known as the gas giants, since they are much larger than the terrestrial planets.
- All the Jovian planets have extensive and active atmospheres, with large storms that may persist at one location for many years.
- Each gas giant has a large system of satellites similar to a miniature Solar System. Another feature common to all of them is their system of rings, composed of ice and rock fragments.

PLUTO: THE NINTH PLANET

- Pluto is the farthest known planet from the Sun, and the only one that has not been visited by spacecraft.
- Pluto is so distant that even the world's most powerful telescopes, including the Hubble Space Telescope, can reveal only vague surface markings.
- The surface of Pluto and its large moon, Charon, are thought to be covered with methane ice.
- Many planetary scientists consider Pluto as belonging to the **Edgeworth–Kuiper belt**, a group of small icy bodies which orbit the Sun beyond Neptune.

ASTEROIDS, COMETS, METEORS AND METEORITES

- As well as the nine planets, the Solar system contains a huge assortment of minor bodies – asteroids and comets as well as smaller fragments of debris.
- To date, astronomers have identified about 10,000 asteroids or minor planets but countless others are thought to exist. The majority are found in the **asteroid belt**, between the orbits of Mars and Jupiter. Many more are thought to exist in a band called the Edgeworth–Kuiper belt, beyond the orbit of Pluto, halfway to the nearest star.
- Asteroids are typically a few tens of kilometres in diameter and are irregular in shape. Those in the inner Solar System are composed of rock and metal while those in the distant Edgeworth–Kuiper belt are icy, like comets. They represent the oldest material left over from the formation of the Solar System.
- Comets are composed of frozen gases and dust and often have eccentric orbits that bring them near the Sun. As they approach the Sun, they heat up and often develop multiple tails as ice vaporizes from their surfaces.
- While most of the debris which enters the atmosphere burns up, producing meteors or shooting stars, larger fragments may survive and reach the surface. These are called meteorites. Meteorites have told us much about the age and environment of the early Solar System.

◀ **Galileo discovered** *the four major satellites orbiting the giant planet Jupiter in 1610. This discovery lent weight to Copernicus's previously unaccepted theory of a Sun-centred Solar System. From left they are Io, Europa, Ganymede and Callisto, all photographed by the space probe named after Galileo.*

OUR SOLAR SYSTEM

WATCHING THE SKY

The night sky is a fascinating place. From our viewpoint on Earth, we have a window into the Universe, full of diverse and extraordinary objects: from tiny, immensely dense black holes to enormous galaxies containing billions of stars. Since the beginning of the twentieth century, professional astronomers have been exploring the sky in radiation other than ordinary light. Branches of astronomy such as infrared, X-ray and gamma ray have opened a Pandora's Box of delights for the professional, allowing them to observe events otherwise hidden from us, such as star formation occurring in the middle of thick clouds of dust and gas.

Even in ordinary light, larger telescopes and advances in technology have opened the optical window wider, and with the launch of the Hubble Space Telescope above Earth's turbulent atmosphere we have enjoyed immeasurably clearer views of even familiar objects.

But the night sky is not the sole domain of the professional astronomer: it is open to everyone. Amateur astronomers have also enjoyed technological advances allowing them to make observations of professional quality and thus contribute to the science of astronomy.

Learning your way around the night sky is the first step into a most enjoyable world. The star charts in this chapter are accompanied by short descriptions which will allow you to learn the major signposts of the sky. With only your eyes and a reasonably dark sky away from city lights, you, too, can explore some of Nature's finest sights.

Watching the Sky
INTRODUCTION

Astronomy is a fascinating hobby which can be enjoyed on many levels. The first step in observing is learning your way around the sky. Just as you become familiar with a new part of town by taking note of prominent landmarks, so you can start to explore the night sky by looking for the more prominent patterns in the constellations, and making a note of the brighter stars. You do not need special equipment to enjoy the sky – just your eyes, a star chart and a dark sky – but even more celestial delights are revealed with a pair of ordinary binoculars. Progressing to a telescope is a bigger step, but many local astronomical societies possess a telescope through which they are always pleased to show people some of the jewels of the night sky.

▼ *A wide range of equipment is available for the amateur astronomer, from small, portable telescopes such as this simple refractor to larger instruments that may need a permanent housing. For many purposes, such as learning the constellations, a pair of binoculars will be adequate.*

A PERSONAL WINDOW

If you are lucky enough to live away from city lights then your window on the Universe lies close at hand. The farther away from any light you travel, the more stars and objects you will be able to see in the night sky. In very dark areas you can see thousands of stars, including the faint misty trail of the **Milky Way** meandering across the sky. This is the plane of our own Galaxy which is composed of billions of stars, dust and gas.

Your view of the night sky is also affected by your **latitude** and the time of year. If you live in the northern hemisphere, for example, then you will not be able to see stars that lie near the **South Celestial Pole**, and the farther north you live, the fewer stars you will be able to see south of the **celestial equator**. In addition, as the Earth orbits the Sun, the night side looks out at different parts of the Universe. This is why different **constellations** are on view at different times of the year.

THE FIRST STEPS

Choose a site as dark as possible and as far from trees and buildings as you can get. This gives you a better all-round view. If you are observing in the summer, you will have to wait until quite late in the evening for the sky to get dark, as the Sun does not drop as far below the horizon as in the winter.

Make sure you are warm; it is not much fun observing when you are shivering. It is surprising how cold you can get, especially during winter evenings, and you need to stay out long enough for your eyes to become dark adapted. Your eyes can take up to half an hour to achieve maximum sensitivity in the dark, and this dark adaption is quickly lost if a light is switched on. Instead of using an ordinary torch to read your star chart, use one with a red bulb as this helps you retain your dark adaption.

Try to gain an idea of where north is, as this will help you orientate yourself on the star charts. If you know the rough direction of north, then you can point the northernmost part of the star chart in this direction, but do not forget that you have to hold the chart above your head at the same time. This is confusing at first, but you will soon

▲ *Even in the brightly lit skies of towns and suburbs it is possible to spot planets and the brightest stars with the naked eye, if you know when and where to look.*

master your own technique of relating the star chart to the sky.

The chart you will need depends on both your latitude and the time of year. For northern-hemisphere dwellers, use the Index I map to select the correct part of the sky. The part of the sky in any chart is best visible around 10 p.m. during the month labelled on the rim, but of course the constellations will still be visible before and after this time during the preceding and following months.

FAMILIAR PATTERNS

Once you have selected the correct chart, choose a prominent constellation or asterism (pattern of stars) on the chart and try and recognize it in the sky. You may be surprised to find how far apart stars are from each other in the sky; some patterns which appear obvious on the star chart can be very spread out and not quite so obvious when you are looking at them in reality. You may also find that some of the fainter stars are not visible, if for example you live in a very light-

INTRODUCTION

WATCHING THE SKY

POLLUTION IN THE SKY

With the spread of urban light, a phenomenon termed 'light pollution' has become the bane of astronomers everywhere. With artificial light reflecting off surfaces or even shining directly into the sky, the night sky starts to glow, often with an orange hue because of the type of lamp used in street lighting. Any light shining into the sky reflects off water droplets and dust particles, causing this 'sky glow'. As the background sky becomes lighter, the fainter stars are drowned out, just as it is more difficult to see fainter magnitude stars in summer twilight. In some cities, the spread of light pollution is so bad that only the very brightest of celestial objects can be seen. This is a type of pollution to which there is a solution, and for years astronomers have campaigned for better lighting control. Astronomers do not expect street lights to be turned off – they wish to be as safe as the next person on the streets – but they ask that the minimum amount of light necessary should be used, and that the light should be directed where it is needed, not up into the sky.

While the dark skies of even 50 years ago will never return, astronomers have had some success in educating lighting engineers and the general public about the problem. After all, the night sky is an area of outstanding natural beauty and should be protected for future generations.

---

polluted area, and this changes some of the stellar patterns. Conversely, if you have a very dark sky, you may be overwhelmed by the number of stars which can confuse the stellar patterns.

Starting from an obvious pattern – and there are many – follow the chart and try and lead yourself to another obvious pattern. This is the first step in star-hopping, a technique used by amateurs at all levels. You can star hop with the naked eye, with binoculars, and through the eyepiece of a powerful telescope, although with increasing magnification the distances you star hop become smaller, and your star charts become more detailed, showing objects of fainter and fainter **magnitude**.

CELESTIAL DELIGHTS

Many fine sights can be seen with just the naked eye, but do not expect to see the detail and colour of images produced by powerful telescopes. Even with a large amateur telescope, you will not see the colours revealed in long-exposure photographs, but there is colour in the night sky, and just looking with the naked eye will reveal that stars are not all the same. When you actually stand under the dome of the **celestial sphere** and look at the faint fuzzy white patch of the Andromeda Galaxy (see Chart 2), or the faint glow of the Orion Nebula (see Chart 9), remember that the **photons** reaching the back of your retina have actually travelled the **light years** through space from the object itself. No wonderful image in a book can match that experience.

Use the star charts and accompanying descriptions in this chapter to pinpoint some of the more recognizable objects. You may want to progress to binoculars to see more details, or to split some of the **double stars** listed, but just knowing that you are looking directly at a **Cepheid variable** (e.g. Polaris in Chart 1), the remnants of a **supernova** that blew up nearly 1,000 years ago (see M1 on Chart 3) or a **globular cluster** (see ω Centauri on Chart 17) is extremely satisfying.

◀ *Bright clusters of stars, such as M7 in Scorpius (see also Chart 18), are attractive targets for binoculars and small telescopes.*

ESTIMATING DISTANCES

Astronomers use angular measurements to describe distances on the celestial sphere. Thus the full Moon is half a degree across, and the constellation of Hydra (see Charts 10, 16 and 17) is the largest in the sky, over 100° long. Estimating such distances in the sky can be difficult, but your own hand can help. For example, a finger, held at arm's length, is somewhat wider than the full Moon. When the Moon is next full, try this trick and see. Select which of your fingers covers the Moon best, and use this when estimating distances between stars. In addition, your fist, again held at arm's length, is about 7° across. This can be very helpful when star hopping large distances across the sky.

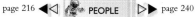

Watching the Sky
PROFESSIONAL ASTRONOMY
OPTICAL & INFRARED ASTRONOMY

▲ *The European Very Large Telescope (VLT)* of the European Southern Observatory in Chile consists of an array of four 8.2-m (323-in) reflectors, each housed in a separate dome. They can be used individually or interlinked to create the equivalent collecting area of a telescope of 16.4-m (645-in) aperture.

The Earth's atmosphere prevents many types of radiation reaching the ground. Radiation with wavelengths from about 300 to 1,000 nm can penetrate to the ground and this part of the spectrum is known as the optical window. It includes the visible spectrum – those wavelengths from about 400 to 700 nm to which the human eye is sensitive – with the near-ultraviolet at the shorter-wavelength end and the near-infrared at the longer-wavelength end. Telescopes operating at these wavelengths are known as optical telescopes and can work from ground-based observatories. In the infrared there are several narrow 'windows' between about 1 and 30 micrometres, allowing some infrared observations to be made from the ground.

THE GREAT TELESCOPES

The largest refractor is the 40-in (1-m) telescope at Yerkes Observatory in Wisconsin, USA, built in 1897. Refracting telescopes use lenses, but large lenses are difficult to make and support, and all the major telescopes of the twentieth century have been reflectors, which use mirrors instead.

The first really large reflector was the 72-in (1.8-m) Leviathan, built by William Parsons, the third Earl of Rosse (1800–67), in the grounds of his castle at Birr, Ireland. It remained the world's largest telescope until 1917, when the 100-in (2.5-m) Hooker reflector came into operation at Mount Wilson, California in the US. This, in turn, was superseded in 1947 by the 200-in (5-m) Hale telescope on Palomar Mountain, California. The Hale telescope remained the largest (and finest) telescope in the world for decades.

At the time of writing, the biggest telescopes are the twin 10-m (390-in) Keck telescopes in Hawaii, whose mirrors are made from a mosaic of hexagonal segments rather than a single piece of glass. The 9.2-m (360-in) Hobby-Eberly telescope, with specially constructed restricted steering, and designed primarily for **spectroscopy**, opened in 1997 at the McDonald Observatory in Texas and a similar instrument is being built in South Africa. The European Southern Observatory in Chile has just completed a group of four 8.2-m (323-in) telescopes, known as the Very Large Telescope, which will act together with an **aperture** equivalent to 16.4 m (645 in). Several other reflectors in the 8-m (315-in) range are either operating or under construction.

IMAGING DETECTORS

Images of astronomical objects have been traditionally recorded by photography, although in professional astronomy the use of photographic plates is now largely outmoded. Emulsions used for astrophotography have very fine grain and can capture large amounts of information quickly. Photography has been particularly suitable for survey work where a large area of sky was to be covered. The telescope used for the Palomar Sky Survey was the first large Schmidt to be used (see below) and had a clear aperture of 1.2 m (48 in). It used photographic plates that covered a 6° × 6° field. Photographic survey plates already in existence continue to have uses when scanned and digitized.

The drawback of photography is that the emulsions capture no more than a few per cent of the light falling on them. More efficient and now more common are astronomical cameras based on the **charge-coupled device (CCD)**, a silicon chip a few millimetres across covered with an array of light-sensitive elements. Typical CCDs are available as 'mosaics' with 1,000 × 1,000 elements or more. CCD cameras are sensitive to a wider range of **wavelengths** than

BERNHARD SCHMIDT

Bernhard Schmidt (1879–1935) was an Estonian telescope maker who designed a wide-field telescope that made possible large-scale photographic surveys of the night sky. Despite losing his right arm in a childhood experiment with gunpowder, Schmidt developed an interest in optics and, in 1904, set up a small business in the German town of Mittweida to manufacture mirrors and lenses for amateur astronomers. In World War I he was interned as a pacifist. In 1927 he was invited to set up a workshop at the Hamburg Observatory. Encouraged by the eminent astronomer Walter Baade (1893–1960), Schmidt came up with a design for a wide-angle telescope.

One drawback of conventional reflecting telescopes is that they produce sharp images over only a narrow patch of sky. Schmidt's idea was to place a thin glass corrector plate in front of a deeply curved spherical mirror. The glass was subtly shaped to correct the distortions that the mirror would produce, resulting in a distortion-free image over several degrees of sky. Schmidt produced his first telescope in 1930.

Although he did not live to see his invention adopted, large Schmidt cameras with apertures of a metre or more have for many years been in routine use in photographic sky surveys, and modified versions are popular medium-sized telescopes for amateur astronomers.

WATCHING THE SKY

THE HUBBLE SPACE TELESCOPE

Ground-based telescopes are limited in **resolution** by atmospheric turbulence, which degrades the quality of the image. One solution is to place telescopes in space. The Hubble Space Telescope (HST) was launched into a 600-km (370-mile) high orbit around the Earth in April 1990. Named after the pioneering US astronomer **Edwin Hubble** (1889–1953), the HST features a 2.4-m (94-in) primary mirror and several cameras and spectrographs to cover the **ultraviolet**, visible and near-infrared bands. It is operated jointly by NASA and ESA. Soon after launch, it was discovered that the primary mirror had been incorrectly figured and could not produce sharp images. Visiting astronauts inserted correcting optics in December 1993 and upgraded the instruments further in 1997 and 1999.

Although the 2.4 m (94-in) primary mirror is modest by modern standards, the freedom from atmospheric distortion allows the HST to achieve a resolution of better than 0.1 **arcsecond**, a tenfold improvement on most ground-based telescopes. With further servicing missions, the HST is expected to continue in operation until 2010.

NASA is pressing ahead with plans for the Next Generation Space Telescope (NGST) which may be as large as 8 m (315 in) in diameter. Concentrating on the near-infrared part of the spectrum, the NGST is due to be launched in 2009.

photographic emulsions and the resulting image can be enhanced by computer. They are now in routine use at all major observatories.

SPECTROGRAPHS

One of the most valuable tools for the astronomer is the **spectrograph**, a device used to split light from an object into its component wavelengths. Spectral analysis can give information about chemical composition, temperature, radial and rotational velocities, and magnetic fields.

A slit placed in the **focal plane** of the telescope is used to isolate the object, such as a **star** or **galaxy**. Light passing through the slit falls on a **diffraction** grating (occasionally a prism), where it is spread out to form a **spectrum** of the object. The image of the spectrum is recorded by a CCD. Modern spectrographs sometimes use optical fibres to channel the light from many objects into the slit side by side.

▶ *The European Infrared Space Observatory (ISO) carried a 0.6-m (24-in) telescope. From orbit it observed infrared radiation that does not penetrate Earth's atmosphere coming from objects such as regions of star formation, dying stars and galactic nuclei.*

BUILDING BIGGER TELESCOPES

While the HST has shown the value of placing telescopes above the blurring effects of the atmosphere, some astronomers question whether the high costs of space-borne telescopes can continue to be justified. Many argue that space observatories should concentrate on the parts of the spectrum that are blocked by the atmosphere, and that visible and near-infrared work is best done from the ground.

Several groups are working on designs for ground-based telescopes very much bigger than the record-holding 10-m (33-ft) Keck Telescopes.

INFRARED TELESCOPES AND DETECTORS

Most **infrared** radiation is absorbed by Earth's atmosphere, although there are several 'windows' where infrared at certain wavelengths can penetrate through to high **altitudes**. Most objects glow brightly at infrared wavelengths, including Earth's atmosphere and the telescope itself. Trying to observe an astronomical infrared source is like looking for a needle in a messy haystack. Sophisticated detectors subtract all the background noise, and have to be cooled to reduce their own emissions. They are cooled by liquid nitrogen or even liquid helium.

Amongst the largest purpose-built infrared telescopes are the 3.8-m (150-in) UK Infrared Telescope (UKIRT) and the 8.2-m (323-in) Subaru Telescope, both sited at high altitude at the Mauna Kea Observatory in Hawaii, where the atmosphere is clear and dry.

The key that makes this feasible is the development of adaptive optics, a variety of techniques which can compensate for atmospheric turbulence, effectively de-blurring the image in real time and realizing the full potential of a large aperture.

All the proposals adopt the segmented mirror approach used in the Keck Telescopes. Among those at an advanced stage are the US 30 m (99 ft) CELT (California Extremely Large Telescope) and the European 100-m (328-ft) OWL (Overwhelmingly Large Telescope). These projects will be expensive, but not prohibitively so. It has been estimated that three OWLs could be constructed for the cost of one Hubble Space Telescope.

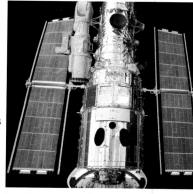

▶ *The Hubble Space Telescope (HST) contains a main mirror of 2.4 m (94 in) diameter. HST is seen here in the cargo bay of the Space Shuttle during one of its service visits by astronauts. Launched in 1990, it is expected to continue operation until 2010.*

Infrared telescopes are used to observe relatively cool objects, such as dust clouds and newly forming stars. Since infrared waves are longer than those of ordinary light, they do not bounce off the particles in the dusty regions, but instead pass through, revealing stars that are hidden at visible wavelengths.

INFRARED SATELLITES

The far-infrared region of the spectrum was first explored by high-altitude aircraft and balloon-borne telescopes in the 1970s, but the first comprehensive surveys awaited the arrival of infrared telescopes on board satellites. The first of these, the Infrared Astronomical Satellite (IRAS), was a Netherlands-UK-US spacecraft that surveyed the whole sky in 1983, detecting a quarter of a million sources. The European Infrared Space Observatory (ISO) followed in 1995. All these spacecraft had short lifetimes because of the need to carry supplies of volatile liquid helium to cool their detectors. Several more infrared spacecraft are planned for the first decade of the new millennium, including the US Space Infrared Telescope Facility (SIRTF) and the 3.5-m (138-in) Herschel Space Observatory.

The region upwards of 0.3 mm is known as the submillimetre and millimetre band and is now usually considered a province of radio astronomy.

Hubble ▶ p. 122

WATCHING THE SKY

RADIO ASTRONOMY

Longwards of the infrared region lies the radio window, where wavelengths from 2 cm up to about 20 m pass unhindered through Earth's atmosphere. Longer waves are reflected by the ionosphere and shorter waves are absorbed by molecules of water vapour and carbon dioxide except in a few relatively clear windows down to about 0.3 mm. At radio wavelengths, observations from the ground are possible 24 hours a day in most weather conditions. Major sources of radio waves include the Sun, interstellar hydrogen clouds, nebulae, supernova remnants, pulsars, galaxies and quasars.

▶ *A Radio Interferometer*
In a simple two-element interferometer, radio waves from a distant source arrive at one telescope (B) slightly before the other (A). The interferometer effectively measures this tiny delay and uses it to calculate the precise direction of the source in the sky. With more telescopes and extended observations a detailed picture of the source can be built up.

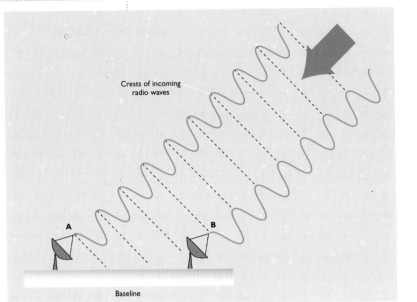

Crests of incoming radio waves

A   B

Baseline

DISCOVERY OF THE RADIO SKY

The beginning of radio astronomy is usually dated from 1932, when Karl Jansky (1905–50), a US electrical engineer investigating interference on transatlantic radio telephone links, discovered that radio waves were coming from the **Milky Way**. In 1936 engineer Grote Reber (b. 1911), who is now regarded as the world's first radio astronomer, built a 31-ft (9.4-m) parabolic dish at his home in Illinois, with which he made the first map of the radio sky. During World War II, J. Stanley Hey (1909–2000), a British physicist working on military radar, discovered radar echoes from **meteor** trails, radio emissions coming from the Sun and later an intense source of radio waves which proved to be the **radio galaxy** Cygnus A. After the war, radio astronomy research groups were set up in several countries, notably the US, the UK, Australia and the Netherlands, and today radio astronomy is firmly established as one of the most powerful techniques for investigating the cosmos.

◀ *The 76-m Lovell Telescope at the Jodrell Bank Observatory in Cheshire, UK, is an example of a fully steerable parabolic dish for radio astronomy. Completed in 1957, the telescope has been used on its own and also as part of interferometer networks.*

BIG DISHES

The most familiar kind of radio telescope has a parabolic dish. The dish collects radio waves and brings them to a focus in much the same way as the primary mirror of an optical telescope. The smoother the surface, the shorter the wavelength at which the telescope can operate. Because radio waves are measured in centimetres, the surface of the dish does not have to be particularly smooth, and can even be a wire mesh rather than a continuous surface. Currently, however, much of the interest in radio astronomy is focused on shorter wavelengths (higher **frequencies**), where surfaces accurate to a fraction of a millimetre are required.

⧗ EARTH ROTATION

Astronomers have traditionally monitored the rotation of the Earth by measuring the time at which certain stars cross the **meridian** at selected observatories. Until the 1950s, this was the basis of **Greenwich Mean Time (GMT)**. Because of the much greater precision possible with radio interferometry, Earth rotation is now monitored primarily by VLBI. Each year about 35 radio telescopes around the world make a quarter of a million observations of about 500 galaxies and quasars whose positions are accurately known. By timing the arrival of the radio waves at each telescope, the orientation of the Earth in space can be measured to a precision of about 0.0002 arcseconds, equivalent to less than one centimetre on the surface of the Earth. The International Earth Rotation Service (IERS), based in Paris, co-ordinates the measurements to form UT1, a variant of **Universal Time (UT)**, that indicates the precise angle through which the Earth has turned on its axis.

MARTIN RYLE

Sir Martin Ryle (1918–84) was one of the pioneers of radio astronomy. He worked on radar during World War II, where he became aware of the potential of this new technology for radio astronomy. Taking up a post at the Cavendish Laboratory in Cambridge in 1945, he had, within a few years, laid the foundations of one of the world's leading radio astronomy centres.

In the 1950s, Ryle discovered how to use several small radio telescopes to imitate the performance of a larger instrument, a technique he called aperture synthesis. He built two major aperture synthesis telescopes at Cambridge, the One-Mile Telescope and the Five-Kilometre Telescope (now known as the Ryle Telescope).

Ryle was a leading proponent of the Big Bang theory for the origin of the Universe, based upon his own surveys of distant radio sources which indicated that radio galaxies had been more powerful in the past and that the Universe must have been evolving. He was knighted in 1966 and served as Astronomer Royal for England from 1972 to 1982. Along with his Cambridge colleague Antony Hewish (b. 1924), Ryle received the Nobel Prize for physics in 1974. It was the first time a Nobel Prize had been awarded for contributions to astronomy.

PROFESSIONAL ASTRONOMY

frequency ▶ p. 339 Greenwich Mean Time (GMT) ▶ p. 339 interferometer ▶ p. 340 meridian ▶ p. 341 meteor ▶ p. 341 Milky Way ▶ p. 341 pulsar ▶ p. 342 quasar ▶ p. 342
radio galaxy ▶ p. 343 redshift ▶ p. 343 spectral line ▶ p. 343 Universal Time (UT) ▶ p. 344 very long baseline interferometry (VBLI) ▶ p. 344

▲ *The Very Large Array (VLA) in New Mexico is an aperture synthesis telescope made up of 27 dishes each 25 m in diameter. They can be moved along railway tracks to form baselines ranging from 33 m to 36 km.*

The largest single-dish telescope in the world is at Arecibo in Puerto Rico and measures 1,000 ft (305 m) in diameter. It is constructed in a natural hollow in the ground and points straight upwards, allowing the rotation of the Earth to move it across the sky. Large steerable dishes can be found at Green Bank in West Virginia (measuring 110 m/360 ft), Effelsberg in Germany (100 m/328 ft), Jodrell Bank in the UK (250 ft/76 m) and Parkes in Australia (210 ft/64 m).

Large dishes are most useful for **very long baseline interferometry (VLBI)** studies, spectroscopy and **pulsar** work, where collecting as much energy as possible is more important than resolution.

INTERFEROMETERS

Since radio waves are typically hundreds of thousands of times longer than waves of visible light, the resolution of even the very largest parabolic dishes is no better than that of the human eye. Radio astronomers get around this limitation by using two or more telescopes together to form an **interferometer**, which can achieve a resolution similar to that of a single dish of diameter equal to the maximum spacing (or baseline) between the telescopes. Signals from each telescope are carried by cable or microwave links to a central point, where they are combined. Interferometry can form detailed images of radio sources in a way that is not possible with single telescopes.

In the technique known as aperture synthesis, two or more telescopes are pointed at the same source for 12 hours as the Earth rotates. They thus sweep out curved paths around each other – strips of a much larger radio dish. By moving the telescopes to change the baselines, data is obtained as if a complete larger radio dish has

been used. The Very Large Array (VLA) in New Mexico is an outstanding example of an aperture-synthesis telescope, with 27 moveable parabolic dishes arrayed in a Y-shape with baselines up to 36 km (22 miles).

In very long baseline interferometry, telescopes separated by hundreds or thousands of kilometres take part in synchronized observations of the same objects. Radio signals are recorded on magnetic tapes at each site, which are later brought together and replayed. VLBI routinely achieves resolutions a thousand times better than the best optical telescopes and is able to discern complex structures in the hearts of radio galaxies and **quasars**.

RECEIVERS

A receiver at the focus of the dish converts the very weak radio waves collected by the telescope into an electrical signal which is amplified before being sent to a control room. There it is further amplified and recorded for later analysis.

Unlike everyday radio sets, radio astronomy receivers are designed to work in a relatively narrow band of wavelengths. A telescope will usually have several receivers, at least one for each of the bands of interest.

Where a spectrum is required, the band of wavelengths received from the telescope may be divided by filters into a large number of channels and each one detected separately. Alternatively, the radio signal across the band may be converted into digital samples and a specialized computer known as an auto-correlator used to form a spectrum by a technique known as Fourier analysis.

MILLIMETRE WAVES

The last part of the radio spectrum to be explored has been the so-called millimetre and submillimetre band, with wavelengths in the range 0.3–10 mm. Although this region is of great astrophysical interest due to its wealth of **spectral lines** from interstellar molecules, it has been difficult to exploit and shares many of the problems associated with infrared astronomy. The atmosphere is only partly transparent at these wavelengths and telescopes designed to observe in them have to be sited on high, dry, mountains where there is a minimum of atmospheric water vapour. Receivers designed for the submillimetre region use techniques borrowed from both radio and infrared astronomy.

Because the waves are so short, the dishes of millimetre-wave telescopes have to be very accurately figured. Notable examples include the

RADIO INTERFERENCE

Radio emissions from distant galaxies are extraordinarily weak: an old estimate was that all the energy received by all the radio telescopes in the world is no more than that of a single falling snowflake, although this is now generally accepted to be an understatement. To ensure that these faint signals are not swamped by interference, a number of narrow bands in the radio spectrum, especially those containing the wavelengths of important spectral lines, are allocated to radio astronomy by international agreement. Other transmissions in these bands are forbidden or tightly controlled.

Two problems are threatening the future of radio astronomy. One is that as astronomers detect ever-more distant galaxies, their high **redshifts** move their spectral lines out of the protected bands. The second is the increasing pressure from commercial users of the radio spectrum. Communications satellites frequently pass through or near the beams of radio telescopes. Such satellites often emit low levels of radio noise away from their allotted frequencies and even the smallest leakage can wipe out the faint signals from a distant galaxy.

Astronomers are campaigning for better controls on emissions from satellites and more protected bands for radio astronomy, especially at millimetre wavelengths where there are numerous spectral lines from interstellar molecules.

45-m (148-ft) dish at Nobeyama in Japan and the 30-m (98-ft) dish on Pico Veleta in Spain. Telescopes for submillimetre waves, such as the 15-m (49-ft) James Clerk Maxwell Telescope and the 10.4-m (34-ft) Caltech Submillimeter Observatory, both in Hawaii, have highly polished dishes and are housed in environmentally controlled enclosures.

Organizations from the US and Europe are planning to build a 64-element interferometer in northern Chile, called the Atacama Large Millimeter Array, that will extend the power of aperture synthesis to these very short waves.

▶ *An artist's impression of one of the 64 12-m telescopes that will make up the Atacama Large Millimetre Array (ALMA) to be constructed in the Atacama desert of northern Chile by astronomers from Europe and the US.*

WATCHING THE SKY

HIGH-ENERGY ASTRONOMY

T he vast expanse of the electro-magnetic spectrum beyond the violet end of the visible range has been difficult for astronomers to explore. The Earth's atmosphere is opaque to all wavelengths shorter than about 300 nm and the only way to detect these high-energy photons is to place instruments on high-altitude balloons, rockets or satellites. The high-energy region can conveniently be divided into ultraviolet (10–400 nm), X-rays (0.01–10 nm) and gamma rays (shorter than 0.01 nm). X-rays and gamma rays are more usually described in terms of photon energy: X-rays have energies in the range 0.1–100 keV and gamma rays have higher energies still.

ULTRAVIOLET ASTRONOMY

The longer waves of ultraviolet light, between 310 and 400 nm, can be observed from the ground by conventional telescopes, but the region between 10 and 310 nm can only be observed from above the atmosphere.

Early observations of ultraviolet radiation were made from rockets and high-flying balloons. The first satellite observations were made with the US

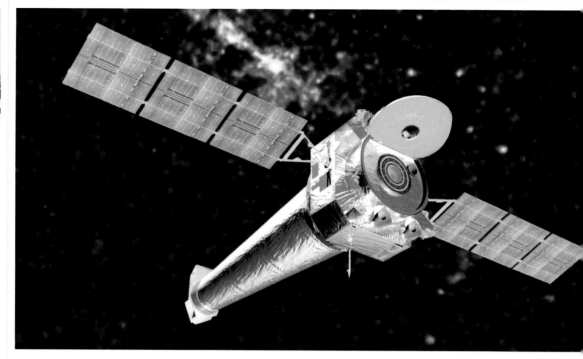

▲ *The Chandra X-ray observatory, launched in 1999, is one of the most powerful instruments ever built for X-ray astronomy. It was named after Indian astrophysicist Subrahmanyan Chandrasekhar.*

Orbiting Astronomical Observatory-2 (OAO-2) spacecraft in 1968. The most famous UV satellite, the International Ultraviolet Explorer, was launched in 1978 and worked for 18 years – the longest-lived of any scientific satellite.

The extreme ultraviolet (XUV or EUV), covering 10–91 nm, was not systematically studied until the early 1990s, when the satellites Rosat and the Extreme Ultraviolet Explorer (EUVE) conducted the first all-sky surveys. Interstellar hydrogen absorbs strongly at these wavelengths, restricting the distance that can be seen within the Galaxy. For this reason the EUV is often considered to be a

spectral region in its own right, bridging the gap between ultraviolet and X-rays.

Detectors and instruments for the ultraviolet are similar to those for visible light at the longer wavelengths, but resemble X-ray technology at EUV.

Major sources of ultraviolet radiation include hot, massive stars, the cores of active galaxies and, in the EUV, newly formed **white dwarfs**.

X-RAY ASTRONOMY

In the 1930s, physicists realized that the Earth's ionosphere must be heated by high-energy radiation from the Sun. Using a V-2 rocket in 1949, Richard Tousey (1908–97) of the US Naval Research Laboratory discovered the Sun is a source of X-rays. Using a Vanguard 3, launched in 1959, Herbert Friedman (b. 1916) and his team investigated solar X-rays, determining that only **prominences** and the **corona** are hot enough to emit them. No one expected that X-rays from other stars – being so far away – would be strong enough to detect.

In 1962 a group in Cambridge, Massachusetts, led by Riccardo Giacconi, launched a rocket to search for X-rays created by solar radiation impacting on the Moon. Instead they found a powerful X-ray source in the **constellation** of

RICCARDO GIACCONI

Riccardo Giacconi (b. 1931) is an Italian astronomer who discovered the first cosmic X-ray source and built two X-ray satellites.

Trained as a cosmic-ray physicist, in 1959 Giacconi joined American Science and Engineering, a small research company set up by scientists from the Massachusetts Institute of Technology. There he led the team that discovered the first source of cosmic X-rays, Scorpius X-1, in 1962. His group went on to build the Uhuru satellite, launched in 1970, which made the first complete survey of the X-ray sky.

In 1973 Giacconi moved his team to the Harvard-Smithsonian Center for Astrophysics, where he was associate director in charge of a new space mission – this became the Einstein X-ray observatory. Launched in 1978, Einstein was the

first X-ray satellite to employ a grazing incidence telescope to make images of X-ray sources, an idea proposed by Giacconi in 1960. It made more than 5,000 observations before it expired in 1981.

Giacconi's experience in operating Einstein as an orbiting observatory led to his appointment in 1981 as the first director of the Space Telescope Science Institute, which operates the Hubble Space Telescope. From 1993 to 1999 he was director general of the European Southern Observatory, his first posting in ground-based astronomy, and he is now president of Associated Universities, the consortium that operates the US National Radio Astronomy Observatory.

◀ *Riccardo Giacconi, one of the pioneers of X-ray astronomy, led the team that discovered the first cosmic X-ray source in 1962. He was responsible for the Uhuru and Einstein spacecraft and went on to direct the Space Telescope Science Institute and the European Southern Observatory.*

WATCHING THE SKY

THE UHURU SURVEY

By the end of the 1960s only about 30 X-ray sources were known, and most of them had been discovered during brief flights by sounding rockets or by satellites designed for other purposes. The need for a systematic survey of the X-ray sky was all too apparent. It finally came in 1970 when NASA launched SAS-1, the first Small Astronomical Satellite.

Launched from a disused oil rig off the coast of Kenya, SAS-1 was renamed Uhuru after the Swahili word for 'freedom'. It carried two sets of proportional counters, one on either side of the cylindrical spacecraft. As Uhuru rotated on its axis five times an hour, the counters swept across the sky picking up X-ray sources. It was also used as an observatory to study the characteristics of individual objects. The mission lasted just over two years and the final Uhuru catalogue, published in 1978, contained 339 sources.

One of the most important discoveries from Uhuru was that many X-ray sources are binary stars, in which one is spilling gas on to the other. It also discovered strong X-ray emission coming from hot gas in the centres of rich clusters of galaxies.

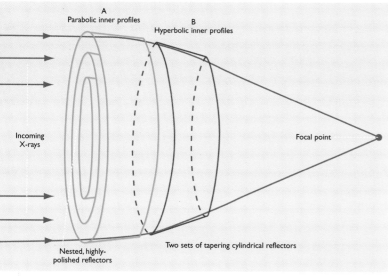

▲ A Grazing-Incidence Telescope
A grazing-incidence telescope focuses X-rays by bouncing them off two tapering sets of nested, highly polished reflectors. The inner surfaces of set B have hyperbolic profiles.

Scorpius, which became known as Scorpius X-1. Further discoveries followed from rockets and high-altitude balloons and, since the 1970s, several surveys of the sky from orbiting satellites have revealed tens of thousands of cosmic sources of X-rays.

Most ordinary stars emit only weak X-rays and the X-ray sky looks very different from the sky at other wavelengths. X-rays come from extremely hot gas (at 10^6–10^8 K), and the most important sources are interacting **binary stars**, **supernova** remnants, active galaxies and rich clusters of galaxies.

X-RAY TELESCOPES

Because of their high energy, X-rays cannot be focused by conventional mirrors and lenses. It is therefore not possible to build an X-ray telescope to collect and image X-rays in the same way as visible or even ultraviolet radiation.

An ingenious solution was proposed in 1952 by German physicist Hans Wolter who was attempting to design an X-ray microscope. He saw that X-rays could be reflected and focused by metal mirrors if they impinged at a glancing angle – perhaps only one or two degrees. Although the microscope was never built, the idea was taken up in the 1960s to design telescopes for X-ray astronomy. A typical grazing incidence telescope contains pairs of parabolic and hyberbolic mirrors that direct incoming X-rays to a focus. The collecting area of each pair of mirrors is small, so two or more pairs are normally nested inside each other.

A prototype X-ray telescope flew on a sounding rocket in 1963, and two large X-ray telescopes for solar astronomy were operated by astronauts on the US Skylab space station in 1973–74. The first telescope designed to detect the much weaker cosmic X-rays flew on the Einstein satellite in 1978 and produced the first detailed images of X-ray sources. The most important X-ray telescopes in use today are on the US Chandra and the European XMM-Newton satellites, both launched in 1999.

GAMMA RAY ASTRONOMY

Gamma rays are **photons** with energy higher than about 100 keV. Like X-rays, they are stopped by the Earth's atmosphere, although the very highest energies can be detected on the ground by their effects on the upper atmosphere.

The first satellite to carry a gamma-ray telescope was the American SAS-2, launched in 1972. In 1975, the ESA COS-B satellite was launched,

💡 **PARTICLE DETECTORS**

At X-ray wavelengths, photons behave more like particles than waves and special techniques are used to detect them. For lower-energy X-rays, the usual detector is a proportional counter, a gas-filled box that detects X-rays by the **ionization** they produce on colliding with gas molecules. For higher-energy X-rays and gamma rays, scintillation counters are used, where the photon causes a flash of light in a crystal, often sodium iodide, which can be detected by a photomultiplier tube. More recently, semiconductor detectors have become available.

The more energetic gamma rays require a device called a spark chamber, in which the passing gamma ray causes electrical discharges within a stack of charged plates. The positions of the sparks on each plate can be used to estimate the direction of travel of the photon.

All these devices are also sensitive to even higher-energy **cosmic-ray** particles, so they are often enclosed by an anti-coincidence system designed to filter out events of higher than expected energy.

Gamma rays above 100,000 MeV are so powerful that they cause flashes of light (Cerenkov radiation) in the upper atmosphere which can be detected by sensitive instruments on the ground.

and it operated for several years. One of the most significant spacecraft dedicated to the study of gamma rays was the US Compton Gamma Ray Observatory, which operated from 1991 to 2000 and observed over 300 gamma-ray sources for each year of its operation.

Major sources of cosmic gamma rays include solar flares, pulsars, supernova remnants, quasars and active galaxies. Thousands of sudden bursts of gamma rays have been detected from all parts of the sky, but their origin is still unclear.

▼ The Compton Gamma Ray Observatory, *pictured here being launched from the Space Shuttle Atlantis in 1991, increased the known number of gamma-ray sources tenfold before burning up in the Earth's atmosphere nine years later.*

BEYOND THE ELECTROMAGNETIC SPECTRUM

The whole of the electromagnetic spectrum is now open for observational astronomy, but electromagnetic waves are not the only means of learning about the cosmos. Three other kinds of phenomena also carry information to Earth about events in space. The earliest to be discovered, the cosmic rays, are high-energy charged particles whose origin is still not understood. Neutrinos are tiny subatomic particles created in nuclear reactions inside the Sun and elsewhere. And gravitational waves, yet to be detected directly, promise to open a new window on stellar collisions and other violent events in the Universe.

COSMIC RAYS

Cosmic rays are highly energetic charged particles that constantly bombard the Earth from space. Their energies range from 10^8 to 10^{19} eV, far in excess of anything that can be produced in particle accelerators. The majority are **protons** (hydrogen nuclei) and most of the remainder are alpha particles (helium nuclei), with a few per cent being **electrons**. These primary cosmic

▶ *One of the* **1,600** *cosmic-ray detectors that will form part of the Pierre Auger Observatory. Each self-contained detector is solar-powered and contains 11,000 litres (3,000 gallons) of water.*

rays collide with **molecules** in the atmosphere to produce large numbers of other particles (secondary cosmic rays) which in turn decay to produce an extensive air shower which may cover several square kilometres of the Earth. A single energetic primary particle can produce up to a half a million secondary particles.

The lower-energy primary cosmic rays can be studied from rockets and satellites using the same kind of detectors as those used for X-rays and gamma rays. The more powerful and rare primaries can only be studied from the ground by detecting secondary particles produced in the extensive air showers.

A new international cosmic ray observatory is under construction. The Pierre Auger Observatory, named after the French physicist who discovered extensive air showers, will consist of two arrays of detectors, one in the US and one in Argentina. Each array will contain 1,600 detectors spread over an area of 3,000 sq km (1,158 sq miles).

WHERE DO COSMIC RAYS COME FROM?

Although cosmic rays were discovered almost 100 years ago, there is still no firm agreement on where they originate. All indications are that they come equally from all directions and that there are no obvious sources of cosmic rays as there are for other kinds of

radiation. This is because, being charged particles, the paths of the rays are bent by the magnetic fields of interstellar space and any trace of their original direction of motion has been lost.

Space physicists have calculated that cosmic rays can survive only a few million years before escaping from the Galaxy, so they must be regularly replenished. Some of the very low-energy cosmic rays are emitted by the Sun and it is possible that other lower-energy rays are emitted in supernova explosions, but as yet the origin of the higher-energy rays remains a mystery. They may even come from outside the Galaxy.

VICTOR HESS

Victor Hess (1883–1964) was an Austrian physicist who discovered cosmic rays in 1912. In 1910 Hess joined the Institute of Radium Research at the University of Vienna. One problem under investigation at the time was the origin of background radiation that appeared in **radioactivity** experiments, even within lead-shielded containers. All attempts to trace the source of these rays had failed, and there was speculation that they were coming from the sky.

Hess made several hazardous ascents in balloons up to heights of 5.3 km (3.3 miles) to discover whether the background radiation varied with height. He found that the radiation gradually decreased at first but then rose again until the intensity at 5 km (3 miles) was twice that at ground level. From these and later studies, Hess demonstrated that

the background radiation was due to extremely powerful cosmic rays coming from space.

The importance of Hess's discovery was not merely a new astrophysical phenomenon, it also showed that the cosmic rays were extremely energetic – far stronger than could be generated in laboratories. The rays were so penetrating that they could be detected through a metre of lead or a 500 m (1,640 ft) depth of water. Investigation of the nature of cosmic rays led directly to the discovery of the **positron** by Carl Anderson, and he shared with Hess the Nobel Prize for physics in 1936.

▶ *Austrian physicist Victor Hess flew balloons to heights of more than 5 km (3 miles) to identify the source of the mysterious radiation he believed was coming from the sky.*

THE HOMESTAKE NEUTRINO EXPERIMENT

The principle of a neutrino detector based on chlorine was worked out as early as the 1940s. If a neutrino collides with the nucleus of a chlorine atom, it is converted into an atom of radioactive argon which can then be detected when it decays.

In the mid-1960s a team led by Ray Davis (b. 1914) at the Brookhaven National Laboratory constructed a neutrino detector almost 1,500 m (5,000 ft) below ground at the Homestake gold mine in South Dakota. At these depths, it is not affected by cosmic-ray particles. It consists of a tank of 400,000 litres (100,000 US gallons) of perchloroethylene (C_2Cl_4), commonly used as dry cleaning fluid. Neutrinos from the Sun occasionally collide with one of the chlorine atoms in the fluid, turning it into argon. The argon is flushed from the tank by helium gas, collected and measured.

The experiment has been running since 1967 and the results have been confirmed by independent studies – only about one-third of the expected number of neutrinos are being detected. This finding poses a continuing challenge to physics.

NEUTRINOS

Neutrinos are tiny, fundamental particles that are emitted in certain kinds of nuclear reaction. They possess little or no **mass**, interact extremely weakly with other matter and always move at, or very close to, the speed of light.

Their interest to astronomers is that they are generated in the nuclear reactions that occur in the cores of stars like the Sun. They stream un-hindered out of the Sun and pass straight through the Earth. Several large detectors have been constructed to search for solar neutrinos. In one type, chlorine atoms are turned into argon on the very rare occasions on which they collide with a neutrino. In another, gallium atoms are turned into germanium. In a third type, neutrinos occasionally collide with water molecules to produce flashes of light. Results so far indicate that fewer neutrinos are being caught than theory predicts, but this is more likely due to subtleties in the nature of the neutrino than failings in our understanding of the solar interior.

Other possible sources of neutrinos are supernova explosions (neutrinos from Supernova 1987A were detected by chance) and active galaxies. Neutrinos created in the Big Bang may form part of the **dark matter** that cosmologists believe accounts for a large part of the mass of the Universe.

GRAVITATIONAL WAVES

Gravitational waves are ripples in space–time predicted by **Albert Einstein**'s (1879–1955) general theory of relativity. In the 1960s, experiments in the US appeared to show that gravitational waves were coming from the centre of the Galaxy. Other experiments failed to confirm this and it is now thought that the claims were mistaken.

Several research groups are nonetheless constructing highly sensitive instruments which they hope will be the first to catch gravitational waves. The most ambitious project is the Laser Interferometer Gravitational Wave Observatory (LIGO) which is being built on two sites in Louisiana and Washington in the US. Each installation is an L-shaped structure with 4-km (2.5-mile) arms. Masses suspended at the corner of the 'L' and at the ends of the arms will move in response to a passing gravitational wave. Laser beams directed along the arms will sense these tiny motions of no more than one thousandth the diameter of a proton. Similar, though smaller-scale, observatories include VIRGO and GEO in Europe.

An international proposal called LISA (Laser Interferometer Space Antenna) envisages a space-borne interferometer consisting of three spacecraft forming an equilateral triangle with sides measuring 5 million km (3 million miles) long. Lasers shining between the spacecraft will monitor their separation and so detect passing gravitational waves.

BINARY PULSARS

Although gravitational waves have not yet been detected directly, astronomers have good reason to believe they exist. In 1974 two US astronomers discovered a pulsar in orbit around a **neutron star**. By making careful measurements of the pulse times, they could calculate the orbits of the stars with great precision. General relativity predicts that the two stars, orbiting each other in less than eight hours, will be radiating energy in the form of gravitational waves, and this should be revealed by the orbits slowly contracting as the stars spiral in towards each other with increasing speed. Within a few years, the orbits were found to be shrinking at precisely the rate predicted by general relativity. So compelling was this evidence for gravitational waves that the two astronomers, Joseph Taylor (b. 1941) and Russell Hulse (b. 1950), were awarded the Nobel Prize for physics in 1993.

▼ *The Sudbury Neutrino Observatory* in Ontario is located 2,000 m (6,800 ft) below ground in a nickel mine. It consists of a 12-m (40-ft) spherical tank containing 1,000 tonnes of heavy water. Neutrinos colliding with the water molecules emit flashes of light which are picked up by an array of 9,600 photomultiplier tubes.

SOURCES OF GRAVITATIONAL WAVES

What sources of gravitational waves might exist in the Universe? According to general relativity, any accelerated mass will radiate energy in the form of gravitational waves which travel through space at the speed of light. Such waves would be extremely weak, so astronomers are interested in situations where high masses are given very large accelerations.

One possibility is a supernova explosion, where most of the mass of a star is suddenly blasted into space. But to generate large amounts of gravi-tational radiation, the explosion must be lopsided. If the explosion ejected matter equally in all directions, no gravitational waves would be emitted. Because the details of supernova explosions are not well understood, theorists cannot predict easily how much gravitational radiation might be produced.

The most reliable source of gravitational waves may be colliding neutron stars. Astronomers have discovered several pairs of neutron stars in orbit around each other. As the orbit decays, the stars are expected to spiral in towards each other until they collide, generating a massive pulse of gravitational waves. The outburst would be so powerful that a detector like LIGO would be able to 'see' any such collisions in galaxies up to 70 million light years from Earth.

Einstein ▶ p. 88

Watching the Sky
TOOLS & TECHNIQUES FOR AMATEURS
BINOCULARS & TELESCOPES

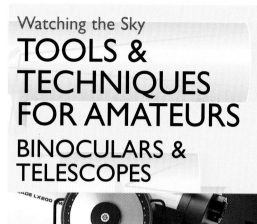

A telescope opens up our beautiful Universe to everyone's scrutiny, whether they are a professional astronomer or casual sky gazer. It is the most important tool available to the astronomer. A telescope collects a wider beam of light than is possible with our small eyes. Collecting more light allows us to see fainter objects and hence enables us to detect objects at much greater distances than our eyes can ever see. The magnificent, but sometimes bewildering, array of telescope types, binoculars, eyepieces and computerized telescopes provide the amateur astronomer with excellent tools for pleasure or research.

BINOCULARS

Binoculars represent the most portable of all astronomical instruments and are an excellent instrument with which to begin. The technical parameters for binoculars consist of two numbers, the **aperture** of each of the main lenses, and the magnification. As an example, a pair of 7×50 binoculars has a seven times magnification and a pair of 50 mm lenses.

The exit pupil is another factor that should be taken into consideration, found by dividing the aperture by the magnification. For our example of 7×50 binoculars, the exit pupil equals 50/7, about 7 mm. The pupil of an average eye, when fully dark-adapted, measures about 7 mm across, so a pair of binoculars such as these is perfectly suited to astronomy. Since many observers live in brightly lit cities, their pupils expand to only about 5 mm in diameter at best, so a pair of 10×50 binoculars might be better suited.

REFRACTORS AND REFLECTORS

There are two principal types of telescope, a refractor and a reflector. A refractor was the kind of telescope used by **Galileo Galilei** (1564–1642) and uses a lens to collect and focus the light. Light from an object enters the lens, which focuses the light and is then magnified by an eyepiece. Single lenses focus different colours of light to different points, resulting in **chromatic aberration**. Using a combination of lenses made from glass of different refractive indices, so light through each layer is bent by different amounts, chromatic aberration can be almost removed. Such lenses are called achromatic lenses, and all good astronomical telescopes use this type.

A reflector uses a concave mirror with a parabolic curve. Reflectors do not suffer from chromatic aberration. In the Newtonian design, invented by **Isaac Newton** (1642–1727), light travels down a tube to the mirror, which focuses the light on to a small diagonal mirror (called the secondary). This diverts the light to the side of the tube where an eyepiece is placed to magnify the image. Another common design of reflector is the Cassegrain. In place of the flat diagonal, a convex secondary sends light back down the tube and through a hole in the main mirror. This kind of telescope has a **focal length** double that of the tube. Further variations on the Cassegrain design exist, the most popular and affordable of which is the **Schmidt-Cassegrain**, a catadioptric type that employs both lenses and mirrors. Typically a spherical mirror with a central hole forms the primary mirror. A weak correcting lens is placed at the front of the telescope tube to correct the aberrations intrinsic in the spherical mirror.

TELESCOPE MOUNTS

A telescope is only as good as the mount which supports it, so a quality mount should contain substantial components. Most amateur telescope mounts are one of three basic types: **altazimuth**, German **equatorial** and equatorial fork. A frequently included option is an electric drive which automatically tracks celestial objects when the telescope is aligned properly.

- Altazimuth: this is the most basic of mounts. A horizontal and vertical axis allows the telescope to be moved up and down and from left to right. Movement in both axes is required to follow a celestial object. Dobsonian telescopes utilize an altazimuth mount. More recently, advanced computerized telescopes use this design in which computer software calculates motions in both axes to track objects across the sky.

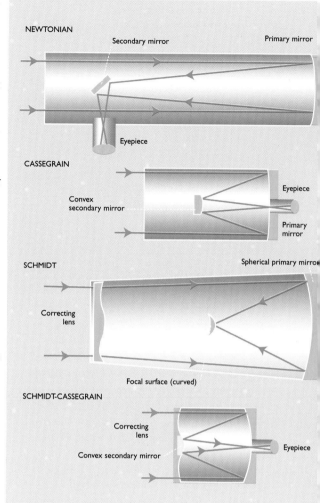

▲ Types of Reflecting Telescope
Light rays are directed along different paths in the four principal types of reflecting telescope: Newtonian, Cassegrain, Schmidt and Schmidt-Cassegrain.

- German equatorial: contains a polar axis that is aligned to the Celestial Pole and a declination axis at 90° to the polar axis. Rotation around the polar axis alone will allow the telescope to track celestial objects.
- Equatorial fork: a variation on the German mount in which the telescope lies inside a U-shaped frame. The two sides of the fork provide the declination axis, and the base of the fork is mounted at the upper end of the polar axis. The popular, short-tube, Schmidt-Cassegrain telescopes use this mount.

EYEPIECES

Selecting an eyepiece is an important consideration for any telescope user and there are many different designs. Beginners will find that a collection of three eyepieces of low, medium and high magnification will be suitable for most observations.

WATCHING THE SKY

Modern eyepieces are made up of several elements to provide excellent colour correction plus a wide field of view and easy eye relief. Good eye relief means that the viewer does not have to place their eye right up to the eyepiece to see the image. The most popular designs are the Orthoscopic, Plössl, Erfle and UltraWide. Most modern eyepieces come with antireflection coatings to avoid significant light loss. The Plössl is most popular today, with a four-element design giving a fine combination of good eye relief and a wide apparent field of view.

COMPUTERIZED TELESCOPES

Once you have a telescope, one of the principal challenges is finding the many faint galaxies and star clusters across the sky. With ever-cheaper computers and miniaturized electronics, many telescope manufacturers now offer instruments with automated targeting of objects. The positions of thousands of galaxies and **nebulae** are stored in the computer memory, often located in a large handset supplied with the telescope. Computerized telescopes require initial set-up by acquiring three stars spread across the sky. Once achieved, a handset allows the observer to select any object of known **right ascension (RA)** and **declination**, and the telescope will slew to the object, placing it in the centre of the field of view.

Instruments with on-board computer systems such as these are tremendously versatile, allowing the observer to view hundreds of objects in a single night. This would be very beneficial to observers interested in searching many galaxies for potential **supernova** explosions. Some telescopes do offer an RS-232 port, allowing a connection to be made to a laptop or personal computer. Planetarium software can interface to these telescopes, displaying the current night sky. Clicking on an object on the screen will cause the telescope to automatically slew to the correct object.

▶ **1. Portable** *computer-controlled Maksutov; 2. 20-cm (8-in) Schmidt-Cassegrain; 3. 40-cm (16-in) Schmidt-Cassegrain; 4. Newtonian reflector; 5. Dobsonian reflector; 6. 60-mm refractor.*

USING THE EYE

Understanding some basics about how the eye sees fine details is as important as the quality of the telescope you use when it comes to observing.

The eye has rods and cones in the retina that detect light. Rods are very sensitive to light and can only detect shades of grey, even in near complete darkness. Rods can detect outlines and silhouettes of objects and are 20 times more numerous than cones. Cones are sensitive to colour and are concentrated in the central region of the retina but they do not work well in low light levels. Consequently, when viewing a dim galaxy through a telescope, the object often becomes more prominent if you divert your eye off to one side a little, throwing the light on to the more numerous light-sensitive rods. Faint objects might appear to blink on and off as you glance away and at the faint object. This technique is widely used by amateur astronomers and is called averted vision.

Adapting your eyes to darkness is an essential prerequisite to an observing session. Rods can become saturated by high light levels. If you walk outside at night from a brightly lit room you will notice it takes some time to adjust your sight to the darkness. It takes at least 10 minutes for the rods to adjust to low light levels. Once dark-

adapted, very faint objects can be viewed through your telescope that would be invisible to the casual viewer. It is also essential to maintain dark adaption by avoiding any bright light. Since the rods are least sensitive to red light, a red flashlight is the most useful accessory for reading star charts and consulting references.

CHOOSING A TELESCOPE

Selecting a telescope for your own use is not easy. The wide array of equipment that comes in very different sizes, designs and prices can be daunting to the beginner. By following a few guidelines, the process can be made much easier. First of all, decide what kind of objects you want to observe. As a general rule, refractors tend to be better for planets and reflectors are ideal for deep-sky objects. Also consider that, aperture for aperture, refractors are significantly more expensive than reflectors.

A telescope is a precision optical instrument, and as such is not cheap. If you find a cheap telescope, then a cheap telescope is often what you will get. Many telescopes are advertised in terms of magnification – and such adverts are misleading. Simply changing the eyepiece can vary magnification. Try to avoid telescopes that are advertised on the basis of magnification, since the claims are usually beyond the normal usefulness of the telescope being offered for sale. The most important aspect of a telescope is aperture – the larger the better – since you will be able to see finer detail (higher **resolution**) and fainter objects.

MAGNIFICATION

The magnification of a telescope is dependent on two physical characteristics: the focal length of the main lens or mirror, and the focal length of the eyepiece.

The focal length is often printed on the side of a telescope. The magnification of a particular eyepiece is found by dividing the focal length of the telescope by the focal length of the eyepiece. For example, if your telescope has a focal length of 2,000 mm (6.6 ft), a 20-mm eyepiece will give a 100 times magnification, and a 10-mm eyepiece will produce a 200 times magnification.

Remember that as the magnification increases, the limited amount of light from an object is spread over a larger area, making it dimmer. Increasing magnification also increases the effect of turbulence in the atmosphere, making it difficult to focus on the object in the eyepiece. In addition, high magnification increases the demands on your telescope mount, since the object will move through the field of view faster and any imperfections in your telescope mount will also be magnified. The maximum magnification a telescope can normally handle is 2.5 times the aperture in mm (60 times the aperture in inches).

Galileo ▶ p. 40 Newton ▶ p. 42

WATCHING THE SKY

FINDING YOUR WAY

T here are many tools to help you find your way around the sky. Learning the constellations is rather like discovering your way around a new city – learn the major routes first. The first item you will need is a map. Concentrate on the principal constellations; then, when you are familiar with these, add in the details of the smaller and dimmer constellations. A good place for northern observers to start is with the Plough (or Big Dipper). Its seven bright stars are useful as a guide to finding other constellations.

STAR MAPS AND PLANISPHERES

Star maps provide a positional layout of **constellations** in the same way that maps of the Earth show roads. The co-ordinate system is analogous with that used for the Earth. On a star map, **longitude** is called right ascension and **latitude** is called declination. Locations of stars, **galaxies**, star clusters and nebulae are indicated by special symbols. Some maps will show only naked-eye stars while others contain hundreds of charts down to very faint **magnitudes** showing thousands of stars.

A planisphere is a unique star map showing the sky on a circular format, synonymous with the entire sky. Two circular disks rotate over each other with a cutaway revealing the constellations above the horizon for any day and time of the year. Simply dial up the current time and date for the current sky view. Remember, though, that a particular planisphere works only in a narrow range of latitude.

PLANETARIUMS AND PUBLIC OBSERVATORIES

The best way to learn your way around the sky is to have someone guide you for an initial orientation. To do this you can join a local astronomical society or visit your local planetarium. Many modern planetariums concentrate on telling the story of discoveries in astrophysics, with spectacular effects, but occasionally they also perform constellation shows. Most planetariums utilize sophisticated star projectors to show exceptional simulations of the night sky, making them excellent venues for learning your way around the sky. Some computer-graphic star projectors produce three-dimensional views of constellations revealing the true distances of the stars, and can even fly you around the **Solar System**, showing how the plane of **orbits**, when viewed from Earth, define the **ecliptic** in relation to the constellations.

Some planetariums, local universities or local astronomical societies operate public observatories and offer open nights for the public to view through large telescopes. Often manned by local amateurs, these are excellent places to become acquainted with the night sky, and a fine opportunity to ask questions of experienced amateurs. These venues, in addition to monthly astronomy magazines, show where the planets are in the current night sky, an important prerequisite to observing them with telescopes.

ATMOSPHERIC TURBULENCE

The finest of telescopes are rendered useless on nights when the atmosphere is so turbulent that it blurs the fine details visible on the surfaces of planets. The level of turbulence is called 'seeing' by astronomers. Good seeing indicates that the atmosphere is very steady and stable. Periods of good seeing are rare in most populated areas of the world. Contrary to what you might expect, the clearest nights are not always times of best seeing. Very clear nights often occur after a period of heavy rainfall that clears the air of dust, making the atmosphere more transparent, yet turbulent

due to the collision of hot and cold air masses. Good transparent nights aid finding dim galaxies but do not assist spotting fine planetary detail. Times of good seeing can occur during a long period of high pressure, since the atmosphere is dense and less likely to be turbulent.

OBSERVING THE PLANETS

Recording the detail that is revealed by small telescopes on the planets is best done by drawing what you see. This requires patience and practice since the image is affected by turbulence of the atmosphere and rarely is the observer in a comfortable position. Begin observing with a moderate power of around ×100. Record the details that you see using a soft pencil on a predrawn disk. Some amateur astronomical organizations issue standard disk blanks for observers to use. Once you have drawn the major features, move to a higher power of ×150 to ×200 by changing the eyepieces. Note the details on the planetary surface during brief moments of steady seeing. Take care with Jupiter, since the rotation of the planet will begin to be noticed within 10 to 15 minutes. Always record the date and time of the observation, plus details of telescope, magnification and viewing conditions. A logbook will act as a permanent record of your observations.

By following a particular planet night after night, the observer will become familiar with the typical appearance of the surface and be able to pick out abnormalities. The giant planet Jupiter is the most rewarding of all the planets because of its ever-changing cloud patterns. The Great Red Spot – in addition to other bright and dark spots, plumes of dark material and bright hollows – makes regular appearances. The other major planets offer equally attractive sights to record.

▼ *Two types of transient object regularly observed by amateurs: a comet (in this case Halley's Comet, seen from Australia at its last return in 1986) and the narrow streak of a meteor burning up in our atmosphere to its right, caught by chance. The fuzzy ball in the upper right is the galaxy NGC 5128.*

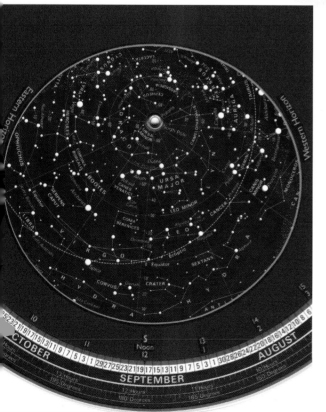

◄ *A planisphere with date and time settings. When buying one, make sure the planisphere works for the latitude in which you live.*

coma ▶ p. 337 comet ▶ p. 337 constellation ▶ p. 337 ecliptic ▶ p. 338 emission nebula ▶ p. 338 galaxy ▶ p. 339 ionization ▶ p. 340 latitude ▶ p. 340
longitude ▶ p. 340 magnitude ▶ p. 341 meteor ▶ p. 341 nova ▶ p. 341 orbit ▶ p. 341 planetary nebula ▶ p. 342 radiant ▶ p. 343 Solar System ▶ p. 343

TRANSIENT PHENOMENA

Some of the fascinating events in the sky happen quickly, and cannot be predicted: **meteors** streaking across the sky, **novae** (faint stars which suddenly flare up in brightness), supernovae (entire stars exploding in distant galaxies), and ghostly **comets** which grace our skies for a few weeks at a time. It is most important to record an accurate time, preferably to the second, of any transient event in the sky for the observation to be scientifically valuable.

Detailed drawings using a pencil are often the best way to capture details of a comet. Subtle

✷ NEBULA FILTERS – BEAT LIGHT POLLUTION

Many towns and cities suffer significantly from light pollution. Streetlights are not shielded properly and their glow illuminates the night sky, robbing millions of the view of a starry sky. However, technology is on the side of the telescope user from such towns and cities. Special filters designed to block light pollution and enhance the contrast of the glow from **planetary nebulae**, galaxies and **emission nebulae** are available, and simply screw into modern eyepieces. These light pollution filters, also called nebula filters, allow observers to see emission nebulae that are otherwise completely invisible from most urban areas.

Unlike stars, nebulae emit light in very narrow wavelength ranges. Nebula filters allow the maximum transmission of wavelengths most commonly emitted by nebulae: hydrogen-alpha, hydrogen-beta and doubly **ionized** oxygen (OIII).

Nebula filters come in two principal varieties, broadband and narrow band. Broadband filters are best used in areas of moderate light pollution and for astrophotography. They suppress light entering your eyepiece from the glow of sodium and mercury vapour lights, and the glow of ionized oxygen in our atmosphere.

Narrow band filters are best used in highly light-polluted regions, or if you want to maximize the contrast for nebulae that are exceedingly faint. They allow maximum transmission of hydrogen-beta and doubly ionized oxygen, typically emitted by planetary nebulae and some emission nebulae. They are not useful for nebulae that emit the predominantly red glow of hydrogen-alpha, since these filters cut its transmission to zero. Experienced observers use both types of filter, as the improved viewing is dependent on the amount of light pollution as well as on the type of object being observed. In dark observing locations, some objects, such as the Veil Nebula in Cygnus, become quite obvious when a nebula filter is employed.

features are often lost when comets are photographed. The **coma** of a comet can contain fans and jets. The tail of a comet can reveal wisps blown off the nucleus days earlier. Comets change daily and sometimes hourly.

Meteors are best observed with the naked eye. Meteor showers are great events to plan a group activity. Each observer can select a region of the sky to observe, and the group can provide whole-sky coverage during the event. Capturing a supernova visually requires a good detailed knowledge of the star field surrounding a galaxy. Photographic surveys are the most productive. Many images of galaxies can be recorded on a single night and checked the following day for a new object.

As soon as a new comet, nova or supernova is discovered, a report is sent to the International Astronomical Union's Central Bureau for Astronomical Telegrams at Harvard University in the USA. This central clearing house for astronomical discoveries disseminates the news via circulars to professional astronomers worldwide.

NEBULAE AND GALAXIES

Nebulae and galaxies have large angular size with low surface brightness. Consequently, low magnification is best to view them. To find them,

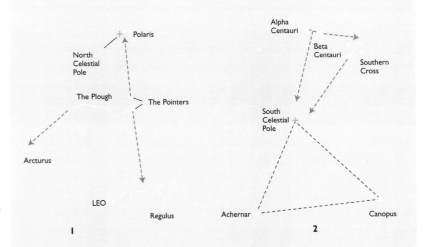

▲ **Signposts in the Sky**
1. Two stars in the bowl of the Plough act as pointers to the north pole star, Polaris. Extending the pointers in the opposite direction brings you to Leo the Lion. The Plough's curving handle points to the bright star Arcturus.
2. The bright stars Alpha and Beta Centauri act as pointers to Crux, the Southern Cross. The long axis of Crux, and a line at right angles to Alpha and Beta Centauri, points to the South Celestial Pole. The bright stars Canopus and Achernar form a large triangle with the pole.

you can use your telescope's setting circles or star-hop to each object using brighter stars as guides. Either way you will need a good-quality star map showing the detailed position of the objects.

Once you have found an object, try and record its appearance permanently in your logbook. This can be achieved by drawing the object using a soft pencil. Draw the principal stars first and then try to accurately represent the appearance of the nebula or galaxy. Since these are diffuse objects, practice smudging the drawing with an Indian rubber eraser or with the tip of your finger.

⎮ METEOR SHOWERS

During a meteor shower it is best to observe a region of the sky about 50° altitude and 60° to 90° away from the **radiant**. The simplest observation is to record the number and magnitude of each meteor for a 10-minute interval. To determine the magnitude of a meteor, compare the brief glimpse of the meteor's brightness to that of major stars in the region. Become familiar with their magnitudes so that a rough estimate of the meteor's brightness can be recorded.

More challenging is to record the path of each meteor on a star chart. At the moment a meteor appears, hold up a pencil to the sky along the path of the meteor. Identify major stars along the meteor's path and record this path on your star chart, along with a record of the exact time it appeared.

Try photographing meteors. Mount a camera on a tripod and point at a region of sky about 30–50° away from the radiant. It is best to use a fast film of at least ISO 800. Use a camera with a 'B' setting and hold the shutter open for 10 minutes per frame using a camera cable release. Make a note of the start and end time of each exposure and note any meteors that pass through the region of sky you are photographing. Expect to use plenty of film before you catch your first meteor.

More detailed observations would include recording anything unusual about the meteor, such as luminous trains remaining after the meteor has passed, or unusually slow or fast meteors.

Submit your observations to meteor observation societies such as the British Astronomical Association's Meteor section, the Association of Lunar and Planetary Observers, or the American Meteor Society. Other countries may also have local meteor societies.

IMAGING THE SKY

Taking photographs of the night sky is one of the most satisfying activities open to the amateur astronomer. Although digital cameras proliferate for everyday use, film remains the leader for amateurs when it comes to capturing wide areas of the sky since film retains speed and high resolution. Equip yourself with a simple camera, tripod and a cable release that allows you to lock the shutter open, and you are ready to begin photographing star trails or planetary conjunctions. More advanced techniques can produce stunning images of planets and galaxies by mounting the camera piggyback on a telescope or photographing through the telescope.

DONALD PARKER

Donald Parker has elevated charge-coupled device (CCD) imaging by amateurs to a level that is equivalent to many professional observatories. His exquisite images of Mars and Jupiter taken through his trusty 16-in (40.6-cm) telescope from his home in Coral Gables, Florida in the US are marvels of patience and dedication.

Parker began his planetary observation career in the 1950s and has been an active contributor to the Association of Lunar and Planetary Observers ever since.

In 1986, prior to the era of CCD imaging, he won the prestigious Walter A. Haas Award for his contributions to planetary photography. Mars was at opposition in the same year and Parker visited the Cerro Tololo Inter-American Observatory, the University of Missouri at St Louis, and the National Geographic Society to photograph the planet. He has continued to excel by developing techniques of planetary CCD imaging during the 1990s which have revealed detail on Mars previously unattained with ground-based imaging. Imaging Mars through three different colour filters, he now achieves extraordinary colour images. Parker has written dozens of scientific papers about solar-system astronomy and astrophotography. He is generally considered to be the world's finest amateur planetary photographer.

CAMERAS

There is a wide variety of cameras, both digital and conventional. Most of the commercially available cameras which dominate the market are unsuitable for most astrophotography applications. The best astronomical cameras are the 35-mm, single-lens reflex variety that use interchangeable lenses and have manual controls for aperture and exposure, including a 'B' setting that allows the shutter to be held open using a cable release. Avoid cameras whose shutter depends on a battery – long exposures on cold nights will drain the power rapidly. Some of the newer digital cameras are good for taking night scenes at star parties, for example, but images tend to be contaminated by digital noise.

When photographing through a telescope, specialized CCD cameras can record extremely faint images that are impossible with conventional film. They have many advantages and a variety of models are commercially available for astronomy. CCD cameras use a **charge-coupled device**, a light-sensitive electronic chip that records an image as an electronic signal. The image can be viewed immediately or stored on a computer and it can be enhanced using image-processing software. A CCD camera includes all the electronics to control the chip, plus a thermo-electric cooler to bring the chip down to -40°C (-40°F).

▼ *An 80-mm refracting telescope. There are two methods of taking atronomical photographs utilizing your telescope: piggyback and prime focus. Piggyback involves mounting the camera on an equatorially-mounted telescope; prime focus means attaching the camera to the telescope in place of the eyepiece as shown here.*

FILM

The wide range of film choices, from black and white or colour, slide or print, fast or slow emulsions, makes choosing a film rather complex. The best place to start is your local camera store, since supermarkets rarely carry the best films for astrophotography. The film you select will depend on the type of photography you want to do. Some basics will help you decide. Film comes in a variety of speeds, indicating its sensitivity to light. Slow films are rated ISO 100 to ISO 200. Faster films have speeds such as ISO 400, 800 or 1,600. The faster the film, the shorter the exposures required to capture a faint object, but the trade-off is often an increase in the graininess of the image. The best way to select a film is to experiment for yourself; doing so is a valuable learning process that will certainly benefit your future astronomical photography exploits.

VIDEO CAMERAS

Commercially available video cameras are capable of recording stars down to 2nd magnitude with little difficulty, and this has spawned a wide range of ways to record astronomical events. In addition, specialized video cameras designed for astronomical use are available. A video camera is an excellent way to record images of the Moon or planets through a telescope. Suitable adapters are available to attach a camera with a removable lens to a telescope. The images can be recorded on videotape and played back later. The video records every turbulent shift and shake of the image. Frame grabbers can select individual

WATCHING THE SKY

video frames that exhibit the best seeing. These can be combined by computer, and the resulting image captures the high resolution of the telescope that is normally not achieved in still photography. Pioneers of this technique have produced outstanding results, including the best Earth-based images of Mercury.

SKY TRACKING

A short exposure of 5–10 seconds on 800 ISO film with a standard 50 mm lens set at its widest aperture and focused to infinity will record most stars visible to the naked eye. The stars will not trail during such short exposures. Longer exposures produce wonderful star trails due to the fact that the Earth is moving, rotating 1° (two Moon diameters) every four minutes. Star trail images produce dramatic scenes, especially if combined with foreground objects.

In order to obtain crisp star images, it is necessary to place the camera on its own equatorial mount or to put the camera piggyback on a telescope. A simple alternative to an expensive equatorial mount is to build a barndoor drive. Tracking the camera and exposing the film for only two minutes will dramatically increase the number of stars recorded on film. You could even create your own photographic atlas of the sky. It is worth making it a habit of taking a couple of

▼ *A CCD camera contains a charge-coupled device, or CCD, an array of tiny electronic light sensors which, when activated, convert light signals into a pattern of electric charges. Each charge is converted to a digital read-out and a computer can reproduce the pattern of light that fell on the CCD array.*

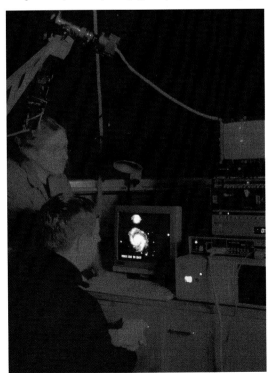

daylight shots at the beginning of each roll of film. This lets your local photolab align the film cutter on a 'normal' image and avoids them mistakenly cutting your wonderful astrophotographs in half.

PIGGYBACK AND PRIME FOCUS

High-quality astronomical photographs can be obtained by mounting a camera piggyback on an equatorially mounted telescope. Many telescope manufacturers offer special mounting adapters for most cameras. Pointing the telescope at a bright star and tracking it removes any problems of stars trailing. However, before the picture is taken, the telescope must be accurately aligned on the pole.

🪐 **STATIC CAMERA EXPOSURE TIMES**

These are the maximum exposure times for a fixed camera photographing stars near the celestial equator in order to prevent stars trailing. Exposure times can be increased slightly for other declinations.

Focal Length of Lens	Exposure
24 mm	29 sec
28 mm	25 sec
35 mm	19 sec
50 mm	13 sec
80 mm	8 sec
135 mm	5 sec

Any misalignment will show up as field rotation, streaking star images near the edge of the picture. Exposures of up to an hour may be obtained with piggybacking, but the limiting factor will be how quickly the film fogs due to light pollution. Only experimentation will determine the best exposures for your location.

The most demanding astronomical photography is with a camera attached directly to the telescope in place of the eyepiece. Detailed photographs of planets and galaxies are possible with this

▲ *A barndoor drive offers a means of taking marvellous constellation photographs and is a system you can build yourself.*

technique. Excellent alignment of the equatorial mount is the foremost requirement, followed by very precise guiding. Guiding can be managed by using an off-axis guider, a device mounted between the camera and telescope. A small prism picks off a slice of light from the telescope and sends it to the guiding eyepiece. Many modern telescopes with sophisticated electronic drives allow CCD cameras to be hooked into the drive system to provide automatic guiding.

THE BARNDOOR DRIVE

A Barndoor drive is simple to build and a most effective drive system for taking constellation photographs. A camera is mounted on top of a hinged board that is moveable. A second fixed board is held at an angle equivalent to your latitude and holds a bolt that, when turned, pushes against the upper board and moves it slowly upwards. The key to the system is to use a 20 threads per inch bolt placed at precisely 290 mm (11.5 in) from the hinge and to align the hinge to the North Celestial Pole. When the bolt is turned clockwise at one revolution per minute (use the second hand of a watch as a guide), the upper board moves at the correct rate to counteract the Earth's rotation. A small lever attached to the bolt will aid turning. The drive will be satisfactory for exposures up to about 10 minutes without trailing.

🧪 **CCD: HOW IT WORKS**

A charged-coupled device, or CCD, is an array of tiny electronic light sensors. Light falls on the array and produces a pattern of electric charges. Each one is measured, converted to a number and transmitted consecutively like a conveyor belt to be stored in a computer. When the computer reads the data, a reproduction of the brightness across the chip is displayed. Each sensor is of finite size, on average 15 microns square, and makes up one picture element or

pixel of an image. A large chip, for example, has an array of 1,024 by 1,024 pixels.

The size of each pixel determines how full each pixel can become with charge, since the greater the capacity, the larger the dynamic range. Many modern CCDs can handle a 13-magnitude difference in brightness. In addition, CCDs respond to light in a linear fashion, meaning that if double the amount of light falls on the chip, the object is double in brightness. This proves invaluable for monitoring variable stars.

TOOLS & TECHNIQUES

COMPUTER-BASED ASTRONOMY

▲ **Image processing**. *This faint digital image of M13 (1) was taken by a CCD camera mounted on a 30-cm (12-in) Schmidt-Cassegrain telescope. M13 is made faint by the substantial traces of light pollution present. The grey background is electronically cut out and replaced with a black background (2). Finally, the limited range of brightness is taken and stretched across a much wider brightness band to make the objects in the image more visible due to greater contrast (3).*

T he way amateurs carry out their hobby is changing due to a revolution in home-based and web-based computing. Detailed star maps can be downloaded from the World Wide Web. Up-to-the-minute coverage of some active meteor showers and eclipses of the Sun and Moon are relayed worldwide via the Internet. Live images of the Sun from the Solar and Heliospheric Observatory (SOHO) keep us up to date with hourly multi-wavelength images. Sophisticated image-processing tools provide the equipment for the amateur wishing to enter the once-exclusive domain of the professional astronomer. Planetarium software can now track the paths of the planets, print out sky charts and plan observing programmes, all with relative ease.

PLANETARIUM SOFTWARE

A planetarium computer program is a software package that is capable of displaying the view of the night sky from any location on Earth, and at any time of the year. The daily positions of planets, the rising and setting of the Sun and Moon and the location of thousands of deep-sky objects are displayed, often with colour coding to identify different types of objects. Position, magnitude, distance, along with rise and set times for your location are displayed on screen at the click of a button. Displays of objects in the **New General Catalogue (NGC)**, **Messier's catalogue** and the Hubble Guide Star Catalogue (GSC) are included in many packages.

Imagine viewing the motion of the planets from high above the plane of the Solar System. You can create time-lapse movies that reveal the elegant interplay of planetary motions as seen from any point. Simple graphical interfaces allow the operator to input basic orbital data of objects, such as a newly discovered comet, and plot their paths many weeks or months in advance. You can even hook the computer up to a telescope, control its pointing, and interface to a CCD camera using the planetarium program.

IMAGE PROCESSING

Astronomical imaging with CCD cameras has created new opportunities for amateurs. Image-processing tools feature a simple user interface that allows the enhancement of digital images by virtually anyone.

Faint images recorded on film are very difficult to print, but they can be readily extracted from the background by contrast stretching – a simple technique that reassigns brightness levels to the exposed **pixels** in the image. In light-polluted skies where the background noise would normally fog the image, multiple, short-exposure images can be added together simulating a longer exposure. Using this technique, images of faint galaxies from highly light-polluted areas in major cities are possible.

CCD cameras record relative light levels only, rendering each image in black and white. Multiple images taken through red, green and blue filters can be combined into a single colour image.

Sequences of images taken of the same region of sky can be readily stored and compared at a later date so that **variable stars** or roving **asteroids** can be detected. Monitoring of known variable stars using image analysis to determine accurate magnitudes is well within the amateur's grasp.

ASTRONOMY ON THE WEB

Modern amateur astronomy is changing so fast that keeping up with current trends can be difficult. However, the Internet has enabled the world of amateur astronomy to become a global village. If you wish to build an observatory, begin CCD imaging, track a new comet, learn how to observe the planets, or read about new discoveries by professional astronomers, you will be able to find the answer somewhere on the World Wide Web. Astronomy magazine web sites, national observatories and amateur societies usually have up-to-date links to many web sites on all topics. Alternatively, you can use one of the popular search engines to find what you want.

A significant advance in recent years is the offering of degree courses in astronomy over the Internet. The courses are targeted at the motivated, non-traditional students who have a passion for astronomy, and are capable of handling independent study through distance learning. On-line tutorials explain concepts in modern astronomy, and on-line exams enable students to take tests from their own home. The variety of on-line courses will expand as more universities develop remote learning tools.

REMOTE-ACCESS TELESCOPES

Professional astronomers use telescopes remotely and have done so for decades. Amateurs began gaining access to remote telescopes with the trailblazing work of the Mount Wilson Observatory's 'Telescopes in Education' project, begun in 1993. By combining a planetarium program with CCD camera control software on your home personal computer, a simple click of a button on the star map can slew the telescope to an object, take a CCD image and download the image to your own computer.

With the global expansion of the Internet, a bold venture will open up the Universe to anyone with an Internet connection. The Global Telescope Network plans for a system of sophisticated telescopes and CCD cameras which will allow anyone with an Internet connection to view and photograph the heavens, 24 hours a day, from anywhere in the world. The current state of the project can be found at www.globaltelescope.com.

▼ **Remote Access Telescopes** *such as the 0.6-m (24-in) Mount Wilson Telescope enable teachers and amateur astronomers to use telescopes from great distances via modem. Operators can swivel the telescope on to the desired object with the click of a button, take images of the sky using a CCD camera and download the result right into their own personal computers.*

WATCHING THE SKY

SETI@HOME – MILLIONS SEARCH FOR ET

SETI@home is an ambitious undertaking that uses the processing power of millions of Internet-connected computers to analyze stacks of data from the radio telescope at Arecibo, Puerto Rico, looking for possible signals from extraterrestrial civilizations.

Following its launch on 17 May 1999, SETI@home attracted over one million participants in the first three months, overwhelming the expectations of the project's designers. Following some major expansion of web-servers for the project, the number of participants reached two million in the first year, making it the largest network of computers dedicated to one project, and the largest computational task in history.

Using a smart screensaver program, each participant's computer downloads blocks of data each day, and while the computer sits unused, a background program sifts the data looking for suspect signals. The program halts any time the computer is in use, so the participant's regular computer use is unaffected by it. At the end of the run, the results are emailed back to the University of California at Berkeley for further analysis and another block of data is sent to each computer.

▲ **Students are shown** *how to link up to the 0.6-m (24-in) Mount Wilson Telescope from the Carnegie Science Center's observatory control room in Pittsburgh, Pennsylvania. Remote telescopes allow students to conduct an observing project and receive the results without having to travel to an observatory.*

CELESTIAL EVENTS ON THE WEB

A new concept in astronomical observation has been developed through the Internet. It introduces almost real-time viewing of rare astronomical events such as eclipses from isolated regions of the world. Video cameras located at widely spaced locations along an eclipse track, for example, are connected to a 'frame grabber' device that digitizes images about every 30 seconds and sends them to the image servers on the Internet. The speed with which images are broadcast to the world depends on the bandwidth of the network. For live events, web servers are set up in a number of central locations with mirror sites around the world to allow for additional viewers.

Today's live broadcasts of such events often get tied up due to oversubscribed networks. As the bandwidth of the Internet increases, higher quality images with few interruptions will be possible. Attempts at live webcasting have included a multi-aircraft observation campaign to observe the 1999 Leonid meteor shower co-ordinated by NASA's Ames Research Center. Most total eclipses of the Sun are now featured in live web casts. The Internet effectively provides the opportunity for a worldwide observatory, accessible from your own home, opening up the Universe to millions of people.

WHAT'S UP TONIGHT?

If you are planning a night's observation, or are casually surfing the Web, many astronomical sites offer current information about events in the night sky. An example of such a site is www.heavens-above.com. After entering your latitude and longitude, the site will offer daily predictions of all bright satellites, including the International Space Station, indicating when and where they can be seen. Small outline charts of all 88 constellations and the positions of **minor planets** (asteroids), planets, Sun and Moon, can be viewed. Nightly events such as when the Moon passes in front of a star (called an **occultation**) are fascinating to observe. Occultation highlights and predictions of future events are provided by the International Occultation Timing Association at www.occultations.org. For the greatest occultation of them all, a total **solar eclipse**, details of visibility and observation hints for solar and **lunar eclipses** that occur during the year, can be found at the web site sunearth.gsfc.nasa.gov.

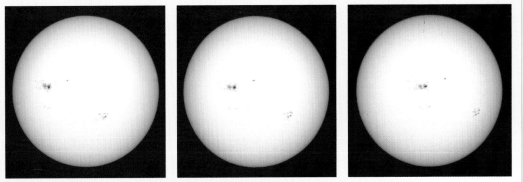

▲ *The surface of the Sun is marked by temporary dark patches known as sunspots, which move across the disk as the Sun rotates. These images were taken at the Big Bear Solar Observatory in California over a period of five days. Regular observation reveals that the number of sunspots rises and falls in a cycle lasting approximately 11 years.*

SOLAR ROTATION PERIOD

An excellent project for home or school is to determine the rotation rate of the Sun. Many solar observatories around the world offer instant access to daily images of the Sun. Two of the best known solar observatories are the Big Bear Solar Observatory in California, USA (www.bbso.njit.edu), and the Solar and Heliospheric Observatory (www.nascom.nasa.gov), a space-based observatory that never suffers from cloudy days!

The Sun frequently displays **sunspots**, cooler, darker regions on the solar surface. Tracking their motion reveals the solar rotation. Access a number of these images over a few days, and print out each image. Draw a horizontal line across the centre of the Sun's disk to represent the equator. Use a simple protractor to determine the angle the sunspots have moved by marking the rim of the solar disk in 10° intervals and extending a perpendicular line from each interval to the equator. Using the time interval and the angle the sunspots have moved, a simple division will determine the rotation rate of the Sun (one full solar rotation represents 360°). Use sunspots near the centre of the disk for better accuracy.

TOOLS & TECHNIQUES

Watching the Sky
STAR CHARTS
INTRODUCTION

Think of the stars as being on the inside of a sphere which rotates around the Earth once every day. Until a few hundred years ago people believed that this was the true situation – and the celestial sphere remains a useful concept for describing the sky, even though we now know that stars are all at different distances from us and that their rising and setting is due not to their own movement, but to the daily rotation of the Earth.

THE CELESTIAL SPHERE

The **celestial sphere** has a north and a south pole, which lie directly overhead at the Earth's poles, and an equator, directly above the Earth's equator. However, there is an additional circle on the celestial sphere which is of considerable importance: it is called the **ecliptic** and it marks the Sun's yearly path against the background stars. Actually, of course, the Sun doesn't move, but

▼ *The rotation of the Earth is demonstrated by star trails around the celestial pole, traced out as the Earth spins during a time exposure. Stars that make a complete circuit of the pole without setting are termed circumpolar.*

the Earth orbits around it, and so we see the Sun projected against different parts of the celestial sphere from month to month. The ecliptic is tilted at 23½° to the celestial equator – that is the amount by which the Earth's axis is tilted from the perpendicular relative to its orbit around the Sun.

Looking at the ecliptic on the celestial sphere, we can begin to understand the significance of terms such as the **solstices** and the **equinoxes**. The ecliptic cuts the celestial equator at two points: these are the equinoxes. When the Sun is at these positions, in spring and autumn, day and night are equal in length the world over. The points where the Sun reaches its farthest **declination** north or south of the celestial equator are known as the solstices. It reaches its most northerly point in June, when the northern hemisphere of Earth experiences its longest day and the southern hemisphere its shortest. At the Sun's southerly extreme of declination, in December, the lengths of day are reversed.

COORDINATES

Positions of stars and other objects on the celestial sphere are specified by coordinates similar to **longitude** and **latitude** on Earth, but called **right ascension (RA)** and declination (Dec.). Declination is the easier of the two to understand since, like latitude, it runs in degrees from 0° at the equator to 90° at either pole.

Right ascension seems more puzzling at first. It is expressed not in degrees but in hours, because of the connection between the Earth's

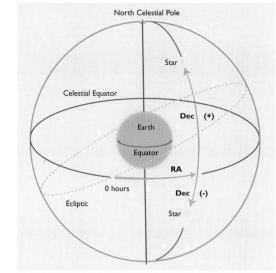

▲ *Celestial Coordinates*
Objects in the sky are plotted using two coordinates comparable with longitude and latitude, called right ascension and declination. For example, a star's right ascension is measured from west to east around the equator, in hours, minutes and seconds from 0 to 24 hours. The star's declination is the angular distance north (+) or south (-) of the celestial equator, given in degrees.

rotation and time – the Earth rotates through 15° in one hour, so one hour of right ascension is equivalent to 15° of angle. As with terrestrial longitude, right ascension is taken to start from an agreed zero **meridian**. On the celestial sphere, the zero meridian runs through the position of the March equinox, which lies in the **constellation** Pisces (Chart 8). Right ascension runs from west to east, which is the direction the Earth turns.

CONSTELLATIONS

The sky is divided into 88 sections known as constellations, a combination of the 48 figures known to the ancient Greeks and others introduced more recently. Their names and boundaries are officially laid down by the International Astronomical Union. Constellations originated in the distant past as easy-to-remember patterns of stars, but nowadays astronomers regard them merely as areas of sky convenient for locating and naming celestial objects. The patterns on which constellations were originally based are purely the product of human imagination – with very few exceptions, the stars in a constellation are unrelated and lie at widely differing distances from us.

NAMES OF THE STARS

Stars bear a variety of names. Some have traditional names dating from antiquity, such as Sirius, Arcturus, Betelgeuse and Rigel. Others are exceptional specimens named after astronomers who discovered or studied them, such as Barnard's Star. But all bright stars are identified by either a

WATCHING THE SKY

letter (usually Greek) or a number, along with the name of the constellation that contains them. For example, Sirius is also Alpha [α] Canis Majoris, meaning 'alpha of Canis Major' – the genitive (possessive) case of the constellation name is always used in this context. In 1603 a German astronomer, Johann Bayer (1572–1625), published an atlas in which the stars were identified by Greek letters, and so these are usually termed Bayer letters. generally they were allocated by contellation in decreasing order of brightness. A supplementary system identifies stars by numbers – called Flamsteed numbers, after the first Astronomer Royal of England. An example is 61 Cygni. **Variable stars**, if they are not already identified on one of the preceding systems, have a different style of nomenclature. Names of the brighter ones consist of one or two Roman letters, e.g. P Cygni, R

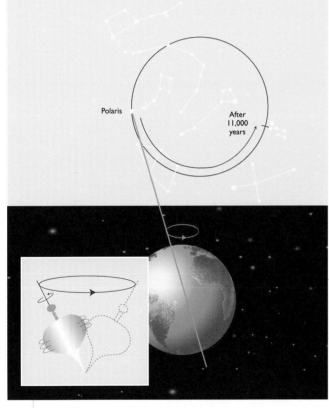

Polaris

After 11,000 years

✹ PRECESSION AND THE POLES

Under the combined gravitational pulls of the Sun and Moon, the Earth executes a slow wobble in space like a spinning top. This motion is called **precession**. As the Earth's orientation relative to the stars changes, the celestial poles (which are defined by the position of the Earth's pole) trace out a complete circle against the stars approximately every 26,000 years. Not only are

💡 TWINKLING AND TRANSPARENCY

Stars seem to twinkle, but this is nothing to do with the stars themselves – it is a result of turbulent air currents in the Earth's atmosphere which cause the star's incoming light to dance around and flash colourfully. The effect varies from night to night, depending on atmospheric conditions, and is most noticeable near the horizon, where we are looking through the thickest layer of atmosphere. Seen from space, stars are steady

Leonis, VV Cephei. Others are called V (for variable) with a three- or four-digit number.

Star clusters, **nebulae** and **galaxies** may have individual popular names, but are generally identified by catalogue numbers. Many of the brighter ones have M, NGC or IC as prefixes. These relate to the objects' listing in either **Charles Messier's catalogue** (the so-called Messier objects, of which there are 110), the **New General Catalogue** (NGC, 7840 entries) and the Index Catalogues (IC), two supplements to the NGC adding over 5,000 more objects.

THE CHANGING SKY

How much of the celestial sphere a person sees depends where they are located on Earth. To take an extreme case, someone at one of the Earth's poles would only ever see half the celestial sphere – the celestial pole would be directly overhead while the celestial equator would coincide with the horizon. All stars would circle the pole without rising or setting as the Earth turned. If

◀ *The Precession of the Earth*
The Earth's rotation axis describes a slow conical motion in space, similar to the wobble of a spinning top. This is caused by the gravitational pull of the Sun and Moon. The motion means that a straight line through the poles traces out a full circle over a period of about 26,000 years. Currently the North Celestial Pole lies near Polaris, the North Star. However, after about 11,000 years, the North Celestial Pole will be near Vega.

the positions of the celestial poles affected: the coordinates of all objects on the celestial sphere gradually change. Think of the coordinate system as a transparent overlay which slides around the celestial sphere every 26,000 years, while the stars themselves remain fixed. The steady march of precession means that, for consistency, the positions of all stars in a catalogue, or the coordinates on a chart, have to be referred to a set date (known as an epoch), which is currently the year 2000.

points of light. Astronomers term the steadiness of the atmosphere the 'seeing' – an unsteady night produces bad seeing, which blurs fine details on planets and makes it difficult to separate the components of close **double stars**. Observations are also affected by the transparency of the atmosphere, which is a different matter from the seeing. For example, a misty night may have very steady seeing but poor transparency. Good transparency is necessary for the best views of diffuse objects such as nebulae and galaxies.

the observer set out towards to the Earth's equator, more and more of the opposite celestial hemisphere would come into view. At the equator itself, the observer would see the celestial poles on the northern and southern horizon and all stars would rise and set as the Earth turned, bringing every part of the celestial sphere into view.

In practice, most people experience something between these two extremes. The angle of the celestial pole above your horizon depends on your latitude on Earth, as navigators have long been aware. As the Earth spins, all stars circle the pole but those closest to the pole will do so without setting – they are termed **circumpolar**. Other stars, closer to the equator, rise in the east and set in the west.

🧪 OBSERVING DEEP-SKY OBJECTS

Anything beyond the **Solar System** that is not a star is lumped under the catch-all term 'deep-sky object' by amateur observers. Deep-sky objects encompass star clusters, nebulae and galaxies. Because their light is spread out, they need clear, dark skies to be seen at their best. Before beginning to observe, one should always allow time for the eyes to become adapted to the dark, and this is particularly important in the case of deep-sky objects. Do not expect to go straight from a brightly lit room to the eyepiece and immediately see faint detail – full dark-adaptation takes at least 20 minutes. Another useful technique for seeing elusive faint objects is averted vision, in which the observer looks slightly to one side of the object of interest. In doing so, the object's light falls on to the outer part of the retina, which is more sensitive than the centre. Measuring the brightness of deep-sky objects is more difficult than that of a star. The **magnitude** figures given for clusters, nebulae and galaxies in the data tables that accompany each star chart are the brightness that the object would appear if all its light were concentrated into a starlike point.

WATCHING THE SKY

THE NORTHERN & SOUTHERN HEMISPHERES

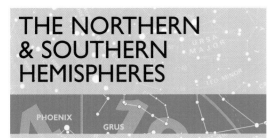

The entire sky from the north celestial pole to the south is illustrated in 20 charts on the following pages. Each chart is accompanied by descriptions of the best sights that the sky has to offer the celestial tourist – double and variable stars, star clusters, nebulae and galaxies – and the equipment needed to observe them, often no more than binoculars.

The two Index charts on these pages provide a key to the 20 individual charts. As a guide to visibility, the months around the rim of the northern hemisphere Index chart show when that part of the sky lies due south at around 10 p.m. – for example, the stars in Chart 10 such as Leo will be well-displayed on evenings in March.

In the case of the southern hemisphere Index chart, the dates show when that part of the sky lies due north at around 10 p.m.

The various sections of the sky are depicted with projections chosen to minimize distortion of the constellation shapes. Each chart slightly overlaps its neighbours, so some objects may appear on more than one chart. Black lines on the charts are celestial co-ordinates (right ascension and declination), while dotted white lines are the constellation boundaries. Other lines mark the ecliptic

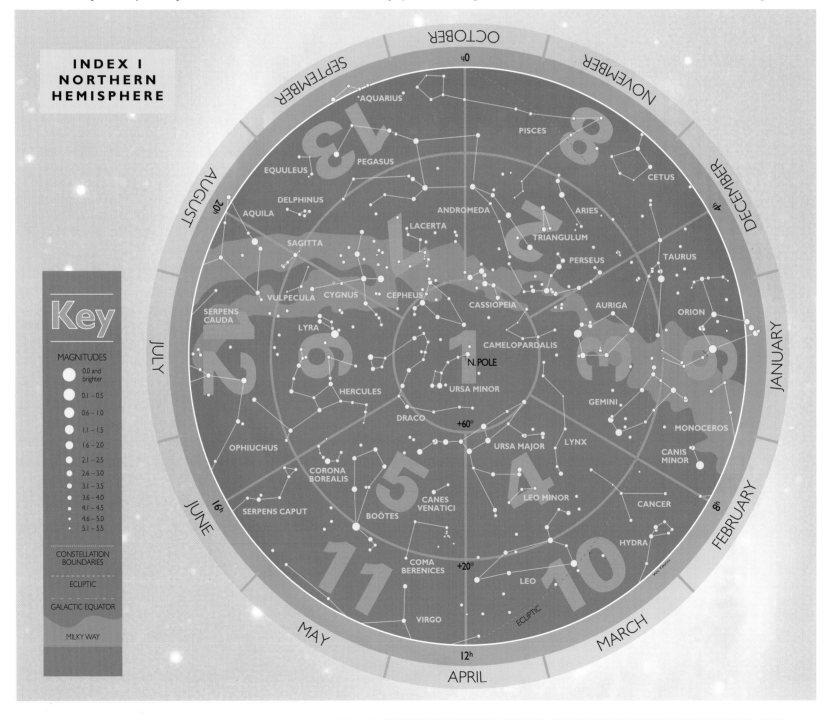

INDEX I NORTHERN HEMISPHERE

Key

MAGNITUDES
- ○ 0.0 and brighter
- ○ 0.1 – 0.5
- ○ 0.6 – 1.0
- ● 1.1 – 1.5
- ● 1.6 – 2.0
- ● 2.1 – 2.5
- ● 2.6 – 3.0
- ● 3.1 – 3.5
- • 3.6 – 4.0
- • 4.1 – 4.5
- • 4.6 – 5.0
- · 5.1 – 5.5

CONSTELLATION BOUNDARIES

ECLIPTIC

GALACTIC EQUATOR

MILKY WAY

WATCHING THE SKY

(a dashed yellow line) and the galactic equator (a dot-and-dash pink line).

The charts include all stars down to magnitude 5.5, which is about as faint as you can expect to see with unaided eyesight under ordinary conditions. The sizes of the star dots are graded in half-magnitude steps – the smaller the dot, the fainter the star. Special symbols are used for particular classes of star: a line bisecting a dot identifies a double or multiple star, while variables have a dot-and-circle combination that indicates their brightness range.

As well as stars, the charts depict the brightest deep-sky objects. These have their own symbols: a green patch for a nebula, a pink ellipse for a galaxy, a dotted white ring for an **open cluster** and a solid white ring for a **globular cluster**. The **Milky Way** is shown a lighter blue.

Planets do not appear on these charts because they are, of course, constantly on the move. When they do appear they will be found near the ecliptic, disturbing the familiar constellation patterns. A few rules of thumb will help you to identify a planet: a brilliant 'star' in

the evening or morning twilight is most likely to be Venus; a bright white star at night will be Jupiter (easily confirmed by binoculars, which readily show its rounded outline); a fainter, yellowish star will be Saturn (confirmable by a small telescope, which reveals the globe and rings); while a bright reddish star will be Mars. Another way to distinguish between a planet and a star is that planets hardly twinkle – unlike stars they are not point sources of light and so are less affected by the air currents in the Earth's atmosphere that cause twinkling.

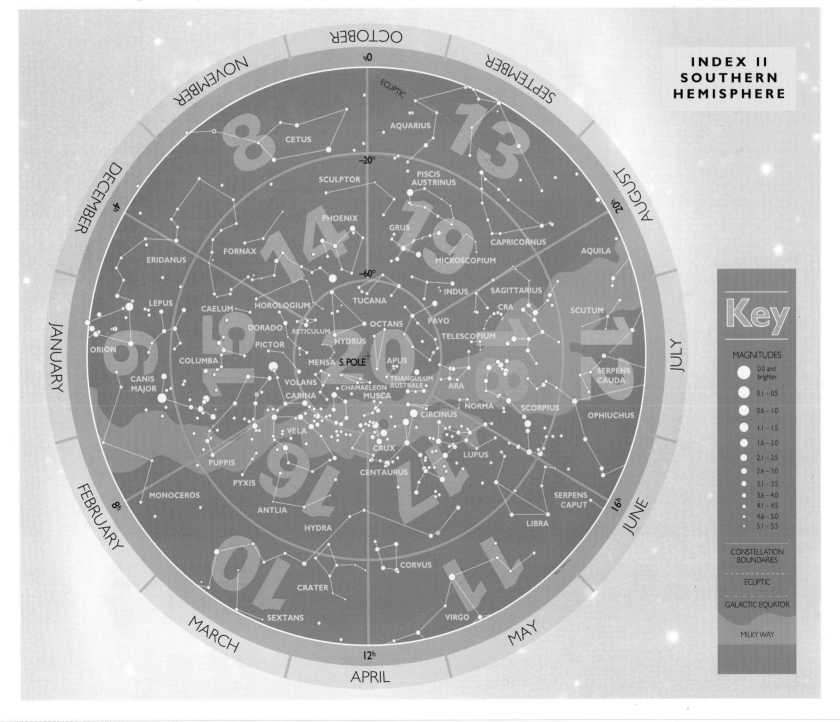

**INDEX II
SOUTHERN
HEMISPHERE**

Key

MAGNITUDES

- 0.0 and brighter
- 0.1 – 0.5
- 0.6 – 1.0
- 1.1 – 1.5
- 1.6 – 2.0
- 2.1 – 2.5
- 2.6 – 3.0
- 3.1 – 3.5
- 3.6 – 4.0
- 4.1 – 4.5
- 4.6 – 5.0
- 5.1 – 5.5

CONSTELLATION BOUNDARIES

ECLIPTIC

GALACTIC EQUATOR

MILKY WAY

CHART I

Our tour of the sky begins in the north polar region. Almost at the centre of this chart is the North Pole Star, Polaris, the pivot around which the sky seems to turn as the Earth spins daily on its axis. On one side of the pole are the seven stars in the constellation Ursa Major that make up the familiar saucepan-shape of the Plough, or Big Dipper, shown at the far left on this map; on the other side of Polaris is a large W composed of five stars in the constellation Cassiopeia. Polaris itself lies at the end of the 'handle' of another dipper, this one the Little Dipper, outlined by the main stars of Ursa Minor. The other constellations in this north polar region of sky – Cepheus, Draco and Camelopardalis – are fairly faint.

OFF TO THE POLE

Some people expect Polaris to be a prominent star but in fact it is fairly ordinary looking, of second magnitude; its importance stems only from the fact that it lies coincidentally close to the northern celestial pole. Currently Polaris lies three-quarters of a degree from the pole but is slowly getting closer because of the effect of precession. Polaris will be at its closest to the pole around the year 2100, less than half a degree away, but will then move away again.

NAME	RA, DEC	MAG(S)	TYPE OF OBJECT	VISIBILITY	
Polaris	2h 32m +89.3°	2.0, 8.2	double star	👀	🔭
γ Ursae Minoris	15h 21m +71.8°	3.0, 5.0	double star	👀 👓	
β Cephei	21h 29m +70.6°	3.2, 7.9	double star		🔭
VV Cephei	21h 57m +63.6°	4.8–5.4	variable star	👓	
NGC 457	1h 19m +58.3°	6.4	open cluster	👓 🔭	
M103	1h 33m +60.7°	7.4	open cluster	👓 🔭	
NGC 663	1h 46m +61.2°	7.1	open cluster	👓 🔭	
NGC 1502	4h 08m +62.3°	5.7	open cluster	👓 🔭	
M52	23h 24m +61.6°	6.9	open cluster	👓 🔭	
NGC 6543	17h 59m +66.6°	8.8	planetary nebula		🔭
M81	9h 56m +69.1°	6.9	galaxy	👓 🔭	
M82	9h 56m +69.7°	8.4	galaxy	👓 🔭	

Polaris is a yellowish-coloured **supergiant** nearly 50 times larger than the Sun and giving out some 2,500 times as much light. It is a **Cepheid variable** – the closest one to us, 430 **light years** away. But its brightness variations are so small, only a few hundredths of a magnitude, they are not noticeable to the eye. However, it is also a double star and small telescopes show its companion star of 8th magnitude. Look around this area with a telescope or binoculars and you should glimpse a ring of faint stars on which Polaris appears to hang like a pearl on a necklace.

VARIABLES AND DOUBLES

For an even more remarkable star, look for VV Cephei, in the top right of the chart. This is one of the largest stars known: a red supergiant perhaps as much as 1,000 times the diameter of the Sun, so huge that it is unstable. Fluctuating irregularly in size and brightness, it can appear between magnitudes 4.8 and 5.4. In this same area of sky is 3rd-magnitude Beta [β] Cephei, a double star with an 8th-magnitude companion detectable through small telescopes.

The stars Beta [β] and Gamma [γ] Ursae Minoris, in the top left of the chart, are popularly known as the Guardians of the Pole. Through binoculars, or with good eyesight, 3rd-magnitude Gamma appears to be double, but the fainter companion is unrelated, lying about 100 light years closer to us.

CLUSTERS IN CASSIOPEIA

Cassiopeia, which also appears on Charts 2 and 7, lies in the Milky Way and contains several star clusters. Prime among them is M52, a large open cluster that can be found with binoculars as a misty, somewhat elongated patch of light. Individual stars, notably an 8th-magnitude orange giant at one edge, can be picked out through telescopes of about 75-mm (3-in) aperture. Another open cluster easily located with binoculars is NGC 457, which includes the 5th-magnitude supergiant Phi [φ] Cassiopeia. As in many such clusters, NGC 457's members seem to form long chains. Look between Delta [δ] and Epsilon [ε] Cassiopeiae for NGC 663, a prominent cluster appearing half the width of the full Moon. Although not on Messier's list, NGC 663 is a better object for binoculars than nearby M103, which needs a small telescope to be fully appreciated. More difficult to find is NGC 1502 in the barren wastes of adjoining Camelopardalis. Although not an impressive cluster through binoculars, it lies next to an unheralded treasure: a chain of stars flowing towards Cassiopeia like water droplets, known as Kemble's Cascade.

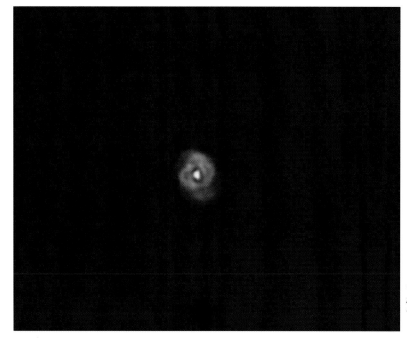

◀ *An amateur CCD image of the Cat's Eye Nebula, NGC 6543, in Draco. This is a so-called planetary nebula, produced when a former red giant star lost its outer layers, leaving its hot core exposed at the centre. The nebula appears as a blue-green ellipse when seen through a small telescope.*

Cepheid variable ▶ p. 337 light year ▶ p. 340 planetary nebula ▶ p. 342 supergiant ▶ p. 344

WATCHING THE SKY

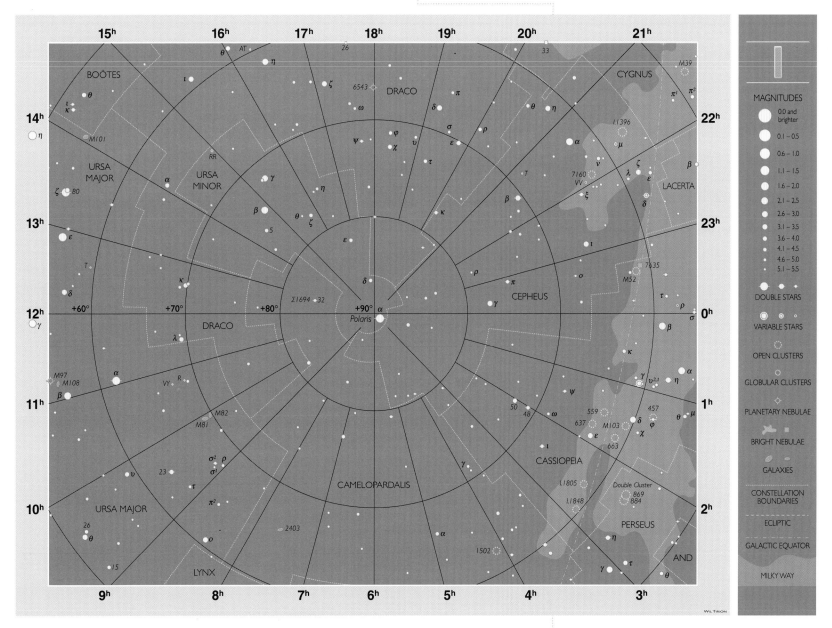

MAGNITUDES

⬤	0.0 and brighter
⬤	0.1 – 0.5
●	0.6 – 1.0
●	1.1 – 1.5
●	1.6 – 2.0
●	2.1 – 2.5
•	2.6 – 3.0
•	3.1 – 3.5
•	3.6 – 4.0
·	4.1 – 4.5
·	4.6 – 5.0
·	5.1 – 5.5

DOUBLE STARS

VARIABLE STARS

OPEN CLUSTERS

GLOBULAR CLUSTERS

PLANETARY NEBULAE

BRIGHT NEBULAE

GALAXIES

CONSTELLATION BOUNDARIES

ECLIPTIC

GALACTIC EQUATOR

MILKY WAY

WIL TIRION

NGC 6543 IN DRACO

At the top of our chart lies NGC 6543, the type of object known as a **planetary nebula**, composed of gas thrown off from a dying star. It achieved fame recently when photographed by the Hubble Space Telescope which revealed the nebula's complex structure, consisting of overlapping loops of gas like an eye – hence its popular name the Cat's Eye Nebula. But these subtle details are beyond the reach of amateur instruments, through which it appears as merely a blue-green disk like an out-of-focus star, of similar apparent size to the disk of Saturn.

GALAXIES M81 AND M82

A celebrated pair of galaxies lie in northern Ursa Major. M81 is a beautiful spiral which in clear,

dark skies can be glimpsed with binoculars as a rounded patch of light. A small telescope shows it more clearly, although it may at first be over-looked because it is larger than expected, half the apparent width of the full Moon. Half a degree to the north lies M82, smaller and elongated in shape since we see it almost edge-on. Long-

exposure photographs of M82 show it has a disturbed appearance due to clouds of gas and dust falling into the galaxy and sparking a burst of star formation. This activity is thought to result from a close encounter with M81 about 200 million years ago. Both galaxies lie about 14 million light years away.

🧪 **FINDING NORTH**

Observers find their way around the sky by using a technique called star hopping. It's as easy as it sounds – you simply jump from star to star until you arrive at your celestial destination. The best-known star hop in the sky uses the familiar pattern of the Big Dipper (or Plough). There are three stars in the handle and four in the bowl of the Dipper. The two end stars of

the bowl are marked Alpha [α] and Beta [β] – their popular names are Dubhe and Merak, respectively. If you hop from Merak (Beta) to Dubhe (Alpha) and then hop about five times as far again to the centre of the chart you'll land at the pole star, Polaris. When you have found it, you know you are facing north. Dubhe and Merak are actually known as the Pointer stars because they point the way to Polaris.

WATCHING THE SKY

CHART 2

NAME	RA, DEC	MAG(S)	TYPE OF OBJECT	VISIBILITY
η Cassiopeiae	0h 49m +57.8°	3.5, 7.5	double star	
γ Andromedae	2h 04m +42.3°	2.3, 4.8	double star	
ρ Persei	3h 05m +38.8°	3.3–4.0	variable star	
Algol (β Persei)	3h 08m +41.0°	2.1–3.4	variable star	
NGC 752	1h 58m +37.7°	5.7	open cluster	
Double cluster	2h 20m +57.1°	4.3, 4.4	open clusters	
M34	2h 42m +42.8°	5.2	open cluster	
α Persei cluster	3h 24m +49.9°	1.2	open cluster	
Pleiades (M45)	3h 47m +24.0°	1.2	open cluster	
M31	0h 43m +41.3°	3.4	galaxy	
M33	1h 34m +30.7°	5.7	galaxy	

This area of sky contains the two nearest large galaxies to us: M31 in Andromeda and M33 in Triangulum, both members of our Local Group and both spirals. Our own Galaxy is also prominent on this chart in the form of the Milky Way, dotted with star clusters; it flows from Cygnus and Lacerta at the upper right via Cassiopeia (whose clusters are dealt with on Chart 1) and Perseus, continuing towards Auriga at the left. One of the most famous variable stars, the eclipsing binary Algol, lies at the centre left of the chart in Perseus, with the best-known and prettiest star cluster in the sky, the Pleiades in Taurus, to the south of it.

GALAXIES IN OUR GANG

M31, the Andromeda Galaxy, is the most distant object you can glimpse with the naked eye – but you will need clear, dark skies to find it. Start by locating Alpha Andromedae, the star at the top left corner of the Square of Pegasus; from there, hop two stars to Beta Andromedae, then two stars up to find M31's misty outline. Being tilted at an angle to us, M31 appears elongated in shape, with a long axis spanning the width of several full Moons – but its full extent is traceable only in binoculars and telescopes. M31 is a **spiral galaxy**, much like our own, lying some 2.5 million light years away, which means the light

we see from it has been on its way since before human beings walked the Earth. As you gaze upon its stellar throng, it is only natural to wonder whether someone is staring back. To spot its small, elliptical companion galaxies M32 and M110 you will need a telescope. Less easy to see than the Andromeda Galaxy is galaxy M33 on the borders of Triangulum and Pisces; for this you will need binoculars and clear skies. Photographs show its spiral shape but visually it appears only as an elliptical cloud of pale light, similar in size to the disk of the Moon. It is depicted to scale on the chart, as is M31.

DOUBLE STARS

One of the delights of double stars is their contrasting colours, and a prime example, Gamma [γ] Andromedae, is to be found in the centre of Chart 2. To the naked eye it appears as a single star of 2nd magnitude, but small telescopes bring

into view a 5th-magnitude companion whose blue colour contrasts gloriously with the orange tone of its giant companion. Another easily divided pair is Eta [η] Cassiopeiae, the brighter component of which appears yellow-white, while the cooler companion appears orange or red. This pair forms a true binary with an **orbital period** of 480 years.

ALGOL AND OTHER VARYING STARS

Some close double stars periodically pass in front of each other as they move along their orbits, temporarily blocking light from the other star. The prototype of these so-called eclipsing binaries (see The Stellar Zoo – Double Stars, page 170) is Algol in the constellation Perseus, also known as Beta [β] Persei. Every 2 days 21 hours this 2nd-magnitude star dips to one-third of its usual brightness, remaining there for 10 hours before returning to normal. This behaviour was first noted in 1782 by a young English amateur astronomer, John Goodricke (1764–86), and now amateurs everywhere hone their observational skills by following the variations of Algol. As you watch Algol, notice that two degrees to the south is a variable star of a different kind: Rho [ρ] Persei, a pulsating **red giant** that changes in brightness between magnitudes 3.3 and 4.0 every seven weeks or so.

SPARKLING STAR CLUSTERS

Scan the Milky Way in Perseus with binoculars and you will encounter some glittering star clusters. One of them is so large that it may not

◀ **The Double Cluster** in the constellation Perseus, a pair of young open clusters lying side by side and visible to the naked eye as a brighter knot in the Milky Way. NGC 869 (at left) is the more condensed of the two, while NGC 884 contains several red giants, evidence that its most massive stars are already growing old. Both clusters fit comfortably into the field of view of a pair of binoculars.

WATCHING THE SKY

STAR CHARTS

orbital period ▶ p. 341   red giant ▶ p. 343   spiral galaxy ▶ p. 344

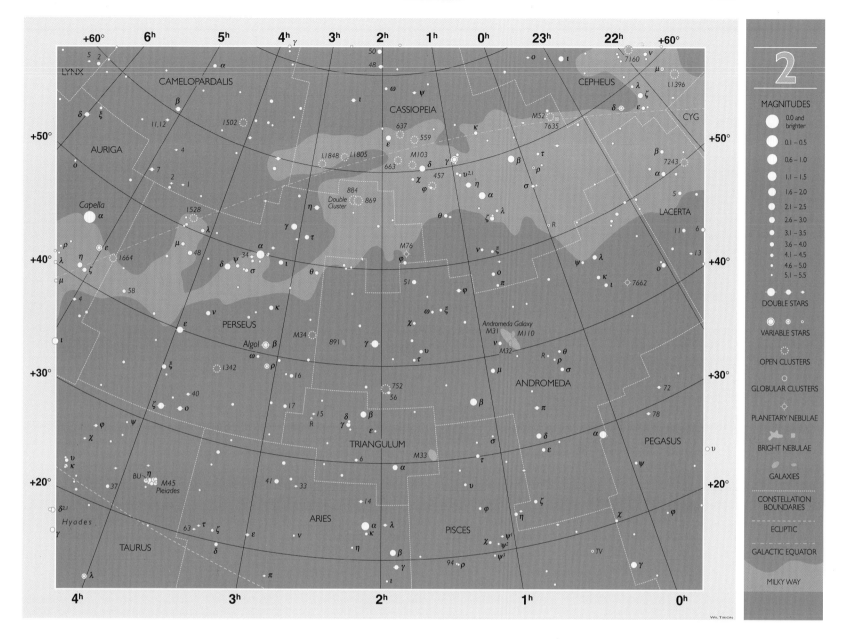

MAGNITUDES

⬤	0.0 and brighter
⬤	0.1 – 0.5
⬤	0.6 – 1.0
⬤	1.1 – 1.5
⬤	1.6 – 2.0
⬤	2.1 – 2.5
⬤	2.6 – 3.0
•	3.1 – 3.5
•	3.6 – 4.0
·	4.1 – 4.5
·	4.6 – 5.0
·	5.1 – 5.5

DOUBLE STARS

VARIABLE STARS

OPEN CLUSTERS

GLOBULAR CLUSTERS

PLANETARY NEBULAE

BRIGHT NEBULAE

GALAXIES

CONSTELLATION BOUNDARIES

ECLIPTIC

GALACTIC EQUATOR

MILKY WAY

even look like a cluster at first – the area around 2nd-magnitude Alpha [α] Persei. Binoculars show a scattering of stars covering an area six Moon diameters wide, known as the Alpha Persei cluster. This widespread group lies about 600 light years away.

Shift your gaze further north and you will encounter two open clusters side by side, NGC 869 and 884, also known as h and Chi [χ] Persei or simply the Double Cluster. Each is of similar apparent diameter to the full Moon and under clear skies can be seen with the naked eye as brighter patches in the Milky Way. Each cluster contains well over 100 stars, the brightest of which can be picked out in small telescopes or even good binoculars. Both clusters formed a few million years ago and still lie adjacent to each other, about 7,300 light years from us in the Perseus spiral arm of our Galaxy.

Now head south to find another open cluster, M34, of similar size to NGC 869 or 884 but with only about half as many stars; the brightest of them can be made out in binoculars. M34 is much closer to us than the Double Cluster, about 1,500 light years away, and older, with an estimated age of 100 million years. Compare M34 with NGC 752 across the border in Andromeda, an even larger cluster but with fainter stars – small telescopes will be needed to make them out individually. NGC 752 is slightly closer than M34, about 1,300 light years away, but is much older, over one billion years, which is unusually ancient for an open cluster.

SEVEN STARRY SISTERS

The prettiest star cluster on this chart, and indeed in the whole sky, lies in the lower left corner – the Pleiades or M45. The group is popularly termed the Seven Sisters, although you will need sharp eyes to see more than its six brightest members which are arranged in the shape of a mini dipper spanning two Moon diameters of sky. The most prominent member is Alcyone, magnitude 2.9. Binoculars show dozens of stars in the cluster and small telescopes bring even more into view. All the brightest stars in the Pleiades are hot, bluish-coloured giants no more than a few million years old; in fact, the entire cluster has come into being since the dinosaurs died out on Earth. The Pleiades lies 380 light years away.

WATCHING THE SKY

STAR CHARTS

CHART 3

NAME	RA, DEC	MAG(S)	TYPE OF OBJECT	VISIBILITY
Castor	7h 35m +31.9°	1.9, 3.0	double star	
ζ Aurigae	5h 02m +41.1°	3.7–4.0	variable star	
ε Aurigae	5h 02m +43.6°	3.0–3.8	variable star	
RT Aurigae	6h 29m +30.5°	5.0–5.8	variable star	
UU Aurigae	6h 37m +38.4°	5.0–7.0	variable star	
η Geminorum	6h 15m +22.5°	3.1–3.9	variable star	
ζ Geminorum	7h 04m +20.6°	3.6–4.2	variable star	
M38	5h 29m +35.8°	6.4	open cluster	
M36	5h 36m +34.1°	6.0	open cluster	
M37	5h 52m +33.5°	5.6	open cluster	
M35	6h 09m +24.3°	5.1	open cluster	
NGC 2392	7h 29m +20.9°	9.0	planetary nebula	
Crab Nebula (M1)	5h 35m +22.0°	8.4	supernova remnant	

At the centre of this chart is the constellation Auriga, whose leading star is Capella, sixth-brightest in the sky. Two other prominent stars with famous names, Castor and Pollux, are found to the lower left in the constellation Gemini. The Milky Way flows from Cassiopeia and Perseus at upper right through Auriga and southwards to Chart 9. Near the edge of the Milky Way in Taurus, at the bottom centre of the chart, is one of the most celebrated objects in the entire sky: the Crab Nebula, the remains of a supernova explosion seen from Earth nearly 1,000 years ago. The two large, bright star clusters at the bottom right, the Pleiades and Hyades, are described on Charts 2 and 9 respectively.

STAR PERFORMERS IN AURIGA

South of sparkling Capella lies a thin triangle of stars, two of which are exceptional eclipsing binaries. Zeta [ζ] Aurigae is an orange giant 150 times larger than the Sun that is orbited by a blue star of about 4 solar diameters. Every 2 years 8 months the smaller blue star passes behind the giant and the brightness of Zeta Aurigae drops by a third; it remains at minimum brightness for over 5 weeks until the blue star emerges again. You will have to wait even longer to see an eclipse of Epsilon [ε] Aurigae, to the north of Zeta, since it has the longest known period of any eclipsing binary, 27 years. Epsilon is an immensely luminous supergiant, some 200 times larger than the Sun and shining over 100,000 times as brightly. The nature of the orbiting companion remains in doubt: it is most likely a close binary enveloped in a disk of dust which actually causes the eclipses during which Epsilon Aurigae halves in brightness, remaining dim for over a year. The next eclipse begins at the end of 2009.

For quicker results, keep an eye on RT Aurigae in the lower left of the constellation, a Cepheid variable with a period of 3.7 days and a range between magnitude 5.0 and 5.8, easily followed in binoculars. Less regular is UU Aurigae, a cool giant with strong lines of carbon in its **spectrum** giving it a deep red colour, noticeable in binoculars; it varies between 5th and 7th magnitudes in eight months or so.

A TRIO OF STAR CLUSTERS IN AURIGA

A chain of three open clusters is draped along the Milky Way in Auriga, all of them visible in binoculars although small telescopes will be needed to resolve their individual stars. M36, in the middle of the chain, is the most prominent despite being the smallest. South of it is M37, the largest and richest of the trio; however, M37's individual stars are the faintest of all and hence it is the most difficult of the three clusters to resolve in small telescopes. Northernmost of the chain is M38, whose brightest stars appear to be arranged in a cross-shape when seen through a telescope.

CASTOR AND POLLUX

Castor and Pollux sit side by side in the sky like the mythological twins after whom they are named, but the stars themselves are not truly related. Pollux, slightly the brighter of the two despite being labelled with the second letter of the Greek alphabet, Beta [β], is an orange giant 34 light years distant. Binoculars emphasize its colour-contrast with blue-white Castor to the north, 52 light years from us. Telescopes of only 60-mm (2.4-in) aperture can split Castor into a dazzling binary, the components of which orbit each other every 470 years or so. Castor was one of the first binary stars in which orbital motion was detected, by the English astronomer **William Herschel** (1738–1822) 200 years ago – although he had to watch it for over 20 years to be sure. Castor also has a faint **red dwarf** companion. All

three of these visible stars are themselves **spectroscopic binaries**, making Castor an astounding sextuplet system.

Gemini also contains two variable stars of note: Zeta [ζ] Geminorum is a Cepheid variable that ranges from magnitude 3.6 to 4.2 and back again every 10.2 days, while Eta [η] Geminorum is a red giant that fluctuates between magnitudes 3.1 and 3.9 every eight months or so.

▶ *The open cluster M35* contains about 200 stars and lies to the north of Eta [η] Geminorum. It is just at the limit of naked-eye visibility but is well seen with binoculars and small telescopes.

STAR CHARTS

pulsar ▶ p. 342 red dwarf ▶ p. 343 spectroscopic binary ▶ p. 344
spectrum ▶ p. 344 white dwarf ▶ p. 344

WATCHING THE SKY

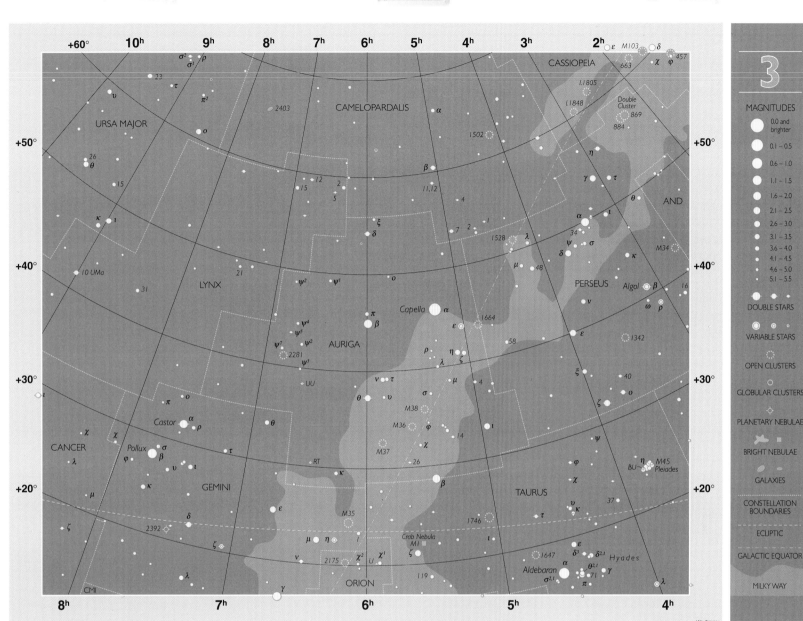

MAGNITUDES

0.0 and brighter
0.1 – 0.5
0.6 – 1.0
1.1 – 1.5
1.6 – 2.0
2.1 – 2.5
2.6 – 3.0
3.1 – 3.5
3.6 – 4.0
4.1 – 4.5
4.6 – 5.0
5.1 – 5.5

DOUBLE STARS

VARIABLE STARS

OPEN CLUSTERS

GLOBULAR CLUSTERS

PLANETARY NEBULAE

BRIGHT NEBULAE

GALAXIES

CONSTELLATION BOUNDARIES

ECLIPTIC

GALACTIC EQUATOR

MILKY WAY

OPEN CLUSTER M35

Close to the border with Taurus lies this major open cluster, nearly as large as the full Moon and a sight to outrank even the clusters in Auriga. M35 is just visible to the naked eye as a brighter patch in the Milky Way under dark skies, but even for those with less favourable conditions it is easily spotted through binoculars, which show it to be elongated in shape. Seen through small telescopes, its brightest stars seem to form sprays of sparks like the explosion of some distant celestial firework.

ESKIMO IN GEMINI

Planetary nebulae, the shells of gas thrown off by dying stars, can assume strange shapes as they expand into space, although the details are usually so subtle that they are visible only with larger apertures. This planetary nebula, NGC 2392 in Gemini, has been christened the Eskimo Nebula because on photographs it resembles a face with a faint outer fringe like a parka. You will not be able to make out the 'face' through a small telescope, but the nebula should show up easily as a blue-green ellipse about the same size as the disk of Saturn; you may also spot its central **white dwarf**, of 10th magnitude.

THE CRAB NEBULA

Two Moon diameters from Zeta [ζ] Tauri lies this legendary object – M1, the first object in Messier's catalogue, which we now know to be the wreckage of a supernova explosion that astronomers on Earth saw in 1054. When the nineteenth-century Irish astronomer Lord Rosse (1800–67) looked at it through his telescope he thought it resembled a crab's claw, so it became known as the Crab Nebula. It can be seen in small to moderate-sized telescopes, appearing as a faint, elongated smudge, with none of the bright filaments that are brought out on long-exposure photographs. Observers can miss it at first because it is bigger than they expect – you should be looking for something midway in size between the disk of Jupiter and the full Moon. The **pulsar** at its centre is far too faint to see through amateur telescopes. The Crab Nebula is about 6,500 light years away.

CHART 4

NAME	RA, DEC	MAG(S)	TYPE OF OBJECT	VISIBILITY
ζ Cancri	8h 12m +17.6°	5.0, 6.2	double star	🔭
ι Cancri	8h 47m +28.8°	4.0, 6.7	double star	🔭
ξ Ursae Majoris	11h 18m +31.5°	4.3, 4.8	double star	🔭
ν Ursae Majoris	11h 19m +33.1°	3.5, 9.9	double star	🔭
Beehive Cluster (M44)	8h 40m +20.0°	3.1	open cluster	👁 👓 🔭
M97	11h 15m +55.0°	c.11	planetary nebula	🔭
M108	11h 12m +55.7°	10.0	galaxy	🔭
M109	11h 58m +53.4°	9.8	galaxy	🔭

Chart 4 covers an area of sky that is visible to observers in the northern hemisphere in the evening during late winter and early spring. It is a region largely bereft of bright stars, although at the upper left is the familiar figure of the Plough or Big Dipper, also featured on Chart 5. As the chart shows, the Plough shape occupies only a small part of the total area of the constellation Ursa Major, the great bear. Sharing this region of sky are three faint constellations, Lynx, Leo Minor and Cancer, the latter containing a major star cluster, the Beehive or M44. The more prominent figure of Leo, the lion, is off to the south and is dealt with on Chart 10.

A STELLAR BEEHIVE

For beginners, the best object to look for in this part of the sky is the open cluster M44 in Cancer, the crab, visible in the lower right of our chart. This cluster is popularly named the Beehive but is also known as Praesepe, Latin for 'manger'. In clear, dark skies it can be glimpsed with the naked eye as a hazy patch, while one glance through binoculars shows why it merits its popular name – a swarm of stars comes into view, buzzing over an area three times the diameter of the full Moon and lying around 580 light years away. To the north and south of the cluster are two fairly faint stars, Gamma [γ] and Delta [δ] Cancri, which astronomers of old regarded as two donkeys feeding at the starry manger.

TWO DOUBLES IN CANCER

In the northern part of Cancer lies Iota [ι] Cancri, a yellow giant of magnitude 4.0. This has a 7th-magnitude companion star that can be glimpsed in good binoculars and is easily seen in small telescopes. Tucked into the bottom right of this chart is another double star divisible in small telescopes: Zeta [ζ] Cancri, a yellow pair of 5th and 6th magnitudes, forming a genuine binary with an orbital period of 1,100 years or so. Larger apertures – about 150 mm (6 in) – further divide the brighter of the two into a close pair of 5th- and 6th-magnitude stars with an orbital period of only 60 years.

MORE DOUBLES IN URSA MAJOR

Much of this chart is occupied by Ursa Major, the great bear, the third-largest of all the constellations – only Hydra and Virgo cover a greater area of sky. The 2nd-magnitude stars Alpha [α] and Beta [β] Ursae Majoris, named Dubhe and Merak, are the Pointers that guide the way to the north pole star, Polaris (see Chart 1). Using the Pointers in the opposite direction guides you to the most southerly naked-eye star in this constellation, Xi [ξ] Ursae Majoris, which is a binary with an orbital period of 60 years. Telescopes of 75-mm (3-in) aperture can separate the two stars, which will become progressively farther apart (and hence easier to separate) until 2035 due to their orbital motion. It was in fact the first binary to have its orbit calculated, in 1827. Three Moon diameters to the north of Xi lies Nu [ν] Ursae Majoris, also a double but the companion is only of 10th magnitude and unimpressive in small apertures.

AN OWLISH NEBULA

Return now to Beta [β] Ursae Majoris: just over four Moon diameters from it lies the Owl Nebula, M97, one of the closest planetary nebulae to us, some 1,500 light years away. One of the faintest objects on Messier's list, it will be difficult to see in the smallest apertures except under ideal conditions, appearing as a featureless circle some three times larger than the disk of Jupiter. Through larger apertures you can start to make out the owlish

▼ **The Beehive Cluster, M44,** in Cancer, is a widespread open cluster visible to the naked eye as a hazy patch under clear, dark skies but better seen through binoculars and small telescopes using low power.

STAR CHARTS

barred spiral galaxy ▶ p. 336 ecliptic ▶ p. 338

WATCHING THE SKY

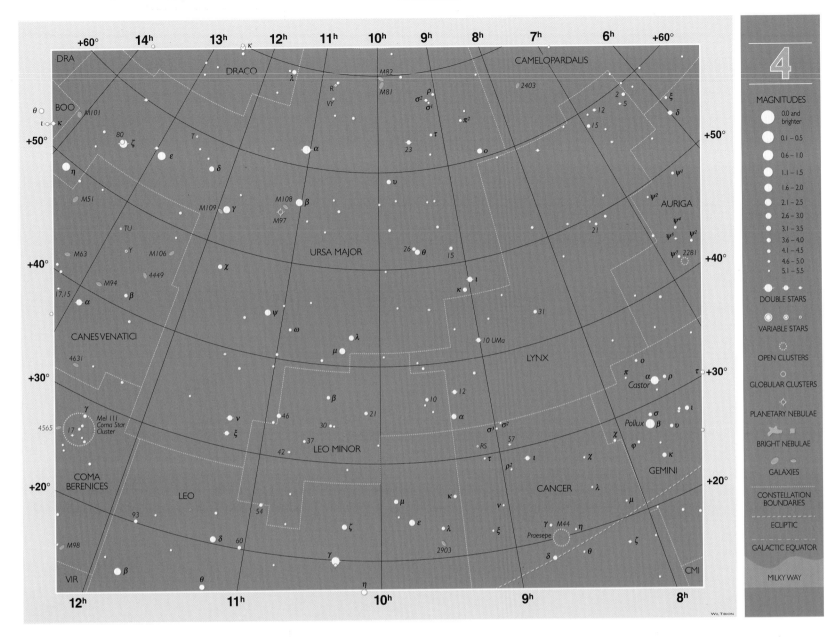

countenance, resulting from two dark hollows, like eyes, that gives this wisp its name.

While in this area, try looking for the galaxy M108, a spiral seen edge-on, with a patchy appear-ance caused by dark clouds of dust. On the other side of the Dipper's bowl, near Gamma [γ] Ursae Majoris, lies another faint Messier galaxy, M109, this one a **barred spiral**. Both are detectable in

100-mm (4-in) apertures, but are not impressive sights. Far better galaxies for small apertures are M81 and M82 in the north of Ursa Major, dealt with under Chart 1.

✳ THE TROPIC OF CANCER

The Sun's yearly path against the stars, known as the **ecliptic**, is depicted on these sky charts as a dashed yellow line, which can be seen passing through the constellation of Cancer in the lower right-hand corner of this chart. About 2,000 years ago the Sun was in Cancer on the June solstice each year, the date on which it reached its most northerly point in the sky and the Earth's northern hemisphere had its longest day. On the June solstice, the midday Sun lies overhead at latitude 23½° north on Earth, a

latitude that came to be known as the Tropic of Cancer (and its southern equivalent the Tropic of Capricorn). These names have since become out of date, due to the motion known as precession. The effect of precession over the past 2,000 years means that the Sun now lies on the Taurus–Gemini border at the June solstice and does not reach Cancer until late July. Similarly, when the Sun reaches its southernmost point in the sky on the December solstice each year it is now in Sagittarius, not Capricorn as was the case 2,000 years ago.

A MISSING ALPHA

Below centre of the chart is the small and faint constellation of Leo Minor, the little lion. Looking at the chart you might think there was a mistake – there is a star labelled Beta in Leo Minor but none labelled Alpha. In fact the mistake was made over 150 years ago, when astronomers were cataloguing the constellation's stars and forgot to allocate the letter Alpha to its brightest star, which is actually 46 Leonis Minoris, on the border with Ursa Major.

WATCHING THE SKY

CHART 5

NAME	RA, DEC	MAG(S)	TYPE OF OBJECT	VISIBILITY
Cor Caroli	12h 56m +38.3°	2.9, 5.6	double star	
Mizar	13h 24m +54.9°	2.2, 4.0, 4.0	multiple star	
κ Boötis	14h 13m +51.8°	4.5, 6.6	double star	
ι Boötis	14h 16m +51.4°	4.8, 7.5	double star	
ε Boötis	14h 45m +27.1°	2.5, 4.6	double star	
μ Boötis	15h 24m +37.4°	4.3, 6.5	triple star	
Y Canum Venaticorum	12h 45m +45.4°	5.0–6.5	variable star	
Coma star cluster	12h 25m +26.0°	1.8	open cluster	
M3	13h 42m +28.4°	6.4	globular cluster	
The Black Eye Galaxy (M64)	12h 57m +21.7°	8.5	spiral galaxy	
The Whirlpool (M51)	13h 30m +47.2°	8.1	spiral galaxy	
M101	14h 03m +54.3°	7.7	spiral galaxy	

Getting your bearings in this part of the sky is easy because of the presence of the Big Dipper, or Plough. This is one of the easiest star patterns to find and every part of it guides you to some other region of sky with interesting objects of its own. For example, if you extend the curve of the Dipper's handle you will end up at the bright star Arcturus, a trick that can be memorized by the phrase 'arc to Arcturus'. Above left of Arcturus lies a distinctive arc of stars that makes up Corona Borealis, the northern crown. There is no need to hop too far from the Dipper on this chart because an impressive range of double stars, clusters and galaxies lies within easy jumps.

MIZAR AND ALCOR

Our first stopping point is the bend of the Big Dipper's handle. With good eyesight you can spot a pair of stars here, and binoculars make them crystal clear. They are Mizar and Alcor, one of the easiest and brightest double stars in the sky. Although not a true binary, they are at similar distances from us, about 80 light years, and move through space together as part of a widely spread cluster that includes several other stars in Ursa Major. A small telescope shows a 4th-magnitude star closer to Mizar, and these probably do form a genuine binary – if so, the orbital period must be measured in thousands of years. Study of the light from Mizar, its companion and Alcor reveals that all three are spectroscopic binaries, thus adding to the complexity of the group.

GALAXIES GALORE

The Big Dipper and Canes Venatici are good places to go galaxy hunting. Start near the last star in the handle of the Dipper, Eta [η] Ursae Majoris. North of it lies M101, large and bright enough to be visible through binoculars as a rounded patch of light in dark skies. Small telescopes pick out its bright central regions, but its full magnificence is apparent only on long-exposure photographs which show that here we are looking at a face-on spiral with far-flung arms. Incidentally, the object that Messier

listed as entry number 102 in his catalogue of nebulous-looking objects was a duplicated observation of M101 with the wrong position. Hence you will not find any object labelled M102 on a star chart.

Look south, over the border into Canes Venatici, and you will encounter the famous Whirlpool Galaxy, M51, a spiral that is interacting with a smaller companion galaxy NGC 5195, which brushed past it a few hundred million years ago. Binoculars show the Whirlpool as a smudgy patch and small telescopes reveal the bright cores of M51 and its companion, but apertures of 250 mm (12 in) will be needed to distinguish the spiral structure. The Whirlpool is of historical interest for it was the first galaxy in which spiral structure was detected, by the nineteenth-century Irish astronomer Lord Rosse – although at the time neither he nor anyone else appreciated its true nature as a separate system of stars far beyond our own Galaxy. Current estimates place it at a distance of about 27 million light years.

For another celebrated galaxy, continue south into Coma Berenices where lies M64, known as the Black Eye Galaxy because a dark cloud of

dust is silhouetted against its bright **nucleus**. Small telescopes will detect M64 as an elliptical haze, but you probably will not see the black eye feature unless you have a telescope of about 150-mm (6-in) aperture or more.

DOUBLES IN BOÖTES

In central Boötes lies a difficult but glorious double star, Epsilon [ε] Boötis. To the eye it appears as an orange star of magnitude 2.5, but telescopes show a 5th-magnitude companion of a beautiful contrasting blue. The closeness of the pair means that high magnification on a telescope of at least 75-mm (3-in) aperture, combined with steady air, will be needed to separate them, but the resulting sight is worth the effort. Further

▶ *A close-up of Mizar (Zeta [ζ] Ursae Majoris) and Alcor (80 Ursae Majoris), the famous pair of stars in the constellation of Ursa Major in the middle of the Big Dipper's handle. A small telescope will reveal that Mizar (at right) is itself a double star with a 4th-magnitude companion, Mizar B.*

WATCHING THE SKY

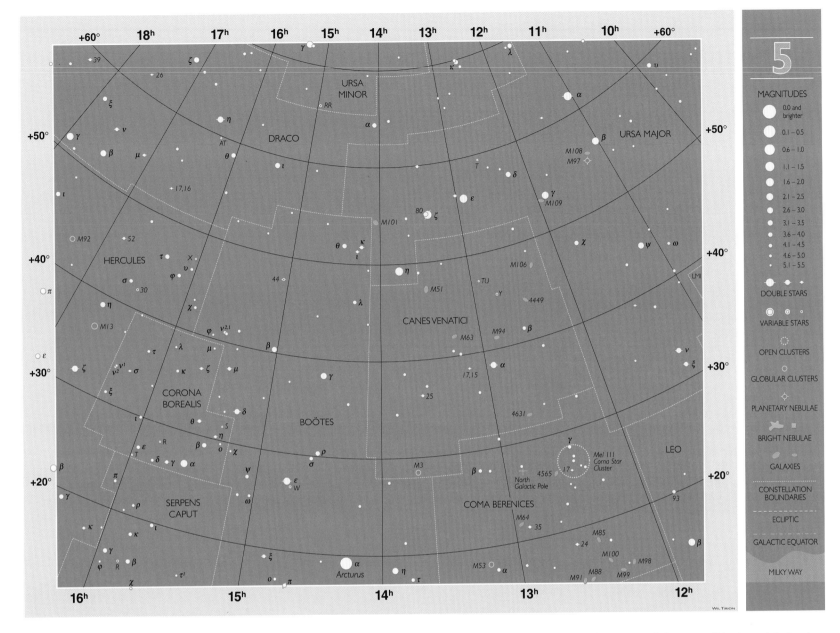

north, near the border with Corona Borealis, is Mu [μ] Boötis, a wide double separable with binoculars; apertures of 75 mm (3 in) or more show the fainter companion to be a close binary, making this a triple star. In the north of the constellation, near the end of the Dipper's handle, lie Kappa [κ] and Iota [ι] Boötis, two easy doubles for small telescopes.

CHARLES'S HEART
Another notable double lies in Canes Venatici, south of the Dipper's handle. Alpha [α] Canum Venaticorum is its proper title, but it also bears the popular name Cor Caroli, meaning Charles's Heart, given to it by British royalists to commemorate King Charles I who was beheaded by his political opponents. Small telescopes easily divide it into

two stars, of 3rd and 5th magnitudes. Both are supposedly white, but some observers have reported subtle colorations. See what you think.

Farther north in Canes Venatici is a variable star worth watching, bearing the letter Y (variables often have Roman letter designations rather than Greek ones). It is a supergiant with a deep red colour, noticeable in binoculars, that varies somewhat irregularly between about magnitudes 5.0 and 6.5; its nickname is La Superba.

CLUSTERS SMALL AND LARGE
Most globular clusters are found in the part of the sky around the centre of the Galaxy, but a few stragglers exist. One such is M3, on the southern border of Canes Venatici — you can find it by dropping down from the tip of the Big Dipper's

handle. It is regarded as one of the best globulars in northern skies, but its distance of some 30,000 light years dims the brilliance of its stars and you will need binoculars to pick up this hazy ball. The brightest of its individual stars can be resolved with telescopes of 100-mm (4-in) aperture or larger.

Far more widespread is the Coma star cluster, containing a few dozen stars strewn across 10 Moon diameters of sky in Coma Berenices. This is an open cluster that is drifting apart, just under 300 light years from us, forming an arrowhead shape easily traceable with binoculars or even the naked eye in dark skies — ancient astronomers knew this group and regarded it as representing either the tresses of Queen Berenice, after whom this constellation is named, or the swishing tail of neighbouring Leo the lion.

CHART 6

Brilliant Vega, fifth-brightest in the sky, is the most noticeable star on this chart. It is the leading star of Lyra, the lyre, a constellation of modest size yet packed with fascinating objects. Much of this chart is taken up by Hercules, representing the strong man of mythology, a constellation with few prominent stars but two globular clusters of note. To the north is part of Draco, representing the dragon that Hercules slew. At left of the chart, the Milky Way flows through Cygnus, the swan. Part of the treasures of Cygnus are dealt with on this chart, the remainder on Chart 7.

NAME	RA, DEC	MAG(S)	TYPE OF OBJECT	VISIBILITY
μ Draconis	17h 05m +54.5°	5.6, 5.7	double star	📷
ν Draconis	17h 32m +55.2°	4.9, 4.9	double star	📷
ε Lyrae	18h 44m +39.7°	4.7, 4.6	multiple star	∞ 📷
ζ Lyrae	18h 45m +37.6°	4.4, 5.7	double star	∞ 📷
β Lyrae	18h 50m +33.4°	7.2, var.	double star	📷
δ Lyrae	18h 54m +37.0°	4.2, 5.6	double star	👀 ∞ 📷
Albireo (β Cygni)	19h 31m +28.0°	3.1, 5.1	double star	∞ 📷
β Lyrae	18h 50m +33.4°	3.3–4.4	variable star	👀
χ Cygni	19h 51m +32.9°	3.3–14	variable star	👀 ∞
M13	16h 42m +36.5°	5.9	globular cluster	∞ 📷
M92	17h 17m +43.1°	6.5	globular cluster	∞ 📷
Brocchi's Cluster	19h 25m +20.2°	3.6	open cluster	∞
Ring Nebula (M57)	18h 54m +33.0°	9.0	planetary nebula	📷
Dumbbell Nebula (M27)	20h 00m +22.7°	8.1	planetary nebula	∞ 📷
Blinking Planetary (NGC 6826)	19h 45m +50.5°	9.8	planetary nebula	📷

DOUBLE STAR DELIGHTS

This is an area filled with outstanding double stars, none better for beginners than Albireo, the popular name for Beta [β] Cygni, on the southern border of Cygnus. The smallest of telescopes – and even good binoculars, if mounted steadily – divides it into two components whose contrasting colours of amber and blue-green resemble two lamps in a celestial traffic light.

In the same binocular field of view as Vega is a wide, white pair, Epsilon [ε] Lyrae. What makes this pair special is that, through telescopes of 75-mm (3-in) aperture under high magnification, each star breaks up into a tight binary, hence its popular name the Double Double. Families of four stars are rare, and this is the finest example. While in the vicinity of Vega look also at Delta [δ] Lyrae, a wide pair of unrelated stars divisible with good eyesight and easy in binoculars, and Zeta [ζ] Lyrae, just within the capacity of good binoculars and easily split with small telescopes.

We cannot leave Lyra's doubles without a mention of Beta [β] Lyrae, easily resolved through small telescopes into a pleasing partnership of cream and blue stars. The brighter star (the cream one) is a fascinating eclipsing binary that changes by a full magnitude every 12.9 days, easily noticeable even to the untrained eye.

If you have not had your fill of doubles, look north into Draco where you will find Nu [ν] Draconis, one of the best pairs for binoculars. For a telescopic challenge turn to nearby Mu [μ] Draconis, a tight binary with an orbital period of some 670 years that should just be divisible with an aperture of 75 mm (3 in).

CELESTIAL SMOKE RING

Midway between Beta [β] and Gamma [γ] Lyrae sits one of the most-photographed planetary nebulae in the sky – M57, the Ring Nebula. You will need a telescope to find it, but its visual appearance may be disappointing compared with photographs in books. Small apertures show it as an elliptical disk larger than Jupiter; larger telescopes are needed to make out the ring shape that gives the nebula its popular name. Do not expect to see the beautiful colours captured on photographs – to the human eye it appears merely grey or bluish.

▼ **The Coathanger,** also known as Brocchi's cluster in Vulpecula, has a charming configuration consisting of 10 stars, six of which form a bar running east–west with the other four creating a hook on the southern side. Remarkably, the stars are not actually related but lie in the same line of sight by chance. This is an excellent object for binoculars.

GLOBULARS IN HERCULES

Four stars arranged in the shape of a keystone – Pi [π], Eta [η], Zeta [ζ] and Epsilon [ε] Herculis – mark the lower part of the body of Hercules. On one side of the keystone, a third of the way from Eta to Zeta, you will find the most impressive globular cluster in northern skies, M13, consisting of some 300,000 stars crowded into a ball over 100 light years across. Even so, at a distance of 25,000 light years all that starlight has faded to the limit of naked-eye visibility. For a definite sighting of M13 binoculars are needed, while small telescopes will pick out the brightest giant stars in the cluster. Less easy to spot is another globular, M92, to the north of the keystone. Being smaller than M13 it can be mistaken for a star, even through binoculars, and telescopes are needed to see it properly. M92 contains some of the oldest stars in our Galaxy, formed an estimated 13 billion years ago.

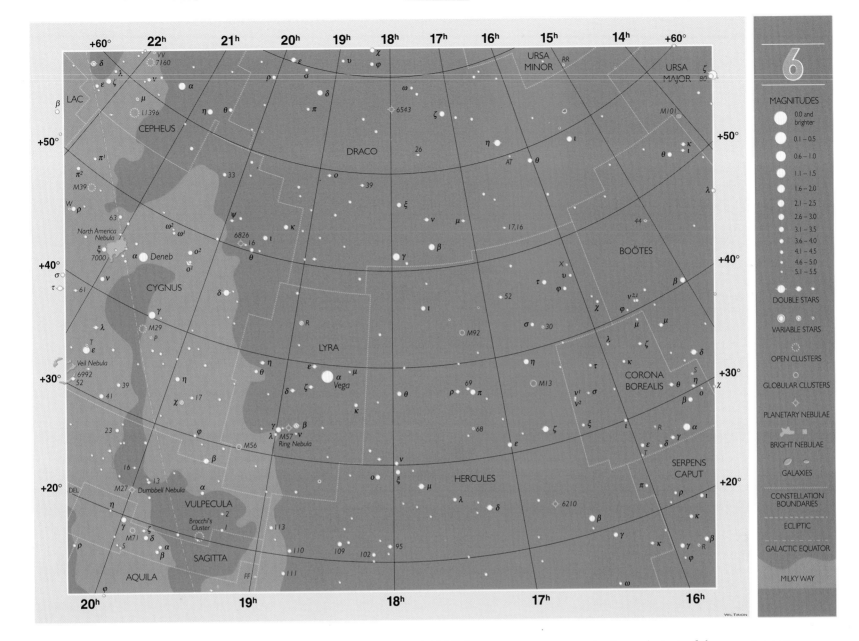

COATHANGER CLUSTER

Vulpecula, the fox, at lower left on this chart, is not one of the best-known constellations, but it harbours two classic objects for small instruments. The first is an attractive grouping, Brocchi's Cluster, more popularly known as the Coathanger because of its unmistakable shape – a line of six stars forming the hanger, with four more completing the hook, spread over three Moon diameters of sky and ideal for binocular users. It turns out that this is not a true cluster at all, but a chance alignment of unrelated stars at different distances from us.

GASEOUS DUMBBELL

Vulpecula's other treasure is M27, the Dumb-bell Nebula, one of the closest planetary nebulae to us, some 1,000 light years away. Being relatively close it also appears quite large, about one-quarter the diameter of the full Moon, and paradoxically this makes it more difficult to see since its light is spread over a greater area. Through binoculars it can be glimpsed as a somewhat elongated wisp, but moderate sized telescopes will be needed to see its shape, supposedly resembling a body-builder's dumbbell – although a better comparison would be a figure eight or an hourglass.

SOME SIGHTS OF CYGNUS

We finish our tour of this part of the sky where we began – in Cygnus. Binocular users should keep an eye on Chi [χ] Cygni, a pulsating red giant of the same type as Mira (see Chart 8). At its brightest, which it reaches every 13 months or so, it is easily visible to the naked eye but at minimum drops to as low as 14th magnitude, difficult to pick out from the background stars even in a telescope.

Farther north is one of the most intriguing planetary nebulae within range of amateur telescopes, NGC 6826. It can be spotted as a round, bluish disk and you may also be able to see the 10th-magnitude central star. Herein lies a peculiar optical effect: look alternately towards the central star and away again, and the nebula appears to blink on and off. This remarkable characteristic has led to its nickname the Blinking Planetary.

WATCHING THE SKY

STAR CHARTS

CHART 7

NAME	RA, DEC	MAG(S)	TYPE OF OBJECT	VISIBILITY
P Cygni	20h 18 m +38.0°	4.8	variable star	👀
o¹ Cygni	20h 14m +46.7°	3.8, 4.8	double star	∞ 🔭
61 Cygni	21h 07m +38.7°	5.2, 6.1	double star	🔭
δ Cephei	22h 29m +58.4°	var., 6.3	double star	👀
δ Cephei	22h 29m +58.4°	3.5–4.4	variable star	👀
μ Cephei	21h 44m +58.8°	3.4–5.1	variable star	∞
β Pegasi	23h 04m +28.1°	2.3–2.7	variable star	👀
ρ Cassiopeiae	23h 54m +57.5°	4.1–6.2	variable star	∞
M29	20h 24m +38.5°	6.6	open cluster	🔭
M39	21h 32m +48.4°	4.6	open cluster	∞ 🔭
N. America Nebula (NGC 7000)	20h 59m +44.3°	6.0	nebula	👀 ∞
Veil Nebula	20h 56m +31.7°	–	supernova remnant	🔭
NGC 7662	23h 26m +42.6°	9.2	planetary nebula	🔭

T he starry band of the Milky Way crosses this chart diagonally from Cassiopeia at top left into Cygnus, the swan, a constellation most easily visualized as a large cross – Deneb is at the top and Beta [β] Cygni at its foot. A dark rift runs along the Milky Way through Cygnus and southwards into Charts 6 and 12, caused by a ribbon of dark dust that blocks starlight from behind it. One of the major features of the northern summer sky is a large triangle marked out by Deneb and two other bright stars, Vega and Altair (shown in the chart's margins). Cepheus, at the top of the chart, contains two classic variable stars, including the original Cepheid. At the bottom left of the chart is Pegasus, noted for its large square of stars, adjoining Andromeda at the left; for more about the Great Square of Pegasus, see Chart 13.

DENEB AND THE NORTH AMERICA NEBULA

Deneb is a highly luminous supergiant, as bright as 250,000 Suns, and the most distant of the 1st-magnitude stars, over 3,000 light years from Earth. Seemingly near it in the sky – although in reality only about half as far away – is a remarkable cloud of gas and dust, NGC 7000, known as the North America Nebula because of its striking resemblance to the shape of that continent. Visually it is elusive; dark skies are needed if you are to see it through binoculars as a brighter patch three times wider than the full Moon against the rich Milky Way background (it is shown to scale on the chart). Under the best conditions it can be detected with the naked eye as a brighter patch in the Milky Way, but its full magnificence is apparent only on photographs.

DOUBLES IN CYGNUS

For an easy introduction to the double stars in this area, turn binoculars on the star Omicron-1 (o¹) Cygni, an orange star which has a wide companion of a pleasantly contrasting blue colour. If the binoculars can be supported steadily, for example on a windowsill or by wedging them against a wall, you may be able to make out a fainter star closer to Omicron-1

Cygni which also appears light blue. Now move to the opposite side of Deneb where lies 61 Cygni, consisting of a pair of orange dwarfs easily separable with small telescopes. A milestone in astronomy is associated with 61 Cygni: it was the first star to have its distance accurately measured by **parallax**, in 1838 (see Properties of Stars). Another well-known double in Cygnus, Albireo or Beta [β] Cygni, is described on Chart 6.

CLUSTERS IN CYGNUS

Two open star clusters in Cygnus appear on Charles Messier's list. The more northerly of them, and the more attractive in binoculars, is M39, consisting of about 20 stars arranged in a triangular shape and covering an area similar to that of the full Moon. The other cluster, M29, lies near the star Gamma [γ] Cygni in the centre of the constellation. Much smaller than M39, it consists of a clutch of stars that are best viewed through small telescopes. Near M29 lies 5th-magnitude P Cygni, one of the rare luminous blue variables, stars of extremely high **mass** and **luminosity** (see The Stellar Zoo – Single Stars, page 168).

VEIL NEBULA

A famous but elusive feature in Cygnus is the Veil Nebula, the filamentary remains of an ancient supernova explosion. While the nebula looks impressive on photographs, it is hard to track down visually. This is one instance in which a nebula filter will definitely help, but even so a fair-sized telescope will be necessary to trace its full extent. The brightest part of it is an arc

labelled NGC 6992. Today the debris is thinly spread and hard to find, but when the star exploded it must have shone like a beacon, brighter than any other object in the night sky.

THE ULTIMATE CEPHEID VARIABLE

John Goodricke, the eighteenth-century English amateur astronomer who discovered

▼ *The North America Nebula (NGC 7000), a softly glowing cloud of gas named for its resemblance to the shape of the continent of North America. It lies in the constellation Cygnus near the 1st-magnitude star Deneb, Alpha Cygni. Deneb is one of the three stars that comprise the so-called Summer Triangle in the northern hemisphere.*

WATCHING THE SKY

luminosity ► p. 340 mass ► 341 parallax ► p. 342
standard candles ► p. 344

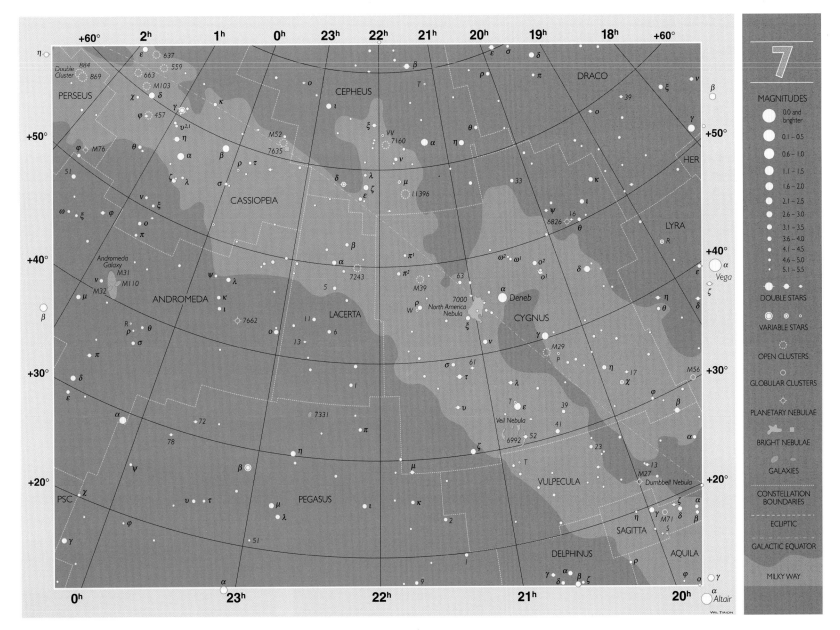

the variations of Algol, also detected the variations of Delta [δ] Cephei. At the time, no one could have foreseen the importance to astronomy of this star, which turned out to be the prototype of a class of regularly pulsating stars whose periodicity is directly linked to their inherent brightness, a valuable characteristic that allows them to function as **standard candles** for measuring distances in space. The term 'Cepheid variable' is now applied to all stars of this type. Delta Cephei itself varies every five days nine hours between magnitudes 3.5 and 4.4, easily visible to the watchful eye as Goodricke demonstrated. Binoculars and small telescopes show that this yellow supergiant is also an attractive double, with a bluish companion star.

THE GARNET STAR AND OTHERS

Near Delta Cephei is another notable variable, Mu [μ] Cephei, nicknamed the Garnet Star by William Herschel on account of its ruddy hue which is readily noticeable through binoculars. A pulsating red supergiant, it is one of the best-known examples of a semiregular variable, stars which have less well-defined periods than regular pulsators such as Delta Cephei. Mu Cephei has a magnitude range from about 3.4 to 5.1 and a rough period of two years.

Look across to Cassiopeia and you will find a similar variable of extraordinary luminosity: Rho [ρ] Cassiopeiae, blazing with the intensity of half a million Suns – it is a hypergiant (see Old Age and Death of High-Mass Stars, page 164). Rho Cassiopeiae rises and falls between

4th and 6th magnitude every 10 or 11 months, although like Mu Cephei the variations do not repeat exactly with each cycle.

A red giant that pulsates entirely irregularly is Beta [β] Pegasi, the star at the top right of the Square of Pegasus, which can be found at any brightness between magnitudes 2.3 and 2.7.

A SNOWBALL IN ANDROMEDA

One of the easiest planetary nebulae to see with small telescopes lies north of the Square of Pegasus, in Andromeda – NGC 7662, nicknamed the Blue Snowball by imaginative observers. Even low magnification shows it as a hazy, blue-green blob, while under higher powers it becomes noticeably elliptical. Its distance is about 4,000 light years.

CHART 8

This is a fairly barren area of sky with no eye-catching stars, although there are several note-worthy objects for those willing to hunt for them. Much of the chart is taken up by the constellation Cetus, representing the sea monster that threatened to devour Andromeda before she was rescued by Perseus. It is one of several constellations in which the star marked Alpha [α] is not the brightest (another example is neighbouring Pisces). In the case of Cetus the honour goes to Beta [β] Ceti, magnitude 2.0, half a magnitude brighter than Alpha Ceti. Cetus contains the prototype of the red giant variables of long period, Mira. Also on this chart are two of the most Sun-like of the nearby stars, one of which is now known to possess a planet. The pretty Pleiades star cluster at top left is described on Chart 2.

NAME	RA, DEC	MAG(S)	TYPE OF OBJECT	VISIBILITY
τ Ceti	1h 44m -15.9°	3.5	Sun-like star	👀
ε Eridani	3h 33m -9.5°	3.7	Sun-like star	👀
ζ Piscium	1h 14m +7.6°	5.2, 6.4	double star	🔭
ρ Piscium	1h 26m +19.2°	5.3, 5.5	double star	∞ 🔭
γ Arietis	1h 54m +19.3°	4.7, 4.6	double star	🔭
α Piscium	2h 02m +2.8°	4.2, 5.2	double star	🔭
γ Ceti	2h 43m +3.2°	3.5, 6.6	double star	🔭
α Ceti	3h 02m +4.1°	2.5, 5.6	double star	∞ 🔭
32 Eridani	3h 54m -3.0°	4.8, 6.1	double star	🔭
Mira (ο Ceti)	2h 19m -3.0°	3–9	variable star	👀 ∞ 🔭
M74	1h 37m +15.8°	9.2	spiral galaxy	🔭
M77	2h 43m 0.0°	8.8	spiral galaxy	🔭

AMAZING MIRA

Astronomers in the seventeenth century were astounded to see the star labelled on their maps as Omicron [ο] Ceti gradually fade from view and then, some months later, reappear again. It was the first star ever seen to behave in such a way and it was named Mira, 'the amazing one'. Since then many thousands of similar stars have been discovered and they are all now known as Mira stars. They are stars with similar masses to the Sun that have swollen up late in their lives, becoming pulsating red giants, a stage that the Sun itself will reach billions of years from now (see Old Age and Death of Low-Mass Stars, page 162). Mira is easily visible to the naked eye when at its brightest, which it achieves every 11 months or so, but in between fades to 9th or 10th magnitude, requiring a small telescope to follow. It is about 400–500 times the diameter of the Sun, but of course is still visible as nothing more than a point of light through even the largest telescopes.

TWO SUN-LIKE STARS

Before turning to some double stars, take the chance to identify the two nearby single stars that are most like the Sun (the most Sun-like of all the nearby stars are the brightest components of Alpha [α] Centauri, but they form a binary). In Eridanus, at the left of this chart, lies Epsilon [ε] Eridani, a slightly cooler star than the Sun with a **spectral type** of K2, appearing somewhat orange in colour (you will need binoculars or a small telescope to discern the colour). It lies a mere 10.5 light years away. Recently, astronomers have found that it is orbited by a planet of similar mass to Jupiter, and it is possible that an entire planetary system is forming around this star. At magnitude 3.7, Epsilon Eridani is easily visible to the naked eye. Slightly hotter and brighter, but also slightly more distant, is Tau [τ] Ceti, whose spectral type is G8. Through binoculars and small telescopes it shows a yellow colour, slightly deeper in tone than the Sun would appear when viewed from afar. At present it is not known whether Tau Ceti has any planets.

DOUBLE-STAR ROUNDUP

An easy binocular double is Alpha [α] Ceti, a red giant lying in almost the same line of sight as a 6th-magnitude background star. To its right is Gamma [γ] Ceti, a much closer pairing that will require at least a 60-mm (2.4-in) telescope and high magnification to divide.

▲ *Two notable galaxies* are within range of amateur instruments in this region of sky. At left is M77 in the constellation Cetus, a spiral with an active nucleus that is the brightest member of the class known as Seyfert galaxies. At right is M74 in Pisces, a beautiful face-on spiral although fairly large apertures are needed to trace its widespread arms.

WATCHING THE SKY

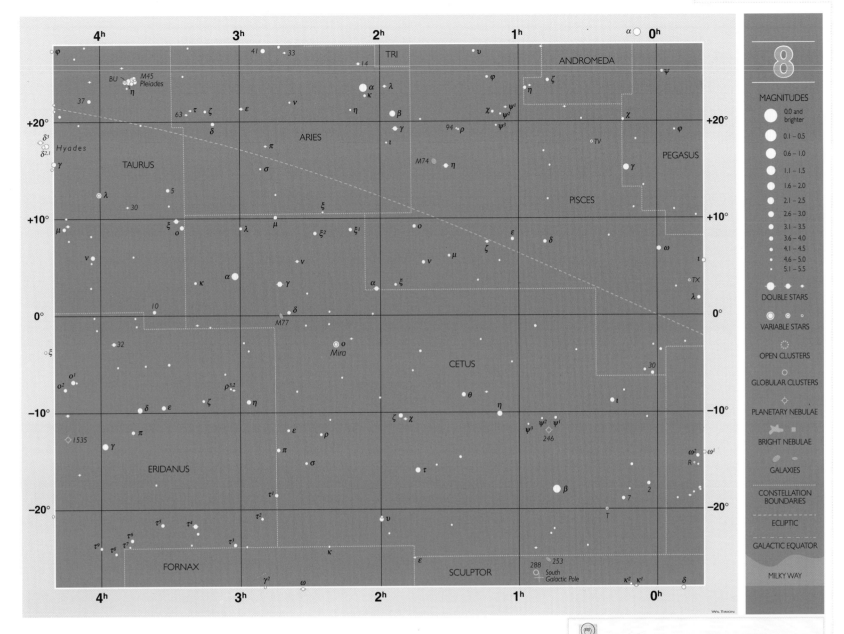

Wil Tirion

MAGNITUDES

0.0 and brighter
0.1 – 0.5
0.6 – 1.0
1.1 – 1.5
1.6 – 2.0
2.1 – 2.5
2.6 – 3.0
3.1 – 3.5
3.6 – 4.0
4.1 – 4.5
4.6 – 5.0
5.1 – 5.5

DOUBLE STARS
VARIABLE STARS
OPEN CLUSTERS
GLOBULAR CLUSTERS
PLANETARY NEBULAE
BRIGHT NEBULAE
GALAXIES
CONSTELLATION BOUNDARIES
ECLIPTIC
GALACTIC EQUATOR
MILKY WAY

Farther to the right again is Alpha [α] Piscium, another example of an Alpha star that is not the brightest in its constellation. This duo form a genuine binary but are close together and a moderate-sized telescope is needed to separate them. Those equipped with small telescopes should instead turn to Zeta [ζ] Piscium, whose 6th-magnitude companion is easily seen, while binocular users should look north to Rho [ρ] Piscium which forms a wide pair with 94 Piscium, an unrelated background star. Now hop to the left, over the border into Aries, to find Gamma [γ] Arietis, an attractive pair of almost identical white stars easily divisible in small telescopes. Rounding out our tour of the doubles in this area is 32 Eridani, at the left of the chart, a beautifully coloured double of yellow and blue-green divisible in small apertures.

TWO SPIRAL GALAXIES

Above left of Mira lies M77, a face-on spiral galaxy visible as a rounded smudge through small telescopes. Its most obvious feature is its nucleus – it is in fact the brightest of the so-called **Seyfert galaxies**, spirals with active centres like subdued versions of **quasars**. For another face-on spiral head north to Pisces and M74, two-and-a-half Moon diameters from Eta [η] Piscium – which is actually the brightest star in Pisces, although you would not think so from its Greek-letter designation. Through small telescopes M74 has a somewhat spotted appearance, but this is caused by faint foreground stars in our Galaxy. Photographs bring out its spiral arms, but these will not be detectable with small apertures.

THE AGE OF AQUARIUS

The celestial equivalent of the Greenwich meridian is the 0h line of right ascension. By definition, it passes through the point where the Sun crosses from south to north of the celestial equator each year, which marks the March equinox. But this crossover point gradually moves against the stars, due to the slow wobble of the Earth in space termed precession. Over 2,000 years ago, when the constellations were first devised, the cross-over point lay in Aries. Subsequently it moved into Pisces, where it still lies, as this chart shows – look at the right, where the dotted yellow line of the Sun's path, the ecliptic, intersects the celestial equator. It will continue to move, eventually reaching Aquarius nearly 600 years from now – which is when the much-heralded Age of Aquarius will in reality start.

WATCHING THE SKY

CHART 9

Orion, the hunter, dominates this rich area, along with his dogs – the constellations Canis Major and Canis Minor, picked out respectively by the bright stars Sirius (the brightest star of all in the night sky) and Procyon. Straddling the celestial equator, Orion is visible equally well from the northern and southern hemispheres on Earth. Three stars in a row comprise the hunter's belt, while below this is the famous Orion Nebula and some star clusters, which together represent his sword. To Orion's upper right is Taurus, the bull, sporting the V-shaped cluster known as the Hyades (the Pleiades cluster, in the top right corner, comes under Chart 2).

NAME	RA, DEC	MAG(S)	TYPE OF OBJECT	VISIBILITY
Rigel	5h 15m −8.2°	0.2, 6.8	double star	telescope
δ Orionis	5h 32m −0.3°	2.2, 6.9	double star	binoculars, telescope
ζ Orionis	5h 41m −1.9°	1.7, 3.9	double star	telescope
ι Orionis	5h 35m −5.9°	2.8, 6.9	double star	telescope
σ Orionis	5h 39m −2.6°	3.8, 6.6	multiple star	telescope
		6.8, 9.0		
Trapezium (θ² Orionis)	5h 35m −5.4°	5.1, 6.7	multiple star	telescope
		6.7, 8.0		
θ² Orionis	5h 35m −5.4°	5.1, 6.4	double star	binoculars, telescope
β Monocerotis	6h 29m −7.0°	4.6, 5.0, 5.4	multiple star	telescope
o² Eridani	4h 15m −7.7°	4.4, 9.5, 11.0	triple star	telescope
Betelgeuse	5h 55m +7.4°	0.0–1.3	variable star	naked eye
Hyades	4h 27m +16°	0.5	open cluster	naked eye, binoculars
NGC 1981	5h 35m −4.4°	4.6	open cluster	telescope
NGC 2232	6h 27m −4.7°	3.9	open cluster	telescope
NGC 2244	6h 32m +4.9°	4.8	open cluster	telescope
NGC 2264	6h 41m +9.9°	3.9	open cluster	binoculars, telescope
M50	7h 03m −8.3°	5.9	open cluster	binoculars, telescope
Orion Nebula (M42)	5h 35m −5.4°	4.0	nebula	naked eye, binoculars, telescope
M43	5h 36m −5.3°	9.0	nebula	binoculars, telescope
NGC 1977	5h 36m −4.9°	–	nebula	binoculars, telescope

ORION NEBULA

We start by visiting one of the most famous stellar birthplaces in the sky: the Orion Nebula. When you explore this region with binoculars or a small telescope you stumble across a wealth of wispy nebulosity and stars, many of them newly born. What is referred to as the Orion Nebula breaks down into several components. The main one is M42, a cloud of gas glowing with the light of a cluster of young stars within it. It is more than two Moon diameters wide and is faintly visible to the naked eye even over a distance of 1,500 light years. Binoculars show

▼ *The V-shaped star cluster called the Hyades, visible to the naked eye, forms the head of Taurus, the bull. The bright red giant star Aldebaran appears to be a member of the cluster but is in fact a foreground object at less than half the distance.*

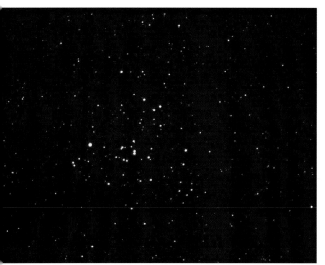

its hazy form more clearly while telescopes bring further details into view. The brightest star within it is actually a quadruple called the Trapezium, or Theta-1 [θ¹] Orionis, and is easily divided in small telescopes. Nearby lies Theta-2 [θ²] Orionis, a wide double star separable with binoculars.

To the north of M42 is M43, an extension of the same cloud. Further north still is NGC 1977, an elongated patch of nebulosity surrounding two stars. Completing the northernmost extent of this complex is NGC 1981, a loose open star cluster visible in binoculars.

South of M42, representing the glinting point of his sword, is the star Iota [ι] Orionis, a neat double divisible in small telescopes. Binoculars will show a wider pair nearby, which does not have a Greek-letter designation.

BELT OF ORION

A row of three 2nd-magnitude stars comprises Orion's jewelled belt. The northernmost of them, Delta [δ] Orionis, has a wide but faint companion detectable in binoculars and small telescopes. At the other end of the line is Zeta [ζ] Orionis, a tight binary that can be split with apertures of 75 mm (3 in) and above. Look below it and to the right to find one of Orion's best-kept secrets: Sigma [σ]

Orionis, a star which bursts into a row of four when seen through small telescopes (the faintest one may be difficult to spot at first), while in the same field of view is a narrow triangle of stars.

Betelgeuse

This red supergiant lies at the top left of Orion, marking his shoulder. It is the most obviously variable of all 1st-magnitude stars, ranging by about a magnitude in brightness with no firm period. Betelgeuse forms a large triangle with Sirius (at its southern apex) and Procyon.

Rigel

In the diametrically opposite corner of Orion from Betelgeuse is a contrasting blue supergiant, Rigel. It has a 7th-magnitude companion, although you will need good conditions and high magnification on your telescope to see it because it is drowned in Rigel's glare. You might expect Rigel and Betelgeuse to lie at similar distances, but in fact Rigel is nearly twice as far away, some 770 light years from us, still only half the distance to the Orion Nebula.

Hyades

Reluctantly leaving Orion, we look above right to find a V-shaped cluster of stars easily visible to the naked eye, called the Hyades. To

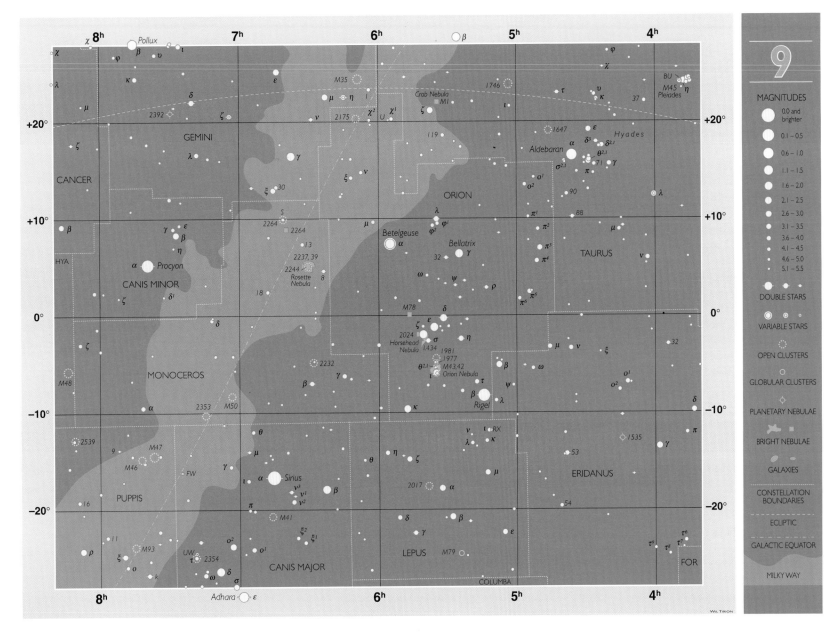

MAGNITUDES

●	0.0 and brighter
●	0.1 – 0.5
●	0.6 – 1.0
●	1.1 – 1.5
●	1.6 – 2.0
●	2.1 – 2.5
●	2.6 – 3.0
●	3.1 – 3.5
●	3.6 – 4.0
●	4.1 – 4.5
·	4.6 – 5.0
·	5.1 – 5.5

DOUBLE STARS

VARIABLE STARS

OPEN CLUSTERS

GLOBULAR CLUSTERS

PLANETARY NEBULAE

BRIGHT NEBULAE

GALAXIES

CONSTELLATION BOUNDARIES

ECLIPTIC

GALACTIC EQUATOR

MILKY WAY

WIL TIRION

ancient astronomers this group formed the face of Taurus the bull, with the red giant Aldebaran as its angry eye – yet we now know that Aldebaran is not a true member of the cluster at all, but a superimposed foreground star. Aldebaran's ruddy colour is more apparent through binoculars, the ideal instrument for sweeping over the widespread stars of the Hyades, which is the nearest large cluster to us, 150 light years away.

RED AND WHITE DWARFS

Now drop south into Eridanus to find Omicron-2 [o²] Eridani, a star similar to the Sun that is accompanied by a white dwarf, the easiest of these faint, old stars to see with a small telescope. More remarkably still, the white dwarf forms a binary with an accompanying red dwarf, which is even fainter but also within reach of small telescopes.

HIDDEN TREASURES IN MONOCEROS

Often overlooked among the sparkling glories of this part of the sky is the constellation Monoceros, the unicorn. Lying in the Milky Way, it contains several objects of note, among them Beta [β] Monocerotis, perhaps the finest triple star in the sky, consisting of an arc of three stars that can be made out with small telescopes.

Lying farther north is the celebrated Rosette Nebula, a loop of gas with a cluster of stars at its centre. Although the nebula shows up beautifully on photographs it is too faint for small apertures unless a nebula filter is used. But the central cluster, NGC 2244, can easily be picked out with binoculars. Another combination of nebula and cluster, NGC 2264, lies farther north still. The cluster can be seen with binoculars but the nebulosity, which includes a dark intrusion known as the Cone Nebula, shows up well only on photographs.

Now sweep south down the Milky Way until you come to M50, an open cluster visible in binoculars as a hazy patch about half the size of the full Moon. A larger but sparser cluster for binoculars, NGC 2232, lies to the north of Beta Monocerotis. To continue the binocular sweep into Puppis and farther along the Milky Way, turn to Chart 15.

WATCHING THE SKY

CHART 10

 NAME	RA, DEC	MAG(S)	TYPE OF OBJECT	VISIBILITY
ε Hydrae	8h 47m +6.4°	3.4, 6.7	double star	
Regulus	10h 08m +12.0°	1.4, 7.7	double star	∞
ζ Leonis	10h 17m +23.4°	3.4, 6.0	double star	∞
γ Leonis	10h 20m +20.0°	2.3, 3.6	double star	
ι Leonis	11h 24m +10.5°	4.1, 6.7	double star	
R Leonis	9h 48m +11.4°	6–10	variable star	∞
U Hydrae	10h 38 -13.4°	4.2–6.6	variable star	∞
M48	8h 14m -5.8°	5.8	open cluster	∞
M67	8h 50m +11.8°	6.9	open cluster	∞
Ghost of Jupiter (NGC 3242)	10h 25m -18.6°	8.6	planetary nebula	
M95	10h 44m +11.7°	9.7	barred spiral	
M96	10h 47m +11.8°	9.2	spiral galaxy	
M105	10h 48m +12.6°	9.3	elliptical galaxy	
M65	11h 19m +13.1°	9.3	spiral galaxy	∞
M66	11h 20m +13.0°	9.0	spiral galaxy	∞

O ne of the most distinctively shaped constellations is Leo, clearly resembling the crouching lion that it is supposed to represent – unlike most constellations which bear only passing resemblance to the creatures or mythological characters after which they are named. Leo's most distinctive feature is a hook shape of six stars called the Sickle, visible at the top of this chart. But Leo is fairly isolated. To its right is Cancer, the faintest of the 12 constellations of the zodiac, while south of it are the relatively barren constellations of Crater, Sextans and Hydra. This latter figure, representing a water snake, is the largest of all 88 constellations – its body trails through Chart 16 and ends on Chart 17 – yet its brightest star, Alpha [α] Hydrae, seen at centre right on the chart, is only of magnitude 2.0.

GAMMA LEONIS AND OTHER DOUBLES

In the arc of the Sickle of Leo lies one of the finest double stars for small telescopes, Gamma [γ] Leonis. A pair of binoculars shows a nearby star of 5th magnitude, but this is unrelated. The real thrill comes when you turn a small telescope on the main star, which divides into two golden-yellow stars, both giants. Long-term monitoring of their motions has revealed that they orbit each other about every 600 years. To the north of it in the Sickle is Zeta [ζ] Leonis, which is flanked by two fainter stars visible in binoculars, although none of them are related.

At the foot of the Sickle is the constellation's brightest star, 1st-magnitude Regulus, also known as Alpha [α] Leonis (of all the 1st-magnitude stars, it is actually the faintest). Regulus, being hotter than the Sun, shows a bluish colour, most noticeable through binoculars and small telescopes. Careful inspection should reveal that it has a much fainter companion star, of 8th magnitude.

Much more challenging is Iota [ι] Leonis, a close binary with an orbital period of 183 years, which you will find near the hind quarters of Leo. A telescope of 100-mm (4-in) aperture will be needed to resolve this pair.

TWO VARYING RED GIANTS

Some way to the right of Regulus lies R Leonis, a variable red giant that never gets very bright – you will probably need binoculars to locate it even at maximum – but is notable for its strong red colour. It is a pulsating star of similar type to Mira in Cetus, hundreds of times larger than the Sun, that rises and falls in brightness over a period of 10 months or so. Much farther south, in Hydra, lies another red giant variable, this one not so regular in its behaviour: U Hydrae, which fluctuates between 4th and 7th magnitudes every four months or so. Through binoculars you will notice its strong red colour, caused by the presence of carbon in its cool outer layers.

GALAXIES IN LEO

Charles Messier catalogued five galaxies in Leo, although none of them are particularly prominent. Start under the hind quarters of Leo where, with care on a dark night, you will find M65 and M66, two spiral galaxies aligned at an angle and hence appearing elongated. A small telescope should pick them out, or even large binoculars.

To their right, under the body of Leo, lie three more Messier galaxies in a little cluster. The two most prominent are M95 and M96, the first a barred spiral – one of the few such in Messier's catalogue – and the other a normal spiral. About a degree to the north of M96 you may be able to make out M105, a smaller **elliptical galaxy**. All five of these Leo galaxies seem to be part of the same cluster, about 35 million light years from us.

CLUSTER IN CANCER

The showpiece star cluster in Cancer is of course the Beehive, M44, dealt with on Chart 4, but don't overlook another open cluster to the south

▼ **M48 is a large** open cluster in Hydra. Its brightest stars can just be made out with binoculars and small telescopes show them clearly. They are arranged into a narrow wedge-shape like the point of an arrow.

WATCHING THE SKY

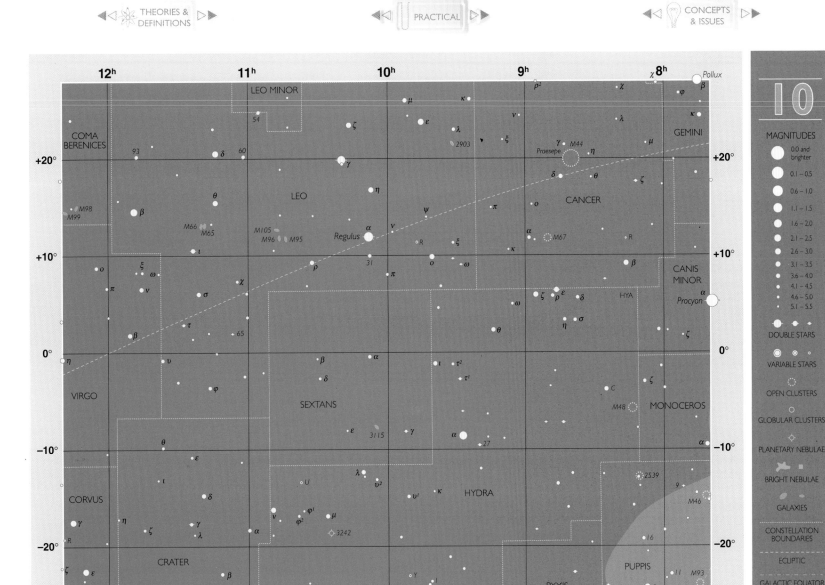

MAGNITUDES

⬤	0.0 and brighter
●	0.1 – 0.5
●	0.6 – 1.0
●	1.1 – 1.5
●	1.6 – 2.0
•	2.1 – 2.5
•	2.6 – 3.0
•	3.1 – 3.5
·	3.6 – 4.0
·	4.1 – 4.5
·	4.6 – 5.0
·	5.1 – 5.5

DOUBLE STARS

VARIABLE STARS

OPEN CLUSTERS

GLOBULAR CLUSTERS

PLANETARY NEBULAE

BRIGHT NEBULAE

GALAXIES

CONSTELLATION BOUNDARIES

ECLIPTIC

GALACTIC EQUATOR

MILKY WAY

WIL TIRION

of it, M67. Smaller and more densely packed than its more celebrated neighbour, M67 is still quite large as clusters go – almost the size of the full Moon and appearing as a misty ellipse when seen through binoculars and small telescopes. Some 200 or so faint stars are crowded in here, and some impression of the cluster's true richness can be obtained through telescopes of 75-mm (3-in) aperture. Because they are quite tightly packed, the stars in this open cluster have stayed together over billions of years without drifting apart.

MONSTER'S HEAD
South of M67 is a loop of half a dozen stars which mark the head of Hydra, the water snake, which represents the multi-headed monster (called the Hydra) that was slain by Hercules.

The head of Hydra is about the only recognizable feature in this entire faint and rambling constellation. Being of 3rd and 4th magnitudes the members of this group are visible to the naked eye but appear more impressive when seen through binoculars. The most northerly of them, Epsilon [ε] Hydrae, is a binary star, but a telescope of at least 75-mm (3-in) aperture will be needed to separate the components, which show an attractive colour contrast of yellow and blue.

OPEN CLUSTER M48
Moving on south, by the border with Monoceros, we come to an impressive open cluster, M48, easily visible in binoculars. It consists of about 80 stars and spans the width

of the full Moon, but it is not round – its stars are arranged in a narrow triangle, resembling an arrowhead. The brightest stars may just be resolved in binoculars, and small telescopes show it well.

GHOST OF JUPITER
Near the bottom of this chart in Hydra, south of the empty area known as Sextans, lies NGC 3242, a planetary nebula popularly termed the Ghost of Jupiter from its similarity in appearance to that planet as seen through a small telescope, although the nebula is bluer in colour. Its 11th-magnitude central star may also be visible. Larger telescopes show a more complex structure to the nebula, consisting of a brighter core within a fainter outer halo.

WATCHING THE SKY

STAR CHARTS

CHART 11

NAME	RA, DEC	MAG(S)	TYPE OF OBJECT	VISIBILITY
δ Corvi	12h 30m −16.5°	2.9, 9.0	double star	
γ Virginis	12h 42m −1.4°	3.5, 3.5	double star	
π Boötis	14h 41m +16.4°	4.9, 5.8	double star	
ξ Boötis	14h 51m +19.1°	4.7, 7.0	double star	
α Librae	14h 50m −16.0°	2.7, 5.2	double star	∞
β Librae	15h 17m −9.4°	2.6	coloured star	∞
M5	15h 19m +2.1°	5.8	globular cluster	∞
M100	12h 23m +15.8°	9.4	spiral galaxy	
M84	12h 25m +12.9°	9.3	elliptical galaxy	
M86	12h 26m +12.9°	9.2	elliptical galaxy	
M49	12h 30m +8.0°	8.4	elliptical galaxy	
M87	12h 31m +12.4°	8.6	elliptical galaxy	
The Sombrero (M104)	12h 40m −11.6°	8.3	edge-on spiral	
M60	12h 44m +11.5°	8.8	elliptical galaxy	

Two bright stars are prominent in this area: Arcturus and Spica. On Chart 5, we saw that the curve of the Big Dipper's handle points to Arcturus in Boötes – a starhop known as 'arc to Arcturus'. Arcturus, the brightest star north of the celestial equator, is classified as a red giant, although to the eye it seems a warm orange colour. Its appearance in the evening sky is a welcome sign to residents of the northern hemisphere that spring is on its way. After you arc to Arcturus, continue the curve southwards for another starhop and you 'speed to Spica', the brightest star in Virgo. Most of this chart is occupied by Virgo, the largest of the 12 constellations of the zodiac. Virgo is home to the Virgo Cluster, the richest nearby cluster of galaxies, whose brightest members are detectable through small telescopes.

DOUBLES – EASY AND DIFFICULT

Before heading off into the depths of space to the Virgo Cluster, we will survey some stellar sights in our own surroundings, starting with some double stars at upper centre of the chart, in Boötes. The best of them for small telescopes is found to the left of Arcturus: Xi [ξ] Boötis, an attractive binary in which a yellow and an orange star orbit each other every 150 years. Somewhat below and to the right you will find Pi [π] Boötis, a sparkling pair of blue-white stars. Turning to the little-known constellation Corvus at the bottom right of the chart, look for Delta [δ] Corvi, a wide pair separable with small telescopes but with a considerable brightness difference between the two stars, which are 3rd and 9th magnitudes.

Many binary systems within range of amateur telescopes have orbital periods so long that the motions of the stars are scarcely detectable over a lifetime of watching. But in a few cases the orbital motions can be obvious over only a few years, a prime example being Gamma [γ] Virginis, above right of Spica. This pair of near-white stars orbit each other every 170 years. When closest together as seen from Earth, around the year 2005, they will be impossible to separate in all but the largest amateur telescopes – but, as they progress around their orbits, they will seem to move apart again, becoming progressively easier to divide. Even small telescopes will be able to split them after about 2012. Watch this pair move over the coming years.

For an easier target, look across to Libra, where Alpha [α] Librae forms a wide double with a 5th-magnitude companion, visible in binoculars. Finally, for a stellar oddity, point your binoculars at 3rd-magnitude Beta [β] Librae. This is reputed to be one of the few bright stars to show a greenish colour tinge, presumably due to some spectral peculiarity. Do you see any colour in it?

THE VIRGO CLUSTER

Most galaxies are to be found in clusters. Our own Local Group is an example of a small cluster, but it is dwarfed by the Virgo Cluster, which is the dominant feature of our corner of the Universe – over 2,000 galaxies are thought to reside in it. The cluster is no respecter of boundaries: part of it spills northwards into neighbouring Coma Berenices, as the chart shows.

The most prominent members of the Virgo Cluster are all giant ellipticals, among them M84 and M86 near the cluster's heart on the northern border of Virgo; these two are close enough together to fit into the same telescopic field of view. Small telescopes should also show M60 to the left of this pair and M49 to the south. Most fascinating of all is M87, an active galaxy with a loud radio voice that is spouting out a high-speed jet of gas from a supermassive **black hole** at its centre. You will not see that jet when you turn a telescope towards M87, but once you know what is going on there, just peering at this fuzzy patch becomes that much more interesting.

The brightest spiral in the Virgo Cluster is M100, across the border in Coma Berenices,

▼ **The Sombrero galaxy, M104,** *is a spiral galaxy seen edge-on, with a prominent dark lane of dust in the plane of its spiral arms. It gets its name from its supposed resemblance to a wide-brimmed Mexican hat. M104 lies in Virgo but is closer than the Virgo cluster.*

STAR CHARTS

black hole ▶ p. 336

WATCHING THE SKY

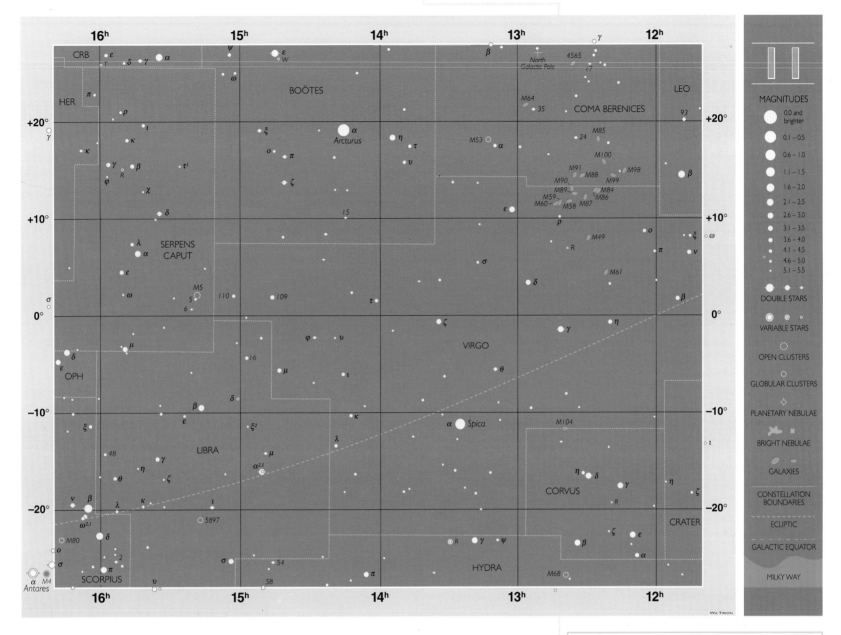

which probably looks much the way our own Galaxy would appear at a similar distance. The Hubble Space Telescope has spied Cepheid variables within M100, from which we know that it – and hence the Virgo Cluster – lies about 55 million light years away. Scan this area with a telescope and you will see many more faint members of this vast cluster.

LOOK FOR A HAT

We end our galaxy quest with M104, a spiral in the south of Virgo that is not a member of the Virgo Cluster, but lies about 10 million light years closer to us. Tilted almost edge-on to us, this galaxy looks rather like a wide-brimmed Mexican hat in photographs, hence its popular name the Sombrero, although small telescopes

show it as only an elongated smudge. Larger apertures will bring into view a dark lane of dust crossing the galaxy's nucleus.

GLOBULAR CLUSTER M5

Returning now to the fringes of our own Galaxy, we take a look at one of the finest globular clusters visible from mid-northern latitudes: M5 in Serpens, left of centre on the chart. Serpens, representing a serpent, is a unique constellation, being split into two, one half either side of Ophiuchus. The half containing M5 is called Serpens Caput, referring to its head. Binoculars show M5 as a hazy glow, just north of a 5th-magnitude foreground star. Apertures of 100 mm (4 in) will resolve this cluster's brightest stars.

KEEPING A SKY LOG

Just as travellers keep a journal of places visited, so do astronomers keep a log of what they have seen in the sky. What it contains is up to you, but always start with the date and time, then note the instrument and magnification used and the sky conditions (nebulous objects will be more difficult to see on hazy nights, while fine details will be more difficult to distinguish on unsteady nights). If you are artistic you may wish to add a sketch. Looking back through the log will remind you of the constellations you have visited, the stars and deep-sky objects you have located and phenomena such as comets and eclipses that you have witnessed.

WATCHING THE SKY

CHART 12

NAME	RA, DEC	MAG(S)	TYPE OF OBJECT	VISIBILITY
α Herculis	17h 15m +14.4°	var., 5.4	double star	
70 Ophiuchi	18h 05m +2.5°	4.2, 6.0	double star	
θ Serpentis	18h 56m +4.2°	4.6, 5.0	double star	
α Herculis	17h 15m +14.4°	3.0–4.0	variable star	👀 ○○
R Scuti	18h 48m -5.7°	4.2–8.6	variable star	○○
η Aquilae	19h 52m +1.0°	3.5–4.4	variable star	👀 ○○
IC 4665	17h 46m +5.7°	4.2	open cluster	○○
M23	17h 57m -19.0°	5.5	open cluster	○○
M24	18h 18m -18.5°	4.5	star cloud	👀 ○○
Eagle Nebula (M16)	18h 19m -13.8°	6.0	open cluster	○○
NGC 6633	18h 28m +6.6°	4.6	open cluster	○○
M25	18h 32m -19.2°	4.6	open cluster	○○
IC 4756	18h 39m +5.4°	5.4	open cluster	○○
M26	18h 45m -9.4°	8.0	open cluster	○○
Wild Duck Cluster (M11)	18h 51m -6.3°	5.8	open cluster	○○
M12	16h 47m -1.9°	6.6	globular cluster	○○
M10	16h 57m -4.1°	6.6	globular cluster	○○
Omega Nebula (M17)	18h 21m -16.2°	7.0	nebula	○○

A s we head back towards the Milky Way, the number of clusters increases. A dark lane of dust in the Milky Way called the Cygnus Rift, first encountered on Chart 7, continues through this area, broadening out in Ophiuchus, the large constellation at centre right. Ophiuchus represents a man entwined by a serpent, the constellation Serpens, its tail on one side (Serpens Cauda) and its head (Serpens Caput) on the other. Although divided in two, Serpens still counts as one constellation. Towards the bottom of this chart we are approaching the centre of our Galaxy, and so the Milky Way is exceptionally thick with stars here. The stars of Scorpius, peeking in at the bottom right, are dealt with on Chart 18.

GIANTS IN THE SKY

Two giants stand head-to-head in the sky: Hercules and Ophiuchus. The head of Hercules is marked by the star Alpha [α] Herculis, sometimes known as Rasalgethi, which is fittingly enough a red giant star, around 400 times the Sun's diameter. Like many red giants Alpha Herculis pulsates in size, fluctuating from 3rd to 4th magnitude but with no detectable period. It is also a binary, with a blue-green companion visible through small telescopes.

A BRIGHT CEPHEID

Cepheid variables are even bigger and brighter than giants – they are supergiants – and one of the easiest to see is Eta [η] Aquilae, at left centre of this chart. It rises and falls between magnitudes 3.5 and 4.4 in a cycle lasting a week, easily followed with the naked eye. In fact, its variability was discovered over 200 years ago by an English amateur astronomer, Edward Pigott, shortly after his colleague and neighbour John Goodricke had recognized the variability of the original member of this class, Delta [δ] Cephei.

DOUBLE STAR SEARCH

Prominent double stars are in surprisingly short supply in this area, but one notable specimen is 70 Ophiuchi, almost central on the chart. The smallest of telescopes shows that it consists of a warmly coloured yellow and orange duo; these

orbit each other every 88 years. Around 70 Ophiuchi is an attractive little clutch of stars, to the north of which lies the faint red dwarf Barnard's Star, the second-closest star to the Sun, after the Alpha Centauri system. Farther to the left on the chart is another easy pair, Theta [θ] Serpentis, both components of which are white.

MILKY WAY STAR CLOUDS

Nowhere are the billowing star clouds of the Milky Way more prominent than in the lower half of our chart. One large patch, the Scutum Star Cloud, spans 12 Moon diameters across the border from Scutum into Aquila. It is the brightest part of the Milky Way outside Sagittarius. In Sagittarius itself, one starry cloud even found its way into Charles Messier's catalogue: M24, measuring four Moon diameters long by two wide, just north of Mu [μ] Sagittarii. In dark skies it can be glimpsed with the naked eye but to appreciate its full glory it should be viewed through binoculars.

A HOST OF OPEN CLUSTERS

An outstanding open cluster for small apertures is M11 in Scutum, visible as a hazy ball in binoculars near the

end of a distinctive arc of stars that straddles the Aquila–Scutum border. Small telescopes are needed to appreciate why it is nicknamed the Wild Duck cluster – its stars are arranged in a fan shape, like migrating ducks. At the fan's apex lies an orange giant, somewhat brighter than the rest. In the same binocular field of view lies R Scuti, a pulsating red supergiant that can be found shining at anything between 4th and 8th magnitude. South of M11

▼ **M11, the Wild Duck Cluster,** in Scutum appears through binoculars as a hazy patch but in small telescopes breaks up into a dense group of stars with a shape reminiscent of a flight of migrating birds.

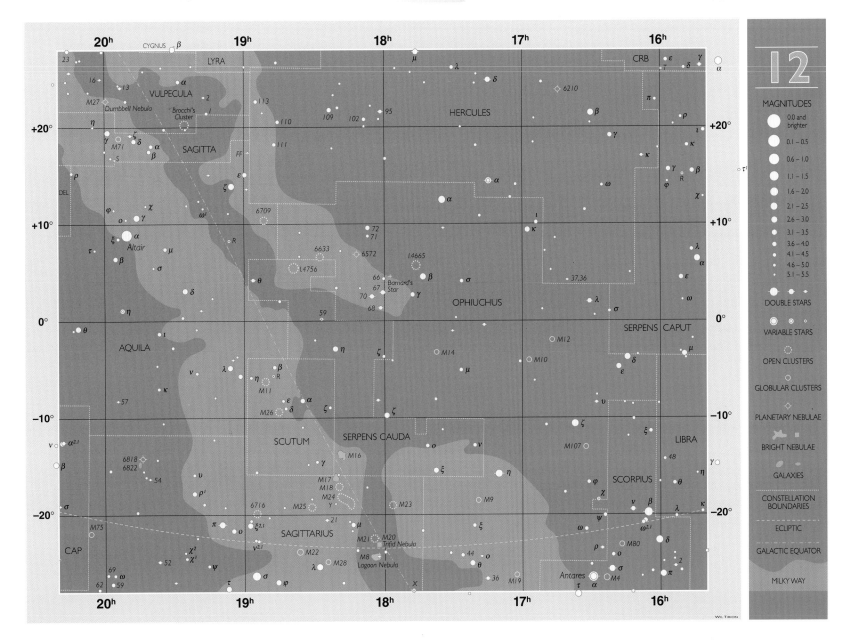

lies M26, a much more sparsely populated cluster best seen with small to moderate-sized telescopes.

Now look north into Serpens and Ophiuchus, where three clusters suitable for binoculars will be found. IC 4756 is the largest of them, but its stars are not particularly bright. NGC 6633 is more compact, but best of all for binoculars is IC 4665, consisting of two dozen or so stars covering an area larger than that of the full Moon. To complete our tour of the finest open clusters on this chart, drop south into Sagittarius, wherein lie M23 and M25. The first of these is notably elongated in shape, while the second is more prominent, its brightest member being a Cepheid variable, U Sagittarii, which ranges from 6th to 7th magnitude.

GLOBULARS IN OPHIUCHUS

Ophiuchus, the serpent-holder, also holds seven globular clusters listed in Messier's catalogue: M9, M10, M12, M14, M19, M62 (on chart 18) and M107. The two best are M10 and M12, at centre right of this chart, of similar size and brightness. Both are visible in binoculars but M12 is more difficult to resolve into stars with telescopes because it is somewhat more distant, around 18,000 light years against 14,000 for M10.

AN EAGLE AND AN OMEGA

We end with a visit to two star-forming regions, below centre of the chart. M16, in Serpens Cauda, appears through binoculars as a hazy patch of light, while small telescopes break it into a ragged cluster with a hint of surrounding mist. On long-exposure

photographs its full magnificence becomes apparent: it spreads wings of nebulosity that give rise to the popular name of the Eagle Nebula. Photographs taken by the Hubble Space Telescope have spied stars that are still coming into being from the gas in this nebula. Just over the border in Sagittarius lies M17, an irregularly shaped patch of nebulosity whose shape will depend on the instrument you are using. Through binoculars it appears elongated, while through larger telescopes it has an arch shape that William Herschel likened to a Greek capital letter Omega – hence the popular name of the Omega Nebula. There is no obvious cluster associated with M17; the illuminating stars must be hidden from our view. About two Moon diameters to the south, small telescopes will show M18, a sparse, open cluster.

WATCHING THE SKY

CHART 13

NAME	RA, DEC	MAG(S)	TYPE OF OBJECT	VISIBILITY
α Capricorni	20h 18m -12.5°	3.6, 4.3	double star	👀 ᗝᗝ
β Capricorni	20h 21m -14.8°	3.1, 6.1	double star	ᗝᗝ 📷
γ Delphini	20h 47m +16.1°	4.3, 5.1	double star	📷
ζ Aquarii	22h 29m 0.0°	4.3, 4.5	double star	📷
TX Piscium	23h 46m +3.5°	4.8–5.2	variable star	👀 ᗝᗝ
M15	21h 30m +12.2°	6.4	globular cluster	ᗝᗝ 📷
M2	21h 34m -0.8°	6.5	globular cluster	ᗝᗝ 📷
NGC 7009	21h 04m -11.4°	8.3	planetary nebula	📷

Some areas of sky have more to offer than others. After the riches of Chart 12, with its Milky Way starfields and numerous clusters, the area covered here seems relatively barren. Nevertheless in the top left corner it contains one of the landmarks of the sky, the Great Square of Pegasus, well displayed on evenings in northern autumn. To the right of the Square is a reminder of warmer evenings now past – the stars of the Summer Triangle (one of them, Altair, appears at the edge of this chart). Pegasus itself contains a major globular cluster, while the zodiacal constellation of Aquarius to the south is the home of two famous planetary nebulae.

SQUARE IN THE SKY

Four stars of 2nd and 3rd magnitudes mark the corners of an immense box in the sky, the Square of Pegasus. Despite the name, only three of the stars actually belong to Pegasus – the brightest, at top left (in the upper margin of our chart), actually belongs to Andromeda, although in the days before hard-and-fast constellation boundaries were drawn up it was shared by the two constellations. The Square is so large that 30 full Moons could be lined up across it, yet this expanse of sky is almost devoid of naked-eye stars (the brightest, Upsilon [υ] Pegasi, is magnitude 4.4). See how many stars you can see within the Square of Pegasus – the number is a good guide to how clear your sky is.

A STAR WITH A PLANET

Halfway down the right-hand side of the Square of Pegasus is a faint star that you might well dismiss as insignificant: 51 Pegasi, magnitude 5.5. Yet look at it and wonder, for in 1995 it became the first star discovered to have a planet. This planet is about half the mass of Jupiter, but its orbit is unlike that of any planet in our Solar System – it whirls around 51 Pegasi every 4 days at only one-twentieth the Earth's distance from the Sun. Hence the planet would be boiling hot and unable to support life. The star itself lies 50 light years away and has a spectral type that identifies it as being almost identical to that of the Sun.

GLOBULARS IN PEGASUS AND AQUARIUS

Few globular clusters are easily visible in small instruments north of the celestial equator, but among the best is M15 in Pegasus. It can be found without difficulty in binoculars, next to a 6th-magnitude star. Small telescopes show it as a misty patch rather like a comet (Charles Messier began to list objects such as this to prevent false alarms while searching for real comets). It lies over 30,000 light years away, and telescopes of at least 100-mm (4-in) aperture will be needed to resolve its individual stars.

Only slightly inferior is M2 in Aquarius, due north of Beta [β] Aquarii. Binoculars and small telescopes will show it as a rounded haze, although since it is more distant than M15 its individual stars are more difficult to resolve.

THE CIRCLET OF PISCES

South of the Great Square of Pegasus is a ring of stars forming the body of one of the two fish of Pisces. Termed the Circlet, this ring consists of seven stars of 4th and 5th magnitudes. At left of the Circlet is TX Piscium, a so-called carbon star, i.e. a red giant with outer layers so cool that sooty particles of carbon form, endowing it with

a strong red colour that is readily apparent through binoculars. This star pulsates irregularly in size, leading to changes in brightness between magnitudes 4.8 and 5.2.

THE WATER JAR OF AQUARIUS

Another feature of this area of constellations with watery associations can be found squarely on the celestial equator: a group of four stars arranged in a Y shape, forming an attractive sight in binoculars. These stars mark the water jar held by Aquarius, from which water flows southwards towards Piscis Austrinus. Zeta [ζ] Aquarii, the central member of the Water Jar group, is an outstanding binary, divisible with small telescopes into twin white stars that orbit each other every 760 years.

THE SATURN NEBULA

In the lower right of Aquarius, near the border with Capricornus, lies a famous planetary nebula, NGC 7009, better known as the Saturn Nebula. This name was given it by the nineteenth-century Irish astronomer, Lord Rosse, because of its

▼ *M15 in Pegasus is one of the globular clusters that swarm in a halo around our Galaxy. It is visible through binoculars but telescopes will be needed to pick out the brightest of its individual stars. It lies some 30,000 light years from us.*

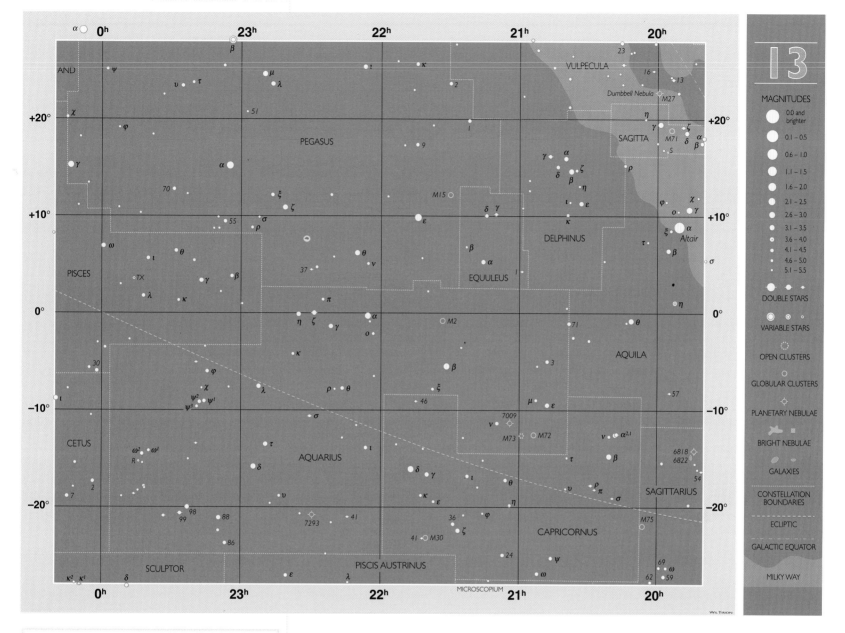

CONSTELLATIONS AND ASTERISMS

As well as the officially defined areas of the constellations, astronomers refer to asterisms, which are unofficial patterns that can consist of stars from one or more constellations. The Square of Pegasus is an example of an asterism composed of stars from two constellations (Pegasus and Andromeda), as is the False Cross (Carina and Vela). The Plough, the Sickle of Leo, the Circlet of Pisces, the Water Jar of Aquarius and the Teapot of Sagittarius are all asterisms within a given constellation. As you scan your way around the sky, feel free to make up, and name, star patterns of your own. Why be hidebound by the traditional constellations?

resemblance to the planet Saturn, an impression amply confirmed by long-exposure photographs. Small telescopes show NGC 7009 as an elongated ellipse similar in size to the globe of Saturn. But the 'rings' – actually long jets of gas ejected by the central star that give it the Saturn-like appearance – become visible only in larger apertures. Another famous planetary nebula in Aquarius, NGC 7293, the Helix, is described on Chart 19.

SWIM WITH A DOLPHIN

Continuing the watery theme of this part of the sky, look for the attractive little constellation of Delphinus, the dolphin, in the upper right of the chart not far from Altair. In this kite-shaped grouping it is easy to imagine a playful dolphin leaping from the water. For users of small telescopes the most interesting star is Gamma [γ] Delphini, in the dolphin's nose, a beautiful double of golden and yellowish stars, sharing the field of view with a fainter 8th-magnitude pair.

DOUBLES IN CAPRICORNUS

Finally, head south into Capricornus, yet another constellation with watery associations, this time representing a fish-tailed goat. Binoculars, or even good eyesight, will reveal that Alpha [α] Capricorni is a wide double, although the two stars (a yellow giant and a yellow supergiant) are at widely differing distances from us and hence their apparent partnership is merely a line-of-sight effect. South of this pair is Beta [β] Capricorni, a double with a considerable brightness difference divisible by small telescopes or good binoculars.

WATCHING THE SKY

CHART 14

The southern sky, glorious for much of its extent, has the reputation of being padded out with small, faint constellations of little interest, and nowhere is this more apparent than in the area covered by Chart 14. Most of this area 'south of the border' is a mystery to northern observers, and the names of the constellations have an unfamiliar ring: Fornax, Sculptor, Caelum, Pictor, Reticulum, Horologium, Phoenix, Hydrus, Tucana and Grus. One welcome exception is Eridanus, the river, which meanders from the borders of Taurus and Orion deep into the southern sky, ending at Achernar, the bright star at the lower centre of this map. The name Achernar means 'river's end', but it is more appropriate to think of it as the wellspring at the head of the river. Far grander sights beckon on all sides, but let us set out with binoculars and small telescope to see what we can find in this often-overlooked 'empty quarter' of the southern sky.

DOUBLES IN ERIDANUS

To prove that good objects for observation can be uncovered in the most unpromising areas, start by turning a small telescope towards Theta [θ] Eridani, left of centre on the chart. It divides neatly into a striking pair of blue-white stars. In ancient Greek times, this star was regarded as marking the end of Eridanus, since it was about as far south as Greek astronomers could see. Since then, Eridanus has been extended to the 1st-magnitude star known as Achernar (Alpha [α] Eridani), which was below the horizon from ancient Greece. Hence observers in high northerly latitudes will not be able to see this star.

Two Moon diameters north of Achernar lies another excellent double. It is identified not by a Greek letter but a lower-case Roman letter: p Eridani. In contrast to the icy glints of Theta Eridani, this is a warm-toned duo of orange dwarfs with an orbital period of some 500 years.

FORNAX AND PHOENIX

There are few signposts to guide the visitor in this area south of Cetus and adjoining the meandering Eridanus, but by careful navigation

NAME	RA, DEC	MAG(S)	TYPE OF OBJECT	VISIBILITY
κ¹ Sculptoris	0h 09m -28.0°	6.1, 6.2	double star	
β Tucanae	0h 32m -63.0°	4.4, 4.5	double star	∞
ζ Phoenicis	1h 08m -55.2°	var., 8	double star	
p Eridani	1h 40m -56.2°	5.8, 5.9	double star	
ε Sculptoris	1h 46m -25.1°	5.3, 8.6	double star	
θ Eridani	2h 58m -40.3°	3.2, 4.3	double star	
α Fornacis	3h 12m -29.0°	3.9, 6.5	double star	
ζ Reticuli	3h 18m -62.5°	5.2, 5.5	double star	👀 ∞
ζ Phoenicis	1h 08m -55.2°	3.9, 4.4	variable star	👀 ∞
NGC 288	0h 53m -26.6°	8.1	globular cluster	
NGC 55	0h 15m -39.2°	8.2	edge-on galaxy	
NGC 253	0h 48m -25.3°	7.1	edge-on galaxy	∞

you should come to Alpha [α] Fornacis, a yellow-white star with a binary companion that orbits it every 300 years, visible through small telescopes.

For observers in mid-northern latitudes the stars of the constellation Phoenix barely scrape above the horizon, but those farther south will be able to find Zeta [ζ] Phoenicis, above right of brilliant Achernar. This star offers two attractions in one, for it is both an eclipsing variable and a visual double. To the naked eye it ranges in brightness by half a magnitude every 1.7 days, while through small telescopes a fainter companion can be seen – this star is not involved in the eclipses.

GALAXIES IN SCULPTOR

The main reason there are so few stars in this area is because here we are looking at right angles to the Milky Way. Hence there is little to block our view of deep space in this direction and numerous galaxies can be seen, although only two are within easy reach of amateur instruments. More prominent of the two is NGC 253 near the constellation's northern edge. An edge-on spiral, it can be detected with binoculars under clear skies as an elongated smudge, although telescopes are required to see it well. Photographs show its spiral arms to be mottled with dust patches, but these become

▶ *The spiral galaxy NGC 253 in the constellation of Sculptor is seen nearly edge-on and so appears elliptical in outline. It is one of the brightest galaxies outside the Local Group. Its arms appear patchy on photographs such as this because they are mottled with dark dust clouds.*

visible only through larger apertures. Nearby in the sky, although actually much closer to us, is NGC 288, a fair-sized globular cluster but too faint to be impressive in small apertures. On the southern border of Sculptor is NGC 55, another edge-on spiral that appears elliptical in small instruments; one half of it is brighter than the other.

Other objects of interest for amateur observers in Sculptor include two double stars: Epsilon [ε] Sculptoris is a binary divisible in small telescopes, while Kappa-1 [κ¹] Sculptoris is a greater challenge, requiring apertures of 75 mm (3 in) or more.

TWO EASY PIECES

Two easy but attractive doubles can be found in the lower reaches of this chart. Zeta [ζ] Reticuli is a pair of yellow stars similar to the Sun, far enough apart to be distinguishable in binoculars

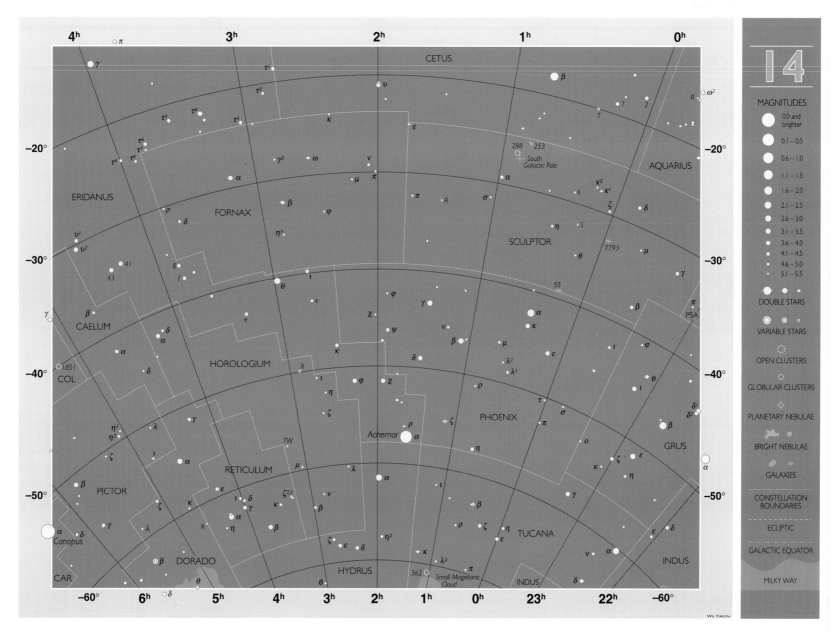

or even with sharp eyesight. Beta [β] Tucanae is more complex, consisting of a well-matched pair of stars divisible with binoculars or a small telescope, with a slightly fainter star nearby. All three move through space together, suggesting they were born from the same cloud of gas even though they do not orbit one another.

FORMING PLANETARY SYSTEM?

Anticipating the wonders to come on Chart 15, we finish in the left-hand corner with Beta [β] Pictoris, not too far from brilliant Canopus. At 4th magnitude, Beta Pictoris is easily overlooked, but it has a secret that astronomers uncovered in the 1980s: it is surrounded by a disk of dust and gas from which a planetary system may currently be forming in an action replay of the birth of our own Solar System. This disk, which is presented edge-on to us, can only be seen with specialized equipment; a photograph of it is shown in Planet Formation, page 188.

💡 ORIGIN OF THE SOUTHERN CONSTELLATIONS

There are no constellations of ancient Greek origin in the far southern sky, for this area was permanently below the horizon from Mediterranean latitudes. Only when European explorers first made their way south on voyages of discovery did they begin to catalogue the southern stars. First to create new constellations in this hitherto uncharted area of sky were two Dutch navigators, Pieter Dirkszoon Keyser and Frederick de Houtman, around 1600. They introduced 12 constellations, most of which represented creatures they had seen on their travels; those depicted on Chart 14 are Grus, the crane, Tucana, the toucan, Hydrus, a small water snake and Dorado, a goldfish, as well as the mythical Phoenix. The remaining gaps were filled in 150 years later by a French astronomer, Nicolas Louis de Lacaille (1713–62), who created 14 new constellations, mostly representing instruments of science and the arts. Constellations on this map of Lacaille's invention are Fornax, a chemical furnace, Sculptor, a sculptor's studio, Horologium, a pendulum clock, Caelum, an engraver's chisels, Reticulum, a set of cross-wires in his telescope's eyepiece and Pictor, a painter's easel.

WATCHING THE SKY

CHART 15

 NAME	RA, DEC	MAG(S)	TYPE OF OBJECT	VISIBILITY
R Doradus	4h 37m –62.1°	4.8–6.6	variable star	👓👓
β Doradus	5h 34m –62.5°	3.5–4.1	variable star	👀 👓👓
L² Puppis	7h 14m –44.7°	3.0–6.0	variable star	👀 👓👓
UW Canis Majoris	7h 19m –24.6°	4.8–5.3	variable star	👀 👓👓
V Puppis	7h 58m –49.2°	4.4–4.9	variable star	👀 👓👓
M41	6h 47m –20.7°	4.5	open cluster	👀 👓👓 🔭
NGC 2362	7h 19m –25.0°	4.1	open cluster	🔭
M47	7h 37m –14.5°	4.4	open cluster	👓👓 🔭
M46	7h 42m –14.8°	6.1	open cluster	👓👓 🔭
M93	7h 45m –23.9°	6.2	open cluster	👓👓 🔭
NGC 2451	7h 45m –38.0°	2.8	open cluster	👓👓 🔭
NGC 2477	7h 52m –38.5°	5.8	open cluster	👓👓 🔭🔭
NGC 2516	7h 58m –60.9°	3.8	open cluster	👀 👓👓 🔭

O n this chart we begin to encounter some of the richest areas of the entire night sky, towards the centre of our Galaxy. By chance, the two brightest stars in the night sky both fall within the boundaries of this chart – Sirius, at the very top, and Canopus, below centre – but both are much closer to us than the densely populated starfields of the Milky Way that run down the left of the chart. At the bottom the Large Magellanic Cloud makes an appearance, looking like a detached portion of the Milky Way but in reality a small nearby galaxy in its own right; it is dealt with on Chart 20. Objects of interest in the fainter constellations to the right of this chart are described under Chart 14.

CANIS MAJOR CLUSTERS

We start with two glorious open clusters that are within view of observers in mid-northern latitudes as well as southern ones. The first, M41, is easily found due south of Sirius, and was known to the astronomers of ancient Greece who could see it with the naked eye. This feat may be difficult in modern urban skies, particularly in more northerly latitudes where M41 remains close to the horizon, but binoculars will show it clearly as a field of stars, covering an area similar to that of the full Moon. The stars seem to be distributed in curving chains – a common characteristic of many open clusters, but probably more to do with the way the eye links up unconnected points rather than a real physical effect.

A second, smaller cluster will require a telescope to be seen at its best: NGC 2362, a tight swarm of a few dozen stars that can be picked out by small telescopes, surrounding the 4th-magnitude blue supergiant Tau [τ] Canis Majoris which is by far the brightest member. To the north of this cluster lies the notable variable star UW Canis Majoris (see below).

VARIABLE STAR ROUND UP

Two pulsating red giants, a Cepheid variable and two eclipsing binaries provide interesting fare for variable-star observers in this area. Most northerly of this quintet is UW Canis Majoris, a remarkable pair of supergiants that orbit each other more closely than Mercury orbits the Sun – so close that their mutual gravitational attraction distorts them into egg shapes. As they orbit every 4.4 days they eclipse each other in turn, leading to a periodic increase and decrease in their combined light of about 50%. UW Canis Majoris can be found just north of the cluster NGC 2362, but it is uncertain whether it is an outlying member of the cluster.

Moving south into Puppis we find L² Puppis, a red giant semiregular variable that can appear as bright as 3rd magnitude and as faint as 6th, although usually its range is not as great as that. Its fluctuations and colour are made more noticeable by the fact that just south of it lies a blue–white star, L¹ Puppis, whose brightness is constant; both stars, which are unrelated, fit comfortably into the same binocular field of view. Tucked into a corner of Puppis, near the boundaries with Vela and Carina, lies V Puppis, an eclipsing binary that varies in brightness by 50% every 35 hours.

We now venture into the deep south to find Beta [β] Doradus, one of the brightest Cepheid variables. Like all members of the Cepheid family it has a regular pulse, in this case rising and falling between magnitudes 3.5 and 4.1 in just under 10 days. Less predictable is the red giant R Doradus, on the border with Reticulum, that ranges between about magnitudes 4.8 and 6.6 every 11 months or so. Watch all these stars for a few cycles to get a feel for their different rhythms.

▼ *The open star cluster M41 in Canis Major lies four degrees due south of Sirius, the brightest star in the sky. Known to the ancient Greeks, it can be spotted with the naked eye as a hazy patch under clear, dark skies and is easily found with binoculars. Small telescopes show its stars to be grouped in bunches and chains.*

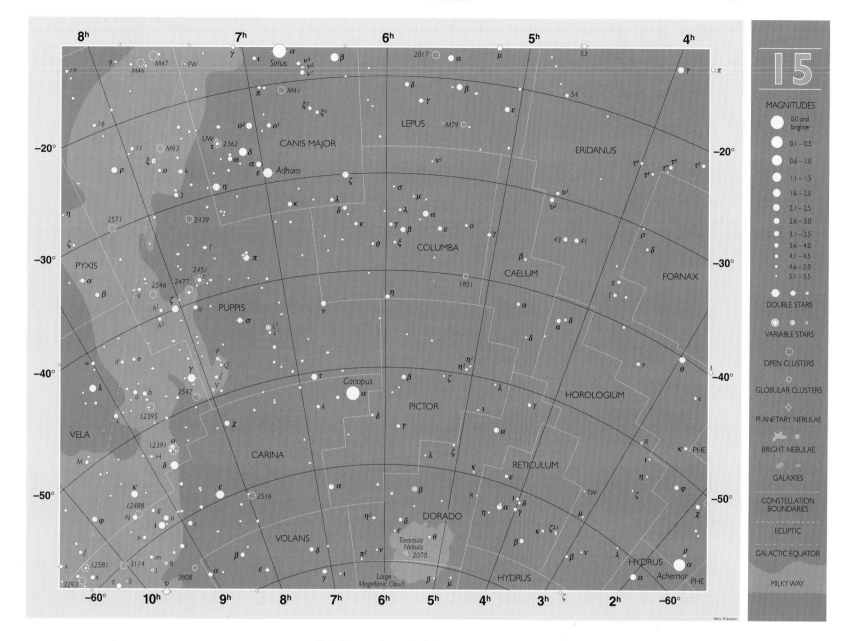

MAGNITUDES

○	0.0 and brighter
○	0.1 – 0.5
○	0.6 – 1.0
●	1.1 – 1.5
●	1.6 – 2.0
●	2.1 – 2.5
●	2.6 – 3.0
•	3.1 – 3.5
•	3.6 – 4.0
•	4.1 – 4.5
·	4.6 – 5.0
·	5.1 – 5.5

DOUBLE STARS

VARIABLE STARS

OPEN CLUSTERS

GLOBULAR CLUSTERS

PLANETARY NEBULAE

BRIGHT NEBULAE

GALAXIES

CONSTELLATION BOUNDARIES

ECLIPTIC

GALACTIC EQUATOR

MILKY WAY

OPEN HOUSE ON OPEN CLUSTERS

An observer could spend a good part of an evening in the Milky Way in this region, skipping with binoculars from cluster to cluster. Start at the top left in Puppis with M46 and M47, two prominent open clusters with contrasting character that are visible in the same binocular field of view. Under dark skies they can be glimpsed with the naked eye as brightenings in the Milky Way. Although they appear close neighbours in the sky, in reality M46 lies about three times farther from us than its companion. In binoculars, M46 shows up as a smudgy patch and small telescopes are required to pick out its faint individual stars, while the members of M47

are brighter and resolvable with binoculars. To the south of these lies M93, visible with binoculars as an elongated patch of light. When seen through small telescopes M93 takes on a notable V shape, with two orange stars near the point.

Equally fine sights, but just too far south to have featured on Messier's lists, are NGC 2451 and 2477, a contrasting pair of open clusters similar to M46 and M47. As with that more northerly duo, they lie at widely differing distances and are not related. The closer of the two, NGC 2451, is easier to spot, being centred on a 4th-magnitude orange giant which adds a touch of colour. But it is a much sparser grouping than NGC 2477, which binoculars show as a hazy patch about

the same apparent size as the full Moon, much like a comet or a loose globular cluster; as with M46, a telescope is needed to see individual stars within it. While in this area, look across at 2nd-magnitude Zeta [ζ] Puppis, a supergiant whose spectral type of O5 distinguishes it as being the hottest – and hence bluest – of all naked-eye stars.

Delving farther south, into Carina, we come to NGC 2516, a scattered cluster of similar apparent size to the full Moon and visible to the naked eye. Its brightest stars, which seem to form a large figure X, are easily picked out in binoculars, most notably a 5th-magnitude red giant. Even richer Milky Way sights in Carina and Vela, await on Chart 16.

CHART 16

NAME	RA, DEC	MAG(S)	TYPE OF OBJECT	VISIBILITY
δ Velorum	8h 10m -47.3°	1.8, 4.3	double star	◯◯ 📷
η Carinae	10h 45m -59.7	-0.8–7.9	variable star	👁 ◯◯ 📷
NGC 2547	8h 11m -49.3°	4.7	open cluster	◯◯ 📷
IC 2391	8h 40m -53.1°	2.5	open cluster	👁 ◯◯
IC 2395	8h 41m -48.2°	4.6	open cluster	◯◯ 📷
NGC 3114	10h 03m -60.1°	4.2	open cluster	◯◯ 📷
NGC 3532	11h 06m -58.7°	3.0	open cluster	👁 ◯◯
NGC 3766	11h 36m -61.6°	5.3	open cluster	📷
NGC 3132	10h 08m -40.5°	8.2	planetary nebula	📷
NGC 3918	11h 50m -57.2°	8.4	planetary nebula	📷
Eta Carinae Nebula (NGC 3372)	10h 44m -60°	2.5	nebula	👁 ◯◯

T here has been a shipwreck in this part of the sky. The ancient Greek constellation of Argo Navis, representing the vessel in which Jason and the Argonauts sailed in search of the golden fleece, was considered so large and unwieldy by the eighteenth-century French astronomer Nicolas Louis de Lacaille that he divided it into three: Vela, the sails, Puppis, the stern, and Carina, the keel. Thus trisected, the ship now floats on the Milky Way's river of stars. Yet the stars in these three constellations retain the Greek letters assigned to them when they were still part of Argo Navis, and it so happens that the stars labelled Alpha [α] and Beta [β] lie in Carina. To the lower left is the famous Southern Cross, Crux, described on the following chart – but there is another southern cross in this area, termed the False Cross because it can fool the unwary into mistaking it for the real one. The False Cross consists of Kappa [κ] and Delta [δ] Velorum and Iota [ι] and Epsilon [ε] Carinae. Compare the two crosses on this chart.

A SHOW-OFF STAR

One of the most remarkable stars in the sky lies within a large nebula in Carina, located at the bottom centre of this chart. Called Eta [η] Carinae, this star drew attention to itself by brightening to magnitude -1 in 1843, outshining even Canopus (at the right of the chart). Subsequently it faded below naked-eye visibility, but brightened again to 5th magnitude during the twentieth century. Through small telescopes you will not see the star itself but the small nebula thrown off during its past outbursts. Hubble Space Telescope photographs reveal that this nebula consists of two lobes of gas and dust expanding at high speed in opposite directions. Eta Carinae is the type of star known as a luminous blue variable (see The Stellar Zoo – Single Stars, page 162). Future variations in brightness may be expected, and it is predicted to explode as a supernova some time within the next 100,000 years.

A KEYHOLE IN A NEBULA

Surrounding Eta Carinae is a much larger nebula known as NGC 3372 or, more usually, the Eta Carinae Nebula. To the naked eye this is the brightest spot in the southern Milky Way, spanning four Moon diameters. Binoculars show its true nature as a glowing gas cloud, similar to the Orion Nebula but about five times as distant, with newborn stars and clusters scattered throughout it. A V-shaped lane of dust divides the nebula into two main parts, with Eta Carinae itself embedded in the brighter half. Nearby is a dark cloud of gas forming a bulbous silhouette against the brighter background. The nineteenth-century astronomer John Herschel (1792–1871) likened it to the shape of a keyhole, a name it has retained even though the appearance of this region has since changed as Eta Carinae itself has faded. To the south of the nebula is IC 2602, a brilliant cluster known as the Southern Pleiades, dealt with on Chart 20.

TOURING OPEN CLUSTERS

To resume our tour of the open clusters along this stretch of the Milky Way that we started on the previous chart, look south of Gamma [γ] Velorum for NGC 2547, a large and scattered cluster consisting of a handful of stars bright enough to be seen in binoculars, with dozens more becoming visible through small telescopes. Now scan farther into the Milky Way to find IC 2395, a smaller group that seems to incorporate a much brighter star which may, in reality, be a superimposed foreground object. Now drop south to find one of the most glittering clusters in the sky, IC 2391, whose main members can be seen with the naked eye, most notably Omicron [o] Velorum, magnitude 3.6. The cluster covers an area greater than that of the full Moon.

Heading on down the Milky Way into Carina we come to NGC 3114, a sprawling cluster whose brightest members can be picked out in binoculars.

Passing to the opposite side of the Eta Carinae Nebula, a region thronged with small groups of stars, we come to an elongated patch of brightness visible to the naked eye which is the cluster NGC 3532, containing well over 100 stars splashed across the width of the full Moon. Binoculars give the best view – they show a star-free lane across its centre, giving a profile like a hot dog. At the cluster's edge is the 4th-magnitude star x Carinae, which is not a member but much more distant. Finally, moving into Centaurus we find a much tighter cluster, NGC 3766, lying in a rich Milky Way starfield and best seen through small telescopes.

TWO PLANETARY NEBULAE

Turning away from clusters at last, there are two planetary nebulae worth hunting out in this area of sky. NGC 3132, on the northern border of Vela with Antlia, is sometimes known as the Southern Ring because of its similarity in appearance to the better-known Ring Nebula in Lyra, although it is not as large. Small telescopes show it as an elliptical disk about the same size as Jupiter. On photographs it appears to consist of loops of gas forming a series of interlocking figures of eight, from which comes its popular name of the Eight-Burst Nebula. Farther south, on the border of Centaurus with Crux, lies NGC 3918, which John Herschel dubbed the Blue Planetary. Small telescopes show it as a bluish disk reminiscent of the planet Uranus, only larger.

THE BRIGHTEST WOLF-RAYET STAR

Finally, return to the centre-right of this chart to find the brightest star in Vela, 2nd-magnitude Gamma [γ] Velorum, near the border with Puppis. Nothing special about it at first glance, you would

WATCHING THE SKY

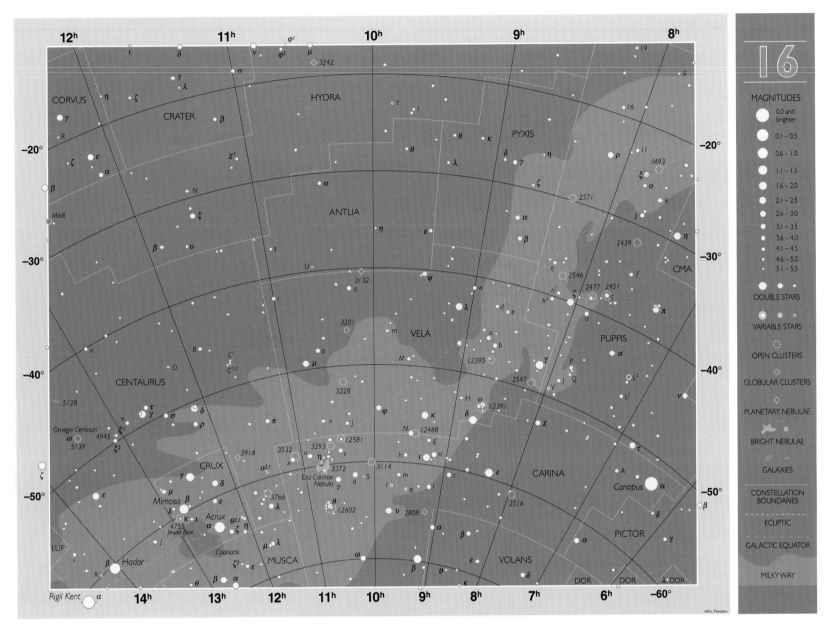

16

MAGNITUDES

⬤	0.0 and brighter
⬤	0.1 – 0.5
⬤	0.6 – 1.0
⬤	1.1 – 1.5
●	1.6 – 2.0
●	2.1 – 2.5
●	2.6 – 3.0
●	3.1 – 3.5
•	3.6 – 4.0
•	4.1 – 4.5
·	4.6 – 5.0
·	5.1 – 5.5

DOUBLE STARS ⊙

VARIABLE STARS ⊚

OPEN CLUSTERS ○

GLOBULAR CLUSTERS ⊕

PLANETARY NEBULAE ◇

BRIGHT NEBULAE ▨

GALAXIES ▬

CONSTELLATION BOUNDARIES

ECLIPTIC

GALACTIC EQUATOR

MILKY WAY

WIL TIRION

think – but analysis of its light through a spectro-scope has revealed that it is in fact a **Wolf-Rayet star**, a rare species of high-mass stars with ultra-hot surfaces named after the two French astronomers who discovered them over a century ago (see Old Age and Death of High-Mass Stars, page 164). Gamma Velorum is in fact the brightest example of such a star visible from Earth and one day it could become a great deal brighter, since such stars are expected to explode as supernovae. Small telescopes or even good binoculars show that it has a 4th-magnitude companion star, but this is in fact a background object and is not truly related.

▶ **The vast Eta Carinae nebula** *is a glowing gas cloud that appears four times wider than the Moon. In its brightest portion, just above centre, lies the peculiar eruptive variable star Eta Carinae, from which the entire nebula takes its name.*

WATCHING THE SKY

289

CHART 17

NAME	RA, DEC	MAG(S)	TYPE OF OBJECT	VISIBILITY
α Crucis	12h 27m -63.1°	1.3, 1.8	double star	
γ Crucis	12h 31m -57.1°	1.6, 6.5	double star	∞
μ Crucis	12h 55m -57.2°	4.0, 5.1	double star	∞
α Centauri	14h 40m -60.8°	0.0, 1.4	double star	
R Hydrae	13h 30m -23.3°	4.0–10.0	variable star	∞
The Jewel Box (NGC 4755)	12h 54m -60.3°	4.2	open cluster	∞
ω Centauri	13h 27m -47.5°	3.7	globular cluster	👀 ∞
Coalsack	12h 53m -63°	–	dark nebula	👀 ∞
NGC 5128	13h 25m -43.0°	7.0	elliptical galaxy	
M83	13h 37m -29.9°	7.6	spiral galaxy	

Two bright stars dominate this area of sky: Alpha [α] and Beta [β] Centauri, below centre of the chart. Alpha is the third-brightest star in the sky but, more importantly, it is the closest naked-eye star to the Sun and an attractive double for small telescopes. A line drawn from Alpha to Beta Centauri directs your gaze to Crux, the Southern Cross. Although the smallest of all 88 constellations, its distinctive shape and bright stars make Crux easy to identify. The dense Milky Way starfields around Crux contain a famous dark nebula and a sparkling open cluster. Most of this chart, though, is taken up by Centaurus, which is filled with celebrated objects, including the brightest globular cluster and an active galaxy. Many of the bright stars in this area lie about 400 to 600 light years from us. That is not coincidence, since they are all members of a vast association of stars, the Scorpius-Centaurus association, which stretches from Scorpius via Lupus into Centaurus.

THE SUN'S NEAREST NEIGHBOURS
To the naked eye Alpha [α] Centauri is the third-brightest star in the sky, but even the smallest telescopes divide it into a glittering double. The brighter of these is almost identical in size and temperature to the Sun while the other, which orbits it every 80 years, is smaller and yellower. The pair lie 4.4 light years from us. Although Alpha Centauri is often said to be the closest star to the Sun, that honour strictly belongs to a third member of this multiple system: Proxima Centauri, a red dwarf 4.2 light years away. Despite its closeness to us, Proxima is so dim that it appears only of 11th magnitude and cannot be seen without a telescope. A good chart is also needed to identify it since Proxima is separated from Alpha by fully four Moon diameters and is not even in the same telescopic field of view. Beta [β] Centauri, incidentally, is not related to Alpha – it is part of the much more distant Scorpius-Centaurus association.

STARS OF THE CROSS
The brightest and most southerly star in the Southern Cross, Alpha [α] Crucis, is an attractive double for small telescopes, easily divisible into two components whose blue-white colour is in clear contrast to the warmer hues of Alpha Centauri. Turn now to the most northerly star in the Cross, Gamma [γ] Crucis, which is a red giant that forms an apparent double with an unrelated 6th-magnitude background star visible through binoculars. Finally, find Mu [μ] Crucis; good binoculars, held steadily, may be able to divide this into two, but it is the easiest of targets for a small telescope.

BLOT ON THE SKYSCAPE
One of the major sights of this area of sky requires no optical aid at all: the Coalsack Nebula, a pear-shaped blot of dust about 14 Moon diameters long and 10 Moon diameters wide, whose inky darkness hides a portion of the Milky Way from our view. It is not entirely black – a few stars do lie in front of it, the brightest being of 5th magnitude. The Coalsack seems to lie next to Alpha and Beta Crucis but, at a distance of 600 light years, is really almost twice as far away as them.

A CASK OF JEWELS
On the northern rim of the Coalsack, near Beta Crucis, lies the spectacular open cluster NGC 4755, better known as the Jewel Box because the nineteenth-century astronomer John Herschel described it as resembling a cask of precious stones. To the naked eye NGC 4755 might be mistaken for a 4th-magnitude star, but binoculars and small telescopes show it is a tightly packed group of dozens of brilliant stars, mostly blue-white but with one prominent orange supergiant. Three of the brightest stars form a line like a miniature version of Orion's Belt.

GREAT GLOBULAR
The finest globular cluster in the heavens lies almost centrally on this chart, in Centaurus. Called Omega [ω] Centauri (or NGC 5139) it can be spotted with the naked eye as an out-of-focus star of 4th magnitude. Turn binoculars towards it and it looks more like the head of a comet. What you are seeing is the combined

▼ *The glittering Jewel Box (NGC 4755) open star cluster in the constellation Crux, the Southern Cross, appears to the naked eye as a hazy star of 4th magnitude. Binoculars resolve its brightest stars, of 6th and 7th magnitudes, and small telescopes bring many more into view. Most of them appear blue-white but there is a red supergiant near the centre. The brightest member is Kappa [κ] Crucis, and the entire group is sometimes referred to as the Kappa Crucis cluster.*

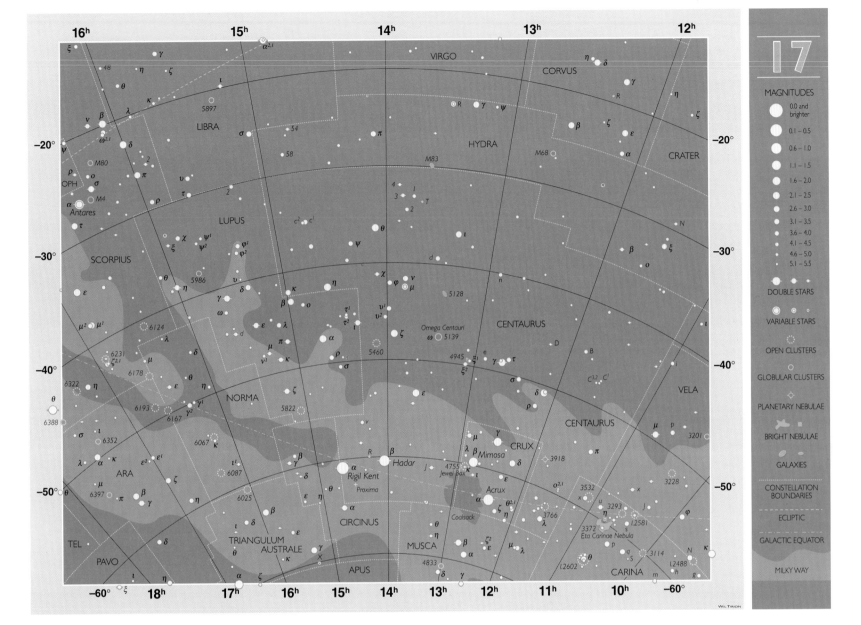

glow of hundreds of thousands of stars, concentrated into an elliptical ball covering more than a Moon diameter of sky; its actual diameter is over 150 light years. Individual stars can be seen through a small telescope, particularly around its periphery. At a distance of 17,000 light years it is among the closest globulars, as well as being the most luminous one known in our Galaxy.

SEE A GALACTIC MERGER

Leaving behind the local sights, we now go in search of larger prey at greater distances. One of the most astounding galaxies within range of modest amateur instruments can be found to the north of Omega Centauri: NGC 5128, also known as the strong radio source Centaurus A.

Binoculars may show it on a clear night, but telescopes will provide a better view. On long-exposure photographs it appears as a giant elliptical galaxy crossed by a lane of dust, which can be seen visually through apertures of 100 mm (4 in) and larger. The puzzle here is that elliptical galaxies should have no dust, so astronomers have concluded that NGC 5128 results from a merger between an elliptical and a spiral galaxy, which has also caused the strong radio emission. As you gaze upon this galaxy, some 16 million light years away, ponder its violent history.

A GALAXY AND
A VARIABLE IN HYDRA

To find a more sedate galaxy, keep moving north into Hydra. Here lies M83, a face-on galaxy

whose picture often appears in books. Although classified as an ordinary spiral, on photographs it displays a noticeable central bar. Small telescopes will show the galaxy as an elliptical cloud, but larger apertures dig deeper to make out the spiral arms of which, unusually, there appear to be three – the usual number is two.

Tonight's concluding target, which returns us to our own Galaxy, lies on the northern border, not far from 3rd-magnitude Gamma Hydrae. The star we are looking for is R Hydrae, a red giant that is one of the brightest of the Mira-type variables. R Hydrae reaches 4th magnitude at its best, but drops to as faint as 10th magnitude at minimum, taking about 13 months to go from one maximum to the next.

WATCHING THE SKY

STAR CHARTS

CHART 18

NAME	RA, DEC	MAG(S)	TYPE OF OBJECT	VISIBILITY
ξ Scorpii	16h 04m -11.4°	4.2, 7.3	multiple star	🔭
		7.4, 8.0		
β Scorpii	16h 05m -19.8°	2.6, 4.9	double star	🔭
ν Scorpii	16h 12m -19.5°	4.3, 5.4	multiple star	🔭
		6.7, 7.7		
Antares	16h 29m -26.4°	1.2, 5.4	double star	🔭
ζ Scorpii	16h 54m -42.4°	3.6, 4.8	double star	👁 ∞
X Sagittarii	17h 48m -27.8°	4.2–4.9	variable star	∞
W Sagittarii	18h 05m -29.6°	4.3–5.1	variable star	∞
NGC 6231	16h 54m -41.8°	2.6	open cluster	∞ 🔭
M6	17h 40m -32.2°	4.2	open cluster	∞ 🔭
M7	17h 54m -34.8°	3.3	open cluster	👁 ∞
M4	16h 24m -26.5°	5.9	globular cluster	∞ 🔭
M22	18h 36m -23.9°	5.1	globular cluster	∞ 🔭
Trifid Nebula (M20)	18h 03m -23.0°	8.5	nebula	🔭
Lagoon Nebula (M8)	18h 04m -24.4°	5.8	nebula	👁 ∞ 🔭

I n this direction we are looking towards the hub of our Galaxy – its exact centre lies just north of where the borders of Sagittarius, Ophiuchus and Scorpius meet. The sky here teems with objects of interest, more than we have room to mention – there are 15 Messier objects in Sagittarius alone. To a casual glance, perhaps the most distinctive feature in this area is the curving chain of stars that mark the body and tail of Scorpius, the scorpion. Another interesting shape, a teapot, is outlined by eight stars in Sagittarius, with Lambda [λ] Sagittarii marking the tip of its lid and Gamma [γ] Sagittarii the spout.

A CLOUDY LAGOON

One of the finest of all nebulae is M8, the Lagoon Nebula in Sagittarius, at top centre of this chart. This elongated cloud of gas, spanning the width of three full Moons, is just detectable with the naked eye under good conditions and its luminous mists are well seen with binoculars. As befits a stellar nursery, one half of the nebula contains a cluster of stars recently formed from the surrounding gas. In the other half is a 6th-magnitude blue supergiant, also newly born, which is one of the main stars that illuminates the nebula. A dark rift running down the centre of M8 gave the impression of a lagoon to some imaginative observers of the past. The Lagoon Nebula lies about 5,000 light years away in the Sagittarius spiral arm of our Galaxy.

A CELESTIAL TRIFID

Also in our Galaxy's Sagittarius arm, to the north of the Lagoon, is M20, known as the Trifid Nebula because it is trisected by dark lanes of dust. Smaller and fainter than the Lagoon, it appears as only a pale patch of light through a small telescope – the dark lanes which give it its name are nowhere near as prominent as on photographs. Near the nebula's centre is a faint double star, born from the gas and now lighting up its surroundings.

GLOBULAR M22

M22 in Sagittarius is the third-best globular cluster in the sky, beaten only by Omega

Centauri and 47 Tucanae. Under ideal conditions it can be glimpsed by the naked eye as a hazy star and is impressive through binoculars, appearing almost as large as the full Moon; small telescopes will resolve its brightest members. M22 has an elliptical profile, a characteristic it shares with Omega Centauri.

ANTARES AND M4

Ruddy-coloured Antares pinpoints the heart of the scorpion, Scorpius. Its name, of Greek origin, refers to its similarity in colour with Mars, the red planet. Antares is a colossal supergiant, around 400 times the diameter of the Sun – capable, ironically enough, of filling Mars's orbit. Antares varies in brightness by a few tenths of a magnitude, but this is scarcely noticeable to the eye. Orbiting Antares every 900 years or so is a much smaller and hotter companion, but you will need a telescope of at least 75-mm (3-in) aperture to pick it out from Antares' glare.

Just over two Moon diameters to the right of Antares lies the globular cluster M4, covering a Moon's breadth of sky. Since its light is so spread out you will need dark skies to be able to detect it with binoculars. Telescopes show that its brightest stars form a bar running north–south across its centre. M4 is one of the closest globulars to us, 7,000 light years away.

DOUBLES AND VARIABLES

As a respite from nebulae and clusters, take a look at some double and variable stars, starting in the

upper right of the chart with Beta [β] Scorpii, an unrelated pairing that can be divided with a small telescope. Next to Beta lies Nu [ν] Scorpii, a quadruple star similar to the famous Double Double in Lyra but not as easily split. Small telescopes, or even strong binoculars, show it as a double; a 75-mm (3-in) telescope will divide the fainter star, while a 100-mm (4-in) telescope confirms that the brighter star, too, is a close pair. As if that were not enough, in the far north of the constellation, just in the top margin of the chart, lies another quadruple combination, Xi [ξ] Scorpii. Small telescopes separate it into two components, with a fainter and wider pair visible in the same field of view.

Variable star observers will want to look out for two Cepheids, W and X Sagittarii, lose enough together to fit into the same binocular field of view. Both have pulsation cycles of a week.

M7

For the remainder of this chart we concentrate on open clusters, starting with M7 in the tail of Scorpius. It is the most southerly Messier object, and it is a stunner – not merely the finest cluster in Scorpius but one of the best in the entire sky. The naked eye will show it as a brighter spot in the Milky Way, and through binoculars it breaks up into a starry field over an area twice the width of the full Moon. To add to the richness of the view, the cluster M6 fits into the same binocular field.

WATCHING THE SKY

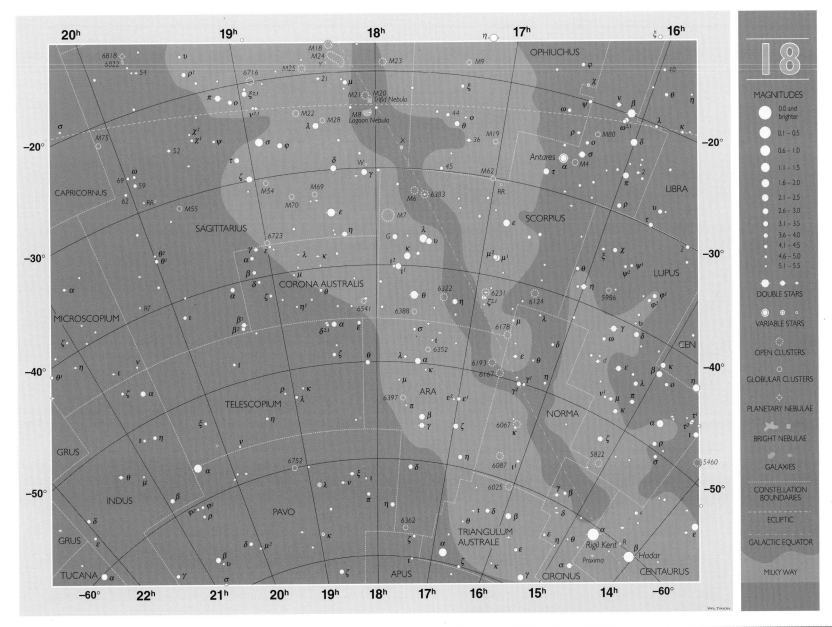

MAGNITUDES

⬤	0.0 and brighter
⬤	0.1 – 0.5
⬤	0.6 – 1.0
●	1.1 – 1.5
●	1.6 – 2.0
•	2.1 – 2.5
•	2.6 – 3.0
•	3.1 – 3.5
·	3.6 – 4.0
·	4.1 – 4.5
·	4.6 – 5.0
·	5.1 – 5.5

DOUBLE STARS

VARIABLE STARS

OPEN CLUSTERS

GLOBULAR CLUSTERS

PLANETARY NEBULAE

BRIGHT NEBULAE

GALAXIES

CONSTELLATION BOUNDARIES

ECLIPTIC

GALACTIC EQUATOR

MILKY WAY

M6

Somewhat overshadowed by its larger and brighter neighbour M7, this cluster is still a remarkable sight in its own right. Through binoculars and small telescopes its stars are seen to be displayed in the shape of a bird or insect with outstretched wings, hence its popular name the Butterfly Cluster. Its brightest member, on one of the wings, is a variable orange giant that can reach 5th magnitude at best. M6 lies 1,600 light years away, nearly twice as far as M7.

NGC 6231

Among all these grand sights do not overlook NGC 6231, an open cluster about half the size of the full Moon and just too far south to have been catalogued by Messier. Binoculars and small telescopes show it as a sparkling group like a miniature version of the Pleiades. To the south lies the wide double star Zeta [ζ] Scorpii. The fainter of this pair, Zeta-1, may be an outlying member of NGC 6231, but the brighter Zeta-2 is an unrelated orange giant much closer to us.

▶ *The Lagoon Nebula, M8, in Sagittarius is a cloud of gas divided by a dark lane. One half of the nebula contains a star cluster, NGC 6530, recently born from the surrounding gas. The entire nebula spans three times the width of the full Moon.*

CHART 19

NAME	RA, DEC	MAG(S)	TYPE OF OBJECT	VISIBILITY
δ Telescopii	18h 32m -45.9°	4.9, 5.1	double star	👓👓
β Sagittarii	19h 23m -44.5°	4.0, 7.2	multiple star	👀 👓👓 📷
		4.3		
μ Gruis	22h 16m -41.3°	4.8, 5.1	double star	👀 👓👓
π Gruis	22h 23m -45.9°	var., 5.6	double star	👓👓
δ Gruis	22h 29m -43.5°	4.0, 4.1	double star	👀 👓👓
σ Gruis	22h 37m -40.6°	5.9, 6.3	double star	👓👓
β Tucanae	0h 32m -63.0°	4.4, 4.5	multiple star	👓👓
		5.1		
NGC 6752	19h 11m -60.0°	5.4	globular cluster	👓👓 📷
M55	19h 40m -31.0°	7.0	globular cluster	👓👓 📷
M30	21h 40m -23.2°	7.5	globular cluster	👓👓 📷
NGC 7293	22h 30m -20.8°	–	planetary nebula	👓👓 📷

fter the dazzling show of the past few charts we return to a quieter area. There is no Milky Way here to enrich the view and only a few bright stars, notably Fomalhaut in Piscis Austrinus, the southern fish (above centre), and Achernar in Eridanus at lower left. Many of the constellations on this chart have been added to the sky since Greek times. Among the constellations that would have been unknown to the ancient Greeks are Telescopium (the telescope), Microscopium (the microscope), Grus (the crane), Indus (the Indian) and Pavo (the peacock). More of these new-style constellations can be found on Charts 14 and 20. For now, let us set out into this little ocean of tranquillity to find what we can catch with simple equipment.

TWO BY TWO

Looking at the constellation Grus, in the centre of the chart, you might think you were seeing double: the stars Mu [μ] and Delta [δ] Gruis are both wide naked-eye pairs. In both cases, the stars are unrelated and their apparent closeness is due purely to a chance line-of-sight effect. The 4th-magnitude components of Delta Gruis are a yellow giant and a red giant, presenting attractive colours for binocular observers, while the Mu Gruis pair, both 5th magnitude, are yellow giants. To the south of Delta the duplicity continues with Pi [π] Gruis, an unrelated pair of stars divisible with binoculars: one is white, the other a red giant that varies in brightness by more than a magnitude. On the opposite side of Delta is yet another binocular pair, Sigma [σ] Gruis, although in this case the stars do lie at the same distance from us.

Turning to the right of the chart, Beta [β] Sagittarii presents another double image: two unrelated stars of 4th magnitude, both visible with the naked eye. The brighter of these has a 7th-magnitude companion, also unrelated, that is detectable through small telescopes. An even richer sight for binoculars is found farther south in Tucana, near the bottom left of this chart, where Beta [β] Tucanae offers a tight double of closely matched blue-white stars; a somewhat fainter third star lies nearby. Although all three

stars have a similar motion through space, they are too far from each other to be genuinely connected by gravity. Astronomers term such an arrangement a common proper motion (c.p.m.) double or multiple. To conclude this feast of double vision look at Delta [δ] Telescopii, near the right-hand edge of the chart, a binocular pair of 5th-magnitude stars that, yet again, are unrelated to each other.

HELIX NEBULA

Many astronomy books carry beautiful colour pictures of our next object, the Helix Nebula (NGC 7293) in southern Aquarius at the top of the chart. Being the closest planetary nebula to us, only about 300 light years away, it is also the largest, half the apparent diameter of the full Moon. Perversely, this makes it more difficult to see, since its light is so spread out. Dark skies will be needed to find it either with binoculars or a small telescope with low power; a nebula filter should help its pale

glow stand out. It is known as the Helix because the gaseous ring appears to be shaped like two overlapping turns of a coil, an effect probably caused by two lobes of gas thrown off in opposite directions, which is a common effect in planetary nebulae.

GLOBULAR CLUSTER HUNT

There are three globular clusters worth hunting out in this area, although none of them really rivals the magnificence of those found on Chart 18. Start at the right of the chart in Sagittarius, where binoculars will show M55 as a hazy patch of weak light. Small telescopes dissect it into a cloud of faint stars. Careful study may reveal that the cluster appears to have a darker notch in one quadrant, as though there is a dearth of bright stars in this region. Our second globular, M30, lies in Capricornus, at top centre of the chart. This is smaller and fainter than M55 but is at least easy enough to locate, lying adjacent to a 5th-magnitude star; at first, the cluster itself may be

▼ *The Helix Nebula, NGC 7293,* is the closest planetary nebula to us. It consists of a doughnut-shaped shell of gas thrown off from a former red giant star. The Helix is not as striking visually as it appears on long-exposure photographs; a broadband light-pollution filter will make it easier to see through a small telescope.

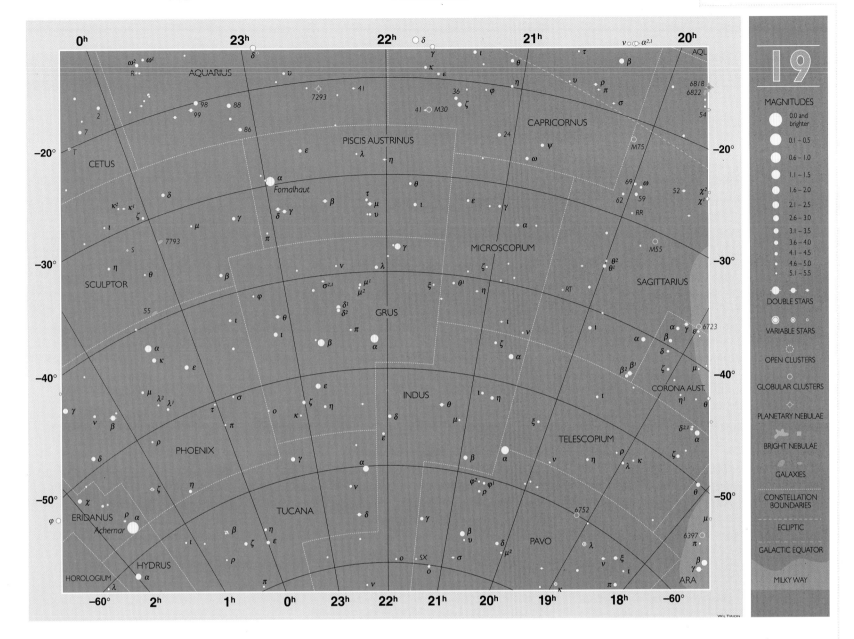

mistaken for another star. Small telescopes show that it is more condensed at the centre than M55, but apertures of 100 mm (4 in) will be needed to resolve individual stars. The best of the bunch, but too far south to have been catalogued by Messier, is NGC 6752 in Pavo, at bottom right of the chart. Binoculars show it as

a small hazy patch, while individual stars, many of them red giants, are resolved by telescopes of 75-mm (3-in) aperture.

SOLAR NEIGHBOUR

Before leaving this area, take a look at one of the nearest Sun-like stars that is visible to the naked

eye: Epsilon [ε] Indi, magnitude 4.6, at lower centre of the chart. Of spectral type K5, it is an orange dwarf somewhat smaller and cooler than the Sun, lying just under 12 light years away. It is similar to Epsilon Eridani, a nearby orange dwarf that was recently discovered to possess planets, but it is not yet known whether Epsilon Indi also has planets.

💡 COLOUR IN THE SKY

Astronomers talk freely of 'red' and 'blue' stars, but in reality such colours are elusive to the human eye. At best, red giants such as Betelgeuse, Antares and Aldebaran will appear deep orange, while hot stars such as Castor, Rigel and Vega will appear blue-white. Binoculars and telescopes will make the colours more obvious, since they will collect

and focus more light from the star into your eye. Perhaps the aspect that disappoints newcomers most of all is the colour of nebulae. Even through optical equipment they appear as nothing more than an ethereal green, or usually just a misty grey, with none of the vibrant colours displayed by photographs in books. It is not that those photographs are misleading (although some may

be processed in false colour to emphasize certain details); rather, it is the human eye that does not respond well to the faint trickle of light that it receives from such diffuse objects. Photographs have an advantage over the eye: they can build up an image with a time exposure, disclosing details that would otherwise have forever been beyond our sight.

CHART 20

T he area within 10 degrees of the south celestial pole is notably bare – there is no equivalent of the northern Polaris to guide your way here. But riches galore can be found not far afield. To one side of the pole is the band of the Milky Way studded with bright stars including Alpha [α] and Beta [β] Centauri (at lower left on this chart) along with the symbol of the southern sky, Crux, the Southern Cross. On the opposite side of the pole hover the Large Magellanic Cloud and its companion the Small Magellanic Cloud, two irregularly shaped satellite galaxies of ours that enchant first-time and experienced observers alike. The Large Cloud contains an exceptional nebula, the Tarantula. Beside the Small Cloud nestles the second-best globular cluster in the sky, although it is actually a member of our own Galaxy.

NAME	RA, DEC	MAG(S)	TYPE OF OBJECT	VISIBILITY
γ Volantis	7h 09m -70.5°	3.8, 5.7	double star	🔭
υ Carinae	9h 47m -65.1°	3.0, 6.0	double star	🔭
θ Muscae	13h 08m -65.3°	5.6, 7.6	double star	🔭
δ Apodis	16h 20m -78.8°	4.7, 5.3	double star	👓
R Carinae	9h 32m -62.8°	4.0–10.0	variable star	👁 👓 🔭
l Carinae	9h 45m -62.5°	3.3–4.2	variable star	👁 👓
κ Pavonis	18h 57m -67.2°	3.9–4.8	variable star	👁 👓
IC 2602	10h 43m -64.4°	1.9	open cluster	👁 👓 🔭
47 Tucanae	0h 24m -72.1°	4.0	globular cluster	👁 👓 🔭
NGC 362	1h 03m -70.8°	6.6	globular cluster	👓 🔭
Tarantula Nebula (NGC 2070)	5h 39m -69.1°	–	nebula	👁 👓 🔭
SMC	0h 53m -73°	2.3	galaxy	👁 👓 🔭
LMC	5h 24m -69°	0.1	galaxy	👁 👓 🔭

SOUTHERN POLE STAR

The nearest naked-eye star to the south celestial pole is Sigma [σ] Octantis – but at a mere magnitude 5.4 it is scarcely prominent even on clear nights. With its equally faint wingmen Tau [τ] and Chi [χ] Octantis, Sigma forms an arrow-head that points to the pole. A more prominent polar pointer lies farther away in the shape of the Southern Cross, Crux, whose long axis is aligned towards the south celestial pole. Currently, Sigma Octantis lies about 1 degree from the pole, and is slowly moving away due to precession. The next naked-eye star to come close to the south pole will be 4th-magnitude Delta [δ] Chamaeleontis over 1,500 years from now.

THE MAGELLANIC CLOUDS

Two satellite galaxies of our Galaxy, called the Magellanic Clouds after the Portuguese explorer Ferdinand Magellan (*c.* 1480–1510), appear to the naked eye like stray Milky Way star clouds directly opposite the south celestial pole from Alpha [α] Centauri. The Large Magellanic Cloud (LMC), which is the closer of the two at a distance of about 170,000 light years, contains roughly one-tenth as many stars as our Galaxy. To the naked eye it spans at least 12 Moon diameters, but becomes even more extensive

when viewed through optical equipment. Binoculars and small telescopes show that the LMC is liberally scattered with star clusters and nebulae. The LMC's major feature is NGC 2070, a glowing gas cloud popularly known as the Tarantula Nebula because of the spidery shape created by its extensive loops of gas. This cloud, visible to the naked eye, is far larger than the greatest nebulae known in our own Galaxy; it is lit up by a cluster of supergiant stars that have recently formed within. In 1987, a supernova bright enough to be visible to the naked eye (Supernova 1987A) erupted in the LMC near the Tarantula Nebula – see Old Age and Death of High-Mass Stars, page 164. By comparison with its larger sibling, the Small Magellanic Cloud is relatively uninteresting, with little conspicuous nebulosity, although a sweep across it with binoculars and small telescopes will reveal various star clusters. Both the LMC and SMC are interacting with our Galaxy (see Galaxy Contents, page 178).

THE SOUTHERN PLEIADES

IC 2602 in Carina, at the upper left of this chart, is an exceptionally rich open cluster popularly termed the Southern Pleiades, although (being more distant than the Pleiades) it appears somewhat smaller than that classic grouping. At least half a dozen members of IC 2602 are visible to the naked eye, splashed over two Moon diameters of sky, and binoculars bring over a dozen more into view. The most prominent member of the cluster is Theta [θ] Carinae,

magnitude 2.7. The large Eta [η] Carinae Nebula to the north of it is described on Chart 16.

TWO BRIGHT CEPHEIDS AND SOME EASY DOUBLES

Carina contains the brightest Cepheid whose variations are large enough to be easily noticed by the naked eye (there are brighter Cepheids, such as Polaris, but their range is tiny). The star in question is l Carinae – note that it is identified not by a Greek letter but a lower-case Roman letter. It rises and falls by nearly a full magnitude in a cycle lasting five weeks. In the same binocular field of view is R Carinae, a red giant variable of Mira type that can reach 4th magnitude at maximum, which comes every 10 months or so. Another bright Cepheid is to be found near the bottom of the chart: Kappa [κ] Pavonis, also with a brightness range of nearly a full magnitude but this time a shorter period of nine days.

South of l Carinae is an attractive double star, Upsilon [υ] Carinae, which small telescopes divide into blue–white partners of 3rd and 6th magnitudes. Even easier to split is Delta [δ] Apodis, below left of centre, a binocular pair of unrelated giant stars, one orange and one reddish. Fainter and more difficult to find is Theta [θ] Muscae, left of centre on the chart. Small telescopes will divide it, but its main point of interest is that the fainter of the pair is a Wolf-Rayet star, the second-brightest such star in the sky (Gamma [γ] Velorum is the brightest of all). Perhaps the most beautiful double for small

WATCHING THE SKY

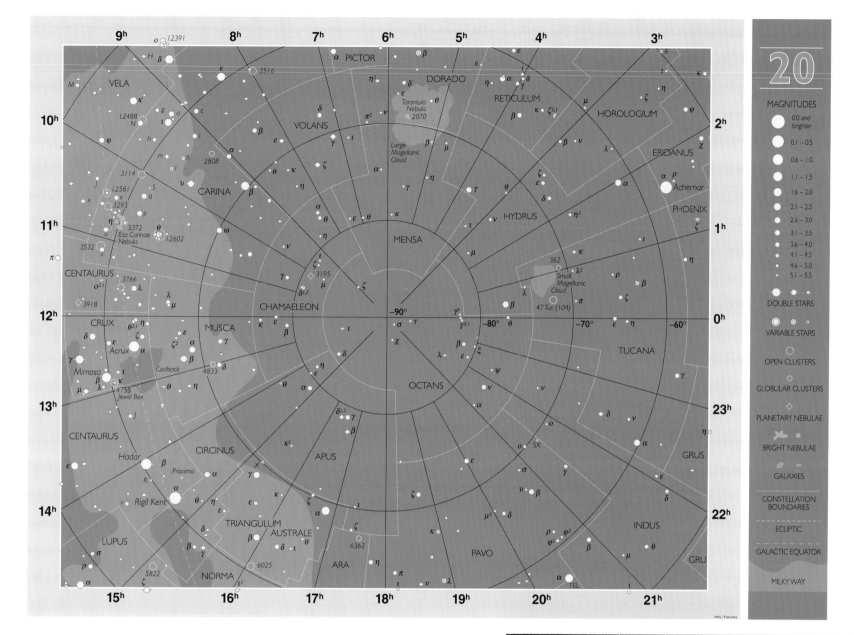

instruments in this part of the sky is Gamma [γ] Volantis, above centre on the chart, which appear gold and cream in colour when seen through a small telescope.

TWO TUCANA GLOBULARS

An object that appears to the naked eye as a fuzzy 4th-magnitude star, designated 47 Tucanae, turns out to be a superb globular cluster second only to Omega Centauri in size and brightness – you can find it at centre right on the chart, next to the Small Magellanic Cloud. Covering more than half the width of the full Moon, it shines with the concentrated light of some 100,000 stars, although telescopes of at least 75 mm (3 in) will be needed to resolve the brightest of them. Lying some 16,000 light years away, 47 Tucanae

is 10 times closer than the Small Magellanic Cloud which it coincidentally appears to adjoin. Binoculars will show another globular cluster near the SMC, this one called NGC 362, but it, too, is a foreground object in our Galaxy, at almost twice the distance of 47 Tucanae and appearing less than half the size.

▶ *Appearing to the naked eye as a fuzzy 4th-magnitude star, 47 Tucanae is in reality the second-best globular cluster in the entire sky, overshadowed only by Omega Centauri. In the sky the cluster appears close to the Small Magellanic Cloud but it is not related to it; 47 Tucanae is actually one of the globular clusters that form a halo around our own Galaxy. Small telescopes are needed to resolve its individual stars, which become increasingly crowded towards the cluster's centre.*

Watching the Sky
SUMMARY

Celestial objects can be viewed through many different kinds of eyes across the whole of the electro-magnetic spectrum. In addition to observing in the radiation to which our human eyes respond, professional astronomers observe in radiation that ranges from the long-wave, low-energy radio waves, to the very high-energy, short-wavelength X-rays and gamma rays. Looking at objects in these different parts of the electromagnetic spectrum reveals much more information than just looking in ordinary light. Each type of radiation produces its own problems for observers, and to observe in much of the electromagnetic spectrum, special telescopes have to be sent into space. However, today's amateurs are also well equipped to observe the fine celestial sights on view. You don't need sophisticated equipment to enjoy the night sky, indeed many objects can be enjoyed using just the naked eye, but some amateurs use CCD cameras and spectro-scopes to produce observations of value to the professionals. Even viewing the night sky without a telescope can produce useful observations, and knowing your way around the glittering constellations gives great personal satisfaction.

TELESCOPES ACROSS THE SPECTRUM

- The world's largest single dish radio telescope is built into a natural hollow at Arecibo, Puerto Rico; it is 1,000 ft (305 m) across.
- Some **infrared wavelengths** penetrate the atmosphere and can be observed by ground-based telescopes on high mountains, but most infrared observations have been made by satellites operating above the Earth's atmos-phere, such as IRAS (the Infrared Astronomical Satellite) and ISO (the Infrared Space Observatory).
- The largest optical telescopes are the twin 10-m (33-ft) Keck telescopes on Mauna Kea, Hawaii, whose mirrors are composed of segments.
- **Ultraviolet** astronomy is done above the Earth's atmosphere and has used satellites such as Rosat, the Extreme Ultraviolet Explorer (EUVE) and the Far Ultraviolet Spectroscopic Explorer (FUSE).
- The two latest satellites launched (in 1999) to explore in the X-ray part of the electromagnetic spectrum are Chandra (formerly called AXAF: the Advanced X-ray Astrophysics Facility) and Newton (formerly called XMM: the X-ray Multi-Mirror Mission).
- Satellites launched to observe **gamma rays**, the most energetic part of the electromagnetic spectrum, include the Compton Gamma Ray Observatory.

▼ *The Compton Gamma Ray Observatory was launched from the Space Shuttle Atlantis in 1991. It increased the number of known gamma-ray sources tenfold before the end of its life in 2000.*

▲ *A planisphere with date and time settings. When buying one, make sure the planisphere works for the latitude in which you live.*

AMATEURS EQUIPPED TO OBSERVE

- Star maps are used to find your way around the sky; different versions map the sky to fainter **magnitudes**, useful when using more powerful telescopes.
- Planispheres show the area of sky visible throughout the year for different **latitudes**.
- Binoculars are described by two numbers: for example a pair of 7 × 50 binoculars has magnification of 7 and a pair of 50 mm lenses.
- Reflecting telescopes use mirrors to collect light from an object and are preferred by some observers for deep-sky observing.
- Refracting telescopes use lenses to bring light from an object to a focus and are preferred by some observers for planetary observation.
- Schmidt-Cassegrain telescopes use both mirrors and lenses and combine larger **aperture**/longer focal length with portability.
- Single lens cameras with the facility of manual focus and exposure setting will allow the user to photograph the sky either by short exposure static shots, piggy-backing on to a telescope or at prime focus of a telescope.
- **Charge-coupled devices (CCDs)** are light-sensitive electronic detectors that allow imaging of celestial objects with significantly shorter exposure times than that required in conventional photography.

▲ *The vast Eta Carinae nebula is a glowing gas cloud that appears four times wider than the Moon. In its brightest portion, just above centre, lies the peculiar eruptive variable star Eta Carinae, from which the entire nebula takes its name.*

STORIES IN THE STARS

Ancient civilizations used the stars for navigation and time keeping. They also discerned patterns in the stars and used them to represent stories which were handed down through the generations. This is the origin of some of the constellations (from the Latin meaning 'groups of stars'), and these myths and legends are beautifully represented on some old star charts where perhaps the art takes precedence over scientific accuracy!

Many constellation names used today have their origin from the ancient Greeks, whereas constellations too far south for the Greeks to observe were named later, between the sixteenth and nineteenth centuries. Major Greek myths represented in the sky include characters of the story where Queen Cassiopeia offended the god of the sea, Poseidon, by boasting her daughter, Andromeda, was more beautiful than any sea nymph. Poseidon sent a sea monster, Cetus, to ravage the land. To appease Poseidon, King Cepheus chained Andromeda to a rock as a sacrifice to Cetus, but she was rescued by a passing hero, Perseus. Cassiopeia and Cepheus are the only married couple to be represented by constellations.

CELESTIAL SIGHTS

◯ There are five planets which can be seen easily with the naked eye at various times: Mercury, Venus, Mars, Jupiter and Saturn, while Uranus and Neptune require binoculars to be seen, and tiny Pluto is difficult to discern even with a powerful telescope.

◯ Some **meteors** (shooting stars) are visible each night as dust burns up in the Earth's atmosphere, while meteor showers occur when the Earth travels through more concentrated debris left by a passing **comet**.

◯ There are usually a few comets visible although they are often very faint and require a telescope to be seen, but some comets become bright enough to be seen with the naked eye.

◯ Stars vary in brightness and colour depending on their size, temperature and distance from us.

◯ Some stars are **double stars**, often presenting contrasting colours through binoculars or a telescope; double stars can be totally unconnected and just appear close together in the sky as they lie in the same direction as viewed from the Earth, but some are gravitationally bound.

◯ **Variable stars** vary in their light output and their changing magnitude can be followed, sometimes with the naked eye. They vary for a variety of reasons; some are binary stars and eclipse each other, some are unstable and pulsate or eject matter from their outer atmosphere.

◯ When lower-mass stars use up all the hydrogen fuel in their core, they expand and lose some of their outer atmosphere which form **planetary nebulae**.

◯ Young stars embedded in the gas and dust that lies in our Galaxy can heat the material, making it glow as an emission nebula; the brightest can be seen with the naked eye.

◯ Massive stars end their lives as **supernovae** when they blow off all their atmosphere in one gigantic explosion. Remnants of supernovae can be seen as faint misty patches through telescopes.

◯ Young stars, still physically close together, can be seen as beautiful open clusters.

◯ Old, spherical clusters called **globular clusters** can be observed and are most numerous in the direction towards the centre of the Milky Way.

◯ In a dark sky, the plane of our Galaxy containing dust, gas and thousands of stars can be seen meandering across the sky as the **Milky Way**.

◯ There are billions of other galaxies in the Universe and many can be observed with an amateur telescope.

MESSIER MARATHON

Some amateur astronomers attempt to observe all the objects listed in Messier's catalogue during one night. Obviously you have to select your observing site so it is physically possible to see all objects, and the time of year when all the objects will be on view at some time during the hours of darkness. This is early spring, and Messier Marathons are generally held on clear, moonless nights in late March and early April. Observers start sweeping the western horizon at dusk and hop from one Messier object to another until observing the last object in the east before dawn. A Messier Marathon is quite a challenge, but a less demanding task is to try and observe all the Messier objects over a much longer period of time. Keep a diary of your observations, and eventually you will have your own Messier catalogue.

WATCHING THE SKY

SUMMARY

SPACE EXPLORATION

Over four decades of planetary exploration have brought stunning surprises. We have sent robust little robots into the electromagnetic cauldron of giant Jupiter and soared with awesome precision through the rings of Saturn. Our Pioneers, Pathfinders, Mariners, Vikings and Voyagers have probed, imaged, sampled and analyzed the farthest reaches of our Solar System. Through their glass eyes and metal arms we have ridden on Mars, stared across Venus's lava fields, soared over iceberg-strewn moonscapes near Jupiter and sailed towards the edge of the Solar System. Competition between rival political systems – capitalism and communism – sent 12 privileged Americans from Earth to the surface of the Moon, where they spent a few precious days on its dusty lava plains and in the foothills of its ancient mountains. Automated probes have explored many worlds beyond the Earth–Moon system but, despite indications of dried-up rivers and lakes on Mars and subsurface oceans on Europa, we have yet to find any conclusive evidence for life. Our home planet seems more precious and unique than we could have ever conceived when we first tried to slip from it and reach for the stars.

Space Exploration
INTRODUCTION

In 1908, just five years after the Wright brothers made their first powered flight, the Russian visionary Konstantin Tsiolkovsky calculated the speed needed to orbit Earth. There was no opportunity at that time to develop this idea, as the two World Wars focused technological advance on weaponry. By the time World War II ended in 1945, though, the developments that had been made in this arena opened up new opportunities – and space travel was one of them.

▲ **A captured German V-2,** *the forerunner of the modern space rocket, veers off course shortly after launch during a test firing in 1946 from White Sands, New Mexico. Such failures were common during early rocket tests.*

REACHING FOR THE SKIES

Eight hundred years ago, the Chinese used the first crude rockets to propel flaming weapons at their enemies. Warfare fuelled their development and evolution from crude giant fireworks filled with gunpowder to sophisticated multi-stage rockets fuelled by super-cooled combustible liquids. Stumbling out of World War II and into a Cold War, which placed a freeze on direct conflict, the Superpowers were forced to dispatch their missiles to other worlds to prove their might. The resulting race for technological supremacy spawned fire-breathing, multi-stage rockets which hurled our first satellites into **orbit** and turned our nearest worlds into objects with their own geology and meteorology – not just blurred disks viewed through ground-based telescopes.

The Soviet Union won this first heat of the race into space in October 1957, when the bleeping signals of Sputnik 1 welcomed the world to the Space Age. This shiny metal ball was still free-falling around Earth when the first animals followed, contained in simple, sealed space capsules perched on top of modified missiles. Humans were keen to follow these pioneering dogs, apes, monkeys and mice, and both Super-powers were already looking further, to the Moon and planets, to demonstrate the power of their rockets.

RACE TO THE MOON AND PLANETS

The Moon was the nearest target for rocket tech-nology and robotic probes. While the Russians showed off by photographing its far side in 1959 and crashing rockets on to its surface, America was still struggling to get out of the starting blocks; but the biggest coup was yet to come. In April 1961 the Russian cosmonaut Yuri Gagarin (1934–68) became the first human to orbit Earth.

In May 1961, with astronaut Alan Shepard (1923–98) back on Earth after America's first suborbital flight, US President John F. Kennedy publicly announced his intention of sending an American to the Moon and returning him safely to Earth by the end of the decade. Robotic exploration of the rest of the **Solar System** became a 1960s sideshow: essentially, the race was on to reach our nearest neighbour.

MOONSTRUCK

Russia was still ahead in this race to the Moon, soft-landing a robotic craft there in 1966 and beaming the first pictures back to Earth. America's Surveyors were close behind and their Orbiter program was hard at work providing detailed maps for the selection of human landing sites.

In anticipation of reaching lunar orbit, both sides were making regular manned flights into Earth orbit to conduct walks in space and refine their techniques for docking and manoeuvring spacecraft.

With hundreds of hours of successful human spaceflight experience in the bag, both Super-power programs were well on course to deliver that first footprint on the lunar surface. Then, in 1967, just as all seemed to be going well, disaster struck both space programs. In January, a fire inside the capsule during a routine ground test on Apollo 1 killed the entire crew: Gus Grissom, Ed White and Roger Chaffee. Four months later, on its maiden flight, the Soyuz 1 capsule crashed when the parachutes failed, killing cosmonaut Vladimir Komarov.

KONSTANTIN TSIOLKOVSKY

The visionary prophet of rocket travel, Konstantin Eduardovich Tsiolkovsky (1857–1935) was born in Kaluga, south-west of Moscow. Childhood illness caused him to lose his hearing at the age of 10, but he went on to qualify as a teacher and continued to dream of spaceflight. In 1883 he published an explanation of how rockets could fly in the vacuum of space, and in *Dream of the Earth* (1895), he predicted artificial satellites positioned 300 km (190 miles) from Earth.

In 1903 selected chapters of his thesis, *The Exploration of Space Using Reaction Devices*, began to appear. They described multi-stage, liquid-fuelled rockets, stabilized by gyroscopes and steered by tilting rocket nozzles. The spacecraft was tear-shaped, with a passenger cabin in the rocket nose housing life-support systems and protected from extremes of temperature and the threat of **meteoroids** by a double skin. Tsiolkovsky even described details of the missions such as **escape velocity**, re-entry and the idea of a spacewalk, although he prophesied that these would not happen until the twenty-first century. **Alexei Leonov**'s (b. 1934) first spacewalk was only 62 years later. In his 78 years, Tsiolkovsky never built a rocket, preferring to write and sketch his ideas, but he never relinquished the belief that mankind's destiny lay among the stars.

SPACE EXPLORATION

THE APOLLO LEGACY

Between 1969 and 1972, 12 men spent a total of 80 hours on the surface of the Moon. The pictures, the rocks, the new knowledge and experience they returned with have in part helped transport all of us to that privileged vantage point. Watching those familiar human forms in that unfamiliar, stark landscape a quarter of a million miles from Earth, carried with it a respect for the genius and heroism which had placed them there. Apollo was a technologically premature chapter of human history. As Eugene Cernan (b. 1934), the last man to walk on the Moon, put it, 'the president had plucked a decade from the twenty-first century and inserted it into the sixties and seventies'. The supercomputer simplicities of our age did not exist then. Those pioneering pilots and navigators had charted their course to the Moon with sextants and computed their trajectories with slide rules. Such talents oozed optimism for the future.

Robotic missions have now taken us to every planet in our Solar System except Pluto. But no amount of photography of these other worlds could prepare us for the sight of our own planet rising above the surface of the barren Moon. Apollo 8 astronaut Bill Anders (b. 1933) was among the first men to look back at such a view of Earth. Those sharp, colour images of our suddenly tiny planet enveloped in a black infinity, awoke our awareness of our insignificance in the vastness of the Universe. 'We came all this way to explore the Moon,' remarked Anders, 'and the most important thing that we discovered was the Earth.'

POST-APOLLO BLUES

After an unsuccessful attempt to put its own crew on the Moon in 1971, Russia modified its Soyuz moonships for use in Earth orbit and established the first small communities in space with a series of ever-lengthening, long-duration orbital missions. With America's last three Apollo flights axed, NASA also established the space station Skylab in Earth orbit and, under diminishing budgets, continued the robotic exploration of the Solar System.

The two Superpowers conducted a rare collaboration in 1975, when their Apollo and Soyuz spaceships linked up for a few hours and the crews exchanged gifts. This was to be the last US human flight until 1981.

DIFFERENT WORLDS

Russia was still finding time to send many probes towards Mars and Venus. They had more success with Venus, soft-landing on its scorching surface in the early 1970s and beaming back a couple of pictures of one of the most extreme environments in the Solar System. Meanwhile, NASA had sent a Mariner mission to orbit Mars and two Pioneer probes to trailblaze a route through the **asteroid belt** to the outer planets.

While Russia perfected the art of surviving the cauldron of Venus and concentrated on the development of its Salyut space stations, NASA landed on Mars in 1976 and the Viking landers tested the soil for signs of life. The end of the 1970s saw a rare planetary alignment in the outer Solar System and, after much lobbying and persuasion, NASA dispatched two Voyager spacecraft on the agency's grandest tour yet –

to visit all four giant gas planets. By the end of the 1980s we had visited every planet in our system except Pluto. In the following years we returned to compile better maps of their moons than ever before.

NOWHERE TO GO

Back on Earth, American astronauts were off the ground again. The arrival of the world's first reusable spacecraft, the Space Shuttle, in 1981, allowed crews to take regular but short trips into low Earth orbit. But in 1986, the Challenger Space Shuttle exploded during its launch, killing all seven of its crew and severely damaging future US plans. The Russians, relying on their tried-and-tested, expendable but simple Soyuz spacecraft, were flying to their giant space stations Salyut 7 and Mir, spending up to one year in space. The US would have to wait some years before they could travel to Mir or claim a stake in the planned International Space Station (ISS). There are no current funded plans to send humans back to the Moon, or on to Mars. For the foreseeable future, we will continue to orbit around our home planet, relying on robots for our reconnaissance of neighbouring worlds, ever hopeful of seeing or hearing the signs which could indicate evidence of life elsewhere in the **Galaxy**.

▼ *The Earth, taken from Apollo 11. The sight of our small planet in the vast expanse of space gave us a new perspective on our place in the Universe.*

▲ *Edwin 'Buzz' Aldrin and Neil Armstrong on a training exercise for the lunar landing in Apollo 11 – the mission that won the Space Race for the US.*

APOLLOS 7 TO 11

The fatal Apollo 1 fire had grounded the program for 18 months, and the next human flight, Apollo 7 in October 1968, was a test of the command and service modules in Earth orbit. A Soyuz capsule containing turtles and other biological specimens sailed around the far side of the Moon in 1968.

With news that Russia was ready to send a crew around the Moon, the American space agency, NASA, made the most significant decision of its history: the next scheduled flight, Apollo 8, would be dispatched to orbit the Moon. The three crewmen would become the first humans to leave the relative safety of Earth orbit and head through that dark, gaping vacuum to experience what lay beyond. Seven months later, Apollo 11 settled down in the Sea of Tranquillity and the Space Race was declared to have been won.

Leonov ▶ p. 320

Space Exploration
ROCKETS

S ince our ancestors first stared up at the sky humans have dreamt of what lies beyond the blue. But the laws of physics declared that we would need to travel at over 40,300 km/h (25,000 mi/h) to slip the shackles of Earth's gravity. Although the reaction principle had been understood for many years, it would take the development of rocket motors to turn theory into any kind of reality. And before we embarked on missions to space – to the Moon, planets and beyond – our reaction engines would be put to more primitive uses: transporting our weapons of destruction across continents.

ROBERT GODDARD

Robert Hutchings Goddard (1882–1945) pioneered the use of liquid fuels to power rockets. He was born in Worcester, Massachusetts and was speculating about spaceflight by the age of 17. During his late twenties – then studying at Clark University, Worcester, Massachussetts – he began working on rocket motors, combustion chambers, propellant delivery systems and multi-stage rockets. In 1915 he was the first to show experimentally that the rocket could work in a vacuum, and hence outer space. His key publication, *A Method of Reaching Extreme Altitudes* (1919), was a turning point for his science, pointing out the possibility of sending a small rocket to the Moon. In 1921, recognizing that liquid fuels carried more energy potential, Goddard began to try and solve problems such as fuel injection, ignition and engine cooling.

On 16 March 1926, Goddard launched the world's first liquid-propelled rocket. In a flight lasting 2.5 seconds, the liquid-oxygen- and gasoline-powered machine flew just over 60 m (200 ft), reaching a height of 13 m (43 ft) at an average speed of 96 km/h (60 mi/h). Goddard continued to refine his liquid-fuelled rockets throughout his life, reaching a maximum height of 2,530 m (8,300 ft) and a speed of 500 km/h (300 mi/h) with his A-series liquid rockets before his death in 1945. NASA's Goddard Spaceflight Center near Washington is named after him.

FIRST RECORDED USE OF ROCKETS

As early as AD 62, the Greek writer Hero described the 'reaction' principle of the rocket motor, conveying it with a practical example. Hero's original reaction engine was a water-filled hollow metal sphere, pivoted like a bicycle wheel on two free-rolling joints. A fire lit beneath would heat the water – pushing steam from two opposite-facing right-angled pipes in the sphere – causing it to spin.

By the seventh century, Chinese reaction devices had become much more rocket-like – spitting fire from explosive chemicals such as gunpowder. Within 600 years such weapons had spread through the rest of Asia and into Europe and were being refined for more precise flight. During those same six centuries, accounts of battles across the world reported an increasing use of rockets as effective weapons.

REACHING FOR THE SKIES

By the early twentieth century, rocketry had become more sophisticated than the glorified gunpowder-driven fireworks which had become so widespread in wars during the preceding centuries. Instrumental in this evolution was the Russian prophet of space travel, **Tsiolkovsky**. His first thesis on rocket propulsion in 1903 suggested that super-cooled liquid hydrogen and liquid oxygen would provide the best reactive forces.

Tsiolkovsky's challenge was not met until the mid-1920s, when the American engineer Robert Goddard launched a series of rockets from his research base in New Mexico. Other groups, inspired by Tsiolkovsky's ideas and Goddard's proof of concept, were also con-

▲ *Wernher von Braun (centre) with one of his V-2 missiles. These were originally developed as weapons and were used during World War II; they were a turning point in rocket technology.*

◄ *Robert Goddard (left), the pioneer of liquid-fuelled rocketry in the US, checking one of his most sophisticated models with his team in 1940.*

vinced of the merits of liquid rockets. Among them was an East Prussian aristocrat, **Wernher von Braun** (1912–77). The German army watched his early work with interest and, in 1932, finally provided him with funding to work on what was to become the world's first long-range ballistic missile. It was von Braun's resulting liquid-fuelled V-2 missiles, on their arching supersonic trajectories, that would touch the edge of space.

ROCKET MOTORS

All rocket motors work by burning fuels to produce a lot of gas that blows out in one direction, pushing the rocket forward in the opposite direction. The force does not come from pushing on air, but from the combined momentum of all the **molecules** in the gas moving in one direction. According to **Isaac Newton**'s (1642–1727) third law of motion, the rocket's momentum in the opposite direction will match this. Typically, the gas molecules are

SPACE EXPLORATION

ion ▶ p. 340 molecule ▶ p. 341 orbit ▶ p. 341

HITLER'S VENGEANCE WEAPONS

Early in the evening of 8 September 1944, the West London suburb of Chiswick was shaken by a blast of high explosives and then a second boom as if an echo of the first blast. There was no sign of a German bomber, nor had there been an air-raid siren. All that told of its origin was a wispy vapour trail that hung above, pointing at the resulting crater. The papers reported it as a gas explosion. In fact it was far more sinister: the first of a new breed of supersonic *Vergeltungswaffe*, or Nazi 'vengeance' weapons. The first was the V-1 buzz bomb, a pilot-less jet plane carrying 850 kg (1,900 lb) of high explosives, designed to teach the Allies a lesson for their blanket bombing of Germany. This new V-2 weapon that had struck Chiswick at 3,220 km/h (2,000 mi/h) had been launched six minutes previously from a park on the outskirts of The Hague, over 160 km (100 miles) away. Fuelled by alcohol and liquid oxygen it had climbed rapidly to a height of 37 km (23 miles) and then arched down towards West London. The 13 people killed and injured by the explosion were casualties of the precursor to the world's first space vehicles.

blown out the back of a rocket at between 8,000 and 16,000 km/h (5,000 and 10,000 mi/h), but to escape Earth's gravity and reach **orbit**, a rocket needs to travel at around 28,000 km/h (17,400 mi/h). To attain this speed, multi-stage rockets are used with smaller, lighter rockets piggybacking on larger ones. Each stage is jettisoned as its fuel runs out, but adds its own velocity cumulatively, so that the final speed of the last stage is several times that of a single-stage rocket, propelling it into orbit.

The 1960s saw a move to add extra thrust on to existing rocket designs by strapping extra solid-fuel powered rocket motors to their sides. Such rocket configuration launched the Viking spacecraft to Mars in 1975 and the two Voyager spacecraft to the outer planets in 1977. Perhaps the most well-known example of strap-on-booster rockets are those mounted on the sides of the Space Shuttle's main external fuel tank. Such boosters are jettisoned once their fuel is spent and tumble back into the ocean to be recovered and flown again. Today there are over 25 expendable commercial launch vehicles in the world, many of which make use of strap-on boosters to place extra-heavy payloads into orbit. The launches of new GPS satellites in the summer of 2000 required nine separate strap-on solid rockets fixed to the Boeing Delta II rockets launching them.

ROCKET FUELS

There are two types of rocket fuels. Solid rockets run on solid powdered fuels which can be chemically designed to burn very quickly without exploding, giving an even thrust. Solid rockets are relatively cheap, but they have one major drawback: the thrust they produce cannot be adjusted during flight and, once ignited, they cannot be stopped or restarted.

For its thrust to be controllable, a rocket needs to burn liquid fuel. Most liquid rockets use a fuel and a source of oxygen, which are pumped together into a combustion chamber. There, like the solid rockets' fuel, they burn, producing lots of hot gases which are forced down, driving the rocket in the opposite direction. The fuel and the oxidizer are often super-cooled liquefied gases such as liquid hydrogen (LH_2) or liquid oxygen (LOX). A complex system of pumps often circulates these cryogenic liquids round the hot parts of the rocket, such as the nozzle and the combustion chamber, to keep the engine cool.

FUTURE DEVELOPMENTS

Any object that throws material out of one end will be propelled in the opposite direction. That material does not need to be burning, but it does need to produce a strong force. Three types of electric rocket engines have already flown in space: electrothermal, using electric elements or plasma (electrically charged or ionized gas) to heat up a propellant; electrostatic which ionize the propellant and then accelerate it through an electric field; and electrodynamic systems which generate a plasma and then accelerate it with an electric or magnetic field. Each can provide high exhaust velocities, but only very gradual accelerations which are only suitable once the craft is in space. The first practical **ion** engine to fly in space was tested on NASA's Deep Space 1 spacecraft, which was launched in October 1998.

Thermonuclear rocket engines have been developed and tested in the US, but not flown in space. They pass the propellant through a nuclear reactor to heat it. In the distant future, nuclear fusion engines and even matter/antimatter systems have been suggested as ways of generating large amounts of energy which could be used to heat a propellant.

▼ **A** *Liquid-fuelled Rocket*
Liquid-fuelled rockets usually combine two liquids, e.g. liquid oxygen and liquid hydrogen. These are pumped through valves into a combustion chamber. The resulting exhaust gases are forced out of the nozzle to drive the rocket in the opposite direction.

▼ **B** *Solid-fuelled Rocket*
A solid-fuelled rocket works by combustion inside a hole bored through the centre of a tube of solid fuel. Viewed in cross-section, the hole often has a star-shaped profile to maximize the surface area available to burn. The exhaust gases driven out of the nozzle give the rocket its thrust.

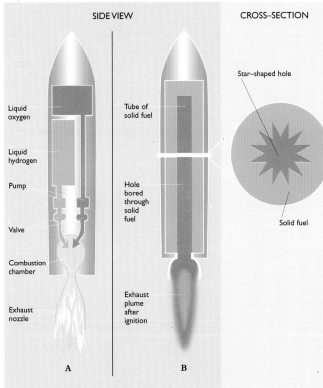

SIDE VIEW — CROSS-SECTION

Liquid oxygen
Liquid hydrogen
Pump
Valve
Combustion chamber
Exhaust nozzle

Tube of solid fuel
Hole bored through solid fuel
Exhaust plume after ignition

Star-shaped hole
Solid fuel

A — B

FUELS, SOLIDS AND LIQUIDS

At the heart of the rocket motor is the energy provided by the fuel or propellant. How easily these fuels ignite, how fast they burn and how much energy they release dictate the efficiency of the motor. Burning is just a chemical reaction and requires oxygen. Because rockets work in the absence of air they must carry their oxygen with them as another chemical. Solid fuel rockets tend to use a compound called ammonium perchlorate to provide the oxygen. Almost 70% of the propellant in the Space Shuttle's solid rocket boosters is made of this. The rest comes

from the fuel itself (aluminium), an iron-oxide catalyst and a polymer which binds it all together.

Liquid propellants are more chemically complex, but can be broadly divided into two categories: monopropellants and bipropellants. A monopropellant fuel carries the oxygen source with it in the same molecule. A bipropellant needs a separate oxygen source to burn. Some combinations (called diergolic fuels) need to be ignited separately after they have mixed, e.g. liquid oxygen and ethanol. Other combinations (called hypergolic fuels) are more potent and ignite spontaneously when they mix together, e.g. red fuming nitric acid and aniline.

SPACE EXPLORATION

Space Exploration

THE SPACE RACE

PRE-ASTRONAUTS

T he ripples of World War II rocket development would be felt through the rest of the twentieth century, and the legacy of the V-2 was to take us to worlds beyond von Braun's dreams; but history could have been very different. With the defeat of the Third Reich in Germany, von Braun manoeuvred his team to be captured by the Americans. They then crossed the Atlantic to continue their work in a country that was already nurturing its rocket designers. The Russians arrived at von Braun's German V-2 labs a few weeks after he had left. Although they did not have von Braun or his designers, they had a number of excellent technicians, and remaining hardware and blueprints. They also had another secret weapon: a brilliant chief rocket designer called Sergei Korolev.

ONCE AROUND THE SUN

On 29 July 1955, the US announced that one of the goals of the International Geophysical Year (IGY) – a global collaboration to study the physics of Earth from 1 July 1957 through to 31 December 1958 – would be to place an artificial satellite into Earth orbit. Both the US Army and the Navy submitted plans for launch vehicles. Despite the Army's careful plan, and the obvious potential of von Braun's rockets, which had already reached

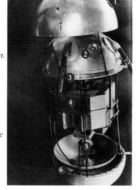

▶ *Inside Sputnik 1*
The spherical Sputnik 1 was powered by chemical batteries. The cone inside the upper part of the satellite contained a radio transmitter. This altered the pitch of its beeps according to the temperature registered by the thermometer.

1. Protective outer shell (top half)
2. Cone containing radio transmitter
3. Sockets to hold (four) whip antennae
4. Chemical batteries
5. Thermometer
6. Protective outer shell (bottom half)

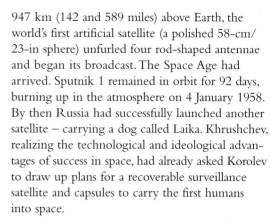

altitudes of 1,098 km (682 miles), it was the Navy's Vanguard project that was approved.

Soviet Premier Nikita Khrushchev had paid little attention to the Russian rocket-building program and the project's chief designer, Sergei Korolev, was left alone to make do as he could. Korolev had helped to build Russia's first liquid-fuelled rocket in 1933 and by 1953, with help from the German team, the V-2 heritage had helped to turn it into a full-size, liquid-fuelled, two-stage intercontinental ballistic missile. By 2 August 1955, the Soviet Union had also declared its desire to launch an artificial satellite during the IGY.

SPUTNIK I

By March 1957, Korolev had a rocket which was far more powerful than anything America had built at that time: the R-7. His plan for the IGY was to launch a massive 1.4-tonne satellite, but it was not ready in time and so Korolev quickly developed Sputnik 1 – built without even a drawing – as a replacement.

On 4 October 1957, after six unsuccessful attempts to orbit a satellite, and disobeying Khrushchev's orders to cancel the project, Korolev's perfected R-7 rocket rose from its pad at Baikonur, carrying Sputnik 1. In orbit between 228 and

947 km (142 and 589 miles) above Earth, the world's first artificial satellite (a polished 58-cm/23-in sphere) unfurled four rod-shaped antennae and began its broadcast. The Space Age had arrived. Sputnik 1 remained in orbit for 92 days, burning up in the atmosphere on 4 January 1958. By then Russia had successfully launched another satellite – carrying a dog called Laika. Khrushchev, realizing the technological and ideological advantages of success in space, had already asked Korolev to draw up plans for a recoverable surveillance satellite and capsules to carry the first humans into space.

THE EXPLORER SERIES AND JAMES VAN ALLEN

Horrified by Korolev's achievement, the US came under strong public pressure to keep up with the Russians. With a small test satellite attached to the top of its third stage, the Vanguard TV-3 rocket was rushed to the launch pad at Cape Canaveral. America was taking a big gamble: Vanguard's second stage had not even been test-flown, but if all three stages worked, they could claim to have launched a satellite. On 6 December 1957, only 1.2 m (4 ft) off the pad, and one second after lift-off, it fell back in a ball of flame, throwing its satellite radio beacon – still transmitting – clear of the explosion.

▼ *Bill Pickering (left), James Van Allen (centre) and Wernher von Braun (right) with the first US satellite, Explorer 1.*

WERNHER VON BRAUN AND SERGEI KOROLEV

Wernher von Braun (1912–77) was an avid follower of the writings of the Romanian engineer Herman Oberth, and an advocate of the use of rockets for spaceflight. He was convinced that he could build liquid-fuelled rockets large enough to carry humans to the Moon and beyond. Joining an amateur rocketry club with like-minded friends, he started building and testing tiny solid rocket missiles in disused army storage compounds outside Berlin. In July 1932, they demonstrated their Repulsor rocket to the German Army, who hired von Braun to work at its experimental rocket test

range. Despite the success of his liquid rockets as World War II weapons, von Braun saw them as space vehicles and designed larger rockets capable of orbiting a 30-tonne payload. Ensconced in America after the war, he told his team, 'Let's not forget that it was our team that first succeeded in reaching outer space. We have never stopped believing in satellites, voyages to the Moon and interplanetary travel'. Over 15 years of testing and flying for the US Army turned his V-2 into the rockets that would launch America's first artificial satellite and, ultimately, take the first American into space.

By his mid-twenties, Sergei Pavlovich Korolev (1906–66) was an active member of the Moscow

Group for the Study of Rocket Propulsion. With this group, Korolev helped to develop the first liquid-fuelled rockets launched in the USSR. He was imprisoned by Stalin in 1938 and forced to work in a scientific labour camp during World War II. Once freed, Korolev returned to rockets, working to improve the design of the captured V-2 missile and turning it into the first Soviet intercontinental ballistic missile, which carried the Sputnik satellites into orbit. With renewed political interest in space and rockets, Korolev went on to mastermind the development of manned Vostok and Voskhod spacecraft and a series of robotic missions to the near planets.

SPACE EXPLORATION

SPUTNIK'S VICTORY

In a blatant piece of political point-scoring, Sputnik's bleeps were broadcast loud and wide on radio frequencies that could be received by any radio operator around the world, rather than on the special frequencies that had been reserved for satellite telemetry for the IGY. Reaction to this triumph in the West was dramatic. Americans were horrified at the Soviet success and in an attempt to unify the US rocket program, the government quickly established the National Aeronautics and Space Administration (NASA) to catch up in what was fast turning into a race for space.

'Flopnik' and 'Kaputnik', as the press dubbed them, spurred US President Dwight Eisenhower to authorize von Braun to resurrect the Army's Project Orbiter plans. On 1 February 1958, Project Orbiter lived up to its name, placing Explorer 1 into a 360 × 2,534-km (224 × 1,575-mile) orbit. James Van Allen (b. 1914), a physicist from Iowa University, had argued for placing a Geiger-Müller counter inside Explorer to detect **cosmic rays**, and the region of trapped **radiation** it revealed was later named the **Van Allen belt**. Explorer 1 beamed back its discoveries until 23 May, but remained in orbit until March 1970.

MORE ANIMALS IN SPACE

The next major milestone in the Space Race was to orbit a human. The largest hurdle to this was establishing the effect of such a journey on life. In the 1940s the US Army had launched two monkeys into space on a ballistic trajectory in the nose cones of captured German V-2 rockets; they did not survive. It was not until 20 September 1951 that a monkey and 11 mice reached the edge of space and returned safely to Earth on an Aerobee rocket.

In November 1957, within a month of launching the first Earth satellite, the Soviet Union had another space first with which to taunt the Americans. The first living creature to make an orbit of Earth was a stray dog from the Moscow streets, called Laika. She was housed in a pressurized compartment and showed no discomfort from the weightlessness of space. The Soviets made no attempt to retrieve her, however, and she died in orbit. More dogs followed between 1957 and 1966.

Animal tests of NASA's new Mercury capsule began in December 1959, when a monkey called Sam flew a suborbital flight on a Redstone rocket. A second primate, Miss Sam, flew in January 1960. A chimp called Ham made a third Mercury suborbital flight in January 1961. In their drive to

▲ *From the mid-1950s through to the early 1960s, American X-15 planes pushed the boundaries of suborbital flight, reaching heights of over 51 km (32 miles) above Earth, allowing the first uninterrupted views of the edge of the Earth.*

rigorously test the capsule, astronaut Alan Shepard's flight was postponed and NASA opted for a fourth chimp flight in November 1961, sending Enos on a two-orbit flight and perhaps costing America the glory of sending the first human to space.

THE PRE-ASTRONAUTS

In the mid-1950s, the American X-planes had reached an altitude of 30 km (19 miles), lifting them above 99% of the atmosphere. Flying at more than twice the speed of sound at this height was reported to be smooth, although the glare from the Sun was disconcerting. In the years when space was still reserved for inanimate bleeping silver balls and hapless dogs and primates, a group of brave men made their own, more leisurely, trips to the edge of the atmosphere. Protected only by the first prototype silver space suits and suspended beneath large silver concert-hall sized balloons which

ONE COLD WAR, FOUR CONTENDERS

In the 1950s, ballistic missile technologies were booming on both sides of the Atlantic and the best way of proving to the world which nation had the most powerful rockets was to use them to reach for space. The quest to fly highest, fastest and carry the most weight into space was being pursued by four teams: the US Army, the Air Force, the Navy and the Soviets. Without a coherent unified American missile program, a 'booster gap' emerged between the nations, but individual American military triumphs initially masked this. In 1956, Redstone nose cones were reaching record altitudes of over 1,000 km (620 miles) into space and, just a day before Sputnik 1's flight, a small US 'Farside' rocket was launched from underneath a balloon and climbed to an impressive 6,400 km (4,000 miles) above Earth. But pooled Russian resources were ensuring that the missiles which had launched Sputnik 1 and 2 were three times more powerful than America's best rockets.

swelled slowly with height, they soared to over 33 km (21 miles) above Earth. From this height they got the first uninterrupted views of the planet's curved horizon, unrivalled until we started to walk in space years later, unencumbered by spacecraft. By the early 1960s, X-15 pilots almost doubled these balloon height records – soaring to 51 km (32 miles), before Al Shepard's suborbital foray to 187 km (116 miles) in 1961.

▼ *Ham the chimpanzee accepts an apple after returning safely from the third Mercury suborbital flight in 1961. The success of tests such as this on primates demonstrated that human ventures into space were a possibility.*

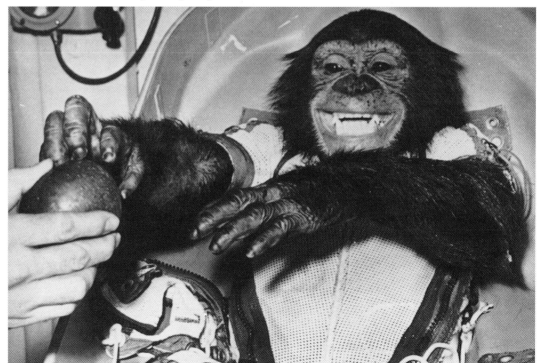

THE FIRST PROBES

Within six months of the first artificial Earth satellite launches, both Superpowers were engaged in ambitious missions to more distant worlds in an expression of their rocket might. The Moon was the first target and voyages to Venus and Mars were not far behind. Early Soviet success with the Moon spurred America on to a more organized space effort with the foundation of NASA. But these first attempts to fly past, orbit and touch down on our planetary neighbours were wracked with problems and pitfalls for both sides. The political drive to reach for the stars was outrunning the technological ability to succeed.

BERNARD LOVELL

Sir Alfred Charles Bernard Lovell (b. 1913) was born in Gloucestershire, England. He received a PhD from the University of Bristol in 1936 and joined the cosmic ray research team at the University of Manchester, conducting radar research during World War II. In December 1945 he installed ex-military radar equipment at the university's botanical research station at Jodrell Bank in the Cheshire countryside, and from there he began to study radar echoes from daytime **meteor** showers, and cosmic rays from the **Milky Way** and beyond.

Radio astronomy was growing in importance as a way of investigating the Universe and Lovell lobbied hard for construction of a vast 76-m (250-ft) steerable dish. The project gained momentum with the imminent birth of the Space Age and was rushed to completion by 1957 – just in time to track Sputnik 1's carrier rocket. The Russians were delighted when Lovell used the telescope to confirm that their Luna 2 probe had reached the Moon, and NASA used it to contact their first deep-space probe, Pioneer 5. Lovell directed the observatory until 1981 and the telescope has since been named after him. Refitted in the 1990s with a new reflective surface, it is still among the most powerful radio telescopes in the world.

LUNA MISSIONS 1–8

The Soviet lunar program was conceived to keep them ahead in the Space Race. By 2 January 1959, their first moonship Luna 1 ('Little Moon') was ready for launch on a trajectory which would take it straight to the Moon. It missed its target and went into orbit around the Sun. The first artificial satellite to leave Earth's gravity was renamed Mechta ('The Dream'). Luna 2 lifted off on 12 September 1959 and hit the Moon two days later – the first human object to land on the Moon. On 4 October 1959 Luna 3 was launched for a flight around the Moon, photographing the far side for the first time.

The next first – a soft-landing – would prove more difficult to achieve. Luna 4 took off on 2 April 1963 and missed the Moon a few days later. Luna 5 crashed while attempting to land; Luna 6 missed the Moon completely; and Lunas 7 and 8 also both crashed on to the lunar surface.

LUNA 9: FIRST PICTURES OF THE LUNAR SURFACE

The multiple Russian attempts to reach the Moon safely finally paid off on 4 February 1966, when Luna 9 touched down in the Ocean of Storms. Luna 9's main rocket fired to brake its speed and it then ejected a lander sphere, which bounced on to the surface and rolled to a halt before unfurling to start its work. In Britain, Bernard Lovell had been tracking the mission with the giant Jodrell Bank dish. When the signal changed on landing, Lovell suspected it might be a picture streaming back and immediately sent for a teletype machine from the *Daily Express* offices in Manchester. Connecting it up to the signal, Lovell and his colleagues were astonished to see the first images of the Moon's surface. They were published in the *Express* the next day, before many of the Russian mission scientists had seen them.

▲ *After five failed attempts to soft-land on the Moon, the Russians finally succeeded with Luna 9, which touched down in the Ocean of Storms on 4 February 1966; this is part of the photographic panorama it sent back.*

PIONEER PROBES 1–9

Announced in 1958, NASA's Pioneer program was to send five spacecraft towards the Moon. After one launch failure, Pioneer 1 lifted off on 11 October 1958, reaching 113,854 km (70,749 miles) before falling back to Earth. None of the next four Pioneers would reach their destination either – the closest missed the Moon by 60,000 km (37,000 miles).

With such an embarrassing start, NASA decided to change tack. The second generation of Pioneer missions would orbit the Sun, monitoring conditions in space between the planets. On 11 March 1960, Pioneer 5 lifted off to become NASA's first successful planetary mission, voyaging 60 million km (37 million miles) from Earth and testing communications in deep space. Pioneers 6–9, launched between December 1965 and November 1968, did equally well, flying to within 118 million km (73 million miles) of the Sun and mapping interplanetary conditions in the inner **Solar System**.

MARINER MISSIONS 1 AND 2

Undaunted by the troubled early Pioneer program, NASA developed a parallel series of planetary missions called Mariner. With great prudence, two identical spacecraft were built to meet a tight launch window to Venus in 1962. Mariner 1 was destroyed during a launch attempt on 22 July, but its twin, Mariner 2, escaped from Earth successfully just over a month later and, on 14 December 1962, became the first spacecraft to fly past Venus. It revealed that the planet's surface was a searing 400°C (750°F) on both the night and the day sides – much hotter than anyone had predicted. With this Venus flyby, America at last had a space 'first' to claim.

◀ *Sir Bernard Lovell, one of the pioneering fathers of radio astronomy, in front of the Jodrell Bank radio dish, which he campaigned to have built in the 1950s.*

▶ *Craters were shown on the best image of Mars taken by the US Mariner 4, the first probe to reach the red planet.*

SPACE EXPLORATION

THE FAR SIDE OF THE MOON

When Luna 3 embarked from the Russian launch site at Baikonur for a flight around the Moon, it carried no electronic memory to remember what it had seen. The designers had to be inventive if they were going to record mankind's first glimpse at the hidden far side of the Moon. The cameras would photograph what they saw and a miniature darkroom carried on board would develop the pictures. Once the little probe flew out from behind the Moon, the negatives would be scanned with a TV camera and the signal beamed back to Earth. To everyone's amazement, the cumbersome system worked and, in early October 1959, the first fuzzy black-and-white pictures of the far side of the Moon appeared on the screens at mission control, creating great excitement. To everyone's surprise, the far side lacked the vast smooth volcanic plains of the familiar Earth-facing side.

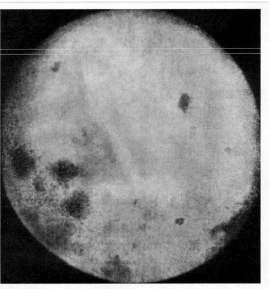

▲ *The first photograph* taken by the Russian lunar probe Luna 3 of the far side of the Moon. Luna 3 was launched in October 1959 and completed a flight around the Moon.

MARINERS 3 AND 4: A FIRST BRUSH WITH MARS

Another pair of Mariners was ready for lift-off in 1964, this time bound for Mars. Once again the first attempt was unsuccessful, but Mariner 4 set off on a historic voyage to the red planet on 28 November. Seven months later and just 9,600 km (6,000 miles) from Mars, humankind got its first close views of a planet which held the promise of life. Carrying no rocket to slow it down, however, Mariner 4 had time to snap just 21 pictures before rushing on into space. As the spacecraft passed behind Mars, its signals were bent through the planet's atmosphere which, to everyone's surprise, turned out to be ultra-thin – just 1/100th of the sea-level pressure on Earth. Each precious picture

took eight hours to trickle back to Earth. Everyone had high hopes for Mars. Centuries of speculation about the seasonal changes on its surface had led many to believe we would glimpse a world not dissimilar to Earth, but the images revealed another apparently dead, cratered world more like the Moon.

MARINERS 5–7

After the success of Mariner 4, mission scientists persuaded NASA to modify the back-up craft to visit Venus. On 19 October 1967, Mariner 5 sailed past Venus, using the same **occultation** techniques as Mariner 4 to measure the planet's surface temperature and pressure. Riding high on the US successes in planetary missions, and with the race to the Moon reaching a climax, Mariners 6 and 7 flew past Mars in the summer of 1969. Their images revealed a southern polar ice cap of solid carbon dioxide and some curious geological features, but the discovery of Mars's most amazing landscapes would have to wait for a later Mariner.

SPUTNIK 3 ONWARDS: MORE ANIMAL EMISSARIES

The Soviet Union continued to build on the successes of Sputniks 1 and 2, launching the massive Sputnik 3 on 15 May 1958. The polished metallic cone, covered in antennae and solar panels, spent 691 days in space, recording data on cosmic rays, the Van Allen belts and micrometeorite collisions.

In preparation for the first human spaceflight, later Sputnik missions flew a variety of biological cargoes into Earth orbit. These missions carried

THE BIRTH OF NASA

America's divided approach to rocket-powered flight through the three branches of its military had cost the country the first round of the Space Race. President Eisenhower was severely criticized for his penny-pinching policies which had lost America its technological leadership. Unifying US rocket research and development into a single program was essential if they were to regain the lead.

In early 1958, a Space Act was drawn up by Congress and the National Advisory Committee for Aeronautics (NACA), which had already been secretly researching the feasibility of orbital flight, became the National Aeronautics and Space Administration (NASA).

On 1 October 1958, NASA officially opened for business, with an annual budget of $340 million. Project Mercury – to place the first American in space – began within a week, and NASA took delivery of its first Mercury capsule on 1 April 1960. After a long fight, the Army's Ballistic Missile Agency was finally transferred to NASA on 1 July of that year, bringing with it the million-pound thrust engine of the Saturn rocket that was to transfer the booster power lead to the Americans and, ultimately, win them the race to the Moon.

mainly dogs. After a day in orbit, Pchelka ('Little Bee') and Mushka ('Little Fly'), on board Sputnik 6, re-entered the atmosphere at too steep an angle and were killed. On 9 March 1961, Sputnik 9 successfully orbited a dog called Chernushka ('Blackie') with a guinea pig, some mice and a dummy cosmonaut. Zvezdochka ('Little Star') on Sputnik 10 was the last dog to fly before Yuri Gagarin lifted off in Vostok 1.

EARLY COSMOS AND ZOND SERIES

Cosmos was the name given to the series of Russian satellites that followed the early Sputniks. The first Cosmos was launched on 16 March 1962, followed by two more over the next two weeks. Since then, some 2,500 Cosmos satellites have been launched, conducting everything from military reconnaissance, space weapons research and communications to astronomy, Earth observations and biological research. The name has also been used as a cover for flights in other series that failed.

Zond began as a name given to some of the early Soviet lunar and planetary missions. In 1968 the Zond name was applied to unmanned test flights for the Russian manned lunar program.

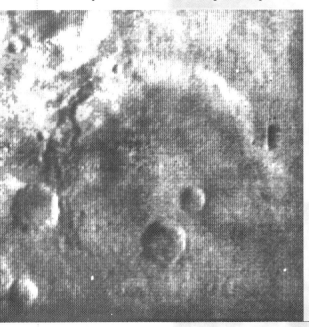

SPACE EXPLORATION

Space Exploration
SATELLITES & PROBES

SURVEYING THE MOON

T he 1960s were characterized by an intensive and ambitious study of the Moon, culminating in a first human visit and safe return to Earth in July 1969. Before that could be accomplished without danger, though, there was much to discover about our nearest neighbour: how to go into a predictable orbit around it; how to slow down enough to land softly; and what the astronauts would be stepping out on to when that long-awaited moment eventually arrived.

THE RANGER AND SURVEYOR SERIES

Success with the Moon had so far eluded NASA and, keen to catch up with the Soviet successes, the Ranger program was conceived to get a close look at its surface. After two unsuccessful test flights, NASA developed Ranger 3 to carry a small TV camera and a landing capsule containing a seismometer to record moonquakes. Neither Ranger 3 nor its clones 4 and 5 succeeded in their objectives and NASA dropped the lander capsule for the final four Ranger missions.

Early in 1964, Ranger 6 was on target for the Moon and all looked well, until the cameras were switched on and no pictures appeared. Still blind, the little probe hit the Moon 20 minutes later. Things picked up for the final three Ranger missions. More than 4,000 pictures were sent back by Ranger 7 in the last 13 minutes of its kamikaze flight, which ended with it crashing into the Sea of Clouds at about 9,300 km/h (5,800 mi/h) in July 1964. Rangers 8 and 9, launched in February and March 1965, repeated this success, revealing objects as small as 1 m (3 ft) across and contributing immensely to the selection of the first Apollo landing sites.

SURVEYORS ON THE MOON

Despite the success of the final three Rangers, a soft lunar landing eluded the Americans and many believed that the millennia of **meteorite** bombardment had left a deep layer of dust at the surface, into which any lander would sink without trace. This debate was still raging when, in June 1966, Surveyor 1 touched down gently in the Ocean of Storms. Six more Surveyors attempted to land on the Moon, the final one in January 1968. Four of them succeeded, beaming back

▲ **Fuelled by the success** of the Soviet Luna missions, the American Ranger program was devised by NASA to take a closer look at the lunar surface and image its topography. The pictures sent back to Earth by the Ranger probes aided the selection of landing sites of the first Apollo missions.

thousands of pictures and carrying out a chemical and physical analysis of the lunar surface, providing invaluable data for the Apollo design team. The next craft to soft-land on the Moon would carry a human.

LUNAS 10–14

Russia's 1960s assault on the Moon combined the success of their soft landers with a flotilla of orbital missions. On 3 April 1966, Luna 10 became the first spacecraft to go into orbit around the Moon; it revealed that the **radiation** field in lunar orbit would not be harmful to humans. Three more orbiters began detailed mapping of the lunar surface to help select sites for a manned Russian landing, chemically analyzing the lunar crust and measuring its density and complex gravity fields. By 1968, Russian knowledge of the Moon was sufficient for a manned mission.

LUNAR ORBITERS

With the Apollo program well underway, America also required detailed images of the Moon to help select scientifically stimulating yet safe landing sites. In May 1964, NASA began working with the Boeing company to create a lunar orbiter. In just over one year, five lunar orbiters were successfully flown, relaying high-resolution pictures of potential Apollo landing sites, detailed enough to pick

EUGENE SHOEMAKER

A geologist by training, Eugene Shoemaker (1928–97) is best known for discovering – with his wife Carolyn and his colleague David Levy – a **comet** that hit Jupiter in July 1994. Shoemaker's signature work, however was his research on the nature and origin of the Barringer Meteor Crater in Arizona, which helped provide a foundation for research into cratering on the Moon and planets.

While still in his teens, Shoemaker realized that one day astronauts would walk on the Moon, and from that point on his whole professional life was directed towards becoming one of them. A medical condition prevented him from ever being selected for the Apollo program, however. 'Not going to the Moon and banging on it with my own hammer has been the biggest disappointment in life,' he later admitted. Although he never realized his dream of doing field geology on the surface of the Moon, Shoemaker was an inspiration behind America's lunar exploration. He took part in the Ranger lunar robotic missions, was principal investigator for the television experiment on the Surveyor lunar landers, and led the geology field investigations team for the first Apollo lunar landings. He later conceived the 1994 Clementine lunar mission, but in 1997 at the age of 69, while doing field work in Central Australia with Carolyn, he was killed in a car crash. As a tribute to the father of lunar geology, some of Shoemaker's ashes were carried to the lunar surface on board the Lunar Prospector mission which impacted there in 1999.

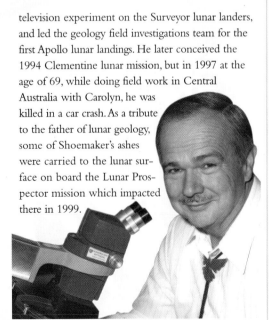

▲ **A trained geologist,** Eugene Shoemaker specialized in the study of the Moon, asteroids and comets. His name is associated with a comet which hit the planet Jupiter in July 1994, jointly discovered by him, his wife and David Levy.

SPACE EXPLORATION

SURVEYOR AND THE MOON'S BASALTS

The first soft-landings by Luna 9 and Surveyor 1 brought the Moon within the realm of geophysicists. Surveyor 3 in April 1967 carried the first shovel to dig in the dust, while on board Surveyor 5, which landed in September 1967, was an instrument which provided the first clues to the composition of the rocks which make up the lunar maria.

Surveyor 5 landed 25 km (16 miles) from the future Apollo 11 landing site in the Sea of Tranquillity. First results suggested the rock was volcanic, probably a kind of basalt. Surveyor 6 found more basalt in the Central Bay region. A predominantly basaltic lunar crust meant that the Moon had been heated strongly from the inside and had become chemically differentiated, with the heavy elements settling into its interior and the lighter elements forming these rocks at the surface.

The first six Surveyor flights had been scouts for possible Apollo landing sites. For the Surveyor 7 mission, the last of its type, NASA finally agreed to send the probe somewhere Apollo would never go: the rough terrain of crater Tycho in the southern highlands. Here, it encountered a variety of other igneous rock types – further evidence for widespread heating of the Moon.

out the Surveyor 1 spacecraft that had landed a year before. These missions together photographed the entire surface of the Moon, including the poles, returning global pictures of the Moon which were not improved on for nearly 30 years.

LUNAS 15–22 AND THE LUNAKHODS

In the late 1960s, sensing that America was further along with their manned lunar program, the USSR tried to develop an unmanned craft that could return a sample of lunar material to Earth before the first American astronauts returned with theirs. Rushing to beat Apollo to a return trip, the first few launch attempts failed and it was not until 13 July 1969 that Luna 15 took off successfully for the Moon, just a few days ahead of Apollo 11. On 21 July, just as Apollo 11 was preparing to lift off, Luna 15 crashed into the Sea of Crises.

Over a year went by before Luna 16 attempted the same mission: this time returning with 101 g (4 oz) of lunar dust from the Sea of Fertility on 24 September 1970. Just six weeks later, Lunakhod 1 – the first unmanned rover vehicle to operate on the surface of another world – arrived on board Luna 17. The small car-sized, eight-

wheeled wagon roamed around the Moon for 321 Earth days, travelling around the Bay of Rainbows where it had landed, while operators watched through its cameras from Earth.

A second, faster Lunakhod arrived near the crater Lemonnier in January 1973, exploring mountainous terrain to the south. A final sample return mission, Luna 24, was launched on 9 August 1976 and recovered samples from 2 m (6.6 ft) below the surface of the Sea of Crises. Eighteen years passed before another spacecraft was sent to explore the Moon.

CLEMENTINE

The Clementine spacecraft was built by the US Naval Research Laboratories to investigate laser performance in space. After a month-long trip from Earth on a circuitous route to conserve fuel, Clementine arrived in lunar orbit on 20 February 1994 and spent the next 10 weeks mapping the whole Moon, recording information about global mineral composition and topography for the first time. An improvised experiment over the lunar south pole, bouncing radio waves into an area of permanent shadow, hinted at the possible existence of water ice. Clementine spun out of control on its way to explore the **asteroid** Geographos after

▼ *Lunakhod 1,* the first unmanned rover vehicle, which arrived on board the Luna 17 craft in 1970 and roamed the lunar landscape for 321 days while operators monitored its progress through its cameras from Earth.

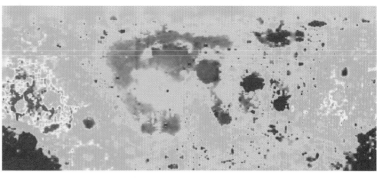

▲ *The moonprobe Clementine* assessed the lunar surface at various wavelengths. The purple areas are basins, green indicates areas of average height and red marks the highest features, those over 7,500 m (25,000 ft).

leaving the Moon. On 20 July 1994, exactly 25 years after mankind had first visited it, Clementine swung by the Moon again and went into solar orbit.

LUNAR PROSPECTOR

The tantalizing hint of ice – perhaps water ice at the lunar south pole – rekindled an interest in the Moon and four years after Clementine's visit, NASA returned to the Moon with a mission called Lunar Prospector. The mission carried no cameras, but instead studied chemical signatures of the lunar crust. Concentrations of hydrogen were detected at the poles, which some scientists have interpreted as further evidence of water ice. In another improvised experiment, at the end of its design life, Lunar Prospector crashed into this southern polar region and the world's telescopes were trained on the impact cloud that was thrown up, scrutinizing it for evidence of water vapour. None was found.

TO ROCKY WORLDS

The Soviets had already been the first to launch a satellite, an animal and a man into orbit, as well as the first to fly around the Moon and crash land on it. They also wanted to be the first to reach the planets and, in autumn 1960, missions were planned for Venus and Mars. The US was closer behind this time and was busy building a network of massive listening dishes around the Earth – a deep-space network (DSN) to keep in touch with their missions.

MARS 1, 2 AND 3

The two first Soviet attempts to send a spacecraft to Mars in 1960 both failed to reach Earth orbit. Mars 1, launched on 1 November 1962, failed just three months from its target. Many other mission failures had befallen their Mars program by 1969. However, in 1971 three Soviet craft were ready to leave for Mars and, after one more launch failure, Mars 2 and 3 succeeded in setting off, just ahead of the US Mariner 9 mission. The race to Mars was on.

Mars 2 and 3 were designed to land on Mars, with the upper part of the spacecraft going into orbit. But when all three missions arrived in November 1971, a dangerous global dust storm was raging on the planet. On 27 November Mars 2 vanished without trace into the storm and probably became the first spacecraft to hit Mars. Mars 3 followed on 2 December. It quickly began to transmit a picture, but after 20 seconds the first Martian broadcast stopped and Mars 3 was never heard from again.

Above the planet, three orbiters were still waiting for the dust to clear. The Russian craft transmitted data and some photographs for nine months, but they were trumped by Mariner 9, which was already beaming back high-quality images of the entire planet.

THE VENERA SERIES

The Russian Venera program to Venus began badly, with a launch failure on 4 February 1961. Four more attempts were made to reach Venus before Venera 4 parachuted into its cloud tops on 18 October 1967, beaming back data for 94 minutes. Pressures of 22 Earth

atmospheres and temperatures of 280°C (536°F) had crushed the capsule 25 km (16 miles) above the surface. Despite redesigns, Veneras 5 and 6 failed to reach the surface as well. In 1970, Venera 7 survived the pressures of 90 Earth atmospheres and 475°C (887°F) to transmit the first short signal from the surface of another planet. Venera 8 lasted 50 minutes in the inferno and revealed that there was as much light down there as a dull day in Moscow. It was time to send a camera.

It was June 1975 before Veneras 9 and 10 snapped the first pictures from the surface of another planet. Veneras 11 and 12 measured the chemistry of the atmosphere during descent and recorded possible lightning flashes. Then, in 1982, Veneras 13 and 14 returned the first colour pictures and drilled into the rocky ground to measure its chemistry. The following year, Veneras 15 and 16 went into orbit around cloud-shrouded Venus and mapped the planet by radar, covering 120 million sq km (46 million sq miles).

MARINERS 8 AND 9

Mariners 8 and 9 were designed to go into orbit around Mars. On 9 May 1971, Mariner 8 failed to reach Earth orbit and fell back into the Atlantic. Mariner 9 was launched successfully three weeks later and reached Mars in November. After sitting out a dust storm, it began to map the planet – discovering a series of giant volcanoes and a vast valley stretching 5,000 km (3,000 miles) across the planet. Everywhere on the surface there were old, dry river channels and gullies. With this evidently wet past, Mars was once again a promising place to hunt for life and determination to land on the planet grew.

MARINER 10

After following a novel **sling-shot** trajectory past Venus, Mariner 10 reached Mercury in March 1974, photographing 40% of the surface. Mercury looked very like our Moon, with a grey, crater-coated surface. Looping around the Sun on its own orbit, Mariner 10 re-encountered

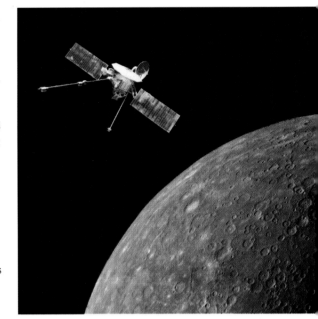

▲ *Mariner 10 shown against Mercury's cratered surface. Using the novel sling-shot technique, Mariner 10 first flew past Venus on its journey to Mercury; its images revealed a grey, crater-filled planet, very much like our Moon.*

Mercury twice, flying over the planet's south pole at 48,000 km (29,800 miles) on the second flyby and only 327 km (203 miles) above the night-time side on the third. Mariner 10 found a more powerful magnetic field than was anticipated – a curious puzzle considering how geologically dead the surface looked. This last Mariner flight remains the only mission to Mercury.

VIKING/PATHFINDER

Vikings 1 and 2 were both orbiters and landers. The orbiters carried television cameras to hunt for potential landing sites and the landers carried a suite of instruments to monitor weather conditions, photograph the surface and, most importantly, search for life. In the summer of 1976, both landers descended to the surface of Mars, relaying dramatic views of their rubble-strewn landing sites beneath a dust-soaked, pink sky. Robotic arms gathered a soil sample that was tested for evidence of life. Initial results proved positive but they were later attributed to chemical reactions in the soil rather than any biological activity. The Viking landers survived on the Martian surface for four years, far longer than their 90-day design life. The Viking orbiters acted as relay stations for the landers, returned over 52,000 reconnaissance photographs and monitored atmospheric circulation.

◀ *The Russian Venera missions represented an attempt to reach and photograph the forbidding surface of Venus. Venera 13 was the first mission to return colour images of the planet's surface (pictured) after penetrating its dense gaseous atmosphere.*

HEROIC FAILURES AND THE GREAT GALACTIC GHOUL

By 1973, Russia had lost numerous spacecraft at varying distances from Earth. Even America had its failures en route to the red planet and the jinx that befell so many Mars missions became known as the Great Galactic Ghoul, which seemed to lurk halfway between the two planets. In 1993 NASA's Mars Observer mission, a sophisticated orbiter, went quiet just as it reached Mars. Russia's ambitious international Mars 96 mission ended up in the

Andes. After the success of Mars Pathfinder and Global Surveyor in the late 1990s, the Ghoul was at work again, putting a stop to the next two US missions – Mars Polar Orbiter and Mars Polar Lander. The orbiter burnt up in the atmosphere when it went into too low an orbit round Mars due to confusion between imperial and metric measurements between NASA and the builders of the satellite, Lockheed Martin. The Polar Lander went silent after entering the atmosphere and, despite a prolonged effort to contact it, the spacecraft was never heard of again.

It was over 21 years before another spacecraft returned to Mars. On 4 July 1997, Mars Pathfinder bounced down. Inside squatted a six-wheeled Martian rover called Sojourner. The little buggy gained global fame when it rolled successfully on to the red surface and spent 10 weeks probing the chemical make-up of the nearby boulders.

PIONEER VENUS AND THE VEGA BALLOONS

The final flight of the US Pioneer series was an ambitious double spacecraft mission to Venus. The first craft went into orbit around Venus on 4 December 1978, carrying a radar to construct a global map. The other spacecraft dropped a cluster of four probes on to the planet a few days later, relaying data on the temperatures, pressures and chemistry of the atmosphere. Although the probes were not designed to survive a landing, one of them did and continued to send back data for over an hour.

The primary purpose of the two Russian Vega missions was to visit Halley's Comet. Using Venus for a gravity sling-shot, in June 1985 each dropped off a Venera-style landing capsule and a French atmospheric balloon. Both floated around the planet carrying instruments to record temperatures, pressures and lightning frequencies. They drifted over 10,000 km (6,000 miles) in Venus's powerful winds.

MAGELLAN

Mapping Venus with the same thoroughness and success as Mars was more difficult. The planet's thick cloud cover meant that little was known about the geology of the planet other than what had been gleaned from the Pioneer Venus inspection, and an advanced radar mapper was called for. Funding and launch delays postponed Magellan's arrival at Venus until 10 August 1990. To save money, the spacecraft had to record the data from each polar orbit on a tape recorder, then turn round to face Earth and use the same mapping antenna to beam back its discoveries to Earth. Within one Venusian day (243 Earth days), 84% of the planet had been mapped.

Magellan's radar maps reflected the roughness of the surface, the brighter areas being rougher than the darker areas. The images revealed that Venus was a vulcanologist's dream – smothered in vast sheets of lava flows, peppered by volcanic craters of all shapes and sizes, and dominated by gigantic country-sized volcanic features called coronae. Sinuous lava channels 4,000 km (2,500 miles) long meandered over the surface and tectonic fractures and cracks criss-crossed the plains and formed intensely wrinkled regions called tesserae. Despite such intense volcanism and clues that the whole surface was relatively young, no eruptions were witnessed. In October 1994, Magellan burnt up in the Venusian atmosphere.

COMET AND ASTEROID MISSIONS

The return of Halley's Comet to the inner Solar System in the mid-1980s provided the first opportunity to mount a cometary mission. Vega, a Russian contribution, reached Halley via Venus, while a European mission, Giotto, flew straight to Halley, recording dust and the comet's chemistry. The two Vegas and Giotto found that the comet's **nucleus** was much darker than had been expected but, just as predicted, it was composed mainly

THE GLOBAL SURVEYOR AND WATER ON MARS

THE GLOBAL SURVEYOR AND WATER ON MARS

In March 1999, NASA's Mars Global Surveyor (MGS) orbiter achieved its planned orbit after a risky period of aerobraking, dipping its solar panels repeatedly into the top of Mars's atmosphere. The most detailed mapping of Mars's landscapes and monitoring of the Martian atmosphere began. MGS's cameras can resolve features on the Martian surface the size of a car, a great improvement on the Mariner 9 and Viking orbiter cameras. In June 2000, mission scientists announced that a small number of high-resolution images revealed gullies on cliff and crater walls which implied that liquid water had seeped out on to the surface very recently.

of ice. Giotto's camera was disabled as it flew through the comet tail, but enough instruments survived to make it worth sending Giotto on to a second comet, Grigg-Skjellerup, in 1992. Giotto is still following a 10-month orbit around the Sun.

Not to be outdone, but without funding for a special comet mission, NASA came up with the International Cometary Explorer (ICE), an existing spacecraft that utilized the Moon's gravity to flip it into a new orbit. ICE rendezvoused with comet Giacobini-Zinner in 1985.

Several asteroid encounters have also been accomplished. NASA's Galileo spacecraft flew past two asteroids, Ida and Gaspra, en route to Jupiter in the early 1990s. NASA's Near-Earth Asteroid Rendezvous (NEAR) mission flew past Mathilde and eventually went into orbit around asteroid Eros in February 2000.

▼ *The barren truth about Venus's surface was revealed through radar data sent back to Earth by the Magellan planetary probe. Computer simulations of the planet's surface were compiled from the data beamed back.*

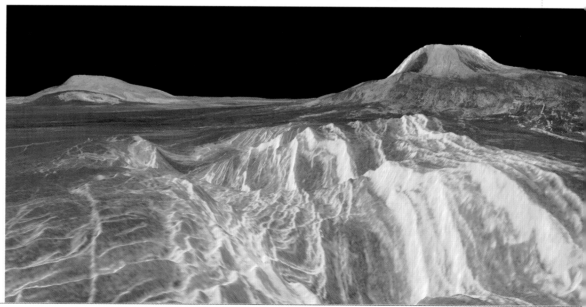

TO GAS WORLDS

Beyond our nearest neighbours the giant gas planets of the Solar System beckoned. If we were ever to journey farther than Mars we would have to prove that the uncharted asteroid belt could be crossed by a space probe. But the extreme distances to the farthest planets on the edge of our Solar System would require more than precise navigation to travel. It was a mathematical break-through in the mid-1960s that opened up a fast-track route to their realms. Trail-blazing journeys to the outer Solar System were flown in the 1970s and 80s and by the 1990s sophisticated robots were returning to the gas planets to map their moons and sample their vast atmospheres.

MICHAEL MINOVITCH AND GARY FLANDRO

It was known that by flying close past a planet a spacecraft could be massively accelerated and redirected on to a new course without using any fuel. But a solution to the difficult three-body problem, which lay at the heart of such a manoeuvre, had remained elusive. The English scientist Derek Lawden is credited with first solving it in 1954, but it was not until 1962 that the technique came to prominence in the US when vacation student Michael Minovitch, working at the Jet Propulsion Laboratory, independently discovered a solution. These breakthroughs gave birth to a powerful 'sling-shot' tool that would open up the Solar System to robotic exploration.

In summer 1964, another vacation student, Gary Flandro (b. 1934), came to JPL. He was interested in planetary positions and computed a series of theoretical sling-shot missions. One stood out from the rest. Flandro found that at the end of the 1970s, a 175-year planetary alignment would enable a spacecraft to visit all the outer planets in one 12-year flight. This rare conjunction demanded a mission, but NASA officials were slow to respond, questioning the mathematics, uninspired by Flandro's discovery. Eventually they accepted the concept, however, and the 'Grand Tour' of the Solar System emerged, later becoming a successful Voyager mission.

PIONEERS 10 AND 11

The aptly named Pioneers 10 and 11 began humankind's first ventures through the **asteroid belt** and on to the realms of the giant gas planets. Pioneer 10 was launched on 3 March 1972, and reached a new record speed of 50,240 km/h (31,214 mi/h). It passed through the asteroid belt after 11 months and reached Jupiter on 4 December 1973, charting the planet's powerful magnetic fields and radiation belts. Spinning gently, the spacecraft scanned Jupiter, recording images better than anything previously seen.

Pioneer 11, launched on 5 April 1973, flew past Jupiter on 4 December 1974, and used the planet's gravitational field to throw it towards Saturn, arriving on 1 September 1979. As well s returning the first spacecraft images of Saturn, Pioneer 11 scouted a route through the planet's rings for the Voyager missions that were already on their way. The last communication from Pioneer 11 was received in November 1995. Without power, it flies on towards the star Lambda Aquilae. Heading in the opposite direction at 12 km/s (7 mi/s) Pioneer 10, currently over 10.88 billion km (6.76 billion miles) from Earth, will pass the star Aldebaran in two million years. Both probes, although drained of power, will outlive the Earth, circling the **Milky Way** for billions of years.

VOYAGERS 1 AND 2

Conceived to take advantage of a rare alignment of the outer planets, NASA planned its most ambitious robotic planetary expedition, and Voyagers 1 and 2 were launched in the late summer of 1977. In the spring of 1979, Voyager 1 reached the realms of Jupiter and beamed back the first clear, close-up views; these were used to make

▲ *Voyager 2 returned numerous fascinating images of Saturn, including this view looking back after it had passed the planet. Initially flying past Jupiter, the probe used the planet's gravitational field to catapult it towards Saturn.*

time-lapse movies of the planet's complex weather patterns. A thin ring around Jupiter was discovered and detailed images of the planet's four largest moons revealed an active sulphurous volcano erupting on Io, the smooth ice world of Europa, the complex ridged and grooved terrain of Ganymede and the crater-coated surface of Callisto. Voyager 2 flew past four months later and recorded more observations of the Jovian system.

Voyager 1 reached Saturn in November 1980 and revealed intricate details of the planet's rings and weather patterns. Swinging past Saturn's largest moon, Titan, Voyager 1 was thrown out of the plane of the Solar System and away from the planets.

When Voyager 2 reached Saturn in August 1981, it successfully imaged more details of the rings, moons and cloud patterns, and permission

▼ *Gravitational Slingshot*
At point A the spacraft is on a trajectory that will take it just 'behind' Jupiter as the planet hurtles on its orbit around the Sun. At B the spacecraft accelerates as it is pulled into the planet's gravity well. At C the spacecraft shoots past the planet on a hyperbolic path. Its trajectory is changed and its speed relative to the Sun is increased as it takes energy from Jupiter's orbital motion. The massive planet's corresponding loss of energy from the encounter is extremely small.

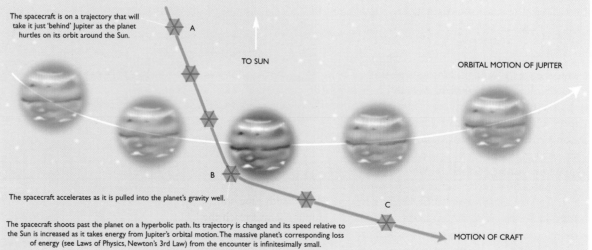

THE GRAND TOUR

In the 1960s a unique window of political will, mathematical know-how, technological ability and planetary conjunction opened up. The giant, outer, gas planets would line up in the 1980s so that a single mission could use the newly proven sling-shot technique to fly past them all. NASA's first proposal for this grand tour was a craft called the Thermoelectric Outer Planets Spacecraft (TOPS). As the concept grew more complex, however, costs soared to over one billion dollars and the project was cancelled in 1972. With only five years left before the launch window closed, JPL officials hurriedly redesigned the

mission. Their cheaper Mariner Jupiter-Saturn (MJS) mission consisted of two spacecraft launched to arrive at Jupiter in 1979 and Saturn in 1981. Mission designers secretly hoped at least one spacecraft would continue on to Uranus and Neptune and budgeted for enough fuel and power just in case these opportunities arose. Renamed Voyager and launched in 1977, the two spacecraft used Jupiter's gravity field to hurl them on to Saturn, thus gaining 57,120 km/h (35,489 mi/h) in speed and slowing Jupiter by 30 cm (12 in) per trillion years. Additional sling-shots at Saturn and Uranus reduced travel time to Neptune by nearly 20 years.

PALE BLUE DOT PICTURE

On 14 February 1990, Voyager 1's camera platform moved into action for the last time. Looking back from its bird's-eye view 5 billion km (3 billion miles) above the Solar System, it snapped 60 frames which snaked a path across the Solar System. Mars and Pluto were too small to be seen, and Mercury was caught up in the glare of the Sun. But six other tiny dots stood out from the black and together became known as 'the family portrait'. In JPL's photographic darkroom, the mosaic was blown up to 6 m (20 ft) across, before the single pixel planets were visible. One pale blue dot was Earth.

was granted to extend the mission. In January 1986, with the Sun over 5 billion km (3 billion miles) away, Voyager 2 imaged the giant planet Uranus and its moons. At Neptune, three-and-a-half years later, an Earth-sized dark storm spot was observed in detail and nitrogen geysers on its moon Triton were photographed. Voyager 2 continues heading out of the Solar System. Power on both spacecraft should last until 2015, although the signals reaching Earth are 20 billion times weaker than the power output from a digital watch.

GALILEO ORBITER AND PROBE

Initially scheduled for launch in the early 1980s, the Galileo mission eventually drifted out of the Shuttle Atlantis cargo bay on 18 October 1989 to embark on a tortuous six-year route to Jupiter via multiple Venus and Earth sling-shots. Unable to unfurl its main antenna properly, the mission initially looked doomed, but re-programming the craft from Earth salvaged many of the mission's goals and after flying close by two asteroids en route, Galileo neared Jupiter in June 1995. A probe was released on 13 June and on 7 December it became the fastest man-made object as it hurtled towards the giant planet at 170,700 km/h (106,000 mi/h). Decelerating to 200 km/h (124 mi/h) in less than two minutes as it hit the top of Jupiter's atmosphere, the coffee-table sized probe briefly weighed as much as a Jumbo Jet. Surviving this phase of its approach, it descended into Jupiter, relaying data on temperatures, pressures and wind speeds to the orbiter for 56.7 minutes until rising temperature caused it to fail. As luck would have it, the Galileo probe had entered a dry spot in the planet – thought to be rather untypical. Galileo has been in orbit around the Jovian system since then, reporting back more details of the planet and its four main moons.

MISSIONS TO THE SUN

The early American Explorer and Pioneer missions first studied interactions between the Sun's radiation and the Earth's magnetic field. In the USSR at this time, the Electron satellite program was carrying out similar studies. Flown from 1962 to 1975, NASA's Orbiting Solar Observatories (OSO) monitored the Sun in **ultraviolet**, **X-ray** and **gamma-ray wavelengths** for its entire 11-year cycle. The US Skylab and Soviet Soyuz flights of the early 1970s flew solar telescopes. In December 1974, the first of two joint German–US solar spacecraft called Helios was propelled to within 46.4 million km (28.8 million miles) of the Sun to study the **solar wind**, magnetic fields and cosmic radiation. In February 1980, NASA's Solar Maximum Mission (SMM) was sent to study the Sun during its most active period. France and Japan also flew their own solar satellites during the 1970s and 1980s. The 1990s saw a revolution in our solar observations with the Solar and Heliospheric Observatory (SOHO) that studies the Sun 24 hours a day and records images in a variety of wavelengths. Launched in 1990, the joint ESA/NASA Ulysses mission used Jupiter's gravity to sling-shot out of the plane of the Solar System into a solar polar orbit, and provide our first observations of the Sun from above and below. And after a failed first launch in 1996, Cluster, four identical spacecraft flying in formation to map Earth's magnetic field in three dimensions, was re-launched in June 2000.

CASSINI-HUYGENS

Cassini-Huygens was launched on 15 October 1997. With sling-shots from Venus, Earth and Jupiter, it will reach the Saturnian system in 2004, firing its main rocket engine for over two hours to brake into orbit. The mission will later

jettison ESA's Huygens probe, which will parachute into the orange cloud tops of Titan, descending to an unknown landing site where it might splash down into an ethane–methane ocean, or touch down on a rocky, icy shore. The orbiter will continue its charting of Saturn's rings and moons for over four years.

FUTURE MISSIONS

If funding is forthcoming and flights are flawlessly executed, then over 20 new missions could open up the Solar System in the next decade. In 2004 Stardust will snatch a sample from comet Wild 2 and return it to Earth. Two new NASA landers will ride more robotic wheels over Mars and ESA's Beagle II will drill into its surface in a bid to find evidence of life. Other European and Japanese missions will return to lunar orbit and Pluto Express may one day head out to the last planet in our Solar System.

▼ *The Voyager missions were launched in the late 1970s and flew past the outer planets. They continue on out of the Solar System.*

MISSIONS TO EARTH

I n tandem with the race to reach the Moon and planets, both Superpowers were aware of the communication, commercial and espionage opportunities of Earth orbit. Missions to Earth orbit were dispatched to chart unmappable parts of our planet, watch the weather, monitor environmental disasters, guide our travels, bounce our conversations round the globe, and to spy on each other. What started as a costly Cold War race would turn into a multi-billion dollar industry which would open up space to everyday use and drive down costs of getting there.

COMMUNICATIONS SATELLITES

The concept of a communications satellite was first proposed by a young electrical engineer named **Arthur C. Clarke** (b. 1917) in 1945. He calculated that a satellite in orbit 35,880 km (22,295 miles) above the Equator would circle the globe in exactly one day, appearing to hover over a point on Earth. With just three satellites in this geostationary orbit it would be possible to bounce signals to almost any point on the planet.

Clarke's orbit proved difficult to reach during the early days of the space program and the first communication attempts were conducted from a lower Earth orbit. In 1958 a US satellite called SCORE transmitted a tape-recorded Christmas message from President Eisenhower.

Communication experiments continued with Telstar, which was launched on 10 July 1962. Although it was not in a geostationary orbit, it was able to transmit live TV pictures and sound between opposite sides of the Atlantic.

◄ *A communications satellite is launched from the cargo bay of the Space Shuttle.*

GEOSYNCHRONOUS ORBITS

The first attempt at a geosynchronous orbit was NASA's Syncom series, first launched in 1963. An international organization called Intelsat was established the same year to manage geosynchronous orbit and launch the world's first commercial communications satellite, Early Bird, in April 1965. The Early Birds could relay 240 telephone calls or one TV channel. By 1969 they covered the whole globe and transmitted worldwide television coverage of the first Moon landing. The current Intelsat-7 generation, built by Ford Aerospace, can simultaneously relay 18,000 telephone calls and three TV channels.

In the USSR the geostationary orbit was considered less useful because of the nation's extreme northern **latitudes**, so they pioneered an alternative communications satellite system using an orbit inclined at 65° to the Equator. As one satellite vanished over the horizon, the ground-tracking dishes could lock on to one another. Their Molniya constellation became the world's biggest domestic satellite network.

NAVIGATION SATELLITES – GPS

The Global Positioning Satellites (GPS) were developed by the US Department of Defense and launched in the early 1990s. Twenty-four satellites in six circular 20,200-km (12,550-mile) orbits are spaced so that at any one time a minimum of six are in view from the ground. Ground stations spread around the world keep in constant contact with the fleet of GPS satellites, computing extremely precise orbital information and relaying this back to each satellite so that they continually know where they are and can act as precise positional reference points. Each GPS satellite transmits its position and a synchronized time signal.

A hand-held GPS receiver can calculate its location anywhere on Earth by measuring the time delay of the signals reaching it from several satellites. The signals arrive at the GPS receiver at slightly different times depending on how far away the satellite is. Once the receiver has used these times to calculate the distances to at least four GPS satellites it can calculate its position in three dimensions on Earth's surface.

EARTH RESOURCES

Following successful land-use mapping from colour pictures of Earth taken by the Apollo crews, NASA launched their first Earth Resources Technology Satellite (ERTS) on 23 July 1972. Its 900-km (560-mile) Sun-synchronous polar orbit provided a constant Sun angle on the Earth to aid comparison of images taken on different days and in different wavelengths of light. ERTS 1 was completely successful and the program, renamed Landsat, continues today.

In 1977 the USSR modified a couple of Vostok capsules to take pictures of the Earth. After a flight of two to four weeks, RESURS-F1 and F2 returned to Earth with the exposed film. Improved versions, which could transmit their pictures, were later launched into two-year polar orbits.

With support from Belgium and Sweden, France joined the Earth observation industry in 1986, with its polar-orbiting SPOT series (Satellite Pour l'Observation de la Terre). Four SPOT satellites have been launched, the latest providing colour 20-m (65-ft) resolution stereoscopic ground views.

Russia, western Europe and the US have all launched other remote-sensing satellites for oceanographic and polar observations. Space Shuttle flights and satellites such as ERS and Radarsat have carried radar for Earth mapping.

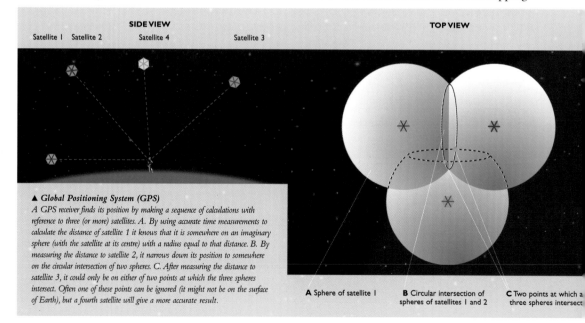

▲ *Global Positioning System (GPS)*
A GPS receiver finds its position by making a sequence of calculations with reference to three (or more) satellites. A. By using accurate time measurements to calculate the distance of satellite 1 it knows that it is somewhere on an imaginary sphere (with the satellite at its centre) with a radius equal to that distance. B. By measuring the distance to satellite 2, it narrows down its position to somewhere on the circular intersection of two spheres. C. After measuring the distance to satellite 3, it could only be on either of two points at which the three spheres intersect. Often one of these points can be ignored (it might not be on the surface of Earth), but a fourth satellite will give a more accurate result.

SIDE VIEW
Satellite 1 Satellite 2 Satellite 4 Satellite 3

TOP VIEW

A Sphere of satellite 1 **B** Circular intersection of spheres of satellites 1 and 2 **C** Two points at which a three spheres intersect

SATELLITES & PROBES

latitude ▶ p. 340 longitude ▶ p. 340 pixel ▶ p. 342

▲ *A Landsat image of the Nile Delta; the Landsat program maps the surface of Earth, allowing comparisons of images taken at different times and light frequencies.*

WEATHER SATELLITES

The US launched the first dedicated weather-watching satellite network called TIROS (Television and InfraRed Observation Satellite) in the first half of the 1960s. These first missions were polar satellites and it was not until 1974 that the first US geostationary weather satellite series was put up. Originally called SMS 1 and 2, they were renamed GOES (Geostationary Operational Environmental Satellites).

Europe's geosynchronous weather satellites are the Meteosat series. Launched on 23 November 1977, they continue to provide round-the-clock observations of Europe and Africa's atmosphere. Russia's main weather satellite program is polar orbiting, but in October 1994 they also launched a geosynchronous series called Geostationary Operational Meteorological Satellite (GOMS). Today the whole planet's weather is covered by a network of five geostationary weather satellites: Meteosat 0° **longitude** (Africa and Europe); GOES-EAST 75° west (Americas); GOES-WEST 135° west (Pacific); GMS Japanese Space Agency (140° east – covering Australasia); and GOMS 76° east (Central Asia). All of them watch the atmosphere in the visible frequencies (to see as the human eye sees), the thermal infrared (to see differences in temperature), and the VW (to see differences in the water vapour absorption).

THE COMMERCIALIZATION OF SPACE

Although the quest to reach space began with government space agencies demonstrating the power of their rockets, the importance of what was being hurled into space soon overtook the size of the rocket in significance. The first commercial satellite was for communications and the business of space soon caught on, driving the need to get into Earth orbit as cheaply and reliably as possible.

But at $24,000 per kg, reaching Earth orbit for commercial purposes requires a strong business plan. In 1998, satellite launches produced revenues of $100 billion, but some estimates suggest the sector will be worth tens of trillions of dollars by 2020: a big prize for those who can find the cheapest and most reliable routes to space. Attempts to bring costs down include Sea Launch – a consortium of companies from the US, Russia and Europe who have converted an oil rig to act as a launch platform in the Pacific, on the Equator. Here, where the Earth spins faster, they can put 30% more payload into space for the same price as launching it from elsewhere. Another American launch system, Pegasus, flies the rocket beneath a plane to 10,000 m (32,800 ft) where it is launched, reducing the amount of thrust needed to overcome Earth's gravity by 10–15%.

RECONNAISSANCE SATELLITES

The first attempts to spy from orbit were conducted by the US Air Force project Corona, which aimed to parachute a film capsule back from space a day or so after launch. After a number of failures, the first successful Corona film capsule was snatched from the air on 19 August 1959. It contained a series of fuzzy images of part of the Soviet Union previously unreachable by U2 spy planes. Details smaller than 10–12 m (33–39 ft) were not clear. By 1972, Corona had mapped 1,200 million sq km (470 million sq miles) of the Earth's surface, mostly over the Soviet Union and China, and with improvements to its cameras, it eventually resolved features 2 m (7 ft) across. In turn these countries also had their own photo-reconnaissance satellite programs to spy on America. Such satellites have also been used for more positive means: their use in treaty verification significantly helped stabilize the world during the Cold War.

In the 1980s film return satellites were discontinued. Today, spy satellite images are usually transmitted back to Earth. In the mid-1990s the US launched its KH-12 Crystal spy satellite – a kind of 18-tonne Hubble telescope pointing back at Earth and relaying real-time images. It can out-manoeuvre anti-satellite interceptors and uses image intensifiers for night observing, and infrared detectors to see buried structures and to penetrate camouflage. To penetrate clouds the US has also developed the Lacrosse satellites, which use radar to resolve features less than a metre in size beneath the clouds.

▼ *A satellite image of San Diego, showing the detail visible from orbit. These images can be processed to emphasize features such as vegetation.*

HOW SATELLITE PICTURES OF EARTH ARE MADE

The early Earth remote-sensing satellites would photograph the planet below on to film and then eject the capsule to be retrieved back on the ground. Later satellites developed the photographs on board, scanned them with TV cameras and then beamed the signals back to Earth. As optics and electronics improved, this intermediate photo-chemical stage became redundant and satellites were able to focus images at a wide spectrum of wavelengths on to light-sensitive electronics, producing digital images made up of tiny picture-element dots or **pixels**.

Infrared wavelengths are particularly helpful for vegetation monitoring. By comparing the colours of these pixels to measurements of colour or spectra on the ground, satellite images can be classified into different land-use areas. Taking this type of image classification further, subtle differences in the colour of vegetation under different environmental conditions can be used to detect stressed plants which could be suffering from adverse environmental conditions or disease. In unvegetated regions different rock types also have different spectral characteristics, allowing geological mapping and mineral prospecting. Using other wavelengths, details of atmospheric chemistry – such as ozone measurements, water vapour content and ocean or land surface temperatures – can also be imaged.

SPACE EXPLORATION

Space Exploration
HUMAN SPACEFLIGHT
THE PIONEERS

The first recorded attempt at human flight by rockets was in 1500, when the Chinese pioneer Wan-Hu attached 47 black powder rockets to a chair suspended by two large kites. He lit all the rockets simultaneously and disappeared in the resulting explosion. Johannes Kepler resurrected the idea of rocket travel in his book *Dream*, published posthumously in 1634. In subsequent centuries many science-fiction writers followed in Kepler's footsteps, describing manned rockets reaching escape velocity and orbiting rotating space stations to counteract weightlessness. Riding ballistic missiles to dizzying heights would eventually become routine, but it would always remain a dangerous and expensive government-sponsored pursuit, reserved for highly trained and dedicated individuals.

▼ *Valentina Tereshkova, the first woman in space, was 26 years old when she was launched on board Vostok 6 on 16 June 1963.*

VALENTINA TERESHKOVA
Valentina Tereshkova (b. 1937) was the first woman in space. On 16 June 1963, the 26-year-old was launched into orbit aboard Vostok 6 and in the next

GAGARIN'S ORBITAL FLIGHT: VOSTOK I
The first spacecraft to take a human into orbit was a spherical capsule called Vostok: 2.3 m (8 ft) in diameter and coated in protective material so that it could re-enter the atmosphere at any orientation without burning up. Five robotic Vostok flights from May 1960 paved the way for the first human mission. Just a few weeks before the first proposed American manned space flight – on 12 April 1961 – a 27-year-old Russian pilot named Yuri Gagarin climbed into his Vostok 1 spacecraft and at 04:10 UT was blasted into **orbit** by **Korolev**'s SL3 rocket. Seventy-eight minutes into the flight, mission control fired the retro-rockets and Gagarin began his descent. The Vostoks had no retrorockets for a soft-landing, and so 8,000 m (26,250 ft) above the ground Gagarin ejected from the capsule and parachuted to Earth, landing on the banks of the Volga not far from Engels. It would be over 10 months before the Americans could improve on this pioneering flight.

ALAN SHEPARD'S SUBORBITAL MERCURY FLIGHT
Alan Shepard was selected as America's first astronaut. He named his spacecraft Freedom 7. The launch from Cape Canaveral in Florida, was originally scheduled for 2 May 1961 but it was postponed until 5 May because of bad weather. At 5.15 a.m. EST, Shepard entered Freedom 7 and over four hours later the Redstone rocket propelled him to 8,214 km/h (5,103 mi/h) before shutting down on schedule 142 seconds after lift-off and separating from the capsule. Shepard manoeuvred the craft about all three axes and tested the reaction control systems. Freedom 7 reached a peak altitude of 187 km (117 miles), before Shepard prepared for re-entry, splashing down in the Atlantic Ocean 15 minutes and 22 seconds after launch.

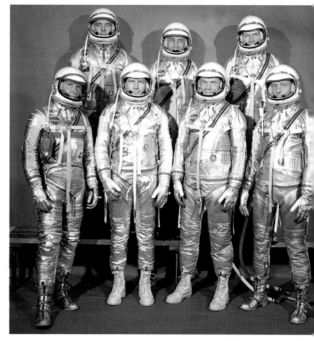

▲ *The Mercury 7 astronauts, who took part in the first US program for human spaceflight. The missions' objectives were to place manned spacecraft in orbital flight around Earth and investigate man's performance capabilities in a space environment.*

OTHER MERCURY FLIGHTS
On 21 July 1961 Virgil 'Gus' Grissom made the second suborbital Mercury flight in a capsule called Liberty Bell 7. All went well until splash-down in the Atlantic after a 15-minute flight. The capsule's hatch burst off and water flooded in, nearly drowning Grissom and sinking Liberty Bell. It rested there until 1999, when it was finally brought to the surface.

After booster problems on a Mercury chimp test flight, rocket designer **von Braun** insisted on a further unmanned booster test flight, and against the wishes of Shepard and others at NASA, an empty Mercury capsule was launched on a flawless test on 24 March. If NASA had overruled von Braun, Shepard would have been the first man in space (though not in orbit), beating Gagarin's flight by three weeks.

three days circled the Earth 48 times, more than the six American Mercury astronauts combined.

Tereshkova was born in the village of Masslenikovo and, at the age of 18, joined her mother and sister to work at the Red Canal textile mill. She studied at the same time and joined a club for parachutists, eventually making over 120 jumps.

In September 1961, inspired by the Vostok flights, Tereshkova wrote a letter to the space centre asking to join the cosmonaut team. With her application helped by her parachuting background she was selected in

March 1962 and, along with four other women, reported to the training centre. Soviet Premier Krushchev personally picked her out for the Vostok 6 flight because of her working-class background.

Famed as a heroine of the women's movement in Soviet society, she went on to a career in politics and married fellow cosmonaut Andrian Nikoleyev. Tereshkova later earned a Candidate of Technical Sciences degree (in 1976) and was eventually promoted to the rank of major general, retiring in March 1997.

THE RACE TO THE MOON

In April 1961 not only was US President Kennedy being humiliated by Soviet triumphs in space, but back on Earth the Soviet-supported communist leader of Cuba, Fidel Castro, also offered a threat – on America's doorstep. Searching for a science stunt to trump the Russians, Kennedy asked his vice-president Lyndon Johnson what move to make next. There were four options: to launch a laboratory into Earth orbit; a trip around the Moon; a rocket landing on the Moon; or a mission to take a man to the Moon and bring him back. America was still only sending apes into space, whereas Russia had orbited a man and sent probes to crash on the Moon and photograph its far side. Only with the last option did the inferior American space program stand a chance of beating the superior Soviets. A Moon race would tax both Superpowers to their limits – and Johnson recommended exactly this. On 25 May 1961, in a 47-minute address to a joint session of Congress, Kennedy declared his belief in landing a man on the Moon and returning him safely to Earth. 'No single space project in this period will be more exciting, or more impressive to mankind, or more important for the long-range exploration of space; and none will be so difficult or expensive to accomplish'. With only 15 minutes of suborbital flight experience in space, NASA was committed to its greatest challenge ever.

JOHN GLENN'S ORBITAL MERCURY FLIGHT

The first Mercury flight to complete an Earth orbit took place on 13 September 1961, carrying a simulated astronaut. After Enos the chimpanzee's flight in November, NASA decided that the Mercury spacecraft was finally ready for a manned flight, and on 20 February 1962 John H. Glenn Jr (b. 1921), became the first American to orbit Earth. On the day of the launch, Glenn boarded his Mercury capsule Friendship 7 at 6.06 a.m. EST. Minor problems delayed the launch several times, but the countdown was completed at 9.47 a.m. and the engines of the Atlas rocket ignited. Five minutes later Glenn was in orbit. Near the end of his first orbit, Friendship 7 was drifting slowly to the right and Glenn switched to manual to correct the problem. On the third orbit, mission control noticed a signal suggesting that the heat shield and landing bag were loose. Glenn was told to keep the retrorocket package on after firing so that its straps would hold the heat shield in place. By the time the pack burned away, aerodynamic pressure would keep the shield from slipping. The retrofire began after four hours

◀ *John Glenn, the first American to orbit Earth. He orbited in his Mercury capsule, and returned to space on the Shuttle in 1999.*

and 33 minutes and during re-entry Glenn observed large chunks of the retrorocket package ablating. Friendship 7 splashed down 22 minutes later.

On 24 May 1962, Scott Carpenter (b. 1925) flew three orbits in a capsule named Aurora 7 and, after a successful scientific mission, landed dangerously short of fuel, 350 km (216 miles) off target. He had to wait almost three hours to be recovered. On 3 October 1962 Walter Schirra (b. 1923) completed a six-orbit flight in Sigma 7, and finally the following year, on 15 May 1963, Gordon Cooper (b. 1927) flew the last Mercury flight in the bold 22-orbit mission of Faith 7.

FURTHER VOSTOK FLIGHTS

Vostok 2 was intended to spend a whole day in space and the 36-year-old jet pilot Gherman Titov (1935–2000) was selected for the mission. He launched on 6 August 1961 and was able to eat, sleep and fly the spacecraft himself. The next day Titov returned – the first to spend over a day in space. A year later, on 11 August 1962, Andrian Nikoleyev (b. 1929) on Vostok 3 and Pavel Popovich (b. 1930) on Vostok 4 were launched a day apart and passed within a few kilometres of each other, before both returning to Earth on 15 August. The following year, Valery Bykovsky (b. 1934) on Vostok 5 was launched on 14 June 1963, and two days later the first woman cosmonaut – 26-year-old Valentina Tereshkova – joined him in Earth orbit just 5 km (3 miles) apart. Tereshkova remained in space for three days, Bykovsky for five: a new record which was never beaten for a single-seater spacecraft.

▼ *Angle of Re-entry*
The angle 'window' through which a space capsule can successfully re-enter Earth's atmosphere is extremely small. Apollo 11, for example, had to enter at between 5.2° and 7.2° to the horizontal to ensure the survival of the crew and the recovery of the craft.

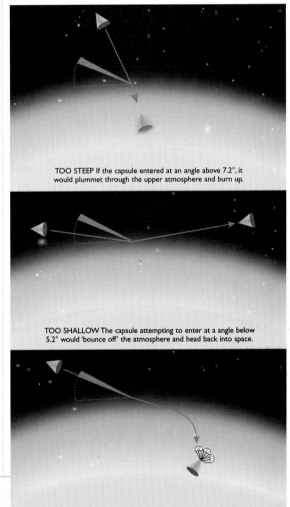

TOO STEEP If the capsule entered at an angle above 7.2°, it would plummet through the upper atmosphere and burn up.

TOO SHALLOW The capsule attempting to enter at a angle below 5.2° would 'bounce off' the atmosphere and head back into space.

CORRECT ANGLE The capsule entering at between 5.2° and 7.2° will be capable of manoeuvring in the atmosphere, altering its trajectory and reducing its speed to allow it to splash down in the ocean with the aid of parachutes.

ORBITS AND RETURNING TO EARTH

The principles of flight and spaceflight dictate that the faster you are travelling the higher you can climb, and at slower speeds you lose height. To orbit Earth at a height of around 320 km (200 miles), a spacecraft needs to travel at 28,000 km/h (17,396 mi/h), so that it drops at the same rate as the Earth is curved. If a spacecraft slows down below this speed, it drops at a steeper angle and no longer stays in orbit, falling back towards Earth. This is the principle of re-entry – using a retrorocket to slow the spacecraft below 28,000 km/h and causing it to drop out of orbit and back into the atmosphere. Hitting the top of the air, the capsule gets heated by friction, slowing it to 480 km/h (298 mi/h) and heating the shield to 3,000°C (7,000°F) – enveloping the spacecraft in plasma and causing a radio blackout. Returning from the Moon straight back to Earth, without first entering orbit, the angle of re-entry is more crucial and should be between 5.2° and 7.2°. Any steeper and you risk falling too fast and burning up; too gentle and you risk bouncing off the top of the atmosphere and back out into space.

SPACE EXPLORATION

MULTI-PERSON MISSIONS

With single-seater spacecraft now making multiple Earth orbits for several days at a time, and the gauntlet of going to the Moon having been laid down, the manned Space Race had entered its second phase with the challenge of launching multi-crewed spacecraft. Russia was struggling to stay ahead and compromised the safety of its crew to cling to their lead. Missions became almost routine in the mid-1960s, but the first casualties of the race to space were imminent and both Superpowers were to be cursed with fatalities in their quest to reach the final finishing post – the Moon.

VOSKHOD: THREE MEN IN A ONE-MAN CAPSULE

With the delay of their next-generation manned spacecraft, Soyuz, Russia's only chance of beating the two-man American Gemini program into orbit was to modify the existing one-man Vostok craft so it could take three cosmonauts. Three couches replaced the single ejection seat and, after one unmanned test flight, three brave cosmonauts squeezed into the tiny Voskhod 1 capsule without pressure suits, and were blasted into orbit on 12 October 1964. Vladimir Komarov (1927–67) commanded the flight, which lasted just one day. His two cramped colleagues were medic Boris Yegorov (1937–94) and design engineer Konstantin Feoktistov (b. 1926). With barely any space to

move, little of scientific value was accomplished, but Russia, it seemed, was still ahead in the race.

Voskhod 2 was launched on 18 March 1965 with just two pressure-suited cosmonauts – commander Pavel Belyayev (1925–70) and Alexei Leonov. About 90 minutes after launch, Leonov passed through an airlock wearing a prototype moonwalking space suit and, tethered to Voskhod on a 5-m safety line, became the first person to 'walk' in space. On trying to get back inside after 20 minutes his suit ballooned up, trapping him outside. He could only rejoin Belyayev by letting air out of the suit. After a 24-hour flight, trouble with re-entry put them 1,000 km (620 miles) off course and they landed in a snowy forest, spending a night with the wolves before being rescued.

THE GEMINI PROGRAM

Undeterred by the Russian Voskhod stunts, the American space program was more meticulous in its advances. After two unpiloted Gemini flights in early 1964, 10 manned missions followed, pioneering complex techniques which would overtake the Russians and carry NASA to the Moon. The first manned flight, Gemini 3 on 23 March 1965, was commanded by Mercury veteran Gus Grissom (1926–67) – the first man to fly to space twice (in this case with co-pilot **John Young**, b. 1930). In

◀ *American astronaut Edward White,* floating free in space during his flight which lasted four days on board Gemini 4 in June 1965.

▼ *The Gemini 7 capsule* as seen from the windows of the Gemini 6 spacecraft. Gemini 6 had been sent out to meet up with Gemini 7; the two crews were able to see each other through the spacecraft windows.

another first for the mission, Gemini 3 used its manoeuvring engines to change orbit.

The Gemini 4 flight on 3 June 1965 kept Jim McDivitt (b. 1929) and Edward White (1930–67) in orbit for a record four days and White became the first American to walk in space. Gemini 6 carried Walter Schirra and Thomas Stafford (b. 1930) to orbit in December, where they manoeuvred to within 2 m (7 ft) of Gemini 7, which had been launched 11 days earlier. The two craft flew in formation for 20 hours – the crews in visual contact through the spacecraft's windows. Gemini 6 returned to Earth after almost 26 hours, leaving Gemini 7's Frank Borman (b. 1928) and James Lovell (b. 1928) to set a new duration record of almost 14 days – a record which would stand for five years.

GEMINIS 8–12

Gemini 8 came close to disaster in March 1966 when a faulty attitude thruster began to misfire shortly after docking with an Agena rocket. The resulting spin nearly caused the crew to black out, but **Neil Armstrong** (b. 1930) managed to control the spin using the re-entry control system. In June 1966 Stafford and Cernan tried again to dock with an Agena rocket on board Gemini 9, but this time problems with the Agena stopped the docking. During a two-hour spacewalk Cernan became dangerously overheated and had trouble returning to the spacecraft.

Geminis 10–12 perfected the spacewalking technique and achieved several more perfect practice dockings with the Agena rocket boosters. On Gemini 11 Charles Conrad (1930–99) and Richard Gordon (b. 1929) used the Agena to set a new altitude record of 1,369 km (851 miles) from Earth. On his second flight, on board Gemini 12, Lovell became the most travelled man in space – notching up a total of 265 orbits. By the end of 1966, and for the first time since Sputnik 1's pioneering flight, America had the edge in space – with enough experience to fulfil Kennedy's dream of reaching the Moon and returning safely to Earth before the decade was out.

ALEXEI LEONOV

Alexei Leonov (b. 1934) became the first man to walk in space when he floated outside the Voskhod 2 spacecraft for 10 minutes on 18 March 1965. Born in the Siberian village of Listvyanka, he came from a large family and decided to become a pilot after one of his older brothers became an air-force mechanic. After attending an air-force school he served as a fighter pilot before being selected as a cosmonaut in March 1960. Initially a candidate for the first Vostok spaceflight, Leonov was discounted in favour of Gagarin and Titov, who were several inches shorter than him.

After his Voskhod 2 flight, Leonov went on to train for the Russian manned circumlunar and lunar

landing missions, and could have become the first man to walk on the Moon. The rest of his cosmonaut career was spent on Earth orbital missions, but he did not fly again until the Apollo–Soyuz test flight in 1975. Leonov served as deputy director of the Gagarin Center for cosmonaut training until his retirement in October 1991. He left military service in March 1992 and is now a principal in the Alfa-Kapital investment fund in Moscow.

▶ *Having been overlooked* from the first Vostok spaceflight in favour of Gagarin and Titov, Alexei Leonov eventually had his moment of glory – making history in March 1965 as the first man to walk in space.

SPACE EXPLORATION

THE EARLY SOYUZ PROGRAMS

Soyuz was conceived as a three-module lunar spaceship by Soviet chief designer Korolev. In 1967 the Apollo program was in trouble, after a serious fire on the launch pad had killed the entire crew of Apollo 1. By April that year the Soyuz program was poised to regain the lead in the race to the Moon. The plan was to dock two Soyuz capsules in orbit and transfer two crew members between spacecraft. Soyuz 2's launch was postponed after Soyuz 1 struck trouble in orbit, eventually crashing when its landing parachutes failed while returning to Earth, killing cosmonaut Komarov. Soyuz 2 eventually flew a year later, as an unmanned target for Georgi Beregovoi (1921–95) flying Soyuz 3. In 1969, Soyuz 4 and 5 were repeats of the first Soyuz missions to practise docking and crew transfer and are thought to have been a rehearsal for their up-coming lunar missions. But by July 1969 Apollo 11 had already reached the Moon and Soyuz 6, 7 and 8 were launched together in October to practise manoeuvring. Soyuz 9, the following summer, set a new endurance record when its crew Andrian Nikoleyev and Vitaly Sevastyanov (b. 1935) remained in orbit for almost 18 days. Further Soyuz flights would be simple ferrying missions to the Salyut and – later – the Mir space stations.

WALKING IN SPACE
EVAs: PERILS AND PRIZES

These days the regularity and apparent ease of spacewalks from the Shuttle's cargo bay and around the International Space Station disguise the difficulties and dangers of such feats, hundreds of kilometres above the Earth. Emerging from the relative safety of a spacecraft while still in orbit places a spacewalker in the line of any debris or micrometeorites and exposes them directly to the hazards of **radiation**. Beyond these threats is the difficulty of manoeuvring with nothing to push against. Although the tether that attached Leonov to his Voskhod capsule had got him into difficulty when trying to re-enter the airlock of the Voskhod spacecraft, Ed White's walk a few months later from Gemini 4 went a lot more smoothly. On later Gemini walks both Cernan and Gordon had found out to their cost how exhausting and potentially dangerous it was to work outside their spacecraft. They both survived their close shaves, but their missions had carried contingency plans for re-entry attempts with a dead crew member abandoned in space.

▶ *A Soviet N-1 rocket booster, designed and built for Russia's manned Moon program. It was similar in size to the US Saturn V, but never flew successfully and was eventually abandoned.*

💡 DICING WITH DEATH: PART 1

Every new Gemini, Voskhod and Soyuz flight entered uncharted territory of spaceflight and the crews that flew these pioneering missions were very aware of the perils. Neil Armstrong and Dave Scott had come close to death when their Gemini 8 capsule went into a severe 60-rpm spin. It was only Armstrong's icy calm that saved their lives and the mission.

Eugene Cernan's Gemini 9 flight made him the second American to walk in space, but he almost paid for this experience with his life. Wrestling the umbilical cord and working against an over-stiff pressure suit, Cernan became dangerously over-heated, steaming up his visor and tripling his heart rate to 180 beats per minute.

Almost as soon as Vladimir Komarov reached orbit in Soyuz 1 things started to go wrong. One of the solar panels failed to deploy and attempts to stabilize the spacecraft failed. He was ordered to come back to Earth after his eighteenth orbit, but during descent the first parachute failed and the reserve chute became tangled. Soyuz 1 smacked into the ground at almost 500 km/h (300 mi/h), killing Komarov instantly.

THE RACE TO THE MOON

With only three years left before President Kennedy's deadline to reach the Moon ran out, both Superpowers' manned space programs were in trouble. A fire during a routine training exercise on the launch pad had killed the three-man crew of Apollo 1 in January 1967 and by April Russia's prestigious Soyuz lunar spacecraft program had killed its sole cosmonaut on its maiden flight. Public and political support for both programs was at an all-time low and the Moon seemed further away than ever. But with only five months to their deadline and spurred on by their competitors and a desire to honour a dead president's words, Apollo finally delivered the first humans to the Moon.

NEIL ARMSTRONG

Neil Armstrong (b. 1930) began flying at the age of 14. He had a pilot's licence by the age of 16, before he could drive. Colleagues said his love of flight bordered on religious devotion. Armstrong apparently even carried a piece of the original Wright Brothers' flier on board his Gemini 8 mission and treated it like a religious relic. Armstrong joined the Navy as a fighter pilot in 1949, flying 78 combat missions in Korea. On one mission he had to bail out when a wire stretched across a valley took the wing off his F9F-2 jet plane.

Back in the US he studied for an aeronautical engineering degree and then joined Edwards Air Force Base as a test pilot in 1955, eventually flying the X-15 rocket plane to a height of 69 km (43 miles) at a speed of Mach 4. NASA selected him in 1962 and he flew on Gemini 8 and Apollo 11. Armstrong stayed with NASA after his pioneering Moon walk, but he never flew into space again and left in 1971 to pursue business and academic interests.

APOLLO 1

On the afternoon of 27 January 1967, the primary crew for Apollo 1 were simulating a countdown on internal power when a spark from the wires set fire to the interior of their capsule. At the high oxygen pressures used, it quickly became an inferno and the inward-opening door prevented their escape. Within minutes the crew had all perished and seemingly with them the chances of reaching the Moon. To prevent a repeat of the disaster, the command module was equipped with more fire-resistant materials and the escape hatch was redesigned to allow it to be opened from the inside. The pure oxygen was replaced with a mix of nitrogen and oxygen, limiting the chances of a fast-spreading fire.

RUSSIANS TO THE MOON

Soviet plans for a manned lunar landing were based on a new super-rocket called the N-1 booster. It had five stages, on top of which sat a lunar landing craft known as the LOK and a two-man Soyuz spacecraft. Once in lunar orbit, a single cosmonaut would make a spacewalk from the Soyuz capsule to enter the lander which would descend to the lunar surface.

A modified Soyuz, known as Zond, was to be used for lunar orbital flights. The Zond's own rockets were not powerful enough to brake the craft in lunar orbit, but Zond could swing around the Moon and return to Earth, allowing the USSR to claim the first flights into deep space. Zond 5 took a cargo of turtles and other biological specimens around the Moon on 18 September 1968, but suffered a severe 10–15 G re-entry. Zond 6 accomplished a second circumlunar biology flight in November and also managed a more gentle re-entry trajectory suitable for a human flight, but its parachutes deployed too early and

▲ **Tragedy struck** the Apollo program on 27 January 1967, when a flash fire flared up in the command module of the Apollo 1 spacecraft during a launch-pad test, killing the entire crew.

were damaged, crushing the capsule on landing. Despite these setbacks, six cosmonauts arrived at Baikonur on 23 November for a human Zond mission around the Moon. NASA decided to make up for the time lost by the Apollo 1 disaster and change the flight plan of Apollo 8.

APOLLO 8

On 21 December 1968, and after only one manned Apollo test flight, Frank Borman, William Anders and James Lovell became the first humans to leave Earth orbit – riding the first Saturn V rocket in Apollo 8 to the far side of the Moon. Their trajectory placed them in lunar orbit on Christmas Eve. During their 20 hours and 10 lunar orbits, the three astronauts beamed back TV pictures of the views and read a Bible passage from Genesis, to a live audience of over a billion on Earth. The three heroes returned on 27 December 1968. The route to the Moon was open, but there was one ingredient left to be tested.

APOLLOS 9 AND 10

At the beginning of 1969, the Lunar Module (LM) which would descend to the Moon's surface and serve as a temporary lunar base had still not flown in space. Nor had the new Apollo space suits. Jim McDivitt, Rusty Schweickart (b. 1935) and Dave Scott (b. 1932) were about to put that right. In Earth orbit, on 3 March that year, Scott and Schweickart climbed into an LM called Spider and flew it, later completing a spacewalk in the new suits. Two months later Apollo 10's LM Snoopy, carrying Stafford and Cernan, descended to only 15 km (9 miles) from the lunar surface in a final dress rehearsal.

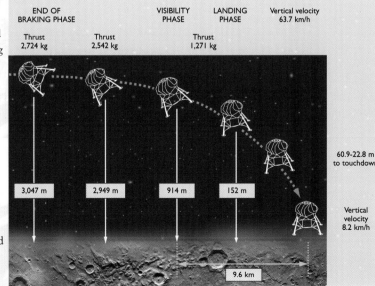

END OF BRAKING PHASE		VISIBILITY PHASE	LANDING PHASE	Vertical velocity 63.7 km/h
Thrust 2,724 kg	Thrust 2,542 kg		Thrust 1,271 kg	
3,047 m	2,949 m	914 m	152 m	60.9-22.8 m to touchdown
				Vertical velocity 8.2 km/h
			9.6 km	

◄ **Lunar Descent Trajectory**
This diagram shows the nominal descent trajectory of an Apollo lunar module. The thrust of its engines is used to reduce speed, to alter its angle with respect to the lunar surface and manoeuvre it into a nearly vertical attitude to make a controlled landing.

solar wind ► p. 343

SPACE EXPLORATION

APOLLO 13: NASA'S FINEST HOUR?

When James Lovell, Jack Swigert (1931–82) and Fred Haise (b. 1933) embarked on the third intended Apollo landing mission on 11 April 1970, the world was not paying much attention. But almost 56 hours into the flight of Apollo 13 a routine operation to stir the liquid-oxygen tanks resulted in an explosion which shut down two fuel cells and began to leak oxygen into space. Using the LM engines to fire them around the Moon and back towards Earth, they shut down the damaged command module and retreated into the LM. The next nerve-wracking 86 hours were survived by switching everything off except the environmental control and communications systems. Temperatures dropped to freezing and ice formed inside the spacecraft, making life dangerous and miserable. When exhaled gases threatened to poison them, mission control helped improvise a carbon dioxide scrubber. On reaching Earth they returned to the command module and splashed down after almost six days in space, surviving the disaster against the odds.

▲ **Astronaut Buzz Aldrin** *steps out on to the 'magnificent desolation' of the Sea of Tranquillity from the lunar module Eagle, during the historic Apollo 11 mission to the Moon. A solar wind experiment is to his right.*

APOLLO 11: 'MAGNIFICENT DESOLATION'

On 20 July 1969 the first Moon landing was poised to touch down on the Sea of Tranquillity. On board, fate had determined that Neil Armstrong and Edwin 'Buzz' Aldrin would be the first. Listening in on their progress in the command module above was Michael Collins. At 20:17 UT, with just 20 seconds of fuel left, Armstrong touched down 19 km (12 miles) south-south-west of crater Sabine D. At the top of the ladder, over six hours later, he pulled a lever that deployed an instrument platform on which a TV camera would film the historic step. Nine rungs down at the foot of the ladder, and now standing on the pad, Neil described the texture of the surface as 'very fine and powdery'. With a TV audience of a billion hanging on his next words Armstrong placed his left foot gingerly on the surface and, still holding on to the ladder, declared 'that's one small step for man, one giant leap for mankind'.

Aldrin followed him down the steps 15 minutes later and the pair busied themselves setting up seismic, **solar wind** and laser-ranging experiments, collecting more samples and taking pictures. They also conducted a brief telephone conversation with President Nixon and unveiled a plaque on the legs of the LM celebrating the achievement of Apollo 'for all mankind'.

After two hours and 31 minutes on the surface they returned to the lander and rested before firing the single rocket motor to blast the upper stage of the LM back into lunar orbit to rejoin Collins. Carrying their precious cargo of 22 kg (49 lb) of lunar samples, they splashed down 21 km (13 miles) from the aircraft carrier Hornet 1,529 km (950 miles) south-west of Hawaii on 24 July. President Nixon called it 'the greatest week since the Creation…'.

APOLLOS 12–14

Just four months later Apollo 12 returned to the Moon, making a precise landing in the Ocean of Storms, close to the unmanned Surveyor 3 spacecraft. Their colour TV camera was damaged and without pictures public interest waned. Curiosity was briefly rekindled when an explosion on Apollo 13 put the lives of the crew at risk in April 1970. A mixture of luck, the ingenuity of the mission controllers and the resilience of the crew eventually returned them safely to Earth. The spacecraft was redesigned before Apollo 14, commanded by America's first astronaut Alan Shepard, touched down in the Fra Mauro highlands 10 months later.

APOLLOS 15–17 AND THE LUNAR ROVER VEHICLE

Budget cutbacks led to the cancellation of Apollos 18–20, leaving only three more chances of visiting the Moon. These 'J class' missions flew an improved lunar module which could carry more equipment and remain on the surface for up to three days. A battery powered lunar rover

LUNAR SCIENCE AND THE ORIGINS OF THE MOON

One of the goals of Apollo was to determine the Moon's origins. Before Apollo, scientists already had a reasonable idea from the Surveyor missions that the lunar surface was made mostly of a volcanic rock called basalt. The 380 kg (840 lb) of samples brought back by Apollo confirmed this and emphasized the importance that heat played in the Moon's formation.

Although oxygen isotopes exist in similar proportions in both Earth and lunar rocks, there were important differences. Some of the lightest elements in the Earth's crust were far less common in Moon rock samples. Water, so abundant in Earth's crust, was absent from the Moon rocks. Two years after Apollo, a planetary scientist called Bill Hartmann, pondering these findings, proposed that a Mars-sized planet had collided with the early Earth, annihilating itself and shaving off a portion of the Earth's crust, which coalesced into a molten Moon. This would account for the lack of water and light elements in the Moon's rocks which had vaporized in the super-heated birth. Chemical similarities with Earth were due to its merged beginnings. We will never know for sure how the Moon came to be, but Hartmann's theory certainly seems to be supported by present information.

vehicle also allowed crews to travel a lot further and be more productive. On the final mission geologist Harrison Schmitt was sent – the only scientist to visit the Moon. Crewmate Eugene Cernan became the last human to date to set foot on the surface. Apollo 17 blasted off from the Taurus Littrow Valley at 5.55 EST (10.55 UT) on 14 December 1972.

▼ **Apollo 17** *in December 1972 was the last manned lunar mission. Here, geologist Harrison Schmitt is dwarfed by a large boulder. In the foreground is the astronauts' electrically powered lunar car.*

POST-APOLLO SPACE STATIONS

With the race to the Moon run and won, both Superpowers began to look for another use for their Moon program technology. While the truncated Apollo lunar program ran its final course, Russia's Soyuz spacecraft were redeveloped into ferries to Earth-orbiting space stations. With America's Space Shuttle program still a decade away from its maiden flight, the challenge of a long-duration human presence in orbit was something to keep the US space program ticking over, too. Unlike the Moon race, this challenge was something that would bring both space-faring nations together with the ultimate handshake of reconciliation on board an Apollo and Soyuz docked together.

SKYLAB

Determined to build on technology developed for Apollo, NASA came up with the Apollo Applications Program, later renamed Skylab. The space station, a converted Saturn rocket stage, would be placed into orbit by the first two stages of a Saturn V rocket. The workshop was divided into two levels by a wire-mesh flooring. The lower section contained the living quarters – a communal area, a toilet, a kitchen area, individual bedrooms and a larger area for exercising. The upper section was the laboratory set aside for scientific experiments. A series of airlocks and docking modules allowed other experiments to be exposed to space and astronauts to climb outside. Skylab 1 was

▲ *Skylab built on the technology developed for the Apollo program. This self-contained space station comprised living areas and a science laboratory for conducting experiments.*

disastrously launched on 14 May 1973. Within seconds a thin meteoroid Sun shield, designed to swing into position after launch, opened too early and was torn off, dragging one solar panel with it and jamming a second panel shut. Once in orbit, four telescope solar panels still deployed and were able to generate enough power to switch all the systems on but, without the shield, temperatures began to rise and the first manned mission scheduled for the next day had to be postponed.

SKYLAB MISSIONS SL2–4

Charles Conrad, who had already visited the Moon on Apollo 12, commanded the first repair mission to Skylab on 25 May 1973. Docking the next day, the crew entered the overheated chamber and pushed a thermal blanket out of one of the air locks to act as a temporary sunshade. During a three-and-a-half-hour spacewalk they finally managed to open the last solar panel and,

with Skylab fully operational, the crew stayed for 28 days, setting a new endurance record.

Alan Bean's (b. 1932) crew arrived in July and undertook solar and stellar astronomy, Earth observations, material science and space medicine studies. During another record-breaking 59-day stay in space they also deployed a more permanent sunshield and undertook several spacewalks to change the film in the solar telescope. The third and final Skylab crew arrived in November 1973 and after a bout of space sickness they gained another space endurance record of 84 days and studied the effect of this long stay on their bodies.

Delays in the Shuttle program, however, and more atmospheric drag than expected meant that Skylab crashed back to Earth. At 12.20 a.m. on 12 July 1979 it lit up the sky. Although most of the 75-tonne space station crashed into the Indian Ocean, some fragments – including a one-tonne oxygen tank – landed in Australia.

DEKE SLAYTON

Donald 'Deke' Slayton (1924–93) was born in Sparta, Wisconsin and grew up on a farm there. At the age of 18 he enlisted in the Army Air Corps and earned his pilot's wings a year later, flying missions over Europe and Japan. Returning to the US after World War II he received a BSc in aeronautical engineering in 1949 and joined the Boeing Company in Seattle, Washington. After a period with the Air Force during the Korean War, Slayton joined the Air Force Test Pilot School at Edwards Air Force Base, California, and was a test pilot there when NASA selected him along with six other Mercury astronauts in 1959. Slayton should

have flown the second American orbital mission, but he was grounded because of an irregular heartbeat. He became head of the Astronaut Office – selecting and assigning astronauts throughout the Gemini and Apollo programs. In 1972 he was returned to flight status and eventually flew on the international Apollo–Soyuz mission in July 1975. A fuel leak during splashdown choked the crew and, luckily for Slayton, during the examination a lung tumour was found and removed. After his Apollo–Soyuz flight Slayton managed the Shuttle testing program until 1981 when he retired from NASA. Slayton died of brain cancer in June 1993.

SALYUT SPACE STATIONS 1–5

Like Skylab, the first Soviet attempt at a space station – Salyut 1 – was a modified upper stage of a rocket containing living and work room, but with only about a quarter Skylab's volume. Salyut 1 was launched by a Proton rocket on 19 April 1971 and the first crew docked in Soyuz 10 on 23 April 1971. A malfunction prevented them entering the station and they returned to Earth. In June a second crew arrived, staying for over three weeks. On their way back to Earth, though, an air leak resulted in their deaths. Salyut 1 was abandoned and burnt up on 11 October 1971.

ORBITAL HANDSHAKES
OVER BOGNOR REGIS

At 12:18 UT on 17 July 1975, 225 km (140 miles) above the coast of southern England a satisfying clunk signalled the docking of what NASA called Apollo-Soyuz and what Russia referred to as Soyuz–Apollo. Stafford called out 'Contact' and a moment later Leonov called back 'Capture … Apollo and Soyuz are shaking hands now'. But the real handshakes and hugs would have to wait a few minutes for the atmospheres between spacecraft to adjust. The five were already good friends and had met many times before to familiarize themselves with each other's equipment and languages.

Stafford and Slayton floated into the Soyuz and Leonov played host as messages were broadcast to the crews from Soviet leader Leonid Brezhnev and US president Gerald Ford. Both spacecraft remained docked together for two days – separating on 19 July to practise re-docking once more before parting permanently to return to Earth. Soyuz landed on the morning of 21 July seven miles from its target. Apollo remained in orbit to conduct observations and experiments before splashing down on 24 July.

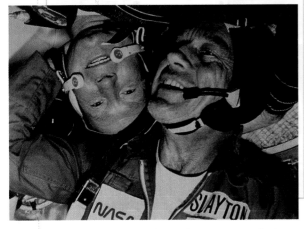

▲ *American astronaut* Deke Slayton and Russian cosmonaut Alexei Leonov celebrate during the Apollo–Soyuz linkup on 17 July 1975.

Salyuts 2, 3 and 5 were military missions, placed into lower orbits for surveillance reasons with unmanned re-entry capsules to return film to the ground. Salyut 4, the next civilian mission, carried a large solar telescope. Alexei Gubarev (b. 1931) and Georgi Grechko arrived in January 1975 for a month-long program of science and set a new Soviet space endurance record. After one more failed attempt to reach Salyut 4 – the last crew arrived in May 1975 and stayed for another record-breaking 63 days. Conditions on board Salyut 4 are reported to have been rather unpleasant, with dampness problems and green mould growing on the walls.

DICING WITH DEATH:
PART 2

Despite many close shaves, no one had been killed in the quest to explore space since 1967, but that would change in 1971. Georgi Dobrovolsky, Vladislav Volkov and Viktor Patsayev had spent a highly successful pioneering stay on board Salyut 1. On 29 June they transferred back to Soyuz 11 and at 18:15 UT they undocked. All three would be dead within four-and-a-half hours. A retrorocket fired to drop them back to Earth and the orbital module separated from the descent module. The shock of separation opened a pressure valve in the descent module, releasing air from the cabin and suffocating all three cosmonauts. The capsule-coffin landed safely less than an hour later.

A second Soviet disaster was averted after a Soyuz launch went badly wrong on 27 September 1983. One minute from lift-off, a fire broke out at the base of the rocket; just as it exploded on the launch pad the escape rocket plucked the Soyuz capsule from the inferno, soaring into the air. Bruised and shaken, Vladimir Titov and Gennadi Strekhalov parachuted back to Earth 4 km (3 miles) from the wrecked launch pad.

SOYUZ 10–18 AND
THE COSMOS MISSIONS

After the launch of Salyut 1, Soyuz had become a kind of space ferry, shuttling crews between Earth and the orbiting labs. But after the Soyuz 11 crew were killed when a valve opened, suddenly releasing their air supply, it was decided that space suits must be worn for complex manoeuvres and for re-entry. To make room for this, the three-man design was changed to a two-man capsule and solar panels were removed. The first test flight of Soyuz 12 was conducted on 27 September 1973. Soyuz 13 flew a dedicated astronomy mission and Soyuz 16 was flown as a practice for the Apollo-Soyuz project.

APOLLO–SOYUZ/SOYUZ–APOLLO

With the main event in the Space Race over, a new spirit of collaboration between the two space-faring nations had developed. At a Superpower summit in May 1972, an agreement was signed to launch a joint mission in July 1975. Soyuz 19 lifted off perfectly at 12:08 UT on 15 July 1975, and was followed seven hours and 22 minutes later by an Apollo spacecraft, boosted to over 160 km (100 miles). After over 50 hours of complex manoeuvres, both spacecraft were in the same orbit and visual contact was established when they were still over 80 km (50 miles) away. Once together the actual docking was completed four minutes ahead of schedule. Astronauts Stafford and Slayton sealed themselves into the docking module and began to adjust the air mix and pressure.

After two minutes, the single remaining hatch separating the two crews was opened and Soyuz commander Leonov's arm stretched through the passageway and clasped Slayton's. Millions of TV viewers around the world were thrilled to witness both commanders embracing each other in space. The historic mission was welcomed as symbolizing both the extension of détente to outer space and the potential of international efforts in space exploration.

It was hoped that this spirit of collaboration in space would develop into further missions, but the chill of the Cold War was about to return. It would be 20 years before the next joint US-Russian docking flight in 1995.

This was also the last flight of a US manned ballistic spacecraft, and the last US manned flight for almost six years. In 31 manned flights, NASA astronauts had notched up 22,503 hours 48 minutes and 58 seconds of spaceflight experience, from launch to splashdown. The next US manned space vehicles would be the reusable Space Shuttles.

▼ *Apollo–Soyuz Docking Procedure*
To allow the Apollo and Soyuz crews to make their historic meeting, it was necessary to equalize the different atmospheric pressures and compositions used by the two craft. Apollo had an atmosphere of pure oxygen at a pressure of 280 mm Hg; Soyuz had oxygen/nitrogen at 520 mm Hg. This diagram shows how it was done. (The hatches are counted left to right).

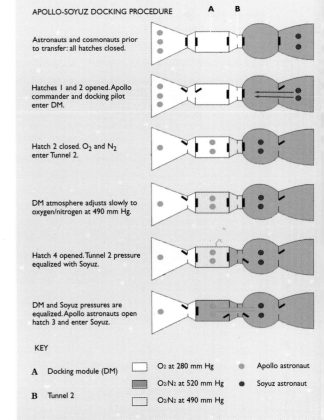

APOLLO-SOYUZ DOCKING PROCEDURE

Astronauts and cosmonauts prior to transfer: all hatches closed.

Hatches 1 and 2 opened. Apollo commander and docking pilot enter DM.

Hatch 2 closed. O_2 and N_2 enter Tunnel 2.

DM atmosphere adjusts slowly to oxygen/nitrogen at 490 mm Hg.

Hatch 4 opened. Tunnel 2 pressure equalized with Soyuz.

DM and Soyuz pressures are equalized. Apollo astronauts open hatch 3 and enter Soyuz.

KEY

A Docking module (DM) | ☐ O_2 at 280 mm Hg | ● Apollo astronaut
B Tunnel 2 | ▨ O_2/N_2 at 520 mm Hg | ● Soyuz astronaut
   | ▨ O_2/N_2 at 490 mm Hg

A PERMANENT PRESENCE IN SPACE

While the American astronauts sat out the late 1970s with their feet firmly on the ground, waiting for the Space Shuttle, Russian cosmonauts were flying into space with impressive regularity, pushing the space endurance records to almost six months on board the later Salyuts. By the time the first Shuttle lifted off in 1981, Russia had leapfrogged ahead of the Americans again, perfecting the art of building laboratories in space and pushing the duration record to 237 days. By 1986 Russia had two major space stations in orbit.

ALMAZ/SALYUT SPACE STATIONS 6 AND 7

Salyut 6 was launched on 29 September 1977, and was visited by a number of short-stay crews dovetailing with the longer-term resident cosmonauts who slowly increased the space endurance record. Cosmonaut Valery Ryumin (b. 1939) completed two six-month stints, making him the most experienced space traveller in history. The last crew left in March 1981 and Salyut 6 was steered into a trajectory which burnt it up on 29 June that year.

Salyut 7 was launched on 19 April 1982, and again served as a home to missions of increasing duration. Female cosmonaut Svetlana Savitskaya (b. 1948), only the second woman to fly in space, visited Salyut 7 in 1982 and became the first woman to spacewalk. Two years later, while the station was uninhabited, Salyut 7 suffered a severe technical failure and in June 1985 Vladimir Dzhanibekov (b. 1942) and Victor Savinykh (b. 1940) were launched on board Soyuz T13 to resurrect the tumbling station. The last crew, on a visit from Mir, arrived on Salyut 7 in March 1986 for a six-week stay before returning to Mir. Salyut 7 crashed back to Earth earlier than anticipated, on 7 February 1991, burning up over South America, and showering parts of Argentina with debris.

LATER SOYUZ 20-TM23 FLIGHTS AND COSMOS MISSIONS

After the Apollo–Soyuz mission had flown, the Soyuz craft continued to act as ferries to the Salyut space stations and performed occasional

▲ *Cosmonauts Piotr Klimuk* and Vitaly Sevastyanov in the orbital space station Salyut 4.

Earth-observation tasks. The next big change to Soyuz design did not come until 1979, by which time miniaturized electronics had created enough room for three suited cosmonauts again. Soyuz T (for Transporter) was primarily a space-station ferry, but it carried two solar panels and was thus able to survive in orbit in the event of a failed docking with a Salyut station. The orbital part of the capsule was also discarded prior to retrofire and re-entry, reducing the amount of fuel required for this manoeuvre.

In 1986 the Soyuz TM series emerged – a modernized T series – with improved power supplies, new parachutes and extra space and weight allowance for equipment. Soyuz has been in service now for over 30 years and more than 180 men and women have flown safely into orbit and back in it.

MIR DEVELOPMENT

Mir (meaning 'a community living in harmony and peace') was the next natural step after Salyut 7's success. The new Russian space station would also be modular, but it would carry more docking ports and a much larger array of solar panels. With this new design came the opportunity for the first permanent human presence in space. The first module was launched on 20 February 1986 and less than a month later Mir's first guests, Leonid Kizim (b. 1941) and Vladimir Solovyev (b. 1946), arrived in Soyuz T15.

The first Soyuz TM craft brought a new crew to Mir in February 1987 and kicked off almost three years of continuous occupancy. The longest stay – a whole year – was completed by Musa Manarov (b. 1951) and Vladimir Titov. Their marathon saw a number of other Progress supply flights and three visits by other teams, including the first Afghan cosmonaut, A. Mohmand, a medical doctor called Valery Poliakov (b. 1942), and a Frenchman called Jean-Loup Chrétien (b. 1938). Chrétien had already spent time on Salyut 7 and on this flight he became the first European to spacewalk.

Chrétien left with Manarov and Titov in December 1988. Poliakov remained on board with the two other TM 7 crew members who had arrived with Chrétien. All three returned to Earth on 27 April 1989 and Mir was left empty for the first time in almost three years.

MIR SECOND PHASE

The first crew to return to Mir arrived on 7 September 1989, starting a second phase of Mir occupation that would settle into a pattern of six-month stays by two-cosmonaut crews. The new two-cosmonaut crews would arrive with a third crew member making a short visit and returning with the old crew. The first new crew, Alexsandr Viktorenko and Alexsandr Serebrov, worked on a new Kvant-2 module which arrived on 6 December 1989. On 1 February 1990 Serebrov flew 33 m (108 ft) from Mir, untethered during

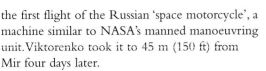

the first flight of the Russian 'space motorcycle', a machine similar to NASA's manned manoeuvring unit. Viktorenko took it to 45 m (150 ft) from Mir four days later.

During these crew exchanges, a number of foreign visitors made short one-week trips to Mir. The first was Japanese journalist Toyohiro Akiyama, who arrived on Soyuz TM11, a rocket which carried advertisements – one from a nappy company. With Akiyama was Musa Manarov, returning to space following his one-year marathon a year before. Manarov went on to spend a total of 500 days on Mir and walked in space for over 20 hours in total. Other cosmonaut passengers included Helen Sharman (b. 1963) on TM12 – the first Briton to go into space. The long-duration spaceflight record was extended to 679 days, spread over two flights by Valery Poliakov.

THE FATE OF MIR

In 1993 NASA, preparing for experience with the upcoming International Space Station project, signed a deal to fly up to 10 Shuttle flights to Mir, taking US astronauts for longer-duration stays in orbit. In the late 1990s, however, a series of serious faults plagued the ageing space station. Fire broke out while British-born astronaut Michael Foale was on board in 1997. Only a few days later, during a practice docking procedure, a Progress supply craft crashed into the Spektr module, puncturing the hapless station and starting off a chain of faults which sent Mir into a dangerous tumble, losing solar power and its main guidance computer.

After the last Shuttle left in June 1998 the plan was to de-orbit Mir at the end of 1999, but a commercial company called MirCorp was set up to save the station, turning it into a space hotel – with tickets for a one-week stay costing $20 million. A number of wealthy businessmen signed up for the trip, but it was not enough to cover the estimated $250 million annual running costs, and Russia demobilised Mir in March 2001, bringing the 137-tonne craft down to Earth.

SPACELAB

Spacelab was conceived in the 1960s when NASA was looking for international collaborations on its post-Apollo space programs. It was eventually decided that the US would develop a reusable

▲ *The Earth surrounded by a swarm of satellites and space debris. As well as the junk, hundreds of artificial satellites circle our planet, feeding back information and enhancing our knowledge of Earth, its atmosphere, and space.*

LONG-DURATION SPACEFLIGHT

From the moment a human body reaches orbit and experiences micro-gravity, biological changes occur. About 60% of astronauts experience space sickness – with headache, nausea and vomiting symptoms which can last from a few hours to a few days as their bodies adjust. Extra calcium appears in the blood and urine almost immediately, as bone mass is reduced. With little work to do in space, muscles are the next things to begin wasting, and the heart can

Space Shuttle and Europe would build a modular laboratory that could be carried on the Shuttle, providing a shirtsleeve environment inside which scientist-astronauts could conduct experiments. The pressurized modules were linked to the main Shuttle cabin with a tunnel. Other experiments and astronomy or Earth-observation equipment could be mounted outside the lab on U-shaped pallets.

The Space Shuttle Columbia on STS-9 carried Spacelab-1 into orbit on 28 November 1983, along with the first European astronaut Ulf Merbold. The mission lasted 10 days and carried 38 different sets of experiments, supporting 73 investigations from 14 countries. Subsequent Spacelabs undertook astronomy, Earth-observation and biological experiments. Its last mission took place in 1998.

SPACE DEBRIS – POLLUTION IN ORBIT

Hurtling round Earth at over 27,000 km/h (17,000 mi/h), even the tiniest fleck of paint can prove lethal to an astronaut or spacecraft. Although no one has yet been killed by one of these space bullets, collisions have already occurred. Shuttle windows have been cracked by impacts and in 1981 a Soviet satellite called Cosmos 1275 suddenly exploded into almost 300 fragments for

weaken. Two hours a day of strenuous exercise can help reduce this muscle deterioration.

Above the Earth's atmospheric and magnetic shields, harmful radiation is a constant threat to astronauts and cosmonauts. Much of this is blocked by the spacecraft walls, but on spacewalks, high doses of radiation from a solar flare could prove fatal. Drugs can help protect against the effects of ionizing radiation on the body, but the best protection comes from proper, dense and radiation-absorbing shielding incorporated into the spacecraft.

◀ *The Space Shuttle about to dock with the Mir space station. Nine Shuttle flights took place to Mir, delivering astronauts for long-duration stays in weightlessness in preparation for the launch of the International Space Station.*

no apparent reason. It is suspected that the satellite collided with an uncatalogued piece of space debris. Somewhere between 23,000 and 26,000 items of space debris are catalogued depending on which catalogue you look in. Many of these objects are large enough to be tracked with radar, but there are countless smaller fragments that are not. Debris comes from objects in orbit that have outlived their usefulness and are floating above the Earth. They include payloads that go wrong, old satellites, run-down batteries, and left-over rocket boosters. Rocket malfunctions have showered exploding debris into orbit on many occasions. The upper stages of seven Delta rockets alone produced over 1,200 fragments, and Titan, Agena and Ariane upper stages have all contributed to this problem.

Although there is a natural drag from the Earth's tenuous upper atmosphere, which pulls many pieces of debris back to burn up, a lot of debris orbits in trajectories which will last for centuries. Although NASA is exploring possibilities of cleaning up Earth orbit, the lack of major disasters so far means that there is little impetus to develop technology that would clean up the debris, and the priority now is to try and make missions less messy.

▼ *Even the tiniest pieces of debris in space have the potential to cause serious damage to craft and death to their occupants. Shuttle windows have been cracked by flying debris and a Soviet satellite has been completely destroyed. Because of their high speeds, even tiny specks of debris can produce impact pits like this one.*

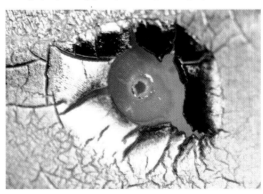

SPACE EXPLORATION

SPACE SHUTTLES

In the 1970s there was still something rather wasteful about human spaceflight. The expendable Saturn V rocket had shed 97% of its mass a few minutes into its flight and the hugely complex space capsules could fly only once, making the journeys extremely expensive. What was needed to bring down the costs of orbital flights was a reusable launch vehicle. Sergei Korolev had proposed the first serious reusable Soviet space plane in the 1960s, but the sudden urgency to reach the Moon had put a stop to further plans for it. In the early 1970s, with the Moon race run, NASA announced its intention to build a reusable Space Shuttle and, not wanting to be left behind, the USSR began its own Shuttle program.

JOHN YOUNG

John Young (b. 1930) was born in San Francisco, California. He attended the Georgia Institute of Technology, receiving a BSc in aeronautical engineering in 1952 before serving aboard the Naval destroyer *USS Laws* for a year and undergoing pilot training from 1955 to 1959. As a pilot he logged over 11,500 hours of flying time. He was one of nine astronauts chosen by NASA in October 1962. Young became the first of the group (which included Charles Conrad, Frank Borman and Neil Armstrong) to be assigned to a spaceflight, when he was named pilot of Gemini 3 in April 1964. He was assigned to flight crews almost continuously for the next nine years: in addition to flights on Gemini 3, Gemini 10, Apollo 10 and Apollo 16, he was backup pilot for Gemini 6, backup command module pilot for Apollo 7, and backup commander for Apollo 13 and Apollo 17.

When many of the early astronauts left the program in the 1970s Young stayed on to work in the astronaut office at Houston and was asked to pilot the first Shuttle flight on 12 April, 1981. In November to December 1983 Young commanded a second Shuttle mission, STS-9, which carried a record crew of six with the first Spacelab. This would be his last flight.

THE BIRTH OF THE US SHUTTLE AND STS 1

The original reasons for developing a Space Shuttle were to service a space station, but plans for that were increasingly delayed, so NASA sold the Space Shuttle to Congress as a versatile flying laboratory, a cheaper reusable launch alternative for the growing commercial satellite market and even a method of re-capturing and repairing satellites once in orbit. The project was approved in 1972 and due to fly in 1978. The design consists of a delta-winged orbiter which sits on a large external liquid oxygen and hydrogen fuel tank (ET) that supplies the Shuttle's reusable main engines at its rear. Strapped to the sides of the ET are two solid rocket boosters used to loft the Shuttle to 44 km (27 miles) high before being jettisoned to parachute back into the sea. Once the main fuel tank is empty it tumbles back to Earth, burning up in the atmosphere.

The orbiter can carry up to eight people in a two-floor crew compartment in the vehicle's nose. The lower level is for sleeping and eating while the upper floor houses the flight deck and a rear work area for operating equipment in the payload bay. The bulk of the Shuttle's 56.1-m (184-ft) length houses the payload bay, which is opened to space by two large doors which also serve as a radiator to keep the Shuttle cool. Delays and budget inadequacies held up the program and although an engine-less Shuttle-shaped glider called Enterprise flew in 1977, the first flight did not take place until 1981.

SHUTTLE FLIGHTS

On 12 April 1981 Commander John Young and pilot Robert Crippen (b. 1937) lay in a reclined position inside Space Shuttle Columbia waiting for the first launch. A textbook launch placed them into a 245-km (152-mile) orbit in eight minutes and 34 seconds. Mission control noticed that a number of silicon re-entry tiles that covered Columbia's fuselage had come off during launch and there were initial

◄ **The spectacular launch** of the Space Shuttle, a versatile reusable transporter capable of launching and recovering large payloads. It glides back to land on a runway. The first Space Shuttle, Columbia, was launched in April 1981, under the command of John Young and piloted by Robert Crippen.

▲ **The flight deck** of the Space Shuttle, showing the seats of the commander and pilot. In all, up to eight astronauts can ride aboard the Shuttle.

fears that this might cause problems later. But in any event everything went smoothly and 54 hours and 20 minutes after launch, the Shuttle glided on to runway 23 at Edwards Air Force base in California.

Seven months later, after a brief launch delay, Columbia roared into orbit on a second test-flight. The five-day mission was cut short when a fuel-cell problem developed, but the world's first reusable space vehicle had made its mark.

Two more developmental flights were followed by a rapid series of satellite launches and a highly publicized series of satellite rescue missions. Civilian launches took place from Cape Canaveral in California, while military missions were meant to be launched from Vandenberg Air Force Base on the west coast of the US. Three Shuttles were added to the fleet and by 1985 launches had become an almost monthly routine.

▼ **Shuttle Flight Profile**

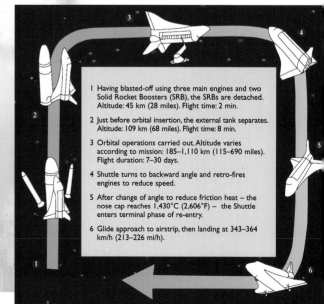

1. Having blasted-off using three main engines and two Solid Rocket Boosters (SRB), the SRBs are detached. Altitude: 45 km (28 miles). Flight time: 2 min.
2. Just before orbital insertion, the external tank separates. Altitude: 109 km (68 miles). Flight time: 8 min.
3. Orbital operations carried out. Altitude varies according to mission: 185–1,110 km (115–690 miles). Flight duration: 7–30 days.
4. Shuttle turns to backward angle and retro-fires engines to reduce speed.
5. After change of angle to reduce friction heat – the nose cap reaches 1,430°C (2,606°F) – the Shuttle enters terminal phase of re-entry.
6. Glide approach to airstrip, then landing at 343–364 km/h (213–226 mi/h).

SHUTTLE-MIR MISSIONS

The NASA Space Shuttle fleet flew to the Russian Mir space station in the mid-1990s in the first collaborative missions between the Superpowers since Apollo–Soyuz. A total of nine link-up flights took place, providing US astronauts with some much-needed long-duration spaceflight experience. Shannon Lucid made the longest stay in a 188-day visit that started on 22 March 1996. The following year Mir started to show its age. During Jerry Linenger's 132-day stay, the space station's main computer was unplugged and Mir began to tumble, losing sunlight on its solar panel array and suffering a major power cut. Mission control and the crew struggled to stabilize the station and after a week of uncertainty and increasingly uncomfortable conditions on board it was stabilized and solar power returned. Despite the dangers, British-born astronaut Michael Foale replaced Linenger on 15 May 1997, but after the Shuttle had left, things started to get worse with a fire on board and a collision with a Progress supply ship. A series of risky spacewalks inside the punctured low-pressure hull to seal it off eventually patched up the space station and two more US astronauts visited. The final Shuttle-Mir docking occurred on 12 June 1998, returning Andrew Thomas from his 140-day flight.

THE STS 51L CHALLENGER DISASTER

After a series of delays, the twenty-fifth shuttle flight – STS-51L – finally rose off the launch pad on 28 January 1986. On board the Shuttle Challenger were commander Dick Scobee, pilot Mike Smith, mission specialists Ellison Onizuka, Judy Resnik and Ron McNair, payload specialist Gregory Jarvis from the Hughes Aircraft Company, and the first teacher to go into space – Christa McAuliffe. She had been selected from 11,000 applicants to broadcast a series of live lessons into US schools from orbit.

It had been a cold night at the Cape and the rubber joining rings on the solid rocket boosters had been weakened. Fifty-eight seconds into the flight, burning fuel began to leak from the bottom of the right booster. Within seconds, the stray flame had burnt through the strut holding the solid booster in place and 72 seconds into the flight it crashed into the top of the external fuel tank and ruptured it. Burning fuel suddenly enveloped the whole Shuttle, tearing it into several large pieces. The crew cabin remained intact, continued briefly upwards under its own momentum with the other fragments and then tumbled back into the Atlantic, impacting with the ocean at 328 km/h (204 mi/h). All seven astronauts were killed.

SHUTTLE FLIGHTS POST CHALLENGER

After a hiatus of over two years, Space Shuttle flights resumed with the launch of STS 26 on 29 September 1988. Routine satellite deployment missions would now only take place on cheaper unmanned rockets and the expensive Shuttle missions were reserved for tasks where a human crew was essential. Satellite repairs were still attempted, including a series of bold repair and maintenance missions to service the Hubble Space Telescope. One of the most high-profile flights of the 1990s was STS 95 which saw 77-year-old John Glenn return to space for the first time since his pioneering Mercury flight in 1962.

As the twenty-first century moves on, there have been over 100 Shuttle flights spanning almost 20 years. In the absence of any alternative reusable launch vehicles the Shuttle fleet has recently been refitted and modernized and will probably fly until at least 2030. Although the Shuttles have proved they can make repeated flights into orbit, launches still cost around $400 million and expendable launch vehicles are proving to be cheaper for reaching space.

BURAN

After a series of atmospheric and suborbital glide and jet-powered flight tests, the Russian Shuttle named Buran ('Snowstorm') was rolled out at the end of the 1980s. Its appearance was very similar

▲ *The Russian Shuttle Buran. Although similar in appearance to the American Space Shuttle, Buran boasted the added functional capacity of being able to fly remotely without being manned by a human crew.*

to the American Shuttle, but it differed in being able to fly remotely without a crew. After one cancelled launch attempt, halted less than a minute before lift-off due to technical problems, Buran finally launched from Baikonur on 15 November 1988. The Shuttle completed two orbits of Earth before re-entering and gliding back to Baikonur three hours and 25 minutes later. Substantial heat damage had occurred during re-entry with an area along the wings' leading edge melted by plasma.

Further flights were planned to visit Mir, but following the severe economic decline of the Soviet Union in the early 1990s and the cheaper alternatives for flying to Mir, the Buran project was abandoned.

▶ *The Manned Manoeuvring Unit (MMU) solved many of the problems encountered with previous types of propulsion backpacks, and allowed a series of daring satellite rescue missions.*

THE MANNED MANOEUVRING UNIT (MMU)

The first Gemini spacewalks in the 1960s had revealed just how difficult it is to manoeuvre accurately in space, and an astronaut propulsion backpack was developed for Gemini 9, but proved hard to test. Skylab's second crew successfully flew a similar prototype system without spacesuits inside the space station's vast main chamber. The third generation backpack called the Manned Manoeuvring Unit (MMU) was built to fly on the tenth Space Shuttle mission in 1984. Bruce McCandless was the first pilot and flew it out to 100 m (330 ft) from the Shuttle's cargo bay during a one-hour 22-minute flight, becoming the first untethered human spacewalker. The success of this trial spurred NASA on to attempt a series of daring satellite rescues using the MMU on the next mission. Astronaut George Nelson attempted to rescue the ailing Solar Maximum Mission in April 1984, but a design flaw with the docking mechanism to grab the satellite meant that in the end the Shuttle's robot arm caught it. In November of that year, astronauts Joe Allen and Dale Gardner used an MMU to successfully capture two communication satellites. The USSR had their own MMU, or 'space motorcycle' which was first flown from Mir in February 1990 by cosmonaut Alexander Serebrov. Both US and Russian versions of this have now been abandoned.

THE FUTURE

With our first International Space Station now in operation, the near future of human spaceflight looks like it will be centred around six-month stays in orbit. At last, NASA's Space Shuttle fleet has somewhere to fly to and, combined with the cheaper Russian Progress and Soyuz supporting missions, human journeys into orbit look likely to settle into a predictable pattern for a while. The promise of single-stage-to-orbit space-planes is still a long way off, and any return to the Moon or human voyages to Mars are only being planned by planetary societies and interested individuals. Government impetus for such ventures is still lacking.

INTERNATIONAL SPACE STATION

Plans for an International Space Station were first announced in 1984, when US president Ronald Reagan declared that NASA would build a large space station in Earth orbit for use by astronauts from the US and other nations for a variety of purposes. After initial setbacks, the former Soviet Union was brought into the program in 1993 and the first Russian component – the 19,320-kg (42,600-lb) Zarya ('Sunrise') module was launched in November 1998. A month later the Shuttle Endeavour attached the Unity docking module – a junction unit allowing other modules to be attached. A second Russian component, Zvezda ('Star'), that contains the living quarters and a propulsion system was docked in July 2000. Other nations, including Brazil, Japan, Canada and the European Space Agency (ESA) have participated in the project. At 37 m (120 ft) in length and with a wingspan of 28 m (95 ft), the ISS is one of the brightest objects in the night sky. The first crew arrived at the beginning of November 2000 for an extended stay of four months. More than 40 flights are required to complete construction by 2005, when it will serve as a permanent orbiting science institute capable of performing long-duration research in a nearly gravity-free environment. By then the ISS will be three times the volume of a two-floored house and able to support a crew of seven.

SPACE TOURISM

The first astronauts were all experienced pilots, but in the late 1970s, Russia's guest cosmonaut program began to open up space to other citizens from friendly nations. The first of these was Vladimir Remek of Czechoslovakia, who visited Salyut 6 in March 1978. Passengers from the other Warsaw Pact countries followed and the program was expanded to include Cuba and Communist countries in Asia.

In the US, non-pilot mission and payload specialists were being recruited for the Space Shuttle and the first – Joseph Allen and William Lenoir – flew on the fifth Shuttle flight in 1982. By the mid 1980s, US politicians and satellite company representatives had flown and the plan was to expand this program to fly journalists, artists and poets into orbit. But the first citizen in space was killed in the Challenger disaster in 1986 and future passenger trips were cancelled.

By then Mir was in orbit and the first fare-paying passenger was not far away. In December 1990 Japanese journalist Toyohiro Akiyama's TV company, TBS, paid $12 million to launch him to Mir on Soyuz TM11 in order to conduct a number of live TV and radio broadcasts from the orbiting station. The Juno project followed – to sponsor a British cosmonaut from industry and the media. Chemist Helen Sharman was selected and flew to Mir for a week in May 1991. An Austrian and a German guest followed. With Mir at the end of its life in 1999, a Dutch-based company called MirCorp was formed to try and turn the vintage space station into a commercial operation, but failure to make enough payments to Russia led to cancellation of this project and the deorbiting of Mir in March 2001.

◀ *How the International Space Station should look when fully assembled in 2005. The first components were launched in 1998 and the first crew arrived in November 2000.*

⧗ **PROBLEMS WITH GOING TO MARS**

There are thousands of problems that must be resolved before a human flight to Mars can become reality. On a journey to the planet a crew could be 40 minutes from a reply from mission control. At these distances, conversations with Earth are impossible and communications become more like letter-writing, even at the speed of light. The shortest return visit to Mars would still require a crew to be away from Earth for up to three years. So far, the longest time a human has spent in space spread over two flights is just over half that period. Although bone and muscle wastage did happen, no permanent damage seemed to result. Thirty years on, no moon-walkers have died from radiation-related illnesses due to doses they received away from Earth, but exactly what the longer-term radiation effect of a three-year voyage to Mars and back would have on the human body is not known.

THE NEXT SPACE-FARING NATION

Almost four decades after Russia and the US hurled their first citizens into space they remain the only countries with this achievement. In February 1978 China confirmed that it was working on a manned space capsule and a Salyut-type space station. Despite subsequent news of Chinese astronaut trainees, however, no manned flights materialized and the program was postponed in 1980. Successful unmanned test flights of their Soyuz-type 'Shenzhou-1' craft began in November 1999. Shenzhou 2 went through a more rigorous test-flight in January 2001 and the first 'taikonaut', or Chinese astronaut, is expected to be launched in the next five years.

MOON BASES AND RETURN TO THE MOON

For four days in December 1972 Eugene Cernan and Harrison Schmitt lived in a lunar base in the Taurus Littrow valley. Their lunar module was a home which they would leave each day to explore and experiment – returning there in the Houston evening to rest. Almost 30 years after Cernan stepped off the Moon, he is still the last human to have trodden on its surface and the post-Apollo promise of lunar bases has completely evaporated.

NASA's first detailed plans for lunar bases were drawn up in the early 1960s, based on Saturn V-sized rockets delivering 11-tonne modules to the Moon's surface. These Lunar Exploration Systems were designed to support up to six astronauts for half a year and, as Apollo hurtled towards its climax, they seemed like the obvious next step. But severe cutbacks to the US space program in the early

◀ *Exercising in space,* an essential daily regime to slow down bone and muscle wastage in zero gravity.

▼ *An artist's impression* of a human base on Mars. Although the first humans might visit Mars in the next 30 years, such extravagant bases are still many centuries away.

1970s put a halt to future plans for a Moon base. Today, the reasons for returning to our nearest neighbour are hard to justify, economically or politically. Governments might only be inspired to return to the Moon if it were a staging post for a human expedition to Mars.

MARS MISSIONS

Dreams of journeying to Mars have filled human minds for as long as we have peered at it through telescopes. But the first realistic study of a human exploration of Mars was not published until 1952, in **von Braun**'s book *The Mars Project*. The book inspired other proposals and when NASA announced its plans to go to the Moon, research into realistic ways for astronauts to reach Mars blossomed. By July 1969, with Apollo 11 en route to the Moon, US vice president Spiro Agnew called for a human mission to reach Mars by the year 2000. More proposals were explored and dates for a mission in the early 1980s were drawn up to take advantage of launch and return windows, via a Venus slingshot, returning to Earth after a 30-day stay on Mars on 14 August 1983.

Only a year after our first footprints on the Moon, with the Apollo program truncated, dreams of reaching Mars were put aside. Three decades on, human spaceflight remains trapped in Earth orbit. The relevant powers have not been idle, but priorities have been assessed and the necessity of maintaining the Shuttle program and, more significantly, getting the ISS off the ground now have the full focus of attention from the deciding parties. The more assured benefits of these goals may well keep missions to Mars just a dream for some time to come.

HUMAN BURIALS IN SPACE

On 31 July 1999 a brief flash on the Moon signalled the first human lunar burial. A small vial of astrogeologist **Eugene Shoemaker**'s ashes had been carried there on board NASA's Lunar Prospector mission. This pioneering funeral had been co-ordinated by Celestis, a Houston-based company in the US. They inaugurated their space memorial business in Earth orbit during a 1996 Pegasus launch, propelling the ashes of 24 people, including those of *Star Trek* creator Gene Roddenberry, into space.

There have been two additional orbital flights since then with a fourth scheduled for 2000. The cremated remains are lipstick-sized samples flown for $5,600 each to burn up in the Earth's atmosphere and scatter on the edge of space. So far over 100 people have been 'buried' in this way.

Celestis are now marketing the lunar option – into orbit or impact on the surface for $12,500, with the first flight expected within the next year. Mareta West, the only female geologist in the Apollo program, and selector of the Apollo 11 landing site, is first on the list.

▲ *The future of spaceflight:* one day it might be possible to build advanced shuttle craft capable of reaching orbit with only one stage. This artist's impression shows how such a single-stage-to-orbit shuttle might look.

REUSABLE LAUNCH VEHICLES

Plans for a totally reusable launch vehicle are still a dream. Although NASA's Space Shuttle is mostly reusable, the giant main fuel tank is not. Attempts to build a space plane which can take off and land on a conventional runway have not progressed far. The X-15 planes, which could fly at Mach 6.7 and soar to over 100 km (60 miles) high, were still launched from a mother ship. Escape velocity at Mach 24 (24 times the speed of sound) has so far remained elusive for a single-stage vehicle. NASA has carried out research on advanced launchers that would reach orbit without dropping rocket motors and fuel tank like the current Space Shuttle, but the ultimate goal of a single-stage-to-orbit vehicle remains beyond present-day technology.

Shoemaker ▶ p. 310 von Braun ▶ p. 306

SPACE EXPLORATION

HUMAN SPACEFLIGHT

Space Exploration
ARE WE ALONE?

From the moment we could build a rocket big enough to leave Earth, we embarked on missions to look for life. Less than five years after Sputnik 1 circled the globe, Russia's Mars 1 mission was dispatched to search for signs of life on Mars. Since then we have scoured the Solar System for evidence that we are not alone: poring over lunar samples returned from Apollo, sifting through the Vikings' Martian dust samples and staring into the cloud tops of the giant planets and on to their moons of fire and ice through Voyager's cameras. Unable to accept that we might be the only intelligent life in this vast Universe, we have built giant dishes and scanned the Galaxy, listening for radio signals from other civilizations. But despite a lot of false starts and a great deal of wishful thinking the only place known to harbour life is here on Earth.

ORIGINS OF LIFE ON EARTH

Even here at home, life's origins are still a puzzle. We can dissect our biochemistry and reduce life to an assortment of organic chemicals, but the moment and the mechanism at play when those elements got together and became 'alive' still elude us. In the 1950s, the favourite explanation, proposed by Harold Urey and his student Stanley Miller working at the University of Chicago, involved subjecting a solution of organic compounds in a sealed glass bottle to high-voltage discharges of electricity and **ultraviolet** light – synthesizing simple amino acids. Urey and Miller argued that such a pre-biotic soup on the primeval Earth might have been subjected to similar conditions creating organic polymers.

The recent discovery of an abundance of primitive subsurface life has led some planetary scientists to suggest that life on Earth might have actually begun below the surface. Iron and nickel sulphide might have collected near hydrothermal vents – acting as a catalyst for pre-biotic chemical reactions forming longer-chain organic molecules. As well as the right chemistry, other factors such as a low rate of fatal **asteroid** impacts, a stabilizing influence of the Moon and the optimum distance from the Sun, appear also to have been vital in transforming Earth into a blooming biosphere.

HABITABLE ZONES AND THE GOLDILOCKS HYPOTHESIS

However and wherever life got started, it seems that Earth was the only place in our **Solar System** where it took such a strong and long-lasting hold. Life as we know it requires the presence of liquid water, and around the Sun today the temperatures to sustain it in abundance at the surface of a planet are only found between 119 and 245 million km (74 and 152 million miles) from the Sun. This is the so-called 'habitability zone' (HZ) where conditions are not too hot and not too cold.

Venus lies outside this zone and the results are a runaway greenhouse effect with any water banished to a vapour in its super-dense atmosphere. Almost twice as far from the Sun as we are, Mars's lower temperatures and tenuous atmosphere render liquid water a rare commodity today. Luckily for us, Earth sits right in the middle of this 'Goldilocks zone' at 149.6 million km (93 million miles) from the Sun and, awash with liquid water, our biosphere blooms. As the Sun grows hotter and larger over the rest of its life this liquid water zone will shift outwards. Within another two billion years the Earth's surface will cease to be habitable as solar ageing transforms us into what Venus is today. Mars should warm up during this period of Solar System history and might come to life, as rivers of liquid water become a possibility once more.

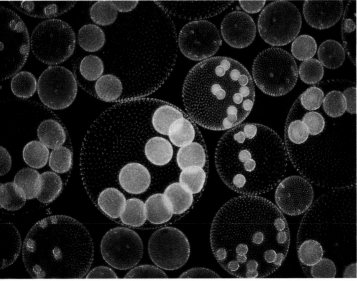

◀ *A microscopic view of groups of simple organisms, living and evolving communally. Scientists propose theories about life on Earth having started as a microbiotic organic soup.*

LIFE ELSEWHERE IN THE SOLAR SYSTEM?

From our ongoing exploration of Earth, we have discovered that life is much hardier than we have ever thought. We find it inside rocks deep beneath our ocean floors, lying patiently in ice blocks on Antarctica and floating in the utter darkness of super-heated sulphurous waters around mid-ocean ridges. However harsh these conditions, though, they all have something in common – water – and our search for life elsewhere in the Solar System must start with a search for liquid water.

The copious dry river channels on Mars were once filled with water. Mars Global Surveyor pictures of possible seepage sites suggest that a deeper liquid-water environment might still exist locked beneath the surface, placing Mars as the strongest contender for finding life near Earth. The robotic search goes on to seek it out.

Further from the Sun, around Jupiter, are two other environments that could harbour liquid water and primitive life. Europa's cracked, wrinkled and refrozen ice crust suggests that a subsurface liquid ocean might exist. Galileo orbiter experiments indicate that Callisto could also contain a salty subsurface ocean. The discovery of abundant subsurface life around Earth's ocean-floor volcanoes has led many astrobiologists to speculate about similar environments teeming with life within Europa or Callisto's subsurface oceans.

Saturn's moon, Titan, is the only moon in the Solar System to be shrouded in cloud and, like the pre-biotic Earth, it is coated in a cocktail of hydrocarbons basking beneath a nitrogen atmosphere. Although its super-chilled temperatures rule out the chance of any advanced life, the right chemistry exists for the building blocks of Earth-like life to be present.

THE SEARCH FOR LIFE

From the moment we had the telescopic power to resolve Mars as more than just a red speck, astronomers began their search for life. **Percival Lowell** (1855–1916) interpreted the networks of channels and seasonal changes he charted on the Martian surface as a tantalizing sign of life. Of course, four decades of spaceflight dashed these hopes and have so far failed to spy out any life elsewhere. But the search continues – homing in on the likely contenders with more landers planned for potentially wet parts of Mars early this century and the Huygens probe already on its way to Saturn's moon, Titan.

ARE WE ALONE?

asteroid ▶ p. 336 Solar System ▶ p. 343 star ▶ p. 344 ultraviolet ▶ p. 344

SETI

The first attempt to search for radio signals from extra-terrestrial intelligence, SETI, was carried out by Frank Drake at the National Radio Astronomy Observatory in Green Bank, West Virginia in 1960. Drake pointed the radio telescope at two nearby Sun-like stars. The only signal they picked up was found to be coming from a military aircraft broadcasting on an unauthorized frequency. Although Drake's experiment had a negative result, the principle of listening out for hypothetical civilizations in other solar systems was established.

As the perception grew that SETI had a reasonable prospect for success, the Americans once again began to observe. During the 1970s, many radio astronomers conducted searches, using existing antennas and receivers. Some of the efforts, employing improved technology, have continued to the present time. Foremost among these are the Planetary Society's Project META, the University of California's SERENDIP project, and Project Phoenix at various observatories worldwide.

NASA had sponsored radio telescope work since 1971 and in 1992 it started on a SETI program, but Congress terminated funding and today the SETI Institute continues with private funding. Using screen savers to process the results over internet downloads, SETI's latest private project has searched through 161 million results, finding about 60 million significant signals. So far none has proved to be extraterrestrial.

▼ *The giant Arecibo radio telescope has been employed by the SETI program to search for signals that might reveal life elsewhere in the Galaxy.*

PANSPERMIA

Although impacts seem to have been responsible for the annihilation of life on Earth, ironically it could have been similar events which seeded our planet with its first life forms. The Panspermia theory claims that in the early Solar System, before the inner planets took on their present environments, life was likely to have started on all of them. The blizzards of asteroids and **comets** in those first few hundred million years ensured that impacts were common and oblique collisions would have blasted rocks housing simple bacteria between planets, cross-contaminating life between worlds. Support for this theory grew in the 1970s with the discovery of some dormant, but surviving bacteria on Surveyor 3's camera brought back by Apollo 12 after three years on the Moon. Over two decades later the discovery of bacteria in a Martian meteorite, although now discredited, kicked off the Panspermia debate again. And experiments to fly unshielded bacteria into Earth orbit and to study the effects of hypervelocity impacts on them are gaining support.

LIFE IN THE GALAXY

Beyond our Solar System, the wobbling of **stars** has revealed that they have their own planetary systems. The number of stars that seem to have their own suites of planets revolving around them grows each month. Although no small rocky planets in the habitability zones of these stars have so far revealed themselves, it is likely that small rocky moons orbit some of the large planets that seem to exist out there. NASA's Terrestrial Planet Finder project plans to resolve Earth-like planets in these new systems and to hunt down tell-tale reactive gases, out of kilter with the chemistry of the planet, which could hint at the presence of life on these very distant balls of rock. Beyond these wobbling stars and the limits of our optical telescopes, giant dishes of the Search for Extra Terrestrial Intelligence (SETI) program are listening out for non-astrophysical sources of radio waves that could come from distant civilizations.

THE DRAKE EQUATION

The Drake Equation, proposed by Frank Drake in 1961, identifies specific factors which might influence the development of technological civilizations and allows us to estimate how many might exist.

$$N = R\star \times fp \times ne \times fl \times fi \times fc \times L$$

Where:

N = The number of communicative civilizations in the Milky Way Galaxy whose radio emissions are detectable.

$R\star$ = The rate of formation of stars with a large enough 'habitable zone' and long enough lifetime to be suitable for the development of intelligent life.

fp = The fraction of sun-like stars with planets; something which is currently unknown, but seems to grow more common with new discoveries made each month.

ne = The number of planets in the stars habitability zone for each planetary system.

fl = The fraction of planets in the habitability zone where life actually develops.

fi = The fraction of life-bearing planets where intelligence develops. Life on Earth began over 3.5 billion years ago. Intelligence took a long time to develop. On other life-bearing planets it may happen faster, it may take longer, or it may not develop at all.

fc = The fraction of planets where technology develops releasing detectable signs of their existence into space.

L = The length of time such civilizations release detectable signals into space. The number of detectable civilizations depends strongly on L. If L is large, then so is the number of signals we might detect.

Lowell ▶ p. 216

ARE WE ALONE?

SPACE EXPLORATION

Space Exploration
SUMMARY

T he invention and refinement of the rocket transformed astronomy from ground-based, atmosphere-blurred observations from afar to a science with the possibility of field research. With every successful mission another distant world was transformed from its flat, two-dimensional disk into a real, three-dimensional object with its own unique vistas. Bodies which previously we had only been able to gaze at from afar we could now touch with robotic arms and test with miniature on-board laboratories. The science of geology was no longer restricted to the Earth. Suddenly three more rocky planets, a handful of asteroids and a suite of moons entered the realms of the geologists. And one lucky scientist got to visit the Moon on Apollo 17.

But the new discipline of planetary science was something that was bolted on to a game of technological trumping, in a ring where two Superpowers were slugging it out with the biggest rockets money could buy. When that game came to an end planetary science had caught on and would not go away. The real-life *Star Treks* were here to stay and the promise of eventually finding other life and proving what we longed to prove – that we were not alone in the Universe – has taken over as the new impetus which now challenges us to reach for the stars.

ROCKET EVOLUTION

- Rocket motors work by ejecting gases in one direction, which push the rocket in the other.
- There are two types of rocket fuels: solid powders which burn quickly without exploding; and liquid fuels which can be pumped into a combustion chamber, regulating the amount of thrust produced.
- The first record of a reaction engine is from AD 62 in the writing of the Greek philosopher Hero.
- By the seventh century, the Chinese were using fire-spitting rockets for warfare.

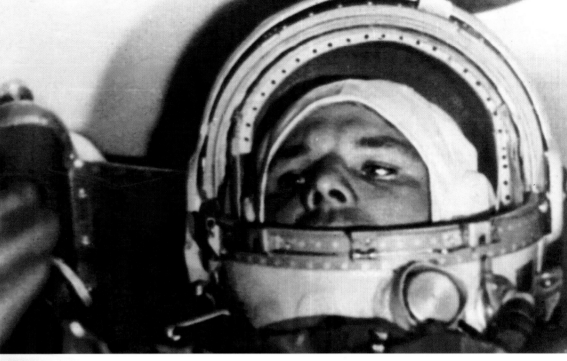

▲ *The Russians* won the race to put the first man in space: Yuri Gagarin orbited Earth in Vostok 1 in April 1961.

- By the mid-twentieth century the liquid-fuelled rocket had emerged and, although still used primarily for war, the possibilities of reaching space with them had emerged.
- Other kinds of reaction engine have been tried – using magnetic or electric fields rather than combustion to accelerate the propellants.

HUMAN SPACEFLIGHT FIRSTS

- Yuri Gagarin became the first man to fly in space orbiting the Earth in his Vostok 1 capsule on 12 April 1961.
- Astronaut Al Shepard was the first American to reach space, in his Mercury capsule Freedom 7 – making a suborbital 15-minute flight on 5 May 1961.
- On 20 February 1962 John Glenn became the first American to orbit the Earth, in Friendship 7.
- **Valentina Tereshkova** became the first woman in space on 16 June 1963, in Vostok 6.
- Voskhod 1 carried the first, cramped three-man crew into orbit on 12 October 1964.
- **Alexei Leonov** became the first man to walk in space – climbing out of Voskhod 2 on 18 March 1965.
- America's two-man spacecraft was called Gemini.

- Ten manned Gemini flights were conducted – perfecting spacewalks, manoeuvres and docking in orbit.
- Ed White became the first American to walk in space on 3 June 1965, on his Gemini 4 flight.
- On 12 April 1981 the Space Shuttle Columbia, the world's first reusable launch vehicle, lifted off. It was 20 years since Gagarin's maiden flight.
- Musa Manarov and Vladimir Titov became the first people to spend a year in space, on Mir, returning to Earth in December 1988.
- The first component of the International Space Station (ISS) was launched in November 1998.
- The first ISS crew arrived in November 2000, for a four-month stay.

FIRST TO THE MOON AND LAST TO LEAVE

- Russia's first Soyuz craft, developed to take a cosmonaut to the Moon, flew in April 1967, but it crash landed back on Earth, killing pilot Vladimir Komarov.
- On 27 January 1967 the crew of Apollo 1 were killed in a capsule fire on the launch pad, grounding America's Moon-shot for over 18 months.
- Despite fatalities in the Russian manned program they sent the first biological cargo around the Moon in their Zond spacecraft in September 1968.

SPACE EXPLORATION

◉ In December 1968 the Apollo 8 flight was sent around the Moon with astronauts Frank Borman, William Anders and James Lovell on board – the first humans to leave Earth orbit.

◉ After two more Apollo test flights in Earth and lunar orbit, Apollo 11 landed safely on the Sea of Tranquillity on 20 July 1969. Astronauts **Neil Armstrong** and Buzz Aldrin spent two and a half hours walking on the surface while Michael Collins orbited in the command module above.

◉ Eugene Cernan and geologist Harrison Schmitt became the last humans to set foot on the Moon in December 1972, during the Apollo 17 mission.

ARE WE ALONE?

◉ The origin of life on Earth is still a mystery and so we are unsure how unique an event it might be.

◉ Despite persistent searching for evidence of life elsewhere in the **Solar System** we have only found it on Earth.

◉ All life on Earth requires liquid water and our search for life elsewhere in the Solar System is focused on a hunt for water.

◉ Around our Sun there is a zone of temperature which allows liquid water to exist at the surface of a planet. This is called the habitability zone – between 119 and 245 million km (74 and 152 million miles) from the Sun.

◉ Outside this zone, subsurface liquid water might still exist and possible evidence for it has been seen on Mars, and under the icy crusts of Jupiter's moons.

▼ *The Gemini and Mercury* space capsules which kicked off America's human flight program.

ALIEN SIGNALS

We have so far been unsuccessful in the search for extra-terrestrial life elsewhere in our Solar System. Looking for it farther afield is more of a listening game – scouring the nearest **stars** in the **Milky Way** to listen for artificial signals. In the summer of 1967, pursuing a PhD at Cambridge, **Jocelyn Bell** was observing **quasars** with a radio telescope when she found an inexplicably regular signal pulsing every second or so and apparently extraterrestrial in origin. Bell and her colleagues couldn't rule out 'little green men' and scribbled the name 'LGM1' on the printout. This time the pulses turned out to be coming from a fast-spinning

neutron star, but on other occasions they have not been so easily explained. After three surveys of the sky by Carl Sagan and Paul Horowitz's late 1980s Mega Channel Extra Terrestrial Assay (META), 11 'events' had been found. Although they satisfied all the criteria for a genuine alien signal, they have never been heard again. Eight of the 11 come from the centre of our Galaxy, an unlikely distribution if they were due to glitches or radio interference from Earth. An as yet unknown astrophysical mechanism might be to blame, as in the 'LGM' case, but without detecting them again it is impossible to conclude that they are from intelligent life.

◉ Beyond the Solar System, the wobble of stars indicates that they have their own planets and the hunt is on for a small rocky one in a star's own habitability zone.

◉ We are also actively listening out for radio signals which have not come from astrophysical objects and might have originated from other technological civilizations.

REMAINING QUESTIONS

◉ How did the Universe and everything in it evolve and what does the future hold for it and us?

◉ How can our understanding of the Universe clarify and enhance our understanding of sciences such as physics and chemistry here on Earth?

◉ Does life really exist elsewhere in the Universe and, if so, in what form?

◉ How can we enhance the revolutionary technological advances of the past few decades to open up space travel to more than just a select few?

◉ How can we develop accurate predictive measures for weather, the environment, climate and natural resources from our current knowledge of the Universe and then apply them to improving our quality and sustainability of life on Earth?

▼ *Space Shuttle Columbia* gliding to a runway landing in 1981. It was the first reusable launch vehicle.

SATELLITE SPOTTING

The International Space Station is already the third brightest object in the night sky after the Moon and Venus. Orbiting the Earth at an angle of 51.6°, its path in the sky plots out a **sinusoidal** route over the turning Earth below. For observers living between **latitudes** of 60°N and 60°S it can be seen in the night sky for a few days every month. Specialist web sites provide the latest positions overhead for the ISS, Shuttles and the thousands of other satellites and spent booster

rockets in Earth orbit. A satellite is visible from the ground when the Sun has set and it is dark where you are, but the satellite, flying high overhead, is still in bright sunlight. What you are seeing is the sunlight reflecting off the satellite, and just how bright it is depends on the orientation of the satellite, how big it is and what it is made of. The best way to find a satellite is to predict when it will pass close to a familiar object or **constellation**. Best and brightest viewing comes when the Sun is to your back.

SPACE EXPLORATION

GLOSSARY

aberration
The small difference between the observed and calculated position of an astronomical object due to the movement of the Earth and the finite velocity of light.

absolute magnitude
The apparent magnitude (brightness) of an astronomical object at 10 parsecs (about 33 light years). The absolute magnitude M can be found from the object's apparent magnitude m and its parallax π in arcseconds using $M = m + 5 + 5\log\pi$. Absolute magnitudes provide a means of comparing the luminosity of one object with that of another.

absorption line
A dark line or band observed in a spectrum caused by the absorption of electromagnetic radiation with a particular wavelength.

absorption spectrum
A spectrum produced when electromagnetic radiation is absorbed. If radiation from a hot source passes through cooler matter, then radiation at certain wavelengths is absorbed, producing dark absorption lines and bands.

accretion
The increase in mass of a celestial body by the addition of matter or smaller objects. Accretion often occurs via an accretion disk.

active galactic nucleus (AGN)
The region at the centre of certain galaxies from which an enormous amount of energy is released. The energy is thought to come from matter accreting via an accretion disk on to a supermassive black hole of greater than 100 million solar masses. *See also* quasars, blazars *and* Seyfert galaxies.

albedo
A measure of the proportion of incident radiation that is reflected by a non-self-illuminating body. The scale runs from 0 for a perfectly absorbing, black object to 1 for a perfectly reflecting, white object. Spherical albedo assumes the object is a perfect sphere, with a diffuse surface reflecting incoming parallel light in all directions. Geometrical albedo is the ratio of the reflectance of the object to that which would be produced by a flat white surface of the same surface area at the same distance. Bond albedo is the fraction of the total incident energy reflected in all directions and calculated over all wavelengths.

altazimuth mounting
A mounting about which a telescope or other instrument can rotate in both the vertical (altitude) and horizontal (azimuth) axes. It offers a firm support but it needs to be moved in both axes to counteract the Earth's diurnal motion in order to observe a celestial object. This can pose problems for the amateur telescope, but sophisticated electronics has led to the altazimuth mount being used in large optical and radio telescopes by professional astronomers.

altitude
The angular distance above an observer's horizon. Measured in degrees with 0° being at the horizon and 90° being the observer's zenith point, altitude is one co-ordinate in the horizontal co-ordinate system. The other co-ordinate in this system is azimuth. If an object is below the observer's horizon, its altitude has a negative value.

angstrom
See Units and Constants p. 368

angular momentum
The momentum associated with the rotational motion of a body. A point-like object of mass m, moving with speed v in a circular orbit of radius r about a point O has orbital angular momentum mvr about the point O. An extended body, such as Earth or the Sun, has a spin angular momentum about its own centre due to the rotation of each part of the body about that centre point.

annular eclipse
A solar eclipse in which the Sun's apparent disk in the sky is larger than the Moon's. In such an eclipse, the Moon's disk does not completely cover the Sun, leaving a ring (Latin: *annulus*) of photosphere around the Moon. The corona is thus not visible. An annular eclipse occurs when the Moon is near apogee, the farthest point from the Earth in its elliptical orbit about the Earth.

anthropic principle
The principle that the Universe is as it is today because life exists to observe it, i.e. that the presence of life today places constraints on the conditions of the early Universe. The weak anthropic principle states that the Universe has properties that have made it possible for intelligent life to evolve. The strong anthropic principle states that the laws of physics which dictate how the Universe has evolved are such that the emergence of intelligent life was inevitable.

aperture
A measure of the effective light (or other radiation) gathering power of a telescope. For optical telescopes it is the clear diameter of the objective lens (for refractors) or primary mirror (for reflectors). The amount of gathered radiation increases with the square of the aperture.

aphelion
The point in an orbit about the Sun which is the farthest from the Sun.

apogee
The point in an orbit about the Earth which is the farthest from the Earth.

apparent magnitude
The brightness of an object as seen from Earth. The apparent magnitude does not take an object's distance into account. The magnitude scale runs backwards in the sense that the brighter an object, the lower the value of its magnitude. The magnitude of an object as it appears to the human eye is known as the apparent visual magnitude.

arcsecond (arcsec or second of arc)
A unit of angular measure, symbol ″, used to measure apparent distances between celestial objects, apparent diameters, or the resolving power of telescopes. There are 60 arcseconds in one arcminute (symbol ′) and 60 arcminutes in one degree.

armillary sphere
An ancient device used to measure and represent the positions of celestial objects.

asteroid (minor planet)
A body smaller than a planet but larger than a meteoroid in orbit about the Sun. Most known asteroids are rocky bodies lying in the asteroid belt, between the orbits of Mars and Jupiter, known as main-belt asteroids. Other asteroids include the Atens which orbit between the Earth and Sun, the Apollos which orbit between Mars and the Sun, the Amors which orbit between the Earth and Mars, the Trojans which orbit at Jupiter's Lagrangian points and the Centaurs which orbit between Saturn and Neptune. Near-Earth asteroids (NEAs) are those which have any chance of colliding with the Earth.

asteroid belt
The region between about 2 and 3.3 AU in which the majority of known asteroids lie. Asteroids lying in the belt are known as main-belt asteroids. Within the belt exist gaps (the Kirkwood gaps) resulting from the gravitational pull of the Sun and Jupiter.

astrometry
The branch of astronomy concerned with the precise measurement of positions of celestial objects.

astronomical unit (AU)
See Units and Constants p. 368

atomic nucleus
The central part of an atom consisting of positively charged protons and uncharged neutrons, thus having an overall positive charge. The nucleus contains most of the mass of an atom.

aurorae (polar lights)
Displays of diffuse, moving, coloured lights in the sky caused by particles from the solar wind spiralling down the Earth's magnetic field at the poles and interacting with the Earth's atmosphere. Known as aurora borealis around the north pole and aurora australis in the south. Aurorae have been observed on both Jupiter and Saturn.

azimuth
The horizontal angular distance measured eastwards from north. Measured in degrees, azimuth is one co-ordinate in the horizontal co-ordinate system. The other co-ordinate in this system is altitude.

Baily's beads
The phenomenon seen at a solar eclipse when the Sun's disk is almost covered by the Moon and photospheric light shines through the lunar hills and valleys at the Moon's limb briefly producing 'beads' of light. If only one bead is visible, it is called the diamond ring. First described by Francis Baily (1774–1844).

barred spiral galaxy
A spiral galaxy in which the spiral arms start from the ends of a bar of material through the nucleus instead of directly from the nucleus. In the Hubble classification they are classified from SBa to SBc with the spiral arms of type SBa being the tightest.

barycentre
The centre of mass of two or more bodies and the point about which they revolve. The barycentre of the Earth–Moon system lies within the Earth as the Earth is so much more massive than the Moon.

baryon
Any class of elementary particles that have a mass greater than or equal to that of a proton, participate in strong interactions and have a spin of one half.

binary star
Two stars that are close enough to be gravitationally bound and to orbit about a common centre of mass. Generally classified according to the means by which their binary nature is observed: in astrometric binaries only one component is observed, but the companion is inferred by perturbations in its proper motion; spectroscopic binaries are revealed by the Doppler shift of lines in their spectrum; in visual binaries both components can be resolved through a telescope. Over 50% of nearby stars appear to be in binary or multiple systems. Binary stars are important for determining information about stellar masses. Some binary stars are so close that transfer of material can occur between the components affecting their evolution and producing variations in their light output. *See also* double star.

black body
A hypothetical body which absorbs all radiation falling on it and which is also a perfect emitter of radiation. The output of radiation from stars mimics a black body sufficiently for their spectra to be interpreted using black-body calculations.

black-body radiation
The radiation emitted by a black body. The radiation is emitted (with differing intensities) across a continuous range of wavelengths, with the output peaking at a wavelength determined by the temperature of the black body. In the case of the Sun (which approximates a black body with a temperature of 5,800 K), the peak occurs at a wavelength of about 500 nm and corresponds to yellow light.

black hole
A massive object so dense that no light or any other radiation can escape from it; its escape velocity exceeds the speed of light. There are thought to be three types: mini black holes created at the Big Bang, stellar black holes created by the collapse of a massive star and supermassive black holes lying at the centres of galaxies. No black hole can be observed directly, but the existence of stellar and supermassive black holes has been inferred by their effect on nearby matter. Matter is pulled into a black hole via an accretion disk; radiation from the disk can be detected.

blazar
A class of active galactic nucleus (AGN) characterized by variations in their visible light by factors as large as 100, strong radio emission and strong optical polarization. In the unified AGN model, a blazar is thought to be produced when an AGN is observed end-on to the emitted jets. *See also* BL Lac objects *and* quasars.

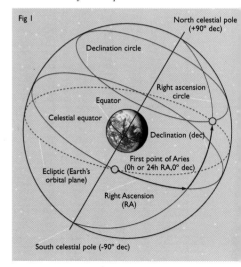
Fig 1

BL Lac objects (BL Lacertae objects)
A class of active galactic nucleus (AGN) typified by the lack of emission and absorption lines in their spectra. In the unified AGN model, a BL Lac object is thought to be produced when the central engine of an AGN is observed end-on.
See also blazars *and* quasars.

blueshift
See redshift.

bolide
A bright meteor (fireball) that produces a sonic boom during its passage through the Earth's atmosphere. Bolides probably originate from asteroids rather than comets and often reach the surface of the Earth as meteorites.

bolometer
An instrument used to measure the total radiation from a celestial source across all wavelengths.

brown dwarf
Stars with mass less than 0.08 solar mass whose core temperatures do not rise high enough to start thermonuclear reactions. Such stars are luminous as they slowly shrink in size and radiate away their gravitational energy. They are known as brown dwarfs as their surface temperatures are below the 2,500 K lower limit for red dwarfs.

cardinal points
The four principal points on an observer's horizon: the north and south points where the celestial meridian crosses the horizon and the east and west points lying 90° from the north and south points.
See also celestial sphere.

celestial equator
The great circle which is the projection of the Earth's equator on to the celestial sphere. The celestial equator is used as the reference plane for the equatorial co-ordinates right ascension and declination. Its orientation changes over the years due to precession.

celestial sphere
An immense imaginary sphere with Earth at its centre on to which all celestial bodies appear to be projected as though they are at a uniform distance. The celestial sphere is used in astronomical co-ordinate systems. Many of the principal astronomical reference points result from the concept of the celestial sphere. (Figure 1).

Cepheid variable
A type of variable star that pulsates in a regular manner with a time-period that is related to its luminosity. Their relatively high luminosity and easy identification makes Cepheids valuable as 'standard candles' that can be used in estimating the distances of galaxies. There are two types of Cepheid variable: the more luminous classical (or type I) Cepheids are massive, young, Population I stars while type II Cepheids (or W Virginis stars) are older Population II stars.

Chandrasekhar limit
The limiting mass of a white dwarf star if it is to avoid collapsing further to become a neutron star or black hole. The limit (around 1.4 solar masses) depends on the white dwarf's composition and rotation rate. Calculated by Subrahmanyan Chandrasekhar (1910–95).

charge-coupled device (CCD)
An imaging device in which photons are collected and a proportional amount of electric charge accumulated. An astronomical CCD usually consists of a two-dimensional array of pixels (picture elements), each pixel responding to the amount of radiation falling on it. The amount of charge in each pixel is coupled to the others, building up a pattern which is recorded in a digital format.

chondrule
Solidified droplets composed mainly of silicates found embedded in chondrites, a type of stony meteorite.

chromatic aberration
A defect in lenses caused by the different wavelengths (colours) of white light being brought to a focus at different distances from the lens. Chromatic aberration causes images to be fringed by different colours, but the effect can be reduced by using an achromatic lens.

chromosphere
The layer about 10,000 km (6,000 miles) thick of the Sun's atmosphere immediately above the photosphere and beneath the corona. Temperatures rise rapidly through the chromosphere from about 4,000 K to 50,000 K at the transition zone. The chromosphere (sphere of colour) is visible as a ring of red light near totality in a solar eclipse.

circumpolar star
A star that remains above the horizon throughout the year. Observers at different latitudes will see different circumpolar stars. For a star to be circumpolar, its polar distance (90° minus its declination) must be less than the observer's latitude.

CNO cycle
A chain of nuclear reactions occurring in main sequence stars slightly more massive than the Sun. The overall effect of the CNO cycle is to fuse four hydrogen nuclei into one helium nucleus with the emission of energy in the form of gamma rays. CNO stands for carbon-nitrogen-oxygen cycle. Nitrogen and oxygen are formed as by-products while carbon is used as a catalyst. Often termed the carbon-nitrogen-oxygen cycle or simply the carbon cycle.

colour index
The difference in apparent magnitude of a star at two standard wavelengths used as a measure of the star's colour and hence temperature. The most commonly used index is the UBV system where the star is measured through different coloured filters, usually blue and yellow giving the B-V index (B for blue, V for visual) or sometimes in the ultraviolet giving the U-B index.

coma
1. A defect in a telescope caused when the incident light falls at an angle and a fan-shaped image is produced. 2. The spherical-shaped cloud of dust and gas that surrounds the nucleus of a comet as it approaches the Sun.

comet
Small icy bodies originating in the Oort cloud and Edgeworth–Kuiper belt which are remnants of material left over from the formation of the Solar System. If perturbed, they can approach the Sun in eccentric, elliptical orbits, often at high inclinations to the ecliptic. As they approach the Sun, heat sublimates the ices, producing the coma which surrounds the nucleus. Near the Sun, dust and gas (ion) tails can be formed, often millions of kilometres long. They are roughly classified into long and short-period comets (with orbital periods greater or less than 200 years).

conjunction
The alignment of two celestial objects (such as two planets or a planet and the Sun) as seen from the Earth. An inferior planet is at inferior conjunction when it lies between the Earth and Sun, and at superior conjunction when it lies on the opposite side of the Sun. A superior planet is at conjunction when it lies on the far side of the Sun.

constellation
One of 88 areas of the sky into which it was divided by the International Astronomical Union in 1930. The areas are bounded by arcs of right ascension and declination (epoch 1875). Stars within each constellation can lie at vastly different distances and are not necessarily physically connected. Many constellation names originate from the ancient Greeks who named patterns in the sky after mythological characters.

Coriolis effect
An effect that influences the motion of objects that are free to move above the surface of a rotating body such as Earth. The effect causes objects moving in the northern hemisphere to be deflected to the right, while those moving in the southern hemisphere are deflected to the left. The Coriolis effect strongly influences the circulation of oceanic and atmospheric currents on Earth, and can also affect the meteorology of any other planet with a substantial atmosphere.

corona
A very tenuous outer layer in the atmosphere of the Sun and other cool main-sequence stars. Coronae are observed to reach incredibly high temperatures; the source of heating is not fully understood. The Sun's corona has a temperature of the order of a million degrees and can be seen at totality during a solar eclipse.

coronagraph
An instrument used to observe the corona of the Sun, at other times than during totality of a solar eclipse, by blocking out the photospheric light of the Sun with an occulting disk.

cosmic background radiation
Electromagnetic radiation observed to come uniformly (to one part in 10,000) from every direction in space. The spectrum of the radiation is that of a black body with a temperature of around 3 K, and therefore peaks in the microwave region. The observed cosmic background radiation is thought to be the cooled, expanded and redshifted remnant of radiation that was emitted about 300,000 years after the Big Bang, when the Universe first became transparent.

cosmic rays
Highly energetic particles moving near the speed of light which enter the Earth's atmosphere from all directions. Mainly protons, they also include electrons, positrons, antiprotons and neutrinos. Primary cosmic rays entering the atmosphere produce cascades of secondary cosmic rays which can be detected at ground level.

cosmology
The study of the Universe as a whole including its origin, evolution and large-scale structure.

dark matter
Matter that is believed to exist because of its gravitational effects, but which has not yet been observed directly. Around 90% of the mass in the Universe may consist of dark matter. Suggestions regarding the nature of dark matter include black holes and exotic fundamental particles.

declination (dec.)
One of the co-ordinates used in the equatorial co-ordinate system, the other being right ascension. The declination (symbol δ) of an object is its angular distance north or south of the celestial equator. Thus declination runs from 0° to 90° and is positive if north of the celestial equator, negative if southwards.

degeneracy pressure
Pressure exerted by matter in a degenerate form due to quantum mechanical effects. When matter becomes so dense that its electrons are packed close together (as in a white dwarf), quantum mechanics does not allow electrons to be in the same energy state and thus a pressure is exerted between the electrons. In neutron stars, the pressure is between neutrons. In degenerate matter, pressure no longer depends on temperature, only on density.

Delta Scuti stars
A type of pulsating variable star that was originally classed as a dwarf Cepheid. They are relatively young stars with masses around 1.5 to 3 solar masses and pulsate with periods of the order of a few hours. They show only slight variation in amplitude and hence are fairly inconspicuous.

density wave
A compressional wave moving through matter causing fluctuations in density. The density wave theory suggests a density wave moving through material in a spiral galaxy causes bursts of star formation producing the spiral structure observed.

differential rotation
The rotation of a non-solid body where different regions rotate at different speeds. Differential rotation is observed in stars and the giant planets where material rotates faster at the equator than the poles. Systems composed of many individual bodies such as a galaxy may also exhibit differential rotation.

diffraction
The spreading of electromagnetic radiation at the edge of obstacles or apertures. The longer the wavelength, the greater the spread. Diffraction is used in diffraction gratings to produce a spectrum. In optical telescopes, diffraction causes diffraction rings and radial diffraction spikes.

diffuse nebula
A luminous cloud of interstellar gas and dust.
See HI and HII regions, emission nebula *and* reflection nebula.

distance modulus
The difference between the apparent magnitude (m) and absolute magnitude (M) of a star giving its distance (d) in parsecs. $d = 10 \times 1.585^{m-M}$.

Doppler shift
The change in frequency of electromagnetic radiation that arises as a result of relative motion between the source of that radiation and the observer. When the source moves away from the observer the frequency is decreased (and the wavelength correspondingly increased) and any spectral lines produced by that source will be redshifted relative to the corresponding lines from a stationary source. The Doppler shift is a very powerful tool that is widely used to determine the radial (line of sight) velocities of stars and other objects. (Figure 2).

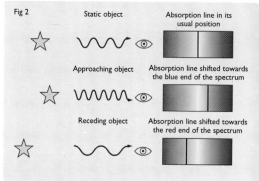

Fig 2

Static object	Absorption line in its usual position
Approaching object	Absorption line shifted towards the blue end of the spectrum
Receding object	Absorption line shifted towards the red end of the spectrum

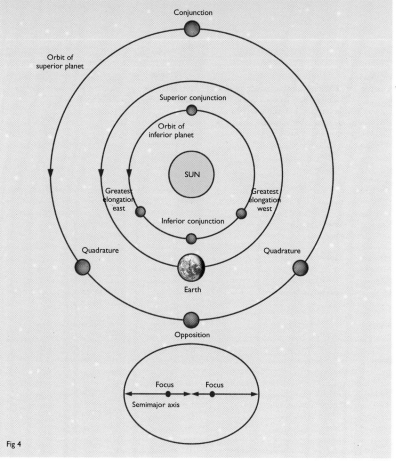

Fig 4

double star
Two stars that appear close together in the sky. Optical doubles are stars that appear to be close due to the line-of-sight effect while physical doubles are gravitationally bound.
See also binary stars.

dwarf star
Another name for a main-sequence star originating from early Hertzsprung-Russell diagrams when stars were classified as either dwarf or giant.

eccentricity
The element of an orbit which describes the extent to which it differs from a circle. The eccentricity, symbol e, is found by dividing the distance between the two foci of the orbit by the major axis. A circle has eccentricity 0, while a parabola has eccentricity 1.

eclipse
The phenomenon that occurs when a celestial body passes into the shadow of another and is totally or partially darkened. A lunar eclipse occurs when the Moon passes into the shadow of the Earth. A solar eclipse is an occultation and occurs when the Moon passes in front of the Sun as seen from Earth.

ecliptic
The projection of the Earth's orbit on to the celestial sphere and thus the apparent path of the Sun in the sky through the year. As the Earth's axis is inclined at an angle of about 23.5°, the ecliptic is inclined by this amount to the celestial equator. This is known as the obliquity of the ecliptic. The equinoxes are the points where the ecliptic crosses the celestial equator and the position of these points changes over the years because of precession.

Edgeworth–Kuiper belt
A region beyond the orbit of Neptune containing rocky/icy debris from which short-period comets are thought to originate. The planet Pluto is sometimes regarded as the largest member of the belt and Triton, the largest satellite of Neptune, was probably once a member.
See also Oort cloud.

electromagnetic radiation
Energy in the form of electromagnetic waves: oscillating electric and magnetic fields at right angles to each other and to the direction of propagation. Light and radio waves are common

examples. Such radiation travels at the speed of light (c = about 300,000 km/s in a vacuum).

electromagnetic spectrum
The full range of electromagnetic radiation from the high-frequency, short-wavelength gamma rays down to the low-frequency, long-wavelength radio waves. Visible light is a very small part of the electromagnetic spectrum. (Figure 3).

electron
A fundamental particle that is a constituent part of every atom. An electron orbiting an atomic nucleus may occupy any one of a series of distinct energy levels dictated by quantum mechanics. Each electron has a negative charge of 1.602×10^{-19} coulomb and a mass of 9.109×10^{-31} kg.

electron degeneracy
The state that exists in matter that is so dense that it is supported by the force exerted by electrons in different quantum states. Electron degeneracy exists in white dwarfs.

elliptical galaxy
An ellipsoidal galaxy composed mainly of older stars and having little gas or dust. In the Hubble Classification scheme, elliptical galaxies run from E0 (almost circular) to E7 (the most elliptical).

elongation
1. The angular distance between the Sun and a Solar System object (usually a planet) measured from 0 to 180° east or west of the Sun. A planet with elongation 0° is at conjunction. An inferior planet has a greatest elongation that is less than 90° east or west of the Sun. A superior planet with elongation 180° is at opposition; with elongation 90° or 270° it is at quadrature. 2. The angular distance between a planet and one of its satellites measured from 0° east or west of the planet. (Figure 4).

emission line
A bright line or band in a spectrum, produced by the emission of electromagnetic radiation at certain wavelengths.

emission nebula
A cloud of interstellar dust and gas that shines by its own light. The nebula is composed primarily of hydrogen that has been ionized. There are three main types of emission nebula: HII regions, planetary nebulae and supernova remnants.

ephemeris
A series of predicted positions of a celestial object such as a planet or comet. The plural is ephemerides.

epicycle
A small circle followed by a planet as it revolved about the Earth in early geocentric models of the motions of the planets. First introduced by Ptolemy.

equant
One of two points equally spaced from the Earth in Ptolemy's geocentric model of the Universe, about which a planet would have uniform motion. The other point is the eccentric, about which the centre of a planet's epicycle revolved.

equator
The great circle on the surface of a near-spherical body such as a planet or star, situated half-way between its poles. The equatorial plane is perpendicular to the body's axis of rotation and passes through the centre of the body.

equatorial mount
A mounting for a telescope or other instrument in which one axis (the polar axis) is parallel to the Earth's axis and points to the Earth's pole. The other axis (the declination axis) is at right angles and is parallel to the Earth's equator. The telescope needs only to be moved around the polar axis to keep an object in the field of view, and this can be done with a motor which turns the polar axis at an appropriate rate as the Earth rotates.

equinox
One of two points on the celestial sphere where the ecliptic crosses the celestial equator. The vernal (or spring) equinox occurs about 21 March when the Sun crosses from south to north, and the autumnal equinox occurs about 23 September when the Sun crosses from north to south. At the equinoxes, the Sun rises due east and sets due west, and the hours of daylight and night are equal. The vernal equinox is also commonly called the First Point of Aries.

equipotential surface
A boundary around a celestial body or system at which the gravitational field is constant. In a close binary system the equipotential surface forms two Roche lobes surrounding each star. Material from either star can pass through either of two Lagrangian points. Mass transfer takes place at the inner Lagrangian point, while mass is lost from the system at the outer Lagrangian point.

escape velocity
The minimum velocity that a body needs to achieve to escape from the gravitational pull of a larger body. The escape velocity of a body of mass M, radius R is $\sqrt{(2GM/R)}$ where G is the gravitational constant. The escape velocity of the Earth is 11.2 km/s and the escape velocity of a black hole is greater than the speed of light.

Fig 3

Type of radiation	Gamma rays		X-rays		Ultra-violet	Visible	Infrared		Radio Waves			
Wavelength	0.0001 0.001	0.01 0.1	1	10	100	1000 nanometres	10 100	1000 micrometres				
								10 millimetres				
								1 10	100 centimetres			
								1 10	100 1000 metres			
Radiation by objects at a temperature of...	100,000,000		10,000,000									
			1,000,000									
				100,000								
					10,000	1000	100 10	1 degree K				

GLOSSARY

event horizon
The largest surface surrounding the singularity of a black hole, from within which no signal can escape. In the case of non-rotating black holes, the event horizon is a sphere with radius equal to the Schwarzschild radius. At the event horizon the escape velocity from the black hole is equal to the speed of light.

excitation
A process by which an atomic electron is given enough energy to move to a higher energy level without escaping entirely from the atom. The energy can be gained during a collision or by absorbing radiation.

exclusion principle
The principle that no two particles can exist in exactly the same quantum state. Often called the Pauli exclusion principle after the physicist who first proposed it.

exoplanet
A planet orbiting another star.

fireball
A meteor brighter than about magnitude −5.

First Point of Aries
Another name for the vernal equinox where the ecliptic crosses the celestial equator and the Sun moves from the south to north celestial hemisphere. It is used as the reference point for right ascension. The vernal equinox used to lie in Aries but, due to precession, it now lies in Pisces.

flare star
A star, usually a faint, cool, red dwarf, that occasionally brightens by a few magnitudes in a short period of time. Also called UV Ceti stars, they are thought to be young stars with strong magnetic fields producing flares similar to those on the Sun. Some flare stars are in close binary systems.

focal length
The distance between a reflecting surface (mirror) or a refracting medium (lens) and its focus or focal point where the image is formed.

focal plane
The plane through the focus in a refracting or reflecting system at right angles to the optical axis.

Fraunhofer lines
Dark absorption lines in the Sun's spectrum caused by the presence of various elements in the Sun's outer atmosphere.

frequency
The number of oscillations per unit time. Symbol ν; frequency in electromagnetic radiation is related to wavelength, λ, by $\nu\lambda = c$, where c is the speed of light.

galactic halo
The approximately spherical distribution of old stars and globular clusters that surrounds the nucleus of a spiral or barred spiral galaxy. The dark halo is a larger distribution of dark matter whose existence is implied by the rotation curve of spiral galaxies.

galaxy
An immense system of stars, dust and gas. The galaxy to which the Sun belongs is the Milky Way. *See also* elliptical, barred spiral, spiral, lenticular *and* irregular galaxies.

gamma rays
The most energetic radiation with the shortest wavelengths: shorter than 0.01 nm. Gamma rays

are absorbed high in the atmosphere, so satellites such as the Compton Gamma Ray Observatory are used to do gamma ray astronomy.

geocentric
Any system which uses the centre of the Earth as its reference point. Compare with heliocentric.

giant branch
The region above and to the right of the main sequence on the Hertzsprung–Russell diagram. This region contains stars in the later stage of evolution that have finished burning hydrogen in their cores and which are burning other elements in concentric shells in their atmospheres.

giant stars
Stars with radii between 10 and 100 solar radii which lie above the main sequence on the Hertzsprung–Russell diagram. A giant star will have higher luminosity than a main-sequence star of comparable radius.
See also red giant *and* supergiant.

gibbous
A phase of the Moon or a planet when more than half, but less than the whole, of its disk is illuminated as seen from Earth.

glitch
A temporary speeding up in the rotation of a pulsar. Pulsars gradually slow down over time and these glitches are believed to be due to adjustments occurring in the core or crust of the neutron star.

globular cluster
A near-spherical cluster of stars. Globular clusters contain mainly old, Population II stars and are situated in the galactic halo. They were probably formed early in the life of the Galaxy.

granulation
The small granular markings observable on the Sun's photosphere. They are the tops of convection cells where hot material is being brought to the surface by convection. Cells are around 1,000 km in size and are part of a larger supergranulation pattern.

gravitation
The universal force of attraction between particles of matter. Newton's law of gravitation states the force F between two particles of mass M_1 and M_2 at distance d apart is given by $F = (GM_1M_2)/d^2$ where G is the gravitational constant. In the presence of very strong gravitation, Einstein's theory of relativity needs to be applied in which gravity changes the geometry of space–time.

gravitational constant (G)
One of the fundamental universal constants of physics. The gravitational constant appears in the mathematical expression of Newton's law of universal gravitation, $F = (GM_1M_2)/d^2$, and has the value $G = 6.673 \times 10^{-11}$ Nm^2 kg^{-2}.

gravitational lens
The bending of the paths of rays of light and other radiation by a massive body such as a black hole or massive galaxy, causing the background object to be split into multiple images and magnified. If the background object is an extended object like a galaxy, the image can be spread into arcs or occasionally into a complete Einstein ring.

greatest elongation
See elongation.

Greenwich Mean Time (GMT)
Mean solar time (in which time is measured with reference to a hypothetical mean sun which moves

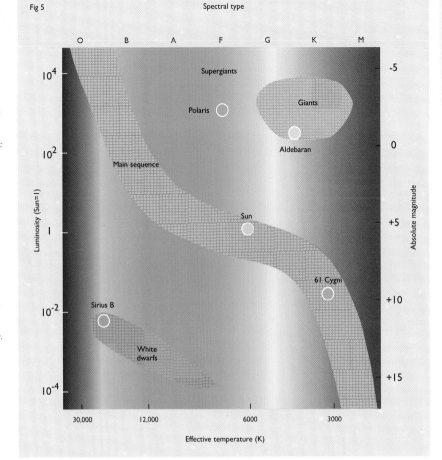

at a constant rate) at Greenwich, UK, on longitude 0°. Renamed Universal Time (UT) for scientific purposes in 1928.

HI and HII regions
Clouds of hydrogen in interstellar space. HI regions are composed of neutral atomic hydrogen while HII regions are composed of primarily ionized hydrogen. The hydrogen in HII regions is ionized by ultraviolet radiation from hot young stars. The Orion Nebula is a prime example of an HII region.
See also emission nebula.

Hawking radiation
Radiation which is emitted from a black hole when a virtual pair of particles (a particle and its antiparticle) is created near the event horizon of a black hole. If one particle lies on the inside and one on the outside, the outside particle would be free to radiate away. A process proposed by Stephen Hawking.

heavy elements
In astronomy, a heavy element is any element heavier (with a greater atomic number) than either hydrogen or helium. Also termed metals.

Heisenberg Uncertainty Principle
The principle that states the position and momentum of a subatomic particle cannot be determined simultaneously with arbitrary precision.

heliocentric
Any system which uses the centre of the Sun as its reference point. Compare with geocentric.

helium flash
An event whereby helium burning starts

explosively in the core of a lower mass star of around 1 to 2 solar masses. When the core of a star has finished burning hydrogen, it collapses and becomes degenerate, causing the temperature to rise rapidly and the helium flash to occur. In stars of greater mass, higher temperatures allow helium fusion to commence more gradually.

Hertzsprung–Russell diagram
Any graph on which a parameter measuring stellar brightness (e.g. magnitude or luminosity) is plotted against a parameter related to a star's surface temperature (e.g. colour index or spectral type). Developed independently by Ejnar Hertzsprung and Henry Norris Russell in the early 1900s, it is a powerful diagram which shows almost all stages in stellar evolution. (Figure 5).

Hohmann orbits
See transfer orbit.

horizontal branch
The horizontal strip on the Hertzsprung–Russell diagram to the right of the main sequence and to the left of the red giant branch, on which stars are burning helium in their core.

hour angle (HA)
The angle measured westwards (opposite to right ascension) along the celestial equator from an observer's meridian to the hour circle of a celestial object.

hour circle
A great circle on the celestial sphere along which declination is measured. The hour circle of an object thus passes through the object itself and both celestial poles.

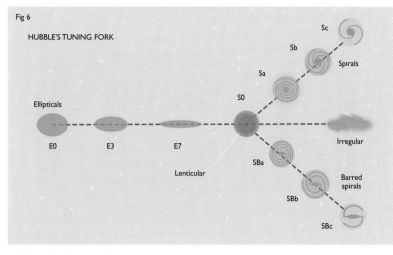

Fig 6

HUBBLE'S TUNING FORK

Ellipticals — E0, E3, E7

Lenticular — S0

Spirals — Sa, Sb, Sc

Barred spirals — SBa, SBb, SBc

Irregular

Hubble Classification scheme

A classification scheme for galaxies proposed by Edwin Hubble in 1925. Also known as the tuning fork diagram, it classifies galaxies according to their appearance. Ellipticals, classified by their ellipticity as seen on the sky, run from E0 (almost spherical) to E7. Spiral galaxies are split into ordinary spirals and barred spirals, and run from Sa (or SBa for barred spirals) with tightly wound arms and large nuclei to Sc (SBc) with loosely wound arms and small nuclei. Lenticular galaxies, S0, lie between the spirals and ellipticals. Any galaxy not falling into this scheme was termed irregular. (Figure 6).

Hubble constant
See Units and Constants p. 368

Hubble law
See Laws and Formulas p. 370

hyperbolic orbit
The orbit of a celestial body which passes another body but is not captured gravitationally and thus does not enter into an orbit around the second body.

inclination
1. The angle between the orbital plane of a body and a reference plane, for example the angle between the orbital planes of the planets and the ecliptic. Inclination, symbol i, is one of the orbital elements. See also orbit. 2. Axial inclination is the angle between the equatorial plane of a near-spherical body and its orbital plane.

inferior conjunction
See conjunction.

inferior planet
A planet (Mercury or Venus) that orbits closer to the Sun than Earth.

inflation
A rapid hypothetical expansion of the Universe which occurred very soon after the Big Bang.

infrared
Electromagnetic radiation with wavelengths from about 0.35 mm down to 700 nm. Observing in the infrared allows astronomers to see through dust clouds that obscure at optical wavelengths.

instability strip
The narrow region of the Hertzsprung–Russell diagram in which pulsating variable stars (e.g. Cepheid variables, RR Lyrae stars etc.) are located. Most stars pass through this region at some time in their lives, the type of pulsating variable the star becomes depending on the star's mass.

intercalation
The insertion of extra days or months into the calendar in order to harmonize it with the solar year.

interference
The phenomenon whereby two sets of waves interact to produce peaks and troughs of intensity.

interferometer
An instrument which uses two or more detectors to collect radiation from a single source. The signals are combined and the interference pattern analyzed, giving more resolution than is possible with one detector.

interstellar matter
The matter lying between the stars of the Galaxy, composed mainly of clouds of hydrogen with some helium and a small percentage of dust grains (mainly silicates).
See also nebula.

inverse-square law
See Maths Toolbox p. 366

ion
Any atom or molecule which has a charge as the result of losing or gaining electrons. Free-moving electrons are often termed negative ions.

ionization
The process by which an atom or molecule is converted to an ion or by which an ion is changed to another ion by the loss or gain of electrons. An atom is said to be completely ionized when it has lost all of its electrons.

irregular galaxy
Any galaxy which does not fit into the Hubble Classification scheme (i.e. neither an elliptical, barred spiral, spiral or lenticular) and which has no well-defined structure or symmetry.

Kepler's laws
See Laws and Formulas p. 370

Kirkwood gaps
Regions within the asteroid belt where few, if any, asteroids are found. The gaps result from periodic disturbances by Jupiter so no stable orbit is possible. They occur at distances that are a simple fraction of Jupiter's orbit, e.g. 1/4, 1/3, 2/3, 1/2 etc. Named after Daniel Kirkwood (1814–95) who first explained them.

Lagrangian points
Points within a gravitational system where the gravitational potential between the bodies in the system is effectively zero. 1. In a system of two massive orbiting bodies, the Lagrangian points are positions at which a smaller body can lie in equilibrium e.g. the Trojan asteroids reside at two of the Lagrangian points in the Jupiter–Sun system. 2. In a system of close binary stars, the Lagrangian points are where the equipotential surfaces (Roche lobes) meet and where material can pass through. (Figure 7).

latitude
1. On a near-spherical body such as planet, the angular distance north or south of the equator. 2. Celestial (or ecliptic) latitude is the angular distance of a body north or south from the ecliptic. 3. Galactic latitude is the angular distance north or south of the galactic equator.

lenticular galaxy
A galaxy which resembles a spiral in morphology, but which has no spiral arms. Designated S0 in the Hubble Classification scheme.

light curve
A graph showing the variation of an object's brightness with time.

light year
The distance travelled by light or any other form of electromagnetic radiation in a vacuum over one year. Symbol l.y., one light year is about 9.4605×10^{12} km; 0.3066 parsecs; 63,240 astronomical units.

Local Group
The small cluster of galaxies to which our galaxy, the Milky Way, belongs. About 40 members are known, most of which are dwarfs; the largest member is the spiral galaxy in Andromeda.

long-period variable
Variable stars with periods of over about 100 days. Most are pulsating red giants or supergiants.
See also Mira stars.

longitude
1. On a near-spherical body such as planet, the angular distance east or west of a reference meridian. 2. Celestial (or ecliptic) longitude is the angular distance of a body east of the vernal equinox. 3. Galactic longitude is the angular distance east of the galactic centre.

luminosity
The total (or absolute) brightness of an object given by the total energy radiated per second. The luminosity (L) of a body over all wavelengths is the bolometric luminosity which is related to the body's effective (or surface) temperature (T_{eff}) by the Stefan–Boltzmann law: $L = 4\pi R^2 \sigma T_{eff}^4$ where R is the radius of the body and σ is the Stefan–Boltzmann constant.

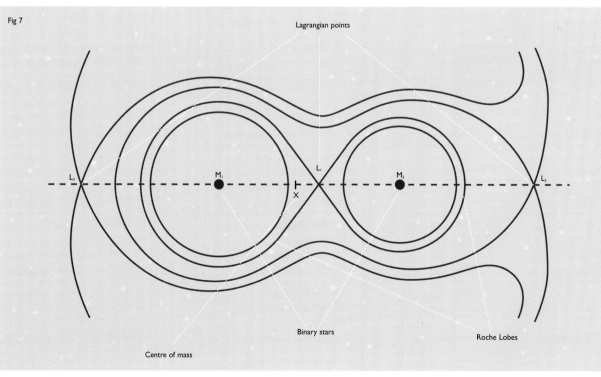

Fig 7

Lagrangian points

L_3 M_1 L_1 M_2 L_2

X

Binary stars

Roche Lobes

Centre of mass

lunar eclipse
An eclipse of the Moon by the Earth, occurring when the Moon passes into the Earth's shadow. If the Moon is completely in the umbra, it is a total lunar eclipse. If the Moon is partially in the umbra, it is a partial lunar eclipse. If the Moon is in the penumbra, it is a penumbral eclipse.

Magellanic Clouds
The two relatively close irregular satellite galaxies of the Milky Way visible from the southern hemisphere. The Large Magellanic Cloud lies about 170,000 light years away, the Small Magellanic Cloud lies about 200,000 light years away. Both galaxies are thought to be gravitationally disrupted by the Milky Way.

magnitude
A measure of the brightness (observed or intrinsic) of an object.
See also apparent magnitude *and* absolute magnitude.

main sequence
The diagonal band on a Hertzsprung–Russell diagram on which most stars lie while they are burning hydrogen in their cores. Stars spend the greatest proportion of their lives on the main sequence, the time depending on their mass: the greater the mass, the higher the temperature and the quicker it burns its core hydrogen and evolves off the main sequence. A star's position on the main sequence also depends on its mass according to the mass–luminosity relation.

mass
The property of a body, typically measured in kilograms, that determines the relative difficulty of accelerating that body. The mass of a body also plays a role in determining the gravitational pull it will exert on other bodies.
See also gravitation.

mass–luminosity relation
A relation between mass (M) and luminosity (L) for stars on the main sequence: L is proportional to M^n where n varies between 2 and 5 according to the mass: the higher the mass, the greater the value of n. For solar-mass stars, n is about 4.

Maunder minimum
A period from about 1645 to 1715 when few sunspots were visible on the Sun. Identified by E. Walter Maunder (1851–1928), the Maunder minimum is believed to be a genuine period of low solar activity.
See also solar cycle.

meridian
1. An imaginary great circle on a near-spherical body such as a planet running perpendicular to its equator through the north and south poles. 2. Abbreviation of celestial meridian which is the great circle on the celestial sphere passing through an observer's zenith, nadir and the two celestial poles. It intersects the observer's horizon at the north and south points.

Messier catalogue
A catalogue of non-stellar astronomical objects first compiled by Charles Messier (1730–1817) to distinguish them from comets for which he was hunting. With additions by others, the total number of objects is 110. Objects are often known by their Messier number, e.g. the Orion Nebula is M42.

meteor
A streak of light observed in the night sky caused by a small particle of interplanetary dust (a meteoroid) burning up in the upper atmosphere. Also commonly known as shooting or falling stars. Sporadic meteors occur throughout the year, whereas meteor showers

occur at specific times when the Earth travels through concentrated meteoroid debris typically left by the passage of a comet.
See also fireball *and* bolide.

meteorite
Interplanetary material reaching the surface of the Earth after falling through the Earth's atmosphere. Roughly classified by their composition into three types: stony, iron and stony–iron. Most meteorites are thought to originate from asteroids, although some material from the Moon and Mars has reached the Earth's surface.

meteoroid
Interplanetary debris smaller than asteroids. If meteoroids pass into the Earth's atmosphere, they burn up as meteors.

Milky Way
1. The galaxy to which the Sun belongs. It is believed to be a barred spiral and the Sun lies in the Orion arm, two thirds of the way out from the centre. 2. The faint band of light visible in a dark sky composed of thousands of stars lying in the plane of the Galaxy.

minor planet
See asteroids.

Mira stars
Long-period pulsating variable stars, usually red giants or supergiants named after the prototype, Mira Ceti. With periods ranging from about 80 to 1,000 days, they pulsate due to instabilities in their atmospheres.

molecular cloud
Cool, dense regions of interstellar matter in which the gas (mainly hydrogen) is in the form of molecules. Ranging in size from small, solar-mass clouds to giant molecular clouds (GMCs) of over a million solar masses, they are often sites of star formation.

molecule
An atom or group of atoms which are capable of independent existence.

momentum
A measure of a body's tendency to continue in its existing state of motion. The linear momentum of a body with mass m and velocity v is mv. The angular momentum of a body of mass m and speed v, moving in a circle of radius r about a point is mvr.

nebula
A cloud of interstellar gas and dust. From the Latin meaning 'cloud', plural is nebulae. There are three main types of nebula: dark, emission and reflection. Composed mainly of hydrogen (see HI and HII regions), emission and reflection nebulae are bright while dark nebulae are observable only because they are silhouetted against a bright background.

neutrino
A fundamental particle with no charge and possibly no mass. They interact only very weakly with matter.

neutron
A fundamental particle that is a constituent part of all atoms except those of common hydrogen. Neutrons are part of the nucleus with no charge and a mass of 1.6749×10^{-27} kg.

neutron star
The remnant of a star in which nuclear fusion has stopped which is held up by neutron degeneracy pressure. Neutron stars lie between the Chandrasekhar limit of 1.4 solar masses up to 2–3 solar masses. Above this, gravity overcomes neutron degeneracy pressure and the object collapses into a black hole. Spinning neutron stars can be detected as pulsars.

New General Catalogue (NGC)
A catalogue of nebulous objects (nebulae, star clusters and galaxies) compiled by J. L. E. Dreyer and published in 1888. Objects are often known by their NGC number, e.g. the Orion Nebula is NGC 1976. The NGC contains around 8,000 entries with about 5,000 additions in two subsequent Index Catalogues, identifying objects by their IC number.

Newton's laws
See Laws and Formulas p. 370

node
One of two points where two orbits intersect e.g. the orbit of a planet with the ecliptic. The ascending node is where the body moves from south to north and the descending node is where it moves from north to south. The line joining the nodes is the line of nodes.

North Celestial Pole (NCP)
The point on the celestial sphere which would be cut by extending the Earth's rotation axis through the north pole. At present it is near the star Polaris in Ursa Minor but, due to precession, the position of the NCP moves in a circle over about 25,800 years.

nova
A star which suddenly brightens by about 10 magnitudes or more in a period of a few days, then slowly declines to its original brightness over a longer period. Classical novae are close binary systems in which one component is a white dwarf that accretes mass from its companion.

nuclear bulge
The ellipsoidal concentration of stars and interstellar matter at the centre of the Galaxy.

nucleus (comet and galaxy)
Generally the inner part of an object. 1. In a comet, the nucleus is the solid body composed mainly of water ice which lies within the coma as it approaches the Sun. 2. In a spiral or barred spiral galaxy, the nucleus is the concentration of material (stars, gas and dust) at the centre.

OB association
Loose groups of young massive stars of spectral types O and B thought to be the result of comparatively recent star formation.

obliquity of the ecliptic
The angle at which the ecliptic is inclined to the celestial equator. This angle, symbol ε, is equal to the tilt of the Earth's axis (currently about 23.5°)

and changes as the Earth's tilt varies, currently by about 47 arcseconds per century.

occultation
When a celestial body passes behind another and its light is totally or partially obscured. A solar eclipse is an occultation when the Moon passes in front of the Sun as seen from Earth. Grazing occultations occur when a body skims the limb of another.

Oort cloud
A reservoir of icy debris surrounding the entire Solar System from which long-period comets are thought to originate. The Oort cloud is thought to extend out to about 50,000 astronomical units but has not been observed directly.
See also Edgeworth–Kuiper belt.

opacity
A measure of the ability of a body to absorb or scatter radiation.

open cluster
Regions where there is a concentration of stars that lie close together in space. The stars are thought to share a common origin and generally travel together through space. Sometimes referred to as galactic clusters, open clusters are less dense than globular clusters, containing no more than a few hundred stars and usually only a few dozen. The Hyades and Pleiades are examples of open clusters.

opposition
The position or time at which a planet or other Solar System body lies directly opposite the Sun in the sky (with an elongation of 180°). It is a favourable time to observe the body as it is then at its closest to the Earth and crosses the meridian at around midnight.

orbit
The path of a celestial object moving in a gravitational field. The planets follow elliptical orbits about the Sun with the Sun at one focus. Seven orbital elements (six for a planet) are needed to define an orbit precisely: the eccentricity, e; the semi-major axis, a; the inclination of the orbital plane to a reference plane, i; the longitude of the ascending node, Ω; the longitude of perihelion, ω; the epoch (time of perihelion passage), T; and the period, P. (Figure 8).

orbital period
The time taken for a body to complete one revolution about another body or the centre of mass of a system.

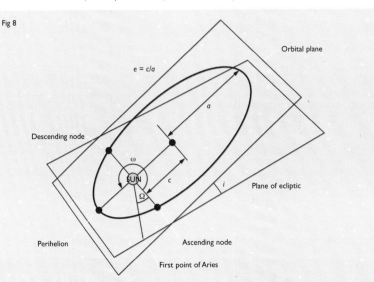

Fig 8

Orbital plane

$e = c/a$

a

Descending node

ω

SUN

c

Ω

i

Plane of ecliptic

Perihelion

Ascending node

First point of Aries

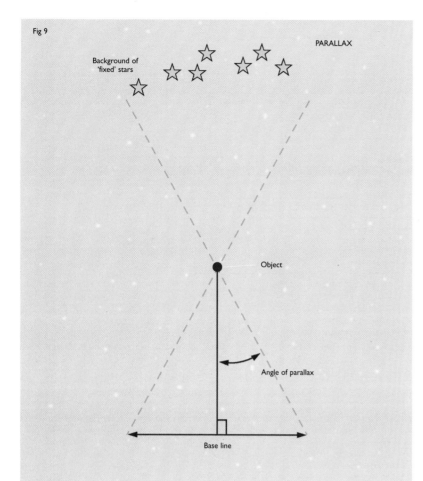

Fig 9

PARALLAX

Background of 'fixed' stars

Object

Angle of parallax

Base line

The energy (*e*) of the photon is directly related to the frequency (ν) of the electromagnetic radiation: $e = h\nu$, where *h* is the Planck constant.

photosphere
The layer in the Sun from which radiation with wavelengths in the visible is emitted. Often termed the Sun's 'surface', it is a layer around 500 km thick below the chromosphere and corona with a temperature of about 5,800 K.

pixel
A 'picture element', the smallest element into which the grid of an electronic imaging device such as a CCD is divided.

Planck constant
See Units and Constants p. 368

planetary nebula
A cloud of gas that has been emitted by a star, often the envelope of a red giant. The material is ionized by the radiation from the stellar core and can be twisted into beautiful shapes by magnetic fields.

planetesimal
A small body from less than a millimetre to several kilometres in size composed of dust, rock or ice from which the planets formed in the early Solar System by the process of accretion.

polarization
The orientation of electromagnetic radiation in a plane perpendicular to its direction of propagation. Polarization occurs by scattering due to interstellar dust grains, or the presence of strong magnetic fields.

position angle (PA)
The direction in the sky of one celestial body from another measured eastwards from north, 0 to 360°.

positron
The antiparticle of the electron with the same mass as the electron (9.109×10^{-31} kg) but the opposite (positive) charge (1.602×10^{-19} coulomb).

precession
The change in direction of the axis of any rotating body. In astronomy it is generally used for the slow change in direction of the Earth's axes. The celestial poles trace out circles with radius equal to the Earth's tilt (about 23.5°) over a period of about 25,800 years (about 50.3 arcseconds a year). This precession causes the positions of the stars to change, and thus any co-ordinate is quoted correct to a standard epoch (currently 2000). Precession of the equinoxes is the movement of the equinoxes along the ecliptic caused by the Earth's precession.

prominence
A huge cloud of solar material extending outwards from the Sun's chromosphere at the limb. If viewed against the Sun's disk, prominences appear as dark filaments. Quiescent prominences can last for many weeks, whereas active prominences have lifetimes of only a few days or hours.

proper motion
The apparent angular motion of a star on the celestial sphere composed of its own motion through space and its motion relative to the Sun. Barnard's star has the greatest known proper motion of 10.3 arcseconds per year.

proplyd
A shortened form of 'protoplanetary disk'.

protogalaxy
An inhomogeneous cloud of gas created by gravitational fluctuations in the primordial matter

formed at the beginning of the Universe from which a galaxy may form.

proton
A fundamental particle that is a constituent part of every atom. Protons are part of the nucleus with a positive charge equal to that of an electron (1.602×10^{-19} coulomb) and a mass of 1.6726×10^{-27} kg.

proton-proton chain
A series of nuclear fusion reactions occurring in the cores of main sequence stars with mass similar to the Sun or less. The overall effect of the proton-proton chain (pp chain) is to fuse four hydrogen nuclei into one helium nucleus with the emission of neutrinos and energy in the form of gamma rays.

protoplanetary disk
A disk of dust and gas around a young star from which planets may form. Over 100 have been observed in the Orion Nebula by the Hubble Space Telescope.

protostar
A stage reached by an embryonic star after it has fragmented from its parent cloud of dust and gas but before temperatures have become high enough for nuclear fusion to occur in its core. Still embedded in their parent cloud, protostars can be imaged in the infrared.

pulsar
A rapidly rotating neutron star detected by pulses of radiation, usually at radio wavelengths. The pulses are believed to be produced by beams of radiation produced by the neutron star along its magnetic axis and detected if the beams are emitted in a direction which sweeps them past the Earth.

pulsating variable
A star which varies in brightness due to pulsations. *See also* Cepheid variable *and* RR Lyrae stars.

quantum electrodynamics
A relativistic quantum theory of electromagnetic interactions.

quantum mechanics
A theory of mechanics of particles at microscopic level in which the Heisenberg Uncertainty Principle and the Pauli exclusion principle dictate the behaviour of matter.

quark
A type of fundamental particle that is also a constituent of protons, neutrons and various other types of particle (but not electrons). There are thought to be six types of quark: up, down, strange, charm, bottom, and top. Quarks have never been observed as free individual particles, only in a combination with other quarks or antiquarks.

quasar
A highly luminous extragalactic compact source originally termed a quasi-stellar object or QSO. Quasars are believed to be extreme examples of active galactic nuclei. In the unified AGN model, a quasar is thought to be produced when the central engine of an AGN is observed obliquely through the disk of matter surrounding the supermassive black hole.
See also Seyfert galaxies, BL Lac objects *and* blazars.

r-process
A method of creating heavy stable nuclei within the interiors of stars by successive capture of neutrons. In the r-process neutrons are added rapidly before the nuclei have time to decay, a process which occurs in supernovae.

P Cygni lines
Double lines in a spectrum consisting of absorption lines adjacent to an emission feature, indicative of an outflow of absorbing gases from a central star.

parabolic orbit
An open orbit whereby a celestial object follows the path of a parabola around another body, such as that followed by some comets around the Sun. In a parabolic orbit, the object does not return to pass the other body again.

parallax
The angular distance an object appears to move against a background of more distant, apparently 'fixed' objects due to being observed from two widely spaced points. The distance of an object can be determined from its parallax provided the distance between the observation points is known. Trigonometric parallax uses a base line such as the Earth's radius (diurnal parallax) or radius of Earth's orbit about the Sun (annual parallax). Other methods of using parallax to determine distance include spectroscopic parallax, moving cluster parallax and statistical parallax. (Figure 9).

parsec
The distance at which an object would have a parallax of one arc second. Symbol pc. One parsec is about 30.857×10^{12} km; 3.2616 light years; 206,265 astronomical units.

penumbra
1. The lighter area surrounding the dark central part of a sunspot. 2. The lighter outer region of a shadow cast by a celestial object. An observer in the penumbral shadow of the Moon sees a partial solar eclipse. *See also* umbra.

perigee
The point in an orbit about the Earth which is the nearest to the Earth.

perihelion
The point in an orbit about the Sun which is the nearest to the Sun.

periodic comet
A short-period comet (period less than 200 years) that has been observed to orbit the Sun more than once. Periodic comets are prefixed with P/ as in P/Halley.

period-luminosity relation
The relationship between the period of light variation of a Cepheid variable and its luminosity. The luminosity (absolute magnitude) of a Cepheid increases with its period of variation. The period-luminosity relation enables Cepheid variables to be used as standard candles, as observing the period gives an indication of the absolute luminosity, and comparison with the apparent luminosity gives an estimate of distance.

perturbation
A small disturbance causing an object or system to deviate slightly from equilibrium. Gravitational effects can perturb celestial bodies in their orbits. The existence of unseen companions to some stars can be inferred from perturbations in the star's proper motion.

photon
A discrete 'packet' of electromagnetic radiation travelling at the speed of light. Electromagnetic radiation travels as waves but it interacts with matter as if it were composed of particles (photons).

GLOSSARY

radial velocity
The component of velocity of an object in the line of sight, either towards or away from the observer. The radial velocity can be calculated from the Doppler shift of spectral lines.

radiant
The point on the celestial sphere from which the members of a meteor shower appear to originate. Usually the meteor shower is named after the constellation in which the radiant lies, e.g. the Leonids have their radiant in the constellation Leo.

radiation
See electromagnetic radiation *and* photon.

radiation laws
See Laws and Formulas p. 370

radio galaxy
A galaxy that emits around a million times more strongly at radio wavelengths than an ordinary galaxy. Usually elliptical galaxies, they are a type of active galaxy similar to quasars. The source of the radio waves is thought to be synchrotron radiation emitted during the accretion of matter on to a central black hole.

radioactivity
Spontaneous disintegration of certain heavy elements producing alpha particles (helium nuclei), beta particles (electrons) and gamma rays (high-energy photons).

radiometric dating
Estimating the age of material by measuring the relative amounts of a radioactive element whose decay rate (half-life) is known and the daughter isotope into which it decays.

red giant
A star which has finished burning hydrogen in its core and which is expanding due to burning hydrogen in a series of shells. As it expands it cools, making it appear red in colour. Red giants have diameters 10 to 1,000 times that of the Sun.

red dwarf
A star at the lower (cool) end of the main sequence with a surface temperature between 2,500 K and 5,000 K and a mass in the range 0.08 to 0.8 times that of the Sun. Red dwarfs are the most common type of star in the Galaxy, comprising at least 80% of the stellar population.

redshift
The increase in wavelength of electromagnetic radiation emitted from a source, due either to the source's own movement away from the observer (Doppler shift) or to the expansion of the Universe (cosmological redshift). The opposite effect is blueshift.

reflection nebula
A cloud of interstellar dust and gas made visible by the reflection of starlight from the dust grains that it contains. Reflection nebulae often appear blue in photographs because the blue light is scattered more efficiently than red.

refraction
The bending of light or any other type of electromagnetic radiation when it crosses from one medium into another. The change of direction is caused by the difference of wave speed in the two media.

resolution
1. The size of the smallest detail made visible by an imaging system. 2. The resolving power of an imaging system. For a telescope, the maximum resolution is given by Dawes' limit: $116/D$

arcseconds where D is the aperture of the telescope in millimetres.

retrograde motion
Motion in the opposite direction to that considered positive in a frame of reference. Positive motion is direct motion. 1. The apparent backwards (east to west) motion of a superior planet in the sky due to the Earth overtaking it on its orbit. 2. Orbital or rotational motion opposite to that of the Earth. (Figure 10).

reversing layer
An old term for the region in the Sun's atmosphere above the photosphere where the spectrum appears almost the reverse of the Fraunhofer spectrum of the photosphere.

right ascension (RA)
One of the co-ordinates used in the equatorial co-ordinate system, the other being declination. Symbol α, right ascension is the angular distance measured eastwards along the celestial equator from the First Point of Aries (the vernal equinox). It is expressed in hours, minutes and seconds (h m s) from 0 to 24h.

Roche limit
The minimum distance at which a planetary satellite can remain in orbit without being pulled apart by the planet's gravitational field. A planet's Roche limit is about 2.5 times its radius.

Roche lobes
See equipotential surface.

rotation curve
A graph of orbital velocity of stars and interstellar matter as a function of radius in a galaxy. The rotation curves of spiral galaxies do not drop off at large distances as would be expected for the amount of luminous matter observed. This implies the existence of dark matter in such systems.

RR Lyrae variables
A type of pulsating variable star that varies over about 0.2 to 2 magnitudes in a period of less than one day, generally found in globular clusters.

Schmidt-Cassegrain telescope
A short-focus telescope combining features of the Schmidt and Cassegrain optical systems. The design includes a Schmidt-type correcting plate with a convex secondary mirror mounted behind it that reflects the light through a hole in the primary mirror to a Cassegrain focus.

Schwarzschild radius
The radius of the event horizon of a black hole of mass M. For a non-rotating black hole with no charge, the Schwarzschild radius is given by $2GM/c^2$ where G is the gravitational constant and c is the speed of light.

Seyfert galaxy
A galaxy with a very luminous, compact nucleus. They are active galactic nuclei (AGNs) with strong emission lines in their spectra. In the unified model of AGNs, Seyfert galaxies are thought to be produced when the central engine is partially obscured (type 1 Seyfert or quasars) or totally obscured (type 2 Seyfert) by the disk of matter surrounding the supermassive black hole. Named after Carl Seyfert (1911–60) who first studied them. See also blazars and BL Lac objects.

sidereal time
Time based on the rotation of the Earth with respect to the stars. Local sidereal noon occurs when the vernal equinox crosses the observer's meridian.

When a celestial object is on the observer's meridian, its right ascension equals the local sidereal time.

singularity
A mathematical point where the laws of physics break down. Calculations predict a singularity exists at the centres of black holes. The Big Bang theory proposes the Universe started from a singularity.

sinusoidal
A quantity that varies as a sine wave.

sling-shot
A term applied to the trajectory of a spacecraft which uses the gravitational field of a larger body, such as a planet, to increase its speed.

solar constant
The total energy radiated by the Sun that passes perpendicularly through a unit area in unit time at a specified distance. The solar constant at the top of the Earth's atmosphere is about 1.37 kilowatts per square metre.

solar cycle
The periodic variation in solar activity. The sunspot cycle has an approximately 11-year period. The magnetic solar cycle is about 22 years as the magnetic polarity in each solar hemisphere reverses after each sunspot cycle.

solar day
The apparent solar day is the time between successive meridian crossings by the Sun. As the motion of the Sun is not constant through the year, the mean solar day uses a hypothetical mean sun which moves at a constant rate through the year.

solar eclipse
An occultation of the Sun's disk by the Moon. An observer in the penumbral shadow sees a partial eclipse, while an observer within the umbra sees a total eclipse in which the complete photosphere is covered and the Sun's corona becomes visible. If the Moon is at a distance such that it does not cover the Sun's disk entirely, an annular eclipse is seen.

Solar System
The Sun and all bodies that are gravitationally bound to it including the nine planets and their moons, the asteroids and the comets.

solar wind
The continuous but gusty outflow of charged particles (mainly electrons and protons) from the Sun.

solstice
Either of the two times during the year when the Sun reaches its most northerly or southerly declination (23.5° north or south). The summer solstice occurs about 21 June, while the winter solstice occurs about 22 December. *See also* tropic.

South Celestial Pole (SCP)
The point on the celestial sphere which would be cut by extending the Earth's rotational axis through the south pole. Unlike the North Celestial Pole, it is

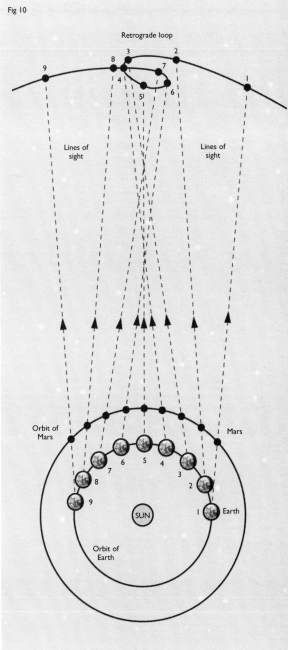

Fig 10

not marked by any nearby star at present. Due to precession, the position of the SCP traces out a circle over about 25,800 years, like the NCP.

spectral line
Bright and dark lines in a spectrum corresponding to the emission and absorption of radiation of a particular wavelength by different elements.

spectral type
The different groups into which stars are classified according to their spectra which are an indication of temperature. The main spectral types are O B A F G K M, O stars being the hottest down to the coolest, M. Each main class is further subdivided and additional lower case letters used to indicate special characteristics.

spectrograph
An instrument used to record the spectrum of a celestial object.

spectroscope
An instrument used to produce spectra of celestial objects using a narrow slit and collimator to produce a narrow beam of radiation which is then spread into a spectrum using a prism or diffraction grating.

spectroscopic binary
A system of two stars showing its binary nature by the Doppler shift of its spectral lines. In a double-lined spectroscopic binary, spectral lines from both components are seen, otherwise it is known as a single-lined spectroscopic binary. Spectroscopic binaries are usually close binaries.

spectroscopy
The science of producing, recording and analyzing spectra.

spectrum
The distribution of intensity of electromagnetic radiation with wavelength. The spectrum of white light through a prism reveals the 'rainbow' band of colours. A continuous spectrum is an unbroken distribution, whereas a line spectrum contains emission and/or absorption lines.

spicules
Short-lived (around 5 to 10 minutes) narrow jets of solar material on the chromosphere, seen at the limb. They have temperatures in the region of 10,000 to 20,000 K and velocities of 20 to 30 km/s. They are located at the edges of super-granulation cells.

spin-orbit coupling
An effect due to tidal interactions between two bodies that causes the less massive body to match its spin period to that of its orbital period. When this occurs (as with the Moon), the body is in synchronous rotation and always shows the same hemisphere towards the more massive body.

spiral galaxy
A galaxy in which spiral arms radiate from the nucleus. In the Hubble classification they are classified from Sa to Sc with the spiral arms of type Sa being the tightest and the nucleus of type Sa being the loosest.
See also barred spiral.

standard candles
Astronomical objects of known luminosity which are therefore useful for determining astronomical distances. Cepheid variables are an example, since the easily observed period of brightness variation is related to the luminosity. Comparisons of the luminosity (absolute magnitude) of any given Cepheid with its apparent magnitude reveals the distance of that Cepheid. Supernovae of Type Ia are another widely used class of standard candle.

star
A gaseous body that emits radiation generated within itself by nuclear fusion.

Stefan–Boltzmann constant
See Units and Constants p. 370.

subdwarf
Stars that are less luminous by 1 to 2 magnitudes than main sequence stars of the same spectral type. Generally old, Population II stars with low metal content that lie below the main sequence on the Hertzsprung–Russell diagram.

sunspot
Dark markings on the photosphere of the Sun often appearing in pairs, but they can grow to become very complex groups. They are regions of cooler solar material; the dark umbra can be up to 2,000 K cooler than the surrounding photosphere, while the penumbra can be about 500 K cooler. They are regions of concentrated magnetic flux.

supercluster
Grouped clusters of galaxies about 100 million light years across. The superclusters appear to be elongated around voids, like soap bubbles. The Local Group belongs to the Local Supercluster which is centred on the Virgo cluster.

supergiant
The largest and most luminous stars known, with spectral types from O to M. Red (M-type) supergiants have the largest radii, of the order of 1,000 times that of the Sun. Supergiants lie above the main sequence and giant region on the Hertzsprung–Russell diagram and are often variable, due to instabilities created by radiation pressure.

superior conjunction
See conjunction.

superior planet
A planet that orbits farther from the Sun than Earth (i.e. Mars, Jupiter, Saturn, Uranus, Neptune or Pluto).

superluminal velocity
A velocity which appears to be faster than light due to geometrical effects.

supernova
A star that explodes, throwing off most of its material and brightening by a factor of about a million. Type I supernovae have no hydrogen in their spectra, Type II do. Core-collapse supernovae (Types Ib, Ic and II) are massive stars whose cores are above the Chandrasekhar limit so they collapse further until neutron degeneracy pressure sets in, causing the explosion. Type Ia are thought to be cores of old massive stars in binary systems sent over the Chandrasekhar limit by mass transfer.

supernova remnant
Material left over from the explosion of a supernova. These are a type of emission nebula that are heated either by interaction with the interstellar medium or by radiation from the pulsar formed in the supernova explosion.

superstring
A theoretical entity associated with families of fundamental particles. Superstring theory allows all the fundamental forces to be unified at the Planck time.

synchrotron radiation
Electromagnetic radiation emitted when charged particles (usually electrons) spiral around magnetic field lines with speeds that are a significant fraction of the speed of light.

synodic period
The period between two successive repetitions of a configuration of a celestial body as viewed from the Earth, e.g. between two successive oppositions of the same planet.

T association
Groups of T Tauri stars (young stars with mass similar to or less than that of the Sun), often still enveloped in material from which they have formed.
See also OB association.

T Tauri star
Very young stars with mass similar to or less than that of the Sun still settling down on to the main sequence. They often show evidence of strong stellar winds.

terraforming
The hypothetical process of altering a planet's surface to become Earth-like.

terrestrial planet
Any of the four small rocky planets, Mercury, Venus, Earth and Mars, situated in the inner Solar System.

thermal radiation
Electromagnetic radiation produced by the interactions of atoms and molecules in a hot, dense medium such as an ionized gas.

total eclipse
When a celestial body passes into the shadow of another and its light is totally obscured. The term is also commonly used to describe those occasions when the Moon completely covers (occults) the solar disk, allowing the Sun's corona to be seen visually from Earth.

transfer orbit
The orbit followed by a satellite or space probe when moving from one orbit to another, possibly around another planet. The transfer requiring the minimum amount of energy is called a Hohmann orbit.

transit
1. The passage of an inferior planet (Mercury or Venus) across the disk of the Sun. 2. The passage of a planetary satellite or its shadow across the disk of its parent planet. 3. The passage of a celestial body across an observer's meridian.

triple-alpha process
The nuclear fusion reaction which converts three alpha particles (helium nuclei) into carbon with a subsequent release of energy. The triple-alpha process occurs at temperatures above 100 million K after stars have consumed all their hydrogen. Also called the Salpeter process, it is the dominant energy-producing process occurring in red giants.

tropic
One of two latitude lines which are the limits reached by the Sun north and south of the equator at the solstices. The Tropic of Cancer (23.5° north) is reached by the Sun at the summer solstice on about 21 June, while the Tropic of Capricorn (23.5° south) is the winter solstice on about 22 December.

tuning-fork diagram
See Hubble classification scheme.

ultraviolet
Electromagnetic radiation with wavelengths from 400 nm down to 10 nm. Ultraviolet radiation with wavelengths longer than 310 nm penetrates the Earth's atmosphere and can be imaged with conventional telescopes, but the rest is absorbed high in the atmosphere.

umbra
1. The dark central part of a sunspot. 2. The dark inner region of a shadow cast by a celestial object. An observer in the umbral shadow of the Moon sees a total solar eclipse. *See also* penumbra.

Universal Time (UT)
Standard time used for scientific purposes, formerly Greenwich Mean Time. UT0 is based on the observed rotation of the Earth with respect to the stars, UT1 is UT0 corrected for the slight variation of the Earth's geographical poles and UTC (Coordinated Universal Time) is based on time given by atomic clocks.

Van Allen belts
Two torus-shaped regions within the Earth's magnetosphere in which charged particles are trapped. The outer belt contains mainly electrons captured from the solar wind. The inner belt contains both protons and electrons. Within the lower belt is a radiation belt containing particles produced by interactions between the solar wind and cosmic rays.

variable star
A star that varies in light output over time. For intrinsic variables the variations are caused by some process occurring in the star itself such as pulsations. Extrinsic variables vary because of some outside effect such as being a member of an eclipsing binary.

vernal equinox
The point on the celestial sphere where the ecliptic crosses the celestial equator and the Sun passes from south to north on about 21 March.

very long baseline interferometry (VLBI)
A technique in radio astronomy that combines the signals from radio telescopes positioned at great distances from each other. The long baselines give greater resolution.

wavelength
The distance over which a periodic wave makes a complete oscillation; symbol λ. Wavelength in electromagnetic radiation is related to frequency, ν, by $\nu\lambda = c$, where c is the speed of light.

white dwarf
The remnants of lower-mass stars in which nuclear fusion has stopped. A white dwarf stops collapsing when electron degeneracy pressure sets in. A white dwarf above the Chandrasekhar limit of about 1.4 solar masses will overcome the electron degeneracy pressure and will collapse further to become a neutron star or black hole. The Sun will probably end up as a white dwarf.

Wien's displacement law
See Laws and Formulas p. 370

Wolf–Rayet stars
Very luminous, hot stars with strong stellar winds and bright emission lines. Often found at the centres of planetary nebulae, they are thought to be massive stars stripped of their outer atmosphere.

wormhole
A hypothetical structure that allows different regions of space–time to be connected by tunnel-like shortcuts. A concept originally implied by solutions to equations arising in Einstein's general theory of relativity, modern cosmology suggests that space–time has, on very small scales, a foam-like structure pervaded by wormholes.

X-rays
Electromagnetic radiation with wavelengths from 10 nm down to 0.01 nm. Most X-rays from space are absorbed before reaching 100 km above the Earth's surface, so X-ray astronomy has to be carried out on high-altitude balloons or satellites.

zenith
The point on the celestial sphere immediately above an observer. The nadir is the opposite point below an observer, 180° away.

zero-age main sequence
Stars on the main sequence which have achieved a stable state with core temperatures sufficiently high for nuclear fusion to begin but have not yet undergone any substantial evolution.

zodiac
A band running about 9° either side of the ecliptic in which the Sun, Moon and planets (except Pluto) are found. The name is also given to the 12 constellations used as astrological signs.

GLOSSARY

Reference
SYMBOLS

Å — Angstrom, a unit of length formerly used by astronomers to measure wavelengths of light; $1 \text{ Å} = 10^{-10}$ m

c — Standard symbol for the speed of light

erg — Symbol used to denote units of energy

G — Standard symbol for the gravitational constant

H_0 — Symbol for the constant in the Hubble expansion law

h — Standard symbol for the Planck constant

K — Kelvin, the unit of temperature in the absolute scale

log — Logarithm; a quantity representing the power to which a fixed number must be raised to produce a given number

m — Standard symbol for mass

N s — Newton second; used to indicate momentum

M_V — Standard symbol for absolute magnitude

m_V — Standard symbol for apparent visual magnitude

v — Standard symbol for velocity

z — Symbol commonly used to designate the Doppler shift ($z = \Delta\lambda/\lambda = v/c$ for velocities much less than the speed of light)

′ — Symbol for arcminutes

″ — Symbol for arcseconds

⊕ — Earth units; used to express bulk properties of planets or stars, e.g. $M_\oplus$ expresses Earth masses

⊙ — Solar units; used to express bulk properties of planets or stars, e.g. $R_\odot$ expresses solar radii

α — The Greek letter alpha; sometimes used to designate an alpha particle

Δλ — The Greek letters delta lambda; used to designate a shift in wavelength, as in the Doppler effect

γ — The Greek letter gamma, sometimes used to designate a gamma-ray photon

λ — The Greek letter lambda, usually used to designate wavelength

ν — The Greek letter nu, the standard symbol for frequency; also used to designate a neutrino in nuclear reactions

π — The Greek letter pi, usually used to designate the parallax angle; also used for the ratio of the circumference of a circle to its diameter

σ — The Greek letter sigma, usually used to designate the Stefan-Boltzmann constant

The Greek Alphabet

Many bright stars are identified by a Greek letter along with the constellation that contains them. Generally they appear in decreasing order of brightness, with alpha (α) being the brightest. See page 254 for a complete explanation.

α	Alpha
β	Beta
γ	Gamma
δ	Delta
ε	Epsilon
ζ	Zeta
η	Eta
θ	Theta
ι	Iota
κ	Kappa
λ	Lambda
μ	Mu
ν	Nu
ξ	Xi
ο	Omicron
π	Pi
ρ	Rho
σ	Sigma
τ	Tau
υ	Upsilon
φ	Phi
χ	Chi
ψ	Psi
ω	Omega

THE HISTORY OF ASTRONOMY

Abbott, David (ed.), *Biographical Dictionary of Scientists: Astronomy*, Peter Bedrick, New York, 1984

Bertotti, B. et al, *Modern Cosmology in Retrospect*, Cambridge University Press, Cambridge, 1990

Bettex, Albert, *The Discovery of Nature*, Thames and Hudson, London, 1965

Chapman, Allan, *Dividing the Circle: The Development of Critical Angular Measurement in Astronomy 1500–1850*, New York and London, 1990

Chapman, Allan, *Astronomical Instruments and Their Users: Tycho Brahe to William Lassell*, Ashgate Publishing, 1996

Cornford, Francis MacDonald, *Plato's Cosmology*, Hackett Publishing Co., Inc., Cambridge, Massachusetts, 1997

Dolgov, A. D., *Basics in Modern Cosmology*, Editions Frontieres, 1990

Duhem, Pierre and Roger Ariew (eds.), *Medieval Cosmology*, University of Chicago Press, Chicago, 1985

Evans, James, *History and Practice of Ancient Astronomy*, Oxford University Press, Oxford, 1998

Gadalla, Moustafa, *Egyptian Cosmology*, Tehuti Research Foundation, London, 1997

Galilei, Galileo, *Siderius Nuncius* (1610), facsimile ed.

Galilei, Galileo, *Dialogue Concerning Two New Sciences* (Great Minds Series), Prometheus Books, New York, 1991

Hoskin, Michael (ed.), *Cambridge Illustrated History of Astronomy*, Cambridge University Press, Cambridge, 1997

Hoskin, Michael (ed.), *Concise History of Astronomy*, Cambridge University Press, Cambridge, 1999

Howse, Derek, *Greenwich Time and Longitude*, Philip Wilson Publishers, 1997

Hoyle, Fred, *Astronomy*, Crescent Books, London, 1962

Kahn, Charles H., *Anaximander and the Origins of Greek Cosmology*, Hackett Publishing Co. Inc., Cambridge, Massachusetts, 1994

Kepler, Johannes, *Kepler's Dream*, University of California Press, Berkeley, 1965

Kepler, Johannes, *Epitome of Copernican Astronomy and Harmonies of the World* (Great Minds Series), Prometheus Books, New York, 1995

Leslie, John, (ed.), *Modern Cosmology & Philosophy*, Prometheus Books, New York, 1998

Liddle, Andrew, *An Introduction to Modern Cosmology*, John Wiley & Sons, New York, 1998

McCluskey, Stephen C., *Astronomies and Cultures in Early Medieval Europe*, Cambridge University Press, Cambridge, 2000

Nasr, S. Hossein, *Islamic Science*, World of Islam Festival Publishing, London, 1976

Newton, Isaac, Cohen, Bernard I. (trans.), Whitman, Anne (trans.), Budenz, Julia (trans.), *The Principia*, University of California Press, Berkeley, 1999

Newton, Isaac, Cohen, Bernard I. (trans.), *Mathematical Principles of Natural Philosophy*, Harvard University Press, Cambridge, Massachusetts, 1990

North, John, *Norton History of Astronomy and Cosmology*, Norton, New York, 1995

North, J. D., *The Measure of the Universe: A History of Modern Cosmology*, Dover Publications Inc., London, 1990

Panek, Richard, *Seeing is Believing: How the Telescope Opened Our Eyes and Minds to the Heavens*, Penguin, New York, 1998

Schechner, Sara J., *Comets, Popular Culture and the Birth of Modern Cosmology*, Princeton University Press, New Jersey, 1999

Swerdlow, N. M. (ed.), *Ancient Astronomy and Celestial Divination*, Dibner Institute for the History of Science and Technology, Cambridge, Massachusetts, 2000

Walker, Christopher (ed.), *Astronomy before the Telescope*, British Museum Press, London, 1996

Westfall, Richard, *The Life of Isaac Newton*, Cambridge University Press, Cambridge, 1996

White, Michael, *Isaac Newton: The Last Sorceror*, Fourth Estate, London, 1998

THE LAWS OF PHYSICS

Blecher, M. (ed.), *Low-Energy Tests of Conservation Laws in Particle Physics*, American Institute of Physics, New York, 1993

Bligh, Bernard Ramsay, *Big Bang Exploded!: Cosmology Corrected, a Commentary with Thermodynamics*, Bernard Ramsay Bligh, 2000

Cartwright, Nancy, *How the Laws of Physics Lie*, Oxford University Press, Oxford, 1983

De Podesta, Michael, *Understanding the Properties of Matter*, Hemisphere Publishing, New York, 1996

Einstein, Albert, *Relativity: The Special and General Theory*, Three Rivers Press, California, 1995

Fackler, O., Thanh Van and Jean Tran (eds.), *Tests of Fundamental Laws of Physics*, Editions Frontieres, 1989

Fang, L. Z. and Y. Q. Chu, *From Newton's Laws to Einstein's Theory of Relativity*, World Scientific Publishing, Singapore, 1987

Feynman, Richard and Steven Weinberg, *Elementary Particles and the Laws of Physics*, Cambridge University Press, Cambridge, 1999

Goldstein, Martin and Inge F. Goldstein, *The Refrigerator and the Universe: Understanding the Laws of Energy*, Harvard University Press, Cambridge, Massachusetts, 1995

Guiderdoni, B. (ed.), *Dark Matter in Cosmology, Clocks and Tests of Fundamental Laws*, Editions Frontieres, 1995

Kovalevsky, J. and V. A. Brumberg (eds.), *Relativity in Celestial Mechanics and Astrometry*, Reidel Publishing Company, Dordrecht, 1986

Lightman, Alan P., *Great Ideas in Physics: The Conservation of Energy, the Second Law of Thermodynamics, the Theory of Relativity and Quantum Mechanics*, McGraw, New York, 2000

Mould, R. A., *Basic Relativity*, Springer-Verlag, New York, 1994

Moulton, Forest Ray, *Introduction to Celestial Mechanics (2nd-edition)*, Dover Publications Inc., London, 1984

Poincaré, Henri, *New Methods in Celestial Mechanics*, American Institute of Physics, New York, 1993

Ribaric, Sustersic M., *Conservation Laws and Open Questions of Classical Thermodynamics*, World Scientific Publishing, Singapore, 1990

Rindler, Wolfgang, *Relativity: Special, General and Cosmological*, Oxford University Press, Oxford, 2001

Stenger, Victor J., *The Unconscious Quantum: Metaphysics in Modern Physics and Cosmology*, Prometheus Books, New York, 1995

Szebehely, V. G. and H. Mark, *Adventures in Celestial Mechanics: A First Course in the Theory of Orbits*, John Wiley & Sons, New York, 1998

Tifft, W. G. (ed.), *Modern Mathematical Models of Time and their Application to Physics and Cosmology*, Kluwer Academic Publishers, Norwell, Massachusetts, 1997

Vinti, John P., *Orbital and Celestial Mechanics*, American Institute of Aeronautics and Astronautics (AIAA), 1998

Yamashita, Yasumasa, *Atlas of Representative Stellar Spectra*, University of Tokyo Press, Tokyo, 1978

IN SEARCH OF QUANTUM REALITY

Close, Frank, et al, *The Particle Explosion*, Oxford University Press, Oxford, 1987

Cushing, James T. (ed.), *Bohemian Mechanics and Quantum Theory: An Appraisal*, Kluwer Academic Publishers, Norwell, Massachusetts, 1996

Davies, Paul and John Gribbin, *The Matter Myth*, Viking, London, 1991

Duck, Ian and E. C. G. Sudarshan, *100 Years of Planck's Quantum*, World Scientific Publishing Company, 2000

Edmonds, A. R., *Angular Momentum in Quantum Mechanics*, Princeton University Press, New Jersey, 1996

Feynman, Richard and Steven Weinberg, *Elementary Particles and the Laws of Physics*, Cambridge University Press, Cambridge, 1999

Feynman, Richard, *QED: The Strange Theory of Light and Matter* (Alix G. Mautner Memorial Lectures), Penguin, London, 1990

Feynman, Richard, *Quantum Electrodynamics*, Perseus Books, 1998

Gribbin, John, *In Search of Schrodinger's Cat*, Corgi, London, 1985

Gross, Franz, *Relativistic Quantum Mechanics and Field Theory*, John Wiley & Sons, New York, 1999

Herbert, Nick, *Quantum Reality: Beyond the New Physics*, Anchor, North Carolina, 1987

Herbert, Nick, *Faster than Light: Superluminal Loopholes in Physics*, New American Library Trade, 1985

House, J. E., *Fundamentals of Quantum Mechanics*, Academic Press, 1998

Norris, Christopher, *Quantum Theory and the Flight from Realism: Philosophical Responses to Quantum Mechanics*, Routledge, London, 2000

Yndurain, F. J., *Relativistic Quantum Mechanics and Introduction to Field Theory*, Springer-Verlag, 1996

THE UNIVERSE: PAST, PRESENT & FUTURE

Al-Khalili, Jim, *Black Holes, Wormholes and Time Machines*, Institute of Physics Publishing, London, 1999

Allen, Harold, W. G., *Cosmic Perspective: Evolution and Reincarnation—The Demise of the Big Bang Theory*, Sunstar Publishing, Fairfield, Iowa, 1998

Bernstein, Jeremy, *Kinetic Theory in the Expanding Universe*, Cambridge University Press, Cambridge, 1988

Calder, Nigel, *Afterglow of Creation*, Arrow Books, London, 1993

Cornell, James, *Bubbles, Voids and Bumps in Time: The New Cosmology*, Cambridge University Press, Cambridge, 1989

Danielson, Dennis Richard (ed.), *The Book of the Cosmos: Imagining the Universe from Heralcitus to Hawking: A Helix Anthology*, Perseus Book Group, New York, 2000

Delsemme, Armand H., *Our Cosmic Origins: From the Big Bang to the Emergence of Life and Intelligence*, Cambridge University Press, Cambridge, 1998

Eberhardt, Nikolai, *From the Big Bang to the Human Predicament*, Pentland, Raleigh, North Carolina, 1998

Eddington, Arthur, *The Expanding Universe*, Cambridge University Press, Cambridge, 1988

Gribbin, John and Martin Rees, *The Stuff of the Universe*, Heinemann, London, 1990

Gribbin, John, *In Search of the Big Bang*, Penguin, London, 1998

Gribbin, John, *Case of the Missing Neutrinos and Other Curious Phenomena of the Universe*, Fromm International, 1998

Gribbin, John, *The Omega Point*, Heinemann, London, 1987

Gribbin, John, *The Birth of Time*, Yale University Press, Connecticut, 2000

Harrison, Edward, *Cosmology: The Science of the Universe* (2nd Edition), Cambridge University Press, Cambridge, 2000

Harrison, Edward, *Darkness at Night: A Riddle of the Universe*, Harvard University Press, Cambridge, Massachusetts, 1987

Hawking, Stephen, *Brief History of Time: From the Big Bang to Black Holes*, Bantam, New York, 1998

Hawking, Stephen, *The Large-Scale Structure of Space-Time*, Cambridge University Press, Cambridge, 1973

Hogan, Craig, J., *Little Book of the Big Bang: A Cosmic Primer*, Springer-Verlag, New York, 1998

Hoyle, Fred, *Different Approaches to Cosmology: From a Static Universe Through the Big Bang Towards Reality*, Cambridge University Press, Cambridge, 2000

Kembhavi, Ajit K. and Jayant V. Narlikar, *Quasars and Active Galactic Nuclei: An Introduction*, Cambridge University Press, Cambridge, 1999

Layzer, David, *Constructing the Universe*, Scientific American Library, 1984

Liddle, Andrew, *An Introduction to Modern Cosmology*, John Wiley & Sons, New York, 1999

Mather, John C. and John Boslough, *The Very First Light*, Basic Books, New York, 1996

Moore, Patrick and Iain Nicolson, *Black Holes in Space*, Orbach and Chambers, 1974

Murray, Carl D. et al, *Solar System Dynamics*, Cambridge University Press, Cambridge, 2000

Nelson, P. G., *Big Bang, Small Voice: Reconciling Genesis and Modern Science*, Whittles Publishing, 1999

Rowan-Robinson, Michael, *Universe*, Longman, London, 1990

Silk, Joseph, *The Big Bang* (3rd edition), W. H. Freeman, 2001

Singh, Jagjit, *Modern Cosmology*, Pelican, London, 1970

Weedman, Daniel W., *Quasar Astronomy*, Cambridge University Press, Cambridge, 1986

CONTENTS OF THE COSMOS

Bertin, Giuseppe, *The Dynamics of Galaxies*, Cambridge University Press, Cambridge, 2000

Binney, James and Michael Merrifield, *Galactic Astronomy*, Princeton University Press, New Jersey, 1998

Challoner, Jack, *Equinox: Space*, Channel 4 Books, London, 2000

Danielson, Richard D. (ed.), *The Book of the Cosmos: Imagining the Universe from Heraclitus to hawking*, Perseus Books, 2000

Fanning, A. E., *Planets, Stars and Galaxies*, Dover Publications, London, 1986

Gaustad, John and Michael Zeilik, *Astronomy: The Comsic Perspective*, John Wiley & Sons, New York, 1990

Giacconi, R. and W. Tucker, *The X-Ray Universe*, Harvard University Press, Cambridge, Massachusetts, 1985

Kaufmann, William J., *Universe*, Freeman and Company, 1994

Lederman, Leon and David Schramm, *From Quarks to the Cosmos*, Scientific American Library, New York, 1989

Longair, Malcolm S., *Our Evolving Universe*, Cambridge University Press, Cambridge, 1997

Luminet, Jean-Pierre, *Black Holes*, Cambridge University Press, Cambridge, 1993

Malin, David and Paul Murdin, *Colours of the Stars*, Cambridge University Press, Cambridge, 1984

O'Meara, Stephen (ed.) and David Leary *The Messier Objects Field Guide*, Cambridge University Press, Cambridge, 1999

Peacock, John A., *Cosmological Physics*, Cambridge University Press, Cambridge, 1999

Ridpath, Ian, *Collins Gem: Stars*, HarperCollins, London, 1999

Riordan, Michael and David Schramm, *Dark Matter and the Structure of the Universe*, Oxford University Press, Oxford, 1993

Ronan, Colin, *The Universe Explained*, Thames and Hudson, London, 1994

Sparke, Linda S. and John S. Gallagher, *Galaxies in the Universe*, Cambridge University Press, Cambridge, 2000

Tayler, Roger, *Galaxies: Structures and Evolution*, Cambridge University Press, Cambridge, 1993

Tayler, Roger, *The Hidden Universe*, Ellis Horwood, 1991

Rubin, Vera, *Bright Galaxies Dark Matters*, American Institute of Physics, New York, 1996

Sagan, Carl, *Cosmos*, Abacus, London, 1995

Seward, Frederick D. and Philip A. Charles, *Exploring the X-ray Universe*, Cambridge University Press, Cambridge, 1995

Weedman, Daniel, *Quasar Astronomy*, Cambridge University Press, Cambridge, 1998

OUR SOLAR SYSTEM

Ahrens, Donald C., *Meteorology Today: An Introduction to Weather, Climate, and the Environment*, Pacific Grove, California, 1999

Barnes-Svarney, Patricia, *Asteroid: Earth Destroyer or New Frontier?*, Plenum Publishing Corps, 1996

Beatty, J. Kelly, Carolyn Collins Petersen and Andrew Chaikin (eds), *The New Solar System*, Harvard University Press, Cambridge, Massachusetts, 1999

Burnham, Robert, *Great Comets*, Cambridge University Press, Cambridge, 2000

Calder, Nigel, *Comets: Speculation and Discovery*, Dover Publications, London, 1994

Davidson, Keay, *Carl Sagan: A Life*, John Wiley & Sons, New York, 1999

Evans, James, *The History and Practice of Ancient Astronomy*, Oxford University Press, New York, 1998

Grant, Edward (ed.), *Stars, Planets, & Orbs*, Cambridge University Press, New York, 1996

Guillermier, P. and S. Koutschmy, *Total Eclipses. Science, Observations, Myths and Legends*, Springer, Heidelberg, 1999

Hunt, Garry and Patrick Moore, *Atlas of Uranus*, Cambridge University Press, Cambridge, 1989

Hunt, Garry and Patrick Moore, *Atlas of Neptune*, Cambridge University Press, Cambridge, 1994

Lang, Kenneth R., *Sun, Earth and Sky*, Springer, Heidelberg, 1997

Levy, David, *Comet: Creators and Destroyers*, Touchstone Books, 1998

Levy, David, *Eclipse: Voyage to Darkness and Light*, Pocket Books, 2000

Levy, David, *Impact Jupiter: The Crash of Comet Shoemaker-Levy 9*, Plenum Publishing Corporation, 1995

Levy, David, *Shoemaker by Levy*, Princeton University Press, New Jersey, 2000

Maran, Stephen P., *Astronomy For Dummies*, Foster City, California, 1999

Peebles, Curtis, *Asteroids: A History*, Smithsonian Institution Press, 2000

Phillips, Kenneth J. H., *Guide to the Sun*, Cambridge University Press, Cambridge, 1995

Poundstone, William, *Carl Sagan: A Life in the Cosmos*, New York, 1999

Ridpath, Ian and Wil Tirion, *Pocket Guide to Stars and Planets*, HarperCollins, London, 2001

Sheehan, William and James O'Meara, *Mars: The Lure of the Red Planet*, Prometheus Books, 2001

Steel, Duncan, *Rogue Asteroids and Domesday Comets: The Search for the Million Megaton Menace theat Threatens Life on Earth*, John Wiley & Sons, New York, 1997

Stern, Alan and Jacqueline Mitton, *Pluto and Charon: Ice Worlds on the Ragged Edge of the Solar System*, John Wiley & Sons, New York, 1999

Taylor, Peter O., *Observing the Sun*, Cambridge University Press, Cambridge, 1991

Vanin, Gabrieli, *Cosmic Phenomena: Comets, Meteor Showers, Eclipses*, Firefly Books, 1999

Weissman, Paul et al, *Encyclopedia of the Solar System*, Academic Press, 1998

Wilson, Robert, *Astronomy Through The Ages*, Taylor and Francis, London, 1997

Zeilik, Michael (ed.), *Astronomy: The Evolving Universe*, New York, 1999

WATCHING THE SKY

Audouze, Jean and Guy Israël (eds.), *The Cambridge Atlas of Astronomy*, Cambridge University Press, Cambridge, 1988

Covington, Michael, *Astrophotography for the Amateur*, Cambridge University Press, Cambridge, 1985

Levy, David, *Observing Comets, Astroids, Meteors and the Zodiacal Light*, Cambridge University Press, Cambridge, 1994

Garfinkle, Robert, *Star-Hopping: Your Visa to the Universe*, Cambridge University Press, Cambridge, 1994

Newton, Jack & Philip Teece, *The Cambridge Deep-Sky Album*, Cambridge University Press, Cambridge, 1984

Dickinson, Terence & Jack Newton, *Splendors of the Universe*, Paradise Kay Publications, Kingston, Ontario, 1997

Howard, N. E., *Standard Handbook for Telescope Making*, New York, 1984

Ingalls, A., *Amateur Telescope Making Vols 1-3*, New York, 1964

Mitton, Jacqueline and Simon, *The Young Oxford Book of Astronomy*, Oxford University Press, Oxford, 1995

Levy, David, *Observing Variable Stars: A Guide for the Beginner*, Cambridge University Press, Cambridge, 1998

Ronan, Colin, *The Practical Astronomer*, Marshall Publishing, London, 1981

Ridpath, Ian, *Norton's Star Atlas*, Longman, Harlow, 1997

Ridpath, Ian and Wil Tirion, *Monthly Sky Guide*, Cambridge University Press, Cambridge, 1996

Scagell, Robin, *Astronomy from Towns and Suburbs*, George Phillip, London, 1994

Scott-Houston, Walter and James O'Meara, *Deep Sky Wonders*, Sky Publishing Corporation, 1999

Sidgwick, John Benson, *The Amateur Astronomer's Handbook*, Dover Publications, New York, 1980

Stephenson, Bruce, et al, *The Universe Unveiled: Instruments and Images Through History*, Cambridge University Press, Cambridge, 2000

Zeilik, Michael and Stephen A. Gregory, *Introductory Astronomy and Astrophysics*, Saunders College Publishing, 1997

SPACE EXPLORATION

Aldrin, B. and M. McConnell, *Men From Earth*, Bantam, London, 1989

Allday, Jonathan, *Apollo in Perspective*, Institute of Physics Publishing, London, 1999

Baker, D., *Spaceflight and Rocketry, A Chronology*, Facts on File Inc, 1996

Bennett, Mary D. and David S. Percy, *Dark Moon*, Aulis Publishers, 1998

Cernan, E. and D. Davis, *The Last Man on the Moon: Astronaut Eugene Cernan and America's Race in Space*, St Martins Press, London, 1999

Freeman, Marsha (ed.), *Challenges of Human Space Exploration*, Springer-Verlag, New York, 2000

Gatland, K., *Space Technology, a Comprehensive History of Space Exploration*, Salamander, London, 1980

Godwin, Robert (ed.), *Apollo 13*, Apogee Books, 2000

Harland, David M., *Jupiter Odyssey, the Story of NASA's Galileo Mission*, Springer-Verlag, New York, 2000

Hurt, H., *For All Mankind*, Atlantic Monthly Press, 1988

Jensen, C., *No Downlink: A Dramatic Narrative about the Challenger Accident and Our Time*, Farrar Straus Giroux, 1996

Kranz, Gene, *Failure is not an Option: Mission Control from Mercury to Apollo 13 and Beyond*, Simon and Schuster, New York, 2000

Leverington, David, *New Cosmic Horizons: A History of Space Astronomy from the V2 to the Hubble Space Telescope*, Cambridge University Press, Cambridge, 2000

McNab, David and James Younger, *The Planets*, BBC Worldwide, London, 1999

Montgomery, Scott, et al, *Back in Orbit: John Glenn's Return to Space*, Longstreet Press Inc., 1998

Moore, Patrick, *The Guinness Book of Astronomy*, Fifth Edition, Guinness, London, 1995

Moore, Patrick, *Mission to the Planets: The Illustrated Story of the Exploration of the Solar System*, Cassell, New York and London, 1995

Newton, David E., *U.S. and Soviet Space Programs: A Comparison*, Franklin Watts, New York, 1988

Saga, C., *Pale Blue Dot: A Vision of the Human Future in Space*, Headline, London, 1995

Slayton, D. and M. Cassutt, *Deke! US Manned Space: From Mercury to the Shuttle*, Forge, 1994

Trux, John, *The Space Race*, New English Library, London, 1985

Wilhelms, D., *To A Rocky Moon, A Geologist's History of Lunar Exploration*, Arizona

Wunsch, Ssi Trautmann, *The Adventure of Sojourner: The Mission to Mars that Thrilled the World*, Mikaya Press, 1998

Zubrin, Robert and Richard Wagner, *The Case for Mars: The Plan to Settle the Red Planet and Why We Must*, Touchstone Books, 1997

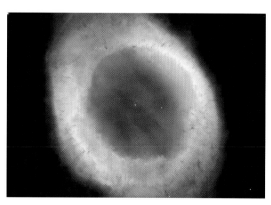

Reference
WEB SITES

THE HISTORY OF ASTRONOMY

http://www.astro.uni.bonn.de/~pbrosche/hist_astr/
ha_pers.html
*One of the most exhaustive web sites of biographies on scientists
who have shaped the history of astronomy.*

http://www.bios.niv.edu/orion/history.html
A brief history of astronomy.

http://webhead.com/WWWWVL/Astronomy/astroweb/
yp_history.html
*Interactive atlas of world astronomy with detailed descriptions of
the astronomy of cultures around the world.*

http://www.gpc.peachnet.edu/~pgore/astronomy/astr10/
ancient.htm
This web site explains basic types of ancient observatories.

http://es.rice.edu/ES/humsoc/Galileo
*Hypertext source of information on the life and work of Galileo
Galilei and the science of his time.*

http://galileoandeinstein.physics.virginia.edu/lectures/
tycho.htm
*Scholarly lectures on the work of Tycho Brahe and his assistant,
Johannes Kepler.*

THE LAWS OF PHYSICS

http://www2.corepower.com:8080/~relfaq/relativity.html
*A good first stop for beginners with non-technical to more
maths-intensive answers to some of the most frequently
asked questions.*

http://math.ucr.edu/home/baez/gr/gr.html
*A site of interconnected web pages that serve as a general
introduction to general relativity. The goal is to demystify what
is often a complicated subject and convey the big ideas without
getting bogged down in complicated calculations.*

http://cosmos.colorado.edu/astr1120/L1S1.html
*Explains why electromagnetic radiation is so crucial for
astronomy.*

http://csep10.phys.utk.edu/astr161/lect/history/newton3
laws.html
*An explanation of how Newton changed our understanding of
the Universe with his three Laws of Motion.*

http://www.herts.ac.uk/astro_ub/a04_ub.html
Explains the importance and function of electromagnetic radiation.

http://www.newton.dep.anl.gov/archive.htm
*Ask a Scientist Archive Directory allows you to select a search
engine using key words or short phrases to search topics through
the available archives.*

http://www.physik.de/jakubows/elmgspec.html
*An in-depth scholarly discussion on the spectrum of
electromagnetic radiation. It also offers a hyperlink to
an introduction to spectroscopy.*

IN SEARCH OF QUANTUM REALITY

http://www.encyclopedia.com/articles/10681.html
*This online encyclopedia provides a broad introduction
to Quantum Mechanics. Hyperlinked text allows you to look up
the following terms and personalities.*

http://antoine.fsu.umd.edu/chem/senese/101/quantum/
faq.shtm
*General chemistry discussions online on topics like radiation and
matter and the Uncertainty Principle.*

http://www.chembio.uoguelph.ca/educmat/chm386/
rudiment/rudiment.htm
Rudiments of quantum theory including Planck's suggestion

*regarding the quantum nature of light and Bohr's and
Rutherford's development of the mathematical model which
became the Old Quantum Theory.*

http://www.colorado.edu/physics/2000/elements_as_
atoms/electron_config.html
*A clear, simple explanation of Wolfgang Pauli's Exclusion
Principle in comic-strip format.*

http://members.aol.com/cclinker/sapfam1.htm
*An explanation of elementary subatomic particle generations
and families.*

http://plato.stanford.edu/entries/qt-idind/
Metaphysical implications of quantum physics.

THE UNIVERSE: PAST, PRESENT & FUTURE

http://www.astro.ucla.edu/~wright/errors.html
*A fun-filled web site for those interested in the errors within
some popular attacks on the Big Bang.*

http://www.astro.ucla.edu/~wight/cosmology_faq.html
*A good discussion forum for topics related to the Big Bang, the
expansion of the Universe, models of the Universe, dark matter,
the collapse of the Universe and much more.*

http://www.image.gsfc.nasa.gov/poetry/ask/a11610.htm
Archive of NASA Image Space Science Question & Answers.

http://www.columbia.edu/cu/newrec/2416/tmpl/
story.2html
*Experiments with tiny particles that could reshape models of
the Universe.*

http://www.ssg.sr.unh.edu/406/Review/rev3.html
Geometrical models of the Universe.

http://www.acuniverse/freeserve.co.uk/history.htm
*Brief summary of traditional theoretical models of
the Universe.*

http://www.time.com/time/time100/scientistprofile/
hubble.html
Astronomers and thinking: a profile of Edwin Hubble.

http://antwrp.gsfc.nasa.gov/diamond_jubilee/d_1996/
sandage_hubble:html
A profile of Edwin Hubble by Allan Sandage.

CONTENTS OF THE COSMOS

http://www.seds.org/messier/
*Images and basic data for the Messier Catalogue of nebulae,
showing many beautiful nearby galaxies.*

http://www.astr.ua.edu/pairs2.html
The University of Alabama atlas of interacting galaxies.

http://www.astro.princeton.edu/~frei/galaxy_catalog.html
*Catalogue of CCD images of a wide variety of galaxies
obtained by Zsolt Frei and James Gunn. In addition to
colour images of the galaxies, the raw data are available for
quantitative analysis.*

http://antwrp.gsfc.nasa.gov/apod/
*Astronomy Picture of the Day. The archive of these pictures and
explanatory texts.*

http://antwrp.gsfc.nasa.gov/apod/lib/aptree.html
Contains an extensive section of items on galaxies.

http://cfpa.berkeley.edu/Bhfaq.html
*A reader-friendly web site tutorial offering a fine non-
technical overview of some of the most important properties of
black holes.*

OUR SOLAR SYSTEM

http://www.sunblock99.org.uk
*This will be the PPARC sponsored site; it was started for
the 1999 eclipse but is being developed into a major
educational site.*

http://umbra.gsfc.nasa.gov/sdac.html
*Contains lots of images and some movies from space and the
ground; has a useful education link.*

http://hesperia.gsfc.nasa.gov/sftheory/
*Mainly about solar flares and related activity, but contains a
useful dictionary of related terms.*

http://www.mssl.ucl.ac.uk/www_solar/sunbasics/index.html
*A nice easy introduction to the Sun with information
on structure, dimensions, the solar cycle, coronal mass ejections,
eclipses; also a short quiz.*

http://www.lmsal.com/YPOP/
*The Yohkoh movie theatre with a number of solar animations,
recent pictures, and some education activities.*

http://sunearth.gsfc.nasa.gov/eclipse/eclipse.html
A wealth of eclipse information for recent and upcoming eclipses.

http://seds.lpl.arizona.edu/nineplanets/nineplanets/
nineplanets.html
The Nine Planets.

http://www.windows.umich.edu/
Windows to the Universe.

http://www.astronomy.com
Astronomy Magazine.

http://helix.nature.com/
Nature Helix.

http://www.jpl.nasa.gov/
NASA/JPL

http://spacescience.com
Science at NASA.

http://exobio.ecsd.edu/Space_Sciences/
Space Sciences Outreach.

http://www.stsci.edu/
Space Telescope Science Institute.

http://sunearth.gsfc.nasa.gov/eclipse.html
*NASA/Goddard Space Flight Center and Fred Espenak's
Eclipse Home Page.*

http://www.lowell.edu/
Lowell Observatory.

http://www.noaa.gov/
National Oceanic and Atmospheric Administration.

http://photojournal.jpl.nasa.gov/
Jet Propulsion Laboratory Photo Journal.

http://www.astronomy.com/
Astronomy magazine.

http://www.stsci.edu/
Space Telescope Science Institute.

WATCHING THE SKY

http://www.skypub.com
Sky & Telescope magazine.

http://cdsweb.u-strasbg.fr/astroweb.html
Astronomy resources on the Internet.

http://cfa-www.harvard.edu/iau/cbat.html
Central Bureau for Astronomical Telegrams.

http://sohowww.nascom.nasa.gov
Solar and Heliospheric Observatory (SOHO).

http://www.bbso.njit.edu
Big Bear Solar Observatory.

http://setiathome.berkeley.edu
SET@home project.

http://www.lpl.arizona.edu/alpo/
Association of Lunar and Planetary Observers.

http://www.occultations.com
International Occultation Timing Association.

http://sunearth.gsfc.nasa.gov
Eclipses.

SPACE EXPLORATION

http://www.spaceport.com
General background to human spaceflight.

http://swampfox.fmarion.edu/web/planet/links.html
Links to the best astronomy sites.

http://www.jpl.nasa.gov/calendar/calendar.html#1099
Great calendar of space events in the coming years.

http://pds.jpl.nasa.gov/planets/welcome.html
Comprehensive planet by planet guide and missions.

http://pds.jpl.nasa.gov/
NASA's planetary data systems nodes — total planetary data.

http://bang.lanl.gov/solarsys/
Solar system pictures and historical astronomy stuff.

http://photojournal.wr.usgs.gov
General tour of the solar system.

http://photojournal.jpl.nasa.gov
A great picture library of solar system shots.

http://www.iki.rssi.ru/Welcome.html
The Russian Space Agency.

http://www.nasa.gov
The American Space Agency.

http://www.ccas.ru/~chernov/vsm/main.html
Russian Space Craft Museum — lots of rare pictures.

http://tele-satellit.com/tse/online/
Encyclopedia of unmanned artificial Earth satellites.

http://nssdc.gsfc.nasa.gov/planetary/chronology.html
Chronology of all missions.

http://solar.rtd.utk.edu:81/~mwade/spaceflt.html
Massive chronology of space flight.

http://leonardo.jpl.nasa.gov/msl/home.html
Mission and spacecraft library.

http://nmp.jpl.nasa.gov/
NASA's New Millennium Programme.

http://quest.arc.nasa.gov/pioneer10/
Pioneer 10 & 11 25th anniversary site.

http://mpfwww.jpl.nasa.gov
NASA's upcoming Mars Missions (including Pathfinder).

http://www.stsci.edu/pubinfo/latest.html
Hubble Space Telescope site.

http://www.jpl.nasa.gov/magellan/
Magellan mission to Venus.

http://www.jpl.nasa.gov/galileo/
Galileo mission to Jupiter and its moons.

http://ccf.arc.nasa.gov/galileo_probe
Galileo probe into Jupiter.

http://ulysses.jpl.nasa.gov/ULSHOME/ulshome.html
Ulysses Solar probe.

http://www.jpl.nasa.gov/mip/voyager.html
Voyager 1 & 2.

http://www.jpl.nasa.gov/cassini/
Cassini mission to Saturn and its moons.

http://www.estec.esa.nl/spdwww/huygens/
Huygens mission to Titan.

http://stardust.jpl.nasa.gov/
Stardust mission to return a comet sample.

http://www.estec.esa.nl/rosetta/html/info0497.html
Rosetta mission to a comet.

http://hurlbut.jhuapl.edu/NEAR/
NEAR mission to an asteroid.

http://lunarprospector.arc.nasa.gov
Lunar Prospector.

Reference
SPACE MISSIONS, CRAFT & PROBES

HUMAN SPACEFLIGHT MISSIONS

The table below lists the important human spaceflight missions to have flown. All missions are included up to the end of 1975, and then only selected Space Shuttle and Soyuz missions are listed, marking further advances in mission complexity or other significant milestones of spaceflight.

SPACECRAFT	SPONSOR	CREW	LAUNCH DATE	DURATION	MISSION
Vostok 1	USSR	1	12–Apr–61	1h:48m	First manned spaceflight 1 orbit
Mercury-Redstone 3	USA	1	5–May–61	15m:22s	First American in space suborbital. Freedom 7
Mercury-Redstone 4	USA	1	21–Jul–61	15m:37s	Second suborbital flight. Liberty Bell 7
Vostok 2	USSR	1	6–Aug–61	1d:1h:18m	First flight longer than 24 hours; 17 orbits
Mercury-Atlas 6	USA	1	20–Feb–62	4h:55m	First American in orbit 3 orbits; telemetry falsely indicated heatshield unlatched. Friendship 7
Mercury-Atlas 7	USA	1	24–May–62	4h:56m	Initiated spaceflight experiments; manual retrofire error caused 400-km (250-mile) landing overshoot. Aurora 7
Vostok 3	USSR	1	11–Aug–62	3d:22h:22m	First twinned flight, with Vostok 4
Vostok 4	USSR	1	12–Aug–62	2d:22h:57m	First twinned flight. On first orbit came within 5 km (3 miles) of Vostok 3
Mercury-Atlas 8	USA	1	3–Oct–62	9h:13m	Developed techniques for long duration missions 6 orbits; closest splashdown to target to date 7 km (4.5 miles). Sigma 7
Mercury-Atlas 9	USA	1	15–May–63	1d:10h:20m	First US evaluation of effects of one day in space 22 orbits; performed manual re-entry after systems failure, landing 6.5 km (4 miles) from target. Faith 7
Vostok 5	USSR	1	14–Jun–63	4d:23h:6m	Second twinned flight, with Vostok 6
Vostok 6	USSR	1	16–Jun–63	2d:22h:50m	First woman in space; passed within 5 km (3 miles) of Vostok 5
Voskhod 1	USSR	3	12–Oct–64	1d:17m	Modified Vostok; first three-man crew in space; first without space suits
Voskhod 2	USSR	2	8–Mar–65	1d:2h:2m	Modified Vostok; first spacewalk 10 min via inflatable airlock
Gemini 3	USA	2	23–Mar–65	4h:53m	First American two-man crew; first piloted spacecraft to change its orbital path; first computer, allowing onboard calculation of manoeuvres
Gemini 4	USA	2	3–Jun–65	4d:1h:56m	First American spacewalk 21 min; first US 4-day flight; manual re-entry made after computer failure
Gemini 5	USA	2	21–Aug–65	7d:22h:56m	First use of fuel cells for electric power; evaluated guidance and navigation system
Gemini 7	USA	2	4–Dec–65	13d:18h:35m	Longest US flight for 8 years 206 orbits, record until Soyuz 9; rendezvous with Gemini 6
Gemini 6	USA	2	15–Dec–65	1d:1h:51m	First manned rendezvous, to within 6 feet of Gemini 7 as planned Agena was lost
Gemini 8	USA	2	16–Mar–66	10h:41m	First docking (with Agena) of one space vehicle with another; emergency re-entry after control malfunction; first Pacific landing
Gemini 9	USA	2	3–Jun–66	3d:21m	127-min EVA rendezvous but no docking with target; landed 0.8 km (0.5 miles) from recovery ship
Gemini 10	USA	2	18–Jul–66	2d:22h:47m	Docked with Agena 10 and used engine to attain record 763-km (474-mile) altitude; rendezvous with Agena 8; 39-min EVA by Collins
Gemini 11	USA	2	12–Sep–66	2d:23h:17m	Used Agena engine to attain record 1,369-km (850-mile) altitude; 163-min EVA by Gordon, connected Gemini and Agena by tether; first automatic computer-guided re-entry
Gemini 12	USA	2	11–Nov–66	3d:22h:34m	Final Gemini mission; Agena docking; record 5.5 hours of EVA by Aldrin, first work carried out during an EVA; automatic computer-guided re-entry
Soyuz 1	USSR	1	23–Apr–67	1d:2h:48m	Retroparachute failed to open; cosmonaut killed in crash landing
Apollo 7	USA	3	11–Oct–68	10d:20h:9m	First piloted flight of Apollo spacecraft, command-service module only; first US 3-man flight; live TV footage of crew
Soyuz 3	USSR	1	26–Oct–68	3d:22h:51m	Rendezvous with unmanned Soyuz 2
Apollo 8	USA	3	21–Dec–68	6d:3h	First manned lunar orbit and piloted lunar return re-entry CSM only; first manned Saturn V; views of lunar surface televised to Earth
Soyuz 4	USSR	1	14–Jan–69	2d:23h:21m	First docking of two piloted spacecraft, with Soyuz 5
Soyuz 5	USSR	3	15–Jan–69	3d:54m	Docked with Soyuz 4; crew transfered by EVA to Soyuz 4
Apollo 9	USA	3	3–Mar–69	10d:1h:1m	First piloted flight of lunar module Earth orbit; 56-min EVA tested lunar suit
Apollo 10	USA	3	18–May–69	8d:3m	First lunar module orbit of Moon, descent to within 15,000 meters (50,000 feet) of Moon's surface; holds manned speed record, 11.0825 km/s (6.8863 mi/s) at atmosphere entry

SPACECRAFT	SPONSOR	CREW	LAUNCH DATE	DURATION	MISSION
Apollo 11	USA	3	16-Jul-69	8d:3h:18m	First lunar landing 20 Jul; 151-min lunar EVA; collected 22 kg (48.5 lb) of soil and rock samples; lunar stay time 21h:36m
Soyuz 6	USSR	2	11-Oct-69	4d:22h:42m	Rendezvous with Soyuz 7/8; first welding of metals in space
Soyuz 7	USSR	3	12-Oct-69	4d:22h:41m	Triple rendezvous with Soyuz 6/8; space station construction test; first time 3 spacecraft, 7 crew members orbited the Earth at once
Soyuz 8	USSR	2	13-Oct-69	4d:22h:51m	Triple rendezvous with Soyuz 6/7; part of space station construction test
Apollo 12	USA	3	14-Nov-69	10d:4h:36m	Second Moon landing; 2 lunar EVAs totalling 465 min; collected 33.9 kg (74.7 lb) of samples; lunar stay time 31h:31m
Apollo 13	USA	3	11-Apr-70	5d:22h:55m	Mission aborted following service module oxygen tank explosion; crew returned safely using lunar module; circumlunar return; holds manned altitude record, 400,187 km (248,665) miles above Earth's surface
Soyuz 9	USSR	2	1-Jun-70	17d:16h:59m	This flight marked the beginning of working in space under weightless conditions; endurance record for solo craft remains
Apollo 14	USA	3	31-Jan-71	9d:42m	Third Moon landing; 2 lunar EVAs totalling 563 min; collected 43.5 kg (96 lb) of lunar samples; lunar stay time 33h:31m
Salyut 1	USSR	0	19-Apr-71	175d	First space station; occupied by Soyuz 11 crew for 23 days; reentered 11-Oct-71
Soyuz 10	USSR	3	23-Apr-71	1d:23h:46m	Adjustment of an improved docking bay between the spacecraft and the orbiting Salyut space station, but no cosmonauts entered the orbiting station
Soyuz 11	USSR	3	16-Jun-71	23d:18h:22m	Docked and entered Salyut 1 space station; orbited in Salyut 1 for 23 days; crew died during re-entry from loss of pressurization
Apollo 15	USA	3	26-Jul-71	12d:7h:12m	Fourth Moon landing; first lunar rover use; first deep spacewalk; 3 lunar EVAs totalling 19h:8m; collected 77 kg (170 lb) of samples; lunar stay time 66h:54m; 38-min Worden EVA; subsatellite released
Apollo 16	USA	3	16-Apr-72	11d:1h:51m	Fifth Moon landing; 3 lunar EVAs totalling 20h:14m; collected 97 kg (213 lb) of lunar samples; lunar stay time 71:14; subsatellite released
Apollo 17	USA	3	7-Dec-72	12d:13h:51m	Sixth piloted lunar landing; 3 lunar EVAs totalling 22h:4m; collected 110 kg (243 lb) of samples; record lunar stay of 74:59; 66-min Evans EVA
Skylab 1	USA	0	14-May-73	2,249d	First US space station; occupied by Skylab 2, 3 & 4 crews; reentered 11-Jul-79
Skylab 2	USA	3	25-May-73	28d:50m	First American piloted orbiting space station; made long-flight tests, crew repaired damage caused during boost; 2 EVAs + SEVA
Skylab 3	USA	3	28-Jul-73	59d:11h:9m	Crew systems and operational tests, exceeded pre-mission plans for scientific activities; 3 EVAs totalling 13:44
Soyuz 12	USSR	2	27-Sep-73	1d:23h:16m	After Soyuz 11 accident, new life support equipment was tested
Skylab 4	USA	3	16-Nov-73	84d:1h:15m	Final Skylab mission; endurance record until Soyuz 29-Salyut 6; 4 EVAs, set then-record spacewalk of 7h:1m 52
Soyuz 19-ASTP	USSR	2	15-Jul-75	5d:22h:31m	First US/USSR joint flight; docked with Apollo 18 for 2 days; conducted experiments, shared meals, and held a joint news conference
Salyut 5	USSR	0	22-Jun-76	412d	4th space station; occupied by 2 crews for 65 days; reentered 8-Aug-77
Soyuz 21	USSR	2	6-Jul-76	49d:6h:23m	First Salyut 5 occupation 48 days; acid fumes forced return
Salyut 6	USSR	0	29-Sep-77	1763d	5th space station; occupied for 676 days by 5 long stay + 11 visiting crews; reentered 28-Jul-82
Soyuz 28	USSR	2	2-Mar-78	7d:22h:16m	First international crew USSR and Czechoslovakia, to Salyut 6
Soyuz 29	USSR	2	15-Jun-78	139d:14h:18m	First 100+ day flight, to Salyut 6; returned in Soyuz 31: Progress 2, 3 & 4 resupply the orbiting complex
Soyuz 30	USSR	2	27-Jun-78	7d:22h:3m	2nd international crew USSR and Poland, to Salyut 6
Soyuz 31	USSR	2	26-Aug-78	7d:20h:49m	3rd international crew USSR and East Germany, to Salyut 6; returned in Soyuz 29
Soyuz 32	USSR	2	25-Feb-79	175d:36m	Cosmonauts board Salyut 6; endurance record; returned in Soyuz 34: Progress 5, 6 and 7 resupply the orbiting complex
Soyuz 33	USSR	2	10-Apr-79	1d:23h:1m	4th international crew USSR and Bulgaria; failed to dock with Salyut 6 after engine failure
Soyuz 35	USSR	2	19-Apr-80	184d:20h:11m	4th Salyut 6 long stay; endurance record; returned in Soyuz 37: Progress 8, 9 and 11 resupply the orbiting complex
Soyuz T2	USSR	2	5-Jun-80	3d:22h:19m	First manned spaceflight of the new spacecraft; manual docking with Salyut 6
STS-1, Columbia	USA	2	12-Apr-81	2d:6h:21m	First space shuttle flight; orbital test flight; some thermal tiles lost
STS-2, Columbia	USA	2	12-Nov-81	2d:6h:13m	First reuse of space shuttle; 2nd orbital test flight; test of Canadian robot arm RMS; 5-day mission halved by fuel cell fault
STS-6, Challenger	USA	4	4-Apr-83	5d:2h:14m	First Challenger flight; first shuttle EVA; Tracking & Data Relay Satellite TDRS
STS-7, Challenger	USA	5	18-Jun-83	6d:2h:24m	First US woman in space; first 5-person crew; 2 COMSATs, German platform SPAS-1
STS-9, Columbia	USA	6	28-Nov-83	10d:7h:47m	First German on US mission; first 6-person crew; first Spacelab Mission SL-1
41-B, Challenger	USA	5	3-Feb-84	7d:23h:16m	First untethered EVA & testing of MMU jetpack; first Kennedy Space Center landing; 2 COMSATs
Soyuz T10B	USSR	3	8-Feb-84	236d:22h:50m	First long-stay triple crew, to Salyut 7; 6 EVAs totalling 22h:56m; returned in Soyuz T11
41-D, Discovery	USA	6	30-Aug-84	6d:56m	First flight of Discovery; first commercial payload specialist; 3 communications satellites
51-A, Discovery	USA	4	8-Nov-84	7d:23h:45m	First satellite retrieval/return; 2 COMSATs
51-L, Challenger	USA	7	28-Jan-86	1m:13s	Exploded during liftoff, all were killed
Mir	USSR	0	20-Feb-86	—	New-generation space station with 6 docking ports; occupied by multiple crews
Soyuz T15	USSR	2	13-Mar-86	125d:1m	First Mir occupation; excursion to Salyut 7 5-May to 26-Jun; two Salyut EVAs totalling 8h:50m; re-used descent module from T-10A abort
Soyuz TM2	USSR	2	5-Feb-87	326d:11h:18m	2nd Mir long stay; new endurance record; return in Soyuz TM-3
Soyuz TM4	USSR	3	21-Dec-87	365d:22h:39m	3rd Mir long stay occupation; new endurance record; 3 EVAs totalling 13h:40m; returned in Soyuz TM6

SPACECRAFT	SPONSOR	CREW	LAUNCH DATE	DURATION	MISSION
Soyuz TM6	USSR	3	29-Aug-88	240d:22h:36m	4th Mir long stay
STS-26, Discovery	USA	5	29-Sep-88	4d:1h	Redesigned shuttle makes first flight; Tracking/Data Relay Satellite TDRS-C
Soyuz TM7	USSR	3	26-Nov-88	151d:11h:9m	5th Mir long stay; returned in Soyuz TM6; Volkov/Krikalev in TM7
STS-30, Atlantis	USA	5	4-May-89	4d:56m	Magellan Venus orbiter launched on IUS stage, arrived Venus Aug 1990
STS-34, Atlantis	USA	5	18-Oct-89	4d:23h:39m	Galileo Jupiter orbiter launched on IUS stage, arrived Jupiter Dec 1995
STS-31, Discovery	USA	5	24-Apr-90	5d:1h:16m	Deployed Hubble Space Telescope HST; set Shuttle altitude record of 619 km (385 miles)
STS-41, Discovery	USA	5	6-Oct-90	4d:2h:10m	Ulysses solar probe launched on IUS stage
STS-47, Endeavour	USA	7	12-Sep-92	7d:22h:30m	50th shuttle mission; 1st black woman in space; 1st Japanese national; Lee & Davis 1st married couple to travel together in space; first Japanese Spacelab SL-J
STS-61, Endeavour	USA	7	2-Dec-93	10d:19h:59m	First Hubble Space Telescope servicing mission; 5 EVAs for 4 crew totalling 35h:28m; Akers set new US EVA duration record 29h:40m
STS-60, Discovery	USA	6	3-Feb-94	8d:7h:9m	First Russian on US shuttle; attempt to deploy the Wake Shield Facility a device to create vacuums in space failed; Spacehab 2
STS-71, Atlantis	USA	7	27-Jun-95	9d:19h:22m	First Mir docking/crew exchange, 100th US human spaceflight, Spacelab carried
STS-82, Discovery	USA	7	11-Feb-97	9d:23h:38m	2nd Hubble Space Telescope servicing mission; 5 EVAs for 4 crew totalling 33h:11m, replaced 10 instruments
STS-95, Discovery	USA	7	29-Oct-98	8d:21h:44m	SpaceHab-SM, SPARTAN-201; John Glenn reflight
STS-88, Endeavour	USA	6	4-Dec-98	11d:19h:18m	ISS assembly flight 2A, Unity Module
STS-96, Discovery	USA	7	27-May-99	9d:19h:13m	ISS assembly flight 2A.1; Starshine
STS-103, Discovery	USA	7	19-Dec-99	7d:23h:12m	3rd Hubble Space Telescope servicing mission; 3 EVAs for 4 crew totalling 24h:33m
Souyuz 2R	USSR	3	30-Oct-00		First crew on ISS – marks start of planned permanent human presence on the station
STS-92, Discovery	USA	7	05-Oct-00		100th Shuttle flight

ROBOTIC SPACE MISSIONS

This table is a list of robotic space missions sent out to flyby, orbit and land on other worlds in our solar system and occasionally return to Earth clutching a precious sample of some exotic extra-terrestrial material. The table is divided into destination bodies, and so missions visiting more than one planet, asteroid or comet appear more than once.

SUN

SPACECRAFT	SPONSOR	LAUNCH DATE	FIRST ENCOUNTER	OUTCOME
OSO 1	USA	07-03-1962	Earth Orbit	Solar flare observations and studies of upper atmosphere
OSO 2	USA	03-02-1965	Earth Orbit	Solar X-ray, UV, gamma ray data, and studies of upper atmosphere
OSO 5	USA	22-01-1969	Earth Orbit	Solar radiation studies and upper atmosphere
OSO 6	USA	09-08-1969	Earth Orbit	Solar radiation studies and upper atmosphere
SMM	USA	14-02-1980	Earth Orbit	UV, gamma ray and X-ray monitoring of solar flares and Sun's atmosphere
MSSTA	USA	13-05-1991	Ballistic trajectory	6 minutes of solar observations in UV and soft X-rays
Yohkoh	Japan	31-08-1991	Earth Orbit	Solar corona and flare observations in X-rays and gamma rays
Ulysses	Eur/USA	06-10-1990	Autumn 1994	Solar polar heliosphere and solar wind measurements
Helios 1	German/USA	10-12-1974	15-03-1975	Solar wind, magnetic field studies – closest flight to the Sun – 45 million km (28 million miles)
Helios 2	German/USA	15-01-1976	17-04-1976	Solar wind and magnetic field studies
Polar	USA	24-02-1996	Earth orbit	Solar wind interactions with Earth
Geotail	Japan/USA	24-07-1992	Earth orbit	Solar wind interactions with Earth
Wind	USA	01-11-1994	Earth orbit	Magnetosphere/solar wind interactions
Interbol 1	Russia & others	02-08-1995	Earth orbit	Solar wind
Interbol 2	Russia & others	29-08-1996	Earth orbit	Solar wind
SOHO	Europe/USA	02-12-1995		Solar corona, wind and interior of Sun
Cluster I	ESA	04-06-1996	Launch Failure	Attempted solar wind monitor
TRACE	US	01-04-1998	Earth orbit	Solar atmospheric observations
ACE	US	25-08-1997	Earth orbit	Solar wind observations
Cluster II	ESA	16-07-2000		Solar wind in 3-D studied by a formation of four satellites

MERCURY

SPACECRAFT	SPONSOR	LAUNCH DATE	FIRST ENCOUNTER	OUTCOME
Mariner 10	USA	04-11-1973	29-03-1974	Triple flyby – mapping 40% of planet – found unexpectedly strong magnetic field

SPACE MISSIONS, CRAFT & PROBES

VENUS

SPACECRAFT	SPONSOR	LAUNCH DATE	FIRST ENCOUNTER	OUTCOME
Venera 1	USSR	12-02-1961	Lost Contact	Attempted Venus flyby
Mariner 1	USA	22-07-1962	Launch Failure	Attempted Venus flyby
Mariner 2	USA	27-08-1962	14-12-1962	Flyby – surface temperature and pressure recorded
Cosmos 27	USSR	02-04-1964	Contact Lost	Flyby at 965,000 km (600,000 miles)
Venera 2	USSR	12-11-1965	27-02-1966	Flyby at 24,000 km (14,900 miles)
Venera 3	USSR	16-11-1965	Contact Lost	Attempted impact
Venera 4	USSR	12-06-1967	18-10-1967	Lander – reached 24km altitude before being crushed
Mariner 5	USA	14-06-1967	19-10-1967	Flyby – surface temperature and pressure
Venera 5	USSR	05-01-1969	16-05-1967	Probe – crushed at 12 km altitude
Venera 6	USSR	10-01-1969	17-05-1967	Probe – crushed at 16 km altitude
Venera 7	USSR	17-08-1970	06-11-1970	Lander – first signals from the surface – 90 atmospheres pressure and 475°C
Venera 8	USSR	27-03-1972	22-07-1972	Lander – survived 50 min on surface
Mariner 10	USA	04-11-1973	05-02-1974	Sling-shot to Mercury
Venera 9	USSR	08-06-1975	20-10-1975	Orbiter and Lander – first pictures of surface
Venera 10	USSR	14-06-1975	25-10-1975	Orbiter and Lander
Pioneer Venus 1	USA	20-05-1978	04-12-1978	Orbiter – radar mapped the planet
Pioneer Venus 2	USA	08-08-1978	09-12-1978	One large probe and three small probes reached surface
Venera 11	USSR	09-09-1978	25-12-1978	Orbiter and lander – 1 hour 35 min on surface
Venera 12	USSR	14-09-1978	21-12-1978	Orbiter and lander – 1 hour 50 min on surface
Venera 13	USSR	30-10-1981	01-03-1982	Lander – first colour pictures of surface and chemistry of rocks
Venera 14	USSR	04-11-1981	05-03-1982	Lander – first colour pictures of surface
Venera 15	USSR	02-06-1983	10-10-1983	Orbiter – radar mapping of entire planet
Venera 16	USSR	06-06-1983	14-10-1983	Orbiter – radar mapping of entire planet
Vega 1	USSR	15-12-1984	10-06-1985	Lander and balloons which flew 10,000 km around planet
Vega 2	USSR	21-12-1984	14-06-1985	Lander and balloons which flew 10,000 km around planet
Magellan	USA	04-05-1989	10-08-1990	Orbiter – radar mapped 84% of the planet revealing weird volcanic terrain
Galileo	USA	18-10-1989	10/02/1990	Sling-shot back to Earth and on to Jupiter
Cassini-Huygens	USA/Europe	15-10-1997	27-04-1998	Sling-shot back to Earth and back to Venus and on to Saturn

MOON

SPACECRAFT	SPONSOR	LAUNCH DATE	FIRST ENCOUNTER	OUTCOME
Pioneer 1	USA	11-10-1958	Launch Failure	Attempted orbit – 113,854 km and fell back to Earth
Pioneer 2	USA	08-11-1958	Launch Failure	Attempted orbiter
Pioneer 3	USA	06-12-1958	Launch Failure	Attempted flyby
Luna 1	USSR	02-01-1959	04-01-1959	Flyby – missed by 5,500km
Pioneer 4	USA	03-03-1959	05-01-1959	Flyby – missed by 60,000 km
Luna 2	USSR	12-09-1959	13-09-1959	First impact – carrying Soviet pendants
Luna 3	USSR	04-10-1959	06-10-1959	Photographed far side for the first time
Ranger 1	USA	23-08-1961	Launch failure	Attempted test flight – fell back to Earth
Ranger 2	USA	18-12-1961	Launch failure	Attempted test flight – fell back to Earth
Ranger 3	USA	26-01-1962	Launch failure	Attempted lander – missed by 36,579 km
Ranger 4	USA	23-04-1962	25-04-1962	Attempted landing – hit far side
Ranger 5	USA	18-10-1962	20-10-1962	Attempted lander – flyby at 720 km
Luna 4	USSR	02-04-1963	06-04-1963	Attempted soft landing – missed Moon by 8,448 km
Ranger 6	USA	30-01-1964	02-02-1964	Impact Sea of Tranquillity – cameras failed
Ranger 7	USA	28-07-1964	31-07-1964	Impact 4,316 pictures of the Sea of Clouds
Ranger 8	USA	17-02-1965	20-02-1965	Impact – Sea of Tranquillity, 7,137 pictures
Ranger 9	USA	21-03-1965	32-03-1965	Impact, 5,814 pictures
Luna 5	USSR	09-05-1965	12-05-1965	Attempted soft landing – crashed into Moon
Luna 6	USSR	08-06-1965	11-06-1965	Attempted soft landing – missed the Moon by 160,000 km
Zond 3	USSR	18-07-1965	21-07-1965	Flyby – photographed far side, on way to Mars
Luna 7	USSR	04-10-1965	07-10-1965	Attempted soft landing – crashed
Luna 8	USSR	03-12-1965	06-12-1965	Attempted soft landing – crashed
Luna 9	USSR	31-01-1966	04-02-1966	First soft landing and pictures from the surface
Luna 10	USSR	31-03-1966	03-04-1966	Orbiter – measured radiation fields
Surveyor 1	USA	30-05-1966	02-06-1966	Landed in the Ocean of Storms, 10,150 pictures returned
Lunar Orbiter 1	USA	10-08-1966	14-08-1966	Orbiter – 413 pictures
Luna 11	USSR	24-08-1966	27-08-1966	Orbiter – studied lunar crust geochemistry
Surveyor 2	USA	20-09-1966	Contact lost	Attempted landing
Luna 12	USSR	22-10-1966	25-10-1966	Orbiter – photographed surface
Lunar Orbiter 2	USA	06-11-1966	10-11-1966	Orbiter – 411 photographs received

MOON

SPACECRAFT	SPONSOR	LAUNCH DATE	FIRST ENCOUNTER	OUTCOME
Luna 13	USSR	21-12-1966	25-12-1966	Lander – selected sites for Russian human Moon missions
Lunar Orbiter 3	USA	05-02-1967	12-02-1967	Orbiter – 275 pictures used to select Apollo landing sites
Surveyor 3	USA	17-04-1967	20-04-1967	Lander – first shovel to trench surface
Lunar Orbiter 4	USA	04-05-1967	08-05-1967	Orbiter – 99% of near side photographed
Surveyor 4	USA	14-07-1967	Contact lost	Attempted landing
Explorer 35	USA	19-07-1967	21-07-1967	Orbiter – studied solar winds and magnetosphere
Lunar Orbiter 5	USA	01-08-1967	05-08-1967	Orbiter – polar orbit photography
Surveyor 5	USA	08-09-1967	11-09-1967	Landed in the Sea of Tranquillity chemical analysis – showed basalt, 19,000 pictures returned
Surveyor 6	USA	07-11-1967	10-11-1967	Landed in Central Bay – more basalt found
Surveyor 7	USA	07-01-1968	10-01-1968	Landed – crater Tycho – southern highlands
Luna 14	USSR	07-04-1968	10-04-1968	Orbiter – monitored lunar environment
Luna 15	USSR	13-07-1969	21-07-1969	Attempted sample return crashed on to Moon
Luna 16	USSR	12-09-1970	24-09-1970	101 g sample returned from the Sea of Fertility to Earth
Luna 17/Lunakhod 1	USSR	10-11-1970	17-11-1970	Rover – travelled 10.5 km around the Bay of Rainbows
Luna 18	USSR	02-09-1971	07-09-1971	Attempted sample return – crashed on to Moon
Luna 19	USSR	28-09-1971	03-10-1971	Orbiter – monitored lunar environment
Luna 20	USSR	14-02-1972	21-02-1972	Sample return from edge of Mare Fecunditatis
Luna 21/Lunakhod 2	USSR	08-01-1973	15-01-1973	Rover explored mountains south of crater Lemonnier
Explorer 49	USA	10-06-1973	15-06-1973	Lunar Orbiter/Radio Astronomy
Luna 22	USSR	29-05-1974	02-06-1974	Orbiter
Luna 23	USSR	28-10-1974	06-11-1974	Attempted core sample return – arm damaged
Luna 24	USSR	09-08-1976	18-08-1976	Sample returned from 2 m below the Sea of Crises
Hiten	Japan	24-01-1990	Contact lost	Attempted flyby/orbiter
Clementine	USA	25-01-1994	20-02-1994	Orbiter – mineral and topography mapping
Lunar Prospector	USA	06-01-1998	11-01-1998	Chemical mapping and search for water ice at lunar south pole

SATURN

SPACECRAFT	SPONSOR	LAUNCH DATE	FIRST ENCOUNTER	OUTCOME
Mars 1	USSR	01-11-1962	Contact Lost	Flyby attempt – failed three months from Mars
Mariner 3	USA	05-11-1964	Launch failure	Flyby failed
Mariner 4	USA	28-11-1964	15-07-1965	Flyby – first close-up pictures sent back and atmospheric pressure
Zond 2	USSR	30-11-1964	Contact Lost	Flyby attempt – failed four months from Mars
Zond 3	USSR	18-07-1965	missed	Attempted flyby – missed
Mariner 6	USA	24-02-1969	31-07-1969	Flyby – southern polar ice cap imaged
Mariner 7	USA	27-03-1969	05-08-1969	Flyby – observations of the south pole
Mariner 8	USA	09-05-1971	Launch failure	Attempted orbiter
Mars 2	USSR	19-05-1971	27-11-1971	Orbiter – failed lander first to hit Mars. Orbiter returned pictures for 9 months
Mars 3	USSR	28-05-1971	02-12-1971	Orbiter and soft lander – dust storm damaged before pictures could be returned Orbiter returned pictures for 9 months
Mariner 9	USA	30-05-1971	16-11-1971	Orbiter first to orbit another planet – mapped whole planet – discovered large volcanoes and giant valley system
Mars 4	USSR	21-07-1973	10-02-1974	Flyby – failed orbiter
Mars 5	USSR	25-07-1973	12-02-1974	Orbiter – pictures returned
Mars 6	USSR	05-08-1973	12-03-1974	Lander – contact lost during descent
Mars 7	USSR	09-08-1973	09-03-1974	Flyby – failed lander
Viking 1	USA	20-08-1975	19-06-1976	Orbiter and lander – first pictures from the surface, searched for life and monitored weather for five years
Viking 2	USA	09-09-1975	07-08-1976	Orbiter and lander – pictures from the surface, searched for life and monitored weather for five years
Phobos 1	USSR	07-07-1988	Contact lost	Orbiter/Phobos lander failed
Phobos 2	USSR	12-07-1988	29-01-1989	Orbiter/Phobos lander failed
Mars Observer	USA	25-09-1992	Contact Lost	Orbiter failed – near Mars
Mars 96	USSR	16-11-1996	Launch failure	Orbiter and lander failed
Mars Pathfinder	USA	04-12-1996	04-07-1997	Lander/rover
Mars Global Surveyor	USA	07-11-1996	11-09-1997	Orbiter – high-resolution mapping of Mars
Mars Climate Orbiter	USA	11-12-1998	23-09-1999	Orbiter – burnt up in Martian atmosphere
Mars Polar Lander	USA	03-01-1999	03-12-1999	Lander – failed
Deep Space 2	USA	03-01-1999	03-12-1999	Penetrators carried on the Polar lander failed
Nozomi (Planet-B)	Japan	03-07-1998	December 2003	Orbiter – failed to reach full acceleration – sent on long route

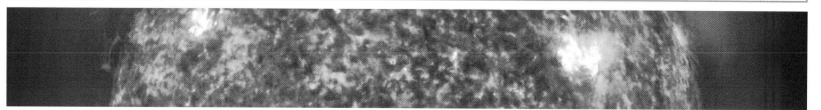

SPACE MISSIONS, CRAFT & PROBES

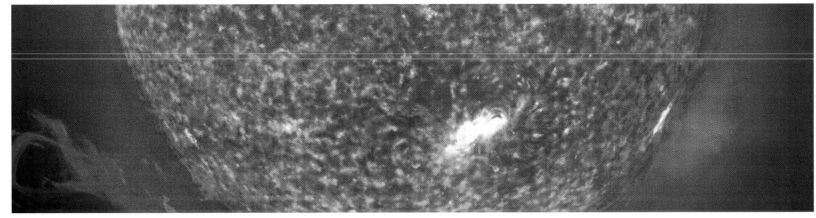

JUPITER

SPACECRAFT	SPONSOR	LAUNCH DATE	FIRST ENCOUNTER	OUTCOME
Pioneer 10	USA	03-03-1972	04-12-1973	Fly-by. First close-up pictures and measurements of magnetic and radiation fields
Pioneer 11	USA	06-04-1973	04-12-1974	Fly-by. Second close-up pictures and radiation measurements
Voyager 1	USA	05-09-1977	05-03-1979	Fly-by. First high-quality close-up pictures of planet and moons
Voyager 2	USA	20-08-1977	09-07-1979	More high-quality close-up pictures
Galileo Orbiter	USA	18-10-1989		Orbit of the Jovian system – multiple flybys of moons
Galileo Probe	USA	18-10-1989	07-12-1995	First entry probe – measured temps/pressures/chemistry of top 200 km of the planet
Ulysses	USA/ESA	06-10-1990	08-02-1992	Sling-shot out of plane of Solar System into solar polar orbit
Cassini-Huygens	USA/ESA	15-10-1997	Dec 2000	Fly-by en route to Saturn

SATURN

SPACECRAFT	SPONSOR	LAUNCH DATE	FIRST ENCOUNTER	OUTCOME
Pioneer 11	USA	06-04-1973	01-09-1979	Fly-by – second close-up pics of planet and rings
Voyager 1	USA	05-09-1977	12-11-1980	Fly-by – through ring plane and sling-shot out of plane of Solar System. First high-quality pictures of planet and moon Titan
Voyager 2	USA	20-08-1977	22-08-1981	Fly-by – close up pics – sling-shot to outer Solar System
Cassini	USA/Europe	15-10-1997	01-07-2004	Orbiter to observe planet, rings and moons close up
Huygens	ESA		Nov 2004	Lander to visit moon Titan

URANUS

SPACECRAFT	SPONSOR	LAUNCH DATE	FIRST ENCOUNTER	OUTCOME
Voyager 2	USA	20-08-1977	24-01-1986	Fly-by – first close-up images of planet, rings and moons

NEPTUNE

SPACECRAFT	SPONSOR	LAUNCH DATE	FIRST ENCOUNTER	OUTCOME
Voyager 2	USA	20-08-1977	25-08-1989	Fly-by – first close-up images of planet, rings and moons

ASTEROIDS AND COMETS

SPACECRAFT	SPONSOR	LAUNCH DATE	FIRST ENCOUNTER	OUTCOME
ICE/ISEE-3	USA	12-08-1978	11-09-1985	Comet P/Giacobini-Zinner flyby
Vega 1	USSR	15-12-1985	06-03-1986	Halley's Comet 9,000 km flyby
Vega 2	USSR	21-12-1985	09-03-1986	Halley's comet flyby
Sakigake	Japan	07-01-1985	10-03-1986	Halley's Comet study from 6.88 million km
Giotto	Europe	02-07-1985	13-03-1986	Halley's Comet flyby and Grigg-Skjellerup 1992
Suisei (Planet A)	Japan	18-08-1985	08-03-1986	Halley's Comet flyby 151,520 km
Galileo	USA	18-10-1989	28-08-1993	Flyby asteroid Ida
Galileo	USA	18-10-1989	29-10-1991	Flyby asteroid Gaspra
NEAR	USA	17-02-1996	27-06-1997	Flyby asteroid 253 Mathilde
NEAR	USA	17-02-1996	14-02-2000	Orbit asteroid 433 Eros
Deep Space 1	USA	24-10-1998	29-07-1999	Asteroid 1992 KD flyby. Comet Borrelly flyby September 2001
Stardust	USA	07-02-1999	Jan 2004	Comet Wild 2 sample return

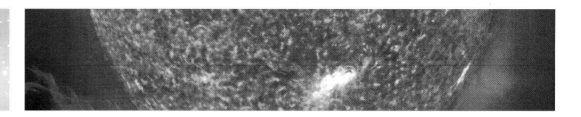

REFLECTORS

NAME	APERTURE	LOCATION	COMPLETION	MOUNT	WEB SITE
Very Large Telescope (VLT)	4 × 8.2 m (323 in)	Cerro Paranal, Chile	2001	Altazimuth	www.eso.org/
Keck Telescopes (Keck I and II)	2 × 9.82 m (387 in)	Mauna Kea, Hawaii, USA	1991 & 1996	Altazimuth	astro.caltech.edu/mirror/keck/index.html
Large Binocular Telescope (LBT)	2 × 8.4 m (331 in)	Mount Graham, Arizona, USA	2004	Altazimuth	medusa.as.arizona.edu/lbtwww/lbt.html
Gran Telescopio Canarias	10.4 m (409 in)	La Palma, Canary Islands	2002	Altazimuth	www.gtc.iac.es
Hobby-Eberly Telescope (HET)	9.1 m (358 in)	Mount Fowlkes, Texas, USA	1997	Azimuth only	www.astro.psu.edu/het/
South African Large Telescope (SALT)	9.1 m (358 in)	Sutherland, South Africa	2003	Azimuth only	www.salt.ac.za/index.html
Subaru Telescope	8.2 m (323 in)	Mauna Kea, Hawaii, USA	1999	Altazimuth	www.naoj.org/
Gemini Telescope (North)	8.1 m (319 in)	Mauna Kea, Hawaii, USA	1999	Altazimuth	www.gemini.edu/
Gemini Telescope (South)	8.1 m (319 in)	Cerro Pachón, Chile	1999	Altazimuth	www.gemini.edu/
MMT 6.5-m Telescope	6.5 m (256 in)	Mount Hopkins, Arizona, USA	2000	Altazimuth	cfa-www.harvard.edu/cfa/oir/MMT/mmt/foltz/mmt.html
Magellan I and II	2 × 6.5 m (256 in)	Las Campanas, Chile	2002	Altazimuth	www.ociw.edu/magellan/
Bolshoi Teleskop Azimutal'ny (BTA)	6.00 m (236 in)	Mount Pastukhov, Russia	1975	Altazimuth	www.sao.ru/
Large Zenith Telescope (LZT)	6.00 m (236 in)	Maple Ridge, B.C. Canada	2000	Fixed	www.astro.ubc.ca/LMT/lzt.html
Hale Telescope (200-in)	5.08 m (200 in)	Mount Palomar, California, USA	1948	Horseshoe yoke	astro.caltech.edu/palomarpublic/index.html
William Herschel Telescope (WHT)	4.20 m (165 in)	La Palma, Canary Islands	1987	Altazimuth	www.ing.iac.es/PR/index.html
SOAR 4-m Telescope	4.20 m (165 in)	Cerro Pachón, Chile	2002	Altazimuth	www.ctio.noao.edu/soar/index.html
Victor M. Blanco Telescope (CTIO 4-m)	4.00 m (157 in)	Cerro Tololo, Chile	1976	Split-ring equatorial	www.ctio.noao.edu/ctio.html
Anglo-Australian Telescope (AAT)	3.89 m (153 in)	Siding Spring, Australia	1974	Split-ring equatorial	www.aao.gov.au/
Nicholas U. Mayall Reflector (Kitt Peak)	3.81 m (150 in)	Kitt Peak, Arizona, USA	1973	Split-ring equatorial	www.noao.edu/kpno/kpno.html
United Kingdom Infrared Telescope (UKIRT)	3.80 m (150 in)	Mauna Kea, Hawaii, USA	1978	English yoke	www.jach.hawaii.edu/JACpublic/UKIRT/
Advanced Electro-optical System Telescope (AEOS)	3.67 m (145 in)	Haleakala, Hawaii, USA	2000	Altazimuth	ulua.mhpcc.af.mil/
Canada-France-Hawaii Telescope (CFHT)	3.58 m (141 in)	Mauna Kea, Hawaii, USA	1979	Horseshoe yoke	www.cfht.hawaii.edu/
Telescopio Nazionale Galileo (Galileo 3.6-m)	3.58 m (141 in)	La Palma, Canary Islands	1998	Altazimuth	www.tng.iac.es/
ESO 3.6-m Telescope	3.57 m (140 in)	La Silla, Chile	1977	Horseshoe fork	www.eso.org/
3.5-m Telescope	3.50 m (138 in)	Calar Alto, Spain	1984	Horseshoe yoke	www.mpia-hd.mpg.de/Public/CAHA/
New Technology Telescope	3.50 m (138 in)	La Silla, Chile	1989	Altazimuth	www.eso.org/
Astrophysics Research Consortium Telescope (ARC)	3.50 m (138 in)	Apache Point, New Mexico, USA	1994	Altazimuth	www.apo.nmsu.edu/
Wisconsin-Indiana-Yale-NOAO Telescope (WIYN)	3.50 m (138 in)	Kitt Peak, Arizona, USA	1994	Altazimuth	www.noao.edu/wiyn/
Starfire Optical Range 3.5-m Reflector	3.50 m (138 in)	Kirtland Air Force Base, New Mexico, USA	1994	Altazimuth	www.de.afrl.af.mil/pa/factsheets/35meter.html
C. Donald Shane Telescope (120-in)	3.05 m (120 in)	Mount Hamilton, California, USA	1959	Fork equatorial	www.ucolick.org/
NASA Infrared Telescope Facility (IRTF)	3.00 m (118 in)	Mauna Kea, Hawaii, USA	1979	English yoke	irtf.ifa.hawaii.edu/
3-m Liquid Mirror Telescope (NODO)	3.00 m (118 in)	Cloudcroft, New Mexico, USA	1996	Fixed	www.astro.ubc.ca/LMT/lmt.html
Harlan J. Smith Telescope (107-in)	2.72 m (107 in)	Mount Locke, Texas, USA	1969	English cross-axis	www.as.utexas.edu/mcdonald/mcdonald.html
Shajn 2.6-m Reflector (Crimean 102-in)	2.64 m (104 in)	Nauchny, Ukraine	1960	Fork equatorial	www.crao.crimea.ua/main.html
Byurakan 2.6-m reflector	2.64 m (104 in)	Mount Aragatz, Armenia	1976	Fork equatorial	bao.sci.am/
Nordic Optical Telescope (NOT)	2.56 m (101 in)	La Palma, Canary Islands	1989	Altazimuth	www.not.iac.es/
Irénée du Pont Telescope (100-in)	2.54 m (100 in)	Las Campanas, Chile	1976	Fork equatorial	www.ociw.edu/lco/
Isaac Newton Telescope (98-in INT)	2.54 m (100 in)	La Palma, Canary Islands	1984	Polar-disk equatorial	www.ing.iac.es/PR/index.html

Hooker Telescope (100-in)	2.50 m (98 in)	Mount Wilson, California, USA	1917	English yoke	www.mtwilson.edu/
Stratospheric Observatory for Infrared Astronomy (SOFIA)	2.50 m (98 in)	Airborne, Boeing 747-SP	2002	Airborne	sofia.arc.nasa.gov/
Sloan 2.5-m Reflector	2.50 m (98 in)	Apache Point, New Mexico, USA	1998	Altazimuth	www.apo.nmsu.edu/
Hubble Space Telescope (HST)	2.40 m (94 in)	Earth orbit, 600 km altitude	1990	3-axis stabilized	oposite.stsci.edu
Hiltner Telescope (2.3-m)	2.34 m (92 in)	Kitt Peak, Arizona, USA	1986	Fork equatorial	www.noao.edu/kpno/kpno.html
Vainu Bappu 2.3-m	2.33 m (92 in)	Kavalur, Tamil Nadu, India	1985	Horseshoe yoke	www.iiap.ernet.in/
Bok Telescope (90-in)	2.30 m (90 in)	Kitt Peak, Arizona, USA	1969	Fork equatorial	www.noao.edu/kpno/kpno.html
Mount Stromlo 2.3-m (ATT)	2.30 m (90 in)	Siding Spring Mountain, Australia	1984	Altazimuth	msowww.anu.edu.au/
Wyoming Infrared Telescope	2.29 m (90 in)	Jelm Mountain, Wyoming, USA	1977	English yoke	faraday.uwyo.edu/wiro/

REFRACTORS

NAME	APERTURE	LOCATION	COMPLETION	MOUNT	WEB SITE
Yerkes 40-in Refractor	1.02 m (40 in)	Williams Bay, Wisconsin, USA	1897	German equatorial	astro.uchicago.edu/yerkes/
36-in Refractor	0.895 m (36 in)	Mount Hamilton, California, USA	1888	German equatorial	www.ucolick.org/
Meudon 33-in (Grande Lunette)	0.83 m (33 in)	Meudon, France	1889	German equatorial	www.obspm.fr/histoire/ meudon/meudon.fr.shtml
Potsdam Refractor	0.80 m (31 in)	Potsdam, Germany	1899	German equatorial	aipsoe.aip.de/refraktor/
Thaw Refractor (30-in)	0.76 m (30 in)	Pittsburgh, Pennsylvania, USA	1914	German equatorial	http://www.pitt.edu/~aobsvtry/
Lunette Bischoffscheim	0.74 m (29 in)	Mont Gros, France	1886	German equatorial	www.obs-nice.fr/
28-in Visual Refractor (Greenwich, RGO)	0.71 m (28 in)	Greenwich, England	1893	English yoke	www.rog.nmm.ac.uk
Grosser Refraktor	0.68 m (27 in)	Berlin, Germany	1896	Hybrid equatorial	home.snafu.de/astw
Vienna Refraktor	0.67 m (26 in)	Vienna, Austria	1880	German equatorial	www.astro.univie.ac.at/ Telescopes.html
McCormick Refractor (26-in)	0.667 m (26 in)	Charlottesville, Virginia, USA	1883	German equatorial	www.astro.virginia.edu/ ~rip0i/jwk/home.html
26-in Equatorial	0.66 m (26 in)	Washington, D.C. USA	1873	German equatorial	www.usno.navy.mil/ USNO26inDome.html
Thompson Refractor (26-in)	0.66 m (26 in)	Herstmonceux, England	1897	German equatorial	closed
Innes Telescope (26-in)	0.66 m (26 in)	Johannesburg, South Africa	1926	German equatorial	no web site
Newall Telescope (25-in)	0.63 m (25 in)	Athens, Greece	1862	German equatorial	www.astro.noa.gr/
Clark Refractor (24-in)	0.61 m (24 in)	Flagstaff, Arizona, USA	1896	German equatorial	www.lowell.edu

SCHMIDT TELESCOPES

NAME	APERTURE	LOCATION	COMPLETION	MOUNT	WEB SITE
2-m Telescope (Tautenberg Schmidt)	1.34 m (52 in)	Tautenberg, Germany	1960	Fork equatorial	www.tls-tautenburg.de/
Oschin 48-in Telescope (Palomar Schmidt)	1.24 m (48 in)	Palomar Mountain, California, USA	1948	Fork equatorial	astro.caltech.edu/ palomarpublic/index.html
UK Schmidt Telescope Unit	1.24 m (48 in)	Siding Spring Mountain, Australia	1973	Fork equatorial	www.aao.gov.au/ukst/
Kiso Schmidt Telescope	1.05 m (41 in)	Kiso, Japan	1975	Fork equatorial	www.ioa.s.u-tokyo.ac.jp/kiso_obs/
3TA-10 Schmidt Telescope (Byurakan Schmidt)	1.00 m (39 in)	Mount Aragatz, Armenia	1961	Fork equatorial	bao.sci.am/
Kvistaberg Schmidt Telescope (Uppsala Schmidt)	1.00 m (39 in)	Kvistaberg, Sweden	1963	Fork equatorial	www.astro.uu.se/history/ Kvistaberg.html
ESO 1-m Schmidt Telescope	1.00 m (39 in)	La Silla, Chile	1972	Fork equatorial	www.eso.org
Venezuela 1-m Schmidt	1.00 m (39 in)	Mèrida, Venezuela	1978	Bent yoke equatorial	www.cida.ve
Télescope Combiné de Schmidt	0.84 m (33 in)	Grasse, France	1981		wwwrc.obs-azur.fr/schmidt/
Schmidt Telescope	0.80 m (31 in)	Riga, Latvia	1968	not available	
Calar-Alto Schmidtspiegel	0.80 m (31 in)	Calar Alto, Spain	1955	Fork equatorial	www.mpia hd.mpg.de/ Public/CAHA/

PLANET & SATELLITE DATA

PLANETS: PHYSICAL DATA

PLANET	AVERAGE DIAMETER[a] (EARTH = 1)	MASS (EARTH = 1)	VOLUME (EARTH = 1)	MEAN DENSITY (WATER = 1)	OBLATENESS	SURFACE GRAVITY[a] (EARTH = 1)	ESCAPE VELOCITY (KM/S)	GEOMETRICAL ALBEDO
Mercury	0.383	0.055	0.056	5.43	0	0.378	4.30	0.11
Venus	0.950	0.815	0.857	5.24	0	0.907	10.36	0.65
Earth	1.000	1.000	1.000	5.52	0.0034	1.000	11.19	0.37
Mars	0.532	0.107	0.151	3.93	0.0065	0.377	5.03	0.15
Jupiter	10.973	317.83	1321.3	1.33	0.0649	2.364	59.5	0.52
Saturn	9.140	95.16	763.6	0.69	0.0980	0.916	35.5	0.47
Uranus	3.981	14.54	63.1	1.27	0.0229	0.889	21.3	0.51
Neptune	3.865	17.15	57.7	1.64	0.0171	1.120	23.5	0.41
Pluto	0.178	0.0021	0.0066	1.75	0	0.059	1.1	0.3

[a] At the 1-bar level in the atmosphere for Jupiter, Saturn, Uranus and Neptune.

Source: National Space Science Data Center, NASA

PLANETS: ORBITAL DATA

PLANET	MEAN DISTANCE FROM SUN (AU)	ECCENTRICITY OF ORBIT	SIDEREAL PERIOD	MEAN SYNODIC PERIOD (DAYS)	INCLINATION OF ORBIT TO ECLIPTIC (DEGREES)	SIDEREAL PERIOD OF AXIAL ROTATION[a]	INCLINATION OF EQUATOR TO ORBIT (DEGREES)
Mercury	0.387	0.206	87.969 d	115.88	7.00	58.646 d	0.01
Venus	0.723	0.007	224.701 d	583.92	3.39	243.019 d (R)	177.36
Earth	1.000	0.017	365.256 d	—	0.00	23.934 h	23.44
Mars	1.524	0.093	686.980 d	779.94	1.85	24.623 h	25.19
Jupiter	5.204	0.049	11.862 y	398.88	1.30	9.925 h	3.13
Saturn	9.582	0.057	29.457 y	378.09	2.48	10.656 h	26.73
Uranus	19.201	0.046	84.011 y	369.66	0.77	17.240 h (R)	97.77
Neptune	30.047	0.011	164.79 y	367.49	1.77	16.110 h	28.32
Pluto	39.236	0.244	247.68 y	366.73	17.16	6.387 d (R)	122.53

[a] R = retrograde. The rotation periods for Jupiter, Saturn, Uranus and Neptune are those of their magnetic fields.

Source: National Space Science Data Center, NASA

PLANETARY SATELLITES

PLANET AND SATELLITE	MASS (10²⁰ KG)	DIAMETER (KM)	DENSITY (WATER = 1)	GEOMETRIC ALBEDO	DISTANCE FROM CENTRE OF PLANET (10³ KM)	PLANETARY RADII	SIDEREAL ORBITAL PERIOD (DAYS)[a]	SIDEREAL ROTATION PERIOD (DAYS)[b]	ORBITAL INCLINATION (DEGREES)	ORBITAL ECCEN-TRICITY
Earth										
Moon	734.9	3,475	3.34	0.12	384.4	60.27	27.322	S	5.15	0.055
Mars										
Phobos	0.000106	27 × 22 × 18	1.90	0.07	9.38	2.76	0.319	S	1.08	0.015
Deimos	0.000024	15 × 12 × 10	1.75	0.08	23.46	6.91	1.262	S	1.79	0.001
Jupiter										
Metis	0.001	40		0.05	127.96	1.79	0.295	S	~0	<0.041
Adrastea	0.0002	26 × 20 × 16		0.05	128.98	1.81	0.298	S	~0	~0
Amalthea	0.072	262 × 146 × 134		0.07	181.3	2.54	0.498	S	0.40	0.003
Thebe	0.008	110 × 90		0.04	221.90	3.11	0.675		0.8	0.015
Io	893.3	3,643	3.53	0.63	421.6	5.91	1.769	S	0.04	0.004
Europa	479.7	3,130	2.99	0.67	670.9	9.40	3.551	S	0.47	0.009
Ganymede	1482	5,268	1.94	0.44	1,070	14.99	7.155	S	0.21	0.002
Callisto	1076	4,806	1.85	0.20	1,883	26.37	16.689	S	0.51	0.007
Leda	0.00006	10		0.07	11,094	155.38	238.72		26.07	0.148

PLANET AND SATELLITE	MASS (10²⁰ KG)	DIAMETER (KM)	DENSITY (WATER = 1)	GEOMETRIC ALBEDO	DISTANCE FROM CENTRE OF PLANET (10³ KM)	PLANETARY RADII	SIDEREAL ORBITAL PERIOD (DAYS)[a]	SIDEREAL ROTATION PERIOD (DAYS)[b]	ORBITAL INCLINATION (DEGREES)	ORBITAL ECCEN-TRICITY
Himalia	0.095	170		0.03	11,480	160.79	250.566	0.4	27.63	0.163
Lysithea	0.0008	24		0.06	11,720	164.15	259.22		29.02	0.107
Elara	0.008	80		0.03	11,737	164.39	259.653	0.5	24.77	0.207
Ananke	0.0004	20		0.06	21,200	297	631R		147	0.169
Carme	0.001	30		0.06	22,600	317	692R		164	0.207
Pasiphae	0.002	36		0.10	23,500	329	735R		145	0.378
Sinope	0.0008	28		0.05	23,700	332	758R		153	0.275
S/1999 J1		10			24,200	338	768R		143	0.125

Saturn

PLANET AND SATELLITE	MASS (10²⁰ KG)	DIAMETER (KM)	DENSITY (WATER = 1)	GEOMETRIC ALBEDO	DISTANCE FROM CENTRE OF PLANET (10³ KM)	PLANETARY RADII	SIDEREAL ORBITAL PERIOD (DAYS)[a]	SIDEREAL ROTATION PERIOD (DAYS)[b]	ORBITAL INCLINATION (DEGREES)	ORBITAL ECCEN-TRICITY
Pan		20		0.5	133.58	2.21	0.575			
Atlas		37 × 34 × 27		0.8	137.67	2.28	0.602		0.3	0.000
Prometheus	0.0014	148 × 100 × 68	0.27	0.5	139.35	2.31	0.613		0.0	0.002
Pandora	0.0013	110 × 88 × 62	0.42	0.7	141.70	2.35	0.629		0.0	0.004
Epimetheus	0.0054	138 × 110 × 10	0.63	0.8	151.42	2.51	0.694	S	0.34	0.009
Janus	0.0192	194 × 190 × 154	0.65	0.9	151.47	2.51	0.695	S	0.14	0.007
Mimas	0.375	418 × 392 × 382	1.14	0.5	185.52	3.08	0.942	S	1.53	0.020
Enceladus	0.73	512 × 494 × 490	1.12	1.0	238.02	3.95	1.370	S	0.00	0.005
Tethys	6.22	1,072 × 1,056 × 1,052	1.00	0.9	294.66	4.88	1.888	S	1.86	0.000
Telesto		30 × 25 × 15		1.0	294.66	4.88	1.888		~0	~0
Calypso		30 × 16 × 16		1.0	294.66	4.88	1.888		~0	~0
Dione	11.0	1,120	1.44	0.7	377.40	6.26	2.737	S	0.02	0.002
Helene		36 × 32 × 30		0.7	377.40	6.26	2.737		0.0	0.005
Rhea	23.1	1,528	1.24	0.7	527.04	8.74	4.518	S	0.35	0.001
Titan	1345.5	5,150	1.88	0.22	1,221.83	20.25	15.945		0.33	0.029
Hyperion	0.2	370 × 280 × 226		0.3	1,481.1	24.55	21.277	C	0.43	0.104
Iapetus	15.9	1436	1.02	0.05 / 0.5	3,561.3	59.03	79.330	S	14.72	0.028
Phoebe	0.004	230 × 220 × 210		0.06	12,952	214.69	550.48R	0.4	175.3	0.163

Uranus

PLANET AND SATELLITE	MASS (10²⁰ KG)	DIAMETER (KM)	DENSITY (WATER = 1)	GEOMETRIC ALBEDO	DISTANCE FROM CENTRE OF PLANET (10³ KM)	PLANETARY RADII	SIDEREAL ORBITAL PERIOD (DAYS)[a]	SIDEREAL ROTATION PERIOD (DAYS)[b]	ORBITAL INCLINATION (DEGREES)	ORBITAL ECCEN-TRICITY
Cordelia		26		0.07	49.77	1.95	0.335		0.08	0.000
Ophelia		30		0.07	53.79	2.10	0.376		0.10	0.010
Bianca		42		0.07	59.17	2.32	0.435		0.19	0.001
Cressida		62		0.07	61.78	2.42	0.464		0.01	0.000
Desdemona		54		0.07	62.68	2.45	0.474		0.11	0.000
Juliet		84		0.07	64.35	2.52	0.493		0.07	0.001
Portia		108		0.07	66.09	2.59	0.513		0.06	0.000
Rosalind		54		0.07	69.94	2.74	0.558		0.28	0.000
S/1986 U10		40			75	2.9	0.62		0.0	0.0
Belinda		66		0.07	75.26	2.94	0.624		0.03	0.000
Puck		154		0.07	86.01	3.37	0.762		0.32	0.000
Miranda	0.66	480 × 468 × 466	1.20	0.27	129.39	5.08	1.413	S	4.22	0.003
Ariel	13.4	1,162 × 1,156 × 1,155	1.67	0.35	191.02	7.48	2.520	S	0.31	0.003
Umbriel	11.7	1,169	1.40	0.19	266.30	10.41	4.144	S	0.36	0.005
Titania	35.2	1,578	1.71	0.28	435.91	17.05	8.706	S	0.14	0.002
Oberon	30.1	1,523	1.63	0.25	583.52	22.79	13.463	S	0.10	0.001
Caliban		60		0.07	7,169	280	579R		139.7	0.082
Stephano					7,942	311	676R		141.5	0.146
Sycorax		120		0.07	12,214	478	1,289R		152.7	0.509
Prospero					16,113	630	1,953R		146.3	0.327
Setebos					18,205	712	2,345R		148.8	0.494

Neptune

PLANET AND SATELLITE	MASS (10²⁰ KG)	DIAMETER (KM)	DENSITY (WATER = 1)	GEOMETRIC ALBEDO	DISTANCE FROM CENTRE OF PLANET (10³ KM)	PLANETARY RADII	SIDEREAL ORBITAL PERIOD (DAYS)[a]	SIDEREAL ROTATION PERIOD (DAYS)[b]	ORBITAL INCLINATION (DEGREES)	ORBITAL ECCEN-TRICITY
Naiad		58		0.06	48.23	1.95	0.294		4.74	0.000
Thalassa		80		0.06	50.08	2.02	0.311		0.21	0.000
Despina		148		0.06	52.53	2.12	0.335		0.07	0.000
Galatea		158		0.06	61.95	2.50	0.429		0.05	0.000
Larissa		208 × 178		0.06	73.55	2.97	0.555		0.20	0.001
Proteus		436 × 416 × 402		0.06	117.65	4.75	1.122		0.55	0.000
Triton	214.7	2,705	2.05	0.77	354.76	14.33	5.877R	S	157.35	0.000
Nereid	0.2	340	1.00	0.4	5,513.4	222.67	360.136		7.23	0.751

Pluto

PLANET AND SATELLITE	MASS (10²⁰ KG)	DIAMETER (KM)	DENSITY (WATER = 1)	GEOMETRIC ALBEDO	DISTANCE FROM CENTRE OF PLANET (10³ KM)	PLANETARY RADII	SIDEREAL ORBITAL PERIOD (DAYS)[a]	SIDEREAL ROTATION PERIOD (DAYS)[b]	ORBITAL INCLINATION (DEGREES)	ORBITAL ECCEN-TRICITY
Charon	19	1,186	2.0	0.5	19.6	17.0	6.39	S	0.0	0.0

[a] R = retrograde [b] S = synchronous rotation, i.e. the same as the orbital period   C = chaotic rotation   *Source: National Space Science Data Center, NASA*

STELLAR & GALAXY DATA

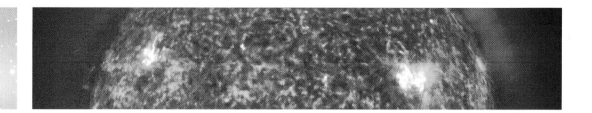

THE BRIGHTEST STARS

STAR	POPULAR NAME	RA 2000.0 (H M)	DEC (° ')	APPARENT MAGNITUDE	SPECTRAL TYPE	PARALLAX (")	DISTANCE (L.Y.)	ABSOLUTE MAGNITUDE
α CMa	Sirius	06 45	-16 43	-1.44	A1V	0.37921	8.60	1.45
α Car	Canopus	06 24	-52 42	-0.62	F0Ib	0.01043	313	-5.53
α Cen	Rigil Kentaurus	14 40	-60 50	-0.28c	G2V + K1V	0.74212	4.39	4.07c
α Boo	Arcturus	14 16	+19 11	-0.05v	K2IIIp	0.08885	36.7	-0.31
α Lyr	Vega	18 37	+38 47	0.03v	A0V	0.12892	25.3	0.58
α Aur	Capella	05 17	+46 00	0.08v	G6III + G2III	0.07729	42.2	-0.48
β Ori	Rigel	05 15	-08 12	0.18v	B8Ia	0.00422	773	-6.69
α CMi	Procyon	07 39	+05 14	0.40	F5IV–V	0.28593	11.4	2.68
α Eri	Achernar	01 38	-57 14	0.45v	B3V	0.02268	144	-2.77
α Ori	Betelgeuse	05 55	+07 24	0.45v	M1Ia-M2Iab	0.00763	427	-5.14
β Cen	Hadar	14 04	-60 22	0.61v	B1III	0.00621	525	-5.42
α Aql	Altair	19 51	+08 52	0.76v	A7V	0.19444	16.8	2.20
α Cru	Acrux	12 27	-63 06	0.77c	B0.5IV + B1V	0.01017	321	-4.19c
α Tau	Aldebaran	04 36	+16 31	0.87	K5III	0.05009	65.1	-0.63
α Vir	Spica	13 25	-11 10	0.98v	B1V	0.01244	262	-3.55
α Sco	Antares	16 29	-26 26	1.06v	M1.5Iab	0.00540	604	-5.29
β Gem	Pollux	07 45	+28 02	1.16	K0III	0.09674	33.7	1.09
α PsA	Fomalhaut	22 58	-29 37	1.17	A3V	0.13008	25.1	1.73
β Cru	Becrux	12 48	-59 41	1.25v	B0.5III	0.00925	353	-3.92
α Cyg	Deneb	20 41	+45 17	1.25v	A2Ia	0.00101	3230	-8.73
α Leo	Regulus	10 08	+11 58	1.36	B7V	0.04209	77.5	-0.52
ε CMa	Adhara	06 59	-28 58	1.50	B2II	0.00757	431	-4.10
α Gem	Castor	07 35	+31 53	1.58c	A2V + A5V	0.06327	51.6	0.59c
γ Cru	Gacrux	12 31	-57 07	1.59v	M3.5III	0.03709	87.9	-0.56
λ Sco	Shaula	17 34	-37 06	1.62v	B1.5IV	0.00464	703	-5.05
γ Ori	Bellatrix	05 25	+06 21	1.64	B2III	0.01342	243	-2.72
β Tau	Alnath	05 26	+28 36	1.65	B7III	0.02489	131	-1.37
β Car	Miaplacidus	09 13	-69 43	1.67	A1III	0.02934	111	-0.99
ε Ori	Alnilam	05 36	-01 12	1.69v	B0Ia	0.00243	1342	-6.38
α Gru	Alnair	22 08	-46 58	1.73	B7V	0.03216	101	-0.73
ζ Ori	Alnitak	05 41	-01 57	1.74c	O9.5 Ib + B0III	0.00399	817	-5.26c
γ²Vel		08 10	-47 20	1.75v	WC8 + O9I	0.00388	841	-5.31
ε UMa	Alioth	12 54	+55 58	1.76v	A0p	0.04030	80.9	-0.21
α Per	Mirphak	03 24	+49 52	1.79	F5Ib	0.00551	592	-4.50
ε Sgr	Kaus Australis	18 24	-34 23	1.79	A0II	0.02255	145	-1.44
α UMa	Dubhe	11 04	+61 45	1.81c	K0III + A8V	0.02638	124	-1.08c
δ CMa		07 08	-26 24	1.83	F8Ia	0.00182	1792	-6.87
η UMa	Alkaid	13 48	+49 19	1.85	B3V	0.03239	101	-0.60
ε Car	Avior	08 23	-59 31	1.86v	K3III + B2V	0.00516	632	-4.58
θ Sco		17 37	-43 00	1.86	F1III	0.01199	272	-2.75
β Aur	Menkalinan	06 00	+44 57	1.90v	A1IV	0.03972	82.1	-0.10
α TrA	Atria	16 49	-69 02	1.91	K2IIb-IIIa	0.00785	415	-3.62
γ Gem	Alhena	06 38	+16 24	1.93	A1IV	0.03112	105	-0.60
δ Vel		08 45	-54 43	1.93	A1V	0.04090	79.7	-0.01
α Pav	Peacock	20 26	-56 44	1.94	B2.5V	0.01780	183	-1.81
α UMi	Polaris	02 32	+89 16	1.97v	F5-8Ib	0.00756	431	-3.64
β CMa	Mirzam	06 23	-17 57	1.98v	B1II-III	0.00653	499	-3.95
α Hya	Alphard	09 28	-08 40	1.99	K3II-III	0.01840	177	-1.69

c = Combined magnitude of double star; v = variable. For the individual components of α Centauri see table opposite. *Sources: The Hipparcos Catalogue and The Astronomical Almanac.*

STARS WITHIN 15 LIGHT YEARS OF THE SUN

STAR	RA 2000.0 (H M)	DEC (° ')	APPARENT MAGNITUDE	SPECTRAL TYPE	PARALLAX (")	DISTANCE (L.Y.)	ABSOLUTE MAGNITUDE
Proxima Centauri (V645 Cen)	14 30	-62 41	11.01 (var)	M5.5Ve	0.7723	4.22	15.45
Alpha Centauri A	14 40	-60 50	-0.01	G2V	0.7421	4.39	4.34
Alpha Centauri B			1.35	K1V			5.70
Barnard's Star	17 58	+04 41	9.54	M5V	0.5490	5.94	13.24
CN Leo (Wolf 359)	10 56	+07 01	13.46 (var)	M6.5Ve	0.419	7.8	16.57
Lalande 21185 (HD 95735)	11 03	+35 58	7.49	M2V	0.3924	8.31	10.46
Sirius A	06 45	-16 43	-1.44	A1V	0.3792	8.60	1.45
Sirius B			8.44	DA2			11.33
UV Ceti A	01 39	-17 56	12.56 (var)	M5.5Ve	0.373	8.7	15.42
UV Ceti B			12.96 (var)	M5.5Ve			15.82
V1216 Sgr (Ross 154)	18 50	-23 50	10.37	M3.5Ve	0.3365	9.69	13.00
Ross 248	23 42	+44 09	12.27	M5.5Ve	0.316	10.3	14.77
Epsilon Eridani	03 33	-09 27	3.72	K2V	0.3108	10.50	6.18
HD 217987 (CoD -36° 15693)	23 06	-35 51	7.35	M2V	0.3039	10.73	9.76
FI Vir (Ross 128)	11 48	+00 48	11.12	M4V	0.2996	10.89	13.50
L789-6 ABC	22 39	-15 17	12.32	M5Ve	0.290	11.2	14.63
61 Cyg A (V1803 Cyg)	21 07	+38 45	5.20 (var)	K5V	0.2871	11.36	7.49
Procyon A	07 39	+05 13	0.40	F5IV–V	0.2859	11.41	2.68
Procyon B			10.7	DF			13.0
61 Cyg B	21 07	+38 45	6.05 (var)	K7V	0.2854	11.43	8.33
HD 173740 (BD +59° 1915 B)	18 43	+59 37	9.70	M4V	0.2845	11.47	11.97
HD 173739 (BD +59° 1915 A)	18 43	+59 38	8.94	M3.5V	0.2803	11.64	11.18
GX And (BD +43° 44 A)	00 18	+44 01	8.09 (var)	M2V	0.2803	11.64	10.33
GX And (BD +43° 44 B)			11.10	M4V			13.34
G51–15	08 30	+26 48	14.81	M6.5Ve	0.276	11.8	17.01
Epsilon Indi	22 03	-56 47	4.69	K4Ve	0.2758	11.83	6.89
Tau Ceti	01 44	-15 56	3.49	G8V	0.2742	11.90	5.68
L372–58	03 36	-44 30	13.03	M4.5V	0.273	11.9	15.21
YZ Cet (L725–32)	01 12	-17 00	12.10 (var)	M5.5Ve	0.2691	12.12	14.25
BD +05° 1668 (Luyten's Star)	07 27	+05 14	9.84	M4V	0.2633	12.39	11.94
VZ Pic (Kapteyn's Star)	05 12	-45 01	8.86 (var)	M1V	0.2553	12.78	10.89
AX Mic (CoD -39° 14192)	21 17	-38 52	6.69	M0Ve	0.2534	12.87	8.71
Krüger 60A (DO Cep)	22 28	+57 42	9.59	M3.5V	0.2495	13.07	11.58
Krüger 60B			11.3	M4Ve			13.3
Ross 614A (V577 Mon)	06 29	-02 49	11.12	M4Ve	0.2429	13.43	13.05
Ross 614B			14				15.9
BD -12° 4523	16 30	-12 40	10.10	M4V	0.2345	13.91	11.95
CoD -37° 15492	00 05	-37 21	8.56	M2V	0.2293	14.23	10.36
Wolf 424A	12 33	+09 03	13.10	M5Ve	0.228	14.3	14.89
Wolf 424 B			13.4				15.2
BD -13° 637B	03 22	-13 16	12.16	--	0.2275	14.34	13.94
van Maanen's Star	00 49	+05 23	12.37	DG	0.2270	14.37	14.15
L1159–16	02 00	+13 01	12.26	M4.5Ve	0.224	14.6	14.01
L143–23	10 45	-61 10	13.87	M4	0.222	14.7	15.60
BD +68° 946	17 36	+68 20	9.15	M3.5V	0.2209	14.77	10.87
LP731–58	10 48	-11 21	15.60	M6.5V	0.221	14.8	17.32
CoD -46° 11540	17 29	-46 53	9.38	M3V	0.2204	14.80	11.10
G208–45	19 54	+44 24	13.99	M6Ve	0.220	14.8	15.70
44AB			13.41	M6Ve			15.12

Source: Adapted from Alan H. Batten, Royal Astronomical Society of Canada Observers' Handbook.

THE CONSTELLATIONS

NAME	GENITIVE	ABBREVIATION	AREA (SQUARE DEGREES)	ORDER OF SIZE	NAME	GENITIVE	ABBREVIATION	AREA (SQUARE DEGREES)	ORDER OF SIZE
Andromeda	Andromedae	And	722	19	Lacerta	Lacertae	Lac	201	68
Antlia	Antliae	Ant	239	62	Leo	Leonis	Leo	947	12
Apus	Apodis	Aps	206	67	Leo Minor	Leonis Minoris	LMi	232	64
Aquarius	Aquarii	Aqr	980	10	Lepus	Leporis	Lep	290	51
Aquila	Aquilae	Aql	652	22	Libra	Librae	Lib	538	29
Ara	Arae	Ara	237	63	Lupus	Lupi	Lup	334	46
Aries	Arietis	Ari	441	39	Lynx	Lyncis	Lyn	545	28
Auriga	Aurigae	Aur	657	21	Lyra	Lyrae	Lyr	286	52
Boötes	Boötis	Boo	907	13	Mensa	Mensae	Men	153	75
Caelum	Caeli	Cae	125	81	Microscopium	Microscopii	Mic	210	66
Camelopardalis	Camelopardalis	Cam	757	18	Monoceros	Monocerotis	Mon	482	35
Cancer	Cancri	Cnc	506	31	Musca	Muscae	Mus	138	77
Canes Venatici	Canum Venaticorum	CVn	465	38	Norma	Normae	Nor	165	74
Canis Major	Canis Majoris	CMa	380	43	Octans	Octantis	Oct	291	50
Canis Minor	Canis Minoris	CMi	183	71	Ophiuchus	Ophiuchi	Oph	948	11
Capricornus	Capricorni	Cap	414	40	Orion	Orionis	Ori	594	26
Carina	Carinae	Car	494	34	Pavo	Pavonis	Pav	378	44
Cassiopeia	Cassiopeiae	Cas	598	25	Pegasus	Pegasi	Peg	1,121	7
Centaurus	Centauri	Cen	1,060	9	Perseus	Persei	Per	615	24
Cepheus	Cephei	Cep	588	27	Phoenix	Phoenicis	Phe	469	37
Cetus	Ceti	Cet	1,231	4	Pictor	Pictoris	Pic	247	59
Chamaeleon	Chamaeleontis	Cha	132	79	Pisces	Piscium	Psc	889	14
Circinus	Circini	Cir	93	85	Piscis Austrinus	Piscis Austrini	PsA	245	60
Columba	Columbae	Col	270	54	Puppis	Puppis	Pup	673	20
Coma Berenices	Comae Berenices	Com	386	42	Pyxis	Pyxidis	Pyx	221	65
Corona Australis	Coronae Australis	CrA	128	80	Reticulum	Reticuli	Ret	114	82
Corona Borealis	Coronae Borealis	CrB	179	73	Sagitta	Sagittae	Sge	80	86
Corvus	Corvi	Crv	184	70	Sagittarius	Sagittarii	Sgr	867	15
Crater	Crateris	Crt	282	53	Scorpius	Scorpii	Sco	497	33
Crux	Crucis	Cru	68	88	Sculptor	Sculptoris	Scl	475	36
Cygnus	Cygni	Cyg	804	16	Scutum	Scuti	Sct	109	84
Delphinus	Delphini	Del	189	69	Serpens	Serpentis	Ser	637	23
Dorado	Doradus	Dor	179	72	Sextans	Sextantis	Sex	314	47
Draco	Draconis	Dra	1,083	8	Taurus	Tauri	Tau	797	17
Equuleus	Equulei	Equ	72	87	Telescopium	Telescopii	Tel	252	57
Eridanus	Eridani	Eri	1,138	6	Triangulum	Trianguli	Tri	132	78
Fornax	Fornacis	For	398	41	Triangulum Australe	Trianguli Australis	TrA	110	83
Gemini	Geminorum	Gem	514	30	Tucana	Tucanae	Tuc	295	48
Grus	Gruis	Gru	366	45	Ursa Major	Ursae Majoris	UMa	1280	3
Hercules	Herculis	Her	1,225	5	Ursa Minor	Ursae Minoris	UMi	256	56
Horologium	Horologii	Hor	249	58	Vela	Velorum	Vel	500	32
Hydra	Hydrae	Hya	1,303	1	Virgo	Virginis	Vir	1,294	2
Hydrus	Hydri	Hyi	243	61	Volans	Volantis	Vol	141	76
Indus	Indi	Ind	294	49	Vulpecula	Vulpeculae	Vul	268	55

STELLAR & GALAXY DATA

MESSIER OBJECTS

M NUMBER NGC	RA (H M)	DEC. (° ′)	SIZE	MAG.	TYPE	M NUMBER NGC	RA (H M)	DEC. (° ′)	SIZE	MAG.	TYPE
1 1952	05 35	+22 01	6 × 4	~8.4	Crab nebula in Taurus	56 6779	19 17	+30 11	7	8.2	Globular cluster in Lyra
2 7089	21 34	-00 49	13	6.5	Globular cluster in Aquarius	57 6720	18 54	+33 02	1	~9.0	Ring nebula in Lyra
3 5272	13 42	+28 23	16	6.4	Globular cluster in Canes Venatici	58 4579	12 38	+11 49	5 × 4	9.8	Spiral galaxy in Virgo
4 6121	16 24	-26 32	26	5.9	Globular cluster in Scorpius	59 4621	12 42	+11 39	5 × 3	9.8	Elliptical galaxy in Virgo
5 5904	15 19	+02 05	17	5.8	Globular cluster in Serpens	60 4649	12 44	+11 33	7 × 6	8.8	Elliptical galaxy in Virgo
6 6405	17 40	-32 13	15	4.2	Open cluster in Scorpius	61 4303	12 22	+04 28	6 × 5	9.7	Spiral galaxy in Virgo
7 6475	17 54	-34 49	80	3.3	Open cluster in Scorpius	62 6266	17 01	-30 07	14	6.6	Globular cluster in Ophiuchus
8 6523	18 04	-24 23	90 × 40	~5.8	Lagoon nebula in Sagittarius	63 5055	13 16	+42 02	12 × 8	8.6	Spiral galaxy in Canes Venatici
9 6333	17 19	-18 31	9	~7.9	Globular cluster in Ophiuchus	64 4826	12 57	+21 41	9 × 5	8.5	Black Eye galaxy in Coma Berenices
10 6254	16 57	-04 06	15	6.6	Globular cluster in Ophiuchus	65 3623	11 19	+13 05	10 × 3	9.3	Spiral galaxy in Leo
11 6705	18 51	-16 16	14	5.8	Wild Duck cluster in Scutum	66 3627	11 20	+12 59	9 × 4	9.0	Spiral galaxy in Leo
12 6218	16 47	-01 57	14	6.6	Globular cluster in Ophiuchus	67 2682	08 50	+11 49	30	6.9	Open cluster in Cancer
13 6205	16 42	+36 28	17	5.9	Globular cluster in Hercules	68 4590	12 40	-26 45	12	8.2	Globular cluster in Hydra
14 6402	17 38	-03 15	12	7.6	Globular cluster in Ophiuchus	69 6637	18 31	-32 31	7	7.7	Globular cluster in Sagittarius
15 7078	21 30	+12 10	12	6.4	Globular cluster in Pegasus	70 6681	18 43	-32 18	8	8.1	Globular cluster in Sagittarius
16 6611	18 19	-13 47	7	6.0	Open cluster in Serpens	71 6838	19 54	+18 47	7	8.3	Globular cluster in Sagitta
17 6618	18 21	-16 11	46 × 37	7	Omega nebula in Sagittarius	72 6981	20 54	-12 32	6	9.4	Globular cluster in Aquarius
18 6613	18 20	-17 08	9	6.9	Open cluster in Sagittarius	73 6994	20 59	-12 38	–	–	Group of four stars in Aquarius
19 6273	17 03	-26 16	14	7.2	Globular cluster in Ophiuchus	74 628	01 37	+15 47	10 × 9	9.2	Spiral galaxy in Pisces
20 6514	18 03	-23 02	29 × 27	~8.5	Trifid nebula in Sagittarius	75 6864	20 06	-21 55	6	8.6	Globular cluster in Sagittarius
21 6531	18 05	-22 30	13	5.9	Open cluster in Sagittarius	76 650–1	01 42	+51 34	2 × 1	~11.5	Planetary nebula in Perseus
22 6656	18 36	-23 54	24	5.1	Globular cluster in Sagittarius	77 1068	02 43	-00 01	7 × 6	8.8	Spiral galaxy in Cetus
23 6494	17 57	-19 01	27	5.5	Open cluster in Sagittarius	78 2068	05 47	+00 03	8 × 6	8	Diffuse nebula in Orion
24 –	18 17	-18 29	90	~4.5	Star field in Sagittarius	79 1904	05 25	-24 33	9	8.0	Globular cluster in Lepus
25 IC4725	18 32	-19 15	32	4.6	Open cluster in Sagittarius	80 6093	16 17	-22 59	9	7.2	Globular cluster in Scorpius
26 6694	18 45	-09 24	15	8.0	Open cluster in Scutum	81 3031	09 56	+69 04	26 × 14	6.8	Spiral galaxy in Ursa Major
27 6853	20 00	+22 43	8 × 4	~8.1	Dumbbell nebula in Vulpecula	82 3034	09 56	+69 41	11 × 5	8.4	Irregular galaxy in Ursa Major
28 6626	18 25	-24 52	11	~6.9	Globular cluster in Sagittarius	83 5236	13 37	-29 52	11 × 10	~7.6	Spiral galaxy in Hydra
29 6913	20 24	+38 32	7	6.6	Open cluster in Cygnus	84 4374	12 25	+12 53	5 × 4	9.3	Elliptical galaxy in Virgo
30 7099	21 40	-23 11	11	7.5	Globular cluster in Capricornus	85 4382	12 25	+18 11	7 × 5	9.2	Elliptical galaxy in Coma Berenices
31 224	00 43	+41 16	178 × 63	3.4	Andromeda spiral galaxy	86 4406	12 26	+12 57	7 × 6	9.2	Elliptical galaxy in Virgo
32 221	0043	+40 52	8 × 6	8.2	Elliptical galaxy in Andromeda	87 4486	12 31	+12 24	7	8.6	Elliptical galaxy in Virgo
33 598	01 34	+30 39	62 × 39	5.7	Triangulum spiral galaxy	88 4501	12 32	+14 25	7 × 4	9.5	Spiral galaxy in Coma Berenices
34 1039	02 42	+42 47	35	5.2	Open cluster in Perseus	89 4552	12 36	+12 33	4	9.8	Elliptical galaxy in Virgo
35 2168	06 09	+24 20	28	5.1	Open cluster in Gemini	90 4569	12 37	+13 10	10 × 5	9.5	Spiral galaxy in Virgo
36 1960	05 36	+34 08	12	6.0	Open cluster in Auriga	91 4548	12 35	+14 30	5 × 4	10.2	Spiral galaxy in Coma Berenices
37 2099	05 52	+32 33	24	5.6	Open cluster in Auriga	92 6341	17 17	+43 08	11	6.5	Globular cluster in Hercules
38 1912	05 29	+35 50	21	6.4	Open cluster in Auriga	93 2447	07 45	-23 52	22	~6.2	Open cluster in Puppis
39 7092	21 32	+48 26	32	4.6	Open cluster in Cygnus	94 4736	12 51	+41 07	11 × 9	8.1	Spiral galaxy in Canes Venatici
40 –	12 22	+58 05	–	8	Double star in Ursa Major	95 3351	10 44	+11 42	7 × 5	9.7	Spiral galaxy in Leo
41 2287	06 47	-20 44	38	4.5	Open cluster in Canis Major	96 3368	10 47	+11 49	7 × 5	9.2	Spiral galaxy in Leo
42 1976	05 35	-05 27	66 × 60	4	Orion nebula	97 3587	11 15	+55 01	3	~11.2	Owl nebula in Ursa Major
43 1982	05 36	-05 16	20 × 15	9	Diffuse nebula in Orion	98 4192	12 14	+14 54	10 × 3	10.1	Spiral galaxy in Coma Berenices
44 2632	08 40	+19 59	95	3.1	Beehive cluster in Cancer	99 4254	12 19	+14 25	5	9.8	Spiral galaxy in Coma Berenices
45 –	03 47	+24 07	110	1.2	Pleiades cluster in Taurus	100 4321	12 23	+15 49	7 × 6	9.4	Spiral galaxy in Coma Berenices
46 2437	07 42	-14 49	27	6.1	Open cluster in Puppis	101 5457	14 03	+54 21	27 × 26	7.7	Spiral galaxy in Ursa Major
47 2422	07 37	-14 30	30	4.4	Open cluster in Puppis	102					Duplicate of M101
48 2548	08 14	-05 48	54	5.8	Open cluster in Hydra	103 581	01 33	+60 42	6	~7.4	Open cluster in Cassiopeia
49 4472	12 30	+08 00	9 × 7	8.4	Elliptical galaxy in Virgo	104 4594	12 40	-11 37	9 × 4	8.3	Sombrero galaxy in Virgo
50 2323	07 03	-08 20	16	5.9	Open cluster in Monoceros	105 3379	10 48	+12 35	4 × 4	9.3	Elliptical galaxy in Leo
51 5194–5	13 30	+47 12	11 × 8	8.1	Whirlpool galaxy in Cances Venatici	106 4258	12 19	+47 18	18 × 8	8.3	Spiral galaxy in Canes Venatici
52 7654	23 24	+61 35	13	6.9	Open cluster in Cassiopeia	107 6171	16 33	-13 03	10	8.1	Globular cluster in Ophiuchus
53 5024	13 13	+18 10	13	7.7	Globular cluster in Coma Berenices	108 3556	11 12	+55 40	8 × 2	10.0	Spiral galaxy in Ursa Major
54 6715	18 55	-30 29	9	7.7	Globular cluster in Sagittarius	109 3992	11 58	+53 23	8 × 5	9.8	Spiral galaxy in Ursa Major
55 6809	19 40	-30 58	19	7.0	Globular cluster in Sagittarius	110 205	00 40	+41 41	17 × 10	8.0	Elliptical galaxy in Andromeda

THE LOCAL GROUP OF GALAXIES

GALAXY	RA 2000.0 (H M)	DEC (° ')	TYPE*	ABSOLUTE MAGNITUDE	VISUAL MAGNITUDE	DISTANCE (MILLION L.Y.)
M31 (NGC 224, Andromeda Galaxy)	00 43	+41 16	Sb I–II	–21.2	3.4	2.48
Milky Way			S(B)bc I–II	–20.9	–	–
M33 (NGC 598, Triangulum Galaxy)	01 34	+30 30	Sc II–III	–18.9	5.7	2.58
LMC	05 20	-69 27	Ir III–IV	–18.5	0.4	0.16
SMC	00 53	-72 48	Ir IV/IV–V	–17.1	2.3	0.20
M32 (NGC 221)	00 43	+40 52	E2	–16.5	8.1	2.48
M110 (NGC 205)	00 40	+41 41	Sph	–16.4	8.1	2.48
IC 10 (UGC 192)	00 20	+59 18	Ir IV	–16.3	10.3	2.15
NGC 6822 (Barnard's Galaxy)	19 45	-14 48	Ir IV–V	–16.0	9	1.63
NGC 185	00 39	+48 20	Sph	–15.6	9.2	2.15
IC 1613	01 05	+02 08	Ir V	–15.3	9.2	2.35
NGC 147	00 33	+48 30	Sph	–15.1	9.5	2.15
WLM (DDO 221)	00 02	-15 28	Ir IV–V	–14.4	10.6	3.10
Sagittarius Dwarf	18 55	-30 29	dSph	–13.8	–	0.10
Fornax Dwarf	02 40	-34 30	dSph	–13.1	7.8	0.46
Pegasus Dwarf (DDO 216)	23 29	+14 45	Ir V	–12.3	12.0	2.48
Cassiopeia Dwarf (And VII)	23 27	+50 42	dSph	–12.0	12.9	2.25
Sagittarius Dwarf Irregular (SagDIG)	19 30	-17 41	Ir V	–12.0	15	3.85
Leo I	10 08	+12 18	dSph	–11.9	10.1	0.82
And I	00 46	+38 00	dSph	–11.8	13.2	2.64
And II	01 16	+33 26	dSph	–11.8	13	2.22
Leo A (DDO 69)	09 59	+30 45	Ir V	–11.5	12.6	2.25
Pegasus II (And VI)	23 52	+24 36	dSph	–11.3	13.3	2.54
Aquarius Dwarf (DDO 210)	20 47	-12 51	Ir V	–10.9	14.1	3.10
Pisces Dwarf (LGS 3)	01 04	+21 54	dIr/dSph	–10.4	15	2.64
And III	00 35	+36 31	dSph	–10.2	13	2.48
Cetus Dwarf	00 26	-11 03	dSph	–10.1	14.4	2.54
Leo II (DDO 93)	11 14	+22 10	dSph	–10.1	11.5	0.68
Sculptor Dwarf	01 00	-33 43	dSph	–9.8	8.8	0.29
Phoenix Dwarf	01 51	-44 27	dIr/dSph	–9.8	–	1.30
Tucana Dwarf	22 42	-64 25	dSph	–9.6	15.2	2.84
Sextans Dwarf	10 13	-01 37	dSph	–9.5	10.3	0.29
Carina Dwarf	06 42	-50 58	dSph	–9.4	11	0.33
And V	01 10	+47 38	dSph	–9.1	15.9	2.64
Ursa Minor Dwarf (DDO 199)	15 09	+67 07	dSph	–8.9	11	0.20
Draco Dwarf (DDO 208)	17 20	+57 55	dSph	–8.6	11	0.26

★ dIr = dwarf irregular, dSph = dwarf spheroidal.   *Source: Adapted from Sidney van den Bergh, Publications of the Astronomical Society of the Pacific, April 2000.*

BRIGHTEST GALAXIES OUTSIDE THE LOCAL GROUP

GALAXY	RA 2000.0 (H M)	DEC (° ')	TYPE	APPARENT MAGNITUDE[a]	DISTANCE[b] (MILLION L.Y.)
NGC 55	00 15	-39 13	Sc	7.9	5.8
NGC 247	00 47	-20 46	Sc	9.1	5.8
NGC 253	00 48	-25 17	Sc	7.2	8.0
NGC 300	00 54	-37 41	Sc	8.1	4.2
NGC 628 = M74	01 37	+15 47	Sc	9.4	43
NGC 1023	02 40	+39 04	SB0	9.4	45
NGC 1068 = M77	02 43	-00 01	Sb	8.9	62
NGC 1097	02 46	-30 17	SBbc	9.5	65
NGC 1291	03 17	-41 06	SBa	8.5	37
NGC 1313	03 18	-66 30	SBc	8.7	13
NGC 1316 = Fornax A	03 23	-37 12	Sa(pec)	8.5	66
NGC 2403	07 37	+65 36	Sc	8.5	8.6
NGC 2841	09 22	+50 58	Sb	9.2	36
NGC 2903	09 32	+21 30	Sc	9.0	24
NGC 2997	09 46	-31 11	Sc	9.4	40
NGC 3031 = M81	09 56	+69 04	Sb	6.9	14
NGC 3034 = M82	09 56	+69 41	Amorphous	8.4	14
NGC 3115	10 05	-07 43	S0	8.9	22
NGC 3368 = M96	10 47	+11 49	Sab	9.3	38
NGC 3379 = M105	10 48	+12 35	E0	9.3	38
NGC 3521	11 06	-00 02	Sbc	9.0	32
NGC 3623 = M65	11 19	+13 05	Sa	9.3	34
NGC 3627 = M66	11 20	+13 00	Sb	8.9	30
NGC 3628	11 20	+13 36	Sb	9.5	36
NGC 4258 = M106	12 19	+47 18	Sb	8.4	26
NGC 4321 = M100	12 23	+15 49	Sc	9.4	55 V
NGC 4374 = M84	12 25	+12 53	E1	9.1	55 V
NGC 4382 = M85	12 25	+18 11	Ep	9.1	55 V
NGC 4406 = M86	12 26	+12 57	S0/E3	8.9	55 V
NGC 4472 = M49	12 30	+08 00	E1/S0	8.4	55 V
NGC 4486 = M87	12 31	+12 23	E0	8.6	55 V
NGC 4594 = M104	12 40	-11 37	Sa/b	8.0	44
NGC 4631	12 42	+32 32	Sc	9.2	31
NGC 4636	12 43	+02 41	E0/S0	9.5	41
NGC 4649 = M60	12 44	+11 33	S0	8.8	55 V
NGC 4697	12 49	-05 48	E6	9.2	52
NGC 4725	12 50	+25 30	Sb/SBb	9.4	59
NGC 4736 = M94	12 51	+41 07	Sab	8.2	17
NGC 4826 = M64	12 57	+21 41	Sab	8.5	18
NGC 4945	13 05	-49 28	Sc	9.3(B)	16
NGC 5055 = M63	13 16	+42 02	Sbc	8.6	28
NGC 5128 = Cen A	13 25	-43 01	S0(pec)	6.8	16
NGC 5194 = M51	13 30	+47 12	Sbc	8.4	27
NGC 5236 = M83	13 37	-29 52	SBc	7.5	16
NGC 5457 = M101	14 03	+54 21	Sc	7.9	17
NGC 6744	19 10	-63 51	Sbc	9.1(B)	33
NGC 6946	20 35	+60 09	Sc	8.8	17
NGC 7331	22 37	+34 25	Sb	9.5	56
NGC 7793	23 58	-32 35	Sd	9.1	8.0

a B = magnitude at blue wavelengths; all others are visual (V) magnitudes.

b Calculated using a Hubble constant of 65 km/s/Mpc. V = member of the Virgo Cluster; for these, the average cluster distance is given.

Sources: A. Sandage & G. Tammann, Revised Shapley-Ames Catalog of Bright Galaxies (1987); The Astronomical Almanac.

MATHS TOOLBOX

Mathematics is the universal language of physical science. It provides a natural medium for rational argument and promotes precise quantitative thinking. Here, gathered together, are some of the basic mathematical tools used in this book and by scientists throughout the world

INDICES AND POWERS OF 10

Indices or *exponents,* as they are sometimes called, are small superscripts used to indicate repeated multiplication. For example, in the equation $2^3 = 8$, the superscript 3 is an index and it indicates *three* twos multiplied together, as in $2 \times 2 \times 2$. Indices provide a valuable shorthand that can be used in association with units or quantities as well as numbers. Thus, the volume of a body can be measured in units of (metre)^3, or m^3 for short. Similarly, a square sheet of paper with sides of length $l = 2$ m will have an area $A = l^2 = (2 \text{ m}) \times (2 \text{ m}) = 4 \text{ m}^2$.

The index associated with a given number or quantity is often referred to as the *power* to which that number is raised. So 2^3 represents 2 raised to the power 3, and l^2 represents l raised to the power 2. Indices of 2 and 3 are especially common, and quantities raised to the power of 2 are usually said to be *squared*, while those raised to the power of 3 are said to be *cubed*.

One of the most frequent uses of indices is in relation to *powers of ten*. These include $10^1 = 10$, $10^2 = 100$ and $10^3 = 1000$, but they also extend to the much larger and smaller numbers that are needed when discussing entities that might be as big as galaxies or as small as atoms. Big numbers simply require the use of large powers, but the use of powers to indicate small numbers involves the following mathematical rules:

Rule 1: Any quantity raised to the power *zero* is equal to 1; in particular, $10^0 = 1$.

Rule 2: A *negative* power indicates division; in particular, $10^{-1} = 1/10 = 0.1$ and $10^{-3} = 1/1000 = 0.001$.

Some important powers of ten, positive and negative, are listed below, together with their equivalent *decimal numbers* and their common names. Note that positive powers of ten indicate the number of zeroes that follow the 1, while negative powers indicate the number of places that the 1 is to the right of the decimal point.

10^9	= 1000 000 000	one billion
10^{-9}	= 0.000 000 001	one billionth
10^6	= 1000 000	one million
10^{-6}	= 0.000 001	one millionth
10^3	= 1000	one thousand
10^{-3}	= 0.001	one thousandth
10^0	= 1	one

SCIENTIFIC NOTATION

Scientific notation provides a standard way of representing numbers and numerical quantities. A number is said to be written in scientific notation when it is expressed as a decimal number between 1 and 10, multiplied by some power of ten. For example, the number 284 is written in scientific notation as 2.84×10^2, and the scientific notation for 0.486 is 4.86×10^{-1}. In the case of quantities that involve units of measurement, such as a length of 24 metre, or a time of 0.060 second, the name of the measurement unit, or the symbol that represents it, is treated as an extra multiplicative factor, and is conventionally written at the end of the expression, as in 2.4×10^1 m and 6.0×10^{-2} s.

Scientific notation offers several advantages over other methods of displaying numerical data. In particular, it helps to avoid the misleading impression of precision given by some decimal numbers. This problem often arises in relation to quantities such as populations, where writing fifty thousand as 50,000 implies that the population is neither as small as 49,999, nor as large as 50,001. Of course, a population is unlikely to be known with this kind of precision, so there is a need for a better method of representing the population than giving the population in scientific notation, as 5.0×10^4, implies only that it is between 49,500 and 50,499 which is more likely to be the case. If the population was actually known with greater accuracy, to the nearest 50 say (rather than 500), then this could be indicated by writing 5.00×10^4. When scientific notation is used in this way all the digits in the decimal part of the number are said to be *significant figures* since they all carry important information about the precision with which the value is known.

ALGEBRAIC NOTATION

Algebra involves the use of symbols, usually letters such as x or M, to represent numbers or quantities. Algebraic quantities may be combined in the same way as numbers, using the basic operations of arithmetic; addition (+), subtraction (-), multiplication (×) and division (÷ or /). In practice, when writing algebraic expressions, multiplication signs are often omitted, so $F = ma$ has the same meaning as $F = m \times a$; and division signs are usually avoided, so $a \div b$ is normally written as a/b or, equivalently, $\frac{a}{b}$. In more complicated expressions, such as $ab + c/d$, it is necessary to know which operations to carry out first. The rule is to perform multiplications and divisions first, then additions and subtractions, so $5 \times 1 + 4/2 = 5 + 2 = 7$. However, despite the rule, it is common practice to use brackets to group together terms that must be worked out first, as in $ab + (c/d)$; this avoids any ambiguity whatsoever.

Apart from the familiar equals sign (=) there are a number of other symbols that are used to indicate specific relationships between expressions or quantities. For example:

$x \approx y$	means x is approximately equal to y
$x \neq y$	means x is not equal to y
$x > y$	means x is greater than y
$x \geq y$	means x is greater than or equal to y
$x < y$	means x is less than y
$x \leq y$	means x is less than or equal to y
$x \propto y$.	means that x is proportional to y, implying $x = ky$ where k is a constant.

Other common mathematical symbols are:
∞ for infinity
± for plus or minus
% for percentage,
the constants $\pi = 3.142$ and $e = 2.718$ (each to four significant figures) and $\sqrt{x}$ for the positive *square root* of x, so $\sqrt{9} = 3$ and $\sqrt{x^2} = x$.
Roots are also indicated by fractional powers; so, for instance $x^{1/2}$ denotes the square root of x, and $x^{1/3}$ denotes the cube root of x.

ANGLES AND TRIGONOMETRY

Angles are often measured in *degrees* (°), with 360° equal to one complete rotation. Fractions of a degree may be expressed in decimal notation, as in 22.5°. Alternatively, fractions of a degree can be expressed using the units *minute of arc* (′) and *second of arc* (″), where

$$1° = 60′ = 3,600″$$

It follows that there are 60 seconds of arc in a minute of arc, and 60 minutes of arc in a degree. Seen from the Earth, the full Moon is about 0.5° across, that is 30 minutes of arc.

Another unit of angle is the *radian*, usually represented by the symbol rad. This is quite a large unit, since, to three significant figures, 1 rad = 57.3°.

More memorable, however, is the exact relation: 2π rad = 360°.

The radian is a 'natural' unit of angle since a *circular arc* of radius R (see Figure 1) will have an *arc length* R when the angle it *subtends* is 1 radian.

More generally, an arc length s measured along the circumference of a circle of radius R will subtend an angle $\theta = s/R$ at the centre of the circle, provided that θ is measured in radians.

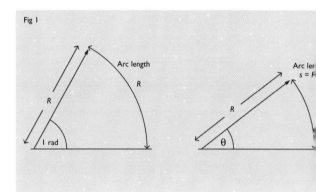

Fig 1

Trigonometry involves the study of *right angled triangles*, that is triangles with an angle of 90° (or $\pi/2$ rad) in one corner. Given a right angled triangle which also contains an angle θ, it is conventional to identify the individual sides as the *opposite*, the *adjacent* and the *hypotenuse*, according to their relation to θ, as shown in Figure 2.

The lengths of the sides are often denoted o, a and h. The hypotenuse is always the longest since, according to *Pythagoras' theorem* $h^2 = o^2 + a^2$.

For a given value of θ the lengths of the sides may be used to define three important quantities known as the *trigonometric ratios*, these are called the *sine*, *cosine* and *tangent* of θ, but they are

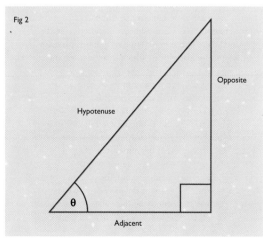

Fig 2

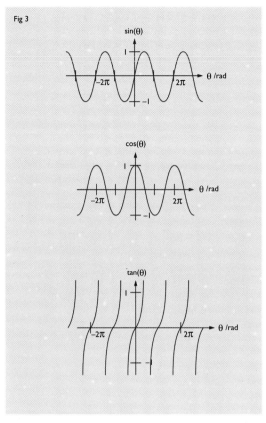

Fig 3

usually represented by their abbreviations sin, cos and tan, and are defined as follows:

$$\sin \theta = o/h$$
$$\cos \theta = a/h$$
$$\tan \theta = o/a$$

These definitions, together with Pythagoras' theorem imply that

$$\tan \theta = \frac{\sin \theta}{\cos \theta} \quad \text{and} \quad (\cos \theta)^2 + (\sin \theta)^2 = 1$$

The trigonometric ratios are only defined for values of θ between 0° and 90°, since the internal angles of any triangle must add up to 180°. However, their values may be generalized to define three *trigonometric functions* – also called sine, cosine and tangent – which each associate a number with any value of θ in a way that repeats itself every time θ is increased by 360°. The repeating graphs of these three functions are shown in Figure 3.

INTERPRETATION OF SIMPLE GRAPHS

Graphs are widely used in science, sometimes, as in Figure 3, to illustrate known relationships, sometimes to display measured data. The simplest kind of graph is probably the familiar rectangular plot in which quantities are measured along simple *linear axes* that meet at a common zero point called the *origin*. There are a number of conventions that are used when plotting such graphs, in order to make them easy to interpret. For instance, the quantities being plotted should always be identified, and so should the relevant units of measurement, if there are any. Another convention is that when one of the plotted quantities is an *independent variable*, the value of which may be chosen at will (such as the θ in Figure 3), then that quantity should be plotted along the horizontal axis. The value of the *dependant variable* – the quantity that is determined by the independent variable (such as sin θ or cos θ in Figure 3) – can then be plotted along the vertical axis.

One of the easiest relationships to recognize from a graph is that of proportionality. If two quantities, x and y say, are proportional, then $y = kx$ where k is a constant, and the graph of y against x will be a straight line that passes through the origin. In such cases, the value of k will be determined by the *gradient* or *slope* of the line, as indicated in Figure 4.

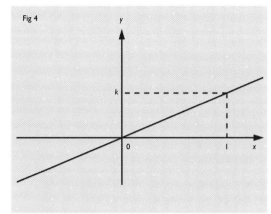

Fig 4

Many natural laws concern proportional relationships. An important example is Hubble's law, which asserts that the recession speed v of a galaxy is proportional to the distance d of that galaxy from the Earth, so $v = H_0 d$. In this case the constant H_0 that determines the slope of the graph is called *Hubble's constant*; its value indicates the rate of expansion of the Universe.

Straight line graphs that do not pass through the origin, such as those shown in Figure 5, indicate a *linear relationship*. Such a relationship is represented by an equation of the form $y = mx + c$, where the constant m determines the slope of the line, while the constant c determines the *intercept* at which the plotted line crosses the vertical axis. Notice that a negative value of m corresponds to a line that slopes down from left to right, so that y decreases as x increases.

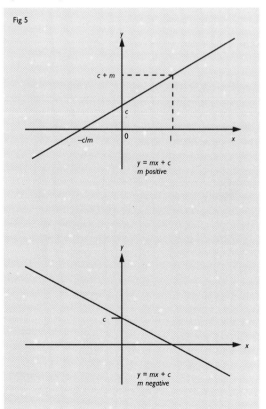

Fig 5

$y = mx + c$
m positive

$y = mx + c$
m negative

More complicated relationships, such as an *inverse square law* of the form $y = k/x^2$, or an *exponential law* of the form $y = ke^x$ (where e =2.718...), lead to graphs that are not straight and are therefore harder to interpret. However, special techniques, such as the use of *logarithmic axes*, can simplify such graphs and thus aid their interpretation.

VECTOR ALGEBRA

Vectors are quantities that have a direction as well as a size or magnitude. An example is wind velocity, which always has a magnitude (the *speed* of the wind) as well as a direction (its compass heading). Vector quantities include forces, accelerations, all kinds of velocity (which should be distinguished from speeds on this basis), and the values of gravitational, electric and magnetic fields at any specified point. All of these quantities may be associated with a direction as well as a magnitude.

In scientific texts, vectors are often indicated by bold symbols, such as $\boldsymbol{F}$ or $\boldsymbol{v}$, and in diagrams they are usually shown as arrows. Like ordinary algebraic quantities, vectors of a similar type, two forces say or two magnetic fields, may be added or subtracted, but in the case of vectors attention must be paid to their directions as well as their magnitudes. One way of achieving this is to add the vectors diagrammatically using a three-step procedure called the *triangle rule*. This is illustrated in Figure 6, where two vectors, $\boldsymbol{v}_1$ and $\boldsymbol{v}_2$, with different magnitudes and directions, are added together. To work out their sum, $\boldsymbol{v}_1 + \boldsymbol{v}_2$, the first step is to represent each vector by an arrow, drawn to scale, so that its length represents the magnitude of the corresponding vector. The next step is to slide the arrows parallel to their original directions until the head of one meets the tail of the other. A third arrow, drawn from the tail of the first to the head of the second, then represents the desired vector sum.

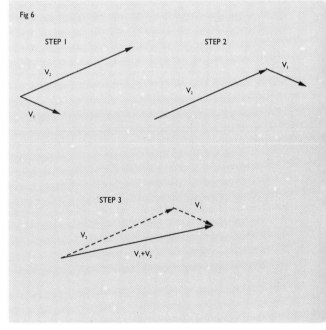

Fig 6

STEP 1

STEP 2

STEP 3

The difference of two vectors, $\boldsymbol{v}_1 - \boldsymbol{v}_2$, can be worked out in almost the same way as their sum. This is because subtracting $\boldsymbol{v}_2$ from $\boldsymbol{v}_1$ is equivalent to adding $-\boldsymbol{v}_2$ to $\boldsymbol{v}_1$, and the vector $-\boldsymbol{v}_2$ is represented by an arrow that has the same magnitude as $\boldsymbol{v}_2$, but points in the opposite direction. Thus, to subtract $\boldsymbol{v}_2$ from $\boldsymbol{v}_1$ simply draw arrows representing the vectors $\boldsymbol{v}_1$ and $-\boldsymbol{v}_2$, and then use the triangle rule to add those two vectors together.

UNITS & CONSTANTS

I n keeping with common scientific practice, this book mainly uses an international system of units known as SI. The letters stand for *Système International d'Unités*, and are themselves an internationally recognized symbol. SI is a well thought out generalization of the old metric system. It is based on seven precisely defined units of measurement, called the *base units*, which include the metre, the kilogram and the second, as well as units of electric current, temperature, amount of substance and luminous intensity. The system gives each of these units a name, and an internationally agreed symbol, such as m for metre, kg for kilogram, and s for second. By combining the base units it is possible to create an unlimited number of *derived units* such as m/s (the metre per second, a unit of speed) or m/s² (the metre per second per second, a unit of acceleration). Some of the most important derived units are given names and symbols of their own; examples include the unit of force, the *newton*, which is defined by the relation 1 N = 1 kg m/s², and the unit of energy, which is called a *joule* and is defined by the relation 1 J = 1 kg m²/s². Note that where SI units are named after scientists, such as Isaac Newton or James Joule, the unit name is always written with a lower case initial letter (newton or joule) even though the unit symbol involves an upper case letter (N or J).

The names of the seven base units, along with the symbols used to represent them and the quantities they measure, are listed in the following table. This is accompanied by simplified versions of the definitions of those units.

Quantity	Base Unit Name	Base Unit Symbol
length	metre	m
mass	kilogram	kg
time	second	s
current	ampere	A
temperature	kelvin	K
amount of substance	mole	mol
luminous intensity	candela	cd

metre: the length of the path travelled by light in a vacuum during a time interval of 1/299 792 458 of a second.

kilogram: the mass of the international prototype kilogram, a manufactured cylinder of platinum–iridium kept under carefully controlled conditions by the International Bureau of Weights and Measures at Sèvres, near Paris.

second: the duration of 9192 631 770 periods of a specific kind of radiation emitted by cesium-133 atoms.

ampere: the electric current which, if maintained in each of two infinitely long, straight, parallel wires placed one metre apart in a vacuum, would cause each wire to be acted on by a force of 2×10^{-7} newtons per metre of its length.

kelvin: the fraction 1/273.16 of the absolute temperature of the triple point of water. (The *triple point* of water determines a unique temperature at which ice, water and water vapour can coexist at a fixed pressure. In SI, this temperature is assigned the value 273.16 K.)

mole: the amount of any pure substance that contains as many basic particles (atoms, molecules, ions etc.) as there are atoms in 0.012 kilogram of carbon-12.

candela: the luminous intensity in a given direction of a source that emits radiation at a frequency of 540×10^{12} hertz with a specified radiant intensity.

The following table includes some of the most important derived units, together with their names (where appropriate) and the quantities they are used to measure.

Quantity	Derived Unit Symbol	Derived Unit Name
speed	m/s	
acceleration	m/s²	
force	kg m/s² = N	newton
energy	kg m²/s² = N m = J	joule
power	kg m²/s³ = J/s = W	watt
pressure	kg/(m s²) = N/m² = Pa	pascal
frequency	1/s = Hz	hertz
charge	A s = C	coulomb
potential difference	kg m²/(s³ A) = J/C = V	volt
magnetic field	kg/(s² A) = T	tesla

Because it is often necessary to use large or small multiples of the base and derived units, SI also includes a number of standard prefixes indicating multiples and submultiples, such as kilo (for 1000) and milli (for 1/1000). Each of these prefixes is represented by an agreed symbol, such as k for kilo and m for milli. Prefix symbols can be combined with unit symbols, just as prefixes can be combined with unit names; this results in combinations such as kilometre (km) and millimetre (mm). Note, however, that when using prefixes the gram (g) is treated as a unit of mass, even though it is the kilogram that is the official base unit. The most important prefixes together with their symbols and meanings are listed in the following table.

Multiple	Prefix	Symbol	Multiple	Prefix	Symbol
10^{12}	tera	T	10^{-3}	milli	m
10^{9}	giga	G	10^{-6}	micro	µ
10^{6}	mega	M	10^{-9}	nano	n
10^{3}	kilo	k	10^{-12}	pico	p
10^{0}			10^{-15}	femto	f

In addition to the SI units already defined, a number of other units, mainly belonging to older systems, are still in widespread use. SI provides exact definitions for some of these units, while others have equivalent SI values that must be determined experimentally. Some of these non-SI units, together with their SI equivalents, exact or experimental, are quoted to four significant figures in the following table.

Quantity	Unit Name	Unit Symbol	SI Equivalent
length	angstrom	Å	10^{-10} m
length	astronomical unit	AU	1.496×10^{11} m
length	light year	ly	9.461×10^{15} m
length	parsec	pc	3.086×10^{16} m
mass	atomic mass unit	u	1.661×10^{-27} kg
mass	tonne	t	10^{3} kg
time	year	yr or a	3.156×10^{7} s
temperature	degree Celsius	°C	1 K
energy	erg	erg	10^{-7} J
energy	electronvolt	eV	1.602×10^{-19} J
pressure	atmosphere	atm	1.013×10^{5} Pa
magnetic field	gauss	G	10^{-4} T

One of the reasons why so many non-SI units are still in use is that many of them have rather natural (and memorable) definitions. Some of these are given below, together with any relevant comments

astronomical unit: the average distance between the centre of the Earth and the centre of the Sun.

light year: the distance that light, travelling in a vacuum, covers in one year.

parsec: the distance at which a length of 1 AU subtends an angle of 1 second of arc.

atomic mass unit: the mass equivalent to one twelfth of the mass of a carbon-12 atom.

year: scientists use many different definitions of the year for different purposes. When used as a unit, the commonest kind of year is a Julian year, which consists of 365.25 days, each containing 86400 seconds.

degree Celsius: SI defines 1 °C = 1 K (exactly), but it should be noted that a temperature T_c measured on the Celsius temperature scale is related to the corresponding temperature T_K on the Kelvin temperature scale by the formula $T_c/°C = (T_K/K) - 273.15$. This implies that absolute zero (0 K) corresponds to -273.15°C.

electronvolt: the energy transferred to an electron when it is accelerated through a potential difference of one volt.

atmosphere: SI defines 1 atm = 101 325 Pa (exactly), but, as a unit, the atmosphere is most easily remembered as being roughly equal to the average pressure of the Earth's atmosphere at sea-level.

PHYSICAL CONSTANTS

The following table shows the names, symbols and currently accepted values of some of the most important physical constants. It should be noted that it is common practice to use quantities such as the mass of the Earth as informal units of measurement. This is done by expressing a given quantity, such as a mass or a radius, as a multiple of the mass or radius of the Earth, as in $M = xM_\oplus$ or $R = yR_\oplus$, where x and y are pure numbers. Thus the mass of Jupiter may be expressed as $M_J = 318\ M_\oplus$, and its radius as $R_J = 11.2\ R_\oplus$.

Constant	Symbol	Value
Speed of light	c	2.99792458×10^8 m/s
Gravitational constant	G	6.673×10^{-11} N m²/kg²
Planck constant	h	6.62607×10^{-34} J s
Electron mass	m_e	9.10938×10^{-31} kg
Proton mass	m_p	1.67262×10^{-27} kg
Stefan-Boltzmann constant	σ	5.67040×10^{-8} W/(m² K⁴)
Wien constant	W	2.89777×10^{-3} m K
Boltzmann constant	k	1.38065×10^{-23} J/K
Solar mass	$M_\odot$	1.989×10^{30} kg
Solar radius	$R_\odot$	6.960×10^8 m
Solar luminosity	$L_\odot$	3.827×10^{26} W
Earth mass	$M_\oplus$	5.974×10^{24} kg
Earth radius	$R_\oplus$	6.378×10^6 m
Tropical year (equinox to equinox)		365.24219 days
Sidereal year (with respect to stars)		365.25636 days = 3.155815×10^7 s

CONVERSIONS OF UNITS

In order to save space in the following tables, powers of ten are indicated by the use of parentheses. Thus, 9.461×10^{17} is written as $9.461(^{17})$

Size and Length

	cm	m	km	$R_\oplus$	$R_\odot$	AU	ly	pc
1 cm =	1	0.01	1(⁻⁵)	1.568(⁻⁹)	1.437(⁻¹¹)	6.685(⁻¹⁴)	1.057(⁻¹⁸)	3.241(⁻¹⁹)
1 m =	100	1	0.001	1.568(⁻⁷)	1.437(⁻⁹)	6.685(⁻¹²)	1.057(⁻¹⁶)	3.241(⁻¹⁷)
1 km =	1(⁵)	1000	1	1.568(⁻⁴)	1.437(⁻⁶)	6.685(⁻⁹)	1.057(⁻¹³)	3.241(⁻¹⁴)
$R_\oplus$ =	6.378(⁸)	6.378(⁶)	6.378(³)	1	9.164(⁻³)	4.263(⁻⁵)	6.742(⁻¹⁰)	2.067(⁻¹⁰)
$R_\odot$ =	6.960(¹⁰)	6.960(⁸)	6.960(⁵)	109.1	1	4.652(⁻³)	7.357(⁻⁸)	2.256(⁻⁸)
1 AU =	1.496(¹³)	1.496(¹¹)	1.496(⁸)	2.346(⁴)	214.9	1	1.581(⁻⁵)	4.848(⁻⁶)
1 ly =	9.461(¹⁷)	9.461(¹⁵)	9.461(¹²)	1.483(⁹)	1.359(⁷)	6.324(⁴)	1	0.3066
1 pc =	3.086(¹⁸)	3.086(¹⁶)	3.086(¹³)	4.838(⁹)	4.433(⁷)	2.063(⁵)	3.262	1

Mass

	g	kg	$M_\oplus$	$M_\odot$
1 g =	1	0.001	1.674(⁻²⁸)	5.028(⁻³⁴)
1 kg =	1000	1	1.674(⁻²⁵)	5.028(⁻³¹)
$M_\oplus$ =	5.974(²⁷)	5.974(²⁴)	1	3.004(⁻⁶)
$M_\odot$ =	1.989(³³)	1.989(³⁰)	3.329(5)	1

Power/Energy

	erg	J	erg/sec	W(= J/s)	$L_\odot$
1 erg =	1	1(⁻⁷)	—	—	—
1 J =	1(⁷)	1	—	—	—
1 erg/sec =	—	—	1	1(⁻⁷)	2.613(⁻³⁴)
1 W (= 1 J/s) =	—	—	1(⁷)	1	2.613(⁻²⁷)
$L_\odot$ =	—	—	3.827(³³)	3.827(²⁶)	1

Wavelength

	Å	nm	µm	mm	cm	m
1 Å =	1	0.1	1(⁻⁴)	1(⁻⁷)	1(⁻⁸)	1(⁻¹⁰)
1 nm =	10	1	0.001	1(⁻⁶)	1(⁻⁷)	1(⁻⁹)
1 µm =	1(⁴)	1000	1	0.001	1(⁻⁴)	1(⁻⁶)
1 mm =	1(⁷)	1(⁶)	1000	1	0.1	0.001
1 cm =	1(⁸)	1(⁷)	1(⁴)	10	1	0.01
1 m =	1(¹⁰)	1(⁹)	1(⁶)	1000	100	1

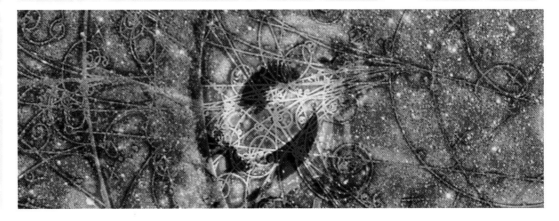

LAWS & FORMULAS

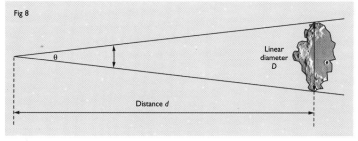

CELESTIAL GEOMETRY

Here are some of the basic formulae used for calculating areas, volumes and densities. Note that the densities are average values; the actual density may vary from one part of a body to another, unless the body is specified as having a uniform density.

Circumference of a circle: $C = 2\pi R$, where R is the radius.

Area of a circle: $A = \pi R^2$.

Surface area of a sphere: $A = 4\pi R^2$.

Volume of a sphere: $V = \pi R^3$.

Average density: $\rho = M/V$, where M is the mass, V is the volume.

Average density of a spherical body: $\rho = 3M/4\pi R^3$.

Given below are brief explanations of some of the fundamental geometric concepts that arise in astronomy and cosmology.

Angular diameter: The angular diameter of an object is the angle between its edges, as measured from a specified observation point (see Figure 7). An example is provided by the Moon, which, observed from the Earth, has an angular diameter of about 0.5°.

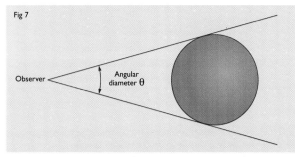

Fig 7

Observer — Angular diameter θ

Unlike the linear diameter of an object (the distance between its edges), the angular diameter is *not* an intrinsic property of the object being observed, since it depends on the distance between the object and the observer. Any increase in the distance of an object of fixed linear diameter will cause a decrease in its angular diameter.

Small angle approximation: this is an approximate relation between an object's angular diameter θ, its linear diameter D, and its distance from the observer d (see Figure 8). It holds true provided the angular diameter is small (less than 5° say), and provided D and d are both measured in the same units (metres say, or light years). When both these conditions are met, the value of θ in radians will be given approximately by:

$$\theta \approx D/d$$

For instance, the Sun has a linear diameter of 1.4×10^6 km, and is at about 1.5×10^8 km from the Earth. It follows that the angular diameter if the Sun is approximately $1.4 \times 10^6/(1.5 \times 10^8) = 0.0093$ rad, which is equal to 0.53°.

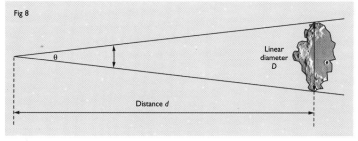

Fig 8

θ — Linear diameter D

Distance d

The small angle approximation may be rearranged to give

$$D \approx \theta \, d$$

which makes it possible to determine the linear diameter of an object from measurements of its distance and angular diameter.

Stellar parallax: stellar parallax is the phenomenon whereby nearby stars, when viewed from the Earth over a period of one year, appear to move in small ellipses relative to the background of more distant stars. The cause of this apparent

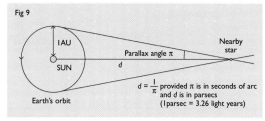

Fig 9

1 AU — SUN — Earth's orbit

Parallax angle π — d — Nearby star

$d = \frac{1}{\pi}$ provided π is in seconds of arc and d is in parsecs (1 parsec = 3.26 light years)

stellar motion is the movement of the observer as the Earth follows its annual orbit about the Sun (see Figure 9).

Parallax provides a simple way of determining the distance to a star. The result of dividing the angular diameter of a star's parallax ellipse by 2 is the parallax angle π, and if this is expressed in seconds of arc, then the distance of the star will be

$$d = 1/\pi$$

where d is expressed in parsecs. For example, a star with a parallax angle $\pi = 0.1$ seconds of arc, would be at a distance $d = 10$ parsecs from Earth (i.e. about 32.6 light years).

Diffraction limit for a telescope: due to the wave nature of light, even a perfectly adjusted telescope of finite diameter will cause a point-like source of light to be seen as a small disk – this is an example of a phenomenon known as diffraction. The diffraction limit of a given telescope is the smallest angular separation between two point sources at which that telescope can still reliably detect the presence of two separate sources, despite diffraction. For a telescope of diameter D, operating at a wavelength λ, the diffraction limit is $\theta = (1.221 \text{ rad}) \lambda/D$, where both λ and D are expressed in the same units (metres say). A modification of this, $\theta = (251 \, 643'') \lambda/D$, gives the diffraction limit in seconds of arc. The diffraction limit of the Hubble Space Telescope, which has a diameter of 2.4 m, when operating at a wavelength of 6.0×10^{-7} m, is 0.063 seconds of arc.

ORBITS

Ever since Johannes Kepler stated his first law of planetary motion – that the planets move in ellipses, with the Sun at one focus (see Figure 10) – there has been good reason for astronomers to be interested in the properties of ellipses. The importance of elliptical orbits was further emphasized when Isaac Newton showed that the gravitational attraction of one body for another inevitably causes the relative motion of two

otherwise isolated bodies to be a conic section, of which an ellipse is a special case. This subsection introduces some of the terminology associated with ellipses and elliptical orbits.

Eccentricity of an elliptical orbit: the eccentricity of an ellipse is a numerical measure of the extent to which the ellipse is flattened compared with a circle. For an ellipse of semi-major axis a, where the distance between the two foci is F_1F_2, the eccentricity is $e = F_1F_2/(2a)$. The eccentricity of a circle is 0.

Perihelion: the point closest to the Sun in the orbit of a body orbiting the Sun. For an orbit of semi-major axis a, and eccentricity e, the distance of the perihelion from the Sun is $P = a (1 - e)$.

Aphelion: the point farthest from the Sun in the orbit of a body orbiting the Sun. For an orbit of semi-major axis a, and eccentricity e, the distance of the aphelion from the Sun is $A = a (1 + e)$.

Relation between synodic and sidereal period for an inferior planet: the synodic period S of a planet is the time that separates identical alignments of that planet with the Sun and the Earth. The sidereal period of a planet T is the time that separates identical alignments of that planet with the Sun and the stars. For an inferior planet (one that orbits closer to the Sun than the Earth) the relation between the synodic and sidereal periods is

$$\frac{1}{S} = \frac{1}{T} - \frac{1}{T_\oplus}$$

where S, T and the sidereal period of the Earth, $T_\oplus$, are all measured in the same units (e.g. years).

Relation between synodic and sidereal period for a superior planet: for a superior planet (one that orbits farther from the Sun than the Earth) the relation between the synodic period S and the sidereal period T is

$$\frac{1}{S} = \frac{1}{T_\oplus} - \frac{1}{T}$$

where S, T and the sidereal period of the Earth, $T_\oplus$, are all measured in the same units (e.g. years).

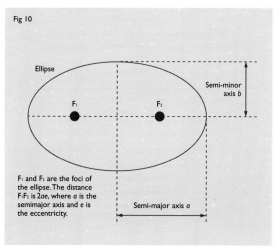

Fig 10

Ellipse

Semi-minor axis b

F_1 — F_2

Semi-major axis a

F_1 and F_2 are the foci of the ellipse. The distance F_1F_2 is $2ae$, where a is the semimajor axis and e is the eccentricity.

Kepler's third law: a law stating that the square of the sidereal period T of a planet is proportional to the cube of the semi-major axis a of its orbit. Expressed in symbols, $T^2 \propto a^3$, which implies

$$T^2 = ka^3$$

where k is a constant. A theoretical analysis of the motion of a planet of mass m_1 around a star of mass m_2, based on Newton's law of gravitation (see below), explains the origin of Kepler's third law and indicates that the constant k is given by

$$k = \frac{4\pi^2}{G(m_1+m_2)}$$

This is a general result that applies to any isolated pair of bodies in a mutual gravitational orbit.

Centre of mass: when a body of mass m_1 is engaged in a mutual gravitational orbit with a second body of mass m_2, the centre of mass of that two-body system is a point located on the line joining the two bodies, at a distance r_1 from the first body and r_2 from the second, such that $m_1r_1 = m_2r_2$. In the absence of any external forces, the centre of mass of a system always moves in a straight line at constant speed.

LAWS OF GRAVITATION

Newton's laws of motion provide a systematic basis for predicting the way in which bodies such as stars, planets and satellites will move in response to known forces. Newton's additional discovery of the law of universal gravitation makes it possible to predict the gravitational forces that act between such bodies, and hence to predict their orbital motion from observations of their positions and velocities at a single time. The powerful combination of laws of motion and a law of gravitation provides a rational explanation of Kepler's laws of planetary motion, and laid the foundations for the detailed study of orbital motion that came to be called celestial mechanics.

Law of universal gravitation: this law asserts that every particle of matter attracts every other particle of matter with a force that is directly proportional to the product of their masses, and inversely proportional to the square of the distance between them. Expressed in symbols, the strength of the force is given by

$$F = \frac{Gm_1m_2}{r^2}$$

where m_1 and m_2 are the masses of the particles, r is the distance between them, and $G = 6.673 \times 10^{-11}$ N m²/kg² is Newton's gravitational constant. Although formulated in terms of particles, the law may also be applied to extended bodies by treating them as assemblies of particles.

Surface gravity: the acceleration due to gravity at the surface of a body. For a spherical body of mass M and radius R, the magnitude of the surface gravity is $g = GM/R^2$. The surface gravity of the Earth varies slightly from place to place, but its average value is 9.81 m/s². The average surface gravity of the Moon is 1.62 m/s².

Escape speed: the minimum speed at which an object must be launched from the surface of a body (ignoring any atmospheric effects) in order for the object to completely escape from the body. For a spherical body of mass M and radius R, the escape speed is $v_{esc} = \sqrt{2GM/R}$ The escape speed from the Earth is about 11.2 km/s.

STELLAR AND GALACTIC PROPERTIES

The general theory of relativity is Einstein's theory of gravity. It regards gravity as a manifestation of space–time distortion. This remarkable insight led to many extensions and corrections to older views based on Newton's law of gravitation. The general theory of relativity has been used to analyze phenomena that range in scale from the cosmic and galactic to the stellar and planetary, and has even been subjected to laboratory tests here on Earth. Here we mention two of its implications for stellar and galactic systems.

Schwarzschild radius: according to general relativity, gravity acts on light, bending the paths of light rays. This effect is small in the neighbourhood of ordinary stars such as the Sun. However, when a star undergoes gravitational collapse to form a *black hole* of mass M it will create a region where the light bending is so severe that no light entering that region is able to escape again. The radius of this light trapping region is known as the Schwarzschild radius and is given by $R = 2GM/c^2$, where G is the gravitational constant and c is the speed of light. In the case of a black hole with the same mass as the Sun, the Schwarzschild radius would be about 3 km.

Hubble expansion law: according to Hubble's law, distant galaxies are receding from the Earth at speeds v that increase in proportion to their distance d. Expressed in symbols, $v = H_0d$, where H_0 is called Hubble's constant and has the approximate value $H_0 = 65$ km/s/Mpc. The simplest models of cosmic geometry, based on general relativity, imply that Hubble's law will hold true for observers anywhere in the Universe, and that the value of H_0 is a measure of the current rate of cosmic expansion. Any changes in that rate over time should cause the more distant galaxies to exhibit departures from Hubble's law.

LIGHT AND RADIATION LAWS

One of the outstanding achievements of nineteenth century science was the theoretical demonstration, by James Clerk Maxwell, that electric and magnetic fields could form waves capable of travelling through empty space at the speed of light. These electromagnetic waves provided an explanation of the nature of light, and implied the existence of other light-like radiations, such as radio waves. However, later investigators, such as Max Planck, discovered that in some circumstances electromagnetic radiation behaved more like a stream of particles, now called photons. The recognition of the dual wave/particle nature of light marked the foundation of the quantum theory that dominates modern physics.

Wavelength and frequency of a wave: The wavelength of a wave is the distance between successive peaks of the wave. The frequency of the wave is the rate (measured in hertz) at which peaks pass a fixed point. If a wave has wavelength λ and frequency f, the speed of the wave will be $v = f\lambda$. For light travelling in a vacuum, this implies $c = f\lambda$.

Energy of a photon: light of a single colour, according to the wave model, has a single frequency f. According to the photon model, such light is a stream of particles each of which has energy $E = hf$, where $h = 6.62607 \times 10^{-34}$ J s is Planck's constant.

Wien's law: a hot body emits electromagnetic radiation with a wide range of wavelengths. However in the case of an ideal emitter (called

a black body, since an ideal emitter is also an ideal absorber) the emissions are distributed across wavelengths in a way that is precisely determined by the surface temperature of the emitter. According to Wien's law, in the case of a black body at temperature T, the emission is a maximum at the wavelength $\lambda_{max} = W/T$, where $W = 2.89777 \times 10^{-3}$ m K is the Wien constant. Wien's law provides a way of determining the surface temperature of a star or other body from observations of the radiation that body emits, provided it can be regarded as a black body.

Stefan's law: this law states that the rate of energy emission per unit area of a black body is $\Phi = \sigma T^4$, where T is the temperature of the black body, and $\sigma = 5.670 \times 10^{-8}$ W/m²/K⁴ is the Stefan-Boltzmann constant.

Stefan–Boltzmann law: the *luminosity* L of a body is the rate at which energy is emitted by that body. In the case of a black body at temperature T, the luminosity is $L = A\Phi = A\sigma T^4$, where A is the surface area of the body. Since a spherical body of radius R has surface area $A = 4\pi R^2$, it follows that the luminosity of a spherical black body is $L = 4\pi R^2\sigma T^4$. This modest extension to Stefan's law is sometimes referred to as the Stefan-Boltzmann law, though that term is also used for Stefan's law itself.

Inverse square law: the intensity I of radiation is a measure of the rate at which the radiation transports energy across a unit area perpendicular to its direction of travel (see Figure 11). According to the inverse square law, the intensity I at a distance d from a source that radiates equally in all directions is given by $I = K/d^2$, where K is a constant. Consequently, if I_1 and I_2 are the intensities at distances d_1 and d_2 from such a source, then $I_1/I_2 = d_2^2/d_1^2$.

Doppler shift: the redshift z of a source of light (or any other electromagnetic radiation) is defined by the relation $z = \Delta\lambda/\lambda$, where $\Delta\lambda$ is the difference between the observed and emitted wavelengths of the light. One possible cause of redshift is the movement of the source away from the observer. When the observed recession velocity v is small compared with the speed of light c, this so-called Doppler shift is given by $z = \Delta\lambda/\lambda = v/c$. (For an emitter that approaches the observer, both v and $\Delta\lambda$ will be negative.)

Relativistic Doppler shift: according to Einstein's special theory of relativity, the Doppler shift of a source with recession velocity v is

$$\frac{\Delta\lambda}{\lambda} = z = \sqrt{\frac{1+v/c}{1-v/c}} - 1$$

This holds true even when v is not small compared with c.

Cosmological redshift: for far-off sources such as distant galaxies, the expansion of the Universe is a significant cause of redshift. The redshift z will increase with the distance d of the galaxy, and, provided z is small, will be given by $z = H_0d/c$, where H_0 is Hubble's constant and c is the speed of light.

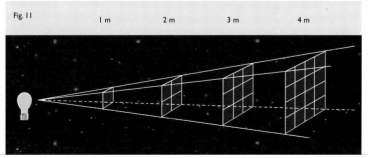

Fig. 11
1 m   2 m   3 m   4 m

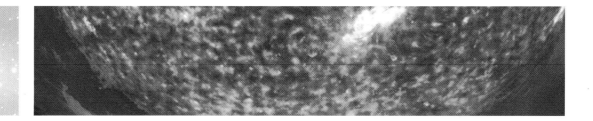

ELEMENT	SYMBOL	ATOMIC NUMBER	ATOMIC MASS	ABUNDANCE SUN	ABUNDANCE METEORITES
Hydrogen	H	1	1.0079	1.0000	1.0000
Helium	He	2	4.0026	0.0977	0.0977
Lithium	Li	3	6.941	1.45×10^{-11}	2.04×10^{-9}
Beryllium	Be	4	9.0122	1.41×10^{-11}	2.63×10^{-11}
Boron	B	5	10.811	3.98×10^{-10}	6.31×10^{-10}
Carbon	C	6	12.011	0.000363	0.000363
Nitrogen	N	7	14.007	0.000112	0.000112
Oxygen	O	8	15.999	0.000851	0.000851
Fluorine	F	9	18.998	3.63×10^{-8}	3.02×10^{-8}
Neon	Ne	10	20.180	0.000123	0.000123
Sodium	Na	11	22.990	2.14×10^{-6}	2.04×10^{-6}
Magnesium	Mg	12	24.305	3.80×10^{-5}	3.80×10^{-5}
Aluminium	Al	13	26.982	2.95×10^{-6}	3.02×10^{-6}
Silicon	Si	14	28.086	3.55×10^{-5}	3.55×10^{-5}
Phosphorus	P	15	30.974	2.82×10^{-7}	3.72×10^{-7}
Sulphur	S	16	32.066	1.62×10^{-5}	1.86×10^{-5}
Chlorine	Cl	17	35.453	3.16×10^{-7}	1.86×10^{-7}
Argon	Ar	18	39.948	3.63×10^{-6}	3.63×10^{-6}
Potassium	K	19	39.098	1.32×10^{-7}	1.35×10^{-7}
Calcium	Ca	20	40.078	2.29×10^{-6}	2.19×10^{-6}
Scandium	Sc	21	44.956	1.26×10^{-9}	1.23×10^{-9}
Titanium	Ti	22	47.88	9.77×10^{-8}	8.51×10^{-8}
Vanadium	V	23	50.942	1.00×10^{-8}	1.05×10^{-8}
Chromium	Cr	24	51.996	4.68×10^{-7}	4.79×10^{-7}
Manganese	Mn	25	54.938	2.45×10^{-7}	3.39×10^{-7}
Iron	Fe	26	55.847	3.47×10^{-5}	3.24×10^{-5}
Cobalt	Co	27	58.933	8.32×10^{-8}	8.13×10^{-8}
Nickel	Ni	28	58.693	1.78×10^{-6}	1.78×10^{-6}
Copper	Cu	29	63.546	1.62×10^{-8}	1.86×10^{-8}
Zinc	Zn	30	65.39	3.98×10^{-8}	4.47×10^{-8}
Gallium	Ga	31	69.723	7.59×10^{-10}	1.35×10^{-9}
Germanium	Ge	32	72.61	2.57×10^{-9}	4.27×10^{-9}
Arsenic	As	33	74.922	–	2.34×10^{-10}
Selenium	Se	34	78.96	–	2.24×10^{-9}
Bromine	Br	35	79.904	–	4.27×10^{-10}
Krypton	Kr	36	83.80	–	1.70×10^{-9}
Rubidium	Rb	37	85.468	3.98×10^{-10}	2.51×10^{-10}
Strontium	Sr	38	87.62	7.94×10^{-10}	8.51×10^{-10}
Yttrium	Y	39	88.906	1.74×10^{-10}	1.66×10^{-10}
Zirconium	Zr	40	91.224	3.98×10^{-10}	4.01×10^{-10}
Niobium	Nb	41	92.906	2.63×10^{-11}	2.51×10^{-11}
Molybdenum	Mo	42	95.94	8.32×10^{-11}	9.12×10^{-11}
Technetium	Tc	43	98.906	–	–
Ruthenium	Ru	44	101.07	6.92×10^{-11}	6.61×10^{-11}
Rhodium	Rh	45	102.91	1.32×10^{-11}	1.23×10^{-11}
Palladium	Pd	46	106.42	4.90×10^{-11}	5.01×10^{-11}
Silver	Ag	47	107.87	8.71×10^{-12}	1.74×10^{-11}
Cadmium	Cd	48	112.41	7.24×10^{-11}	5.75×10^{-11}
Indium	In	49	114.82	4.57×10^{-11}	6.61×10^{-12}
Tin	Sn	50	118.71	1.00×10^{-10}	1.38×10^{-10}
Antimony	Sb	51	121.76	1.00×10^{-11}	1.10×10^{-11}
Tellurium	Te	52	127.60	–	1.74×10^{-10}
Iodine	I	53	126.90	–	3.24×10^{-11}
Xenon	Xe	54	131.29	–	1.70×10^{-10}
Cesium	Cs	55	132.91	–	1.32×10^{-11}
Barium	Ba	56	137.33	1.35×10^{-10}	1.62×10^{-10}
Lanthanum	La	57	138.91	1.66×10^{-11}	1.58×10^{-11}
Cerium	Ce	58	140.12	3.55×10^{-11}	4.07×10^{-11}
Praseodymium	Pr	59	140.91	5.13×10^{-12}	6.03×10^{-12}
Neodymium	Nd	60	144.24	3.16×10^{-11}	2.95×10^{-11}
Promethium	Pm	61	146.92	–	–
Samarium	Sm	62	150.36	1.00×10^{-11}	9.33×10^{-12}
Europium	Eu	63	151.96	3.24×10^{-12}	3.47×10^{-12}
Gadolinium	Gd	64	157.25	1.32×10^{-11}	1.17×10^{-11}
Terbium	Tb	65	158.93	7.94×10^{-13}	2.14×10^{-12}
Dysprosium	Dy	66	162.50	1.26×10^{-11}	1.41×10^{-11}
Holmium	Ho	67	164.93	1.82×10^{-12}	3.16×10^{-12}
Erbium	Er	68	167.26	8.51×10^{-12}	8.91×10^{-12}
Thulium	Tm	69	168.93	1.00×10^{-12}	1.35×10^{-12}
Ytterbium	Yb	70	170.04	1.20×10^{-11}	8.91×10^{-12}
Lutetium	Lu	71	174.97	5.75×10^{-12}	1.32×10^{-12}
Hafnium	Hf	72	178.49	7.59×10^{-12}	5.37×10^{-12}
Tantalum	Ta	73	180.95	–	1.35×10^{-12}
Tungsten	W	74	183.85	1.29×10^{-11}	4.79×10^{-12}
Rhenium	Re	75	186.21	–	1.86×10^{-12}
Osmium	Os	76	190.2	2.82×10^{-11}	2.40×10^{-11}
Iridium	Ir	77	192.22	2.24×10^{-11}	2.34×10^{-11}
Platinum	Pt	78	195.08	6.31×10^{-11}	4.79×10^{-11}
Gold	Au	79	196.97	1.02×10^{-11}	6.76×10^{-12}
Mercury	Hg	80	200.59	–	1.23×10^{-11}
Thallium	Tl	81	204.38	7.94×10^{-12}	6.61×10^{-12}
Lead	Pb	82	207.2	7.08×10^{-11}	1.12×10^{-10}
Bismuth	Bi	83	208.98	–	5.13×10^{-12}
Polonium	Po	84	209.98	–	–
Astatine	At	85	209.99	–	–
Radon	Rn	86	222.02	–	–
Francium	Fr	87	223.02	–	–
Radium	Ra	88	226.03	–	–
Actinium	Ac	89	227.03	–	–
Thorium	Th	90	232.04	1.32×10^{-12}	1.20×10^{-12}
Protactinium	Pa	91	231.04	–	–
Uranium	U	92	238.03	3.55×10^{-13}	3.24×10^{-13}
Neptunium	Np	93	237.05	–	–
Plutonium	Pu	94	239.05	–	–
Americium	Am	95	241.06	–	–
Curium	Cm	96	244.06	–	–
Berkelium	Bk	97	249.08	–	–
Californium	Cf	98	252.08	–	–
Einsteinium	Es	99	252.08	–	–
Fermium	Fm	100	257.10	–	–
Mendelevium	Md	101	258.10	–	–
Nobelium	No	102	259.10	–	–
Lawrencium	Lr	103	262.11	–	–

Source: *Allen's Astrophysical Quantities* (ed. Arthur N. Cox, Springer-Verlag, New York, 2000) Abundances are given in terms of the number of atoms relative to hydrogen, as measured for the solar photosphere and in meteorites. These values are considered as being typical of the composition of the Universe as a whole, and are hence sometimes known as cosmic abundances. The largest difference between the two sets of data is for lithium, which is destroyed in the Sun's outer layers and hence is under-represented in the photosphere.

Reference

ACKNOWLEDGEMENTS & PICTURE CREDITS

CONTRIBUTORS

IAN RIDPATH
General Editor; Star Chart Entries; Mercury; Our Solar System Data Boxes
Ian Ridpath is editor of the world-famous *Norton's Star Atlas*, regarded as the amateur astronomer's 'bible', and of the authoritative *Oxford Dictionary of Astronomy*. He is author of several standard guides to the night sky, including the *Collins Pocket Guide to Stars and Planets, Collins Gem: Stars* and *The Monthly Sky Guide* (CUP).

MARTIN REES
Foreword
Martin Rees is Royal Society Research Professor at Cambridge and a former Director of the Institute of Astronomy. He also holds the honorary title of Astronomer Royal and is a former President of the British Association for the Advancement of Science. He is an international leader in cosmological research and is also well-known as a lecturer and writer for general audiences. His books include *Before the Beginning* and *Just Six Numbers*.

AUTHORS

JIM AL-KHALILI
In Search of Quantum Reality
Jim Al-Khalili is a theoretical nuclear physicist and lecturer in the Department of Physics at Surrey University. He is author of the popular science book *Black Holes, Wormholes and Time Machines* and performed the 1998 Institute of Physics Schools Lecture Tour culminating with a finale at the Royal Institution. He is a recent recipient of an Institute of Physics Public Awareness of Physics Award and has been nominated for the Royal Society Michael Faraday Award in the Public Understanding of Science. His active area of research is the study of properties of exotic types of atomic nuclei. His recent TV and radio appearances have seen him discuss subjects ranging from the nature of Free Will to the science behind *Star Wars*.

RETA BEEBE
Outer Planets
Dr Reta Beebe received her BA in chemistry from the University of Colorado and her PhD in Astrophysics from the University of Indiana. She has worked as a research associate, Full Professor of Astronomy and College Professor of Astronomy at New Mexico State University and as Discipline Scientist for the Planetary Atmospheres Program and Jupiter System Data Analysis Program (1997–99) at the NASA Headquarters Office of Space Science Program in Washington, DC. Beebe was a member of the NASA Voyager Imaging Team and is an associate member of the Galileo Imaging Team. She has made extensive use of the Hubble Space Telescope to observe Saturn storms and weather phenomena on Jupiter. She has also managed the Atmospheric Discipline Node of the Planetary Data System, NASA's planetary data archive (1996–2001).

ALLAN CHAPMAN
The History of Astronomy
Allan Chapman is a professional historian of astronomy at Wadham College, Oxford University. His interests include the relationship between astronomical innovation and key developments in instrumentation, along with wider cultural issues, such as economic patronage, philosophy and religion. Among his books are *Dividing the Circle* (1990, 1995) and *The Victorian Amateur Astronomer* (1998).

DUNCAN COPP
Our Solar System Additional Material
Dr Duncan Copp gained his PhD in astronomy and planetary science, the geology of Venus. He is a writer, researcher, consultant and broadcaster on space and the Universe. He has researched and presented a number of documentaries for the BBC and writes and lectures on the subject.

WYN EVANS
The Milky Way
Dr Wyn Evans is Reader in Physics at the University of Oxford and has published over 70 papers on theoretical astrophysics. His research interests span all areas of dynamical astronomy, including the structure and evolution of solar systems and galaxies, as well as gravitational lensing and microlensing.

BONNIE GORDON
Our Solar System Introduction and Summary
Bonnie Gordon is the chief editor of *Astronomy* magazine, the largest circulation, English-language astronomy magazine in the world. She has worked in science journalism for 25 years and helped write and edit chapters in two Time-Life Books astronomy series. She has authored three books of poetry, countless magazine stories, has three adult children and lives with her husband and dogs in Milwaukee, Wisconsin.

DAVE HAWKSETT
Outer Planets Introduction; Mars
Dave Hakwsett is a science writer, consultant, lecturer and researcher. He is responsible for running the UK Planetary Forum and is working towards a PhD in planetary science. He is also responsible for science at Guinness World Records.

RODNEY HILLIER
The Universe: Past, Present & Future; Black Holes; Quasars (with RICHARD SALE)
Before retiring, Rodney Hillier was a senior lecturer in the physics department of the University of Bristol and his research interests were in the field of gamma ray astronomy. He has more than 30 years' experience of lecturing to the public on astronomy and is a vice-president of the Bristol Astronomical Society.

TONY JONES
The Laws of Physics; Professional Astronomy
Dr Tony Jones did research in astronomy at the universities of London and Manchester and is now a science writer and part-time associate lecturer with the Open University. He has written numerous articles on astronomy, science and technology, contributed to several books and is author of *Splitting the Second: The Story of Atomic Time*.

ROBERT LAMBOURNE
Units and Conversions; Maths Toolbox
Dr Robert Lambourne is Senior Lecturer in the Department of Physics and Astronomy at the Open University, where he is also Director of the Centre for Collaborative Science Teaching. He has written and edited a number of books, including *Basic Mathematics for the Physical Sciences*, and *The Restless Universe*.

MICHAEL MERRIFIELD
Galaxies
Michael Merrifield received his undergraduate degree from Oxford and a PhD from Harvard University. He then went on to astronomical research positions in Canada and the UK. He currently holds the Chair of Astronomy at the University of Nottingham, where his research seeks to understand the formation, evolution and structure of galaxies.

MARTIN RATCLIFFE
Tool & Techniques for Amateurs; Observatories and Telescopes
Martin Ratcliffe is President of the International Planetarium Society (2001–02) and a Fellow of the Royal Astronomical Society. An accomplished writer and monthly columnist for *Astronomy* magazine, he earned his Bachelor of Science degree from University College London (England) in Astronomy. Martin also finds time to teach Astronomy as adjunct faculty of Baker University and at Baker University in Wichita, Kansas and is currently employed as Director of the Boeing CyberDome theater at Exploration Place in Wichita. Martin is a keen observer, owning a Celestron 11-inch telescope in a roll-off roof observatory, and enjoys astrophotography with a 5.5-inch Schmidt camera. He also films total eclipses of the Sun for television. He is a past Council member and Deep-Sky Section director of the British Astronomical Association.

CHRIS RILEY
Space Exploration; Space Missions, Crafts and Probes
Dr Christopher Riley gained his doctorate from Imperial College in 1995 and has worked as a science journalist in broadcasting, print and online for over a decade. He acted as series researcher for the highly acclaimed, award-winning 1999 BBC2 landmark series *The Planets*. He is still a regular broadcaster covering astronomy related subjects for the corporation's growing number of channels and web sites and continues to work for BBC Science as a producer and director.

SARA RUSSELL
Planet Formation; Asteroids, Comets and Meteorites
Sara Russell is a cosmochemist at the Natural History Museum in London. Her research interest is the origin of the Solar System, and she has written over 30 scientific papers. She has contributed planetary science articles to *New Scientist, Ad Astra, Physics World* and many other publications. Dr Russell was co-editor of *Protostars and Planets IV* (University of Arizona Press).

RICHARD SALE
The Universe: Present and Future; Black Holes; Quasars (with Rodney Hillier)
Richard Sale's MSc was in theoretical astrophysics, his PhD in experimental gamma ray astronomy and included the discovery of a gamma ray pulsar. After many years as a research scientist, he took up writing as a full-time career, although he has maintained his interest in all things astrophysical.

PAM SPENCE
Watching the Sky Introduction and Summary; Glossary
Pam Spence gained an MSc in astronomy at the University of Sussex and worked on a research fellowship in solar physics at the Mullard Space Science Laboratory, University College London, until leaving to become Editor of the popular monthly astronomy magazine *Astronomy Now*. Now a freelance astronomer and science writer, she continues to be very involved with the amateur astronomical community and is a former President of the Federation of Astronomical Studies.

DAVE STICKLAND
Stars; The Sun
Dave Stickland gained his degrees at the Universities of St Andrews and Sussex. He joined the RGO at Herstmonceux in 1971 and was the first UK Resident Astronomer in Madrid on the IUE project. Transferred to the RAL in 1984, he has been managing editor of *The Observatory* since 1983.

WIL TIRION
Star Map Cartography
Wil Tirion started making his first star atlas in 1977, in his free hours, with stars down to magnitude 6.5 (the whole sky on five large maps). It was published in the *Encyclopedia of Astronomy*, edited by Colin Ronan, in 1979 and in 1981 as a separate set of maps by the British Astronomical Association (*B.A.A. Star Charts 1950.0*). Still as a hobby, he started working on a larger atlas, *Sky Atlas 2000.0*, showing stars down to magnitude 8.0. Its publication, in 1981, resulted in requests from several publishers for star maps for different purposes. In 1983 he started working as a full-time uran-ographer. Since then he has contributed to many books and magazines on astronomy.

KELLY KIZER WHITT
Terrestrial Planets
Kelly Kizer Whitt is the copy-editor and photo editor for *Astronomy* magazine. She regularly writes observing and science articles for the magazine and edits the monthly *Sky Show* and *Hot Shots* departments. A graduate of the University of Wisconsin-Madison, she studied English and astronomy.

ADVISORY BOARD

DAVID L. GOODSTEIN
The Laws of Physics
Dr David L. Goodstein is vice provost and professor of physics and applied physics at the California Institute of Technology in Pasadena, where he has been on the faculty for more than 30 years. He has served on numerous scientific and academic panels, including the National Advisory Committee to the Mathematical and Physical Sciences Directorate of the National Science Foundation, which he currently chairs. In 1995 he was named the Frank J. Gilloon Distinguished Teaching and Service Professor. Dr Goodstein has recently co-authored the best-seller *Feynman's Lost Lecture* with his wife Dr Judith R. Goodstein.

BRIAN GREENE
In Search of Quantum Reality
Brian Greene gained his PhD in String Theory and Quantum Gravity at the University of Oxford in 1987. He is currently Professor in the Department of Physics at Columbia University. His area of research is superstring theory, in particular quantum geometry. He has contributed many articles on the theory to journals and other publications.

THOMAS HOCKEY
History of Astronomy
Thomas Hockey (University of Northern Iowa) is a member of the International Astronomical Union's Commission on the History of Astronomy. He is presently editing a biographical encyclopedia containing entries on over 1,400 astronomers from antiquity to the twentieth century. Professor Hockey's course may be visited on the World Wide Web.

ROGER D. LAUNIUS
Space Exploration
Roger D. Launius is chief historian of the National Aeronautics and Space Administration, Washington, D.C. His office is responsible for preparing books, monographs and special studies on aerospace history in the United States; managing the NASA Historical Reference Collection of documentary materials about the history of the agency; and providing historical services to both the NASA staff and the public. A graduate of Graceland College in Lamoni, Iowa, he received his Ph.D. from Louisiana State University, Baton Rouge, in 1982. He has written or edited several books on aerospace history, including *Reconsidering Sputnik: Forty Years Since the Soviet Satellite* (Amsterdam: Harwood Academic Publishers, 2000); *Innovation and the Development of Flight* (College Station: Texas A&M University Press, 1999); *NASA & the Exploration of Space* (New York: Stewart, Tabori, & Chang, 1998); and *Frontiers of Space Exploration* (Westport, CT: Greenwood Press, 1998).

ANDREW LIDDLE
General Consultant; The Universe: Past, Present & Future; Contents of the Cosmos
Andrew Liddle is Professor of Astrophysics and Director of the Astronomy Centre at the University of Sussex, where he has been based since January 2000. His main research interest is theoretical cosmology, and he is the author of two textbooks: *An Introduction to Modern Cosmology* and *Cosmological Inflation and Large-Scale Structure*.

LUCY MCFADDEN
Our Solar System
Lucy McFadden is an authority on the small bodies of the Solar System – asteroids and comets – and has participated in many educational and public outreach programmes on the Solar System, serving as the on-camera expert for the MSNBC cable television network during the Mars Pathfinder mission, and is one of the three editors of *The Encyclopedia of the Solar System*, published by Academic Press in 1998. She is a science team member for NASA's NEAR mission to asteroid 433 Eros and the Deep Impact mission to Comet Tempel 1.

JACQUELINE MITTON
Star Charts
Dr Jacqueline Mitton is author of 17 astronomy books. She has also made major contributions to several encyclopedias. She gained her first degree in Physics at the University of Oxford and her PhD at the University of Cambridge. She has been Press Officer of the Royal Astronomical Society since 1989.

PICTURE CREDITS

AKG London: 82 (b), 139 (l), 156 (l, r)

Christie's Images 2001: 32, 34, 56 (l), 88 (l)

Galaxy Pictures/Robin Scagell: 48 (l), 56 (r), 63 (r), 74 (l), 76 (m), 78 (r), 81 (r), 118-119, 120, 121 (t), 122 (t), 124 (t), 125 (r), 127 (t), 130 (t), 133, 135 (t, b), 141 (t, b), 143 (l), 144 (t), 145 (t), 148, 149 (t, b), 157 (t, b), 158 (l, r), 161 (t), 162 (b), 168 (b), 171 (r), 175 (b), 178 (b), 179 (l), 184 (l, r), 197 (b), 200 (l), 203 (b), 204 (t), 209 (t, r), 210 (l), 228 (r), 229 (r), 236 (r), 237, 240 (b), 250, 251 (l), 254 (l), 260, 262, 264, 272, 274, 286, Isaac Newton Group 152, 180 (b), Chris Livingstone 179 (r), Hubble Space Telescope/NASA 153 (l), Pedro Re 276, 282, Don Davis/NASA 231 (t), Carnegie Mellon University 231 (b), Nigel Evans 230 (b), D. Roddy/LPI 230 (t), Calvin J. Hamilton 202 (r), 213, Thierry Legault 212 (t), Eric Hutton 211, Philip Perkins 278, Stephen Fielding 280, George Haig 251 (r), Maurice Gavin 252 (t), 258, Martin Ratcliffe 253 (l), European Southern Observatory 180 (t), 181 (t), 234-235, 238, 241 (b), Adrian Catterall 266, JPL 192-193, 204 (b), 206 (t), 215 (b), 217 (l, r), 218 (l, r), 219 (l), 220 (t), 220 (l), 222 (l,

b), 223 (t, b), 224 (l), 225 (t), 227 (t), 228 (l), 232 (b), 233, 312 (b), 313, 315, Nuffield Radio Astronomy 86, ESA 179 (l), 239 (l), Kipp Teague 303 (r), 323 (l, r), 334 (b), Dave Finley/NRAO/AUI/NFF 241 (t), Ed Hengeveld/NASA 322 (t), Lowell Observatory 216 (l), 227 (l), ESA/NASA 199 (b), NASA 56 (t), 57 (r), 69 (t), 88 (t), 109 (r), 127 (l), 127 (r), 153 (l, r), 174 (b), 187 (b), 191 (b), 194, 200 (r), 206 (b), 207 (l, r), 208 (r), 210 (b), 212 (b), 214 (t, b), 217 (t), 232 (t), 234-235, 239 (r), 242 (t), 243 (r), 298 (l), 306 (l), 308 (t), 309 (t, b), 310 (t), 311 (t), 318 (t), 319 (l), 324, 325 (l), 327 (l, r), 328 (t), 329 (b), 330 (l, r), 331 (r), Hencoup 219 (r), 219 (r), Dr Francisco Diego 199 (r), Swedish Solar Vacuum Telescope 197 (t), Big Bear Solar Observatory 196, 253 (r), Gordon Garradd 182 (l), 290, Michael Stecker 52-53, 62 (t), 66 (t), 85 (r), 157 (r), 161 (b), 164 (l), 167 (b), 175 (b), 177 (t), 178 (t), 268, 270, 284, 289, 293, 294, 297, ESO 210 (t), AURA/NOAO/NSF 182 (b), 184 (t), DSS: 184 (b), Image Select/Ann Ronan 62 (b), Martin Ratcliffe 253 (l), Max-Planck-Institut für Extraterrestriche Physik 183, Howard Brown-Greaves 176 (l), AURA 119 (l), Richard Bizley 171 (b), Malin Space Science Systems 200 (b), STSCI 92-93, 142, 154 (t), 157 (l), 158 (l, r), 162 (t), 164 (r), 165 (r), 166 (l), 167 (t), 171 (l), 172, 173 (t), 186 (t), 187 (b), 188 (t, b), 190 (t), 191 (t), 216 (r), 224 (r), 227 (r, b), 229 (l), 242 (b), Harvard University Archives 154 (b), Meade Instruments 236 (l), 247, George Philip 248 (l), 298 (r), Robert McNaught 248 (r), Pierre Auger Observatory 244 (t), Sudbury Neutrino Observatory 245, Chris Wahlberg 252 (b), MACHO Collaboration 177 (b), John Dubinski, University of Toronto and San Diego Supercomputing Center 187 (t), USGS, Carolyn S. Shoemaker 310 (b), SDSS Collaboration 143 (b), CERN 134 (t), Skyview Technologies 146 (l)

Genesis Space Photo Library/Tim Furness: 61 (t), 64 (t), 300-301, 307 (t), 309 (b), 312 (t), 314 (t), 315, 316 (l), 317 (t, b), 320 (t, r), 321, 322 (t), 327 (t), 328 (l), 329 (t, b), 331 (l), 333, 335 (r)

Mary Evans Picture Library: 26-27, 29 (l), 30 (t), 31 (b), 33 (l, r), 35 (r), 36 (r), 38, 40 (r), 43 (r), 44 (l), 45 (l), 49, 50 (l), 57 (l), 58 (l), 60, 66 (b), 70 (b), 72 (l), 76 (l), 96 (l), 100 (b), 102 (b), 103 (l), 105 (b), 107 (t), 110 (b), 112 (b), 116 (t), 136, 137 (r), 189 (l, r), 244 (b)

Novosti Picture Library: 309 (t), 311 (b), 326

Photodisc Library: 150-151, 159, 163 (l, r), 167 (m)

Science & Society Picture Library: 28, 30 (l), 39 (l), 42, 43 (t), 45 (l), 46 (l, r), 47 (b), 51, 69 (t), 101 (r), 111 (b), 113, 114, Science Museum 82 (t), 124 (b), Department. of Physics, Imperial College 55 (t), CERN 87 (b), 121 (b), NASA 195, 205 (t)

Science Photo Library: 95 (t, b), 96 (r), 97 (t), 100 (t), 116 (l), 122 (b), British Technical Films 77, David Parker 75 (t), Aldred Pasieka 76, (t), 91 (t), Dr Gary S. Settles/Stephen McIntyre 58 (r), David A. Hardy 55 (b), 80 (t), Dept. of Physics, Imperial College 83 (t)

Topham Picturepoint: 29 (r), 33 (b), 35 (l), 36 (r), 39 (r), 40 (t), 41, 43 (l), 44 (r), 45 (r), 47 (b), 48 (r), 50 (r), 54, 57 (l), 64 (b), 67, 68, 70 (t), 71, 78 (l), 79 (l, r), 81 (l), 88 (b), 91 (b), 94 (t), 94 (b), 103 (r), 105 (t), 108, 109 (b), 126, 131, 132 (b), 140 (b), 143 (t), 144 (r, m), 146 (r), 175 (t), 302, 303 (l, b), 304 (t, b), 306 (b), 307 (b), 308 (b), 318 (b), 320 (l), 332, 334 (t), 335 (l), David Gallant 84

All graphics courtesy of Foundry Arts.

Special thanks to the following for graphics references: 105 courtesy of John Gribbin, *In Search of Schrödinger's Cat*; 115 courtesy of Brian Greene, *The Elegant Universe*; 188 Radial Velocity Curve courtesy of Vincent Mannings, Alan P. Boss amd Sara S. Russell (eds), *Protostars and Planets IV*.

gamma decay 108
gamma-ray astronomy 242-3
gamma-ray satellites 298
Gamow, George 126
Ganymede (moon of Jupiter) 193, 220-1, 233
Gaposchkin, Sergei 154
Gardner, Dale 329
Garnet star, observing 271, *271*
Gascoigne, William 41
gases 58, *66*, 66-7, 82-4, 90
　'ideal gas' model 55
　kinetic theory of 66
　see also dust and gas, interstellar
Gebir 33
Geiger, Johannes Wilhelm 100
Gell-Mann, Murray *110*, 110-11
Gemini, observing *262-3*, 262-3
Gemini spacecraft 320-1, 334, *335*
Germer, Lester 102-3
Giacconi, Riccardi 242
giant stars 156-7, *161*, 162-3, 190
Gilbert, William 40
Giotto spacecraft 313
Glenn, John 318, *318*, 329, 334
'glitches' 167
globular clusters 48-9, 122, 125, *125*, 130, *156*, 168,
　177-8, 191
　observing 278, 280-1, 297, *297*, 299
　see also star clusters
gluons *def* 117, 108, 111, *111*
GMCs (giant molecular clouds) 158
Goddard, Robert 304
GOES (Geostationary Operational Environmental
　Satellites) 317
Gold, Thomas 138
GOMS (Geostationary Operational Meteorological
　Satellites) 317
Gordon, Richard 320-1
Goudsmit, Sam 105
Gould's Belt 178
GPS (Global Positioning Satellites) 316, *316*
gravitation 35, 39, 40-51, 46-7, 54-7, 60-1, 62-3,
　86-7, 90, 94, 112, 114, 114-15, 117, 120, 125,
　131-3, 135-7, 140, 142-4, 147-9, *149*, 153, 158,
　165-6, 169, 172, 174, 179-80, 187, 192-3, 232,
　245, 304
　and binary stars 46
　distorting space–time 54-5, 90-1, 120-1, 137,
　172-3
　'dynamo model' 201
　and electricity 71
　existing throughout Universe 60
　gravitational constant 60, 149
　gravitational energy *def* 58, 61
　gravitational lensing 89, 177, 189
　gravitational redshifts 88, 172
　integrating with other forces 114-15
　and mass 172
　singularities 132
　in space 167

and time 172
universality of 60-1
see also general theory of relativity *under* relativity
gravitons 114
'Great Debate' 181
Grechko, Georgi 325
Greece, ancient 29-37, *34*, 50-1, *66*, 71, 100, 136, 304
Greenwich Observatory 43
Grissom, Virgil 'Gus' 302, 317, 320
Grus constellation, observing 294-5, *295*
Gubarev, Alexei 325
Guth, Alan 132
GUTs (Grand Unified Theories) 114-15, 147

H

hadrons *def* 111, 117
Hahn, Otto 109
Hainzel, Paul 38
Haise, Fred 232
Hale, George Ellery 196
half-lives 124
Halley, Edmond 43-5, *44*, 50, 62, 124
　Halley's Comet 44-5, *45*, 62-3, 229, *229*, *248*, 313
Harrison, John 45
Hartle, Jim 133
Hartmann, Bill 323
Harvard Observatory 46, 48, 154
Hawking, Stephen 117, *132*, 132-3, 147, 174, *174*
　Hawking radiation 147, 173-5
Hazard, Cyril 184
heat *see* thermodynamics
heavy hydrogen *see* deuterium
Heisenberg, Werner 102, *102*, 104, 108, *116*
　Heisenberg's Uncertainty Principle 102-3, 105, 112,
　115-17, 133, 143
helium 108-9, *128*, *130*, 130-1, 148-9, 153, 155, *157*,
　160, 162-5, 168, 196, 220-2, 233
　discovery of 83
　helium flash 162-3
Henderson, Thomas 153
Herbig-Haro objects *158*
Hercules Cluster 45
　observing 268, *269*
Hero 304
Herschel, Caroline 44-5
Herschel, Sir John 46, 152
Herschel, Sir William 44-6, 50-1, 78, 136, 140-1, 214,
　218-19, 224-5, 271, 281
Hertz, Heinrich 55, 72, *78*, 78-9, 91, 98
Hertzsprung, Ejnar 156, *156*
　Hertzsprung-Russell (HR) diagram 156-8, 161, 163-5,
　168, 190
　main sequence *156*, 156-8, 160-2, 168, 178, 190, 192
Hess, Victor *244*, 244
Hewish, Antony 166
Hewitt, David 219
Higgs, Peter 113
　Higgs boson 113, 117
Hinduism 29-30, 32
Hipparchus 30-1, *31*, 51

Hipparcos satellite 179, *179*
Hippocrates 33
Holmberg, Eric 186
Hooke, Robert 42-3, 76, 219
'horizon problem' 132, *132*
Horrocks, Jeremiah 34, 39, 41
Hourglass nebula *163*
Hoyle, Sir Fred 138, 204
Hubble, Edwin 48-9, 51, 120, *122*, 122-4, 127, 136-7,
　140-2, 148, 153, 181-2
　Hubble Deep Field Survey 142, *187*
　Hubble flow 123
　Hubble time 124-5
　Hubble's Constant 49, 120, 123-5, 140, 144, *144*, 145,
　149, 178
　Hubble's Law 120, 122-3, *123*, 125, 127
　Hubble Guide Star Catalogue 252
　Hubble Space Telescope (HST) 49-51, *88*, 123, *125*,
　130, 135, 142, *149*, 185-6, *187*, 227, 233, 239, *239*
Huggins, Sir William 47, *47*, 83
Hulse, Russell 245
Humason, Milton 49
Huygens, Christiaan 41-2, 55, 76, *76*-7, 214, 219
Hyades cluster 178
　observing 274-5, *275*
Hydra, observing 276-7, *277*
hydrogen 67, 83, 101, 109, 130, 138, 149, 153-4, *157*,
　158-60, 162-5, 168, 177, 179, 183, 192, 196, 198-9,
　218, 220-2, 224, 233, 244
　'hydrogen flash' 171
hypergiant stars 164
Hyperion (moon of Saturn) 63, *63*, 223

I

Iapetus (moon of Saturn) 223
Ibn Yunis 33
IC (Index Catalogue) 152
ICE (International Cometary Explorer) 313
imaginary numbers 105
impulse *def* 59
India 29, 32-3, 83
inertia *def* 56-7, 51, 61
　rotational inertia *def* 58
infinity 35
infrared waves 73, 78, 90, 155, 159, 176, 182
instability strip 164
interference *see under* light
interferometers 49, 77, 170, *199*, *240*, 240-1
Io (moon of Jupiter) 193, 219-20, *220*, 221, 233
ionization *def* 67, 67, 79, 84, 111, 155, 168
ions 66, 111, 155, 176, 196, 198
Iran 33
IRAS (Infrared Astronomical Satellite) 159
iron 83, 155, 164-5, 200, 203, 233
Irwin, James B *210*
Islam 50
ISO (Infrared Space Observatory) 159
isobars *def* 109
isotones *def* 109
isotopes *def* 109